T0211634

# Lecture Notes in Computer Science 12296

More information about this series at http://www.springer.com/series/7407

Emmanuel Hebrard · Nysret Musliu (Eds.)

# Integration of Constraint Programming, Artificial Intelligence, and Operations Research

17th International Conference, CPAIOR 2020
Vienna, Austria, September 21–24, 2020
Proceedings

 Springer

*Editors*
Emmanuel Hebrard (ID)
LAAS - CNRS
Toulouse, France

Nysret Musliu (ID)
TU Wien
Vienna, Austria

ISSN 0302-9743 ISSN 1611-3349 (electronic)
Lecture Notes in Computer Science
ISBN 978-3-030-58941-7 ISBN 978-3-030-58942-4 (eBook)
https://doi.org/10.1007/978-3-030-58942-4

LNCS Sublibrary: SL1 – Theoretical Computer Science and General Issues

This Springer imprint is published by the registered company Springer Nature Switzerland AG
The registered company address is: Gewerbestrasse 11, 6330 Cham, Switzerland

# Preface

This volume contains the papers that were presented at the 17th International Conference on the Integration of Constraint Programming, Artificial Intelligence, and Operations Research (CPAIOR 2020), exceptionally held online from Vienna, Austria, September 21–24, 2020. The conference received a total of 92 submissions, including 72 regular paper and 20 extended abstract submissions. The regular papers reflect original unpublished work, whereas the extended abstracts contain either original unpublished work or a summary of work that was published elsewhere. Each regular paper was reviewed by at least three Program Committee members. The reviewing phase was followed by an author response period and a general discussion by the Program Committee. The extended abstracts were reviewed for appropriateness for the conference. At the end of the review period, 36 regular papers were accepted for presentation during the conference and publication in this volume, and 12 abstracts were accepted for short presentation at the conference. Among the 36 regular papers, 4 were published directly in the journal via a fast-track review process. The abstracts of these papers can be found in this volume.

In addition to the regular papers and extended abstracts, four invited talks, whose abstracts and/or articles can be found in this volume, were given:

- **Margarida Carvalho** (University of Montreal, Canada): "Algorithmic approaches for integer programming games and a story on policy making"
- **Georg Gottlob** (University of Oxford, UK, and TU Wien, Austria): "Hypertree Decompositions: Questions and Answers"
- **Sebastian Pokutta** (Technische Universität Berlin, Zuse Institute Berlin, Germany): "Restarting Algorithms: Sometimes there is Free Lunch"
- **Peter Stuckey** (Monash University, Australia): "Combinatorial Optimisation for Multi-Agent Path Finding"

The conference program included a Master Class on the topic "Recent Advances in Optimisation Paradigms and Solving Technology":

- **Laurent Perron and Frédéric Didier** (Google Paris, France) "Constraint Programming"
- **Armin Biere** (Johannes Kepler University Linz, Austria): "Satisfiability (SAT)"
- **Günther Raidl and Andrea Schaerf** (TU Wien, Austria, and University of Udine, Italy): "(Meta)Heuristics and Hybridisation"
- **Inês Lynce** (University of Lisbon, Portugal): "MaxSAT, Multi-Objective Optimisation, and Parallelism"
- **Timo Berthold** (Fair Isaac Germany GmbH, Germany): "Mixed-Integer Programming"
- **Marie Pelleau** (Université Nice Sophia-Antipolis, France): "Numerical Constraint Programming"

The COVID-19 pandemic imposed significant hardship on the organization of this conference, which was initially scheduled to May 26–29, but was eventually held online during September 21–24. We want to express our deepest gratitude to the Local Organizing Committee members (Juliane Auerböck, Tobias Geibinger, Lucas Kletzander, Florian Mischek, Mihaela Rozman, and Felix Winter) and the Master Class organizers (Emir Demirović, Andrea Rendl, and Mohamed Siala). We would also like to thank the Program Committee members and external reviewers for their outstanding work in reviewing and discussing – often in great details – every paper, and in particular for providing extra reviews on short notice when required. Of course, we also thank the authors for upholding the high standards of the conference! Finally, we would like to express our gratitude to our sponsors for their support: Vienna Center for Logic and Algorithms (VCLA), *Artificial Intelligence Journal* (AIJ), FICO, Österreichische Post AG, Springer, Vienna Convention Bureau, and TU Wien.

July 2020                                                        Emmanuel Hebrard
                                                                 Nysret Musliu

# Organization

## Program Chairs

Emmanuel Hebrard      LAAS-CNRS, Université de Toulouse, France
Nysret Musliu      TU Wien, Austria

## Conference Chair

Nysret Musliu      TU Wien, Austria

## Master Class Chairs

Emir Demirović      The University of Melbourne, Australia
Andrea Rendl      Satalia, Austria
Mohamed Siala      INSA, LAAS-CNRS, Université de Toulouse, France

## Program Committee

| | |
|---|---|
| Tobias Achterberg | Gurobi, Germany |
| Christian Artigues | LAAS-CNRS, Université de Toulouse, France |
| Chris Beck | University of Toronto, Canada |
| Nicolas Beldiceanu | IMT Atlantique (LS2N), France |
| David Bergman | University of Connecticut, USA |
| Timo Berthold | Fair Isaac Germany GmbH, Germany |
| Hadrien Cambazard | G-SCOP, Grenoble INP, CNRS, Joseph Fourier University, France |
| Andre Augusto Cire | University of Toronto, Canada |
| Mathijs de Weerdt | Delft University of Technology, The Netherlands |
| Sophie Demassey | CMA, MINES ParisTech, France |
| Emir Demirović | Delft University of Technology, The Netherlands |
| Luca Di Gaspero | DPIA, University of Udine, Italy |
| Pierre Flener | Uppsala University, Sweden |
| Graeme Gange | Monash University, Australia |
| Martin Josef Geiger | Helmut Schmidt University, Germany |
| Bernard Gendron | University of Montreal, Canada |
| Tias Guns | Vrije Universiteit Brussel (VUB), Belgium |
| John Hooker | Carnegie Mellon University, USA |
| Matti Järvisalo | University of Helsinki, Finland |
| Serdar Kadioglu | Brown University, USA |
| George Katsirelos | MIA Paris, INRAE, AgroParisTech, France |
| Philip Kilby | Data61 and The Australian National University, Australia |

# Additional Reviewers

Björdal, Gustav
Buchheim, Christoph
Cappart, Quentin
Castro, Margarita
Curry, Timothy
Derval, Guillaume
Gottwald, Leona
Hoppmann, Kai
Houndji, Vinasétan Ratheil
Isoart, Nicolas
Kapoor, Reena
Lam, Edward
Le Bodic, Pierre

Lhomme, Olivier
Libralesso, Luc
Lo Bianco, Giovanni
Lozano, Leonardo
Luce, Robert
Refalo, Philippe
Senderovich, Arik
Stelmach, Fabian
Talbot, Pierre
Tesch, Alexander
Zarpellon, Giulia
Zitoun, Heytem

# Extended Abstracts

The following extended abstracts were accepted for presentation at the conference:

- Sina Aghaei, Andres Gomez, and Phebe Vayanos. "Learning Optimal Classification Trees: Strong Max-Flow Formulations."
- Hassan Anis and Roy Kwon. "Data-Driven Construction of Financial Factor Models."
- Harun Aydilek, Muberra Allahverdi, Asiye Aydilek, and Ali Allahverdi. "Algorithms and Effective Dominance Relations for a No-Wait Flowshop Scheduling with Random Setup Times."
- Roland Braune. "Machine Learning-based Queuing Model Regression – Example Selection, Feature Engineering and the Role of Traffic Intensity."
- Jan Elffers, Stephan Gocht, Ciaran McCreesh, and Jakob Nordström. "Justifying All Differences Using Pseudo-Boolean Reasoning."
- Nikolaus Frohner, Bernhard Neumann, and Günther Raidl. "A Beam Search Approach to the Traveling Tournament Problem."
- Patrick Gerhards. "Using CP and MIP techniques to tackle the Multi-mode Resource Investment Problem."
- Benjamin Hogstad, Jonas Falkner, and Lars Schmidt-Thieme. "A Search Heuristic Guided Reinforcement Learning Approach to the Traveling Salesman Problem."
- Matthias Horn, Marko Djukanovic, Christian Blum, and Günther R. Raidl. "On the Use of Decision Diagrams for the Repetition-Free Longest Common Subsequence."
- Thomas Jatschka, Tobias Rodemann, and Günther Raidl. "A Large Neighborhood Search for Distributing Service Points in Mobility Applications with Capacities and Limited Resources."
- Luc Libralesso, Abdel-Malik Bouhassoun, Hadrien Cambazard, and Vincent Jost. "Solving the Sequential Ordering Problem with anytime tree search."
- Maximilian Moll and Leonhard Kunczik. "A Reinforcement Learning Approach to the Labeled Maximum Matching Problem."

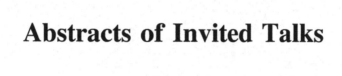

# Abstracts of Invited Talks

# Algorithmic Approaches for Integer Programming Games and a Story on Policy Making

Margarida Carvalho ⓘ

Université de Montréal
carvalho@iro.umontreal.ca

**Abstract.** Integer programming games (IPGs) are a class of problems that can suitably model non-cooperative interactions between decision makers (players). Under such formulations, each player goal in the game is described by a parametric integer program where interactions between players are reflected in their objective functions. This talk will begin with the description of IPGs, the challenge they represent and an algorithmic framework to solve them that integrates ideas of normal-form games [2]. The notion of solution to IPGs will motivate the second part of this talk, where refinements will be discussed, in particular, in the context of policy making for the kidney exchange game [1]. The latter will highlight an opportunity to integrate (patients) fairness which we address as a constraint satisfaction problem.

**Acknowledgments.** The author wishs to thank the support of the Institut de valorisation des données and Fonds de Recherche du Québec through the FRQ–IVADO Research Chair in Data Science for Combinatorial Game Theory, and the Natural Sciences and Engineering Research Council of Canada through the discovery grant 2019-04557.

This research was enabled in part by support provided by Calcul Québec (www.calculquebec.ca) and Compute Canada (www.computecanada.ca).

# References

1. Carvalho, M., Lodi, A.: Game theoretical analysis of kidney exchange programs (2020). arXiv:1911.09207
2. Carvalho, M., Lodi, A., Pedroso, J.P.: Computing Nash equilibria for integer programming games (2018). Technical report: DS4DM-2018-006

# The HyperTrac Project: Recent Progress and Future Research Directions on Hypergraph Decompositions

Georg Gottlob[1] ⓘ, Matthias Lanzinger[2] ⓘ, Davide Mario Longo[2] ⓘ
Cem Okulmus[2] ⓘ, Reinhard Pichler[2] ⓘ

[1] University of Oxford, Oxford, UK
`georg.gottlob@cs.ox.ac.uk`
[2] TU Wien, Vienna, Austria
`{mlanzing,dlongo,cokulmus,pichler}@dbai.tuwien.ac.at`

**Abstract.** Constraint Satisfaction Problems (CSPs) play a central role in many applications in Artificial Intelligence and Operations Research. In general, solving CSPs is NP-complete. The structure of CSPs is best described by hypergraphs. Therefore, various forms of hypergraph decompositions have been proposed in the literature to identify tractable fragments of CSPs. However, also the computation of a concrete hypergraph decomposition is a challenging task in itself. In this paper, we report on recent progress in the study of hypergraph decompositions and we outline several directions for future research.

This work was supported by the Austrian Science Fund (FWF): P30930-N35 in the context of the project "HyperTrac". Georg Gottlob is a Royal Society Research Professor and acknowledges support by the Royal Society for the present work in the context of the project "RAISON DATA" (Project reference: RP\R1\201074). Davide Mario Longo's work was also supported by the FWF project W1255-N23.

# Restarting Algorithms:
# Sometimes There Is Free Lunch

Sebastian Pokutta[1,2]

[1] Technische Universit¨at Berlin, Berlin, Germany
pokutta@zib.de
[2] Zuse Institute Berlin, Berlin, Germany

**Abstract.** In this overview article we will consider the deliberate restarting of algorithms, a meta technique, in order to improve the algorithm's performance, e.g., convergence rates or approximation guarantees. One of the major advantages is that restarts are *relatively* black box, not requiring any (significant) changes to the base algorithm that is restarted or the underlying argument, while leading to *potentially significant* improvements, e.g., from sublinear to linear rates of convergence. Restarts are widely used in different fields and have become a powerful tool to leverage additional information that has not been directly incorporated in the base algorithm or argument. We will review restarts in various settings from continuous optimization, discrete optimization, and submodular function maximization where they have delivered impressive results.

**Keywords:** Restarts · Convex optimization · Discrete optimization · Submodular optimization

# Combinatorial Optimisation for Multi-agent Path Finding

Peter Stuckey ⓘ

Monash University
peter.stuckey@monash.edu

**Abstract.** Multi-Agent Path Finding (MAPF) is a problem that requires one to compute a set of collision-free paths for a team of moving agents. The problem appears in variety of practical applications including warehouse logistics, traffic management, aircraft towing and computer games. The general version of the problem (minimizing makespan or sum of path costs, on graphs with parallel actions and rotations) is known to be NP-hard. One of the leading methods for solving MAPF optimally, employs a strategy known as Conflict-based Search (CBS). The central idea is to plan paths for each agent independently and resolve collisions by branching the current plan. Each branch is a new candidate plan wherein one agent or the other is forced to find a new path that avoids the selected collision. When we examine CBS from an optimisation perspective, it is clearly a form of (Logic-based) Benders Decomposition. This begs the question: can we use combinatorial optimisation techniques to tackle the MAPF problem efficiently? In this talk I will show two approaches: the first uses core-guided search together with a nogood learning Constraint Programming solver [1]; the second uses Branch-and-Cut-and-Price together with a MIP solver [2]. Both methods prove to be highly competitive to previous CBS approaches.

## References

1. Gange, G., Harabor, D., Stuckey, P.J.: Lazy {CBS}: Implicit conflict-based search using lazy clause generation. In: Lipovetzky, N., Onaindia, E., Smith, D. (eds.) Proceedings of the 29th International Conference on Automated Planning and Scheduling. pp. 155–162. AAAI Press (2019). https://www.aaai.org/ojs/index.php/ICAPS/article/view/3471
2. Harabor, D., Lam, E., Le Bodic, P., Stuckey, P.J.: Branch-and-cut-and-price for multi-agent pathfinding. In: Kraus, S. (ed.) Proceedings of the 28th International Joint Conference on Artificial Intelligence. pp. 1289–1296. IJCAI Press (2019). https://doi.org/10.24963/ijcai.2019/179

# Abstracts of Fast-Track
# Journal Papers

# Power of Pre-processing: Production Scheduling with Variable Energy Pricing and Power-Saving States

Ondřej Benedikt[1,2], István Módos[1,2], and Zdeněk Hanzálek[1]

[1] Czech Institute of Informatics, Robotics and Cybernetics, Czech Technical, University in Prague, Czech Republic
[2] Faculty of Electrical Engineering, Czech Technical University in Prague, Czech Republic
{ondrej.benedikt,istvan.modos,zdenek.hanzalek}@cvut.cz

In recent years, the research interest in energy-efficient scheduling has been increasing [3, 4]. Besides the traditional performance-oriented criteria, such as makespan, the authors also consider energy optimization, in order to make the production cost-efficient and environmentally friendly.

In this work, we study a single machine scheduling problem to minimize the total energy cost (TEC) of the production, assuming the power-saving states of the machine as well as time-of-use (TOU) energy pricing.

The integration of the power-saving states and the TOU pricing was initially proposed by Shrouf et al. [5], who designed an integer linear programming (ILP) model for the single machine problem with the fixed order of the jobs. Later, Aghelinejad et al. [1] improved and generalized the existing ILP model to consider even an arbitrary order of the jobs, in which case the problem is $\mathcal{NP}$-hard [2]. However, in both [1] and [5], only small instances of the problem have been solved optimally. One of the reasons for the inefficiency of the models is that the proposed ILP models explicitly formulate the transition behavior of the machine, and optimize it jointly with the scheduling of the jobs. In consequence, the size of the ILP models is large, and only the medium instances can be solved optimally.

We propose a novel pre-processing technique for the single machine scheduling problem with TOU pricing and machine states. Our technique pre-computes the optimal switching behavior of the machine states in time w.r.t. energy costs. Then, the pre-computed costs of the optimal switchings allow us to design efficient exact ILP and constraint programming models called and ILP-SPACES and CP-SPACES.

As shown by the experiments[1], our approach outperforms the existing ILP model ILP-REF [1], which is, to the best of our knowledge, the state-of-the-art among the

This work was supported by the Technology Agency of the Czech Republic under the National Competence Center - Cybernetics and Artificial Intelligence TN01000024.

[1] The source codes are publicly available at https://github.com/CTU-IIG/EnergyStatesAndCosts Scheduling, while all the benchmark instances can be found at https://github.com/CTU-IIG/ EnergyStatesAndCostsSchedulingData.

**Table 1.** Comparison of upper bound *ub*, lower bound *lb* and runtime *t* on MEDIUM dataset with the transitions graph having two standby states. Numbers $n$ and $h$ denote the number of jobs and pricing intervals, respectively. Time-limit is 600 s, and TLR stands for time-limit reached.

| MEDIUM+TWOSBY | | | | | | | | | | | |
|---|---|---|---|---|---|---|---|---|---|---|---|
| Instance | | ILP-REF [1] | | | CP-SPACES | | | ILP-SPACES | | | |
| $n$ | $h$ | $ub$ [-] | $lb$ [-] | $t$ [s] | $ub$ [-] | $lb$ [-] | $t$ [s] | $ub$ [-] | $lb$ [-] | $t$ [s] |
| 30 | 106 | **3 815** | **3 815** | 29.4 | **3 815** | 1 240 | TLR | **3 815** | **3 815** | 1.4 |
| 30 | 129 | **3 804** | **3 804** | 30.7 | 3 815 | 1 220 | TLR | **3 804** | **3 804** | 2.3 |
| 30 | 152 | **3 804** | **3 804** | 42.0 | 3 815 | 1 210 | TLR | **3 804** | **3 804** | 7.0 |
| 30 | 175 | **3 804** | **3 804** | 61.4 | 3 815 | 1 210 | TLR | **3 804** | **3 804** | 9.5 |
| 60 | 254 | **10 863** | **10 863** | 588.1 | **10 863** | 4 190 | TLR | **10 863** | **10 863** | 2.0 |
| 60 | 311 | 10 289 | 10 087 | TLR | 10 401 | 3 860 | TLR | **10 248** | **10 248** | 43.3 |
| 60 | 368 | **9 917** | 9 696 | TLR | 10 104 | 3 470 | TLR | **9 917** | **9 917** | 82.1 |
| 60 | 426 | 20 346 | 9 133 | TLR | 9 954 | 3 340 | TLR | **9 874** | **9 874** | 233.9 |
| 90 | 370 | 17 179 | 14 818 | TLR | 15 401 | 5 900 | TLR | **15 379** | **15 379** | 140.2 |
| 90 | 454 | 22 808 | 12 951 | TLR | 14 973 | 5 680 | TLR | **14 923** | **14 923** | 138.6 |
| 90 | 538 | 25 992 | 11 868 | TLR | 14 729 | 5 500 | TLR | **14 548** | **14 548** | 403.8 |
| 90 | 621 | 29 558 | 11 406 | TLR | 14 900 | 4 620 | TLR | **14 392** | **14 392** | 225.8 |
| Average time [s]: | | | | 412.6 | | | >600 | | | 107.5 |
| Average optimality gap [%]: | | | | 16.02 | | | 0.84 | | | 0.00 |

exact methods for this problem. Our ILP model can solve all the benchmark instances with up to 190 jobs and 1200 pricing intervals within the time-limit. On the other hand, state-of-the-art ILP-REF model from the literature scales only up to instances with 60 jobs and 300 intervals. The results for MEDIUM instances are shown in Table 1.

# References

1. Aghelinejad, M., Ouazene, Y., Yalaoui, A.: Production scheduling optimisation with machine state and time-dependent energy costs. Int. J. Prod. Res. **56**(16), 5558–5575 (2018). https://doi.org/10.1080/00207543.2017.1414969
2. Aghelinejad, M., Ouazene, Y., Yalaoui, A.: Complexity analysis of energy-efficient single machine scheduling problems. Oper. Res. Perspect. **6**, 100105 (2019). https://doi.org/10.1016/j.orp.2019.100105
3. Gahm, C., Denz, F., Dirr, M., Tuma, A.: Energy-efficient scheduling in manufacturing companies: a review and research framework. Eur. J. Oper. Res. **248**(3), 744–757 (2016). https://doi.org/10.1016/j.ejor.2015.07.017

4. Gao, K., Huang, Y., Sadollah, A., Wang, L.: A review of energy-efficient scheduling in intelligent production systems. Complex Intell. Syst. (2019). https://doi.org/10.1007/s40747-019-00122-6

5. Shrouf, F., Ordieres-Meré J., García-Sánchez, A., Ortega-Mier, M.: Optimizing the production scheduling of a single machine to minimize total energy consumption costs. J. Clean. Prod.**67**, 197–207 (2014). https://doi.org/10.1016/j.jclepro.2013.12.024

# Learn to Relax: Integrating 0–1 Integer Linear Programming with Pseudo-Boolean Conflict-Driven Search

Jo Devriendt[1,3] and Ambros Gleixner[2], and Jakob Nordström[3,1]

[1] Lund University, Lund, Sweden
jo.devriendt@cs.lth.se
[2] Zuse Institute Berlin, Berlin, Germany
gleixner@zib.de
[3] University of Copenhagen, Copenhagen, Denmark
jn@di.ku.dk

Conflict-driven pseudo-Boolean (PB) solvers optimize 0–1 integer linear programs by generalizing the conflict-driven clause learning (CDCL) paradigm [3, 12, 14] from SAT solving. Some PB solvers essentially encode the input back to CNF and run CDCL [8, 13, 15], but another approach, which is our focus in this work, is to extend the solvers from CNF to reason natively with linear constraints [5, 10, 11, 16]. Such solvers have the potential to run exponentially faster than CDCL solvers, since the cutting planes method [6] they use is exponentially stronger than the resolution method underlying CDCL [4]. In practice, however, PB solvers can sometimes get hopelessly stuck even in parts of the search space where the linear programming (LP) relaxation of the residual problem is infeasible [9].

Inspired by mixed integer programming (MIP), we address this problem by interleaving incremental LP solving with the conflict-driven pseudo-Boolean search. Our integration is fully dynamic, with the PB and LP solvers communicating continuously during execution. In order to balance resources and avoid that the LP solver starves the PB solver, LP calls are made with a strict time budget and are terminated as soon as this budget is exceeded. If the LP solver detects infeasibility, we use Farkas' lemma to combine existing constraints into a new linear constraint that can serve as the starting point of pseudo-Boolean conflict analysis. When the LP solver instead finds a rational solution, we generate Gomory cuts that prune away this solution and tighten the search space both on the PB and the LP side. The PB solver can also use information from the rational solution to direct the search, e.g., by determining how to assign variables, and we also explore passing constraints learned during conflict analysis from the PB solver to the LP solver. To the best of our knowledge, this is the first time techniques from MIP solving such as LP relaxations and cut generation have been combined with full-blown pseudo-Boolean conflict analysis, which learns new linear inequalities by operating directly on the linear constraints (rather than applying resolution on clauses derived from such constraints, as has been done previously in MIP and constraint programming solvers in, e.g., [1, 7]).

We report on extensive experiments with a combined solver integrating the LP solver *SoPlex* [17] (part of the MIP solver *SCIP* [2]) with the pseudo-Boolean solver *RoundingSat* [10]. Although we believe that there is ample room for further improvements, this hybrid approach already exhibits significantly improved performance on a wide range of benchmarks, approaching a "best of two worlds'" scenario between SAT-style conflict-driven search and MIP-style branch-and-cut.

# References

1. Achterberg, T.: Conflict analysis in mixed integer programming. Discrete Optim. **4**(1), 4–20 (2007)
2. Achterberg, T., Berthold, T., Koch, T., Wolter, K.: Constraint integer programming: a new approach to integrate CP and MIP. In: Proceedings of the 5th International Conference on the Integration of AI and OR Techniques in Constraint Programming for Combinatorial Optimization Problems (CPAIOR 2008), pp. 6–20 (2008)
3. Bayardo Jr., R.J., Schrag, R.: Using CSP look-back techniques to solve real-world SAT instances. In: Proceedings 14th National Conference on Artificial Intelligence (AAAI 1997), pp. 203–208 (1997)
4. Beame, P., Kautz, H., Sabharwal, A.: Towards understanding and harnessing the potential of clause learning. J. Artif. Intell. Res. **22**, 319–351 (2004)
5. Chai, D., Kuehlmann, A.: A fast pseudo-Boolean constraint solver. IEEE Trans. Comput. Aided Des. Integr. Circ. Syst. **24**(3), 305–317 (2005)
6. Cook, W., Coullard, C.R., Turán, G.: On the complexity of cutting-plane proofs. Discrete Appl. Math. **18**(1), 25–38 (1987)
7. Downing, N.R.: Scheduling and Rostering with Learning Constraint Solvers. Ph.D. thesis, University of Melbourne (2016)
8. Eén, N., Sörensson, N.: Translating pseudo-Boolean constraints into SAT. J. Satisf. Boolean Model. Comput. **2**(1-4), 1–26 (2006)
9. Elffers, J., Giráldez-Cru, J., Nordström, J., Vinyals, M.: Using combinatorial benchmarks to probe the reasoning power of pseudo-Boolean solvers. In: Proceedings of the 21st International Conference on Theory and Applications of Satisfiability Testing (SAT 2018), pp. 75–93 (2018)
10. Elffers, J., Nordström, J.: Divide and conquer: Towards faster pseudo-Boolean solving. In: Proceedings of the 27th International Joint Conference on Artificial Intelligence (IJCAI 2018), pp. 1291–1299 (2018)
11. Le Berre, D., Parrain, A.: The Sat4j library, release 2.2. J. Satisf. Boolean Model. Comput. **7**, 59–64 (2010)
12. Marques-Silva, J.P., Sakallah, K.A.: GRASP: a search algorithm for propositional satisfiability. IEEE Trans. Comput. **48**(5), 506–521 (1999)
13. Martins, R., Manquinho, V., Lynce, I.: Open-WBO: a modular MaxSAT solver,. In: Sinz, C., Egly, U., (eds) SAT 2014. LNCS, vol. 8561, pp. 438–445. Springer, Cham (2014). https://doi.org/10.1007/978-3-319-09284-3_33
14. Moskewicz, M.W., Madigan, C.F., Zhao, Y., Zhang, L., Malik, S.: Chaff: engineering an efficient SAT solver. In: Proceedings of the 38th Design Automation Conference (DAC 2001), pp. 530–535 (2001)

15. Sakai, M., Nabeshima, H.: Construction of an ROBDD for a PB-constraint in band form and related techniques for PB-solvers. IEICE Trans. Inf. Syst. **98-D**(6), 1121–1127 (2015)
16. Sheini, H.M., Sakallah, K.A.: Pueblo: a hybrid pseudo-Boolean SAT solver. J. Satisf. Boolean Model. Comput. **2**(1–4), 165–189 (2006)
17. SoPlex—the Sequential object oriented simPlex. https://soplex.zib.de

# The Potential of Quantum Annealing for Rapid Solution Structure Identification

Yuchen Pang[1], Carleton Coffrin[2], Andrey Y. Lokhov[2],
and Marc Vuffray[2]

[1] University of Illinois at Urbana-Champaign, Champaign IL 61801, USA
yuchenp2@illinois.edu
[2] Los Alamos National Laboratory, Los Alamos NM 87545, USA
{cjc,lokhov,vuffray}@lanl.gov

As the challenge of scaling traditional transistor-based computing technology continues to increase, experimental physicists and high-tech companies have begun to explore radically different computational technologies, such as gate-based quantum computers, quantum annealers, neuromorphic computers, memristive circuits, and coherent Ising machines. The goal of all of these technologies is to leverage the dynamical evolution of a physical system to perform a computation that is challenging to emulate using traditional computing technology, a notable motivating application being the simulation of quantum physics. Despite their entirely disparate physical implementations, optimization of quadratic functions over binary variables has emerged as a challenging computational task that a wide variety of these hardware platforms can address. As these technologies mature, it may be possible for this specialized hardware to rapidly solve challenging combinatorial problems, such as Max-Cut or Max-Clique. However, at this time, understanding the computational advantage that these hardware platforms can bring to established optimization algorithms remains an open question. It is unclear if the primary benefit will be dramatically reduced runtimes due to highly specialized hardware implementations or if the behavior of the underlying analog computing model will bring intrinsic algorithmic advantages.

Focusing on quantum annealing, this work provides new insights on the properties of this computing model and identifies problem structures where it can provide a computational benefit over a broad range of established solution methods. Through the careful design of contrived optimization problems, called Corrupted Biased Ferromagnets, this work provides new insights into the computational properties of quantum annealing and suggests that this model has an uncanny ability to avoid local minima and quickly identify the structure of high-quality solutions. A meticulous comparison to a variety of algorithms spanning both complete and local search suggest that quantum annealing's performance on the proposed optimization tasks is unique. This result provides new insights into the time scales and types of optimization problems where quantum annealing has the potential to provide notable performance gains over established optimization algorithms and suggests the development of hybrid algorithms that combine the best features of quantum annealing and state-of-the-art classical approaches.

# A New Constraint Programming Model and Solving for the Cyclic Hoist Scheduling Problem

Mark Wallace[1] and Neil Yorke-Smith[2]

[1] Monash University, Australia
mark.wallace@monash.edu
[2] Delft University of Technology, The Netherlands
n.yorke-smith@tudelft.nl

**Abstract.** The cyclic hoist scheduling problem (CHSP) is a well-studied optimisation problem due to its importance in industry [1]. In its simplest form, the problem requires one to specify the operation of an industrial hoist which operates along a linear track above a set of tanks. The hoist must move a fixed, repeating sequence of items to be processed through the tanks.

When many items are simultaneously in process, the hoist or hoists have to be available to complete all the moves between items. Moreover, each hoist must itself travel from the end of the last move that it performed to the start of the next move. Thus hoist availability is a complex resource constraint. The challenge is to find a feasible schedule that minimises the cycle time, which is termed its *period*. The central disjunctive constraint in the CHSP connects the period with the temporal decisions about hoists [4].

Despite the wide range of solving techniques applied to the CHSP and its variants, the models have remained complicated and inflexible, or have failed to scale up with larger problem instances.

This paper re-examines modelling of the CHSP and proposes a new simple and flexible constraint programming formulation [3]. We compare current state-of-the-art solvers on this formulation, and show that modelling in a high-level constraint language, MiniZinc [2], leads to both a simple, generic model and to computational results that outperform the state-of-the-art previous models. We benchmark on standard and new problem instances against results reported in the literature, using integer programming, constraint programming and lazy clause generation solvers.

We further demonstrate that combining integer programming and lazy clause generation, using the multiple cores of modern processors, has potential to improve over either solving approach alone.

**Acknowledgements.** We thank the CPAIOR reviewers for their comments, and for their recommendation to the *Constraints* journal where the full version of this work is expected to appear. Thanks also to C. Chu, K. Fleszar, S. van der Laan, W. Lei, K. Leo, G. Tack and F. Wimmenauer.

# References

1. Boysen, N., Briskorn, D., Meisel, F.: A generalized classification scheme for crane scheduling with interference. Eur. J. Oper. Res. **258**(1), 343–357 (2017)
2. Nethercote, N., Stuckey, P.J., Becket, R., Brand, S., Duck, G.J., Tack, G.: MiniZinc: Towards a Standard CP Modelling Language. In: Bessière, C. (eds.) CP 2007. LNCS, vol. 4741, pp. 529–543. Springer, Heidelberg (2007). https://doi.org/10.1007/978-3-540-74970-7_38
3. Riera, D., Yorke-Smith, N.: An improved hybrid model for the generic hoist scheduling problem. Annals of Oper. Res. **115**(1–4), 173–191 (2002). https://doi.org/10.1023/A:1021101321339
4. Rodošek, R., Wallace, M.: A generic model and hybrid algorithm for hoist scheduling problems. In: Maher, M., Puget, J.F. (eds.) CP 1998. LNCS, vol. 1520. Springer, Heidelberg (1998). https://doi.org/10.1007/3-540-49481-2_28

# Contents

# Invited Papers

# The HyperTrac Project: Recent Progress and Future Research Directions on Hypergraph Decompositions

Georg Gottlob[1] [ID], Matthias Lanzinger[2] [ID], Davide Mario Longo[2(✉)] [ID], Cem Okulmus[2] [ID], and Reinhard Pichler[2] [ID]

[1] University of Oxford, Oxford, UK
georg.gottlob@cs.ox.ac.uk
[2] TU Wien, Vienna, Austria
{mlanzing,dlongo,cokulmus,pichler}@dbai.tuwien.ac.at

**Abstract.** Constraint Satisfaction Problems (CSPs) play a central role in many applications in Artificial Intelligence and Operations Research. In general, solving CSPs is NP-complete. The structure of CSPs is best described by hypergraphs. Therefore, various forms of hypergraph decompositions have been proposed in the literature to identify tractable fragments of CSPs. However, also the computation of a concrete hypergraph decomposition is a challenging task in itself. In this paper, we report on recent progress in the study of hypergraph decompositions and we outline several directions for future research.

## 1 Introduction

Constraint Satisfaction Problems (CSPs) are arguably among the most important problems in Artificial Intelligence with a wide range of applications including diagnosis, planning, natural language processing, machine learning, etc. [10,21,47,48,53,54]. CSPs provide convenient means to formulate combinatorial problems and are, therefore, also used in many applications in Operations Research spanning scheduling [22,39,45], vehicle routing [9,41,51], all kinds of graph problems such as colouring, matching, and many other areas [12,15,48].

Formally, solving a CSP comes down to model-checking of a first-order formula, where the formula only uses the connectives $\exists, \wedge$ but not $\forall, \vee, \neg$. In this sense, solving CSPs is the equivalent problem to answering Conjunctive Queries (CQs) – one of the most fundamental kinds of queries in the database world, which essentially corresponds to (unnested) SELECT-FROM-WHERE queries in the popular database query language SQL or Basic Graph Patterns (BGPs) in

This work was supported by the Austrian Science Fund (FWF): P30930-N35 in the context of the project "HyperTrac". Georg Gottlob is a Royal Society Research Professor and acknowledges support by the Royal Society for the present work in the context of the project "RAISON DATA" (Project reference: RP\R1\201074). Davide Mario Longo's work was also supported by the FWF project W1255-N23.

E. Hebrard and N. Musliu (Eds.): CPAIOR 2020, LNCS 12296, pp. 3–21, 2020.
https://doi.org/10.1007/978-3-030-58942-4_1

the Semantic Web query language SPARQL. The underlying structure of these problems is best captured by a hypergraph. A hypergraph $H = (V, E)$ consists of a set $V$ of vertices and a set $E$ of edges with $E \subseteq 2^V$. An FO-formula $\phi$ representing a CSP or CQ gives rise to the hypergraph $H = (V, E)$, where $V$ contains the set of variables of $\phi$ and $E$ contains a set $e$ of variables as an edge if and only if there is an atom in $\phi$ whose variables are precisely the ones in $e$.

Solving CSPs and answering CQs are classical NP-complete problems [13]. Therefore, there is a long history of research on finding tractable fragments of these problems. A natural approach to this task is to search for structural properties of the underlying hypergraph which ensure tractability of CSP solving and CQ answering. A key result in this area is that CSP instances whose underlying hypergraph is acyclic can be solved in polynomial time [55]. Several generalisations of acyclicity have been identified by defining various forms of hypergraph *decompositions*, each associated with a specific notion of *width* [14,27]. Intuitively, the width measures how far away a hypergraph is from being acyclic, with a width of 1 describing the acyclic hypergraphs. The most important forms of decompositions are *hypertree decompositions (HDs)* [28], *generalized hypertree decompositions (GHDs)* [28], and *fractional hypertree decompositions (FHDs)* [32]. These decomposition methods give rise to the following notions of width of a hypergraph $H$: the *hypertree width* $hw(H)$, *generalized hypertree width* $ghw(H)$, and *fractional hypertree width* $fhw(H)$, where $fhw(H) \leq ghw(H) \leq hw(H)$ holds for every hypergraph $H$. For definitions, see Sect. 2.

The use of decompositions can significantly speed up CSP solving and CQ answering. In fact, in [1], a speed-up of up to a factor of 2,500 was reported for the CQs studied there. Structural decompositions are therefore already being used in commercial products and research prototypes, both in the CSP area as well as in database systems [1,3,4,33,40]. However, deciding if a given hypergraph $H$ has width $\leq k$ for given $k$ (for one of the width-notions mentioned above) is itself a challenging task. Formally, for a given width-notion *width* and a desired value $k$ of the width, we are thus confronted with the following family of problems:

---

CHECK(*width*, $k$)

*Instance:* A hypergraph $H$.
*Question:* Is $width(H) \leq k$?

---

We are also interested in the functional counterpart of these problems where, in case of a "yes"-answer, a witnessing decomposition of width $\leq k$ should be output as well. However, all decision procedures recalled in this paper also compute an explicit witness and so there is little need to distinguish between the decision variant and function variant of this family of problems.

The CHECK($hw, k$) problem is decidable in polynomial time for any fixed $k$ [28]. In contrast, CHECK($ghw, k$) and CHECK($fhw, k$) are NP-hard already for $k = 2$ [20,29]. Nevertheless, since $ghw$ and $fhw$ are in general smaller than $hw$, using GHDs and FHDs allows, in theory, for even more efficient algorithms for solving CSPs and answering CQs than using HDs. This is due to the fact that

CSP and CQ algorithms using decompositions have a runtime which is exponential in the width. Hence, a smaller width may ultimately pay off even if the search for a GHD or FHD is harder than for an HD. In light of the hardness result, the search for islands of tractability for CHECK($ghw, k$) and CHECK($fhw, k$) has, therefore, evolved as an important research goal. In total, we see the following three main research directions to further increase the applicability of decomposition techniques to CSP solving in AI- and OR-applications:

- *Complexity Analysis.* We need to identify restrictions on hypergraphs that guarantee the tractability of the CHECK($ghw, k$) and CHECK($fhw, k$) problems for fixed $k \geq 1$. Such restrictions should fulfill two main criteria: (i) they need to be *realistic* in the sense that they apply to a large number of CSPs and/or CQs in real-life applications, and (ii) they need to be *non-trivial* in the sense that the restriction itself does not already imply bounded $ghw$ or $fhw$. Trivial restrictions would be, e.g., bounded treewidth or acyclicity.
- *Algorithm Design.* The main motivation for identifying tractable fragments is to lay the foundation for algorithms which perform well on problem instances that fall into these fragments. Consequently, there have been several different approaches to the algorithm design for the CHECK problem, including a top-down construction of the decomposition (as proposed in the original paper on HDs [28]), a parallel approach to constructing a decomposition, and reductions to other problems such as SMT. In addition, preprocessing in the form of simplifications of a given hypergraph plays an important role.
- *From Theory to Practice.* To make sure that the decomposition algorithms work well in practice, extensive empirical evaluation is necessary. Above all, a good understanding of the hypergraphs occurring in real-world applications is required. Of course, in the real world, we do not encounter hypegraphs as such but CSPs and database queries with some underlying hypergraph structure. Especially for database queries (to a lesser extent also for CSPs) it has turned out that extracting these hypergraph structures is a non-trivial task by itself, since the CQs are somehow "hidden" behind the syntax of real-world SQL or SPARQL queries. In this paper, we report on the challenges encountered when setting up a hypergraph benchmark, that has already been used for several validation tasks and in competitions.

The paper is organized as follows: in Sect. 2, we recall some basic notions and results. Sections 3–5 are then devoted to a report on recent developments in the three main research areas mentioned above, i.e., "complexity analysis", "algorithm design", and "from theory to practice". In Sect. 6, we briefly summarize the current state of affairs and outline promising directions for future research.

## 2 Preliminaries

We have already introduced in Sect. 1 hypergraphs as pairs $(V, E)$ consisting of a set $V$ of vertices and a set $E$ of edges. It is convenient to assume that $V$ contains no isolated vertices (i.e., vertices not contained in any edge). We can then identify a hypergraph $H$ with its edge set $E$ and implicitly assume $V = \bigcup E$. A subhypergraph $H' = (V', E')$ of $H$ is then simply obtained by taking a subset $E'$ of $E$ and setting $V' = \bigcup E'$. The *primal graph* $G = (W, F)$ of a hypergraph $H = (V, E)$ is obtained by setting $W = V$ and defining $F$ such that two vertices form an edge in $G$ if and only if they occur jointly in some edge in $E$.

We are interested in the following structural properties of hypergraphs: the *rank* of $H$ is the maximum cardinality of the edges of $H$; the *degree* of $H$ refers to the maximum number of edges containing a particular vertex. A class $\mathcal{C}$ of hypergraphs is said to have *bounded rank* (or *bounded degree*) if there exists a constant $c$ such that every hypergraph in $\mathcal{C}$ has rank (or degree) $\leq c$. In [26], the notion of $(c,d)$-hypergraphs for integers $c \geq 1$ and $d \geq 0$ was introduced: $H = (V, E)$ is a $(c,d)$-hypergraph if the intersection of any $c$ distinct edges in $E$ has at most $d$ elements, i.e., for every subset $E' \subseteq E$ with $|E'| = c$, we have $|\bigcap E'| \leq d$. A class $\mathcal{C}$ of hypergraphs is said to satisfy the *bounded multi-intersection property (BMIP)* [20], if there exist $c \geq 1$ and $d \geq 0$, such that every $H$ in $\mathcal{C}$ is a $(c,d)$-hypergraph. As a special case studied in [19,20], a class $\mathcal{C}$ of hypergraphs is said to satisfy the *bounded intersection property (BIP)*, if there exists $d \geq 0$, such that every $H$ in $\mathcal{C}$ is a $(2,d)$-hypergraph.

For the definition of hypergraph decompositions and their widths, we need the following notions: *edge weight functions* are of the form $\gamma \colon E \to [0,1]$. We define $B(\gamma) = \{v \in V \mid \sum_{e \in E, v \in e} \gamma(e) \geq 1\}$ as the set of all vertices *covered* by $\gamma$ and $weight(\gamma) = \sum_{e \in E} \gamma(e)$ as the *weight* of $\gamma$. The set of edges with non-zero weight is called the *support* of $\gamma$, i.e., $supp(\gamma) = \{e \in E \mid \gamma(e) > 0\}$. We call $\gamma$ a *fractional edge cover* of a set $X \subseteq V$ by edges in $E$, if $X \subseteq B(\gamma)$. For $X \subseteq V$, we write $\rho_H^*(X)$ to denote the minimum weight over all fractional edge covers of $X$. For *integral* edge covers, the edge weight functions are restricted to integral values, i.e., $\gamma \colon E \to \{0,1\}$. We write $\rho_H(X)$ to denote the minimum weight over all integral edge covers of $X$. Clearly, $\rho_H^*(X) \leq \rho_H(X)$ holds for any $H = (V, E)$ and $X \subseteq V$. The ratio $\rho_H(V)/\rho_H^*(V)$ is referred to as the *integrality gap*.

A tuple $(T, (B_u)_{u \in T})$ is a *tree decomposition (TD)* of hypergraph $H = (V, E)$, if $T$ is a tree, every $B_u$ is a subset of $V$, and the following conditions are satisfied:

(1) For every edge $e \in E$, there is a node $u$ in $T$, such that $e \subseteq B_u$, and
(2) for every vertex $v \in V$, $\{u \in T \mid v \in B_u\}$ is connected in $T$.

The vertex sets $B_u$ are usually referred to as the *bags* of the TD. Note that, by slight abuse of notation, we write $u \in T$ to express that $u$ is a node in $T$.

A *fractional hypertree decomposition* (FHD) of a hypergraph $H = (V, E)$ is a tuple $\langle T, (B_u)_{u \in T}, (\gamma_u)_{u \in T} \rangle$, such that $\langle T, (B_u)_{u \in T} \rangle$ is a TD of $H$ and the following condition holds:

(3) For each $u \in T$, $B_u \subseteq B(\gamma_u)$ holds, i.e., $\gamma_u$ is a fractional edge cover of $B_u$.

A *generalized hypertree decomposition* (GHD) is an FHD, where $\gamma_u$ is an integral edge weight function for every $u \in T$. Hence, by condition (3), $\gamma_u$ is an integral edge cover of $B_u$. A *hypertree decomposition* (HD) of $H$ is a GHD with the following additional condition (referred to as the "special condition" in [28]):

(4) For each $u \in T$, $V(T_u) \cap B(\gamma_u) \subseteq B_u$, where $V(T_u)$ denotes the union of all bags in the subtree of $T$ rooted at $u$.

Because of condition (4), it is important to consider $T$ as a *rooted* tree in case of HDs. For TDs, FHDs, and GHDs, the root of $T$ can be arbitrarily chosen or simply ignored. The *width* of an FHD, GHD, or HD is defined as the maximum weight of the functions $\gamma_u$ over all nodes $u \in T$. The fractional hypertree width, generalized hypertree width, and hypertree width of $H$ (denoted $fhw(H)$, $ghw(H)$, and $hw(H)$) is the minimum width over all FHDs, GHDs, and HDs of $H$.

We next recall some notions which are of great importance in most of the current decomposition algorithms. Consider a hypergraph $H = (V, E)$ and let $S \subseteq V$. A set $C$ of vertices with $C \subseteq (V \setminus S)$ is $[S]$-*connected* if for any two distinct vertices $v, w \in C$, there exists a sequence of vertices $v = v_0, \ldots, v_h = w$ and a sequence of edges $e_0, \ldots, e_{h-1}$ $(h \geq 0)$ such that $\{v_i, v_{i+1}\} \subseteq (e_i \setminus S)$, for each $i \in \{0, \ldots, h-1\}$. A set $C \subseteq V$ is an $[S]$-*component*, if $C$ is maximal $[S]$-connected. Such a vertex set $S$ that is used to split a hypergraph into components is referred to as a *separator*. Note that a separator $S$ also gives rise to disjoint subsets of $E$ with $E_C := \{e \in E \mid e \cap C \neq \emptyset\}$. The *size of an* $[S]$-*component* $C$ is defined as the number of edges in $E_C$. We call $S$ a *balanced separator* if all $[S]$-components of $H$ have size $\leq \frac{|E|}{2}$. We say that a TD $(T, (B_u)_{u \in T})$ (analogously for FHD, GHD, or HD) is in *normal form* if every internal node $u$ of $T$ satisfies the following condition: let $u_1, \ldots, u_\ell$ be the child nodes of $u$. Then, for each $i \in \{1, \ldots, \ell\}$, there is a $[B_u]$-component $C_i$ of $H$ with $C_i = V(T_{u_i}) \setminus B_u$, where $V(T_{u_i})$ denotes the union of all bags in the subtree of $T$ rooted at $u_i$.

# 3 On the Complexity of Checking Widths

The search for tractable fragments of CHECK($ghw, k$) and CHECK($fhw, k$) has seen significant progress in recent years. Where there were individual proofs for various properties at first, we now have an overarching theoretical framework and tractability results for highly general properties that unify the current theory of tractable CHECK fragments. In this section we give a brief overview of this uniform theory and the resulting tractable classes for CHECK. The presentation here follows [26] which is the source of all stated results.

The BMIP will play an important role in our discussion. On the one hand, the structure of edge intersections has been identified as an important factor in the complexity of the problem. On the other hand, it can be argued that real-world problems correspond to $(c, d)$-hypergraphs with low $c$ and $d$. An empirical study of these parameters in real-world instances is presented later in Sect. 5.

## 3.1 Decompositions from Candidate Bags

In hypertree decompositions, the special condition implies a kind of lower bound on the bags of the decomposition in the sense that certain vertices need to be included in certain bags. With the generalization to $ghw$ and $fhw$ the special condition is dropped and we lose the lower bound. This then leaves us with exponentially many possible choices, even in trivial hypergraphs: e.g., every subset of an edge is a possible bag. We will see soon that this exponential number of possible bags is in fact the main challenge in the construction of polynomial time algorithms for width checking.

To illustrate this point we consider the computational complexity of constructing a tree decomposition from a set of *candidate bags* which is given to the procedure as an input. Through rather straightforward dynamic programming this is indeed possible in polynomial time.

**Theorem 1.** *Let* $H = (V, E)$ *be a hypergraph and* $\mathbf{S} \subseteq 2^V$. *There exists an algorithm that takes* $H$ *and* $\mathbf{S}$ *as an input and decides in polynomial time whether there exists a tree decomposition in normal form such that for every node* $u$ *it holds that* $B_u \in \mathbf{S}$.

Interestingly, the restriction to normal form decompositions is necessary here. Without it the problem is in fact NP-complete [26].

We call such a tree decomposition where each bag is from $\mathbf{S}$, a *candidate tree decomposition* (w.r.t. $\mathbf{S}$). Theorem 1 thus reduces the check problem to the problem of computing a set of candidate bags $\mathbf{S}$ for hypergraph $H$ such that there exists a candidate tree decomposition w.r.t. $\mathbf{S}$ if and only if $ghw(H) \leq k$. The following example illustrates this idea.

*Example 1.* Let $k$ and $r$ be constant integers and let $H = (V, E)$ be a hypergraph with rank at most $r$. Let $\mathbf{S} \subseteq 2^V$ be the set of all subsets of unions of $k$ edges, i.e.,

$$\mathbf{S} = \bigcup_{E' \in E^k} 2^{\bigcup E'}$$

where $E^k$ contains all $k$ element subsets of $E$. Since there are $\binom{|E|}{k} \leq |E|^k$ combinations of $k$ edges and each edge has rank at most $r$ we see that $\mathbf{S}$ can be computed in time $O(|E|^k 2^{kr})$. Clearly, $\mathbf{S}$ contains all bags that can be covered by $k$ edges. It is then easy to see that there exists a candidate tree decomposition w.r.t. $\mathbf{S}$ if and only if $ghw(H) \leq k$.

Thus, computing $\mathbf{S}$ and then using Theorem 1 gives us a polynomial time procedure for CHECK($ghw, k$) for hypergraph classes with bounded rank.

Note that it is enough to consider only tree decompositions in Theorem 1. In our setting it is always possible to find the respective covers of bags, if they exist, in polynomial time (recall that $k$ is considered constant).

## 3.2   Computing Candidate Bags

Example 1 illustrates the way we can use candidate bags for tractability results. However, the problem becomes much more complex as soon as we abandon bounded rank since we can no longer enumerate all (exponentially many) bags. The problem thus shifts to constructing appropriate sets of candidate bags.

This splitting of the problem into constructing candidate tree decompositions and computing lists of candidate bags then becomes very convenient. It separates the algorithmic considerations for constructing a decomposition from the combinatorial problem of limiting the number of bags. Proving both separately is significantly simpler than doing both at the same time.

Hence, for a polynomial size list of candidate bags we need to consider a more limited set of decompositions. In particular, we focus only on *bag-maximal* GHDs in which every bag is made as large as possible. Bag-maximal GHDs have two important properties: First, there always exists a bag-maximal GHD of $H$ with width $ghw(H)$. Second, for every edge $e$ and node $u$ in the decomposition we can characterize $e \cap B_u$ (assuming it is not empty) by some covers in the decomposition in the following way

$$e \cap B_u = \bigcap_{j=1}^{\ell} e \cap B(\gamma_{u_j})$$

where $(u_1, \ldots, u_\ell)$ is the path from $u$ to the node $u_\ell$ in which $e$ is covered completely, i.e., $e \subseteq B_{u_\ell}$.

However, the length of such paths cannot be bounded in terms of $c$, $d$ and $k$. Instead, we make use of the assumption that we are dealing with a $(c, d)$-hypergraph for constant $c$ and $d$. Intuitively, we distinguish between two cases based on whether the intersection along the path intersects $e$ with many $(\geq c)$ distinct edges. If so, the intersection is small $(\leq d)$ and we can compute all subsets of the intersection. One of them will be $e \cap B_u$. In the other case we can explicitly compute all the intersections of $e$ with up to $c$ edges, which will again contain $e \cap B_u$. A detailed argument can be found in Section 5 of [26].

**Theorem 2.** *Fix constant $k \geq 1$. For every hypergraph class $\mathcal{C}$ that enjoys the BMIP the $\textsc{Check}(ghw, k)$ problem is tractable.*

We see that $\textsc{Check}(ghw, k)$ is tractable in a wide range of cases. The BMIP properly generalizes many important hypergraph properties. A hypergraph with rank $r$ is a $(1, r)$-hypergraph and a hypergraph with degree $\delta$ is a $(\delta + 1, 0)$-hypergraph. Hence, bounded rank and bounded degree (and bounded intersection) are all simply special cases of the BMIP.

## 3.3 One Step Further: Fractional Hypertree Width

The restriction to bounded rank is only one part of what made Example 1 simple. The other part is that we considered only $ghw$. With the step to $fhw$ it is no longer clear of which sets we want to consider the subsets as more than $k$ edges can be necessary to cover a set of vertices with weight $k$. In general, the integrality gap for edge cover in hypergraphs is $\Theta(\log |V|)$ [44]. This means that we would need the union of $k \log |V|$ edges to cover every set of vertices $U$ with $\rho^*(U) \leq k$. If we follow the naive approach in Example 1, we would thus get a time bound of $O(|E|^{k \log |V|} 2^{rk \log |V|})$ which is no longer polynomial.

We see that for CHECK($fhw, k$) we have an additional challenge, bounding the support of fractional edge covers with weight at most $k$. Indeed, in cases where the support is boundable we can, in a sense, reduce the problem to the $ghw$ case. This reduction results in Theorem 3 below.

**Definition 1.** *We say an FHD $\langle T, (B_u)_{u \in T}, (\gamma_u)_{u \in T} \rangle$ is q-limited if for every node $u$ in $T$ it holds that $|support(\gamma_u)| \leq q$. Analogous to $fhw(H)$ we define $fhw_q(H)$ as the minimum width of all q-limited FHDs of $H$.*

**Theorem 3.** *Fix constants $q$ and $k$. For every hypergraph class $\mathcal{C}$ that enjoys the BMIP the CHECK($fhw_q, k$) problem is tractable.*

Theorem 3 abstracts away the computation of the tree decomposition and the computation of the candidate bags. We can show that $fhw$ checking is tractable for some class $\mathcal{C}$ if for each hypergraph $H \in \mathcal{C}$ there exists a constant $q$ such that $fhw_q(H) = fhw(H)$.

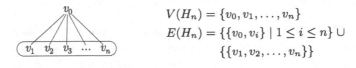

$$V(H_n) = \{v_0, v_1, \ldots, v_n\}$$
$$E(H_n) = \{\{v_0, v_i\} \mid 1 \leq i \leq n\} \cup$$
$$\{\{v_1, v_2, \ldots, v_n\}\}$$

**Fig. 1.** (2,1)-hypergraphs $H_n$ with large support

However, bounding the support of fractional edge covers is a difficult problem. Consider the $(2, 1)$-hypergraphs given in Fig. 1. We have $\rho^*(H_n) = 2 - \frac{1}{n}$ where the optimal cover assigns weight $\frac{1}{n}$ to all the edges incident to $v_0$ and $1 - \frac{1}{n}$ to the big edge. That is, the optimal cover has support of size $n$.

The example demonstrates that it is not possible to bound the support of the *optimal* cover, even in $(2, 1)$-hypergraphs. However, if we were interested in checking $fhw \leq 2$ for such a $H_n$, we do not necessarily need to consider the optimal covers but can instead consider slightly heavier covers (still with weight $\leq 2$) for which we can bound the support. In this case it is easy as we can always cover the whole hypergraph with 2 edges that are assigned weight 1.

While the problem is much more complex in general, the main idea stays the same. One can show that for every cover $\gamma$ of weight at most $k$, there exists another weight assignment $\nu$ such that $B(\gamma) \subseteq B(\nu)$, $weight(\nu) \leq k$, for which the support can be bounded in terms of $k$, $c$, and $d$ if either $c \leq 2$ or $d = 0$. In consequence we arrive at the following theorem.

**Theorem 4.** *Fix constants $q$ and $k$. For every hypergraph class $\mathcal{C}$ that enjoys either bounded degree or bounded intersection the CHECK($fhw, k$) problem is tractable.*

Note that Theorem 4 holds also for classes of bounded rank because it is a special case of bounded intersection.

# 4   Hypergraph Decomposition Algorithms and Systems

In this section we give a brief overview on recent developments in the field of hypergraph decomposition algorithms. We highlight some selected works which implement decomposition algorithms, i.e., released systems which can produce decompositions and are therefore advancing the practical ability to use decompositions within more complex systems, such as a database management system or a CSP solver.

## 4.1   Hypergraph Preprocessing

Since simplifications of the input hypergraph are common across many of the presented systems and algorithms, we present here an overview of the techniques so far developed. The common idea is always to reduce the input in such a way that no valid solutions are lost, thus reducing the effective search space.

Fichte et al. [18] use a number of simplifications, each reducing the size of the SMT encoding: 1) the removal of edges contained in other edges, 2) splitting $H$ into its biconnected components, and working on each of them separately (for a definition of biconnected components we refer to [25]), 3) the removal of vertices of degree one and 4) the removal of simplicial vertices, defined as vertices whose neighbourhood in $H$ forms a clique in the primal graph. This is allowed for the computation of $fhw$, since the fractional cover number of this clique then forms a lower bound on the $fhw$ of H. Additional simplifications from the literature are: 5) the removal of all vertices (bar one) of the same *type* [30], where a *type* of a vertex is the set of all its incident edges and 6) splitting $H$ into its *hinges* [25]. This is a generalisation of simplification 2). A definition of hinge is found in [25].

## 4.2   Top-Down Construction

**HD Computation.** We will briefly recall the basic principles of the DetKDecomp algorithm from [31], which improves significantly on the first implementation, called OptKDecomp [43] of the original HD algorithm from [28]. For a fixed $k \geq 1$, DetKDecomp tries to construct an HD of a hypergraph $H$ in a top-down fashion. Its input is a subhypergraph $H'$ (initially the same as $H$). It produces a new node $u$ (initially serving as the root), then computes the $[B_u]$-components $C_1, \ldots, C_\ell$. We define for each component a new hypergraph $H_i = (V_i, E_i)$, where $E_i = \{e \mid e \cap C_i \neq \emptyset\}$ and $V_i = \bigcup E_i$. Then DetKDecomp recursively searches for an HD of width $\leq k$ for each hypergraph $H_i$. If this succeeds for each $H_i$, then DetKDecomp accepts. If there exists an $H_i$ s.t. no HD of width $\leq k$ can be found, then DetKDecomp backtracks and produces a new node $u$. If all choices for nodes have been exhausted, it rejects.

**Tractable GHD Computation.** Novel algorithms for solving the aforementioned problem of CHECK($ghw, k$) in polynomial time for $(2, d)$-hypergraphs with low $d$ are presented in [20]. Based on these results, implementations of these algorithms are developed in [19], built on the basis of `DetKDecomp` (with the new system aptly named `NewDetKDecomp`). The source code is publicly available under https://github.com/TUfischl/newdetkdecomp.

We proceed to sketch out the tractable *ghw* algorithm used in `NewDetKDecomp`. As mentioned in Sect. 3, the main reason for the NP-completeness of the CHECK(*ghw, k*) problem is the exponential number of candidate bags. Following the construction from [20], `NewDetKDecomp` explicitly computes intersections of up to *k* edges, and considers these subsets of edges (called *subedges* in the sequel) as part of the input. From Theorem 2 it follows that this can be done in polynomial time for hypergraph classes with the BIP (as it is a special case of BMIP). Based on when exactly those subedges are added to the currently considered subhypergraph, two variants were defined, called `GlobalBIP` and `LocalBIP`. We leave out details here and refer readers to [19].

## 4.3  Parallel Approaches

**Balanced Separator Algorithm.** Yet another *ghw* algorithm from [19], `BalSep` builds on an observation from [2]: For any hypergraph $H$ with *ghw* $\leq k$, there exists a balanced separator $S$ with $\rho_H(S) \leq k$. This gives rise to an algorithm which checks for the presence of such separators, and if they cannot be found, can immediately reject. By definition, a balanced separator reduces the size of hypergraphs to be considered by at least half. This means that `BalSep` has logarithmically bounded recursion depth, compared with the linear recursion depth of `DetKDecomp`. This property makes `BalSep` a promising candidate for a parallel approach to computing *ghw*.

**Parallel GHD Computation.** On the basis of the Balanced Separator algorithm, a parallel algorithm for computing GHDs is presented in [30], as well as a series of generally applicable algorithmic improvements for computing GHDs. This system, called `BalancedGo`, is able to decompose nearly twice as many real-world CSP instances within a feasible time, when compared to `NewDetKDecomp`. Written in the programming language Go [16], `BalancedGo` is available under https://github.com/cem-okulmus/BalancedGo. We proceed to detail this parallel approach below.

The following generally applicable improvements are presented in [30]: 1) While existing algorithms, such as `DetKDecomp` make use of heuristics consisting of ordering the edges in such a way as to try out promising separators first, none of the existing methods proved fruitful for speeding up the search for balanced separators. A number of heuristics are considered in [30], ultimately settling on one, which balances out the speed-up of the search against the actual time to compute the heuristic itself, and 2) the existing implementation of the Balanced Separator algorithm from [19] proved to be inefficient w.r.t. considering all relevant subsets of a given separator. Reorganising the way subedges are considered, as well as more effective caching helps to provide significant speed-ups here.

The programming language Go has a model of parallelisation inspired by Communicating Sequential Processes of Hoare [34]. It is based on light-weight

threads, called *goroutines*, which communicate over *channels*. This model minimises, as far as possible, the need for explicit synchronisation. For `BalancedGo`,
there are two areas of parallelism: the search of balanced separators, and the
recursive calls. Each is implemented via goroutines, in such a manner as to reduce
the need for explicit synchronisation, to enable effective backtracking and utilise
existing CPU resources as best as possible. For details we refer to [30].

Finally, a hybrid algorithm is presented [30], which combines the above mentioned parallel Balanced Separator algorithm, with the existing `NewDetKDecomp`
algorithm (extended to compute *ghw* as mentioned above): for a constant number $m$ of recursions, it uses the Balanced Separator algorithm. After recursion
depth $m + 1$ has been reached, it proceeds to use `NewDetKDecomp`. The Balanced
Separator algorithm is effective at initially reducing the size of instances, but
gets slower as it needs to backtrack more and more often. The `NewDetKDecomp`
algorithm is very effective at quickly computing HDs for smaller instances, or
rejecting if no HD of sufficiently low width exists. The hybrid approach therefore
combines the best of both worlds.

### 4.4   Using Established Solvers and Other Approaches

**HD and GHD Computation via SMT Encoding.** A very different approach to compute hypergraph decompositions is utilised by Fichte et al. [18].
Instead of implementing or designing algorithms to compute decompositions,
the aim is instead to encode the problem into SMT (SAT modulo Theory) with
Linear Arithmetic and then use the SMT solver *Z3*. From this result a provably
minimum width FHD can then be constructed. This system, called FraSMT, is
available under https://github.com/daajoe/frasmt.

The basis of the aforementioned encoding is an ordering-based characteri-
sation of *fhw*, similar to the well known elimination ordering for treewidth [7].
As the elimination ordering has already been used successfully for SAT encodings of treewidth [49], it seems natural to investigate a similar approach for
*fhw* computation. To this end, Fichte et al. define, for a given ordering of the
vertices $L = (v_1, \ldots, v_n)$, an extension of the hypergraph $H^i_L$, which iteratively
constructs and adds a new edge $e_i$, covering all such vertices $v_j$, where $i < j \leq n$
and there exists some edge $e$ in $H^{i-1}_L$ such that $v_i, v_j \in e$. The *fhw of H w.r.t.*
*a linear ordering* $L$ is then defined as the largest fractional cover number of the
vertices $e_i \cup v_i$, for any $v_i \in L$, where only edges in $H$ are considered for the
fractional cover. Fichte et al. then prove that the *fhw* of $H$ is exactly the same
as the smallest *fhw* of $H$ w.r.t. to any linear ordering.

The above defined ordering is then translated into a formula $F(H, w)$, where
$F(H, w)$ is true if and only if $H$ has a linear ordering $L$ such that the *fhw* of $H$
is $\leq w$. For symmetry breaking, Fichte et al. consider the *hyperclique* (defined
as cliques in the primal graph), with the highest fractional cover number, and
require the vertices of this hyperclique to appear at the end of the ordering, thus
reducing the search space.

An extension of FraSMT, is presented by Schidler and Szeider [50], called
HtdSMT, available under https://github.com/ASchidler/htdsmt. They extend

the above defined encoding to SMT in order to express the special condition of HDs. They won first prize in the PACE 2019 Challenge [17], in the track for the exact computation of *hw* of up to 100 instances in less than 30 min, beating out a version of `NewDetKDecomp`.

**Approaches Based on the Enumeration of Potential Maximal Cliques.** Korhonen et al. [37] present a novel approach to compute *ghw* based on the Bouchitté-Todinca (BT) algorithm [11], which enumerates so-called *potential maximal cliques* (PMC). It is available under https://github.com/Laakeri/Triangulator. While PMC-based approaches were previously used to define a number of algorithms for solving treewidth, minimum fill-in width and other measures, Korhonen et al. are the first to present a practical implementation of the BT algorithm which computes *ghw*. Based on their evaluation, it compares quite favourably to `DetKDecomp`, despite computing *ghw* instead of *hw*.

# 5   From Theory to Practice

In this section we discuss the problem of evaluating the quality of decomposition algorithms over real-world instances. Indeed, while the theory of hypergraph decompositions is well understood and implementations show promising results, very little was known about the typical instances that these should decompose. To this end, we focus here on the challenges encountered during the development of a benchmark that can be used to reduce the gap between theory and practice.

## 5.1   The Need for Benchmarks

Over the years, the performance evaluation of algorithms and systems tackling variants of the CHECK problem has been conducted against provisional datasets. When `DetKDecomp` [31] appeared, a collection of CSPs was used to show the superiority of `DetKDecomp` w.r.t. `OptKDecomp`. However, since the dataset lacks CQs, no assessment of `DetKDecomp` can be made for usage in databases.

Later on, Scarcello et al. proposed in [23,24] a system for the evaluation of SQL queries using hypertree decompositions. Although their study showed promising results, the dataset used to evaluate the system performance consisted only of a limited set of queries coming from the same source. Thus, the conclusion cannot be generalized to query answering in general.

These examples highlight the need for a comprehensive, easily extensible, public benchmark. This could not only be used to evaluate concrete CSP-solving and database systems, but also to empirically test theoretical properties. Indeed, if a large part of the instances satisfies a certain property, then it is worth developing specialized algorithms exploiting the property.

## 5.2   HyperBench: Challenges and Results

The desire for collecting hypergraphs and comparing algorithm performances led to the implementation of *HyperBench* [19], which is a comprehensive collection of circa 3000 hypergraphs representing CSPs and CQs. The hypergraphs and the experimental results are available at http://hyperbench.dbai.tuwien.ac.at.

**Collecting Instances.** HyperBench altogether contains 3070 hypergraphs divided into three classes: Application-CSPs, Application-CQs, Random. Table 1 shows, for each class, the number of instances and the number of cyclic hypergraphs, i.e., the ones having $hw \geq 2$. Out of the 1172 Application-CSPs, 1090 come from XCSP [5] and 82 were used in previous analyses [6,31]. The 535 Application-CQs have been fetched from a variety of sources. In particular, all the 70 SPARQL queries having $hw \geq 2$ from [8] and all the cyclic SQL queries from SQLShare [35] have been included. The remaining CQs come from different benchmarks such as the Join Order Benchmark (JOB) [42] and TPC-H [52]. The Random class contains 863 random CSP instances from [5] and 500 random conjunctive queries generated with the tool used for [46]. This class has the purpose of comparing real application instances to synthetic ones.

**Table 1.** Overview of the classes of instances contained in HyperBench [19].

| Class | Num. instances | $hw \geq 2$ |
|---|---|---|
| Application-CSPs | 1172 | 1172 |
| Application-CQs | 535 | 81 |
| Random | 1363 | 1327 |
| Total | 3070 | 2580 |

**Obtaining the Hypergraphs.** Collecting CSPs and CQs is only a preliminary step in building a benchmark. The very next task is the translation of instances into a uniform hypergraph format. The details of this phase depend on the language in which the instances are written, thus, here, we briefly go over the translation of two specific sets of instances: the CSPs from [5] and the SQL queries from [35]. While the translation of the first set did not pose any particular challenge, the second one turned out to be rather difficult.

The CSPs fetched from XCSP are encoded in well-structured XML files in which variables and constraints are defined explicitly. Moreover, an extensive library for parsing the instances, in which most of the process is automatized, is available. In this case, it is sufficient to redefine the behaviour of some callback methods so that, whenever the program reads a variable, it adds a vertex to the hypergraph, and, whenever it reads a constraint, it adds an edge containing the vertices corresponding to the variables affected by the constraint.

Given its multifaceted nature, the SQLShare dataset poses numerous challenges for the translation of the queries into hypergraphs. Since it is a collection of databases and handwritten queries by different authors, the original format is highly irregular and requires several refinement phases. Some of them follow:

1. Cleaning the queries from trivial errors that impede parsing.
2. Extracting table definitions from the databases.
3. Inferring the definition of undefined tables from the queries.
4. Resolving ambiguities in the queries semantics, e.g., choosing one definition for the tables that appear with the same name in different databases.
5. Extracting conjunctive query cores from a complex SQL query, i.e., given a single query, producing a collection of simpler conjunctive queries that can be used to compute the result of the original query.

Particular attention has been devoted to views. Indeed, a query that uses views must be expanded first and only afterwards should be translated. In this way, the resulting hypergraph will accurately reflect the query structure.

**Experiments.** We report on some experiments carried out using HyperBench. The results reveal a comprehensive picture of the hypergraph characteristics of the collection. We present aggregate results for the classes in Table 1.

For sake of uniformity, we adapt the terminology of [19] to the one of Sect. 3. A hypergraph $H$ has degree bounded by $\delta$ if and only if $H$ is a $(\delta+1, 0)$-hypergraph. We say $H$ has $c$-multi-intersection size $d$ if $H$ is a $(c, d)$-hypergraph. In the special case of $c = 2$, we talk of intersection size of $H$. If we do not have in mind any particular $c$, we simply speak of multi-intersection size of $H$.

**Table 2.** Percentage of $(c, d)$-hypergraphs with degree $\leq 5$ and $c$-multi-intersection size $\leq 5$, for $c \in \{2, 3, 4\}$. $(6,0)$-hypergraphs are the ones with degree at most 5 [19].

| Class | $(6, 0)$-hgs (%) | $(2, 5)$-hgs (%) | $(3, 5)$-hgs (%) | $(4, 5)$-hgs (%) |
|---|---|---|---|---|
| Application-CSPs | 53.67 | 99.91 | 100 | 100 |
| Application-CQs | 81.68 | 100 | 100 | 100 |
| Random | 10.12 | 76.82 | 90.17 | 93.62 |
| Total | 39.22 | 89.67 | 95.64 | 97.17 |

One of the goals of [19] was to find out whether low (multi-)intersection size is a realistic and non-trivial property. For the purposes of the study, the value $d = 5$ has been identified as a threshold separating low values from high values. Table 2 shows the percentage of instances having low degree and low $c$-multi-intersection size, for $c \in \{2, 3, 4\}$. It can be seen that for each class the amount of instances with low (multi-)intersection is greater than the ones having low degree. Also, the (multi-)intersection size tends to be (very) small for both CSPs and CQs taken from applications, while it is still reasonably small for random instances.

An additional correlation study between the hypergraph properties establishes that there is no correlation between (multi-)intersection size and $hw$, thus low (multi-)intersection size does not imply low $hw$.

After the analysis of structural properties, lower and upper bounds for $hw$ have been computed for the whole dataset. For these experiments a timeout of 1 h was set. The results are summarized in Table 3. It has been determined that 694 of all 1172 Application-CSPs (59.22%) have $hw \leq 5$ and, surprisingly, $hw \leq 3$ for all Application-CQs. In total, considering also random instances, 1849 (60.23%) out of 3070 instances have $hw \leq 5$. For 1778 of them, the bound on $hw$ is tight, while for the others the actual value of $hw$ could be even less. To conclude, for the vast majority of CSPs and CQs (in particular those from applications), $hw$ is small enough to allow for efficient CSP solving or CQ answering, respectively.

**Table 3.** Number and percentage of instances having $hw \leq 5$ [19].

| Class | $hw \leq 5$ | Total | % |
|---|---|---|---|
| Application-CSPs | 694 | 1172 | 59.22 |
| Application-CQs | 535 | 535 | 100 |
| Random | 620 | 1363 | 45.49 |
| Total | 1849 | 3070 | 60.23 |

As computing $ghw$ is more expensive, the algorithms ran on the hypergraphs with small width. Thus, for all the hypergraphs having $hw \leq k$ with $k \in \{3,4,5,6\}$, the check $ghw \leq k - 1$ has been performed. If the algorithm did not timeout and gave either a *yes* or *no* answer, we say the instance is *solved*. Though it is known that, for each hypergraph $H$, $hw(H) \leq 3 \cdot ghw(H) + 1$ holds [2], surprisingly it turns out that 98% of the solved instances, which form 57% of all instances, have identical values of $hw$ and $ghw$.

## 5.3  Further Uses of HyperBench

Since its publication, HyperBench has been used in several ways in the world of decomposition techniques. As already discussed, the restrictions defining tractable fragments of variants of the CHECK problem presented in Sect. 3 have been already investigated in [19]. Moreover, it has been used to gain an understanding of the differences between $hw$ and $ghw$ in real-world CSPs and CQs.

In [36], the edge clique cover size of a graph is identified as a parameter allowing fixed-parameter-tractable algorithms for enumerating potential maximal cliques. The latter can be used to compute exact $ghw$ and $fhw$. An edge clique cover of a graph is a set of cliques of the graph that covers all of its edges. In case of a CSP with $n$ variables and $m$ constraints, the set of constraints is an edge clique cover of the underlying (hyper)graph. Thus, this property can be exploited for CSPs having $n > m$ and HyperBench has been used to verify that it happens in circa 23% of the instances.

HyperBench has also been used in the PACE 2019 Challenge [17] to compare the performance of several HD solvers. The challenge dedicated two tracks to HDs: in the *exact* track, the participants had to compute $hw(H)$ for as many hypergraphs as possible, while in the *heuristic* track, the task was to compute a decomposition with low $hw$ in short time.

## 6    Conclusion and Future Work

In this paper, we have reported on recent progress in the research on hypergraph decompositions. This progress has several facets: we now have a fairly good understanding of the complexity of constructing various kinds of decompositions. In particular, we have seen several structural restrictions on hypergraphs which make the CHECK($ghw, k$) and CHECK($fhw, k$) problems tractable, such as low rank, low degree, small intersection and, in case of $ghw$, also small multi-intersection. On the algorithmic side, several different approaches have been proposed for the computation of concrete hypergraph decompositions – either for some desired upper bound $k$ on the width or with minimum width. The main categories of these decomposition algorithms are the "classical" top-down construction of a decomposition (as suggested in [28]), a parallel approach, and the reduction to other problems such as SMT. Finally, we have recalled the work on the HyperBench benchmark, which has already been used for the empirical evaluation of several implementations of decomposition algorithms and which has also allowed us to get a realistic picture as to which structural properties the hypergraphs underlying CSPs and CQs in practice typically have.

There are several promising directions of future research in this area. As mentioned in Sect. 1, decomposition techniques have already been introduced into research prototypes and first commercial products. Our aspiration is to see decomposition techniques incorporated more widely also into mainstream systems – both in the CSP world and in the database world. Another kind of system where decomposition techniques may have a lot of potential are integer programming solvers. Indeed, integer programs are readily modelled as hypergraphs whose vertices correspond to the variables in the integer program and the non-zero entries in each row are represented by an edge. The application of hypergraph decompositions, in particular, to sparse integer programs seems very promising to us. The topic has been touched on in [38] but a deeper investigation is missing to date. Another important direction for future work is to study the dynamics of decompositions when the corresponding CSP or CQ is slightly modified. In such a case, does the entire decomposition have to be re-computed from scratch or can it be obtained from the "old" one via suitable modifications?

## References

1. Aberger, C.R., Tu, S., Olukotun, K., Ré, C.: EmptyHeaded: a relational engine for graph processing. In: Proceedings of SIGMOD 2016, pp. 431–446 (2016)

2. Adler, I., Gottlob, G., Grohe, M.: Hypertree width and related hypergraph invariants. Eur. J. Comb. **28**(8), 2167–2181 (2007)
3. Amroun, K., Habbas, Z., Aggoune-Mtalaa, W.: A compressed generalized hypertree decomposition-based solving technique for non-binary constraint satisfaction problems. AI Commun. **29**(2), 371–392 (2016)
4. Aref, M., et al.: Design and implementation of the LogicBlox system. In: Proceedings of SIGMOD 2015, pp. 1371–1382 (2015)
5. Audemard, G., Boussemart, F., Lecoutre, C., Piette, C.: XCSP3: an XML-based format designed to represent combinatorial constrained problems (2016). http://www.xcsp.org/
6. Berg, J., Lodha, N., Järvisalo, M., Szeider, S.: Maxsat benchmarks based on determining generalized hypertree-width. In: MaxSAT Evaluation 2017: Solver and Benchmark Descriptions, vol. B-2017-2, p. 22 (2017)
7. Bodlaender, H.L.: Discovering treewidth. In: Vojtáš, P., Bieliková, M., Charron-Bost, B., Sýkora, O. (eds.) SOFSEM 2005. LNCS, vol. 3381, pp. 1–16. Springer, Heidelberg (2005). https://doi.org/10.1007/978-3-540-30577-4_1
8. Bonifati, A., Martens, W., Timm, T.: An analytical study of large SPARQL query logs. VLDB J. **29**, 655–679 (2019). https://doi.org/10.1007/s00778-019-00558-9
9. Booth, K.E.C., Beck, J.C.: A constraint programming approach to electric vehicle routing with time windows. In: Rousseau, L.-M., Stergiou, K. (eds.) CPAIOR 2019. LNCS, vol. 11494, pp. 129–145. Springer, Cham (2019). https://doi.org/10.1007/978-3-030-19212-9_9
10. Booth, K.E.C., Tran, T.T., Nejat, G., Beck, J.C.: Mixed-integer and constraint programming techniques for mobile robot task planning. IEEE Robot. Autom. Lett. **1**(1), 500–507 (2016)
11. Bouchitté, V., Todinca, I.: Treewidth and minimum fill-in: grouping the minimal separators. SIAM J. Comput. **31**(1), 212–232 (2001)
12. Brailsford, S.C., Potts, C.N., Smith, B.M.: Constraint satisfaction problems: algorithms and applications. Eur. J. Oper. Res. **119**(3), 557–581 (1999)
13. Chandra, A.K., Merlin, P.M.: Optimal implementation of conjunctive queries in relational data bases. In: Proceedings of STOC 1977, pp. 77–90. ACM (1977)
14. Cohen, D.A., Jeavons, P., Gyssens, M.: A unified theory of structural tractability for constraint satisfaction problems. J. Comput. Syst. Sci. **74**(5), 721–743 (2008)
15. Dechter, R.: Constraint Processing. Morgan Kaufmann Publishers Inc., San Francisco (2003)
16. Donovan, A.A.A., Kernighan, B.W.: The Go Programming Language. Addison-Wesley Professional, Boston (2015)
17. Dzulfikar, M.A., Fichte, J.K., Hecher, M.: The PACE 2019 parameterized algorithms and computational experiments challenge: the fourth iteration. In: Proceedings of IPEC 2019, Leibniz International Proceedings in Informatics (LIPIcs), vol. 148, pp. 25:1–25:23. Schloss Dagstuhl-Leibniz-Zentrum für Informatik (2019)
18. Fichte, J.K., Hecher, M., Lodha, N., Szeider, S.: An SMT approach to fractional hypertree width. In: Hooker, J. (ed.) CP 2018. LNCS, vol. 11008, pp. 109–127. Springer, Cham (2018). https://doi.org/10.1007/978-3-319-98334-9_8
19. Fischl, W., Gottlob, G., Longo, D.M., Pichler, R.: HyperBench: a benchmark and tool for hypergraphs and empirical findings. In: Proceedings of PODS 2019, pp. 464–480. ACM (2019)
20. Fischl, W., Gottlob, G., Pichler, R.: General and fractional hypertree decompositions: hard and easy cases. In: Proceedings of PODS 2018, pp. 17–32. ACM (2018)

21. Gange, G., Harabor, D., Stuckey, P.J.: Lazy CBS: implicit conflict-based search using lazy clause generation. In: Proceedings of ICAPS 2019, pp. 155–162. AAAI Press (2019)

22. Geibinger, T., Mischek, F., Musliu, N.: Investigating constraint programming for real world industrial test laboratory scheduling. In: Rousseau, L.-M., Stergiou, K. (eds.) CPAIOR 2019. LNCS, vol. 11494, pp. 304–319. Springer, Cham (2019). https://doi.org/10.1007/978-3-030-19212-9_20

23. Ghionna, L., Granata, L., Greco, G., Scarcello, F.: Hypertree decompositions for query optimization. In: Proceedings of ICDE 2007, pp. 36–45. IEEE Computer Society (2007)

24. Ghionna, L., Greco, G., Scarcello, F.: H-DB: a hybrid quantitative-structural SQL optimizer. In: Proceedings of CIKM 2011, pp. 2573–2576. ACM (2011)

25. Gottlob, G., Hutle, M., Wotawa, F.: Combining hypertree, bicomp, and hinge decomposition. In: Proceedings of ECAI 2002, pp. 161–165. IOS Press (2002)

26. Gottlob, G., Lanzinger, M., Pichler, R., Razgon, I.: Complexity analysis of generalized and fractional hypertree decompositions. CoRR abs/2002.05239 (2020). https://arxiv.org/abs/2002.05239

27. Gottlob, G., Leone, N., Scarcello, F.: A comparison of structural CSP decomposition methods. Artif. Intell. **124**(2), 243–282 (2000)

28. Gottlob, G., Leone, N., Scarcello, F.: Hypertree decompositions and tractable queries. J. Comput. Syst. Sci. **64**(3), 579–627 (2002)

29. Gottlob, G., Miklós, Z., Schwentick, T.: Generalized hypertree decompositions: NP-hardness and tractable variants. J. ACM **56**(6), 30:1–30:32 (2009)

30. Gottlob, G., Okulmus, C., Pichler, R.: Fast and parallel decomposition of constraints satisfaction problems. In: Proceedings of IJCAI 2020, pp. 1155–1162 (2020)

31. Gottlob, G., Samer, M.: A backtracking-based algorithm for hypertree decomposition. ACM J. Expe. Algorithmics **13** (2008)

32. Grohe, M., Marx, D.: Constraint solving via fractional edge covers. ACM Trans. Algorithms **11**(1), 4:1–4:20 (2014)

33. Habbas, Z., Amroun, K., Singer, D.: A forward-checking algorithm based on a generalised hypertree decomposition for solving non-binary constraint satisfaction problems. J. Exp. Theor. Artif. Intell. **27**(5), 649–671 (2015)

34. Hoare, C.A.R.: Communicating sequential processes. Commun. ACM **21**(8), 666–677 (1978)

35. Jain, S., Moritz, D., Halperin, D., Howe, B., Lazowska, E.: SQLShare: results from a multi-year SQL-as-a-service experiment. In: Proceedings of SIGMOD 2016, pp. 281–293. ACM (2016)

36. Korhonen, T.: Potential maximal cliques parameterized by edge clique cover. CoRR abs/1912.10989 (2019). https://arxiv.org/abs/1912.10989

37. Korhonen, T., Berg, J., Järvisalo, M.: Solving graph problems via potential maximal cliques: an experimental evaluation of the Bouchitté-Todinca algorithm. ACM J. Exp. Algorithmics **24**(1), 1.9:1–1.9:19 (2019)

38. Korimort, T.: Heuristic hypertree decomposition. Ph.D. thesis, Vienna University of Technology (2003)

39. Laborie, P., Rogerie, J., Shaw, P., Vilím, P.: IBM ILOG CP optimizer for scheduling - 20+ years of scheduling with constraints at IBM/ILOG. Constraints Int. J. **23**(2), 210–250 (2018)

40. Lalou, M., Habbas, Z., Amroun, K.: Solving hypertree structured CSP: sequential and parallel approaches. In: Proceedings of RCRA@AI*IA 2009, CEUR Workshop Proceedings, vol. 589 (2009). CEUR-WS.org

41. Lam, E., Hentenryck, P.V., Kilby, P.: Joint vehicle and crew routing and scheduling. Transp. Sci. **54**(2), 488–511 (2020)
42. Leis, V., et al.: Query optimization through the looking glass, and what we found running the join order benchmark. VLDB J. **27**(5), 643–668 (2018)
43. Leone, N., Mazzitelli, A., Scarcello, F.: Cost-based query decompositions. In: Proceedings of SEBD 2002, pp. 390–403 (2002)
44. Lovász, L.: On the ratio of optimal integral and fractional covers. Discret. Math. **13**(4), 383–390 (1975)
45. Musliu, N., Schutt, A., Stuckey, P.J.: Solver independent rotating workforce scheduling. In: van Hoeve, W.-J. (ed.) CPAIOR 2018. LNCS, vol. 10848, pp. 429–445. Springer, Cham (2018). https://doi.org/10.1007/978-3-319-93031-2_31
46. Pottinger, R., Halevy, A.Y.: Minicon: a scalable algorithm for answering queries using views. VLDB J. **10**(2–3), 182–198 (2001). https://doi.org/10.1007/s007780100048
47. Raedt, L.D., Guns, T., Nijssen, S.: Constraint programming for data mining and machine learning. In: Proceedings of AAAI 2010. AAAI Press (2010)
48. Rossi, F., Van Beek, P., Walsh, T.: Handbook of Constraint Programming. Elsevier, Amsterdam (2006)
49. Samer, M., Veith, H.: Encoding treewidth into SAT. In: Kullmann, O. (ed.) SAT 2009. LNCS, vol. 5584, pp. 45–50. Springer, Heidelberg (2009). https://doi.org/10.1007/978-3-642-02777-2_6
50. Schidler, A., Szeider, S.: Computing optimal hypertree decompositions. In: Proceedings of ALENEX 2020, pp. 1–11. SIAM (2020)
51. Shaw, P.: Using constraint programming and local search methods to solve vehicle routing problems. In: Maher, M., Puget, J.-F. (eds.) CP 1998. LNCS, vol. 1520, pp. 417–431. Springer, Heidelberg (1998). https://doi.org/10.1007/3-540-49481-2_30
52. Transaction Processing Performance Council (TPC): TPC-H decision support benchmark (2014). http://www.tpc.org/tpch/default.asp
53. Tsang, E.: Foundations of Constraint Satisfaction. Academic Press Limited, Cambridge (1993)
54. Verhaeghe, H., Nijssen, S., Pesant, G., Quimper, C., Schaus, P.: Learning optimal decision trees using constraint programming. In: Proceedings of BNAIC 2019. CEUR Workshop Proceedings, vol. 2491(2019). CEUR-WS.org
55. Yannakakis, M.: Algorithms for acyclic database schemes. In: Proceedings of VLDB 1981, pp. 82–94 (1981)

# Restarting Algorithms: Sometimes There Is Free Lunch

Sebastian Pokutta[1,2]([⊠])

[1] Technische Universität Berlin, Berlin, Germany
pokutta@zib.de
[2] Zuse Institute Berlin, Berlin, Germany

**Abstract.** In this overview article we will consider the deliberate restarting of algorithms, a meta technique, in order to improve the algorithm's performance, e.g., convergence rates or approximation guarantees. One of the major advantages is that restarts are *relatively* black box, not requiring any (significant) changes to the base algorithm that is restarted or the underlying argument, while leading to *potentially significant* improvements, e.g., from sublinear to linear rates of convergence. Restarts are widely used in different fields and have become a powerful tool to leverage additional information that has not been directly incorporated in the base algorithm or argument. We will review restarts in various settings from continuous optimization, discrete optimization, and submodular function maximization where they have delivered impressive results.

**Keywords:** Restarts · Convex optimization · Discrete optimization · Submodular optimization

## 1 Introduction

Restarts are a powerful meta technique to improve the behavior of algorithms. The basic idea is to deliberately restart some base algorithm, often with changed input parameters, to speed-up convergence, improve approximation guarantees, reduce number of calls to expensive subroutines and many more, often leading to provably better guarantees as well as significantly improved real-world computational performance. In actuality this comes down to running a given algorithm with a given set of inputs for some number of iterations, then changing the input parameters usually as a function of the output, and finally restarting the algorithm with new input parameters; rinse and repeat.

One appealing aspect of restarts is that they are relatively black-box, requiring only little to no knowledge of the to-be-restarted base algorithm except for the guarantee of the base algorithm that is then amplified by means of restarts. The reason why restarts often work, i.e., improve the behavior of the base algorithm is that some structural property of the problem under consideration is not explicitly catered for in the base algorithm, e.g., the base algorithm might work

© Springer Nature Switzerland AG 2020
E. Hebrard and N. Musliu (Eds.): CPAIOR 2020, LNCS 12296, pp. 22–38, 2020.
https://doi.org/10.1007/978-3-030-58942-4_2

for general convex functions, however the function under consideration might be strongly convex or sharp. Restarts cater to this additional problem structure and are in particular useful when we want to incorporate data-dependent parameters. In fact, for several cases of interest the only known way to incorporate that additional structure is via restarts often pointing out a missing piece in our understanding.

On the downside, restarts often explicitly depend on parameters arising from the additional structure under consideration and obtained guarantees are off by some constant factor or even log factor. The former can be often remedied with adaptive or scheduled restarts (see e.g., [23,39]) albeit with some minor cost. This way we can obtain fully adaptive algorithms that adapt to additional structure without knowing the accompanying parameters explicitly. The latter shortcoming is inherent to restart scheme as due to their black box nature additional structural information might not be incorporated perfectly.

Restarts have been widely used in many areas and fields and we will review some of these applications below to provide context. We would like to stress that references will be incomplete and biased; please refer also to the references cited therein.

*SAT Solving and Constraint Programming.* Restarts are ubiquitous in SAT Solving and Constraint Programming to, e.g., explore different parts of the search space. Also after new clauses have been learned, these clauses are often added back to the formulation and then the solver is restarted. This can lead to dramatic overall performance improvements for practical solving; see e.g., [7,25] and references contained therein.

*Global Optimization.* Another important area where restarts are used is global optimization. Often applied to non-convex problems, the hope is that with randomized restarts different local optima can be explored, ideally one of those being a global one; see e.g., [24] and their references.

*Integer Programming.* Modern integer programming solvers use restarts in many different ways, several of which have been inspired by SAT solving and Constraint Programming. In fact, Integer Programming solvers can be quite competitive for pseudo-Boolean problems [6]. A relatively recent approach [4] is *clairvoyant restarts* based on online tree-size estimation that can significantly improve solving behavior.

Most of the restart techniques mentioned above, while very important, come without strong guarantees. In this article, we are more interested in cases, where provably strong guarantees can be obtained that also translate into real-world computational advantages. In the following, we will restrict the discussion to three examples from convex optimization, discrete optimization, and submodular function maximization. However, before we consider those, we would like to mention a two related areas where restarts have had a great impact not just from a computational point of view but also to establish new theoretical guarantees, but that are unfortunately beyond the scope of this overview.

*Variance Reduction via Restarts.* Usually when we consider stochastic convex optimization problems where the function is given as a general expectation and we would like to use first-order methods for solving the stochastic problem, we cannot expect a convergence rate better than $O(1/\sqrt{t})$ under usual assumptions, where $t$ is the number of stochastic gradient evaluations. However, it turns out that if we consider so-called finite sum problems, a problem class quite common in machine learning, where the expectation is actually a finite sum and some mild additional assumptions are satisfied, then we can obtain a linear rate of convergence by means of variance reduction. This is an exponential improvement in convergence rate. Variance reduction techniques replace the stochastic gradient which is an unbiased estimator of the true gradient with a different, lower variance, unbiased estimator that is formed with the help of a reference point obtained from an earlier iterate. This reference point is then periodically reset via a restart scheme. Important algorithms here are for example Stochastic Variance Reduced Gradient Descent (SVRG) [26] and its numerous variants, such as e.g., the Stochastic Variance Reduced Frank-Wolfe algorithm (SVRFW) [22].

*Acceleration in Convex Optimization.* Restarts have been heavily used in convex optimization both for improving convergence behavior via restarts in real-world computations (see e.g., [37]) but also as part of formal arguments to establish accelerated convergence rates and design provably faster algorithms. As the literature is particularly deep, we will sample only a few of those works in the context of first-order methods here that we are particularly familiar with; we provide further references in the sections to come. For example restarts have been used in [1] to provide an alternative explanation of Nesterov's acceleration as arising from the coupling mirror descent and gradient descent. In [39] it has been shown how restarts can be leveraged to obtain improved rates as the sharpness of the function (roughly speaking how fast the function curves around its minima) increases and these restart schemes have been also successfully carried over to the conditional gradients case in [27]. Restarts have been also used to establish dimension-independent local acceleration for conditional gradients [16] by means of coupling the Away-step Frank-Wolfe algorithm with an accelerated method. As we will see later in the context of submodular maximization, restarts can be also used to reduce the number of calls to expensive oracles. This have been extensively used for lazification of otherwise expensive algorithms in [11,12] leading to several orders of speed-up in actual computations while maintaining worst-case guarantees identical to those of the original algorithms and in [28] a so-called optimal method based on lazification has been derived. Very recently, in [23] a new adaptive restart scheme has been presented that does not require any knowledge of otherwise inaccessible parameters and its efficacy for saddle point problems has been demonstrated.

## Outline

In Sect. 2 we consider restart examples from convex optimization and in Sect. 3 we consider examples from discrete optimization. Finally we consider submodular function maximization in Sect. 4. We keep technicalities to a bare minimum, sometimes simplifying arguments for the sake of exposition. We provide references though with the full argument, for the interested reader.

## 2  Smooth Convex Optimization

Our first examples come from smooth convex optimization. As often, the examples here are (arguably) the cleanest ones. We briefly recall some basic notions:

**Definition 1 (Convexity).** *Let $f : \mathbb{R}^n \to \mathbb{R}$ be a differentiable function. Then $f$ is* convex, *if for all $x, y$ it holds:*

$$f(y) - f(x) \geq \langle \nabla f(x), y - x \rangle .$$

*In particular, all local mimima of $f$ are global minima of $f$.*

**Definition 2 (Strong Convexity).** *Let $f : \mathbb{R}^n \to \mathbb{R}$ be a differentiable convex function. Then $f$ is $\mu$-strongly convex (with $\mu > 0$), if for all $x, y$ it holds:*

$$f(y) - f(x) \geq \langle \nabla f(x), y - x \rangle + \frac{\mu}{2} \|y - x\|^2.$$

**Definition 3 (Smoothness).** *Let $f : \mathbb{R}^n \to \mathbb{R}$ be a differentiable function. Then $f$ is $L$-smooth (with $L > 0$), if for all $x, y$ it holds:*

$$f(y) - f(x) \leq \langle \nabla f(x), y - x \rangle + \frac{L}{2} \|y - x\|^2.$$

In the following let $x^* \in X^* = \operatorname{argmin} f(x)$ denote an optimal solution from the set of optimal solutions $X^*$. Choosing $x = x^*$ and applying the definition of strong convexity (Definition 2) we immediately obtain:

$$f(y) - f(x^*) \geq \langle \nabla f(x^*), y - x^* \rangle + \frac{\mu}{2} \|y - x^*\|^2 \geq \frac{\mu}{2} \|y - x^*\|^2, \tag{1}$$

where the last inequality follows from $\langle \nabla f(x^*), y - x^* \rangle \geq 0$ by first-order optimality of $x^*$ for $\min f(x)$, i.e., the primal gap upper bounds the distance to the optimal solution. This also implies that the optimal solution $x^*$ is unique.

**Smooth Convex to Smooth Strongly Convex: The Basic Case.** Let $f : \mathbb{R}^n \to \mathbb{R}$ be an $L$-smooth convex function. Then using *gradient descent*, updating iterates $x_t$ according to $x_{t+1} \leftarrow x_t - \frac{1}{L} \nabla f(x_t)$, yields the following standard guarantee, see e.g., [21,29,33,35].

**Proposition 1 (Convergence of gradient descent: smooth convex case).**
*Let $f : \mathbb{R}^n \to \mathbb{R}$ be a smooth convex function and $x_0 \in \mathbb{R}^n$ and $x^* \in X^*$. Then gradient descent generates a sequence of iterates satisfying*

$$f(x_t) - f(x^*) \leq \frac{L\|x_0 - x^*\|^2}{t}. \tag{2}$$

Now suppose we additionally know that the function $f$ is $\mu$-strongly convex. Usually, we would expect a linear rate of convergence in this case, i.e., to reach an additive error of $\varepsilon$, we would need at most $T \leq \frac{L}{\mu} \log \frac{f(x_0)-f(x^*)}{\varepsilon}$ iterations. However, rather than reproving the convergence rate (which is quite straightforward in this case) we want to reuse the guarantee in Proposition 1 as a black box and the $\mu$-strong convexity of $f$. We will use the simple restart scheme given in Algorithm 1: in restart phase $\ell$ we run a given base algorithm $\mathcal{A}$ for a fixed number of iterations $T_\ell$ on the iterate $x^{\ell-1}$ output in the previous iteration:

---

**Algorithm 1.** Simple restart scheme
___
**Input:** Initial point $x_0 \in \mathbb{R}^n$, base algorithm $\mathcal{A}$, iteration counts $(T_\ell)$.
**Output:** Iterates $x^1, \ldots, x^K \in \mathbb{R}^n$.
 1: **for** $\ell = 1$ **to** $K$ **do**
 2:   $x^\ell \leftarrow \mathcal{A}(f, x^{\ell-1}, T_\ell)$        {run base algorithm for $T_\ell$ iterations}
 3: **end for**

---

A priori, it is unclear whether the restart scheme in Algorithm 1 is doing anything useful, in fact even convergence might not be immediate as we in principle could undo work that we did in a preceding restart phase. Also note that when restarting vanilla gradient descent with a fixed step size of $\frac{1}{L}$ as we do here the final restarted algorithm is identical to vanilla gradient descent, i.e., the restarts do not change the base algorithm. This might seem nonsensical and we will get back to this soon; the reader can safely ignore this for now.

In order to analyze our restart scheme we first chain together Inequalities (2) and (1) and obtain:

$$f(x_t) - f(x^*) \leq \frac{L\|x_0 - x^*\|^2}{t} \leq \frac{2L}{\mu} \frac{f(x_0) - f(x^*)}{t}. \tag{3}$$

This chaining together of two error bounds is at the core of most restart arguments and we will see several variants of this. Next we estimate how long we need to run the base method, using Inequality (3) to halve the primal gap from some given starting point $x_0$ (this will be the point from which we are going to restart the base method), i.e., we want to find $t$ such that

$$f(x_t) - f(x^*) \leq \frac{2L}{\mu} \frac{f(x_0) - f(x^*)}{t} \leq \frac{f(x_0) - f(x^*)}{2},$$

which implies that it suffices to run gradient descent for $T_\ell \doteq \left\lceil \frac{4L}{\mu} \right\rceil$ steps for all $\ell = 1, \ldots, K$ to halve a given primal bound as there is no dependency on the state of the algorithm in this case. Now, in order to reach $f(x_T) - f(x^*) \leq \varepsilon$, we have to halve $f(x_0) - f(x^*)$ at most $K \doteq \left\lceil \log \frac{f(x_0) - f(x^*)}{\varepsilon} \right\rceil$ times and each of the halving can be accomplished in at most $\left\lceil \frac{4L}{\mu} \right\rceil$ gradient descent steps. All in all we obtain that after at most

$$T \geq \sum_{\ell=1}^{K} T_\ell = K \cdot T_1 = \left\lceil \frac{4L}{\mu} \right\rceil \left\lceil \log \frac{f(x_0) - f(x^*)}{\varepsilon} \right\rceil \tag{4}$$

gradient descent steps we have obtained a solution $f(x_T) - f(x^*) = f(x^K) - f(x^*) \leq \varepsilon$. With this we have obtained the desired convergence rate. Note that the iterate bound in Inequality (4) is optimal for vanilla gradient descent up to a constant factor of 4; see e.g., [21,29,33].

In the particular case from above it is also important to observe that our base algorithm gradient descent is essentially memoryless. In fact, the restarts do not 'reset' anything in this particular case and so we have also indirectly proven that gradient descent *without* restarts will converge with the rate from Inequality (4). This is particular to this example though and will be different in our next one. Also, note that a direct estimation would have yielded the same rate up to the factor 4 discussed above.

**Smooth Convex to Smooth Strongly Convex: The Accelerated Case.** While the rate from Inequality (4) is essentially optimal for vanilla gradient descent it is known that (vanilla) gradient descent itself is not optimal for smooth and strongly convex functions and also Proposition 1 is not optimal for smooth and (non-strongly) convex functions. In fact Nesterov showed in [36] that for smooth and (non-strongly) convex functions a quadratic improvement can be obtained; a phenomenon commonly referred to as *acceleration*:

**Proposition 2 (Convergence of accelerated gradient descent).** *Let $f : \mathbb{R}^n \to \mathbb{R}$ be an $L$-smooth convex function and $x_0 \in \mathbb{R}^n$ and $x^* \in X^*$. Then accelerated gradient descent generates a sequence of iterates satisfying*

$$f(x_t) - f(x^*) \leq \frac{cL\|x_0 - x^*\|^2}{t^2},$$

*for some constant $c > 0$.*

Again, we could try to directly prove a better rate via acceleration for the smooth and strongly case (which is non-trivial this time) or, as before, invoke our restart scheme in Algorithm 1 in a black-box fashion, which is what we will do here. As before we will use an analog of Inequality (3) to estimate how long it takes to halve the primal gap, i.e., we want to find $t$ such that

$$f(x_t) - f(x^*) \leq \frac{2cL}{\mu} \frac{f(x_0) - f(x^*)}{t^2} \leq \frac{f(x_0) - f(x^*)}{2},$$

which implies that it suffices to run accelerated gradient descent for $T_\ell \doteq \left\lceil \sqrt{\frac{4cL}{\mu}} \right\rceil$ steps for all $\ell = 1, \ldots, K$ to halve a given primal gap. With the same reasoning as above we need to halve the primal gap at most $K \doteq \left\lceil \log \frac{f(x_0)-f(x^*)}{\varepsilon} \right\rceil$ times to reach an additive error of $\varepsilon$. Putting everything together we obtain that after at most

$$T \geq \sum_{\ell=1}^{K} T_\ell = K \cdot T_1 = \left\lceil \sqrt{\frac{4cL}{\mu}} \right\rceil \left\lceil \log \frac{f(x_0) - f(x^*)}{\varepsilon} \right\rceil \tag{5}$$

accelerated gradient descent steps we have obtained a solution $f(x_T) - f(x^*) = f(x^K) - f(x^*) \leq \varepsilon$. Note that the iterate bound in Inequality (5) is optimal for strongly convex and smooth functions (up to a constant factor). In contrast to the unaccelerated case, this time the restart actually 'resets' the base algorithm as accelerated gradient descent uses a specific step size strategy that is then reset.

*Remark 1.* Sometimes it is also possible to go backwards. Here we recover the optimal base algorithm for the smooth and (non-strongly) convex case from the strongly convex one. The argument is due to [34] (we follow the variant in [42]). Suppose we know an optimal algorithm for the strongly convex and smooth case that ensures $f(x_T) - f(x^*) \leq \varepsilon$ after $O\left(\sqrt{\frac{L}{\mu}} \log \frac{f(x_0)-f(x^*)}{\varepsilon}\right)$ iterations. Now consider a smooth and convex function $f$ and an initial iterate $x_0$ together with some upper bound $D$ on the distance to some optimal solution, i.e., $\|x_0 - x^*\| \leq D$. Given an accuracy $\varepsilon > 0$, we consider the auxiliary function

$$f_\varepsilon(x) \doteq f(x) + \frac{\varepsilon}{2D^2} \|x - x_0\|^2,$$

which is $\left(L + \frac{\varepsilon}{2D^2}\right)$-smooth and $\frac{\varepsilon}{2D^2}$-strongly convex. It can be easily seen that

$$f(x) - f(x^*) \leq f_\varepsilon(x) - f_\varepsilon(x^*) + \frac{\varepsilon}{2},$$

so that finding an $\varepsilon/2$-optimal solution to $\min f_\varepsilon$ provides an $\varepsilon$-optimal solution to $\min f$. We can now run the purported optimal method on the smooth and strongly convex function $f_\varepsilon$ to compute an $\varepsilon/2$-optimal solution to $\min f_\varepsilon$, which we obtain after:

$$O\left(\sqrt{\frac{L + \frac{\varepsilon}{2D^2}}{\frac{\varepsilon}{2D^2}}} \log 2 \frac{f_\varepsilon(x_0) - f_\varepsilon(x^*)}{\varepsilon}\right) \leq O\left(\sqrt{\frac{2LD^2 + \varepsilon}{\varepsilon}} \log \frac{(L + \varepsilon)D^2}{\varepsilon}\right),$$

iterations, where we used $f_\varepsilon(x_0) - f_\varepsilon(x^*) \leq \frac{(L+\varepsilon)D^2}{2}$. Finally note, ignoring the log factor, $\sqrt{\frac{2LD^2+\varepsilon}{\varepsilon}} \leq T \Leftrightarrow \frac{2LD^2+\varepsilon}{T^2} \leq \varepsilon$, which is the bound from Proposition 2.

The approach used in this section to obtain better rates of convergence under stronger assumptions by means of the simple restart scheme in Algorithm 1 works in much broader settings in convex optimization (including the constrained case).

For example it can be used to improve the $O(1/\sqrt{t})$-rate for general non-smooth convex functions via sub-gradient descent into the $O(1/t)$-rate for the non-smooth strongly convex case. Here the base rate is $f(x_t) - f(x^*) \leq \frac{G\|x_0 - x^*\|}{\sqrt{t}}$, where $G$ is a bound on the norm of the subgradients. We obtain the restart inequality chain (analog to Inequality (3)):

$$f(x_t) - f(x^*) \leq \frac{G\|x_0 - x^*\|}{\sqrt{t}} \leq \frac{G}{\sqrt{t}} \sqrt{\frac{f(x_0) - f(x^*)}{\mu}},$$

and halving the primal gap takes at most $\frac{4G^2}{\mu(f(x_0)-f(x^*))}$ iterations. Following the argumentation from above, we then arrive that the total number of required subgradient descent iterations using Algorithm 1 to ensure $f(x_t) - f(x^*) \leq \varepsilon$ is at most $t \geq \frac{8G^2}{\varepsilon\mu}$ for the non-smooth but $\mu$-strongly convex case, which is optimal up to constant factors.

*Related Approaches.* In a similar way we can incorporate additional information obtained e.g., from so-called *Hölder(ian) Error Bounds* or *sharpness* (see, e.g., [9, 10] and references contained therein for an overview). The careful reader might have observed that the restart scheme in Algorithm 1 requires knowledge of the parameter $\mu$. While this could be acceptable in the strongly convex case, for more complex schemes to leverage, e.g., sharpness, this is unacceptable as the required parameters are hard to estimate and generally inaccessible. This however, can be remedied in the case of sharpness, at the cost of an extra $O(\log^2)$-factor in the rates, via *scheduled restarts* as done in [39] that do not require sharpness parameters as input or when an error bound (of similar convergence rate) is available as in the case of conditional gradients [27]; see also [23] for a very recent adaptive restart scheme using error bounds estimators.

## 3 Discrete Optimization

In this section we consider a prominent example from integer programming: optimization via augmentation, i.e., optimizing by iteratively improving the current solution.

We consider the problem:

$$\max \{cx \mid x \in P \cap \mathbb{Z}^n\}, \tag{6}$$

where $P \subseteq \mathbb{R}^n$ is a polytope and $c \in \mathbb{Z}^n$.

To simplify the exposition we assume that $P \subseteq [0, 1]^n$ and $c \geq 0$ (the latter is without loss of generality by flipping coordinates), however the arguments here generalize to the general integer programming case. Suppose further that we can compute *improving solutions*, i.e., given $c$ and a solution $x_0$, we can compute a new solution $x$, so that $c(x - x_0) > 0$ if $x_0$ was not already optimal; such a step (Line 3 in Algorithm 2) is called an *augmentation step*. Then a trivial and inefficient strategy is Algorithm 2, where we continue improving the solution

---

**Algorithm 2.** Augmentation
***

**Input:** Feasible solution $x^0$ and objective $c \in \mathbb{Z}_+^n$
**Output:** Optimal solution of max $\{cx \mid x \in P \cap \{0,1\}^n\}$
1: $\tilde{x} \leftarrow x^0$
2: **repeat**
3:     **compute** $x \in P$ integral with $c(x - \tilde{x}) > 0$ and set $\tilde{x} \leftarrow x$     {improve solution}
4: **until** no improving solution exists
5: **return** $\tilde{x}$     {return optimal solution}

---

until we have reached the optimum. It is not too hard to see that Algorithm 2 can take up to $2^n$ steps, essentially enumerating all feasible solutions to reach the optimal solution; simply consider the cube $P = [0,1]^n$ and an objective $c$ with powers of 2 as entries.

**Bit Scaling.** We will now show that we can do significantly better by restarting Algorithm 2, so that we obtain a number of augmentation steps of $O(n \log \|c\|_\infty)$, where $\|c\|_\infty \doteq \max_{i \in [n]} c_i$. This is an exponential improvement over base algorithm and the restart scheme, called *bit scaling*, is due to [41] (see also [17,20]). It crucially relies on the following insight: Suppose we decompose our objective $c = 2c_1 + c_0$ with $c_0 \in \{0,1\}^n$ (note this decomposition is unique) and we have already obtained some solution $x_0 \in P \cap \{0,1\}^n$ that is optimal for max $\{c_1 x \mid x \in P \cap \mathbb{Z}^n\}$, then we have for all $x \in P \cap \{0,1\}^n$:

$$c(x - x_0) = 2\underbrace{c_1(x - x_0)}_{\leq 0} + c_0(x - x_0) \leq n, \tag{7}$$

by the optimality of $x_0$ for $c_1$ and $c_0, x, x_0 \in \{0,1\}^n$. Hence starting from $x_0$, for objective $c$, there are at most $n$ augmentation steps to be performed with Algorithm 2 to reach an optimal solution for $c$. Equipped with Inequality (7) the following strategy emerges: slice by the objective $c$ according to its bit representation and then successively optimize with respect to the starting point from a previous slice. We first present the formal bit scaling restart scheme in Algorithm 3, where $\mathcal{A}$ denotes Algorithm 2.

Next, we will show that restart scheme from Algorithm 3 requires at most $O(n \log \|c\|_\infty)$ augmentation steps (Line 3 in Algorithm 2) to solve Problem (6). First observe, that by construction and the stopping criterion in Line 5 of Algorithm 3 it is clear that we call $\mathcal{A}$ in Line 3 at most $\lceil \log C \rceil$ times. Next, we bound the number of augmentation steps in Line 3 executed within algorithm $\mathcal{A}$. To this end, let $\tilde{x}$ and $\mu$ denote the input to $\mathcal{A}$. In the first iteration $c^\mu \in \{0,1\}^n$, so that $\mathcal{A}$ can perform at most $n$ augmentation steps. For later iterations observe that $\tilde{x}$ was optimal for $c^{2\mu} = \lfloor c/(2\mu) \rfloor$. Moreover, we have $c^\mu = \lfloor c/\mu \rfloor = 2c^{2\mu} + c_0$, where $c_0 \in \{0,1\}^n$ as before. Via Inequality (7) we obtain for all feasible solutions $x \in P \cap \mathbb{Z}^n$:

$$c^\mu(x - \tilde{x}) = 2c^{2\mu}(x - \tilde{x}) + c_0(x - \tilde{x}) \leq n,$$

---

**Algorithm 3.** Bit Scaling

---

**Input:** Feasible solution $x^0$
**Output:** Optimal solution to $\max \{cx \mid x \in P \cap \mathbb{Z}^n\}$
1: $C \leftarrow \|c\|_\infty + 1$, $\mu \leftarrow 2^{\lceil \log C \rceil}$, $\tilde{x} \leftarrow x^0$, $c^\mu \leftarrow \lfloor c/\mu \rfloor$    {initialization}
2: **repeat**
3:    **Call** $\tilde{x} \leftarrow \mathcal{A}(\tilde{x}, c^\mu)$
4:    $\mu \leftarrow \mu/2$, $c^\mu \leftarrow \lfloor c/\mu \rfloor$
5: **until** $\mu < 1$
6: **return** $\tilde{x}$    {return optimal solution}

---

which holds in particular for the optimal solution $x^*$ to Problem (6). As each augmentation step reduces the primal gap $c^\mu(x - \tilde{x})$ by at least 1, we can perform at most $n$ augmentation steps. This completes the argument.

**Geometric Scaling.** The restart scheme in Algorithm 3 essentially restarted via bit-scaling the objective function, hence the name. We will now present a more versatile restart scheme that is due to [40] (see also [30] for a comparison and worst-case examples), which essentially works by restarting a regularization of our objective $c$. For comparability we also consider Problem (6) here, however the approach is much more general, e.g., allowing for general integer programming problems and with modifications even convex programming problems over integers.

Again, we will modify the considered objective function $c$ in each restart. Given the original linear objective $c$, we will consider:

$$c^\mu(x, \tilde{x}) = c(x - \tilde{x}) - \mu\|x - \tilde{x}\|_1.$$

Note that $c^\mu(x, \tilde{x})$ is a linear function in $x \in \{0,1\}^n$ for a given $\tilde{x} \in \{0,1\}^n$. In particular we can call Algorithm 2 with objective $c^\mu(\cdot, \cdot)$ and starting point $\tilde{x}$. The restart scheme works as follows: For a given $\mu$ we call Algorithm 2 with objective $c^\mu(\cdot, \cdot)$ and starting point $\tilde{x}$. Then we halve $\mu$ and repeat.

As in the bit-scaling case, the key is to estimate the number of augmentation steps performed in such a call. To this end let $x_0$ be returned by Algorithm 2 for a given $\mu$ and starting point $\tilde{x}$. Then

$$c^\mu(x, x_0) = c(x - x_0) - \mu\|x - x_0\|_1 \leq 0,$$

holds for all $x \in P \cap \mathbb{Z}^n$ and in particular for the optimal solution $x^*$; this is simply the negation of the improvement condition. Now let $x'$ be any iterate in the following call to Algorithm 2 for which an augmentation step is performed with objective $c^{\mu/2}(\cdot, \cdot)$ and starting point $x_0$, i.e., there exists $x^+$ so that

$$c^{\mu/2}(x^+, x') = c(x^+ - x') - \mu/2\|x^+ - x'\|_1 > 0.$$

We can now combine these two inequalities, substituting $x \leftarrow x^*$, to obtain

$$2\frac{c(x^+ - x')}{\|x^+ - x'\|_1} > \mu \geq \frac{c(x^* - x_0)}{\|x^* - x_0\|_1},$$

which implies

$$c(x^+ - x') \geq \frac{1}{2} \frac{\|x^+ - x'\|_1}{\|x^* - x_0\|_1} c(x^* - x_0) \geq \frac{1}{2n} c(x^* - x_0),$$

where $\|x^+ - x'\|_1 \geq 1$ as the iterates are not identical and $\|x^* - x_0\|_1 \leq n$ as $x^*, x_0 \in P \subseteq [0,1]^n$. As such each augmentation step recovers at least a $\frac{1}{2n}$-fraction of the primal gap $c(x^* - x_0)$ and therefore we can do at most $2n$ such iterations before the condition in Line 3 has to be violated. With this we can formulate the geometric scaling restart scheme in Algorithm 4. The analysis now is basically identical to the one as for Algorithm 3, however this time we have $O(\log n \|c\|_\infty)$ restarts, leading to an overall number of augmentation steps of $O(n \log n \|c\|_\infty)$, which can be further improved to $O(n \log \|c\|_\infty)$, matching that of bit-scaling, with the simple observation in [30].

---

**Algorithm 4.** Geometric Scaling

---

**Input:** Feasible solution $x^0$
**Output:** Optimal solution of $\max \{cx \mid x \in P \cap \mathbb{Z}^n\}$
1:  $C \leftarrow \|c\|_\infty + 1,\ \mu \leftarrow nC,\ \tilde{x} \leftarrow x^0,\ c^\mu(x,y) \doteq c(x-y) - \mu \|x-y\|_1$.    {initialization}
2: **repeat**
3:     **Call** $\tilde{x} \leftarrow \mathcal{A}(\tilde{x}, c^\mu)$
4:     $\mu \leftarrow \mu/2$
5: **until** $\mu < 1$
6: **return** $\tilde{x}$    {return optimal solution}

---

*Related Approaches.* Chvátal-Gomory cutting planes, introduced by Chvátal in [13], are an important tool in integer programming to approximate the integral hull $\text{conv}(P \cap \mathbb{Z}^n)$ by means of successively strengthening an initial relaxation $P$ with $\text{conv}(P \cap \mathbb{Z}^n) \subseteq P$. This is done by adding new inequalities valid for $\text{conv}(P \cap \mathbb{Z}^n)$ cutting off chunks of $P$ in each round. A key question is how many rounds of such strengthenings are needed until we recover $\text{conv}(P \cap \mathbb{Z}^n)$. In [14] it was shown that in general the number of rounds can be arbitrarily large. It was then shown in [8] via a restart argument that for the important case of polytopes contained in $[0,1]^n$ the number of rounds can be upper bounded by $O(n^3 \log n)$. The key here is to use basic bounds on the number of rounds, e.g., from [14], first for inequalities with some maximum absolute entry $c$, then doubling up $c$ to $2c$, and restarting the argument. This bound was further improved in [18] to $O(n^2 \log n)$ by interleaving two restart arguments, one multiplicative (e.g., doubling) over the maximum absolute entry $c$ and one additive (e.g., adding a constant) over the dimension, which matches the lower bound of $\Omega(n^2)$ of [38] up to a log factor; closing this gap remains an open problem. As mentioned in the context of the scheduled restarts of [39], it might be possible that the additional log factor is due to the restart schemes itself and removing it might require a different proof altogether.

Another important application is the approximate Carathéodory problem, where we want to approximate $x^0 \in P$, where $P$ is a polytope, by means of a sparse convex combination $x$ of vertices of $P$, so that $\|x_0 - x\| \leq \varepsilon$ for some norm $\|\cdot\|$ and target accuracy $\varepsilon$. In general it is known that this can be done with a convex combination of $O(1/\varepsilon^2)$ vertices. However, it turns out as shown in [31] that whenever $x_0$ lies deep inside the polytope $P$, i.e., we can fit a ball around $x_0$ with some radius $r$ into $P$ as well, then we can exponentially improve this bound via restarts to $O(\frac{1}{r^2}\log\frac{1}{\varepsilon})$. This restart argument here is particularly nice. We run the original $O(1/\varepsilon^2)$-algorithm down to some fixed accuracy and obtain some approximation $\tilde{x}$, then scale-up the feasible region by a factor of 2, and restart the $O(1/\varepsilon^2)$-algorithm on the residual $x_0 - \tilde{x}$ and repeat. The argument in [31] relies on mirror descent as underlying optimization routine. More recently, it was shown in [15] that the restarts can be removed and adaptive bounds for more complex cases can be obtained by using conditional gradients as base optimization algorithm, which automatically adapts to sharpness (and optima in the interior) [27,43].

# 4 Submodular Function Maximization

We now turn our attention to submodular function maximization. Submodularity captures the *diminishing returns* property and is widely used in optimization and machine learning. In particular, we will consider the basic but important setup of maximizing a monotone, non-negative, submodular function subject to a single cardinality constraint. To this end we will briefly repeat necessary notions. A set function $g : 2^V \to \mathbb{R}_+$ is *submodular* if and only if for any $e \in V$ and $A \subseteq B \subseteq V \setminus \{e\}$ we have $g_A(e) \geq g_B(e)$, where $g_A(e) \doteq g(A + e) - g(A)$ denotes the *marginal gain* of $e$ w.r.t. $A$ and $A + e \doteq A \cup \{e\}$, slightly abusing notation. The submodular function $g$ is *monotone* if for all $A \subseteq B \subseteq V$ it holds $g(A) \leq g(B)$ and *non-negative* if $g(A) \geq 0$ for all $A \subseteq V$.

Given a monotone, non-negative submodular function $g$ over ground set $V$ of size $n$ and a budget $k$, we consider the problem

$$\max_{S \subseteq V, |S| \leq k} g(S) \tag{8}$$

It is well known that solving Problem (8) exactly is NP-hard under the value oracle model, however the greedy algorithm (Algorithm 5) that in each iteration adds the element that maximizes the marginal gain yields a $(1 - 1/e)$-approximate solution $S^+ \subseteq V$ with $|S^+| \leq k$, i.e., $g(S^+) \geq (1 - 1/e)\,g(S^*)$, where $S^* = \operatorname{argmax}_{S \subseteq V, |S| \leq k} g(S) \subseteq V$ is an optimal solution to Problem (8) and e denotes the Euler constant (see [19,32]).

The proof of the approximation guarantee of $1 - 1/e$ is based on the insight that in each iteration it holds:

$$g(S^*) - g(S^+) \leq k \cdot \max_{e \in V} g_{S^+}(e). \tag{9}$$

---

**Algorithm 5.** Greedy Algorithm

---

**Input:** Ground set $V$ of size $n$, budget $k$, and monotone, non-negative, submodular
  function $g : 2^V \to \mathbb{R}_+$.
**Output:** feasible set $S^+$ with $|S^+| \leq k$.
 1: $S^+ \leftarrow \emptyset$
 2: **while** $|S^+| \leq k$ **do**
 3:    $e \leftarrow \mathrm{argmax}_{e \in V \setminus S^+} g_{S^+}(e)$
 4:    $S^+ \leftarrow S^+ + e$
 5: **end while**

---

To see that Inequality (9) holds, let $S^* = \{e_1, \ldots, e_k\}$, then

$$g(S^*) \leq g(S^* \cup S^+) = g(S^+) + \sum_{i=1}^{k} g_{S^+ \cup \{e_1, \ldots, e_{i-1}\}}(e_i)$$

$$\leq g(S^+) + \sum_{i=1}^{k} g_{S^+}(e_i) \leq g(S^+) + k \max_{e \in V} g_{S^+}(e),$$

where the first inequality follows from monotonicity, the equation follows from
the definition of $g_S(v)$, the second inequality from submodularity, and the last
inequality from taking the maximizer.

With Inequality (9) the proof of the $(1 - 1/e)$-approximation is immediate.
In each iteration the greedy element we add satisfies $\max_{e \in V} g_{S^+}(e) \geq \frac{1}{k}(g(S^*) - g(S^+))$, therefore after $k$ iterations we have obtained a set $S^+$ with $|S^+| = k$, with

$$g(S^*) - g(S^+) \leq (1 - 1/k)^k (g(S^*) - g(\emptyset)) \leq (1 - 1/k)^k g(S^*) \leq \frac{1}{e} g(S^*),$$

so that the desired guarantee $(1 - \frac{1}{e})g(S^*) \leq g(S^+)$ follows.

Unfortunately, due to Line 3 in Algorithm 5 computing such a $(1 - 1/e)$-
approximate solution can cost up to $O(kn)$ evaluations of $g$ in the value oracle
model, where we can only query function values of $g$. For realistic functions this
is often quite prohibitive. We will now see a different application of a restart
scheme to reduce the total number of function evaluations of $g$ by allowing for
a small error $\varepsilon > 0$. We obtain a total number of evaluations of $g$ of $O(\frac{n}{\varepsilon} \log \frac{n}{\varepsilon})$,
quasi-linear and independent of $k$, to compute a $(1-1/e-\varepsilon)$-approximate solution.
The argument is due to [5] and similar in nature to the argument in Sect. 3 for
geometric scaling. We simplify the argument slightly for exposition; see [5] for
details.

The basic idea is rather than computing the actual maximum in Line 3 in
Algorithm 5, we collect all elements of marginal gains that are roughly maximal within a $(1 - \varepsilon)$-factor, then scale down the estimation of the maximum,
and then restart. We present the restart scheme, the so-called Threshold Greedy
Algorithm in Algorithm 6. This time we present the scheme and the base algorithm directly together. Note that the inner loop in Lines 3 to 7 in Algorithm 6
adds all elements that have approximately maximal marginal gain. The restarts

are happening whenever we go back to the beginning of the outer loop starting in Line 2, with a reset value for $\Phi$.

---

**Algorithm 6.** Threshold Greedy Algorithm

---

**Input:** Ground set $V$ of size $n$, budget $k$, accuracy $\varepsilon$, and monotone, non-negative, submodular function $g : 2^V \to \mathbb{R}_+$
**Output:** feasible set $S^+$ with $|S^+| \leq k$.
1: $S^+ \leftarrow \emptyset$, $\Phi_0 \leftarrow \max_{e \in V} g(e)$, $\Phi \leftarrow \Phi_0$
2: **while** $\Phi \geq \frac{\varepsilon}{n}\Phi_0$ **do**
3:     **for** $e \in V$ **do**
4:         **if** $|S^+| < k$ and $g_{S^+}(e) \geq \Phi$ **then**
5:             $S^+ \leftarrow S^+ + e$
6:         **end if**
7:     **end for**
8:     $\Phi \leftarrow \Phi(1 - \varepsilon)$
9: **end while**

---

We will first show that the gain from any new element $e \in V$ added in Line 5 of Algorithm 6 is at least

$$g_{S^+}(e) \geq \frac{1 - \varepsilon}{k} \sum_{x \in S^* \setminus S^+} g_{S^+}(x).$$

To this end suppose we have chosen element $e \in V$ to be added. Then $g_{S^+}(e) \geq \Phi$ by Line 4 and for all $x \in S^* \setminus (S^+ + e)$ we have $g_{S^+}(x) \leq \Phi/(1 - \varepsilon)$; otherwise we would have added $x$ in an earlier restart with a higher value $\Phi$ already. Combining the two inequalities we obtain

$$g_{S^+}(e) \geq (1 - \varepsilon)g_{S^+}(x),$$

for all $x \in S^* \setminus (S^+ + e)$ and averaging those inequalities leads to

$$g_{S^+}(e) \geq \frac{1 - \varepsilon}{|S^* \setminus S^+|} \sum_{x \in S^* \setminus S^+} g_{S^+}(x) \geq \frac{1 - \varepsilon}{k} \sum_{x \in S^* \setminus S^+} g_{S^+}(x), \tag{10}$$

which is the desired inequality. From this we immediately recover the (approximate) analog of Inequality (9). We have via submodularity and non-negativity

$$\sum_{x \in S^* \setminus S^+} g_{S^+}(x) \geq g_{S^+}(S^*) \geq g(S^*) - g(S^+),$$

and together with Inequality (10)

$$g_{S^+}(e) \geq \frac{1 - \varepsilon}{k}(g(S^*) - g(S^+)).$$

Therefore, as before, after $k$ iterations we obtain a set $S^+$ with $|S^+| = k$, with

$$g(S^*) - g(S^+) \leq (1 - (1 - \varepsilon)/k)^k (g(S^*) - g(\emptyset)) \leq (1 - (1 - \varepsilon)/k)^k g(S^*)$$

$$\leq \frac{1}{e^{(1-\varepsilon)}} g(S^*) \leq \left(\frac{1}{e} + \varepsilon\right) g(S^*),$$

leading to our guarantee $g(S^+) \geq \left(1 - \frac{1}{e} - \varepsilon\right) g(S^*)$. If we do fewer than $k$ iterations, the total gain of all remaining elements is less than $\varepsilon$, establishing the guarantee in that case.

Now for the number of evaluations of $g$, first consider the loop in Line 2 of Algorithm 6. The loops stops after $\ell$ iterations, whenever $(1 - \varepsilon)^\ell \leq \frac{\varepsilon}{n}$, which is satisfied if $1/(1 - \varepsilon)^\ell \geq (1 + \varepsilon)^\ell \geq \frac{n}{\varepsilon}$ and hence $\ell \geq \frac{1}{\varepsilon} \log \frac{n}{\varepsilon}$. For each such loop iteration we have at most $O(n)$ evaluations of $g$ in Line 4, leading to the overall bound of $O(\frac{n}{\varepsilon} \log \frac{n}{\varepsilon})$ evaluations of $g$.

*Related Approaches.* The approach presented here for the basic case with a single cardinality constraint can be applied more widely as already done in [5] for matroid, knapsack, and $p$-system constraints. It can be also used to reduce the number of evaluations in the context of robust submodular function maximization [2,3].

A similar restart approach has been used to 'lazify' conditional gradient algorithms in [11,12,28]. Here is the number of calls to the underlying linear optimization oracle is dramatically reduced by reusing information from previous iterations by solving the linear optimization problem only approximately as done in the case of the Threshold Greedy Algorithm. The algorithm, in a similar vein, is then restarted, whenever the threshold for approximation of the maximum is too large.

**Acknowledgement.** We would like to thank Gábor Braun and Marc Pfetsch for helpful comments and feedback on an earlier version of this article.

# References

1. Allen-Zhu, Z., Orecchia, L.: Linear coupling: An ultimate unification of gradient and mirror descent. arXiv preprint arXiv:1407.1537 (2014)
2. Anari, N., Haghtalab, N., Naor, S., Pokutta, S., Singh, M., Torrico, A.: Structured robust submodular maximization: offline and online algorithms. In: Proceedings of AISTATS (2019)
3. Anari, N., Haghtalab, N., Naor, S., Pokutta, S., Singh, M., Torrico, A.: Structured robust submodular maximization: offline and online algorithms. INFORMS J. Comput. (2020+, to appear)
4. Anderson, D., Hendel, G., Le Bodic, P., Viernickel, M.: Clairvoyant restarts in branch-and-bound search using online tree-size estimation. In: Proceedings of the AAAI Conference on Artificial Intelligence, vol. 33, pp. 1427–1434 (2019)
5. Badanidiyuru, A., Vondrák, J.: Fast algorithms for maximizing submodular functions. In: Proceedings of the Twenty-Fifth Annual ACM-SIAM Symposium on Discrete Algorithms, pp. 1497–1514. SIAM (2014)

6. Berthold, T., Heinz, S., Pfetsch, M.E.: Solving pseudo-Boolean problems with SCIP (2008)
7. Biere, A.: Adaptive restart strategies for conflict driven SAT solvers. In: Kleine Büning, H., Zhao, X. (eds.) SAT 2008. LNCS, vol. 4996, pp. 28–33. Springer, Heidelberg (2008). https://doi.org/10.1007/978-3-540-79719-7_4
8. Bockmayr, A., Eisenbrand, F., Hartmann, M., Schulz, A.: On the Chvátal rank of polytopes in the 0/1 cube. Discrete Appl. Math. **98**, 21–27 (1999)
9. Bolte, J., Daniilidis, A., Lewis, A.: The łojasiewicz inequality for nonsmooth subanalytic functions with applications to subgradient dynamical systems. SIAM J. Optim. **17**(4), 1205–1223 (2007)
10. Bolte, J., Nguyen, T.P., Peypouquet, J., Suter, B.W.: From error bounds to the complexity of first-order descent methods for convex functions. Math. Program. **165**(2), 471–507 (2016). https://doi.org/10.1007/s10107-016-1091-6
11. Braun, G., Pokutta, S., Zink, D.: Lazifying conditional gradient algorithms. In: Proceedings of the International Conference on Machine Learning (ICML) (2017)
12. Braun, G., Pokutta, S., Zink, D.: Lazifying conditional gradient algorithms. J. Mach. Learn. Res. (JMLR) **20**(71), 1–42 (2019)
13. Chvátal, V.: Edmonds polytopes and a hierarchy of combinatorial problems. Discrete Math. **4**, 305–337 (1973)
14. Chvátal, V., Cook, W., Hartmann, M.: On cutting-plane proofs in combinatorial optimization. Linear algebra Appl. **114**, 455–499 (1989)
15. Combettes, C.W., Pokutta, S.: Revisiting the Approximate Carathéodory Problem via the Frank-Wolfe Algorithm. Preprint (2019)
16. Diakonikolas, J., Carderera, A., Pokutta, S.: Locally accelerated conditional gradients. Proceedings of AISTATS (arXiv:1906.07867) (2020, to appear)
17. Edmonds, J., Karp, R.M.: Theoretical improvements in algorithmic efficiency for network flow problems. J. ACM **19**(2), 248–264 (1972)
18. Eisenbrand, F., Schulz, A.: Bounds on the Chvátal rank on polytopes in the 0/1-cube. Combinatorica **23**(2), 245–261 (2003)
19. Fisher, M.L., Nemhauser, G.L., Wolsey, L.A.: An analysis of approximations for maximizing submodular set functions–ii. In: Balinski, M.L., Hoffman, A.J. (eds.) Polyhedral Combinatorics, pp. 73–87. Springer, Heidelberg (1978). https://doi.org/10.1007/BFb0121195
20. Graham, R.L., Grötschel, M., Lovász, L.: Handbook of Combinatorics, vol. 1. Elsevier (1995)
21. Hazan, E.: Lecture notes: Optimization for machine learning. arXiv preprint arXiv:1909.03550 (2019)
22. Hazan, E., Luo, H.: Variance-reduced and projection-free stochastic optimization. In: International Conference on Machine Learning, pp. 1263–1271 (2016)
23. Hinder, O., Lubin, M.: A generic adaptive restart scheme with applications to saddle point algorithms. arXiv preprint arXiv:2006.08484 (2020)
24. Hu, X., Shonkwiler, R., Spruill, M.C.: Random restarts in global optimization (2009)
25. Huang, J., et al.: The effect of restarts on the efficiency of clause learning. IJCAI **7**, 2318–2323 (2007)
26. Johnson, R., Zhang, T.: Accelerating stochastic gradient descent using predictive variance reduction. In: Advances in Neural Information Processing Systems, pp. 315–323 (2013)
27. Kerdreux, T., d'Aspremont, A., Pokutta, S.: Restarting Frank-Wolfe. In: Proceedings of AISTATS (2019)

28. Lan, G., Pokutta, S., Zhou, Y., Zink, D.: Conditional accelerated lazy stochastic gradient descent. In: Proceedings of the International Conference on Machine Learning (ICML) (2017)
29. Lan, G.: First-order and Stochastic Optimization Methods for Machine Learning. Springer, Heidelberg (2020). https://doi.org/10.1007/978-3-030-39568-1
30. Le Bodic, P., Pfetsch, M., Pavelka, J., Pokutta, S.: Solving MIPs via scaling-based augmentation. Discrete Optim. **27**, 1–25 (2018)
31. Mirrokni, V., Paes Leme, R., Vladu, A., Wong, S.C.W.: Tight bounds for approximate Carathéodory and beyond. In: Proceedings of the 34th International Conference on Machine Learning, pp. 2440–2448 (2017)
32. Nemhauser, G.L., Wolsey, L.A., Fisher, M.L.: An analysis of approximations for maximizing submodular set functions–I. Math. Program. **14**(1), 265–294 (1978)
33. Nemirovski, A.: Lectures on modern convex optimization. In: Society for Industrial and Applied Mathematics (SIAM). Citeseer (2001)
34. Nesterov, Y.: How to make the gradients small. Optima. Math. Optim. Soc. Newslett. **88**, 10–11 (2012)
35. Nesterov, Y.: Lectures on Convex Optimization. Springer, Heidelberg (2018). https://doi.org/10.1007/978-3-319-91578-4
36. Nesterov, Y.E.: A method for solving the convex programming problem with convergence rate $O(1/k^2)$. Dokl. akad. nauk Sssr. **269**, 543–547 (1983)
37. O'donoghue, B., Candes, E.: Adaptive restart for accelerated gradient schemes. Found. Comput. Math. **15**(3), 715–732 (2015)
38. Rothvoß, T., Sanità, L.: 0/1 polytopes with quadratic chvátal rank. Oper. Res. **65**(1), 212–220 (2017)
39. Roulet, V., d'Aspremont, A.: Sharpness, restart and acceleration. ArXiv preprint arXiv:1702.03828 (2017)
40. Schulz, A.S., Weismantel, R.: The complexity of generic primal algorithms for solving general integer programs. Math. Oper. Res. **27**(4), 681–692 (2002)
41. Schulz, A.S., Weismantel, R., Ziegler, G.M.: 0/1-integer programming: optimization and augmentation are equivalent. In: Spirakis, P. (ed.) ESA 1995. LNCS, vol. 979, pp. 473–483. Springer, Heidelberg (1995). https://doi.org/10.1007/3-540-60313-1_164
42. Scieur, D., d'Aspremont, A., Bach, F.: Regularized nonlinear acceleration. In: Advances in Neural Information Processing Systems, pp. 712–720 (2016)
43. Xu, Y., Yang, T.: Frank-Wolfe Method is Automatically Adaptive to Error Bound Condition (2018)

# Regular Papers

# Discriminating Instance Generation from Abstract Specifications: A Case Study with CP and MIP

Özgür Akgün, Nguyen Dang$^{(\boxtimes)}$, Ian Miguel, András Z. Salamon, Patrick Spracklen, and Christopher Stone

School of Computer Science, University of St Andrews, St Andrews, UK
nttd@st-andrews.ac.uk

**Abstract.** We extend automatic instance generation methods to allow cross-paradigm comparisons. We demonstrate that it is possible to completely automate the search for benchmark instances that help to discriminate between solvers. Our system starts from a high level human-provided problem specification, which is translated into a specification for valid instances. We use the automated algorithm configuration tool irace to search for instances, which are translated into inputs for both MIP and CP solvers by means of the CONJURE, Savile Row, and MiniZinc tools. These instances are then solved by CPLEX and Chuffed, respectively. We constrain our search for instances by requiring them to exhibit a significant advantage for MIP over CP, or vice versa. Experimental results on four optimisation problem classes demonstrate the effectiveness of our method in identifying instances that highlight differences in performance of the two solvers.

**Keywords:** Instance generation · MIP · Constraint Programming

## 1 Introduction

When developing a model of a combinatorial problem class, a set of representative instances drawn from the class is essential for evaluating the model's performance. Recently, Akgün et al. [1] demonstrated how to generate instances automatically from the ESSENCE[1] specification of a problem class [10]. The instances generated are *graded*: neither too difficult nor too easy relative to a given solver and resource limits. Graded instances are particularly valuable for model evaluation, since they are less likely to be solved trivially with the model under development or to remain unsolved at the expiry of a time budget. Either of these outcomes would reveal little useful information about model performance.

---

[1] ESSENCE is an abstract constraint specification language that supports a formal statement of a problem without committing to detailed modelling decisions.

© Springer Nature Switzerland AG 2020
E. Hebrard and N. Musliu (Eds.): CPAIOR 2020, LNCS 12296, pp. 41–51, 2020.
https://doi.org/10.1007/978-3-030-58942-4_3

In this paper we consider a more complex situation: rather than focus on modelling for a single solving paradigm, it is often the case that we might wish to evaluate two or more solving paradigms for a problem of interest. In this context, it is desirable to generate instances that are not only graded, but also *discriminating*, i.e. which exhibit a pronounced difference in solving performance among the solving paradigms under consideration. Discriminating instances are valuable both for a manual inspection of the instance characteristics that favour one paradigm over the others, and to provide coverage of the instance space when training the selection process for an algorithm portfolio [28]. Our hypothesis is that starting from a single high-level specification of a problem we can generate discriminating instances automatically via synthesising an "instance generator model" and using standard algorithm tuning tools.

We consider two paradigms: Constraint Programming (CP) and Mixed Integer Programming (MIP). Extending the approach of Akgün et al., we employ the automated configuration tool irace [22] to search for graded discriminating instances for the CP solver Chuffed [7] and the MIP solver CPLEX [17]. This search is performed twice, with each solver in turn first considered the *base* solver, and the other the *favoured* solver. irace is guided to search for discriminating instances where the favoured solver performs significantly better than the base solver. The advantage of this approach is that, even when one of the two solvers predominantly performs better, the search for discriminating instances is pushed towards regions of the instance space where the generally weaker solver has the advantage. This provides good coverage of the instance space and a clear picture of relative solver performance, as our empirical results demonstrate.

## 2    Related Work

Benchmarks play an important role in combinatorial optimisation as they are often the main device employed to verify the quality of solvers. For a long time this involved bundling a collection of problem instances, with one or more problem classes, that are then solved and compared by practitioners. In many cases, this has led to the reuse of the same set of instances for several decades causing algorithms to become highly tailored to solve those specific sets and becoming less generally applicable [13]. This is not an ideal practice, as it has been observed that different algorithmic techniques have their own strengths and weaknesses [6,19,30]. An alternative approach is making use of instance generators that can produce a stream of new instances. The two main approaches to generate instances in an automated manner are based on handcrafted programs [9,15,32], where practitioners use their knowledge to specify desired characteristics, and on meta-heuristic approaches where instances are created and selected according to some objective functions [18,29]. In both cases these generators can produce instances only for specific problem classes and making them applicable to new problem classes would require substantial modifications. Another criticism raised [31] is that often only a small number of algorithms are tested on a small set of instances to certify the superiority of one algorithm over another instead of studying the strengths and weaknesses of both.

Belov et al. [6] demonstrated automatic translation of CP specifications expressed in the MiniZinc language to lower-level FlatZinc [24], using knowledge of the target paradigm to guide the translation. Their experiments showed that the MIP and CP solving paradigms have different sets of strengths and weaknesses, and found discriminating instances among the MiniZinc benchmarks.

Generation of discriminating instances is increasingly popular [4], for instance to measure algorithm performance across instances [28] or to improve algorithm selection tools [14,30]. Most studies in this area tackle one problem class at a time. We here extend [1] to automatically produce instances for any problem class. The goal is to automate finding instances that are particularly suited to one algorithm but not another, and to study characteristics of these instances.

## 3   Background

CP and MIP solvers work by solving a *problem instance* model composed of decision variables with associated domains, a set of constraints on the decision variables (and optionally an objective function for optimisation problems). CP modelling languages typically offer a richer language in comparison to MIP modelling languages thanks to having a richer set of constraint types. Modern modelling languages for both formalisms allow models to be written for a *problem class*. A problem class model is instantiated by a modelling tool before it is given to a solver to achieve a problem instance model.

ESSENCE is a problem specification language for combinatorial decision and optimisation problems [10]. ESSENCE supports *abstract* decision variables, such as multiset, relation and function, as well as *nested* types, such as multiset of sets. In addition to language features for specifying decision variables, constraints and the objective function (**find**, **such that**, **min/maximising** respectively) it allows the specification of problem parameters which define problem instances (**given**) and restrictions on values that parameters can take on valid instances (**where**).

Problem specifications written in ESSENCE are converted to class level constraint models by CONJURE [2,3], which are then fed into Savile Row [26] to instantiate the model and convert it into input suitable for a supported solver. Savile Row also applies instance level model improvements automatically [26].

A problem specification can be automatically converted to an *instance generator specification* by CONJURE [1]. First, the **given** statements that declare parameters are converted into **find** statements that declare corresponding decision variables. Second, the **where** statements are converted to **such that** statements. This process is explained in more detail in [1].

The main objective of [1] was to generate *graded* instances (neither too easy nor too hard for a selected backend solver). This is achieved by using a general-purpose automated algorithm tuning tool to search for generator configurations covering the problem instance space, with solving time between the given bounds. In [1], irace was used for this task. irace is an automated algorithm configuration tool that supports tuning parameters of algorithms efficiently. The core idea behind irace is *iterated racing*, an iterative search procedure where at each iteration, statistical tests are used to eliminate configurations with poor performance early, so that the budget is saved for evaluating more promising configurations.

# 4    Experimental Method

## 4.1    Problem Classes

We demonstrate our method on the following four optimisation problem classes. The first three are typically solved by Operations Research methods. We work with existing ESSENCE specifications from CSPlib [12], where available.

**Transshipment (TP):** Given costs of transporting goods from a warehouse to a transshipment point and from a transshipment point to a customer, warehouse stock levels, and customer demand, the objective is to minimise the total transport cost while meeting customer demand. TP is known to have efficient linear programming solutions [27] and we expected CPLEX to dominate Chuffed.

**Progressive Party (PPP,** CSPlib 013): The objective is to minimise the number of boats hosting a party at a yacht club, where some boats (with capacities) are designated as hosts, and the crews of the remaining boats visit the host boats for fixed time periods; two guest crews may meet at most once. PPP is a classic CP problem and we expected Chuffed to dominate CPLEX.

**Warehouse Location (WLP,** CSPlib 034): A central warehouse will supply depots, each with a maintenance cost and a capacity; each store will be supplied from exactly one depot at some cost. The objective is to find a subset of depots to open so as to minimise the sum of the maintenance and supply costs. We had no prior opinion on whether CP would outperform MIP.

**Capacitated Vehicle Routing (CVRP):** The task is to find least cost routes for identical vehicles with capacities, delivering goods from a central depot. Each location is visited once by one vehicle. A route starts at the depot and finishes there [8,20]. We had no prior opinion on whether MIP would outperform CP.

## 4.2    irace's Scoring Function to Find Discriminating Instances

Each *evaluation* during the tuning involves a generator configuration and a random seed, both sampled by irace. The CP solver minion [11] is used to solve the configuration with the given random seed. A solution is returned as an instance of the original problem. That instance is evaluated using the two solvers and a score value (to be minimised) is calculated. The default setting of irace compares configurations based on ranking. Therefore, the absolute difference between score values is not important. Details of the scoring are as follows.

- If the generator configuration is unsatisfiable, then a special infinite score value is returned. irace will discard the configuration immediately.

– If the generator configuration is too large be to translated (Savile Row is out of time or memory), or not solvable by `minion`, then the score is set to 2.
– If the generator configuration is satisfiable and an instance is found:
  • if the instance is unsatisfiable or too large to be translated (Savile Row is out of time or memory for either solver), then a score of 1 is returned,
  • if the instance is too difficult for the favoured solver, or too easy for the base solver, then a score of 0 is returned, or
  • if the instance is solved within the given time and memory limits, then the negation of the ratio between the solving time of the base and the favoured solvers is returned.

## 4.3  Experimental Setup

The memory limit given to each evaluation is 7 GB. The time limit for Savile Row and `minion` is 5 min each. The time limit for the favoured solver is 5 min, while the base solver is allowed between 10 s and 25 min. Chuffed version 0.10.3 and CPLEX version 12.9 are used. Instances are translated to Chuffed directly via Savile Row. CPLEX input is translated to MiniZinc format first using Savile Row, and then to CPLEX input format using MiniZinc [6]. The compilation time required by MiniZinc was never more than a few seconds, as the input MiniZinc files have been pre-processed and optimised by Savile Row.

Solving time on an instance is calculated as the average value across three runs. Each experiment is run on a cluster node with two 2.1 GHz Intel Xeon E5-2695 processors. Since irace supports parallelism, 30 cores are used per experiment. Each tuning is given a budget of 5000 evaluations and 48 h of wall-time, and is stopped when either of the two budgets is exhausted.

## 5  Results

The discriminating instances found[2] for each problem class are plotted in Fig. 1. Table 1 details how many evaluations the tuning spent on each type of instance, with numbers describing what the search space looks like during tuning.

For CVRP and PPP, we found discriminating instances for both CPLEX and Chuffed. However, the number of instances found in the Chuffed-favoured experiment is larger than in the CPLEX-favoured experiment ($\approx$2000 vs. $\approx$50 instances for CVRP, and $\approx$1400 vs. $\approx$600 instances for PPP). Detailed results on the search space during the tuning show that in the Chuffed-favoured experiment, the majority of evaluations is spent on instances solved by Chuffed within 300 s, while in the CPLEX-favoured experiment, the majority is spent on instances where CPLEX timed out (both problem classes), or on instances that are very easy for Chuffed (CVRP only). This indicates that in our current setting, although instances where CPLEX is better than Chuffed on those two problem classes exist (and are found by our tuning), overall Chuffed is better at solving these problems than CPLEX.

---

[2] Code and data are at: https://github.com/stacs-cp/CPAIOR2020-InstanceGen.

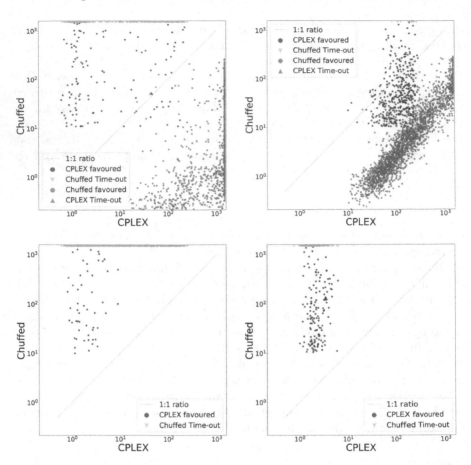

**Fig. 1.** Solving time of Chuffed and CPLEX on the discriminating instances found. We run irace twice with each solver favoured in turn. The plotted time-outs are only for the base solver.

For the two remaining problems, Transshipment and Warehouse Location, we see a different picture. Many CPLEX-favoured instances are found, while there are no Chuffed-favoured instances at all. For Transshipment, in the majority of evaluations in both tuning experiments either Chuffed times out or the generator configuration was not solved in time. Figure 1 also shows that CPLEX's solving time on the discriminating instances of the same problem is quite small ($\leq 10$ s). These observations suggest that CPLEX completely dominates Chuffed on Transshipment, which is exactly what we expected. For Warehouse Location, the story is a bit different. Instances in the CPLEX-favoured experiment are mostly too easy for Chuffed. A more detailed look at those instances reveals that Chuffed is comparable to CPLEX on the instances that are "too easy". Therefore, we conjecture that CPLEX may not completely dominate Chuffed on Warehouse Location, but it does dominate Chuffed on the "more difficult" instances which take at least 10 s to solve by Chuffed in our setup.

## 6   Feature Analysis

To gain more insights into the discriminating instances found, we extract their FlatZinc features using the fzn2feat tool (part of mzn2feat [5]). There are 95 features grouped into 6 categories (variables, constraints, domains, global constraints, objective, and solving features) [5]. For each of the two problem classes where discriminating instances are found, CVRP and PPP, we use the Balanced Random Forest classifier from the Python package imblearn [21] with 200 estimators and 5-fold cross validation. To identify the most important features representing the discriminating property between the two solvers, Mean Decrease Impurity [23] of each feature is calculated across 20 runs. Random Forests have been shown to be the overall best choice for modelling running time of CP and SAT solvers [16]. The Mean Decrease Impurity (MDI) is a widely-used measurement for feature importance analyses in Random Forest models. The MDI of a feature in each tree is the weighted decrease in impurity (using Gini importance) across all tree nodes where the feature is used in the splits. The overall MDI of a feature is calculated by averaging the MDI values across all trees in the forest.

Table 1. Number of runs for each instance type during tuning. Experiment name has the problem class and the favoured solver. Columns: gen failed (unsolved generator configurations), SR-timeout&unsat (Savile Row timed out or the instances are unsatisfiable), base easy (solved by the base solver within 10 s), favoured timeout (the favoured solver timed out), within the range (sat instances solved within 300 s by the favoured solver and not solved within 10 s by the base solver). The four final numbers for each experiment show how discriminating instances are; ratio is solving time of the base to the favoured solver. If the base solver times out, 25 min is used.

| | gen failed | SR-timeout & unsat | base easy | favoured timeout | within the range ratio≤1 | ratio>1 | ratio>2 | ratio>10 |
|---|---|---|---|---|---|---|---|---|
| PPP-chuffed | 85 | 932 | 538 | 554 | 4 | 1474 | 1471 | 1419 |
| PPP-cplex | 67 | 1019 | 301 | 1290 | 14 | 670 | 665 | 597 |
| CVRP-chuffed | 457 | 178 | 1117 | 741 | 4 | 2225 | 2224 | 2080 |
| CVRP-cplex | 686 | 178 | 1768 | 1865 | 330 | 110 | 59 | 8 |
| TP-chuffed | 1762 | 30 | 323 | 2862 | 0 | 0 | 0 | 0 |
| TP-cplex | 2108 | 479 | 50 | 643 | 0 | 1096 | 1096 | 1038 |
| WLP-chuffed | 626 | 58 | 4277 | 34 | 0 | 0 | 0 | 0 |
| WPL-cplex | 819 | 298 | 1174 | 0 | 0 | 519 | 518 | 481 |

To avoid noise in the measurement of solving time, we only consider discriminating instances where the ratio between the solving time of the bad solver and the good solver is larger than 1.5. Each instance is labelled as either Chuffed-favoured or CPLEX-favoured.

Figure 2 show the importance values of the top 10 features for PPP and CVRP. For PPP, the first feature, v_cv_domdeg_vars, shows a much higher importance value compared to the rest. This feature defines the Coefficient of Variance of the ratios between domain size and degree (number of constraints involved) of all variables. A more detailed look into the data indicates that Chuffed-favoured instances tend to have similar ratios between domain size over degree across different variables, while for CPLEX-favoured instances those ratios differ more drastically between variables. For CVRP, there is no clearly distinguished single feature. Moreover, the list of the most important features varies between the two different problem classes. This suggests that the favouring-behaviours of the two solvers depend on the problem.

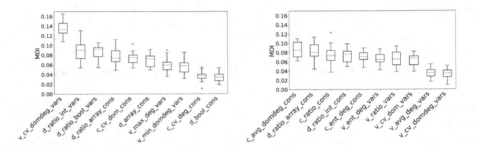

**Fig. 2.** Top 10 FlatZinc features for discriminating instances based on MDI

# 7   Conclusion and Future Work

In this work, we presented an automated instance generation system that can produce discriminating instances between two solvers. We demonstrated our method on four problem classes with the CP solver Chuffed and the MIP solver CPLEX. This revealed the strengths and weaknesses of each solver. A further analysis of the discriminating instances using FlatZinc features [5] suggests that the discriminating behaviour is problem dependent.

In future work, we plan to extend our system for finding discriminating instances for a portfolio of more than two solvers; the open question here is how to define the discriminating property. We also plan more detailed feature analysis investigating the relationship between solver performance and instance space. This can involve defining new instance features based on the high-level types supported by ESSENCE, which may provide more high-level structural information about the instances. Finally, we currently rely on irace's built-in exploration of the generator configuration space to ensure the diversity of the generated instances. This can be improved by investigating more advanced approaches for controlling instance diversity more directly, including incorporating a diversity measurement such as multi-objective indicators [25] into the scoring values of the tuning, or forcing each generator configuration to generate instances far away from the current instance set by adding constraints into the generator model.

**Acknowledgements.** This work is supported by EPSRC grant EP/P015638/1 and used the Cirrus UK National Tier-2 HPC Service at EPCC (http://www.cirrus.ac.uk) funded by the University of Edinburgh and EPSRC (EP/P020267/1).

# References

1. Akgün, Ö., Dang, N., Miguel, I., Salamon, A.Z., Stone, C.: Instance generation via generator instances. In: Schiex, T., de Givry, S. (eds.) CP 2019. LNCS, vol. 11802, pp. 3–19. Springer, Cham (2019). https://doi.org/10.1007/978-3-030-30048-7_1
2. Akgun, O., et al.: Automated symmetry breaking and model selection in CONJURE. In: Schulte, C. (ed.) CP 2013. LNCS, vol. 8124, pp. 107–116. Springer, Heidelberg (2013). https://doi.org/10.1007/978-3-642-40627-0_11
3. Akgün, Ö., Miguel, I., Jefferson, C., Frisch, A.M., Hnich, B.: Extensible automated constraint modelling. In: AAAI 2011: Proceedings of the Twenty-Fifth AAAI Conference on Artificial Intelligence, pp. 4–11. AAAI Press (2011). https://www.aaai.org/ocs/index.php/AAAI/AAAI11/paper/viewPaper/3687
4. Alissa, M., Sim, K., Hart, E.: Algorithm selection using deep learning without feature extraction. In: GECCO 2019: Proceedings of the Genetic and Evolutionary Computation Conference, pp. 198–206. ACM (2019)
5. Amadini, R., Gabbrielli, M., Mauro, J.: An enhanced features extractor for a portfolio of constraint solvers. In: SAC 2014: Proceedings of the 29th Annual ACM Symposium on Applied Computing, pp. 1357–1359. ACM (2014). https://doi.org/10.1145/2554850.2555114. Code available from https://github.com/CP-Unibo/mzn2feat
6. Belov, G., Stuckey, P.J., Tack, G., Wallace, M.: Improved linearization of constraint programming models. In: Rueher, M. (ed.) CP 2016. LNCS, vol. 9892, pp. 49–65. Springer, Cham (2016). https://doi.org/10.1007/978-3-319-44953-1_4
7. Chu, G., de la Banda, M., Stuckey, P.: Exploiting subproblem dominance in constraint programming. Constraints **17**(1), 1–38 (2012). https://doi.org/10.1007/s10601-011-9112-9. Code available from https://github.com/chuffed/chuffed
8. Dantzig, G.B., Ramser, J.H.: The truck dispatching problem. Manag. Sci. **6**(1), 80–91 (1959). https://doi.org/10.1287/mnsc.6.1.80
9. Drexl, A., Nissen, R., Patterson, J.H., Salewski, F.: Progen/πx-an instance generator for resource-constrained project scheduling problems with partially renewable resources and further extensions. Eur. J. Oper. Res. **125**(1), 59–72 (2000)
10. Frisch, A.M., Harvey, W., Jefferson, C., Martínez-Hernández, B., Miguel, I.: Essence: a constraint language for specifying combinatorial problems. Constraints **13**(3), 268–306 (2008). https://doi.org/10.1007/s10601-008-9047-y
11. Gent, I.P., Jefferson, C., Miguel, I.: Minion: a fast scalable constraint solver. In: Proceedings of ECAI 2006, pp. 98–102. IOS Press (2006). http://ebooks.iospress.nl/volumearticle/2658
12. Gent, I.P., Walsh, T.: CSPlib: a benchmark library for constraints. In: Jaffar, J. (ed.) CP 1999. LNCS, vol. 1713, pp. 480–481. Springer, Heidelberg (1999). https://doi.org/10.1007/978-3-540-48085-3_36. http://www.csplib.org
13. Hooker, J.N.: Testing heuristics: we have it all wrong. J. Heuristics **1**(1), 33–42 (1995)

14. Hoos, H.H., Kaufmann, B., Schaub, T., Schneider, M.: Robust benchmark set selection for boolean constraint solvers. In: Nicosia, G., Pardalos, P. (eds.) LION 2013. LNCS, vol. 7997, pp. 138–152. Springer, Heidelberg (2013). https://doi.org/10.1007/978-3-642-44973-4_16

15. Horie, S., Watanabe, O.: Hard instance generation for SAT. In: Leong, H.W., Imai, H., Jain, S. (eds.) ISAAC 1997. LNCS, vol. 1350, pp. 22–31. Springer, Heidelberg (1997). https://doi.org/10.1007/3-540-63890-3_4

16. Hutter, F., Xu, L., Hoos, H.H., Leyton-Brown, K.: Algorithm runtime prediction: methods & evaluation. Artif. Intell. **206**, 79–111 (2014)

17. IBM: IBM ILOG CPLEX Optimization Studio documentation (2019)

18. Julstrom, B.A.: Evolving heuristically difficult instances of combinatorial problems. In: GECCO 2009: Proceedings of the 11th Annual conference on Genetic and Evolutionary Computation, pp. 279–286. ACM (2009). https://doi.org/10.1145/1569901.1569941

19. Kotthoff, L.: Algorithm selection for combinatorial search problems: a survey. In: Bessiere, C., De Raedt, L., Kotthoff, L., Nijssen, S., O'Sullivan, B., Pedreschi, D. (eds.) Data Mining and Constraint Programming. LNCS (LNAI), vol. 10101, pp. 149–190. Springer, Cham (2016). https://doi.org/10.1007/978-3-319-50137-6_7

20. Laporte, G.: The vehicle routing problem: an overview of exact and approximate algorithms. Eur. J. Oper. Res. **59**(3), 345–358 (1992). https://doi.org/10.1016/0377-2217(92)90192-C

21. Lemaître, G., Nogueira, F., Aridas, C.K.: Imbalanced-learn: a python toolbox to tackle the curse of imbalanced datasets in machine learning. J. Mach. Learn. Res. **18**(17), 1–5 (2017). http://jmlr.org/papers/v18/16-365.html

22. López-Ibáñez, M., Dubois-Lacoste, J., Cáceres, L.P., Birattari, M., Stützle, T.: The irace package: iterated racing for automatic algorithm configuration. Oper. Res. Perspect. **3**, 43–58 (2016). https://doi.org/10.1016/j.orp.2016.09.002. http://iridia.ulb.ac.be/irace/

23. Louppe, G., Wehenkel, L., Sutera, A., Geurts, P.: Understanding variable importances in forests of randomized trees. In: Advances in Neural Information Processing Systems, pp. 431–439 (2013)

24. Nethercote, N., Stuckey, P.J., Becket, R., Brand, S., Duck, G.J., Tack, G.: MiniZinc: towards a standard CP modelling language. In: Bessière, C. (ed.) CP 2007. LNCS, vol. 4741, pp. 529–543. Springer, Heidelberg (2007). https://doi.org/10.1007/978-3-540-74970-7_38

25. Neumann, A., Gao, W., Wagner, M., Neumann, F.: Evolutionary diversity optimization using multi-objective indicators. In: GECCO 2019: Proceedings of the Genetic and Evolutionary Computation Conference, pp. 837–845. ACM (2019)

26. Nightingale, P., Akgün, Ö., Gent, I.P., Jefferson, C., Miguel, I., Spracklen, P.: Automatically improving constraint models in Savile Row. Artif. Intell. **251**, 35–61 (2017). https://doi.org/10.1016/j.artint.2017.07.001

27. Orden, A.: The transhipment problem. Manag. Sci. **2**(3), 276–285 (1956). https://doi.org/10.1287/mnsc.2.3.276

28. Smith-Miles, K., Baatar, D., Wreford, B., Lewis, R.: Towards objective measures of algorithm performance across instance space. Comput. Oper. Res. **45**, 12–24 (2014)

29. Smith-Miles, K., van Hemert, J.: Discovering the suitability of optimisation algorithms by learning from evolved instances. Ann. Math. Artif. Intell. **61**(2), 87–104 (2011). https://doi.org/10.1007/s10472-011-9230-5

30. Smith-Miles, K., Lopes, L.: Generalising algorithm performance in instance space: a timetabling case study. In: Coello, C.A.C. (ed.) LION 2011. LNCS, vol. 6683, pp. 524–538. Springer, Heidelberg (2011). https://doi.org/10.1007/978-3-642-25566-3_41
31. Smith-Miles, K., Lopes, L.: Measuring instance difficulty for combinatorial optimization problems. Comput. Oper. Res. **39**(5), 875–889 (2012)
32. Vanhoucke, M., Maenhout, B.: NSPLib-a nurse scheduling problem library: a tool to evaluate (meta-) heuristic procedures. In: Operational Research for Health Policy: Making Better Decisions, Proceedings of the 31st Annual Meeting of the Working Group on Operations Research Applied to Health Services, pp. 151–165 (2007)

# Bilevel Optimization for On-Demand Multimodal Transit Systems

Beste Basciftci[1,2]($\boxtimes$) and Pascal Van Hentenryck[1]

[1] Georgia Institute of Technology, 30332 Atlanta, GA, USA
beste.basciftci@gatech.edu, pascal.vanhentenryck@isye.gatech.edu
[2] Sabancı University, 34956 Istanbul, Turkey

**Abstract.** This study explores the design of an On-Demand Multimodal Transit System (ODMTS) that includes segmented mode switching models that decide whether potential riders adopt the new ODMTS or stay with their personal vehicles. It is motivated by the desire of transit agencies to design their network by taking into account both existing and latent demand, as quality of service improves. The paper presents a bilevel optimization where the leader problem designs the network and each rider has a follower problem to decide her best route through the ODMTS. The bilevel model is solved by a decomposition algorithm that combines traditional Benders cuts with combinatorial cuts to ensure the consistency of mode choices by the leader and follower problems. The approach is evaluated on a case study using historical data from Ann Arbor, Michigan, and a user choice model based on the income levels of the potential transit riders.

**Keywords:** On-demand transit system · Mode choice · Bilevel optimization · Benders decomposition · Combinatorial cuts

## 1 Introduction

On-Demand Multimodal Transit Systems (ODMTS) [13,15] combines on-demand shuttles with a bus or rail network. The on-demand shuttles serve local demand and act as feeders to and from the bus/rail network, while the bus/rail network provides high-frequency transportation between hubs. By using on-demand shuttles to pick up riders at their origins and drop them off at their destinations, ODMTS addresses the first/last mile problem that plagues most of the transit systems. Moreover, ODMTS addresses congestion and economy of scale by providing high-frequency along congested corridors. They have been shown to bring substantial convenience and cost benefits in simulation and pilot studies in the city of Canberra, Australia and the city of Ann Arbor, Michigan.

The design of an ODMTS is a variant of the hub-arc location problem [4,5]: It uses an optimization model that decides which bus/rail lines to open in order to maximize convenience (e.g., minimize transit time) and minimize costs [13]. This optimization model uses, as input, the current demand, i.e., the set of

© Springer Nature Switzerland AG 2020
E. Hebrard and N. Musliu (Eds.): CPAIOR 2020, LNCS 12296, pp. 52–68, 2020.
https://doi.org/10.1007/978-3-030-58942-4_4

origin-destination pairs over time in the existing transit system. Transit agencies however are worried about latent demand: As the convenience of the transit system improves, more riders may decide to switch modes and adopt the ODMTS instead of traveling with their personal vehicles. By ignoring the latent demand, the ODMTS may be designed suboptimally, providing a lower convenience or higher costs. This concern was raised in [3] who articulated the potential of leveraging data analytics within the planning process and proposing transit systems that encourage riders to switch transportation modes.

This paper aims at remedying this limitation and explores the design of ODMTS with both existing and latent demands. It considers a pool of potential riders, each of whom is associated with a personalized mode choice model that decides whether a rider will switch mode for a given ODMTS. Such a choice model can be obtained through stated and revealed preferences, using surveys and/or machine learning [17]. The main innovation of this paper is to show how to integrate such mode choice models into the design of ODMTS, capturing the latent demand and human behavior inside the optimization model. More precisely, the contributions of the paper can be summarized as follows:

1. The paper proposes a novel bilevel optimization approach to model the ODMTS problem with latent demand in order to obtain the most cost-efficient and convenient route for each trip.
2. The bilevel optimization model includes a personalized mode choice for each rider to determine mode switching or latent demand.
3. The bilevel optimization model is solved through a decomposition algorithm that combines both traditional and combinatorial Benders cuts.
4. The paper demonstrates the benefits and practicability of the approach on a case study using historical data over Ann Arbor, Michigan.

The remainder of the paper is organized as follows. Section 2 reviews the relevant literature. Section 3 specifies the ODMTS design problem. Section 4 proposes a bilevel optimization approach for the design of ODMTS with latent demand, and Sect. 5 develops the novel decomposition methodology. The case study is presented in Sect. 6 and Sect. 7 concludes the paper with final remarks.

## 2 Related Literature

Hub location problems are an important area of research in transit network design (see [7] for a recent review). More specifically, the transit network design problem considering hubs can be considered as a variant of the hub-arc location problem [4,5], which focuses on determining the set of arcs to open between hubs, and optimizing the flow with minimum cost. Mahéo, Kilby, and Van Hentenryck [13] extended this problem to the ODMTS setting by introducing on-demand shuttles and removing the restriction that each route needs to contain an arc involving a hub. Furthermore, instead of the restriction of hubs being interconnected in the network design, they consider a weak connectivity within system by ensuring the sum of incoming and outgoing arcs to be equal to each other for each hub. Although these studies provide efficient solutions for a given demand, they neglect the effect of the latent demand which can change the design of the transit systems.

Bilevel optimization is an important area of mathematical programming, which mainly considers a leader problem, and a follower problem that optimizes its decisions under the leader problem's output. Due to this hierarchical decision making structure, this area attracted attention in different urban transit network design applications [8,11] such as discrete network design problems [9] by improving a network via adding lines or increasing their capacities, and bus lane design problems [16] under traffic equilibrium. Another line of research focuses on the pricing aspects of these problems for toll optimization by considering a multi-commodity network flow problem [1], and [2] extends this setting by jointly designing the underlying network. Studies [6,14] provide an overview of various solution methodologies to address these problems including reformulations based on Karush–Kuhn–Tucker (KKT) conditions, descent methods and heuristics. User preferences and the corresponding latent demand constitute important factors impacting the network design. Because of the computational complexity involved with solving bilevel problems, it is preferred to model rider preferences within a single level optimization problem [10]. To this end, our approach provides a novel bilevel optimization framework for solving the ODMTS by integrating user choices, and developing an exact decomposition algorithm as its solution procedure.

## 3   Problem Statement

This section defines the problem statement and stays as close as possible to the original setting of the ODMTS design [13]. In particular, the input consists of a set of (potentially virtual) bus stops $N$, a set of potential hubs $H \subseteq N$, and a set of trips $T$. Each trip $r \in T$ is associated with an origin stop $or^r \in N$, a destination stop $de^r \in N$, and a number of passengers taking that trip $p^r \in \mathbb{Z}_+$. This paper often abuses terminology and uses trips and riders interchangeably, although a trip may contain several riders. The distance and time between each node pair $i, j \in N$ is given by parameters $d_{ij}$ and $t_{ij}$, respectively. These parameters can be asymmetric and are assumed to satisfy the triangular inequality. The network design optimizes a convex combination of convenience (mostly travel time) and cost, using parameter $\theta \in [0,1]$: In other words, convenience is multiplied by $\theta$ and cost by $1 - \theta$. The investment cost of opening a leg between the hubs $h, l \in H$ is given by as $\beta_{hl} = (1 - \theta) \, b \, n \, d_{hl}$, where $b$ is the cost of using a bus per mile and $n$ is the number of buses during the planning period. For each trip $r \in T$, the weighted cost and convenience of using a bus between the hubs $h, l \in H$ is given by $\tau^r_{hl} = \theta(t_{hl} + S)$, where $S$ is the average waiting time of a bus (the bus cost is covered by the investment).

This paper adopts a pricing model where the ODMTS subsidizes part, but not all, of the shuttle costs. More precisely, for simplicity in the notations, the paper assumes that the transit price is half of the shuttle cost of a trip.[1] With this pricing model, the weighted cost and convenience for an on-demand shuttle between $i$ and $j$ for the ODMTS and riders is given by $(1-\theta)\frac{g}{2}d_{ij}+\theta t_{ij}$, where $g$ is the cost of using a shuttle per mile. Moreover, the shuttles act as feeders to bus system or serve the local demand. As a result, their operations are restricted to serve requests in a certain distance. This is captured by a threshold value of $\Delta$ miles that characterizes the trips that shuttles can serve. As a result, it is suitable to define the weighted cost and convenience of an on-demand shuttle between the stops $i, j \in N$ as follows:

$$\gamma_{ij}^r := \begin{cases} (1-\theta)\frac{g}{2}d_{ij}+\theta t_{ij} & \text{if } d_{ij} \leq \Delta \\ M & \text{if } d_{ij} > \Delta. \end{cases}$$

where $M$ is a big-M parameter.

To capture latent ridership, this paper assumes that a subset of trips $T' \subseteq T$ currently travel with their personal vehicles, while the trips in $T \setminus T'$ already use the transit system. The goal of the paper is to capture, in the design of the ODMTS, the fact that some riders may switch mode and use the ODMTS instead of their own cars as the transit system has a better cost/convenience trade-off. Each rider $r \in T'$ has a choice model $C^r$ that determines, given the cost/convenience of the novel ODMTS, whether $r$ will switch to the transit system. For instance, the cost model could be

$$C^r(d^r) \equiv \mathbb{1}(d^r \leq \alpha^r \, d_{car}^r)$$

where $d_{car}^r$ represents the weighted cost and convenience of using a car for rider $r$, $d^r$ represents the weighted cost and convenience of using the ODMTS in some configuration, and $\alpha^r \in \mathbb{R}_+$. In other words, rider $r$ would switch to transit if its convenience and cost is not more than $\alpha^r$ times the cost and convenience of traveling with her personal vehicle. The choice model could of course be more complex and include the number of transfers and other features. It can be learned using multimodal logic models or machine learning [17].

## 4  Model Formulation

This section proposes an optimization model for the design of an ODMTS following the specification from Sect. 3. In the model, binary variable $z_{hl}$ is 1 if there is a bus connection from hub $h$ to $l$. Furthermore, for each trip $r$, binary variables $x_{hl}^r$ and $y_{ij}^r$ represent whether rider $r$ uses a bus leg between hubs $h, l \in H$ and a shuttle leg between stops $i$ and $j$ respectively. Binary variable $\delta^r$ for $r \in T'$ is 1 if rider $r$ switches to the ODMTS. The bilevel optimization model for the

---

[1] The results in this paper generalize to other subsidies and pricing models, and they will be discussed in the extended version of the paper.

ODMTS design can then be specified as follows:

$$\min \quad \sum_{h,l \in H} \beta_{hl} z_{hl} + \sum_{r \in T \setminus T'} p^r d^r + \sum_{r \in T'} p^r \delta^r d^r \tag{1a}$$

$$\text{s.t.} \quad \sum_{l \in H} z_{hl} = \sum_{l \in H} z_{lh} \quad \forall h \in H \tag{1b}$$

$$\delta^r = \mathcal{C}^r(d^r) \quad \forall r \in T' \tag{1c}$$

$$z_{hl} \in \{0,1\} \quad \forall h,l \in H \tag{1d}$$

$$\delta^r \in \{0,1\} \quad \forall r \in T' \tag{1e}$$

where $d^r$ is the cost and convenience of trip $r$, i.e.,

$$d^r = \min \quad \sum_{h,l \in H} \tau_{hl}^r x_{hl}^r + \sum_{i,j \in N} \gamma_{ij}^r y_{ij}^r \tag{2a}$$

$$\text{s.t.} \quad \sum_{\substack{h \in H \\ \text{if } i \in H}} (x_{ih}^r - x_{hi}^r) + \sum_{i,j \in N} (y_{ij}^r - y_{ji}^r) = \begin{cases} 1 & \text{, if } i = or^r \\ -1 & \text{, if } i = de^r \\ 0 & \text{, otherwise} \end{cases} \quad \forall i \in N \tag{2b}$$

$$x_{hl}^r \le z_{hl} \quad \forall h,l \in H \tag{2c}$$

$$x_{hl}^r \in \{0,1\} \quad \forall h,l \in H, y_{ij}^r \in \{0,1\} \quad \forall i,j \in N. \tag{2d}$$

The resulting formulation is a bilevel optimization where the leader problem (Eqs. (1a)–(1e)) selects the network design and the follower problem (Eqs. (2a)–(2d)) computes the weighted cost and convenience for each rider $r \in T$ in the proposed ODMTS.

The objective of the leader problem (1a) minimizes the investment cost of opening legs between hubs and the weighted cost and convenience of the routes in the ODMTS for those riders. Constraint (1b) ensures weak connectivity between the hubs and constraint (1c) represents the rider choice, i.e., whether rider $r \in T'$ switches to the ODMTS.

The follower problem of a given trip minimizes the cost and convenience of its route between its origin and destination, under a given transit network design between the hubs (objective function (2a)). Constraint (2b) ensures flow conservation for the bus and shuttle legs. Constraint (2c) guarantees that only open legs are considered by each trip. The follower problem has a totally unimodular constraint matrix, once the leader problem determines the transit network design decisions $z$. In this case, integrality restrictions (2d) can be relaxed, and the problem can be solved as a linear program.

As specified, the follower problem takes into account all of the arcs between each node pair $i,j \in N$ for possible rides with on-demand shuttles. However, due to the triangular inequality, it is sufficient to consider a subset of the arcs for the on-demand shuttles of each trip. More precisely, the optimization only needs to consider arcs i) from origin to hubs, ii) from hubs to destination, and iii) from origin to destination. This subset of necessary arcs for trip $r$ is denoted by $A^r$. Consequently, the model only needs the following decision variables for describing the on-demand shuttles used in trip $r$:

$$y_{or^r h}^r, y_{hde^r}^r \in \{0,1\} \quad \forall h \in H$$
$$y_{or^r de^r}^r \in \{0,1\}.$$

This preprocessing step significantly reduces the size of the follower problem and provides a significant computational benefit.

# 5 Solution Methodology

This section presents a decomposition approach to solve the bilevel problem (1). The decomposition combines traditional Benders optimality cuts with combinatorial Benders cuts to capture the rider choices. The Benders master problem is associated with the leader problem and considers the complicating variables $(z_{hl}, \delta^r, d^r)$ and the subproblems are associated with the follower problems. The master problem relaxes the user choice constraint (1c). The duals of the subproblems generate Benders optimality cuts for the master problem. Moreover, combinatorial Benders cuts are used to ensure that the rider mode choices in the master problem are correctly captured by the master problem. The overall decomposition approach iterates between solving the master problem and guessing $(\bar{z}_{hl}, \bar{\delta}^r, \bar{d}^r)$ and solving the subproblems to obtain the correct value $d^r$ from which the switching decision $C^r(d^r)$ can be derived. The overall process terminates when the lower bound obtained in the master problem and upper bound computed through the feasible solutions converge.

Section 5.1 presents the master problem and Sect. 5.2 discusses the subproblem along with some preprocessing steps. Section 5.3 introduces the cut generation procedure and proposes stronger cuts under some natural monotonicity assumptions. Section 5.4 specifies the proposed decomposition algorithm and proves its finite convergence. Finally, Sect. 5.5 improves the decomposition approach with Pareto-optimal cuts.

## 5.1 Relaxed Master Problem

The initial master problem (3) is a relaxation of the bilevel problem (1), i.e.,

$$\min \quad \sum_{h,l \in H} \beta_{hl} z_{hl} + \sum_{r \in T \setminus T'} p^r d^r + \sum_{r \in T'} p^r \delta^r d^r$$

$$\text{s.t.} \quad (1b), (1d), (1e). \tag{3a}$$

Each iteration first solves the relaxed master problem (3), before identifying combinatorial and Benders cuts to add to the master problem. These cuts depend on the proposed transit network design and rider choices as discussed in Sects. 5.2 and 5.3. The objective function (3a) involves nonlinear terms and needs to be linearized. Since the mode choice is binary, the nonlinear terms can be linearized easily by defining $\nu^r = \delta^r d^r$ and adding the following constraints to the master problem for each trip $r \in T'$:

$$\nu^r \leq \bar{M}^r \delta^r \tag{4a}$$

$$\nu^r \leq d^r \tag{4b}$$

$$\nu^r \geq d^r - \bar{M}^r(1 - \delta^r) \tag{4c}$$

$$\nu^r \geq 0, \tag{4d}$$

where the constant $\bar{M}^r$ is an upper bound value on the objective function value of the lower level problem of trip $r$.

## 5.2  Subproblem for Each Trip

The subproblems for the decomposition algorithm are the duals of the follower problems (2). Since the follower problems have a totally unimodular constraint matrix for a given binary $\bar{z}$ vector, the integrality condition for variable $x^r_{ij}$ can be relaxed into by $x^r_{ij} \geq 0$ and the bounds $x^r_{ij} \leq 1$ can be discarded since it is redundant due to constraint (2c). Then, the dual of the subproblem for each route $r \in T$ can then be specified by introducing the dual variables $u^r_i$ and $v^r_{hl}$:

$$\max \quad (u^r_{or^r} - u^r_{de^r}) - \sum_{h,l \in H} \bar{z}_{hl} v^r_{hl} \tag{5a}$$

$$\text{s.t.} \quad u^r_h - u^r_l - v^r_{hl} \leq \tau^r_{hl} \quad \forall h, l \in H \tag{5b}$$

$$u^r_i - u^r_j \leq \gamma^r_{ij} \quad \forall i, j \in A^r \tag{5c}$$

$$u^r_i \geq 0 \quad \forall i \in N, v^r_{hl} \geq 0 \quad \forall h, l \in H. \tag{5d}$$

Problem (2) is trivially feasible by using the direct trip between origin and destination (which may have a high cost) and hence the dual problem (5) bounded. The optimal objective value of subproblem (2) under solution $\{\bar{z}_{hl}\}_{h,l \in H}$ is denoted by $SP^r(\bar{z})$. In the following section, this value is utilized to evaluate the rider's mode choice and possibly to generate combinatorial cuts.

## 5.3  The Cut Generation Procedure

The cut generation procedure receives a feasible solution $(\{\bar{z}_{hl}\}_{h,l \in H}, \{\bar{\delta}^r\}_{r \in T'}, \{\bar{d}^r\}_{r \in T})$ to the relaxed master problem. It solves the dual subproblem (5) for each trip $r \in T$ under the network design $\bar{z}$. For any trip $r \in T'$, the cut generation procedure then analyzes the feasibility and optimality of the solution of the relaxed master problem, depending on the value of $SP^r(\bar{z})$. The cut generation first needs to enforce the consistency of the choice model.

**Definition 1 (Choice Consistency).** *For a given trip $r$, the solution values $\{\bar{z}_{hl}\}_{h,l \in H}$ and $\bar{\delta}^r$ are consistent with $SP^r(\bar{z})$ if*

$$\bar{\delta}^r = \mathcal{C}^r(SP^r(\bar{z})).$$

As a result, it is useful to distinguish the following cases in the cut generation process:

1. Solution values $\{\bar{z}_{hl}\}_{h,l \in H}$ and $\bar{\delta}^r$ are *inconsistent* with $SP^r(\bar{z})$
   (a) $\bar{\delta}^r = 1$ and $C^r(SP^r(\bar{z})) = 0$;
   (b) $\bar{\delta}^r = 0$ and $C^r(SP^r(\bar{z})) = 1$.
2. Solution values $\{\bar{z}_{hl}\}_{h,l \in H}$ and $\bar{\delta}^r$ are *consistent* with $SP^r(\bar{z})$.

The first inconsistency (case 1(a)) can be removed by using the cut

$$\sum_{(h,l):\bar{z}_{hl}=0} z_{hl} + \sum_{(h,l):\bar{z}_{hl}=1} (1 - z_{hl}) \geq \delta^r \tag{6}$$

**Proposition 1.** *Constraint* (6) *removes inconsistency 1(a).*

The second inconsistency (case 1(b)) can be removed by using the cut

$$\sum_{(h,l):\bar{z}_{hl}=0} z_{hl} + \sum_{(h,l):\bar{z}_{hl}=1} (1 - z_{hl}) + \delta^r \geq 1 \tag{7}$$

**Proposition 2.** *Constraint* (7) *removes inconsistency 1(b).*

Combinatorial cuts (6) and (7) ensure the consistency between the rider choice model and the transit network design $\bar{z}$. These cuts can be strenghtened under a monotonicity property.

**Definition 2 (Anti-Monotone Mode Choice).** *A choice function $C$ is anti-monotone if $d_1 \leq d_2 \Rightarrow C(d_1) \geq C(d_2)$.*

**Proposition 3.** *Let $r \in T$. If $\bar{z}_1 \leq \bar{z}_2$, then $SP^r(\bar{z}_1) \geq SP^r(\bar{z}_2)$.*

*Proof.* If $\bar{z}_1 \leq \bar{z}_2$, more arcs are available in the network defined by $\bar{z}_2$ than in the network defined by $\bar{z}_1$. Therefore, the length of the optimum shortest path for trip $r$ under $\bar{z}_1$ is greater than or equal to that of $\bar{z}_2$. □

The following proposition follows directly from Proposition 3.

**Proposition 4.** *Let $r \in T$ and $C^r$ be an anti-monotone choice function. If $\bar{z}_1 \leq \bar{z}_2$, then $C^r(SP^r(\bar{z}_1)) \leq C^r(SP^r(\bar{z}_2))$.*

When the choice function is anti-monotone, stronger cuts can be derived.

**Proposition 5.** *Consider an anti-monotone choice function. Then constraint* (6) *for case 1(a) can be strengthened into constraint*

$$\sum_{(h,l):\bar{z}_{hl}=0} z_{hl} \geq \delta^r \tag{8}$$

*Proof.* Consider case 1(a) and network design $\bar{z}$. Let $\tilde{z}$ be a network design obtained by removing some arcs from $\bar{z}$. By Proposition 3, $SP^r(\tilde{z}) \geq SP^r(\bar{z})$ for any trip $r$. Hence, by Proposition 4, $C^r(SP^r(\tilde{z})) \leq C^r(SP^r(\bar{z}))$. Therefore, the right term of cut (6) does not remove the inconsistency and the result follows. □

**Proposition 6.** *Consider an anti-monotone choice function. Then constraint (7) for case 1(b) can be strengthened into constraint*

$$\sum_{(h,l):\bar{z}_{hl}=1} (1 - z_{hl}) + \delta^r \geq 1 \tag{9}$$

*Proof.* Consider case 1(b) and network design $\bar{z}$. Let $\tilde{z}$ be a network design obtained by adding some arcs to $\bar{z}$. By Proposition 3, $SP^r(\tilde{z}) \leq SP^r(\bar{z})$ for any trip $r$. Hence, by Proposition 4, $C^r(SP^r(\tilde{z})) \geq C^r(SP^r(\bar{z}))$. Thus, the left term of cut (7) does not remove the inconsistency and the result follows.           □

Since the dual subproblem (5) is bounded, it is also possible to add an optimality cut to the master problem in both cases of 1 and 2 using the weighted cost and convenience of each obtained route. This cut is the standard Benders optimality cut and it uses the vertex $(\bar{u}^r, \bar{v}^r)$ obtained when solving the dual subproblem as follows:

$$d^r \geq (\bar{u}^r_{or^r} - \bar{u}^r_{de^r}) - \sum_{h,l \in H} z_{hl} \bar{v}^r_{hl}. \tag{10}$$

It is also possible to obtain an upper bound from the solutions to the subproblems. Indeed, the rider choices can be derived from the solutions of the subproblems and used instead of the corresponding master variables for the mode choices.

The experimental results use the choice function $C^r(d^r) \equiv \mathbb{1}(d^r \leq \alpha^r \, d^r_{car})$: A rider $r$ chooses the ODMTS if her weighted cost and convenience is not greater than $\alpha^r$ times the weighted cost and convenience $d^r_{car}$ of using her personal car. This choice function is anti-monotone.

**Proposition 7.** *The choice function* $C^r(d^r) \equiv \mathbb{1}(d^r \leq \alpha^r \, d^r_{car})$ *is anti-monotone.*

*Proof.* By definition, $d^r$ decreases when adding arcs to a network and $d^r_1 \leq d^r_2$ implies $C^r(d^r_1) \geq C^r(d^r_2)$.           □

## 5.4   Decomposition Algorithm

The decomposition is summarized in Algorithm 1. It uses a lower and an upper bound to the bilevel problem (1) to derive a stopping condition. The master problem provides a lower bound and, as mentioned earlier, an upper bound can be derived for each network design by solving the subproblems and obtaining the mode choices for the trips.

**Proposition 8.** *Algorithm 1 converges in finitely many iterations.*

---

**Algorithm 1.** Decomposition Algorithm

---

1: Set $LB = -\infty$, $UB = \infty$, $z^* = \emptyset$.
2: **while** $UB > LB + \epsilon$ **do**
3:    Solve the relaxed master problem (3) and obtain the solution ($\{\bar{z}_{hl}\}_{h,l \in H}$, $\{\bar{\delta}^r\}_{r \in T'}$, $\{\bar{d}^r\}_{r \in T}$).
4:    Update $LB$.
5:    **for all** $r \in T$ **do**
6:       Solve the subproblem (5) under $\bar{z}$, and obtain $SP^r(\bar{z})$.
7:       Add optimality cut in the form (10) to the relaxed master problem (3).
8:    **for all** $r \in T'$ **do**
9:       **if** $\{\bar{z}_{hl}\}_{h,l \in H}$ and $\bar{\delta}^r$ are inconsistent with $SP^r(\bar{z})$ **then**
10:          Add cuts in the form (8) or (9) to the relaxed master problem.
11:       **if** $C^r(SP^r(\bar{z}))$ is 1 **then**
12:          Set $\hat{\delta}^r = 1$.
13:       **else**
14:          Set $\hat{\delta}^r = 0$.
15:    $\widehat{UB} = \sum_{h,l \in H} \beta_{hl} \bar{z}_{hl} + \sum_{r \in T \setminus T'} p^r SP^r(\bar{z}) + \sum_{r \in T'} p^r \hat{\delta}^r SP^r(\bar{z})$.
16:    **if** $\widehat{UB} < UB$ **then**
17:       Update $UB$ as $\widehat{UB}$, $z^* = \bar{z}$.

---

*Proof.* The algorithm generates traditional Benders optimality cuts and, in addition, the consistency cuts of the form (8) or (9). When all the consistency cuts are generated, the algorithm reduces to a standard Benders decomposition. There are only finitely many consistency cuts, because the decision variables $z$ and $\delta^r$ are binary. Since each iteration adds at least one new consistency or Benders cut, the algorithm is guaranteed to converge in finitely many iterations.    □

## 5.5   Pareto-Optimal Cuts

The decomposition algorithm can be further enhanced by utilizing Pareto-optimal cuts [12] through alternative optimal solutions of the subproblems. To this end, the algorithm first solves the follower problem (2) under a given network design, obtains the optimal objective value for the corresponding trip, and then solve the Pareto subproblem, i.e., a restricted version of the dual subproblem (5) under this optimal value.

Observe that, once the transit network design $\bar{z}$ is given, the follower problem of each trip $r$ is equivalent to solving a shortest path problem considering the union of the arcs defined by $\bar{z}$ and the arcs in the set $A^r$. Consequently, this shortest path information can be obtained by solving a linear program and obtaining the objective value $\sigma^r$ for trip $r$. Using this information, the Pareto subproblem for trip $r$ is defined as follows:

$$\max \quad (u^r_{or^r} - u^r_{de^r}) - \sum_{h,l \in H} z^0_{hl} v^r_{hl} \tag{11a}$$

$$\text{s.t.} \quad u^r_h - u^r_l - v^r_{hl} \le \tau^r_{hl} \quad \forall h, l \in H \tag{11b}$$

$$u^r_i - u^r_j \le \gamma^r_{ij} \quad \forall i, j \in A^r \tag{11c}$$

$$(u^r_{or^r} - u^r_{de^r}) - \sum_{h,l \in H} \bar{z}_{hl} v^r_{hl} = \sigma^r \tag{11d}$$

$$u^r_i \ge 0 \quad \forall i \in N, v^r_{hl} \ge 0 \quad \forall h, l \in H, \tag{11e}$$

where $z^0$ is a core point that satisfies the weak connectivity constraint (1b). To obtain an initial core point, it suffices to select a value $\eta \in (0, 1)$, and set $z_{hl} = \eta$ for all $h, l \in H$.

## 6    Computational Results

The computational study considers a data set from Ann Arbor, Michigan with 10 hubs located around high density corridors and 1267 bus stops. The experiments examine a set of trips from 6 pm to 10 pm on a specific day. The studied data set involves 1503 trips with a total of 2896 users, where the origin and destination of each trip are associated with bus stops. The costs and times between the bus stops are asymmetric in the studied data set. The study included a preprocessing step to ensure the triangular inequality with respect to the cost and convenience parameters of the on-demand shuttles between the stops.

To model rider preferences in the formulation, the computational study used an income-based classification. This approach assumes that, as the income level of a rider increases, she becomes more sensitive to the quality of the ODMTS route (convenience). In particular, the study considers three classes of riders: i) low-income, ii) middle-income and iii) high-income, where a certain percentage of riders from each class is assumed to use the ODMTS. The trips are then classified with respect to their destination locations, which can be associated with the residences of the corresponding riders. In particular, in the base scenario, 100% of low-income riders, 75% of middle-income riders, and 50% of high-income users utilize the transit system, whereas the remaining riders have the option to select the ODMTS or use their personal vehicles by comparing the obtained route with their current mode of travel.

The convenience parameter $\theta$ is set to 0.01 for weighting cost and convenience. The cost of an on-demand shuttle per mile is taken as $g = \$2.86$ and the cost of a bus per mile is $b = \$7.24$. The buses between hubs have a frequency of 15 min, resulting in 16 buses during the planning horizon with length of 4 h. As mentioned earlier, the price of a ride in the ODMTS is half the cost of the shuttle legs. The base case of the case study sets $\alpha^r$ to 1.25 and 1 for middle-income and high-income riders respectively. The distance threshold for the on-demand shuttles, $\Delta$, is set to 2 or 5 mi.

## 6.1   Transit Design and Mode Switching

Figure 1 depicts the transit network design between hubs under the proposed approach. The bus stops associated with the lowest income level are red dots, those of the middle-income level are grey boxes, and those of the high-income level are green plus symbols. In the resulting network design, almost every hub is connected to at least another hub ensuring weak connectivity of the network.

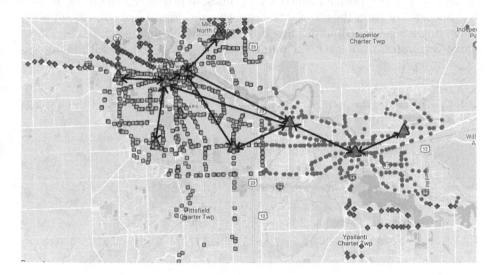

**Fig. 1.** Network design for the ODMTS with 10 Hubs with $\Delta = 2$.

**Table 1.** Adoption rates, average route time and average ride cost for the ODMTS.

| Income level | #trips | %adoption | #riders | %adoption | Avg route time (s) | Avg route cost ($) |
|---|---|---|---|---|---|---|
| Low | 476 | 1.00 | 877 | 1.00 | 901.45 | 2.41 |
| Middle | 784 | 0.96 | 1615 | 0.97 | 553.43 | 2.43 |
| High | 149 | 0.72 | 285 | 0.79 | 583.10 | 2.78 |

Table 1 shows the rider preferences, and the average time and cost of the obtained routes. In particular, columns '#trips' and '#riders' represent the number of trips and riders of the ODMTS. The "%adoption" columns correspond to the adoption rate, i.e., the percentage of trips or riders utilizing the ODMTS. When computing the adoption rate, these numbers include the initial set of riders, i.e., 100% of low-income riders, 75% of middle-income riders and 50%

of high-income riders. The cost and convenience of the ODMTS is sufficiently attractive to exhibit significant mode switching, even for the high-income population. Columns for the average route time and cost represent the averages for the obtained routes regardless of the fact that whether riders adopts the transit system or not. The results highlight the high adoption rates. The average route time is the longest for the low-income riders given their long commuting trips. Similar results are observed for the number of transfers, which include the transfers between on-demand shuttles and buses, and between the buses in the hubs. Specifically, from the set of riders choosing the transit system, 22% of low-income riders, 8% of medium-income riders, and 3% of high-income riders have at least 3 transfers. Moreover, the number of transfers decreases with increases in income level.

**Table 2.** Comparing the average cost and time of the ODMTS trips and those using personal vehicles (Cars).

| | ODMTS trips | | | | Car trips | | | |
|---|---|---|---|---|---|---|---|---|
| | Time | | Cost | | Time | | Cost | |
| Income level | ODMTS | Cars | ODMTS | Cars | ODMTS | Cars | ODMTS | Cars |
| Low | 901.45 | 405.96 | 2.41 | 10.72 | NA | | NA | |
| Medium | 528.94 | 296.95 | 2.38 | 7.14 | 1489.80 | 585.03 | 5.17 | 14.55 |
| High | 529.84 | 326.53 | 2.30 | 7.06 | 93.77 | 31.51 | 0.21 | 0.88 |

Table 2 presents a cost and convenience analysis for the ODMTS trips and those using personal vehicles (cars). The columns corresponding to "ODMTS Trips" represent users who chose the transit system, whereas columns corresponding to "Car Trips" are for those using their personal vehicles, once they observe the transit network design. It also provides the cost and convenience of the other mode, i.e., the convenience and cost of using a personal vehicle for those using the ODMTS and vice-versa. As can be seen, the cost of using the ODMTS is significantly lower, although personal vehicles would decrease the commute time significantly for low-income riders. Note however that the ODMTS has also achieved low commuting times. Riders using personal vehicles do so because the transit times are simply too large for their trips.

The next results examine the effect of the threshold value $\Delta$ on the rides with on-demand shuttles. Figure 2 presents the network design with $\Delta = 5$ mile. This allows for longer shuttle rides from origin to destination of each trip compared to the $\Delta = 2$ case in Fig. 1. As a result, the network design has fewer connections between hubs. Although the investment cost for the network design is lower in this case, the average trip cost increases and the average time of the trips decreases through the adoption of more on-demand shuttles. This highlights the trade-off between the high-frequency buses and on-demand shuttles.

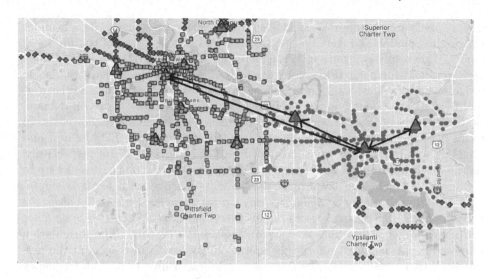

**Fig. 2.** Network design for the ODMTS with 10 Hubs with $\Delta = 5$.

## 6.2   The Benefits of the Formulation

This section compares the novel bilevel formulation with latent demand (L) with the original formulation that ignores the latent demand (O). In other words, the original formulation designs the network with $T \setminus T'$ trips but is evaluated on the complete set $T$ of trips. The two network designs are then compared in terms of cost and convenience. To obtain a realistic setting, the share of public transit is assumed to be 10% for each income level. The results are presented in Table 3.

**Table 3.** Comparison of the proposed (L) and original (O) models with $\Delta = 5$.

| Model | Income | Adoption | Investment ($) | ODMTS trips ($) | Conv. (s) | Cost & Conv. |
|-------|--------|----------|----------------|-----------------|-----------|--------------|
| L | Low | 1.00 | 2482.38 | 17530.24 | 1269263.87 | 32505.13 |
|   | Middle | 1.00 | | | | |
|   | High | 0.84 | | | | |
| O | Low | 1.00 | 861.54 | 20685.56 | 1167457.12 | 33006.20 |
|   | Middle | 0.99 | | | | |
|   | High | 0.82 | | | | |

The results show that both models have similar results in terms of mode switching. However, the new formulation has a higher investment cost and a lower cost for the ODMTS trips compared to the original formulation. The difference between the models is highlighted in Fig. 3, which shows the network designs under the two approaches: The dashed legs represent the design under the original model, and the other legs correspond to the design of the proposed model. This result is intuitive: With more ridership, the ODMTS should open more legs and further reduces congestion. It shows that the novel formulation provides a more robust solution that should reassure transit agencies. As the original formulation opens fewer legs between hubs, users utilize more on-demand shuttles, resulting in trips with more convenience but at much higher costs. In terms of the total investment and trips cost, the results show that the new and original formulations have total costs of $20012.62 and $21547.10, respectively. As this cost improvement corresponds to a planning horizon of 4 h, it scales up to a gain of $1227585 over a yearly plan with 200 days over 16 h. *This is significant for this case study and highlights why transit agencies are worried about the success of ODMTS when they are planned with the existing demand only: They will under-invest in bus lines and sustain higher shuttle costs.* The formulation proposed in this paper remedies this limitation: By taking into account the personalized choice models of riders, the network design invests in high-frequency buses, decreasing the overall cost while maintaining an attractive level of convenience. Note also that, the pricing model adopted in this paper keeps the transit costs low but is also conducive to numerous mode switchings, since the transit system subsidies half the cost. It is also important to report the computational performance of the proposed algorithms. The formulation with latent demand requires 513 s to converge in 8 iterations, whereas the original formulation requires 189 s in 8 iterations for the case study.

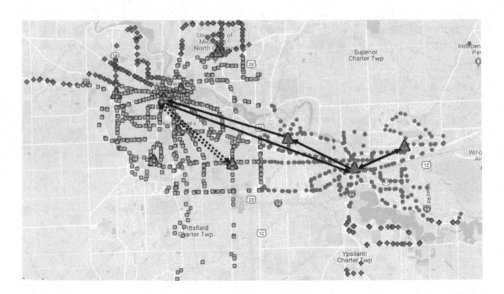

**Fig. 3.** Network designs of the proposed model (L) and the original model (O).

# 7    Conclusion

This study presented a bilevel optimization approach for modeling the ODMTS by integrating rider preferences and considering latent demand. The transit network designer optimizes the network design between the hubs for connecting them with high frequency buses, whereas each rider tries to find the most cost-efficient and convenient route under a given design through buses and on-demand shuttles. The paper considered a generic preference model to capture whether riders switch to the ODMTS based on the obtained route and their current mode of travel. To solve the resulting optimization problem, the paper proposed a novel decomposition approach and developed combinatorial Benders cuts for coupling the network design decisions with rider preferences. A cut strengthening was also proposed to exploit the structure of the follower problem and, in particular, a monotonicity assumption of the choice model. The potential of the approach was demonstrated on a case study using a data set from Ann Arbor, Michigan. The results showed that ignoring latent demand can lead to significant cost increase (about 7.5%) for transit agencies, confirming that these agencies are correct in worrying about customer adoption. This is the case even for a pricing model where the transit agency and riders share the shuttle costs. The new formulation can also be solved in reasonable time.

Current work is devoted to examining the impact of various cost models and different choice models for riders. Applications of the model to the city of Atlanta is also contemplated and should reveal some interesting modeling and computational challenges given the size of the city.

**Acknowledgements.** This research is partly supported by NSF Leap HI proposal NSF-1854684.

# References

1. Brotcorne, L., Labbé, M., Marcotte, P., Savard, G.: A bilevel model for toll optimization on a multicommodity transportation network. Transp. Sci. **35**(4), 345–358 (2001)
2. Brotcorne, L., Labbé, M., Marcotte, P., Savard, G.: Joint design and pricing on a network. Oper. Res. **56**(5), 1104–1115 (2008)
3. Campbell, A.M., Van Woensel, T.: Special issue on recent advances in urban transport and logistics through optimization and analytics. Transp. Sci. **53**(1), 1–5 (2019)
4. Campbell, J.F., Ernst, A.T., Krishnamoorthy, M.: Hub arc location problems: Part ii–formulations and optimal algorithms. Manage. Sci. **51**(10), 1556–1571 (2005)
5. Campbell, J.F., Ernst, A.T., Krishnamoorthy, M.: Hub arc location problems: Part i–introduction and results. Manage. Sci. **51**(10), 1540–1555 (2005)
6. Colson, B., Marcotte, P., Savard, G.: Bilevel programming: a survey. 4OR **3**(2), 87–107 (2005)
7. Farahani, R.Z., Hekmatfar, M., Arabani, A.B., Nikbakhsh, E.: Hub location problems: a review of models, classification, solution techniques, and applications. Comput. Ind. Eng. **64**(4), 1096–1109 (2013)

8. Farahani, R.Z., Miandoabchi, E., Szeto, W., Rashidi, H.: A review of urban transportation network design problems. Eur. J. Oper. Res. **229**(2), 281–302 (2013)
9. Fontaine, P., Minner, S.: Benders decomposition for discrete-continuous linear bilevel problems with application to traffic network design. Transp. Res. Part B: Methodological **70**, 163–172 (2014)
10. Laporte, G., Marín, A., Mesa, J.A., Perea, F.: Designing robust rapid transit networks with alternative routes. J. Adv. Transp. **45**(1), 54–65 (2011)
11. LeBlanc, L.J., Boyce, D.E.: A bilevel programming algorithm for exact solution of the network design problem with user-optimal flows. Transp. Res. Part B: Methodological **20**(3), 259–265 (1986)
12. Magnanti, T.L., Wong, R.T.: Accelerating benders decomposition: algorithmic enhancement and model selection criteria. Oper. Res. **29**(3), 464–484 (1981)
13. Mahéo, A., Kilby, P., Van Hentenryck, P.: Benders decomposition for the design of a hub and shuttle public transit system. Transp. Sci. **53**(1), 77–88 (2019)
14. Sinha, A., Malo, P., Deb, K.: A review on bilevel optimization: from classical to evolutionary approaches and applications. IEEE Trans. Evol. Comput. **22**(2), 276–295 (2018)
15. Van Hentenryck, P.: Social-aware on-demand mobility systems. ISE Mag. (Fall 2019)
16. Yu, B., Kong, L., Sun, Y., Yao, B., Gao, Z.: A bi-level programming for bus lane network design. Transp. Res. Part C Emerg. Technol. **55**, 310–327 (2015)
17. Zhao, X., Yan, X., Van Hentenryck, P.: Modeling heterogeneity in mode-switching behavior under a mobility-on-demand transit system: an interpretable machine learning approach. CoRR abs/1902.02904 (2019). http://arxiv.org/abs/1902.02904

# Local Search and Constraint Programming for a Real-World Examination Timetabling Problem

Michele Battistutta[1], Sara Ceschia[2] (iD), Fabio De Cesco[1], Luca Di Gaspero[2] (iD),
Andrea Schaerf[2(✉)] (iD), and Elena Topan[1]

[1] EasyStaff srl, Via Adriatica, 278, 33030 Campoformido, UD, Italy
{michele,fabio,elena}@easystaff.it
[2] DPIA, University of Udine, Via delle scienze 206, 33100 Udine, Italy
{sara.ceschia,luca.digaspero,andrea.schaerf}@uniud.it

**Abstract.** We investigate the examination timetabling problem in the context of Italian universities. The outcome is the definition of a general problem that can be applied to a large set of universities, but is quite different in many aspects from the classical versions proposed in the literature.

We propose both a metaheuristic approach based on Simulated Annealing and a Constraint Programming model in MiniZinc. We compare the results of the metaheuristic approach (properly tuned) with the available MiniZinc back-ends on a large set of diverse real-world instances.

**Keywords:** Examination timetabling · Simulated annealing · MiniZinc

## 1 Introduction

Examination timetabling (ETT) is one of the classical problems that every university has to deal with on a regular basis. Many formulations of the ETT problem have been proposed in the literature, some of which have received considerable attention [5, 12].

We propose a novel formulation of ETT, which applies to Italian universities. This formulation is quite different from the ones proposed in the literature [15, 16], as it involves many specific constraints and objectives. For example, some exams are composed by separate written and oral part, which must be scheduled at suitable distance and have different overlap acceptability levels in relation to other exams. In addition, the same exam might be repeated more than once in a session, with prescribed minimal distances among rounds. As another quite distinctive feature, exams might require multiple rooms, typically in exclusive use.

In this work, we propose a metaheuristic approach based on a tailored neighborhood structure, a Simulated Annealing procedure, and a statistically-principled parameter tuning. The main motivation for the choice of Simulated

E. Hebrard and N. Musliu (Eds.): CPAIOR 2020, LNCS 12296, pp. 69–81, 2020.
https://doi.org/10.1007/978-3-030-58942-4_5

Annealing is that it already turned out to be capable of obtaining state-of-the-art results in many educational timetabling problems (see, e.g., [1,2,6]).

We also developed a MiniZinc model that allows us to compare the results of the metaheuristic, by testing many available back-ends on this challenging problem.

The outcome is that Simulated Annealing is able to solve real-world instances to a good quality. Conversely, most instances are currently beyond the reach of exact Constraint Programming methods.

As a byproduct of this research, we are collecting many real-world instances, which are made available to the community for future comparisons, and could potentially become a new benchmark. Instances, solutions, and the MiniZinc model are available at https://bitbucket.org/satt/ExamTimetablingUniudData. The repository includes also a Python validator that checks the cost of a solution, so as to provide against possible misunderstanding about the constraints and the objectives.

## 2   Problem Formulation

Our problem consists of scheduling an examination session at the end of a semester for a university campus. The problem is based on the following entities:

**Courses, Exams, & Events:** For each course, we have to schedule one or more exams within the session. Each exam might be a single event (either written or oral) or composed by two events, identified as the written part (first) and the oral part (second), to be handed out in this strict order.

**Rooms & Roomsets:** Some events require one or more rooms, others do not, as they take place in teacher's office or in external rooms. Rooms are classified as small, medium, or large, and for each written event we set the number and the type of rooms requested (mixed requests are not considered). Due to logistic issues, not all combinations of homogeneous rooms can be assigned to a single event. The available ones, called *roomsets*, are explicitly listed in the input data. Oral events might require at most one room (of any size).

**Days, Timeslots, & Periods:** The session is divided in days and each day in divided in timeslots, with the same number of timeslots for each day. Each pair day/timeslot represents a period of the session.

**Curricula:** A curriculum is a set of courses that have students in common, which might enroll in the corresponding exams. The set of courses of a curriculum is split into *primary* courses, that are the ones taught in the current semester, and the *secondary* ones, that have been taught in the previous semester, but such that some students might still have to undertake them. The level of conflict between primary and secondary exams of a curriculum varies, as detailed below.

The problem consists in assigning a period and a *location* (a room, a roomset, or nothing) to each event, satisfying the hard and soft constraints explained in the following paragraphs.

COURSES

| Name | Teacher | #Exams | Exam Distance | Exam Type | $\mathcal{W}/\mathcal{O}$ Distance | Location |
|------|---------|--------|---------------|-----------|--------------------------------|----------|
| Databases | Andrea | 2 | 5 | $\mathcal{W} + \mathcal{O}$ | (4,8) | 2 Large |
| Algorithms | Luca | 2 | 5 | $\mathcal{W}$ | — | 1 Large |
| Program. | Elena | 2 | 4 | $\mathcal{W}$ | — | 2 Large |
| Oper. Res. | Sara | 2 | 4 | $\mathcal{W} + \mathcal{O}$ | (1,1) | 1 Small + 1 Small |
| Analytics | Michele | 1 | — | $\mathcal{O}$ | — | — |

ROOMS

| Name | Type |
|------|------|
| A | Large |
| B | Large |
| C | Large |
| G | Small |
| H | Small |

ROOMSETS

| Roomset | Size | Type |
|---------|------|------|
| {A, B} | 2 | Large |
| {A, C} | 2 | Large |

PERIODS

| 20 |
|----|

CURRICULA

| Name | Primary | Secondary |
|------|---------|-----------|
| Man. Eng. | Databases Oper. Res. | Analytics |
| Ele. Eng. | Databases Algorithms | Oper. Res. Analytics |

**Fig. 1.** A toy instance.

All the data representing a toy instance is shown in Fig. 1. The table COURSES shows the structure of the courses, with teacher, number of exams, distance between exams, exam type ($\mathcal{W}$ for written, $\mathcal{O}$ for oral), and location requested. The other tables show the respective features, including the available roomsets. Preferences and constraints are not shown for the sake of brevity.

We propose a possible solution in Fig. 2. We can see that the 5 courses are "exploded" into 13 events. Each single event has its own conflict and distance constraints, as discussed in the following paragraphs.

The hard constraints are the following:

**H1. RoomRequest:** for each written event, type and number of the rooms assigned must be correct; for oral exams, a single room, of any type, must be assigned, if requested.

**H2. RoomOccupation:** in one period, a room can be used by at most one event.

**H3. HardConflicts:** Two events in *hard conflict* cannot be scheduled in the same period. Two events are in hard conflict in the following cases:
- They are part of courses that are both primary courses of one curriculum.
- They have the same teacher.
- There is an explicit constraint stating that the overlap of the two events is forbidden.

**H4. Precedences:** When two events have a *precedence* constraint, the first must be scheduled strictly before the second. Two events have a precedence constraint in the following cases:
- They are part of two exams of the same course.
- They are part of the same exam (written and oral).

**H5. Unavailabilities:** An event might be explicitly stated as unavailable in a specific period, so that it cannot be assigned to that period. Similarly, a

SOLUTION

| Course | Event | | | Assignment | |
| | Exam | Part | ID | Period | Rooms |
|---|---|---|---|---|---|
| Databases | #1 | W | 0 | 2 | A, B |
| | | O | 1 | 5 | — |
| | #2 | W | 2 | 12 | A, B |
| | | O | 3 | 15 | — |
| Algorithms | #1 | W | 4 | 0 | B |
| | #2 | W | 5 | 10 | A |
| Programming | #1 | W | 6 | 0 | A, C |
| | #2 | W | 7 | 17 | A, B |
| Oper. Res. | #1 | W | 8 | 6 | H |
| | | O | 9 | 7 | G |
| | #2 | W | 10 | 16 | H |
| | | O | 11 | 17 | G |
| Analytics | #1 | O | 12 | 8 | — |

Fig. 2. A solution for the toy instance.

room might be unavailable for a specific period, so that no event can use it in that period.

The objectives (soft constraints) are they following:

**S1. SoftConflicts:** Two events in *soft conflict* should not be scheduled in the same period. Two events are in soft conflict in the following cases:
- They belong to courses that are in the same curriculum, either as primary and secondary or both as secondary.
- There is an explicit constraint stating that their overlap is undesirable.

**S2. Preferences:** Like Unavailabilities, preferences between events and periods and between periods and rooms stating the undesirability of an assignment can be expressed explicitly. For periods, it is also possible to state a positive preference for a specific event, so that in presence of preferred periods for an event, all indifferent ones are assumed undesired (and explicitly undesired one are given a larger penalty).

**S3. Distances:** Among events there might be requested separations in term of periods. Distances can be either *directed*, imposing that one event must precede the other, or *undirected* so that any order is acceptable. The situations that require a distance are the following:
- Same exam (directed): different parts of the same exam have a minimum and a maximum distance, stated specifically for each course (e.g., events 0 and 1 in the example).
- Same course (directed): different exams of the same course must be separated. The separation constraint is applied between the first (or single) part of each of the two exams (e.g., events 0 and 2 in the example).
- Same curriculum (undirected): if two courses belong to the same curriculum, there should be a separation between the exams (as above, for

two-part exams, we consider the first one). The amount of separation and the weight for its violation depend on the type of the two (primary or secondary) memberships.

– Additional requests can be added explicitly.

The weight of the violation of the various types of soft conflicts are set by the end-user. Similarly, all distance limits and the corresponding weights are configurable.

We can see that the solution of Fig. 2 has a few soft constraint violations. For example, the distance between the two parts of Databases is 3 periods for both exams, whereas the minimum is 4. Another violation is related to the curriculum Ele. Eng. for which the written part of the first exam of Databases (at period 2) is too close to the written part of the single exam of Algorithms (at period 0), given that the minimum distance for primary/primary exams is assumed to be 6.

## 3  Related Work

The literature on the examination timetabling problem is rather vast. We refer to Qu et al. [15] for a relatively up-to-date survey.

Among the large set of proposed formulations, two have received considerable attention in the literature. The first is the classical one from Carter et al. [5]. This is a very essential version of the problem, in which each exam is a single event and rooms are not considered. Differently from our formulation, which is curriculum-based, in the work of Carter et al. conflicts are based on student enrollments, so that each student contributes to the penalty of a schedule whenever the exams (s)he takes are too close. Given also that rooms are not involved, the only constraint is HardConflicts (H3) and the only objective is Distances (S3). Distances are penalized in a fixed *exponential* patterns: the cost of scheduling two exams with $k$ common students at distance 1, 2, 3, 4, and 5 periods is $16k$, $8k$, $4k$, $2k$, and $k$, respectively.

This formulation comes with a challenging dataset of 13 real-world instances form North American universities that are still not solved to proven optimality. The dataset has been subsequently extended and generalized by other authors (see, e.g.., [13]). Recent results on this formulation have been obtained by Leite et al. [10].

The other popular formulation is the one coming from Track 1 of the 2nd International Timetabling Competition [12]. This is a much richer formulation, taken mainly from British universities, that includes rooms and many specific constraints. Among others, periods and exams can have different length, so that some periods might be unsuitable for certain exams. Rooms might be used either exclusively or shares among exams (preferably of the same length). Exams might have precedence constraints. The objective function components regard mainly the spreading of exams for students and the assignment of large exams in the initial periods.

For this formulation 12 instances are available which are also quite challenging and not solved to optimality up to today. The contributions on this problem include [1,4,11].

Another notable formulation is the one proposed by Müller [14], which represents a complex real-world problem. The formulation comes along with 9 challenging instances in XML format.

# 4   Local Search Solver

We now introduce our local search method. We first describe the search space, the cost function, and the initial solution (Sect. 4.1), then we introduce the neighborhood relations (Sect. 4.2), and finally we discuss our metaheuristic strategy (Sect. 4.3).

## 4.1   Search Space, Cost Function, and Initial Solution

A state in the search space is represented by two vectors that store for each event the period when it is scheduled and the location where it takes place. As already mentioned, the location can be either a single room, a roomset, or the *dummy* room (no room assigned). Any period can be assigned to an event unless it is unavailable for the event. For the location assignment, instead, we consider only locations of the correct type, according to the requirements.

This means that the constraints H1 and H5 are always satisfied. Conversely, we let the search method visit also infeasible state, by violating constraints H2, H3, and H4.

The cost function is thus the weighted sum of the penalties of the soft constraints S1–S3 and the violations of hard constraints H2–H4 multiplied by a suitably high weight, such that a single hard constraint violation is never preferred to any combination of soft ones.

The initial solution is generated at random, but satisfying constraints H1 and H5. In particular, for the selection of the location, the choice is among the compatible ones. For example, an event that does not require a room is compatible only with the dummy room, so that this is the location that is always selected.

## 4.2   Neighborhood Relations

The typical basic neighborhood relation in examination timetabling is the reposition of a single event. We call this neighborhood MoveEvent:

**MoveEvent (ME).** Given an event $e$, a period $p$ and a location $l$, the move $ME\langle e, p, l \rangle$ repositions $e$ at $p$ in $l$.
    Preconditions: $p$ is available for $e$ and $l$ is compatible with $e$.

In our case, the presence of exams composed by two events suggests that it might be useful to consider a neighborhood that could move the two paired events jointly. However, we should consider the possibility of moving either the single event or the complete exam. This leads us to the following neighborhood:

**MoveEventOrExam (MEE).** Given an event $e$, a period $p$, a location $l$, a Boolean $b$, a period $p'$, and a location $l'$, the move $\text{MEE}\langle e, p, l, b, p', l'\rangle$ repositions $e$ at $p$ in $l$. If $b$ is true, it also repositions the event $e'$ associated with $e$ at $p'$ in $l'$. If $b$ is false, then $p'$ and $l'$ are ignored.

Preconditions: $p$ is available for $e$ and $l$ is compatible with $e$; if $e$ is not part of a composite exam then $b$ is false; if $b$ is true, $p'$ is available for $e'$ and $l'$ is compatible with $e'$.

## 4.3   Simulated Annealing

As a metaheuristic to guide the local search we use Simulated Annealing (SA) [9]. For an up-to-date exhaustive introduction to Simulated Annealing and its variants we refer to the work of Franzin and Stützle [7].

Our SA procedure starts from an initial random state and at each iteration draws a random move in the MEE neighborhood. For the MEE move selection, we first select uniformly the event $e$, the period $p$, and the location $l$. If $e$ is not part of a composite exam, then $b$ is set to false. If instead $e$ is part of a composite exam, we select $b = true$ with probability $p_b$ and $b = false$ with probability $1 - p_b$. If $b$ is true, $p'$ and $l'$ are randomly selected, whereas if $b$ is false $p'$ and $l'$ are ignored.

As customary for SA, calling $\Delta$ the difference of cost induced by the selected move, this is accepted if $\Delta \leq 0$, whereas it is accepted based on time-decreasing exponential distribution (called Metropolis) in case $\Delta > 0$. Specifically, a worsening move is accepted with probability $e^{-\Delta/T}$, where $T$ is the *temperature*. The temperature starts at the initial value $T_0$, and decreases by being multiplied by a value $\alpha$ (with $0 < \alpha < 1$), each time a fixed number of samples $n_s$ has been drawn.

To the basic SA procedure, we add the *cut-off* mechanism that speeds up the early stages. The idea is to decrease the temperature also when a given number of moves has been accepted. That is, we add a new parameter $n_a$, that represents the maximum number of accepted moves at each temperature. The temperature is decreased when the first of the following two conditions occurs: $(i)$ the number of sampled moves reaches $n_s$, $(ii)$ the number of accepted moves reaches $n_a$.

We use the total number of iterations $\mathcal{I}$ as stop criterion. This guarantees that the running time is the same for all configurations of the SA parameters. With respect to a criterion based on a strict time limit, our criterion has the advantage that it is deterministic (given the random seed) and it is not dependent on the environment, so that each run can be reproduced precisely.

In order to keep total number of iterations $\mathcal{I}$ fixed, one of the parameters, namely $n_s$, is not left free, but is computed from $\mathcal{I}$ and from the others using the formula:

$$n_s = \mathcal{I}/\left(\frac{\log(T_f/T_0)}{\log \alpha}\right)$$

where $T_f$ is the *expected* final temperature. Notice that $T_f$ is used to compute $n_s$, but the actual final temperature might fall below, as the potential iterations saved in the early stages, due to the cut-off mechanism, are returned at the end of the search.

Given that $n_s$ is not fixed, $n_a$ is not fixed directly, but is set to be a fraction $\rho$ of the computed $n_s$, where $\rho$ is a new parameter (that replaces $n_a$).

The running time is equal for all configurations on the same instance, but may be different from instance to instance, as the computation of the costs depends on the size of the instance.

## 5    Constraint Programming Model

In order to use MiniZinc, the input format needs to be preprocessed in such a way to flatten all the data (conflicts, distances, ...) at the level of each single event, and write it in terms of a set of arrays.

The decision variables are two vectors of size equal to the number of events, called `EventPeriod` and `EventLocation`, storing the assigned period and the assigned location, respectively. As for the local search solver, the location assigned can be either a single room, a roomset, or the dummy room.

The constraint stating that the same room cannot be used simultaneously by two events is the following, where `LocationOverlap` is a binary input matrix, stating of two locations overlap (1) or not (0).

```
constraint
  forall(e1 in 1..Events-1, e2 in e1+1..Events)
    (EventPeriod[e1] != EventPeriod[e2] \/
      LocationOverlap[EventLocation[e1], EventLocation[e2]] = 0);
```

Obviously, two locations overlap if they have a room in common; the dummy room does not overlap with any room, not even with itself, so that more than one event can be placed in the dummy room at the same time.

This is the most critical constraint, as it involves a disjunction. The other constraints are relatively straightforward, and are not shown here.

The objective function is obtained as the weighted sum of various components. For example, the variable carrying the count of the soft conflict violations is connected to the main variables by the following constraint.

```
constraint
  ConflictCost = sum(e1 in 1..Events-1, e2 in e1+1..Events
                  where Conflicts[e1,e2] > 0)
        ((EventPeriod[e1] = EventPeriod[e2]) * Conflicts[e1,e2]);
```

**Table 1.** Main features of the instances.

| Dept. | #inst | Courses | | Events | | Periods | | Rooms | | Slots | |
|-------|-------|---------|-----|--------|-----|---------|-----|-------|-----|-------|-----|
|       |       | min | max | min | max | min | max | min | max | min | max |
| D1 | 7 | 239 | 281 | 239 | 281 | 26 | 52 | 64 | 65 | 2 | 2 |
| D2 | 3 | 57 | 58 | 61 | 62 | 156 | 204 | 0 | 0 | 6 | 6 |
| D3 | 9 | 76 | 89 | 78 | 177 | 48 | 188 | 14 | 15 | 4 | 4 |
| D4 | 6 | 223 | 240 | 235 | 514 | 38 | 88 | 34 | 34 | 2 | 2 |
| D5 | 5 | 125 | 156 | 132 | 426 | 24 | 136 | 17 | 20 | 2 | 2 |
| D6 | 8 | 189 | 207 | 346 | 539 | 52 | 90 | 29 | 29 | 2 | 2 |
| D7 | 2 | 60 | 63 | 136 | 150 | 155 | 330 | 22 | 22 | 5 | 10 |

The symmetric matrix `Conflicts` carries the value of the penalty of the conflict between two events. It has the conventional value $-1$ in case of hard conflict. The other components have similar structure and they are not shown here.

Finally, for variable and value selections, we use the default strategy (i.e., no search annotation is used).

## 6  Problem Instances

At present, we have collected 40 instances coming from 7 different departments (of 6 different universities), which show a good variety of diverse practical situations. Table 1 summarizes, for each department, the values (minimum and maximum) of the main features of the corresponding instances.

Notice that one department (D2) has 0 rooms, so that all events are assigned to the dummy room. This means that the management has decided to leave outside the system the assignment of the rooms. Notice also that the number of rooms and courses is rather stable within the instances of the same department, whereas the number of events might change considerably. This is due to the fact that in different sessions during the year the exam of the same course might be repeated a different number of times.

## 7  Experimental Analysis

The experiments have been run on an Ubuntu Linux 18.4 machine with 4 cores Intel® i7-7700 (3.60 60 GHz), with a single core dedicated to each experiment.

### 7.1  Tuning

The tuning phase of the local search solver has been performed using the tool JSON2RUN [17], which samples the configurations using the *Hammersley point set* [8] and implements the F-Race procedure [3] for comparing them.

The resulting best configuration is shown in Table 2, which shows also the initial intervals selected based on preliminary experiments.

**Table 2.** Parameter settings.

| Name | Description | Value | Range |
|------|-------------|-------|-------|
| $T_0$ | Start temperature | 118.75 | 100—500 |
| $T_f$ | Final temperature | 0.28 | 0.1—1 |
| $\alpha$ | Cooling rate | 0.99 | 0.8—0.999 |
| $\rho$ | Accepted moves ratio | 0.2 | 0.1—0.3 |
| $p_b$ | Move written/oral together rate | 0.75 | 0.0—1.0 |

## 7.2   Comparative Results

For the MiniZinc model we have tested the back-ends available in the standard distribution (v. 2.3.2), plus `cplex` (v. 12.9). For all of them we set a timeout of 1 h for each run.

The results obtained are shown in Table 3, along with the average and best results out of 30 runs of the Simulated Annealing solver, with $\mathcal{I} = 10^8$. The outcome is that all instances have feasible solutions. In addition, some cases can be easily solved to a *perfect* solution (0 cost), some others are more challenging and result in relatively high costs. For the MiniZinc back-ends, the symbol × means that the solver exhausted the memory, and the symbol — that it has not been able to produce any feasible solution within the time limit. Optimality has been proved only for the 0 cost solutions.

The metaheuristic approach has been able to obtain satisfactory solutions, and it proved to be quite robust, as the gap between the best costs and the average ones is relatively low.

On the contrary, the straightforward CP model in MiniZinc turned out to be unusable for most practical instances, leaving room for search strategies and/or smarter encodings to be developed.

Only for the department D2, `cplex` has been able to provide better results than SA in all three instances, and `coin-bc` in two of them. In particular, in those two instances, both have found the perfect solution, whereas SA is consistently stuck in a solution of cost 22.

**Table 3.** Comparative results.

| Instance | SA | | | MiniZinc | | | |
|---|---|---|---|---|---|---|---|
| | avg. | best | time | chuffed | gecode | coin-bc | cplex |
| D1-1-16 | **180.40** | 180 | 568.1 | × | — | × | × |
| D1-1-17 | **134.00** | 134 | 499.2 | × | — | × | × |
| D1-2-16 | **258.63** | 257 | 541.6 | × | — | × | × |
| D1-2-17 | **352.00** | 351 | 698.9 | × | — | × | × |
| D1-3-16 | **478.37** | 477 | 483.3 | × | — | × | × |
| D1-3-17 | **354.57** | 354 | 493.1 | × | — | × | × |
| D1-3-18 | **80.00** | 80 | 536.6 | × | — | × | × |
| D2-1-18 | 427.77 | 426 | 94.7 | — | 8731 | 906 | **406** |
| D2-2-18 | 22.00 | 22 | 88.7 | 1543 | 4022 | 0 | 0 |
| D2-3-18 | 22.00 | 22 | 95.0 | 1873 | 3985 | 0 | 0 |
| D3-1-16 | **0.00** | 0 | 61.9 | — | 75947 | × | — |
| D3-1-17 | **0.00** | 0 | 83.5 | — | 82948 | × | — |
| D3-1-18 | **0.00** | 0 | 83.9 | — | 82433 | × | — |
| D3-2-16 | **0.00** | 0 | 0.8 | 0 | — | — | 0 |
| D3-2-17 | **0.00** | 0 | 3.1 | 0 | — | — | 0 |
| D3-2-18 | **0.00** | 0 | 3.5 | — | — | — | 0 |
| D3-3-16 | **0.00** | 0 | 1.0 | 0 | — | — | 0 |
| D3-3-17 | **0.00** | 0 | 2.3 | 0 | — | — | 0 |
| D3-3-18 | **0.00** | 0 | 2.2 | 0 | — | — | 0 |
| D4-1-17 | **132.43** | 18 | 312.0 | × | — | × | × |
| D4-1-18 | **567.73** | 563 | 1401.4 | × | — | × | × |
| D4-2-17 | **575.50** | 566 | 1307.3 | × | — | × | × |
| D4-2-18 | **11609.33** | 9685 | 39.3 | × | — | × | × |
| D4-3-17 | **137.03** | 137 | 462.6 | — | — | × | × |
| D4-3-18 | **379.50** | 372 | 555.2 | — | — | × | × |
| D5-1-17 | **7361.03** | 5870 | 2.9 | — | — | × | × |
| D5-1-18 | **38.00** | 36 | 824.2 | — | — | × | × |
| D5-2-17 | **60.20** | 60 | 930.0 | — | — | × | × |
| D5-2-18 | **274.37** | 270 | 1237.8 | × | — | × | × |
| D5-3-18 | **0.00** | 0 | 68.4 | — | — | × | — |
| D6-1-16 | **898.83** | 872 | 1429.1 | × | — | × | × |
| D6-1-17 | **740.87** | 723 | 1385.5 | × | — | × | × |
| D6-1-18 | **881.50** | 873 | 1420.1 | × | — | × | × |
| D6-2-16 | **948.00** | 935 | 1551.3 | × | — | × | × |
| D6-2-17 | **943.13** | 920 | 1428.5 | × | — | × | × |
| D6-2-18 | **692.03** | 683 | 1650.4 | × | — | × | × |
| D6-3-16 | **355.40** | 355 | 882.4 | × | — | × | × |
| D6-3-17 | **381.57** | 381 | 969.0 | × | — | × | × |
| D7-1-17 | **373.20** | 360 | 222.5 | — | 56891 | × | — |
| D7-2-17 | **766.50** | 758 | 219.3 | — | 114385 | — | — |

# 8   Conclusions and Future Work

We have modeled a complex real-world version of the examination timetabling problem.

The metaheuristic solver has found good results on most instances, although the presence of a few results far from the optimum is a clue that further improvements are possible. To this aim, we plan to devise new neighborhood relations and different metaheuristic strategies.

The results show that the problem, in its straightforward modeling, is beyond the reach of MiniZinc back-ends. Smarter encodings are necessary in order to try to improve the performances of all the back-ends.

The future work includes also the extension of the model to features that appear in a few cases, which have been neglected in our current formulation. The most important ones are: events that span over several periods, heterogeneous roomsets, conflicts at the level of the day (not only of the single period), exams to be given in the same day and in the same location, and uniform spreading of the primary courses of a curriculum.

# References

1. Battistutta, M., Schaerf, A., Urli, T.: Feature-based tuning of single-stage simulated annealing for examination timetabling. Ann. Oper. Res. **252**(2), 239–254 (2015)
2. Bellio, R., Ceschia, S., Di Gaspero, L., Schaerf, A., Urli, T.: Feature-based tuning of simulated annealing applied to the curriculum-based course timetabling problem. Comput. Oper. Res. **65**, 83–92 (2016)
3. Birattari, M., Yuan, Z., Balaprakash, P., Stützle, T.: F-Race and iterated F-Race: an overview. In: Bartz-Beielstein, T., Chiarandini, M., Paquete, L., Preuss, M. (eds.) Experimental Methods for the Analysis of Optimization Algorithms, pp. 311–336. Springer, Heidelberg (2010)
4. Bykov, Y., Petrovic, S.: An initial study of a novel step counting hill climbing heuristic applied to timetabling problems. In: Proceedings of the 6th Multidisciplinary International Conference on Scheduling : Theory and Applications (MISTA-13), pp. 691–693 (2013)
5. Carter, M.W., Laporte, G., Lee, S.Y.: Examination timetabling: algorithmic strategies and applications. J. Oper. Res. Soc. **74**, 373–383 (1996)
6. Ceschia, S., Di Gaspero, L., Schaerf, A.: Design, engineering, and experimental analysis of a simulated annealing approach to the post-enrolment course timetabling problem. Comput. Oper. Res. **39**, 1615–1624 (2012)
7. Franzin, A., Stützle, T.: Revisiting simulated annealing: a component-based analysis. Comput. Oper. Res. **104**, 191–206 (2019)
8. Hammersley, J.M., Handscomb, D.C.: Monte Carlo Methods. Chapman and Hall, London (1964)
9. Kirkpatrick, S., Gelatt, D., Vecchi, M.: Optimization by simulated annealing. Science **220**, 671–680 (1983)
10. Leite, N., Fernandes, C., Melício, F., Rosa, A.: A cellular memetic algorithm for the examination timetabling problem. Comput. Oper. Res. **94**, 118–138 (2018)

11. Leite, N., Melício, F., Rosa, A.: A fast simulated annealing algorithm for the examination timetabling problem. Expert Syst. Appl. **122**, 137–151 (2019)
12. McCollum, B., et al.: Setting the research agenda in automated timetabling: the second international timetabling competition. INFORMS J. Comput. **22**(1), 120–130 (2010)
13. Merlot, L.T.G., Boland, N., Hughes, B.D., Stuckey, P.J.: A hybrid algorithm for the examination timetabling problem. In: Burke, E., De Causmaecker, P. (eds.) PATAT 2002. LNCS, vol. 2740, pp. 207–231. Springer, Heidelberg (2003)
14. Müller, T.: Real-life examination timetabling. J. Sched. **19**(3), 257–270 (2014)
15. Qu, R., Burke, E., McCollum, B., Merlot, L., Lee, S.: A survey of search methodologies and automated system development for examination timetabling. J. Sched. **12**(1), 55–89 (2009)
16. Schaerf, A.: A survey of automated timetabling. Artif. Intell. Rev. **13**(2), 87–127 (1999)
17. Urli, T.: json2run: a tool for experiment design & analysis. CoRR abs/1305.1112 (2013)

# Parameterised Bounds on the Sum of Variables in Time-Series Constraints

Nicolas Beldiceanu[1]([✉]), Maria I. Restrepo[1], and Helmut Simonis[2]

[1] TASC (LS2N-CNRS), IMT Atlantique, 44307 Nantes, France
nicolas.beldiceanu@imt-atlantique.fr, mrestrep@uco.fr
[2] Insight Centre for Data Analytics, University College Cork, Cork, Ireland
helmut.simonis@insight-centre.org

**Abstract.** For two families of time-series constraints with the aggregator Sum and features one and width, we provide parameterised sharp lower and upper bounds on the sum of the time-series variables wrt these families of constraints. This is important in many applications, as this sum represents the cost, for example the energy used, or the manpower effort expended. We use these bounds not only to gain a priori knowledge of the overall cost of a problem, we can also use them on increasing prefixes and suffixes of the variables to avoid infeasible partial assignments under a given cost budget. Experiments show that the bounds drastically reduce the effort to find cost limited solutions.

## 1 Introduction

Time series is an increasingly important format of data in many applications, from financial to scientific. Time series are sequences of values taken at successive equally spaced points in time. Two traditional topics are time series forecasting [16] and time series pattern recognition [15,19]. A more recent topic is the generation of time series satisfying a given set of constraints. Indeed, in an industrial or commercial setting, time series are constrained by physical laws or organisational regulations. In this case, when time series correspond to a resource produced or consumed, the question of maximising or minimising the sum of the elements of a time series becomes important. This article focuses on this issue.

*Context and Motivation.* From a constraint perspective work on time-series constraints was introduced in [13] to formalise the notions of exact and approximate similarity between time-series patterns and data. More recently, some authors have proposed quantitative regular expressions [1,2] as a way to *(i)* formalise and identify common types of time-series patterns [9,18], and to *(ii)* express

This publication has emanated from research conducted with the financial support of Science Foundation Ireland under Grant number 12/RC/2289-P2 which is co-funded under the European Regional Development Fund as well as from the Gaspard-Monge program.

E. Hebrard and N. Musliu (Eds.): CPAIOR 2020, LNCS 12296, pp. 82–98, 2020.
https://doi.org/10.1007/978-3-030-58942-4_6

time-series constraints, which are then used to generate constrained time series. To improve propagation, implied constraints and cuts were derived in [3,6,7].

These ideas have been used to solve real-life problems including the analysis of the output of electric power stations over multiple days [11], the solution of a staff scheduling problem in a service company [5], power management for large-scale distributed systems [10] and the generation of typical energy consumption profiles of a data centre [12,14]. Most of these problems require the incorporation of an objective function which is represented by the sum of the decision variables. Hence, computing bounds on such sum is an important issue.

*Time-Series Constraints.* A time-series constraint $\gamma(X, R)$ is a constraint which restricts an integer result variable $R$ to be the result of some computations over a sequence of $n$ integer variables $X$. The components of a time-series constraint we reuse from [9] are a *pattern* $\sigma$, a *feature* $f$, and an *aggregator* $g$. A pattern is described by a *regular expression* over the alphabet $\Sigma = \{$ ' $<$ ',' $=$ ',' $>$ ' $\}$ whose language $\mathcal{L}_\sigma$ does not contain the empty word. For instance, in [4] the Plateau pattern is characterised by the expression ' $<=^*>$ '. A feature and an aggregator are functions over integer sequences.

- A time-series constraint with the aggregator Sum and the feature one restricts $R$ to be the number of occurrences of pattern $\sigma$ in $X$.
- A constraint with the aggregator Sum and the feature width restricts $R$ to be the sum of the widths of the maximal occurrences of pattern $\sigma$ in the integer sequence. The *width* of an occurrence of $\sigma$ is the number of time-series variables included in $\sigma$ minus a constant corresponding to the sum of two integer trimming values. For instance, consider a time series $X = \langle 0, 3, 3, 0 \rangle$ with one occurrence of $\sigma =$ Plateau $=$ ' $<=^*>$ '; the width of the occurrence of $\sigma$ is equal to 2, as the two integer trimming values of $\sigma$ are equal to 1.

*Motivating Example.* Assume we have to generate a time series of size $n = 14$ with $R = 5$ increasing terraces, i.e. $\sigma =$ ' $<=^+<$ ', while maximising the sum of the 14 variables, each restricted to be in $[\ell, u] = [2, 6]$. Ignoring the 5 terraces leads to an upper bound of 84, while, as shown in Part (D) of Fig. 1, considering the 5 terraces gives a sharp upper bound $n \cdot u - p \cdot (2 \cdot t + s + 1) - r \cdot (2 \cdot s + 3) = 59$. The procedure for deriving the formulas $p = \min(R, \lfloor \frac{n-2 \cdot R}{2} \rfloor)$, $s = \lfloor \frac{R}{\max(1, p)} \rfloor$, $t = \frac{s \cdot (s+1)}{2}$, $r = R \mod \max(1, p)$, is presented in Sect. 3. Our goal is to *find a method to derive such formula for different patterns.*

*Focus and Contributions of This Paper.* We focus on the $g\_f\_\sigma(X, R)$ families of time-series constraints with $g$ being Sum, with $f$ being either one or width, and with $\sigma$ being a pattern described by a regular expression over the alphabet $\Sigma = \{$ ' $<$ ',' $=$ ',' $>$ ' $\}$. Our contributions consist of *parameterised sharp upper and lower bounds on the sum of the time-series variables* for the SUM_ONE_$\sigma(X, R)$ (also denoted as NB_$\sigma(X, R)$) and the SUM_WIDTH_$\sigma(X, R)$ families provided all $X$ variables are in the interval $[\ell, u]$. The parameters in the bounds correspond to the sequence length, the values $\ell$ and $u$, and the regular expression $\sigma$. The limits $\ell$ and $u$ are typically given by physical limitations of the system, which are time independent, and therefore apply to all variables. The parameterised bounds are valid provided some condition on the regular expression $\sigma$ holds, which in practice is true for 80% of the 22 regular expressions of [4]. Note that an approach encoding the full problem with an automaton would lead to a pseudo-polynomial algorithm since such automaton would have $O(n^2u^3)$ states: assuming $\ell = 0$, each state would record the values of $R$ (from 0 to $n$), of $X_{i-1}$ (from 0 to $n$), of the partial sum $X_1 + \cdots + X_i$ (from 0 to $u \times n$), and would have up to $u$ outgoing transitions.

*Outline of the Paper.* Section 2 presents a background on time-series constraints. Section 3 introduces our contribution, a unique per family expression that defines upper and lower bounds on the sum of the time-series variables wrt the time-series constraints. Section 4 evaluates the impact of the bounds. Section 5 concludes.

## 2    Background

We present the background to define bounds on the sum of the time-series variables wrt the time-series constraints with aggregator Sum and features one and width. A time-series constraint imposed on a sequence of $n$ integer variables $X = \langle X_1, ..., X_n \rangle$ and a result variable $R$ is described by a feature $f$, an aggregator $g$, and a pattern $\sigma$ as mentioned in the introduction. Let $S = \langle S_1, ..., S_{n-1} \rangle$ be the *signature* of a time series $X$, which is defined by: $(X_i < X_{i+1} \Leftrightarrow S_i = ' < ') \wedge (X_i = X_{i+1} \Leftrightarrow S_i = ' = ') \wedge (X_i > X_{i+1} \Leftrightarrow S_i = ' > ')$ for all $i \in [1, n-1]$. If a sub-signature $\langle S_i, ..., S_j \rangle$ is a maximal word matching $\sigma$ in the signature of $X$, then the subsequence $\langle X_{i+b_\sigma}, ..., X_{j+1-a_\sigma} \rangle$ is called a $\sigma$-*pattern*, and the subsequence $\langle X_i, ..., X_{j+1} \rangle$ is called an *extended $\sigma$-pattern*. The constants $b_\sigma$ and $a_\sigma$ respectively trim the left and right borders of an extended $\sigma$-pattern to obtain a $\sigma$-pattern from which a feature value is computed. They are useful when there is the need to perform computations from only a part of the occurrence of $\sigma$, as shown in Ex. 1. As in [9], we assume $\sigma$-patterns not to overlap.

*Example 1.* Consider the $\sigma = $ IncreasingTerrace $= ' <=^+< '$ regular expression with $a_\sigma = b_\sigma = 1$ and the time series $X$ shown in the figure in the right over the interval $[2, 5]$ and with signature $S = \langle <, =, <, >, <, =, =, < \rangle$. A $\sigma$-pattern called increasing terrace within $X$ is a subset whose signature is a

maximal occurrence of $\sigma$ in the signature of $X$. Time series $X$ contains two increasing terraces, labelled ① and ②, namely $\langle 3, 4, 4, 5 \rangle$ and $\langle 2, 3, 3, 3, 5 \rangle$ with widths 2 and 3, respectively. Hence, the aggregation of the number of occurrences using the aggregator Sum is 2 and the aggregation of their widths using Sum is 5. The corresponding time-series constraints are

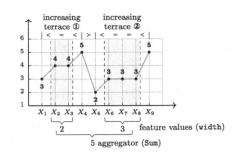

NB_$\sigma(X, R)$ and SUM_WIDTH_$\sigma(X, R)$, respectively.    △

*Regular-expression characteristics* were introduced as a way to parameterise the bounds on the result value of time-series constraints [8] and to derive AMONG implied constraints [6] in a systematic way. We now present a brief definition of the characteristics we reuse in this paper and illustrate them with one example.

- The *size* of $\sigma$, denoted by $\omega_\sigma$, is the length of a shortest word in the language $\mathcal{L}_\sigma$ of the regular expression $\sigma$.
- The *height* of $\sigma$, denoted by $\eta_\sigma$, is the smallest difference between the domain upper and lower limits, i.e. $u - \ell$, such that there is a *ground time series* (all $X_i$ are fixed) over $[\ell, u]$ whose signature has at least one occurrence of $\sigma$.
- The *range* of $\sigma$ wrt $n$, denoted by $\phi_\sigma^{\langle n \rangle}$, is the minimum difference between the maximum and the minimum values in an extended $\sigma$-pattern of width $n$.
- The *set of inducing words* of $\sigma$, denoted by $\Theta_\sigma$, is a subset of $\mathcal{L}_\sigma$ such that for every word $v$ in $\mathcal{L}_\sigma$, there exists a word $w = w_1 w_2 ... w_k$ in $\Theta_\sigma$ such that every $w_i$ is non-empty and every $v$ in $\mathcal{L}_\sigma$ can be represented as $v_1 w_1 v_2 w_2 ... v_k w_k v_{k+1}$ with every $v_i$ being a word in $\{`<', `=', `>'\}^*$.
- The *overlap* of $\sigma$ wrt $\langle \ell, u \rangle$, denoted by $o_\sigma^{\langle \ell, u \rangle}$, is the maximum number of time-series variables that belong simultaneously to two consecutive extended $\sigma$-patterns of a time series among all time series over $[\ell, u]$. If such maximum is not bounded, then $o_\sigma^{\langle \ell, u \rangle}$ is undefined.
- The *smallest variation of maxima* of $\sigma$ wrt $\langle \ell, u \rangle$, denoted by $\delta_\sigma^{\langle \ell, u \rangle}$, corresponds to the smallest difference between the maximum values of two consecutive extended $\sigma$-patterns that have at least one common time-series variable.
- The *set of supporting time series* of a word $v$ in $\mathcal{L}_\sigma$ wrt $\langle \ell, u \rangle$, denoted by $\Omega_\sigma^{\langle \ell, u \rangle}(v)$, is a set of time series where each element of $\Omega_\sigma^{\langle \ell, u \rangle}(v)$ is a time series over $[\ell, u]$ whose signature is $v$.

*Example 2.* Consider the $\sigma = \texttt{IncreasingTerrace} = `<=^+<'$ regular expression and the sequence $X = \langle 3, 4, 4, 5, 5, 6 \rangle$. The figure on the right illustrates

regular-expression characteristics associated with $X$. The common time-series variables of the two consecutive extended $\sigma$-patterns are coloured in grey. The first (resp. second) extended $\sigma$-pattern is shown in blue (resp. red). Points $L_1$ and $L_2$ correspond to the overlap $o_\sigma^{\langle \ell, u \rangle}$. The difference between the $y$-coordinates of points $L_2$ and $L_3$ corresponds to the value of $\delta_\sigma^{\langle \ell, u \rangle}$.    $\triangle$

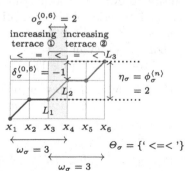

We reuse in Sect. 3 the notions of interval without restart and superposition of two words from [8] that we now recall. An *interval without restart* consists of a subsequence such that every two consecutive extended $\sigma$-patterns within this subsequence have $o_\sigma^{\langle \ell, u \rangle} > 0$ common time-series variables. The intervals without restart are always disjoint. Consequently, two consecutive extended $\sigma$-patterns belonging to distinct intervals without restart do not share any time-series variables. The *superposition of two words* $v$ and $w$ in $\mathcal{L}_\sigma$ wrt $\langle \ell, u \rangle$ is the signature $q$ of some ground time series over $[\ell, u]$ that contains at least two $\sigma$-patterns. For instance, the word $z = $ ' $<=<=<$ ' is the superposition of the two increasing terraces in the figure from Example 2.

## 3    Bounds on the Sum of the Time-Series Variables

Consider a regular expression $\sigma$, an integer interval $[\ell, u]$, and a time series $X = \langle X_1, \ldots, X_n \rangle$, with every $X_i$ ranging over $[\ell, u]$. We present a method to derive upper bounds on the sum of the $X_i$ for NB_$\sigma(X, R)$ and SUM_WIDTH_$\sigma(X, R)$. Wlog lower bounds are obtained in a similar way.

### 3.1    New Regular-Expression Characteristics

We present in this section two new regular-expression characteristics that will be used to maximise the sum of the time-series variables, while at the same time (i) constructing a fixed number of pattern occurrences, or (ii) building a number of pattern occurrences achieving a given total width. We first motivate and give the intuition of such characteristics in the context of the **IncreasingTerrace** = ' $<=^+<$ ' pattern before providing their formal definitions.

- The first characteristic corresponds to the *maximum weight* of the inducing word of a regular expression $\sigma$. For example, given ' $<=^+<$ ' and the domain value $u$, the maximum weight is the maximum value which can be achieved by a supporting time series of the inducing word ' $<=<$ ', i.e. $(u - 2) + (u - 1) + (u - 1) + u = 4 \cdot u - 4$.
- The second characteristic corresponds to the *weight of the overlap of the inducing word* of a regular expression $\sigma$ with itself. We need to know this quantity to evaluate the maximum weight that can be achieved by a supporting time series of a stretch of overlapping inducing words. For example, given ' $<=^+<$ ' and the domain value $u$, the maximum weight of the overlap highlighted in grey in ' $<=<=<$ ' of two consecutive inducing words ' $<=<$ ' is equal to $(u - 2) + (u - 1) = 2 \cdot u - 3$.

**Definition 1 (Maximum weight of $\sigma$).** *Consider a regular expression $\sigma$ with exactly one word $v \in \Theta_\sigma$ with length $\omega_\sigma$, and an integer interval domain $[\ell, u]$. The* maximum weight *of $\sigma$ wrt $\langle u \rangle$, denoted by $\lambda_\sigma^{\langle u \rangle}$, is a function that maps an element of $\mathcal{R}_\Sigma \times \mathbb{Z}$ to $\mathbb{Z}$. It is defined by $\lambda_\sigma^{\langle u \rangle} = u \cdot (\omega_\sigma + 1) - \nu_\sigma$, where $\nu_\sigma$ is the weight variation of $\sigma$. The function $\nu_\sigma$ maps an element of $\mathcal{R}_\Sigma$ to $\mathbb{Z}$,*

$$\nu_\sigma = \min_{t \in \Omega_\sigma^{\langle \ell, u \rangle}(v)} \left[ (\omega_\sigma + 1) \cdot \max_{X_i \in t} X_i - \sum_{X_i \in t} X_i \right],$$

*where $t$ is a supporting time series of $v \in \Theta_\sigma$ wrt $\langle \ell, u \rangle$ denoted by $\Omega_\sigma^{\langle \ell, u \rangle}(v)$, and $\mathcal{R}_\Sigma$ denotes the set of regular expressions over the alphabet $\Sigma$.*

**Definition 2 (Total weight of the overlap of $\sigma$).** *Consider a regular expression $\sigma$ with exactly one word $v$ in $\Theta_\sigma$, and an integer interval domain $[\ell, u]$. The total weight of the overlap of $\sigma$ wrt $\langle u \rangle$, denoted by $\alpha_\sigma^{\langle u \rangle}$, is a function that maps an element of $\mathcal{R}_\Sigma \times \mathbb{Z}$ to $\mathbb{Z}$. It is defined by $\alpha_\sigma^{\langle u \rangle} = u \cdot o_\sigma^{\langle \ell, u \rangle} - \xi_\sigma$, where $\xi_\sigma$ is the weight variation of the overlap of $\sigma$, defined by*

$$
\xi_\sigma = \begin{cases} \min\limits_{t \in \Omega_\sigma^{\langle \ell, u \rangle}(v,v)} \left[ o_\sigma^{\langle \ell, u \rangle} \cdot \max_{X_i \in t} X_i - \sum_{X_i \in t_o} X_i \right], & \text{if } \Gamma_\sigma^{\langle \ell, u \rangle}(v,v) \neq \emptyset \\ 0, & \text{otherwise.} \end{cases}
$$

*where $\Gamma_\sigma^{\langle \ell, u \rangle}(v, w)$ is the shortest superposition of words $v$ and $v$ in $\Theta_\sigma$, $\Omega_\sigma^{\langle \ell, u \rangle}(v, v)$ is the supporting time series set for the shortest superposition between $v$ and $v$ wrt $\langle \ell, u \rangle$, and $t_o$ is a subsequence of $t$ corresponding to the overlap of two consecutive extended $\sigma$-patterns from $\Gamma_\sigma^{\langle \ell, u \rangle}(v, v)$.*

*Example 3.* Consider $\sigma_1 = \texttt{StrictlyDecreasingSequence}$, $\sigma_2 = \texttt{Peak} = $ ' $< (< \mid =)^* (> \mid =)^* > $ ', and $\sigma_3 = \texttt{IncreasingTerrace} = $ ' $<=^+< $ ', and the interval $[0, 3]$. Table 1 presents the values for the weight variation and the total weight regular-expression characteristics of the inducing words and the overlap of $\sigma_1, \sigma_2$ and $\sigma_3$. $\triangle$

## 3.2    Time-Series Constraints with Feature ONE

We show how to derive bounds on the sum of the time-series variables for the $\texttt{NB\_}\sigma(X, R)$ constraint family, provided all variables are in an interval $[\ell, u]$.

**Table 1.** Regular-expression characteristics for $\texttt{StrictlyDecreasingSequence}$, $\texttt{Peak}$, $\texttt{IncreasingTerrace}$; column "length" gives the number of variables in the time series of interest, i.e. the number of filled dots in the column "illustration".

| $\sigma$ | word type | $\Theta_\sigma$ | length | illustration | new characteristics weight variation | total weight |
|---|---|---|---|---|---|---|
| Strictly-Decreasing-Sequence | inducing word | ' $>$ ' | 2 | | 1 | $2u - 1$ |
| | overlap | - | 0 | | 0 | $0u - 0$ |
| Peak | inducing word | ' $<>$ ' | 3 | | 2 | $3u - 2$ |
| | overlap | - | 1 | | 1 | $u - 1$ |
| Increasing-Terrace | inducing word | ' $<=<$ ' | 4 | | 4 | $4u - 4$ |
| | overlap | - | 2 | | 3 | $2u - 3$ |

- First, Example 4 provides the basis for understanding the intuition of the method.
- Second, we list the properties required by a regular expression to use the intuitions we just described for deriving an upper bound.
- Finally, based on these properties, we give a greedy method to construct a time series that maximises the sum of its variables wrt the $\mathrm{NB\_}\sigma(X, R)$ family of time-series constraints.

## From an Intuition to a Methodology

*Example 4. (Intuition for constructing a time series reaching the upper bound).*

Figure 1 gives three examples of how to build a time series that maximises the sum of its variables, while reaching a given number of pattern occurrences. Part (A) gives three constraints of the form $\mathrm{NB\_}\sigma_i$ with $\sigma_1 =$ StrictlyDecreasingSequence $= $ ' $>^+$ ', $\sigma_2 = $ Peak $= $ ' $< (< \mid =)^*(> \mid =)^* >$ ', and $\sigma_3 = $ IncreasingTerrace $= $ ' $<=^+<$ ', respectively enforcing 3 occurrences of $\sigma_1$, 3 occurrences of $\sigma_2$, and 5 occurrences of $\sigma_3$.

- Since strictly decreasing sequences cannot overlap, Part (B) shows a time series with three intervals without restart where each interval corresponds to a strictly decreasing minimum size sequence positioned at its highest level, the remaining variables $X_7$, $X_8$ being set at their maximum value.
- Even if consecutive peaks may overlap, their maximal values may remain at the same level, Part (C) shows a time series with a single interval without restart containing three minimum size peaks positioned at their highest level, the remaining variables $Y_8$, $Y_9$ being set at their maximum value.
- As two consecutive intersecting increasing terraces are necessarily offset in height, Part (D) shows a time series containing the maximum number of possible intervals without restart, given that 5 increasing terraces have to be positioned in a sequence of size 14. The 5 terraces ①, ②, ..., ⑤ are distributed in two intervals without restart in the most balanced way, i.e. 3 and 2 terraces, by placing them at the highest possible level.                                    △

To build a time series whose sum of variables is maximum, while having $R$ maximal occurrences of the pattern $\sigma$, we proceed as follows.

- [MAXIMISING THE NUMBER OF VARIABLES SET TO $u$] We minimise the overall size taken by all $R$ maximal occurrences of $\sigma$ in order to set all remaining variables to their maximum value $u$.
- [POSITIONING PATTERN OCCURRENCES AS CLOSE AS POSSIBLE TO $u$] We try to position the $R$ maximal occurrences of $\sigma$ at their maximum height wrt to $u$. Unfortunately, as shown in Example 4 for the IncreasingTerrace pattern, this is not always possible: in Part (D) of Fig. 1, only the terraces labelled with ① and ② are placed at their highest possible level. This can occur for patterns such that $o_\sigma^{\langle \ell, u \rangle} \neq 0$ and $\delta_\sigma^{\langle \ell, u \rangle} \neq 0$, when $R$ is too large wrt the size of the time series. In this case, the $R$ pattern occurrences are distributed in a balanced way over as many intervals without restart as possible.

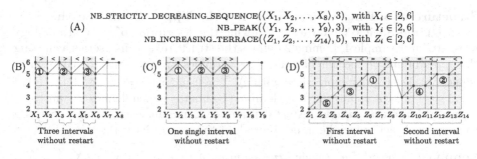

**Fig. 1.** (A) Three constraints and their corresponding time series that maximise the sum of the time-series variables respectively containing (B) three strictly decreasing sequences, (C) three peaks, and (D) five increasing terraces

– [SELECTING EACH PATTERN OCCURRENCE] Finally, each maximal occurrence of $\sigma$ corresponds to a supporting time series $X_1, X_2, \ldots, X_{\omega_\sigma+1}$ of a word $v$ of $\mathcal{L}_\sigma$ verifying simultaneously all the following conditions:
  i  $v$ is a word whose size is as short as possible; hence its size is $\omega_\sigma + 1$.
  ii $X_1, \ldots, X_{\omega_\sigma+1}$ minimises the variation wrt the maximum value of its variables, i.e. $(\omega_\sigma + 1) \cdot \max_{i \in [1, \omega_\sigma+1]} X_i - \sum_{i \in [1, \omega_\sigma+1]} X_i$.

**Required Properties of Regular Expressions.** As shown before, building in a greedy way a time series $t$ that maximises the upper bound on the sum of the time-series variables wrt a time-series constraint with aggregator Sum and feature one, requires finding $R$ maximal words of $\mathcal{L}_\sigma$, such that the superposition of these $R$ words wrt an integer interval domain $[\ell, u]$ *simultaneously optimises* several regular-expression characteristics. To define these properties, we use two regular-expression characteristics presented in Sect. 2 and Sect. 3.1: the set of inducing words and the weight variation of word $v$, denoted by $\Theta_\sigma$ and $\nu_\sigma(v)$.

Prop. 1. The language of $\sigma$ does not include the word ' $=^+$ ', i.e., ' $=^+$ ' $\notin \mathcal{L}_\sigma$.
Prop. 2. Regular expression $\sigma$ has only one inducing word, i.e., $\mid \Theta_\sigma \mid = 1$.
Prop. 3. The weight variation wrt the maximum domain value $u$ of the only inducing word of $\sigma$, denoted by $v$, is lower than or equal to the weight variation of any other word in the language of $\sigma$, i.e., $\nu_\sigma(v) \leq \nu_\sigma(w)$, for each $w \in \mathcal{L}_\sigma : w \neq v$.

Prop. 1 guarantees that when the number of time-series variables included in the $R$ maximal occurrences of $\sigma$ is lower than the sequence length $n$, the time series $t$ can be completed by setting all its reminder variables in the maximal domain value $u$. Prop. 2 guarantees that the smallest possible number of time-series variables is used to include $R$ maximal occurrences of pattern $\sigma$ in time series $t$. Prop. 3 ensures that the weight variation of a $\sigma$ occurrence is minimised. Hence, the upper bound on the sum of the time-series variables associated with the $R$ occurrences of $\sigma$ in $t$ is maximal. We show in Lemma 2 that these three properties give a sufficient condition for getting a sharp upper bound on the sum of time-series variables wrt a NB_$\sigma(X, R)$ constraint.

**Structure of a Time Series Achieving the Upper Bound on the Sum of the Time-Series Variables.** Following the description of the methodology presented in Example 4, Lemma 2 defines the structure of a time series achieving the upper bound on the sum of time-series variables wrt a NB_$\sigma(X, R)$ time-series constraint. For regular expressions with $o_\sigma^{\langle \ell, u \rangle} \neq 0$ and $\delta_\sigma^{\langle \ell, u \rangle} \neq 0$ (e.g., Part (D) of Fig. 1), we present an intermediary lemma (Lemma 1) which defines the *maximal number of intervals without restart* containing $R$ maximal occurrences of $\sigma$ in a time series $X$ achieving the upper bound on the sum of its variables.

**Lemma 1.** *Consider a regular expression $\sigma$, a time series $X = \langle X_1, \ldots, X_n \rangle$ with every $X_i$ ranging over the same integer interval domain $[\ell, u]$, a NB_$\sigma(X, R)$ constraint with $R \geq 0$. When $o_\sigma^{\langle \ell, u \rangle} \neq 0$, $\delta_\sigma^{\langle \ell, u \rangle} \neq 0$ and Prop. 2 holds, the maximal number of intervals without restart, denoted by $p$, is defined by*

$$p = \min \left( R, \left\lfloor \frac{n - R \cdot (\omega_\sigma + 1 - o_\sigma^{\langle \ell, u \rangle})}{o_\sigma^{\langle \ell, u \rangle}} \right\rfloor \right). \tag{1}$$

*Proof.* When $o_\sigma^{\langle \ell, u \rangle} \neq 0$ and $\delta_\sigma^{\langle \ell, u \rangle} \neq 0$ the $R$ $\sigma$-patterns might be contained in one or more intervals without restart. Since each interval without restart contains at least one $\sigma$-pattern, $p$ cannot exceed $R$. Wlog assume that we have only one $\sigma$-pattern in the first $p - 1$ intervals without restart and $R - p + 1$ in the last one; we remark that moving one $\sigma$-pattern from an interval without restart containing more than one $\sigma$-pattern to another interval without restart, does not change the overall number of time-series variables belonging to the $R$ $\sigma$-pattern occurrences. By Prop. 2 we use the only inducing word of $\sigma$, hence:

- In the first $p - 1$ intervals without restart the total number of time-series variables used is $(p - 1) \cdot (\omega_\sigma + 1)$.
- In the last interval without restart the total number of time-series variables used is $(R - p + 1) \cdot (\omega_\sigma + 1) - (R - p) \cdot o_\sigma^{\langle \ell, u \rangle}$.

Since the total number of time-series variables used by the $R$ $\sigma$-patterns must be lower than or equal to $n$ we have:

$$(p - 1) \cdot (\omega_\sigma + 1) + (R - p + 1) \cdot (\omega_\sigma + 1) - (R - p) \cdot o_\sigma^{\langle \ell, u \rangle} \leq n.$$

By isolating $p$, and since $p$ is an integer, we obtain $p \leq \left\lfloor \frac{n - R \cdot (\omega_\sigma + 1 - o_\sigma^{\langle \ell, u \rangle})}{o_\sigma^{\langle \ell, u \rangle}} \right\rfloor$, which is thus the second term inside the min term in Eq. (1). $\square$

**Lemma 2.** *Consider a regular expression $\sigma$ that has Prop. 1, Prop. 2 and Prop. 3. Then for any integer number $n \geq 2$ and given number of occurrences of $\sigma$ $R \geq 0$, there exists a word $z$ with an associated ground time series $t$ of length $n$ over $[\ell, u]$ achieving the upper bound on the sum of the $X_i$ time-series variables.*

*Proof.* We first construct a word $z$ composed by the concatenation of two words, a prefix, denoted by $z$, containing $R$ maximal occurrences of $\sigma$, and a suffix, denoted by $z$, containing zero occurrences of $\sigma$. Second, we prove that there exists a supporting time series wrt $[\ell, u]$ with signature $z$ that maximises the sum of the time-series variables.

**Part A: Construction of the Word $z$.** When building word $z$, if $o_\sigma^{\langle \ell,u \rangle} = 0$, each pair of consecutive $\sigma$-patterns does not share any time-series variables. Hence, each extended $\sigma$-pattern belongs to a different interval without restart and $p = R$. If $o_\sigma^{\langle \ell,u \rangle} \neq 0$ and $\delta_\sigma^{\langle \ell,u \rangle} = 0$, all pairs of consecutive extended $\sigma$-patterns share $o_\sigma^{\langle \ell,u \rangle}$ time-series variables. Hence, time series $t$ has a single interval without restart that contains all $\sigma$-patterns and $p = 1$. By Lemma 1, if $o_\sigma^{\langle \ell,u \rangle} \neq 0$ and $\delta_\sigma^{\langle \ell,u \rangle} \neq 0$, all $\sigma$-pattern occurrences of time series $t$ are contained in $p \geq 1$ intervals without restart. There exists $R$ words of $\mathcal{L}_\sigma$, a prefix word $z$ including the $R$ words, and a concatenation of $z$ with a suffix word $z$ such that all the conditions of Prop. 1, Prop. 2 and Prop. 3 are satisfied. We construct the signature of the time series, denoted by $z$, by first building the signature $z_k$ (with $k \in [1, p]$) of every interval without restart of $t$ by imposing the following conditions:

- **[Structure of each interval without restart]** Each word $z_k$ (with $k \in [1, p]$) has $c_k$ occurrences of $\sigma$ and is defined by

$$
\begin{cases}
z_k = v^{c_k}, & c_k = 1, & \text{if } o_\sigma^{\langle \ell,u \rangle} = 0 \\
z_k = v^{c_k}, & c_k = R, & \text{if } o_\sigma^{\langle \ell,u \rangle} \neq 0 \text{ and } \delta_\sigma^{\langle \ell,u \rangle} = 0 \\
z_k = vw^{c_k-1}, & \begin{cases} c_k = s+1, & \text{if } k \leq p' \\ c_k = s, & \text{otherwise} \end{cases} & \text{otherwise}
\end{cases} \quad (2)
$$

   where $v \in \Theta_\sigma$, $v^k$ denotes the concatenation of $k$ occurrences of $v$, $vw$ is the superposition between $v$ and $v$, $s = \lfloor \frac{R}{\max(1,p)} \rfloor$, and $p' = R \bmod \max(1, p)$.

- **[Combining the intervals without restart: structure of $z$ ]** Word $z$ is defined by

$$
\begin{cases}
w^{R-1}v, & \text{if } o_\sigma^{\langle \ell,u \rangle} = 0 \\
z_1, & \text{if } o_\sigma^{\langle \ell,u \rangle} \neq 0 \text{ and } \delta_\sigma^{\langle \ell,u \rangle} = 0 \\
z_1 \,\text{'} < \text{'}...\text{'} < \text{'}\, z_p, & \text{if } o_\sigma^{\langle \ell,u \rangle} \neq 0 \text{ and } \delta_\sigma^{\langle \ell,u \rangle} > 0 \\
z_1 \,\text{'} > \text{'}...\text{'} > \text{'}\, z_p, & \text{if } o_\sigma^{\langle \ell,u \rangle} \neq 0 \text{ and } \delta_\sigma^{\langle \ell,u \rangle} < 0
\end{cases} \quad (3)
$$

   where $v \in \Theta_\sigma$, and word $w$ belonging to $\{$'$v >$ ', '$v =$ ', '$v <$ '$\}$ is not a proper factor of any word in $\mathcal{L}_\sigma$ and its height is $\eta_\sigma$.

- **[Completing the set of intervals without restart: structure of $z$]** Word $z$ with length $m$ is defined by

$$
\begin{cases}
\varepsilon, & \text{if } m = 0 \\
\text{'} <=^* \text{'}, & \text{if } m > 0 \text{ and ' } > (=|>)^* \text{'} \text{ is a suffix of } v \\
\text{'} =^+ \text{'}, & \text{otherwise}
\end{cases} \quad (4)
$$

**Part B: Proving that There is a Ground Time Series $t$ with Signature $z$ that Maximises the Sum of the Time-Series Variables**. Since we assume that regular expression $\sigma$ has Prop. 1, time-series variables in $z$ can be assigned to the maximal domain value $u$ without creating a new occurrence of pattern. Hence, to prove the maximality on the sum of the $X_i$ variables belonging to $t$, it suffices to show that there exists a ground time series over $[\ell, u]$ obtained with the signature of word $z$ achieving the upper bound on the sum of its variables. For space reasons we only consider the case where $\delta_\sigma^{\langle \ell, u \rangle} \neq 0$. We define two ground time series $t^*$ and $t'$ such that their signatures contain $R$ $\sigma$-pattern occurrences and $p$ intervals without restart:

- $t^*$ corresponds to the ground time series with signature $z$ satisfying Eq. (2) and where the first $\sigma$-pattern occurrence of each interval without restart is at level 0, i.e. the level closest to the maximal domain value $u$.
- $t'$ corresponds to any other ground time series where the number of $\sigma$-patterns located at level 0 is strictly less than $p$.

To obtain the total weight of a ground time series, i.e. the upper bound on the sum of the time-series variables, we first define the maximum weight of a $\sigma$-pattern located at level $e$ by $\underbrace{\lambda_\sigma^{\langle u \rangle}}_{A} - \underbrace{(\omega_\sigma + 1) \cdot \mid \delta_\sigma^{\langle \ell, u \rangle} \mid \cdot e}_{B}$, and the weight of the overlap between two consecutive $\sigma$-patterns located at levels $e$ and $e+1$ by $\underbrace{\alpha_\sigma^{\langle u \rangle}}_{C} - \underbrace{o_\sigma^{\langle \ell, u \rangle} \cdot \mid \delta_\sigma^{\langle \ell, u \rangle} \mid \cdot e}_{D}$. Terms A and C, defined in Sect. 3.1, correspond to the maximum weight of a $\sigma$-pattern and to the total weight of the overlap between two consecutive $\sigma$-patterns, respectively. B and D are two correction terms which respectively adjust the weight of a $\sigma$-pattern and the weight of the overlap between two consecutive $\sigma$-patterns, caused by a change in the level of a $\sigma$-pattern occurrence.

The total weight of a ground time series $t$, denoted by $W_t$, is the sum of the weights of the $R$ $\sigma$-patterns minus the sum of the weights of the $R - p$ overlaps between consecutive pairs of $\sigma$-patterns. Hence, $W_t$ is defined by

$$W_t = \left( R \cdot \lambda_\sigma^{\langle u \rangle} - (\omega_\sigma + 1) \underbrace{\sum_{k=1}^{p} \sum_{e=i_k}^{j_k} \Delta_e}_{B_T} \right) - \left( (R-p) \cdot \alpha_\sigma^{\langle u \rangle} - o_\sigma^{\langle \ell, u \rangle} \underbrace{\sum_{k=1}^{p} \sum_{e=i_k}^{j_k - 1} \Delta_e}_{D_T} \right), \quad (5)$$

where $\Delta_e = \mid \delta_\sigma^{\langle \ell, u \rangle} \mid \cdot e$, and $i_k, j_k$ are the highest and the lowest levels of the $\sigma$-patterns in interval without restart $k \in [1, p]$, respectively. Note that in Eq. (5), the only terms that depend on the level of the $\sigma$-pattern occurrences are the correction terms $B_T$ and $D_T$. Let $i_k = 0$ and $j_k = c_k - 1$ be the levels of the highest and the lowest $\sigma$-pattern occurrence in interval $k \in [1, p]$ for $t^*$. For $t'$ we assume that at least one $i_k > 0$ with $k \in [1, p]$. Therefore, we compare the terms $B_T$ and $D_T$ for $t^*$ and $t'$ in the following way:

$$\sum_{k=1}^{p} \sum_{e=i_k}^{j_k} \Delta_e \geq \mid \delta_\sigma^{\langle \ell, u \rangle} \mid + \sum_{k=1}^{p} \sum_{e=0}^{c_k - 1} \Delta_e \quad (6)$$

$$\sum_{k=1}^{p} \sum_{e=i_k}^{j_k-1} \Delta_e \geq | \delta_\sigma^{\langle \ell, u \rangle} | + \sum_{k=1}^{p} \sum_{e=0}^{c_k-2} \Delta_e \qquad (7)$$

Our objective is to show that $W_{t^*} > W_{t'}$, i.e. the total weight of $t^*$ is strictly greater than the total weight of $t'$. Hence, by using Eq. (5) to define $W_{t^*}$ and $W_{t'}$, by including Inequalities (6) and (7) in $W_{t'}$ and by factorising, we have

$$o_\sigma^{\langle \ell, u \rangle} \sum_{k=1}^{p} \sum_{e=0}^{c_k-2} \Delta_e - (\omega_\sigma + 1) \sum_{k=1}^{p} \sum_{e=0}^{c_k-1} \Delta_e > \qquad (8)$$

$$o_\sigma^{\langle \ell, u \rangle} \left( | \delta_\sigma^{\langle \ell, u \rangle} | + \sum_{k=1}^{p} \sum_{e=0}^{c_k-2} \Delta_e \right) - (\omega_\sigma + 1) \left( | \delta_\sigma^{\langle \ell, u \rangle} | + \sum_{k=1}^{p} \sum_{e=0}^{c_k-1} \Delta_e \right)$$

By factorising Inequality (8), we have

$$\omega_\sigma + 1 > o_\sigma^{\langle \ell, u \rangle} \qquad (9)$$

Since the size of $\sigma$ is always greater than or equal to the overlap of $\sigma$, i.e. $\omega_\sigma \geq o_\sigma^{\langle \ell, u \rangle}$, Inequality (9) holds and $W_{t^*} > W_{t'}$.      □

**Upper Bound on the Sum of the Time-Series Variables.** Consider a NB_$\sigma(X, R)$ family of time-series constraints with every $X_i$ ranging over the same interval $[\ell, u]$. Theorem 1 provides an upper bound on the sum of the time-series variables wrt the time-series constraint.

**Theorem 1.** *Consider a regular expression $\sigma$ satisfying the conditions of Prop. 1, Prop. 2 and Prop. 3. The upper bound on the sum of the time-series variables for the NB_$\sigma(X, R)$ family is defined by*

$$\sum_{k=1}^{p} \sum_{e=0}^{c_k-1} (\lambda_\sigma^{\langle \ell, u \rangle} - (\omega_\sigma + 1) \cdot \Delta_e) - \sum_{k=1}^{p} \sum_{e=0}^{c_k-2} (\alpha_\sigma^{\langle \ell, u \rangle} - o_\sigma^{\langle \ell, u \rangle} \cdot \Delta_e) + m \cdot u, \qquad (10)$$

*where $m$ is defined by:* $m = n - \left[ \sum_{k=1}^{p} \sum_{e=0}^{c_k-1} (\omega_\sigma + 1) - \sum_{k=1}^{p} \sum_{e=0}^{c_k-2} o_\sigma^{\langle \ell, u \rangle} \right].$

*Proof.* It uses the construction of the proof of Lemma 2.      □

This upper bound is valid for all 22 regular expressions of [4], except for Inflexion, Zigzag, Steady and SteadySequence, since the first two regular expressions do not satisfy the condition in Prop. 2 and the last two regular expressions do not satisfy the condition in Prop. 1.

### 3.3  Time-Series Constraints with Feature WIDTH

For patterns $\sigma$ satisfying Prop. 1 and Prop. 2 we sketch a method to derive bounds on the sum of the time-series variables for the SUM_WIDTH_$\sigma(X, R)$ family, provided all $X_i$ (with $i \in [1, n]$) variables are in an interval $[\ell, u]$. To build a time series $t$ whose sum of variables is maximum, while having $R$ as the sum of the widths of the occurrences of the pattern $\sigma$, we use a two-step procedure.

- [STEP 1: NORMALISING THE PATTERN OCCURRENCES] For each $\sigma$ pattern, we define a transformation $\mathcal{T}_\sigma$ whose repeated application from any initial signature $S_{initial}$ leads to a target signature $S_{target}$. $S_{initial}$ and $S_{target}$ have the same value for $R$, and no matter the value of $S_{initial}$, this signature will converge to a signature $S_{target}$ with the same number of $\sigma$-pattern occurrences. A single application of $\mathcal{T}_\sigma$ from a signature $S$ to a signature $S'$ has the following properties:
  - i  $S$ and $S'$ share the same sum of the widths for their $\sigma$ patterns.
  - ii  The largest sum of the $X_i$ variables compatible with $S$ is less than or equal to the largest sum of the $X_i$ variables compatible with $S'$.

  To find the time series with the largest sum of the $X_i$ variables compatible with signature $S$ we first perform generalised arc consistent (GAC) in the induced constraint satisfaction problem. Second, we fix all $X_i$ variables to their respective maximal value. Note that for a binary constraint of the type $<$, $=$ or $>$, we can always set its two variables to their respective maximal values, while satisfying the constraint in question.
- [STEP 2: NORMALISATION OUTSIDE THE PATTERN OCCURRENCES] We modify $S_{target}$ to $S_{final}$ so that all $X_i$ variables that do not belong to an extended $\sigma$-pattern of $S_{final}$ can be set to their maximum value $u$.

We define two transformations, denoted by $\mathcal{T}_\sigma^1$ and $\mathcal{T}_\sigma^2$. For space reasons, we sketch the two transformations but we only illustrate $\mathcal{T}_\sigma^1$ in Example 5.

- $\mathcal{T}_\sigma^1$ transforms $S_{initial}$ into a sequence $S_{target}$ containing the smallest possible words in $\mathcal{L}_\sigma$, i.e. inducing words whose widths are equal to $\pi_\sigma = \omega_\sigma + 1 - a_\sigma - b_\sigma$. $\mathcal{T}_\sigma^1$ works for $\sigma = \texttt{DecreasingSequence}$, $\texttt{IncreasingSequence}$, $\texttt{StrictlyIncreasingSequence}$, and $\texttt{StrictlyDecreasingSequence}$, and for $\texttt{Gorge}$ and $\texttt{Summit}$ when $n \geq p_\sigma^{\langle R \rangle} \cdot (\omega_\sigma + 1) - (p_\sigma^{\langle R \rangle} - 1) \cdot o_\sigma^{\langle \ell, u \rangle}$, i.e. there is enough space to create $p_\sigma^{\langle R \rangle} = \lfloor \frac{R}{\pi_\sigma} \rfloor$ inducing words of $\sigma$. The upper bound on the sum of $X_i$ variables when $\mathcal{T}_\sigma^1$ is used is

$$\underbrace{p_\sigma^{\langle R \rangle} \cdot \lambda_\sigma^{\langle \ell, u \rangle} - (p_\sigma^{\langle R \rangle} - 1) \cdot \alpha_\sigma^{\langle \ell, u \rangle}}_{\text{I}} + \underbrace{(R \bmod \pi_\sigma) \cdot (u - (\eta_\sigma + 1))}_{\text{II}} + \underbrace{m \cdot u}_{\text{III}},$$

where $m = n - (p_\sigma^{\langle R \rangle} \cdot (\omega_\sigma + 1) - (p_\sigma^{\langle R \rangle} - 1) \cdot o_\sigma^{\langle \ell, u \rangle} + R \bmod \pi_\sigma)$. Term I corresponds to the maximum weight of the concatenation of $p_\sigma^{\langle R \rangle}$ occurrences of the only inducing word of $\sigma$. Term II is related to a correction term which is used when it is not possible to obtain a sum of the widths equal to $R$ with $p_\sigma^{\langle R \rangle}$ inducing words. Term III corresponds to the maximum weight of the variables that do not belong to any $\sigma$-pattern occurrence. In Part (C) of Fig. 2 points ●, ◎ and ○ respectively contribute to terms I, II and III.
- $\mathcal{T}_\sigma^2$ transforms $S_{initial}$ into a sequence $S_{target}$ containing one occurrence of $\sigma$. $\mathcal{T}_\sigma^2$ works for 10 other $\sigma$-pattern including $\texttt{IncreasingTerrace}$ and $\texttt{Peak}$. The upper bound on the sum of $X_i$ variables when $\mathcal{T}_\sigma^2$ is used is $\lambda_\sigma^{\langle R, u \rangle} + m \cdot u$, where $m = n - (R + a_\sigma + b_\sigma)$, and $\lambda_\sigma^{\langle R, u \rangle}$ is the maximum weight of the regular expression $\sigma$ where words in $\mathcal{L}_\sigma$ have a fixed length of $R + a_\sigma + b_\sigma - 1$.

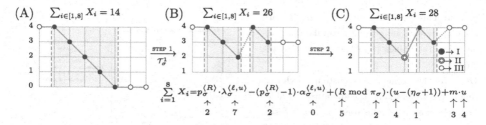

**Fig. 2.** Transforming an initial time series to a final time series that maximises the sum of the $X_i$ variables, where both time series share the same value, i.e. $R = 5$, for the sum of the widths of the strictly decreasing sequences

*Example 5.* Figure 2 gives an example of how to build a time series that maximises the sum of its variables while reaching a given sum of the widths of the pattern occurrences. The constraint used is SUM_WIDTH_$\sigma(\langle X_1, \ldots, X_8 \rangle, 5)$ with $\sigma = $ '$>^+$', $a_\sigma = b_\sigma = 0$, and $X_i \in [0,4]$. Part (A) shows an initial time series with the largest sum of the $X_i$ variables compatible with signature $S_{initial} = \langle =, >, >, >, >, =, = \rangle$. Part (B) presents a time series with the largest sum of the $X_i$ variables compatible with $S_{target} = \langle =, >, >, <, >, =, = \rangle$. $S_{target}$ is obtained after applying $\mathcal{T}_\sigma^1$ to $S_{initial}$ by changing the fourth signature variable from '$>$' to '$<$'. Note that $S_{initial}$ and $S_{target}$ share the same value for $R$ and that the largest sum of the $X_i$ variables compatible with $S_{initial}$ is less than the largest sum of the $X_i$ variables compatible with $S_{target}$. Part (C) shows a final time series with the largest sum of the $X_i$ variables compatible with $S_{final} = \langle =, >, >, <, >, <, = \rangle$, which is obtained by applying STEP 2 to $S_{target}$, i.e. by changing the sixth signature variable from '$=$' to '$<$'. This allows one to obtain a larger value for the sum of the $X_i$, i.e. 28 instead of 26.                    △

## 4   Evaluation

As a test for our procedure, we run all time-series constraints from the NB_$\sigma$ and the SUM_WIDTH_$\sigma$ families for synthetic time series with length between 5 and 60, and for all possible result values (in all 45,835 runs), and find a single optimal solution minimising the sum of the time-series variables. The individual constraints use a state-of-the-art implementation, combining optimised automata [5], bounds on the result variables [8], glue-matrix constraints linking all prefixes and suffixes [5], and selected redundant linear constraints based on Farkas lemma [17]. For the variable assignment, we compare four search methods shown below, while using the bounds obtained for cost variables.

**Search** This is the default search in SICStus, the variables are assigned in natural order, enumerating the values from the smallest to the largest.

**Custom** This implements a custom search routine based on assigning the signature variables first. The same method is used for all test cases.

**Search Impose** This uses the default search in SICStus, but first assigns the cost variable to its smallest value. As the bounds are sharp, the first solution found is optimal.

**Custom Impose** We use the custom search method, but also initially impose the lower bound of our method for each constraint.

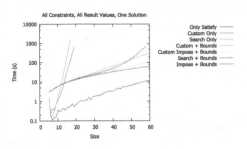

**Fig. 3.** Comparing baseline solutions with different search strategies

Since the conjunction of arithmetic constraints encoding bounds can propagate poorly, which results in some poor performance in the context of optimisation, we do not propagate the bounds directly; we rather use a table constraint to link the cost and result variables with a pre-computed table of all possible pairs. We compare against three baseline solutions. The first one (Only Satisfy) finds a single feasible solution, the second one (Search Only) solves the optimisation problem without bounds on the cost and with the default search routine, the third one (Custom Only) uses the custom search, again without the bounds on the cost variable. All experiments were run with SICStus Prolog 4.3.5 on a single core of a Windows 10 laptop with an Intel Core i7 CPU running at 2.9 GHz and with 64 Gb of memory. We stop the search if, for a given size, the time to run all its instances exceeds 600 s, or if we reach size 60.

As we observed that both families NB_$\sigma$ and SUM_WIDTH_$\sigma$ behave similarly in our benchmarks, the results are shown in Fig. 3, using a log scale for the y axis. We see that without the new bounds on the cost even a custom search routine does not find solutions for all cases if the size exceeds 18. Adding the bounds significantly increases the size of the problem one can handle. The custom search outperforms the default search for larger sizes, and further improvements are possible if we impose the lower bound before starting the search on the time-series variables. The best search combines imposing the lower bound with the default search, which seems to impose only a very limited overhead compared to the Only Satisfy base line, which only finds feasible solutions.

# 5    Conclusion

On the one hand, the theoretical contribution of this paper consists of *parameterised sharp bounds* on the sum of the time-series variables for two families of time-series constraints. Future work may look how to extend this work to any linear cost function, e.g. linear functions where all coefficients are not set to one.

On the other hand, the practical insight of this paper is related to the importance of encoding all *arithmetic constraints* representing a bound as a table constraint in order to get all the benefits from the bounds. An interesting avenue for future research is related to the derivation of bounds for the conjunction of time-series constraints.

# References

1. Abbas, H., Rodionova, A., Bartocci, E., Smolka, S.A., Grosu, R.: Quantitative regular expressions for arrhythmia detection algorithms. In: Feret, J., Koeppl, H. (eds.) CMSB 2017. LNCS, vol. 10545, pp. 23–39. Springer, Cham (2017). https://doi.org/10.1007/978-3-319-67471-1_2

2. Alur, R., Fisman, D., Raghothaman, M.: Regular programming for quantitative properties of data streams. In: Thiemann, P. (ed.) ESOP 2016. LNCS, vol. 9632, pp. 15–40. Springer, Heidelberg (2016). https://doi.org/10.1007/978-3-662-49498-1_2

3. Arafailova, E., Beldiceanu, N., Carlsson, M., Flener, P., Francisco Rodríguez, M.A., Pearson, J., Simonis, H.: Systematic derivation of bounds and glue constraints for time-series constraints. In: Rueher, M. (ed.) CP 2016. LNCS, vol. 9892, pp. 13–29. Springer, Cham (2016). https://doi.org/10.1007/978-3-319-44953-1_2

4. Arafailova, E., et al.: Global constraint catalog, volume II, time-series constraints. arXiv preprint arXiv:1609.08925 (2016)

5. Arafailova, E., Beldiceanu, N., Douence, R., Flener, P., Francisco Rodríguez, M.A., Pearson, J., Simonis, H.: Time-series constraints: improvements and application in CP and MIP contexts. In: Quimper, C.-G. (ed.) CPAIOR 2016. LNCS, vol. 9676, pp. 18–34. Springer, Cham (2016). https://doi.org/10.1007/978-3-319-33954-2_2

6. Arafailova, E., Beldiceanu, N., Simonis, H.: **among** implied constraints for two families of time-series constraints. In: Beck, J.C. (ed.) CP 2017. LNCS, vol. 10416, pp. 38–54. Springer, Cham (2017). https://doi.org/10.1007/978-3-319-66158-2_3

7. Arafailova, E., Beldiceanu, N., Simonis, H.: Generating linear invariants for a conjunction of automata constraints. In: Beck, J.C. (ed.) CP 2017. LNCS, vol. 10416, pp. 21–37. Springer, Cham (2017). https://doi.org/10.1007/978-3-319-66158-2_2

8. Arafailova, E., Beldiceanu, N., Simonis, H.: Deriving generic bounds for time-series constraints based on regular expressions characteristics. Constraints 23(1), 44–86 (2017). https://doi.org/10.1007/s10601-017-9276-z

9. Beldiceanu, N., Carlsson, M., Douence, R., Simonis, H.: Using finite transducers for describing and synthesising structural time-series constraints. Constraints 21(1), 22–40 (2015). https://doi.org/10.1007/s10601-015-9200-3

10. Beldiceanu, N., Feris, B.D., Gravey, P., Hasan, S., Jard, C., Ledoux, T., Li, Y., Lime, D., Madi-Wamba, G., Menaud, J.-M., Morel, P., Morvan, M., Moulinard, M.-L., Orgerie, A.-C., Pazat, J.-L., Roux, O., Sharaiha, A.: Towards energy-proportional clouds partially powered by renewable energy. Computing 99(1), 3–22 (2016). https://doi.org/10.1007/s00607-016-0503-z

11. Beldiceanu, N., Ifrim, G., Lenoir, A., Simonis, H.: Describing and generating solutions for the EDF unit commitment problem with the ModelSeeker. In: Schulte, C. (ed.) CP 2013. LNCS, vol. 8124, pp. 733–748. Springer, Heidelberg (2013). https://doi.org/10.1007/978-3-642-40627-0_54

12. Eeckhout, L., De Bosschere, K., Neefs, H.: Performance analysis through synthetic trace generation. In: 2000 IEEE International Symposium on Performance Analysis of Systems and Software. ISPASS (Cat. No. 00EX422), pp. 1–6. IEEE (2000)

13. Goldin, D.Q., Kanellakis, P.C.: On similarity queries for time-series data: constraint specification and implementation. In: Montanari, U., Rossi, F. (eds.) CP 1995. LNCS, vol. 976, pp. 137–153. Springer, Heidelberg (1995). https://doi.org/10.1007/3-540-60299-2_9

14. Kegel, L., Hahmann, M., Lehner, W.: Template-based time series generation with loom. In: EDBT/ICDT Workshops, vol. 1558. Citeseer (2016)

15. Lin, J., Williamson, S., Borne, K.D., De Barr, D.: Pattern recognition in time series. In: Way, M.J., Scargle, J.D., Ali, K.M., N, S.A. (eds.) Advances in Machine Learning and Data Mining for Astronomy. CRC (2016)

16. Montgomery, D.C., Jennings, C.L., Kulahci, M.: Introduction to Time Series Analysis and Forecasting, 2nd edn. Wiley (2016)

17. Rodríguez, M.A.F., Flener, P., Pearson, J.: Implied constraints for automaton constraints. In: Gottlob, G., Sutcliffe, G., Voronkov, A. (eds.) Global Conference on Artificial Intelligence, GCAI 2015, Tbilisi, Georgia, 16–19 October 2015. EPiC Series in Computing, vol. 36, pp. 113–126. EasyChair (2015)

18. Rodríguez, M.A.F., Flener, P., Pearson, J.: Automatic generation of descriptions of time-series constraints. In: 2017 IEEE 29th International Conference on Tools with Artificial Intelligence (ICTAI), pp. 102–109. IEEE (2017)

19. Shokoohi-Yekta, M., Chen, Y., Campana, B.J.L., Hu, B., Zakaria, J., Keogh, E.J.: Discovery of meaningful rules in time series. In: Cao, L., Zhang, C., Joachims, T., Webb, G.I., Margineantu, D.D., Williams, G. (eds.) Proceedings of the 21th ACM SIGKDD International Conference on Knowledge Discovery and Data Mining, Sydney, NSW, Australia, 10–13 August 2015, pp. 1085–1094. ACM (2015)

# A Learning-Based Algorithm to Quickly Compute Good Primal Solutions for Stochastic Integer Programs

Yoshua Bengio[2,3], Emma Frejinger[2], Andrea Lodi[1,3], Rahul Patel[1,3(✉)], and Sriram Sankaranarayanan[1]

[1] Canada Excellence Research Chair, Polytechnique Montreal, Montreal, Canada
`andrea.lodi@polymtl.ca`, `rahul.polymtl@gmail.com`
[2] Department of Computer Science and Operations Research, University of Montreal, Montreal, Canada
[3] Mila - Quebec Artificial Intelligence Institute, Montreal, Canada

**Abstract.** We propose a novel approach using supervised learning to obtain near-optimal primal solutions for two-stage stochastic integer programming (2SIP) problems with constraints in the first and second stages. The goal of the algorithm is to predict a *representative scenario* (RS) for the problem such that, deterministically solving the 2SIP with the random realization equal to the RS, gives a near-optimal solution to the original 2SIP. Predicting an RS, instead of directly predicting a solution ensures first-stage feasibility of the solution. If the problem is known to have complete recourse, second-stage feasibility is also guaranteed. For computational testing, we learn to find an RS for a two-stage stochastic facility location problem with integer variables and linear constraints in both stages and consistently provide near-optimal solutions. Our computing times are very competitive with those of general-purpose integer programming solvers to achieve a similar solution quality.

**Keywords:** Stochastic integer programming · Machine learning · Heuristics

## 1 Introduction

Two-stage stochastic integer programming (2SIP) is a standard framework to model decision making under uncertainty. In this framework, first the so-called *first-stage* decisions are made. Then, the values of some uncertain parameters in the problem are determined, as if sampled from a known distribution. Finally, the second set of decisions are made depending upon the realized value of the uncertain parameters, the so-called *second-stage* or *recourse* of the problem. The decision maker, in this setting, minimizes the sum of (i) a linear function of the first-stage decision variables and (ii) the expected value of the second-stage optimization problem.

© Springer Nature Switzerland AG 2020
E. Hebrard and N. Musliu (Eds.): CPAIOR 2020, LNCS 12296, pp. 99–111, 2020.
https://doi.org/10.1007/978-3-030-58942-4_7

2SIP is studied extensively in the literature [5,8,12,14–16,18–21,25] owing to its applicability in various decision making situations with uncertainty, like the stochastic unit-commitment problems for electricity generation [19,20], stochastic facility location problems [15], stochastic supply chain network design [22], among others. With the overwhelming importance of 2SIP a wide range of solution algorithms have been proposed, for example, [2,3,7,16,23,24].

In this paper we are interested in using machine learning (ML) to obtain good primal solutions to 2SIP. Along this line, Nair et al. [17] proposed a reinforcement learning-based heuristic solver to quickly find solutions to 2SIP. Given that the agent can be trained offline, the algorithm provided solutions much faster for some classes of problems compared to an open-source general-purpose mixed-integer programming (MIP) solver, in their case, SCIP [10,11]. However, their method is based on the following restrictive assumptions:

a. All first-stage variables are required to be binary. General integer variables or continuous variables in the first stage cannot be handled.
b. Any assignment of the binary variables is required to be feasible for the first stage of the problem, i.e., no constraints are allowed in the first stage.

Assumption a above is intrinsic to the method in [17], as both the *initialization policy* and *the local move policy* of the method involves flipping the bits of the first-stage decision vector. Hence, one cannot easily have general integer variables or continuous variables. Assumption b is again crucial to the algorithm in [17], as flipping a bit in the first stage could potentially make the new decision infeasible and it might require a more complicated feedback mechanism to check and discard infeasible solutions. In fact, if there are constraints, it is $NP$-hard to decide if there exists a flip that keeps the decision feasible. Alternatively, one could empirically penalize the infeasible solutions, but tuning the penalty might be a hard problem in itself.

In contrast, our method does not require either of these two restrictive assumptions. We allow binary, general integer as well as continuous variables in both first and second stage of the problem. We also allow constraints in both stages of the problem. Furthermore, we have a simple and direct approach to handle the first-stage constraints, without requiring any empirical penalties.

We make the following common assumption to exclude pathological cases, where an uncertain realization can turn a feasible first-stage decision infeasible.

**Assumption 1.** The 2SIP has *complete recourse*, i.e., if a first-stage decision is feasible given the first-stage constraints, then it is feasible for all the second stage problems as well.

We make another assumption of uncertainty with finite support, so we can have a proper benchmark to compare our solution against. However, this assumption can be readily removed, without affecting the proposed algorithm.

**Assumption 2.** The uncertainty distribution in the 2SIP has a finite support.

## 2    Problem Definition

We formally define a 2SIP as follows:

$$\min_{x \in \mathbb{R}^{n_1}} \quad c^{\mathsf{T}}x + \mathbb{E}_{\xi}\left[Q(x,\xi)\right] \tag{1a}$$

$$\text{subject to} \quad Ax \leq b \tag{1b}$$

$$x_i \in \mathbb{Z}, \quad \forall i \in \mathcal{I}_1 \tag{1c}$$

where,

$$Q(x,\xi) \quad = \quad \min_{y \in \mathbb{R}^{n_2}}\left\{q_{\xi}^{\mathsf{T}}y_{\xi} : Wy_{\xi} \leq h_{\xi} - T_{\xi}x, y_{\xi} \geq 0; y_i \in \mathbb{Z}\,\forall i \in \mathcal{I}_2\right\}$$

where $x \in \mathbb{R}^{n_1}$ and $y \in \mathbb{R}^{n_2}$ are the first and second-stage decisions respectively, $c \in \mathbb{R}^{n_1}$, $A \in \mathbb{R}^{m_1 \times n_1}$, $b \in \mathbb{R}^{m_1}$, $y_{\xi} \in \mathbb{R}^{n_2}$, $q_{\xi} \in \mathbb{R}^{n_2}$, $W \in \mathbb{R}^{m_2 \times n_2}$, $T_{\xi} \in \mathbb{R}^{m_2 \times n_1}$, $h_{\xi} \in \mathbb{R}^{m_2}$, $\mathcal{I}_1 \subseteq \{1, \ldots, n_1\}$, $\mathcal{I}_2 \subseteq \{1, \ldots, n_2\}$, $\xi \in \varXi$ and where $(\varXi, \mathscr{F}_{\varXi}, p)$ defines a probability space.

When Assumption 2 holds, the 2SIP described above can also be expressed as a single deterministic MIP as follows:

$$\min_{x,y} \quad c^{\mathsf{T}}x + \sum_{\forall \xi \in \varXi} p_{\xi}q_{\xi}^{\mathsf{T}}y_{\xi} \tag{2a}$$

$$\text{subject to} \quad Ax \leq b \tag{2b}$$

$$Wy_{\xi} \leq h_{\xi} - T_{\xi}x \qquad \forall \xi \in \varXi \tag{2c}$$

$$x_i \in \mathbb{Z}, \quad \forall i \in \mathcal{I}_1 \tag{2d}$$

$$y_{\xi i} \in \mathbb{Z}, \quad \forall \xi \in \varXi, \forall i \in \mathcal{I}_2. \tag{2e}$$

where, $\varXi$ is the set of random scenarios and $p_{\xi}$ is the probability of a random scenario $\xi \in \varXi$.

When Assumption 2 does not hold, the formulation (2) could be a finite-sample approximation of (1), which is extensively studied in the stochastic programming literature. Imitating [17], we compare our algorithm against solving (2) with a general-purpose MIP solver.

## 3    Methodology

In this section, we discuss the algorithmic contribution of the paper.

### 3.1    Surrogate Formulation

We first define the objective value function (OVF) $\varPhi : \mathbb{R}^{n_1} \rightarrow \mathbb{R}$, mapping $x \mapsto c^{\mathsf{T}}x + \mathbb{E}_{\xi}\left[Q(x,\xi)\right]$ - the function we are trying to optimize over the mixed-integer set defined in (1).

Given (2), we define *the surrogate problem associated with* $\bar{\xi} = (\bar{q}, \bar{h}, \bar{T})$, as follows:

$$\min_{x,y} \quad c^\mathsf{T}x + \bar{q}^\mathsf{T}y \tag{3a}$$

$$\text{subject to} \quad Ax \leq b \tag{3b}$$

$$Wy \leq \bar{h} - \bar{T}x \tag{3c}$$

$$x_i, y_j \in \mathbb{Z}, \quad \forall i \in \mathcal{I}_1; j \in \mathcal{I}_2 \tag{3d}$$

In other words, should the value that the uncertain parameters are going to take is deterministically known to be $\bar{\xi}$, then the decision maker can solve the surrogate problem associated with $\bar{\xi}$. Now, the idea behind the algorithm proposed in this paper is captured by Conjecture 1.

**Conjecture 1.** Let (2) (and hence (1)) have an optimal objective value of $f^*$. There exists $q^*, h^*, T^*$ in $\mathbb{R}^{n_2}$, $\mathbb{R}^{m_2}$ and $\mathbb{R}^{m_2 \times n_1}$ such that if $(x^\dagger, y^\dagger)$ solves the (much smaller) surrogate problem defined by $(q^*, h^*, T^*)$, then, $f^* = \Phi(x^\dagger)$.

Observe that by construction, $x^\dagger$ is feasible to the original problem in (1). Also, Conjecture 1 asserts that, there exists a realization of the uncertainty $(\xi^* = (q^*, h^*, T^*))$ such that if one deterministically optimizes for that realization $\xi^*$, then its solutions are optimal for the original 2SIP. Each such $\xi^*$ is called a *representative scenario* (RS) for the given 2SIP.

Now, given adequate computing resources, one can solve the following bilevel program to obtain an RS.

$$\min_{\substack{U,v,w \\ x,y}} \quad c^\mathsf{T}x + \sum_{\forall \xi \in \Xi} p_\xi q_\xi^\mathsf{T} y_\xi \tag{4a}$$

$$\text{subject to} \quad (x, w) \in \arg\min_{x,w} \left\{ c^\mathsf{T}x + v^\mathsf{T}w : \begin{array}{ll} Ax \leq b; \\ Ww \leq v - Ux; \\ x_i \in \mathbb{Z} & \forall i \in \mathcal{I}_1 \\ w_i \in \mathbb{Z} & \forall i \in \mathcal{I}_2 \end{array} \right\} \tag{4b}$$

$$Wy_\xi \leq h_\xi - T_\xi x \quad \forall \xi \in \Xi \tag{4c}$$

$$y_{\xi i} \in \mathbb{Z}, \quad \forall \xi \in \Xi, \forall i \in \mathcal{I}_2 \tag{4d}$$

If the optimal value of this problem matches the optimal value of the original 2SIP, then the corresponding values for $(U, v, w)$ form the RS. Note that if $T_\xi$ is the same for all $\xi \in \Xi$, then (4) is a mixed-integer bilevel *linear* program (MIBLP) and can hopefully be solved faster than the general case.

## 3.2    Learning Algorithm

The goal of ML algorithms is to predict an optimal $(U, v, w)$ to (4), given the data for the 2SIP. On the one hand, we are expecting the ML algorithms to predict the solutions of a seemingly much harder optimization problem than the original 2SIP. On the other hand, this is easier for ML since there are no constraints on the predicted variables – $U, v, w$. Supervised learning is a natural tool to achieve this goal.

Supervised learning can be used if there is a training dataset of problem instances and their corresponding RS. The task of predicting RS can be formulated as a regression task as RS is real valued. The algorithm tries to minimize the mean squared error (MSE) between the true and predicted RS. The prediction can also be evaluated on the merits of optimization metrics, comparing the solution and objective value of true and predicted RS.

# 4    Computational Study

This section discusses the computation study performed to support Conjecture 1.

## 4.1    Problem Definition

In this work, we consider a version of two-stage stochastic capacitated facility location (S-CFLP) for computational analysis. The problem is enhanced such that both the first and the second stage of the problem have integer as well as continuous variables. More precisely, the first stage consists of deciding (i) the locations where a facility has to be opened (binary decisions), and (ii) if a facility is open, then the maximum demand that the facility can serve (continuous decisions). There are also constraints which dictate bounds on the total number of facilities that can be opened. The uncertainty in the problem pertains to the demand values at various locations in the second stage of the problem, which are sampled from a finite distribution. Once the demand is realized, the second stage consists in deciding (i) if a given open facility should serve the demand in a location (binary decisions) (ii) if a facility serves the demand in a location, then what quantity of demand is to be served (continuous decisions). These decisions have to ensure that the demand and supply constraints are met. The problem formulation is presented formally in Appendix A.1.

## 4.2    Data Generation

*Generate Instances.* We generate 50K instances of S-CFLP, with 10 facilities, 10 clients and 50 scenarios. We provide details on how the data for these instances are generated in Appendix A.2. The generated instances are solved using Gurobi, running 2 threads, to at most 2% gap or 10 min time limit.

Next, we compute an RS for each of the 50K instances. As stated earlier, one could solve mixed-integer bilevel programs (4) to obtain the RSs. However, due to the computational burden, we use heuristics that work using the knowledge of the (nearly) optimal solutions to the 2SIP already obtained from Gurobi. These heuristics are detailed in Appendix A.3. Out of 50K instances, they find an RS for 49,290 instances. We believe that a more thorough search will enable us to find the RS for all the problems.

### 4.3  Learning Algorithm

We formulate the task of predicting the RS as a regression task. The size of the dataset, which comprises of instances and their corresponding RS, is 49,290. The dataset is split into training and test sets of size 45K and 4,290, respectively. Further, a validation set of 5K is carved out from the training set. We use linear regression (LR) and artificial neural network (ANN) to minimize the MSE between the true and predicted RS.

*Feature Engineering.* It is well known that features describing the connection between variables, constraints and other interaction help ML to perform well rather than just providing plain data matrices [4,6,9,13]. In this spirit, along with the fixed and variable costs to open facilities at different locations, we also provide aggregated features on the set of scenarios. These features give information about each of the potential locations for facilities in S-CFLP as well as the way different locations interact through the demands in adjacent nodes. A detailed set of the features is given in Appendix A.4.

### 4.4  Comparison

In order to evaluate the ML-based prediction of $\xi^\star$, which we refer to as $\hat{\xi}^\star$, we compare the solution obtained by solving the surrogate problem associated with $\hat{\xi}^\star$ against solutions obtained by various algorithms.

We use LR and ANN to predict $\xi^\star$. We compare these predictions against (i) Solution obtained using Gurobi by solving (2) (GRB) (ii) a solution obtained by solving the surrogate associated with the average scenario, namely $\sum_{i=1}^{N} \Xi_i / N$ (AVG) (iii) a solution obtained by solving the surrogate associated with a randomly chosen scenario from the $N$ choices (RND) (iv) a solution obtained by solving the surrogate associated with a randomly chosen scenario from the distribution of the scenario predicted by LR (DIST). Note that GRB produces better solutions (in most cases) than the ML methods, however, taking a significantly longer time. We therefore assess the time it takes GRB to get a solution of comparable quality to LR and ANN. We refer to these as GRB-L and GRB-A, respectively.

## 5  Results

Table 1 reports the *objective value difference ratio* defined as ((Obj val by a method−GRB obj val)/GRB obj val) for each method and Table 2 statistics on computing times. Before analyzing the results in more detail, we note a key finding that emerges. Namely, LR and ANN perform almost as good as GRB (in terms of quality of the objective value) in a fraction of the time taken by GRB. Figure 1 captures the trade off between the quality of solutions obtained by different methods as well as the time taken to obtain these solutions.

We observe from Table 1 that LR and ANN produce decisions that are as good as GRB ones on an average (and by the median value), and in some cases the

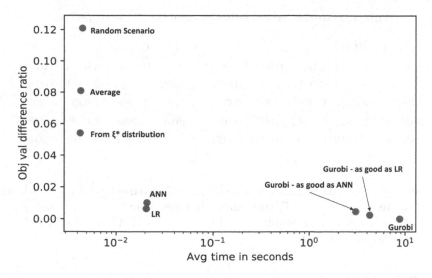

**Fig. 1.** Objective value difference ratio vs. avg time in seconds

ML-based methods even beat GRB, i.e., produce solutions whose objective value is better than that of GRB. This is possible because GRB does not necessarily solve the problem to optimality, but only up to a gap of 2%. Further, even in the worst of the 4,290 test cases, LR is at most 2.64% away from GRB. To show that this is not easily achieved, we also compare GRB against AVG, RND and DIST. We observe that these methods perform much poorer than GRB, unlike LR and ANN.

**Table 1.** Objective value difference ratio, GRB vs. the five other methods (in %)

| GRB vs. | AVG | RND | DIST | LR | ANN |
|---|---|---|---|---|---|
| Min | 3.36 | −0.33 | −0.17 | −0.62 | −0.45 |
| Max | 14.23 | 94.42 | 49.87 | 2.64 | 7.85 |
| Avg. | 8.10 | 12.10 | 5.41 | 0.64 | 1.02 |
| Median | 8.08 | 8.24 | 3.54 | 0.60 | 0.90 |
| Std. dev. | 1.59 | 12.11 | 5.39 | 0.41 | 0.70 |

Analyzing the time improvement in using LR and ANN, we observe from Table 2 that these methods solve the S-CFLP *orders of magnitude* faster than GRB. Indeed, GRB takes over 8 s on an average to solve these problems, while the maximum time is 0.046 s using LR and ANN. We emphasize that the time taken to solve using the ML methods *includes* the time elapsed in computing the values of the features used in ML and the time elapsed in solving the surrogate associated with the corresponding $\hat{\xi}^\star$. Recall that GRB-L and GRB-A denote

**Table 2.** Statistics on computing times of the different methods

| Method | GRB | AVG | RND | DIST | GRB-L | GRB-A | LR | ANN |
|---|---|---|---|---|---|---|---|---|
| Min | 0.4354 | 0.0041 | 0.0041 | 0.0040 | 0.4049 | 0.4268 | 0.0194 | 0.0197 |
| Max | 559.32 | 0.0077 | 0.0081 | 0.0065 | 140.43 | 140.43 | 0.0454 | 0.0457 |
| Avg. | 8.7713 | 0.0043 | 0.0045 | 0.0043 | 4.2650 | 3.0398 | 0.0206 | 0.0209 |
| Median | 2.2621 | 0.0043 | 0.0044 | 0.0042 | 1.4613 | 1.2898 | 0.0200 | 0.0204 |
| Std. dev. | 17.919 | 0.0001 | 0.0003 | 0.0001 | 8.5576 | 6.8362 | 0.0019 | 0.0019 |

the time it takes GRB to produce a solution of comparable quality to LR and ANN. The results show that GRB cannot produce a solution of the same quality as LR and ANN in a comparable time. In fact, LR and ANN are still orders of magnitude faster than GRB.

## 6   Discussion

In this paper, we present an algorithm to solve 2SIP using ML-based methods. The method hinges on the existence of the RS conjectured in Sect. 3.1. Computationally, we see that the methods proposed in this paper consistently provide good quality solutions to S-CFLP orders of magnitude faster.

An important observation we had while training the models is that we were never able to get the training loss close to zero. Naturally, the predicted RS in the test dataset were quite different from the RS estimated using our heuristics. The differences in the predicted values of the components of RS and those obtained using the heuristics are documented by Fig. 2 in the Appendix. Despite this, the solution value to the 2SIP as determined by our algorithm were near optimal as shown in the results and significantly better than those obtained with other methods. The mismatch between the ML metrics and those characterizing discrete optimization problems is a known issue [4] requiring extensive research and, in our context, we believe that exploring this avenue might produce better solutions.

Another interesting observation is that LR beats ANN in this task. We suspect that this is partly caused by the parsimony offered by LR. However, this is also encouraging news that the sample complexity of the learning task might be relatively small in general. We believe that a natural extension to this work is to provide these analyses more formally.

Further, we believe that computational tests assessing the performance of the algorithms in different datasets of 2SIP is crucial to show how much and where our method generalizes. This might involve learning solutions to other forms of 2SIP like the stochastic unit-commitment problem, the stochastic supply chain-network design problem etc. These are cases where we believe that Conjecture 1 still holds, but we do not have computational validation.

Finally, we would also be interested in extending the theory side when Conjecture 1 is not expected to hold at all or holds only with weaker guarantees;

for example, in the case where both the stages are mixed-integer nonlinear programming (MINLP) problems. In such cases, it will be useful to understand the reach of ML-based solution techniques as opposed to traditional MINLP solvers.

## A    Computational Test Details

### A.1    Problem Formulation

We provide below the problem considered in this work for computational study.

$$\min_{b \in \{0,1\}^n, v \in \mathbb{R}^n_+} \sum_{i=1}^{n} \left( c_i^f b_i + c_i^v v_i \right) + \mathbb{E}_\xi (Q(x, \xi)) \tag{5a}$$

$$\frac{n}{10} \leq \sum_{i=1}^{n} b_i \leq \frac{3n}{4} \tag{5b}$$

$$v_i \leq M b_i \tag{5c}$$

where $Q(x, \xi)$ is

$$\min_{u \in \{0,1\}^{n \times n}, y \in \mathbb{R}^{n \times n}_+} \sum_{i=1}^{n} \sum_{j=1}^{n} c_{ij}^{tv} y_{ij} + \sum_{i=1}^{n} \sum_{j=1}^{n} c_{ij}^{tf} u_{ij} \tag{5d}$$

$$\sum_{j=1}^{n} y_{ij} \leq v_i \tag{5e}$$

$$\sum_{i=1}^{n} y_{ij} = d_j(\xi) \tag{5f}$$

$$y_{ij} \leq u_{ij} M \tag{5g}$$

In this problem, we minimize the fixed and variable costs of opening a facility along with the fixed and variable costs of transportation between the facilities and the demand nodes. There are $n$ potential locations where a facility could be opened. A fixed cost of $c_i^f$ is incurred, if a facility is opened in location $i$, and a variable cost of $c_i^v$ is incurred per-unit capacity of the facility opened in location $i$. The binary variable, $b_i$ tracks if a facility is opened in location $i$ and the continuous variable $v_i$ indicates the size of the facility at location $i$. The constraint in (5c), along with the binary constraint on $b$ ensures that the costs are incurred in the right way. Finally, (5b) is a complicating constraint, which says that at least a tenth of the locations must have a facility open and not more than three-quarters of the locations must have a facility open.

In the second stage, $c_{ij}^{tf}$ is the fixed cost incurred in transporting from location $i$ to $j$; $c_{ij}^{tv}$ is the per-unit variable cost incurred in transporting from $i$ to $j$. The binary variable $u_{ij}$ denotes if any transportation happens from $i$ to $j$ and the continuous variable $y_{ij}$ denotes the actual quantity transported from $i$ to $j$.

---

**Algorithm 1:** GENERATEXIHAT

---

**Data:** 2SIP $\mathcal{P}$ with first-stage solution $x^e$ and objective value $o^e$, max iterations
    *iter*, a constant $c > 0$

**Result:** $\xi^*$ or *NULL*

1  $\bar{\xi} \leftarrow \frac{1}{|\Xi|} \sum_{i=1}^{|\Xi|} \xi_i$;

2  **while** *iter* **do**

3  |    Formulate surrogate problem $\mathcal{P}'$ associated with $\bar{\xi}$;

4  |    Solve $\mathcal{P}'$ and extract first-stage solution $x^{\bar{\xi}}$ with objective $o^{\bar{\xi}}$;

5  |    **if** $\frac{o^{\bar{\xi}}}{o^e} \leq c$ **then**

6  |    |    **return** $\bar{\xi}$;

7  |    **else**

8  |    |    Perturb $\bar{\xi}$ using heuristics based on $x^e$ and $x^{\bar{\xi}}$

9  **return** *NULL*

---

The second-stage objective in (5d) minimizes the transportation cost incurred under a random demand scenario parameterized by $\xi$. Then, (5e) ensures that the total quantity transported out of a facility is not greater than the capacity of the facility, while (5f) ensure that the total quantity supplied to a location $j$ equals the (random) demand at $j$. Finally, constraints (5g) link $u$ and $y$ variables appropriately.

## A.2    Data Generation

*Instance Generation.* We generate 50K instances of S-CFLP, with 10 facilities, 10 clients and 50 scenarios. The random parameters $c^f$, $c^v$ and $\Xi = [\xi_1, \ldots, \xi_{50}]$ vary across instances, where as $c^{tf}$ and $c^{tv}$ are fixed across all instances. Moreover, $c^f$ and $c^v$ are sampled from a discrete uniform distribution $[15, 20)$ and $[5, 10)$, respectively. The demand matrix $\Xi$ is generated by first evaluating $\lambda = \lfloor (c^f + 10c^v)/\sqrt{n} \rfloor$. The $i^{th}$ demand scenario ($1 \leq i \leq 50$) is generated by sampling from a Poisson distribution with the mean equal to $\lambda$.

The generated instances are solved using Gurobi, running 2 threads, to optimality (less than 2% gap) or 10 min time limit. We store the objective value, gap closed and master solution $x^* = (b^*, v^*)$. We were able to solve all the instances up to the specified gap, within the specified time limit.

We follow Algorithm 1 for generating $\xi^*$. Then, $|\Xi|$ refers to the cardinality of $\Xi$ and $c = 1.01$ in step 6. The heuristics for updating the RS, based on $x^e$ and $x^{\bar{\xi}}$, are described in Appendix A.3.

## A.3    Heuristics

Let $x^* = (b^*, v^*)$ and $x^{\bar{\xi}} = (b^{\bar{\xi}}, v^{\bar{\xi}})$ be the first-stage optimal and surrogate solution associated with the scenario $\bar{\xi}$, respectively. There are three heuristics that we use in tandem to generate $\xi^*$. The first heuristic is based on the comparison of facilities being open or close in the optimal and surrogate solution. The demand in the $\bar{\xi}$ is zeroed out at clients for which the $b^*$ is close and $b^{\bar{\xi}}$ is open, as suggested by

$$b_i^* = 0 \wedge b_i^{\bar{\xi}} = 1 \implies \bar{\xi}_i = 0 \quad i = 1, \ldots, n. \tag{6}$$

The remaining two heuristics are based on the comparison of capacities installed in facilities in the optimal and the surrogate solution. First, the client with maximum absolute difference between capacities installed in optimal and surrogate solution is identified i.e., $\operatorname{argmax}_i |v_i^* - v_i^{\bar{\xi}}|$. The demand at this client in the $\bar{\xi}$ is updated either by a fixed percentage $p$ of the current demand

$$\bar{\xi}_i = \bar{\xi}_i + \frac{v_i^* - v_i^{\bar{\xi}}}{|v_i^* - v_i^{\bar{\xi}}|} \times p\bar{\xi}_i, \tag{7}$$

or by a fraction $f$ of the difference between the capacities installed in optimal and surrogate solutions

$$\bar{\xi}_i = \bar{\xi}_i + (v_i^* - v_i^{\bar{\xi}})f\bar{\xi}_i. \tag{8}$$

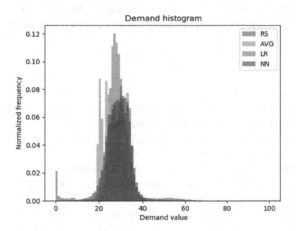

**Fig. 2.** Demand histogram for different methods.

## A.4    Feature Engineering

The inputs of the models are $c^f, c^v$ and $\Xi$. We do not provide $c^{tf}$ and $c^{tv}$ in the input as they are fixed across all instances. We do feature engineering on $\Xi$, instead of providing it as a raw input, to extract information which can be useful in predicting $\xi^*$. Let $\Xi$ be an $m \times n$ matrix, where $m$ is the number of scenarios and $n$ is the number of clients. We calculate minimum, maximum, average, standard deviation, median, $75^{th}$ quantile, and $25^{th}$ quantile of $\Xi_{[:,i]}$ ($i^{th}$ column of $\Xi$) for $i = 1, \ldots, n$.

We also find the percentage of scenarios in which some fraction of demand for a client is greater than and less than the demand on all the other nodes

$$\frac{c * \Xi_{[:,i]} \geq \Xi_{[:,\neq i]}}{m} \quad \text{and} \quad \frac{c * \Xi_{[:,i]} \leq \Xi_{[:,\neq i]}}{m}.$$

We set $c$ to different values (0.9, 1, 1.1, 1.2 and 1.5) and we thus end up with an input vector of size 190, combining $c^f$, $c^v$ and features extracted from $\Xi$. The feature extraction performs an aggregation over the number of scenarios.

### A.5    ML Model Details

For the LR model, we use the vanilla LR implementation provided by Scikit-Learn without regularization.

For the ANN based regression, we consider a fully connected feed-forward neural network with two hidden layers. The configuration of the network was an input of 190 dimensions, two hidden layers with 128 neurons each and an output of 10 dimensions. We used ReLU activation in the hidden layers and trained the ANN using Stochastic Gradient Descent with momentum 0.9. The ANN was implemented using PyTorch. An implementation of the algorithm to solve capacitated facility location problem is available at https://github.com/ds4dm/nectar [1].

## References

1. Nectar: CPAIOR2020 release - code archived using zenodo. https://doi.org/10.5281/zenodo.3690293. Accessed 30 Sept 2010
2. Ahmed, S.: A scenario decomposition algorithm for 0–1 stochastic programs. Oper. Res. Lett. **41**(6), 565–569 (2013)
3. Ahmed, S., Tawarmalani, M., Sahinidis, N.V.: A finite branch-and-bound algorithm for two-stage stochastic integer programs. Math. Program. **100**(2), 355–377 (2004)
4. Bengio, Y., Lodi, A., Prouvost, A.: Machine learning for combinatorial optimization: a methodological tour d'horizon. arXiv preprint arXiv:1811.06128 (2018)
5. Birge, J.R., Louveaux, F.: Introduction to Stochastic Programming. Springer, Heidelberg (2011). https://doi.org/10.1007/978-1-4614-0237-4
6. Bonami, P., Lodi, A., Zarpellon, G.: Learning a classification of mixed-integer quadratic programming problems. In: van Hoeve, W.-J. (ed.) CPAIOR 2018. LNCS, vol. 10848, pp. 595–604. Springer, Cham (2018). https://doi.org/10.1007/978-3-319-93031-2_43
7. Carøe, C.C., Tind, J.: L-shaped decomposition of two-stage stochastic programs with integer recourse. Math. Program. **83**(1–3), 451–464 (1998)
8. Dupačová, J., Gröwe-Kuska, N., Römisch, W.: Scenario reduction in stochastic programming. Math. Program. **95**(3), 493–511 (2003)
9. Gasse, M., Chételat, D., Ferroni, N., Charlin, L., Lodi, A.: Exact combinatorial optimization with graph convolutional neural networks. arXiv preprint arXiv:1906.01629 (2019)

10. Gleixner, A., et al.: The SCIP Optimization Suite 6.0. Technical report, Optimization Online, July 2018. http://www.optimization-online.org/DB_HTML/2018/07/6692.html
11. Gleixner, A.: The SCIP Optimization Suite 6.0. ZIB-Report 18–26, Zuse Institute Berlin, July 2018. http://nbn-resolving.de/urn:nbn:de:0297-zib-69361
12. Kall, P., Wallace, S.W.: Stochastic Programming, John Wiley and Sons, Chichester, (1994)
13. Khalil, E.B., Le Bodic, P., Song, L., Nemhauser, G., Dilkina, B.: Learning to branch in mixed integer programming. In: Thirtieth AAAI Conference on Artificial Intelligence (2016)
14. Linderoth, J., Shapiro, A., Wright, S.: The empirical behavior of sampling methods for stochastic programming. Ann. Oper. Res. **142**(1), 215–241 (2006)
15. Louveaux, F.V., Peeters, D.: A dual-based procedure for stochastic facility location. Oper. Res. **40**(3), 564–573 (1992)
16. Lulli, G., Sen, S.: A branch-and-price algorithm for multistage stochastic integer programming with application to stochastic batch-sizing problems. Manag. Sci. **50**(6), 786–796 (2004)
17. Nair, V., Dvijotham, D., Dunning, I., Vinyals, O.: Learning fast optimizers for contextual stochastic integer programs. In: UAI, pp. 591–600 (2018)
18. Nemirovski, A., Juditsky, A., Lan, G., Shapiro, A.: Robust stochastic approximation approach to stochastic programming. SIAM J. Optim. **19**(4), 1574–1609 (2009)
19. Powell, W.B., Meisel, S.: Tutorial on stochastic optimization in energy-part i: modeling and policies. IEEE Trans. Power Syst. **31**(2), 1459–1467 (2015)
20. Powell, W.B., Meisel, S.: Tutorial on stochastic optimization in energy-part ii: an energy storage illustration. IEEE Trans. Power Syst. **31**(2), 1468–1475 (2015)
21. Prékopa, A.: Stochastic Programming, vol. 324. Springer, Heidelberg (2013)
22. Santoso, T., Ahmed, S., Goetschalckx, M., Shapiro, A.: A stochastic programming approach for supply chain network design under uncertainty. Eur. J. Oper. Res. **167**(1), 96–115 (2005)
23. Sen, S.: Stochastic mixed-integer programming algorithms: beyond benders' decomposition. Wiley Encyclopedia of Operations Research and Management Science (2010)
24. Sen, S., Higle, J.L.: The $C^3$ theorem and a $D^2$ algorithm for large scale stochastic mixed-integer programming: set convexification. Math. Program. **104**(1), 1–20 (2005)
25. Shapiro, A., Dentcheva, D., Ruszczyński, A.: Lectures on Stochastic Programming: Modeling and Theory. SIAM (2009)

# Integer Programming Techniques for Minor-Embedding in Quantum Annealers

David E. Bernal[1,4,5]($\boxtimes$), Kyle E. C. Booth[2,4,5], Raouf Dridi[3], Hedayat Alghassi[3], Sridhar Tayur[3], and Davide Venturelli[4,5]

[1] Department of Chemical Engineering, Carnegie Mellon University, Pittsburgh, PA 15213, USA
bernalde@cmu.edu
[2] Department of Mechanical and Industrial Engineering, University of Toronto, Toronto, ON M5S 3G8, Canada
kbooth@mie.utoronto.ca
[3] Tepper School of Business, Carnegie Mellon University, Pittsburgh, PA 15213, USA
{rdridi,halghass,stayur}@andrew.cmu.edu
[4] Quantum AI Laboratory (QuAIL), NASA Ames Research Center, Moffett Field, CA 94035, USA
davide.venturelli@nasa.gov
[5] USRA Research Institute for Advanced Computer Science (RIACS), Mountain View, CA 94043, USA

**Abstract.** A major limitation of current generations of quantum annealers is the sparse connectivity of manufactured qubits in the hardware graph. This technological limitation has generated considerable interest, motivating efforts to design efficient and adroit minor-embedding procedures that bypass sparsity constraints. In this paper, starting from a previous equational formulation by Dridi et al. (arXiv:1810.01440), we propose integer programming (IP) techniques for solving the minor-embedding problem. The first approach involves a direct translation from the previous equational formulation to IP, while the second decomposes the problem into an assignment master problem and fiber condition checking subproblems. The proposed methods are able to detect instance infeasibility and provide bounds on solution quality, capabilities not offered by currently employed heuristic methods. We demonstrate the efficacy of our methods with an extensive computational assessment involving three families of random graphs of varying sizes and densities. The direct translation as a monolithic IP model can be solved with existing commercial solvers yielding valid minor-embeddings but it is outperformed, overall, by the decomposition approach. Our results demonstrate the promise of our methods for the studied benchmarks, highlighting the advantages of using IP technology for minor-embedding problems.

**Keywords:** Graph minors · Quantum annealers · Integer programming · Decomposition · Algebraic geometry

© Springer Nature Switzerland AG 2020
E. Hebrard and N. Musliu (Eds.): CPAIOR 2020, LNCS 12296, pp. 112–129, 2020.
https://doi.org/10.1007/978-3-030-58942-4_8

# 1    Introduction

Quantum annealing processors have been developed to perform the quantum annealing algorithm, which searches for the minimum of a quadratic unconstrained binary optimization (QUBO) problem equivalent to finding the minimum energy state of an Ising spin system [18]. The most successful implementation of quantum annealers use superconducting quantum bits (qubits), where the interactions between qubits are controlled in order to perform an adiabatic evolution to a state whose energy function represents the objective function to be optimized. This is the case of the quantum annealers manufactured by D-Wave, which is the current largest quantum annealing hardware producer.

The hardware configuration of D-Wave quantum annealers follows a *Chimera* graph topology (Fig. 1). The *Chimera* graph, $C_{L,M,N}$, is a grid of $M \times N$ cells of $K_{L,L}$ biclique graphs connected in a nearest-neighbor fashion by means of nonplanar edges [22]. This architecture is selected due to the advantages it provides, both in terms of the physical implementation (e.g., the ability to incorporate on-chip control circuitry, 2D chip integration, and minimization of noise-to-signal ratio), the chip topology characteristics (e.g., non-planarity, ability to embed complete graphs [3]). Figure 1 presents photographs of the D-Wave Two processor chip, and how this physical topology corresponds to a *Chimera* $C_{4,8,8}$ graph.

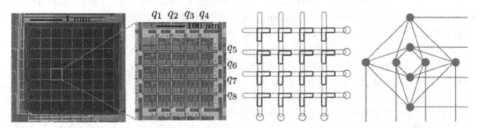

**Fig. 1.** *Chimera* graph with one cell magnified. Each cell contains 8 qubits, whose internal connections can be described by a $K_{4,4}$ graph. Qubits are laid in thin loops (red), and connected with each other in the *Chimera* cell by couplers (blue cells), and outside of it to neighbor cells (blue circles, in this case only to cells located south and east). Images adapted from Supplementary material of [18] and [30,31]. (Color figure online)

Given the connectivity restrictions of the graph defining the processor's architecture, representing an arbitrary QUBO requires the use of graph minor theory (GMT). GMT, the central theme of this work, is prominent across many fields. Mapping a dense *source* graph to a sparse *target* graph can be achieved by constructing connected subgraphs of the target graph from the high degree logical vertices in the source graph. The resulting mapping is called a *minor-embedding* of the source graph inside the target graph. In quantum computing,

GMT is employed to extend the scope of problems that can be represented on current quantum annealing hardware [5,19]. An example of minor-embedding is presented in Fig. 2, where an example source graph is embedded in $C_{4,1,2}$.

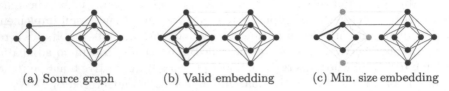

(a) Source graph          (b) Valid embedding          (c) Min. size embedding

**Fig. 2.** Source graph of illustrative example [9], a valid embedding and its minimal size embedding in $C_{4,1,2}$. Grey nodes and edges represent unused nodes and edges in embedding, but present at the target graph. Bold edges represent edges in chains.

Numerous heuristics for finding minor-embeddings have been proposed [1, 4,32], with some work focused on finding embeddings within *Chimera* graphs specifically [14,23,29,32]. While these approaches are generally fast, they do not provide guarantees on the quality of the produced minor-embeddings (e.g., minimal size of the embedding), nor can they prove the nonexistence of a minor-embedding for infeasible problems. An approach that attempts to address these shortcomings was recently introduced by Dridi et al. [9] and uses tools from algebraic geometry to produce an equational formulation, as opposed to a purely combinatorial approach, to the minor-embedding problem.

In this paper, starting from the equational formulation presented by Dridi et al. [9], we propose integer programming (IP) techniques for tackling the embedding problem. Our proposed approaches differ from the computationally demanding Groebner bases computation used previously and are aimed at more efficiently computing embeddings while retaining interesting properties that arise from the equational formulation of the problem. Our first approach, detailed in Sect. 3, directly translates the previous equational formulation to IP, while our second approach decomposes the problem into an assignment master problem and fiber condition checking subproblems, as described in Sect. 4. The proposed methods are able to detect instance infeasibility and provide bounds on solution quality for a specific objective function, capabilities not offered by currently employed heuristic methods. While recent work uses an approach with IP to address the embedding problem based on templates specific to D-Wave quantum annealers [28], the techniques we present in this paper are target graph agnostic.

We conduct an extensive empirical analysis involving a benchmark consisting of three different families of random graphs in Sect. 5. There we present our results in an illustrative and challenging case for heuristics, which motivates the use of IP over computational algebraic geometry (CAG) methods in random structured and unstructured graphs as well as in applications for quantum annealing. The results of the experiments indicate that, while the IP-based methods are slower than currently employed heuristics whenever the heuristics are able to find an embedding, the IP methods provide infeasibility proofs and quality guarantees which the heuristics are unable to provide. Furthermore, our experiments suggest that the decomposition method outperforms the monolithic IP in finding compact embeddings. However, the decomposition approach does not always perform as efficiently as the monolithic IP approach in providing optimality or infeasibility guarantees, especially seen in the illustrative example and small structured graphs. We provide concluding remarks in Sect. 6.

*Notations.* All graphs considered in this paper are simple and undirected. We use $\mathbf{V}(X)$ and $\mathbf{E}(X)$ to denote the vertex and edge sets of a graph $X$, respectively. We also define $n = |\mathbf{V}(X)|$, and $m = |\mathbf{V}(Y)|$. Finally, given a vector $v$, $\mathbf{v}$ denotes the concatenation $\mathbf{v} = (v_1, \ldots, v_{|\mathbf{v}|})$.

## 2  Problem Definition

Let $X$ and $Y$ be a fixed target and source graph, respectively. The problem we consider is finding a minor-embedding of the graph $Y$ inside $X$. As the target graph (often) cannot directly represent all the edges in the source graph, a vertex in the source graph will typically be represented by one or more vertices in the target graph that form a connected subgraph. We call the collection of vertices in the target graph that represent a vertex, $y \in \mathbf{V}(Y)$, in the source graph the *vertex model* of $y$, denoted $\phi(y)$. An embedding, then, is defined as the union of all vertex models of $y \in \mathbf{V}(Y)$. The vertex models are constructed such that, for each edge $\{y_1, y_2\} \in \mathbf{E}(Y)$, there exists at least one edge in $\mathbf{E}(X)$ connecting the two vertex models $\phi(y_1)$ and $\phi(y_2)$. For the sake of a simplicity, we use the term embedding instead of minor-embedding throughout the remainder of the paper. Suppose we have an embedding of the graph $Y$ inside the graph $X$. The embedding is given by subgraph of $X$ that satisfies $\phi(Y) := \cup_{y \in \mathbf{V}(Y)} \phi(y)$, which in GMT is known as a $Y$ minor. In the context of quantum annealers, this embedding represents the quantum processor's representation of the source graph, since it does not distinguish between qubits representing different nodes of the source graph or qubits representing the same node in the source graph. The vertices of $X$ belonging to the same vertex model are known as *chains* in the quantum annealing community and as *fibers* using CAG terminology. In practice, quantum annealers use a strong ferromagnetic coefficient to enforce these replicated values to be equal (i.e., acting as a single qubit).

In the equational approach, embedding the source graph $Y$ inside the target graph $X$ is represented by a map $\pi : X \to Y$ such that: $\pi^{-1}(y) = \phi(y), y \in \mathbf{V}(Y)$. The map $\pi$ is required to be surjective to guarantee that all logical qubits are embedded. The goal, then, is finding an embedding, which is equivalent to finding the mapping $\pi$ given graphs $X$ and $Y$.

Although finding a valid embedding is sufficient for a problem represented by graph $X$ to be implemented in a quantum annealer described by graph $Y$, as in Fig. 2, it is often desirable to find an embedding which optimizes an objective function. In particular, given the limitations on the number of qubits in available quantum hardware, a desired property of an embedding is to have a small *qubit footprint* [7], as shown in Fig. 2(c).

## 3   Problem Formulation

We tackle the problem of determining the mapping $\pi$ using integer programming (IP). IP is a mathematical optimization technique used for problems modeled as a set of decision variables taking on integer values, constrained by linear constraints and with a linear objective function. The standard solution approach to IP models is branch-and-bound tree search. Indeed, due to their many practical applications, the computational capabilities of modern IP solvers have increased tremendously in recent years [2]. These IP solvers are capable of proving instance infeasibility and providing certificates of optimality and bounds on solution quality.

In our first approach, the previously proposed polynomial equations [9] are reformulated such that they represent the original logic and are representable in the IP formalism (i.e., linear constraints involving integer variables). Defining the mapping $\pi$ explained above as:

$$\pi(x_i) = \sum_{j:y_j \in \mathbf{V}(Y)} \alpha_{ij} y_j, \quad \forall x_i \in \mathbf{V}(X) \tag{1}$$

where $\alpha_{ij}$ are binary coefficients. For this map to be well-defined we impose:

$$\sum_{j:y_j \in \mathbf{V}(Y)} \alpha_{ij} \leq 1 \quad \forall x_i \in \mathbf{V}(X), \tag{2}$$

that is, at most one $\alpha_{ij}$ is non-zero for each vertex in the graph $X$. The unique non zero $\alpha_{ij}$ (if any) represents whether the physical qubit $x_i$ embeds $y_j$, i.e., $\phi(y_j) = x_i$. When all the coefficients $\alpha_{ij}$ are zero, we get $\pi(x_i) = 0$ indicating that the physical qubit is not used. In other words, while the domain of definition of $\pi$ is $\mathbf{V}(X)$, its support is only a subset of $\mathbf{V}(X)$. The other conditions included in the definition of the embedding $\phi(Y)$ (e.g., the connectivity of the fibers) can similarly be written in equational form.

### 3.1   Constraints

Here, we present the IP formulation of the polynomial conditions, from Dridi et al. [9], that the coefficients $\alpha_{ij} \in \{0,1\}, \forall x_i \in \mathbf{V}(X), \forall y_j \in \mathbf{V}(Y)$ need to satisfy for $\pi$ to be a valid embedding. This constitutes the first contribution of the paper. Note that, with a slight abuse of notation, our IP approaches redefine $\alpha_{ij}$ as a binary decision variable equal to 1 if $x_i$ belongs to the vertex model of $y_j$, and 0 otherwise.

*1. Minimum and maximum size.* These constraints ensure that the total number of qubits is bounded within the number of variables in the original problem and the total number of qubits $n$.

$$m \leq \sum_{i:x_i \in \mathbf{V}(X)} \sum_{j:y_j \in \mathbf{V}(Y)} \alpha_{ij} \leq n. \tag{3}$$

*2. Well-definition of the map $\pi$.* This is captured by Eq. (2).

*3. Fiber size constraint.* This constraint on the size of the vertex models $|\phi(y_j)|$, known as *fiber size*, is given by:

$$1 \leq \sum_{i:x_i \in \mathbf{V}(X)} \alpha_{ij} \leq k \quad \forall y_j \in \mathbf{V}(Y). \tag{4}$$

where $k$ is the desired maximum size of each fiber $\pi^{-1}(y_j)$. The lower bound ensures that all the logical variables are embedded i.e., the map $\pi$ is a surjection on the set $\mathbf{V}(X)$. We also include the following constraint:

$$1 \geq \alpha_{i_1 j} + \alpha_{i_2 j} \quad \forall x_{i_1}, x_{i_2} \in \mathbf{V}(X), \min d(x_{i_1}, x_{i_2}) > k, \forall y_j \in \mathbf{V}(Y). \tag{5}$$

This additional refinement excludes pairs $(x_{i_1}, x_{i_2})$ from being in the fiber $\pi^{-1}(y_j)$ whenever their distance, $d(x_{i_1}, x_{i_2})$, is larger than $k$, the desired maximum size of the fiber.

*4. Fiber condition.* We require that each fiber to be a connected subtree of $X$:

$$\forall x_{i_1}, x_{i_2} \in \pi^{-1}(y_j) : \alpha_{i_1 j} + \alpha_{i_2 j} + \left( \sum_{c_k(x_{i_1}, x_{i_2}) \in C_k(x_{i_1}, x_{i_2})} (\gamma_{c_k, j}) - 1 \right) \leq 2. \tag{6}$$

The binary variable $\gamma_{c_k, j}$ takes a value of 1 if a fiber $c_k(x_{i_1}, x_{i_2})$ is used in the vertex model of $y_j$, and 0 otherwise. Here $c_k(x_{i_1}, x_{i_2})$ is a fiber of size $\leq k$ connecting the two physical qubits $x_{i_1}$ and $x_{i_2}$, and $\mathcal{I}_{k,i_1,i_2} = \text{int}(c_k(x_{i_1}, x_{i_2})) = c_k(x_{i_1}, x_{i_2}) \backslash \{x_{i_1}, x_{i_2}\}$. We also write $C_k(x_{i_1}, x_{i_2})$ to denote the set of all fibers of size $\leq k$ connecting $x_{i_1}$ and $x_{i_2}$. This condition implies the existence of a unique fiber connecting the pair and completely contained in $\pi^{-1}(y_j)$. This automatically implies that $\pi^{-1}(y_j)$ is connected. The binary $\gamma_{c_k, j}$ is defined using the following IP representable constraints: $\forall c_k(x_{i_1}, x_{i_2}) \in C_k(x_{i_1}, x_{i_2})$ and $\forall y_j \in \mathbf{V}(Y)$:

$$\gamma_{c_k, j} = \prod_{\ell:x_\ell \in \mathcal{I}_{k,i_1,i_2}} \alpha_{\ell j} \Leftrightarrow \begin{cases} \gamma_{c_k, j} \leq \alpha_{\ell j} \quad \forall x_\ell \in \mathcal{I}_{k,i_1,i_2} \\ \gamma_{c_k, j} \geq 1 - (k-1) + \sum_{\ell:x_\ell \in \mathcal{I}_{k,i_1,i_2}} \alpha_{\ell j} \end{cases} \tag{7}$$

The constraint in Eq. (6) does not exclude the cases where 2 variables in the source graph $(y_{j_1}, y_{j_2}) \in \mathbf{E}(Y)$ are mapped to 4 different qubits in a fiber $\{x_{i_1}, \ldots, x_{i_4}\}$, where the vertex models are intercalated, i.e. $\phi(y_{j_1}) = \{x_{i_1}, x_{i_3}\}, \phi(y_{j_2}) = \{x_{i_2}, x_{i_4}\}$. The following constraint ensures that if two nodes in the target graph are in the vertex model of the same logical variable, and are not neighbors in the target graph, then one of the fibers joining them has to be active.

$$\alpha_{i_1 j} + \alpha_{i_2 j} - \sum_{c_k(x_{i_1}, x_{i_2}) \in C_k(x_{i_1}, x_{i_2})} (\gamma_{c_k, j}) \leq 1 \quad \forall y_j \in \mathbf{V}(Y) \tag{8}$$

$$\forall (x_{i_1}, x_{i_2}) \in \mathbf{V}(X), (x_{i_1}, x_{i_2}) \notin \mathbf{E}(X), \min d(x_{i_1}, x_{i_2}) \leq k$$

5. *Pullback condition.* We require that for each edge $(y_{j_1}, y_{i_2})$ in $\mathbf{E}(Y)$, there exists at least one edge in $\mathbf{E}(X)$ connecting the fibers $\pi^{-1}(y_{j_1})$ and $\pi^{-1}(y_{i_2})$. The way we guarantee this is by requiring that the quadratic form of the source graph $y$ vanishes modulo the (pullback along $\pi$ of the) quadratic form of the graph $X$. The details of this are in [9]. The resulting constraint can be written as

$$1 \leq \sum_{i_1,i_2:(x_{i_1},x_{i_2})\in\mathbf{E}(X)} \left( \delta^{\|}_{i_1 i_2 j_1 j_2} + \delta^{\perp}_{i_1 i_2 j_1 j_2} \right) \quad \forall (y_{j_1}, y_{j_2}) \in \mathbf{E}(Y), \qquad (9)$$

where we have introduced the binaries $\delta^{\|}_{i_1 i_2 j_1 j_2}$ and $\delta^{\perp}_{i_1 i_2 j_1 j_2}$ defined $\forall (x_{i_1}, x_{i_2}) \in \mathbf{E}(X)$ and $\forall (y_{j_1}, y_{j_2}) \in \mathbf{E}(Y)$: The binary variable $\delta^{\|}_{i_1 i_2 j_1 j_2}$ is one if $x_{i_1}$ and $x_{i_2}$ are edges of the vertex-models $\phi(y_{j_1}), \phi(y_{j_2})$ respectively, and the binary variable $\delta^{\perp}_{i_1 i_2 j_1 j_2}$ is one if $x_{i_2}$ and $x_{i_1}$ are edges of the vertex-models $\phi(y_{j_1}), \phi(y_{j_2})$ respectively. This conditions are equivalent to $\delta^{\|}_{i_1 i_2 j_1 j_2} = \alpha_{i_1 j_1} \alpha_{i_2 j_2}$ and $\delta^{\perp}_{i_1 i_2 j_1 j_2} = \alpha_{i_1 j_2} \alpha_{i_2 j_1}$. We can then represent these new variables using linear inequalities as follows: $\forall (x_{i_1}, x_{i_2}) \in \mathbf{E}(X), \forall (y_{j_1}, y_{j_2}) \in \mathbf{E}(Y)$:

$$\delta^{\|}_{i_1 i_2 j_1 j_2} = \alpha_{i_1 j_1} \alpha_{i_2 j_2} \Leftrightarrow \begin{cases} \delta^{\|}_{i_1 i_2 j_1 j_2} \leq \alpha_{i_1 j_1} \\ \delta^{\|}_{i_1 i_2 j_1 j_2} \leq \alpha_{i_2 j_2} \\ \delta^{\|}_{i_1 i_2 j_1 j_2} \geq \alpha_{i_1 j_1} + \alpha_{i_2 j_2} - 1 \end{cases}$$

and equivalently for $\delta^{\perp}_{i_1 i_2 j_1 j_2}$. Both variables cannot be one for a single combination of $(i_1 i_2 j_1 j_2)$ simultaneously. This leads to the following constraint.

$$\delta^{\|}_{i_1 i_2 j_1 j_2} + \delta^{\perp}_{i_1 i_2 j_1 j_2} \leq 1 \quad \forall (x_{i_1}, x_{i_2}) \in \mathbf{E}(X), \forall (y_{j_1}, y_{j_2}) \in \mathbf{E}(Y). \qquad (10)$$

## 3.2   Complete IP Model

The feasible region of the IP formulation is defined by:

$$F = \left\{ (\boldsymbol{\alpha}, \boldsymbol{\gamma}, \boldsymbol{\delta}^{\|}, \boldsymbol{\delta}^{\perp}) | (\boldsymbol{\alpha}, \boldsymbol{\gamma}, \boldsymbol{\delta}^{\|}, \boldsymbol{\delta}^{\perp}) \in ((2) \cap \cdots \cap (10)) \right\}. \qquad (11)$$

A constant objective function can be set for this problem such that any solution that lies within the feasible region defined in Eq. (11) optimizes it.

*Embedding Size.* The objective function that minimizes the embedding size is

$$\min \sum_{i:x_i \in \mathbf{V}(X)} \sum_{y_j \in \mathbf{V}(Y)} \alpha_{ij} \quad \text{s.t. } (\boldsymbol{\alpha}, \boldsymbol{\gamma}, \boldsymbol{\delta}^{\|}, \boldsymbol{\delta}^{\perp}) \in F. \qquad (12)$$

Other objective functions such as fiber size minimization, minimal fiber size dispersion, and available edges in the embedding, among others are also IP representable and can be implemented within this framework.

# 4    Decomposition Approach

Implementing all the constraints at once in the IP formulation leads to a model which is often intractable in practice. The fiber conditions require many constraints to be enforced, and only a small fraction of these are active in optimal solutions. We investigate the application of a decomposition approach which iterates between a qubit assignment master problem and fiber condition checking subproblems. The strategy adds strengthened 'no-good' constraints (i.e., cuts) to the master problem when they are found to be violated. Such an approach bears resemblance to decomposition techniques used for scheduling and routing problems, such as classical and logic-based Benders decomposition and branch-and-check [13,17,27].

## 4.1    Master Problem

In the master problem, we relax the fiber conditions, permitting a node in the source graph to be mapped in multiple parts of the target graph without being connected. For our master problem, we introduce a new binary decision variable, $z_{e_x e_y}$ $\forall e_x \in \mathbf{E}(X), \forall e_y \in \mathbf{E}(Y)$, to track the embedding of problem edges in the target graph edges. The variable takes on a value of 1 if edge $e_x = (x_{i_1}, x_{i_2}) : e_x \in \mathbf{E}(X)$ is mapped through the embedding in edge $e_y = (y_{j_1}, y_{j_2}) : e_y \in \mathbf{E}(Y)$, and 0 otherwise. For modeling purposes, we also denote $e_{x,1} = x_{i_1}, e_{x,2} = x_{i_2}, e_{y,1} = y_{j_1}$, and $e_{y,2} = y_{j_2}$. This master problem formulation includes previously expressed mapping constraints, Eq. (2), and size constraints in Eq. (4), in addition to constraints (13) through (15) as follows:

*Assignment of Edges.* Each edge in the source graph has to be assigned to an edge in the target graph.

$$\sum_{e_x \in \mathbf{E}(X)} z_{e_x e_y} = 1 \quad \forall e_y \in \mathbf{E}(Y). \tag{13}$$

*Linking Constraints.* To link the assigned qubit values to the $z_{e_x e_y}$ variables, we use the following set of constraints $\forall e_x \in \mathbf{E}(X), \forall e_y \in \mathbf{E}(Y)$:

$$z_{e_x e_y} \leq \alpha_{e_{x,1} e_{x,2}} \quad z_{e_x e_y} \leq \alpha_{e_{y,1} e_{y,2}}. \tag{14}$$

Together, these constraints ensure that a problem edge can only be assigned to an edge in the target graph if the pair of nodes involved in that edge take on the required values, which are aggregated in the following constraint

$$2 \cdot z_{e_x e_y} \leq \alpha_{e_{x,1} e_{x,2}} + \alpha_{e_{y,1} e_{y,2}} \quad \forall e_x \in \mathbf{E}(X), \forall e_y \in \mathbf{E}(Y). \tag{15}$$

*Subproblem Relaxation.* Although the constraints above already represent the assignment problem to be modeled in the master problem, we can include a relaxation of the subproblem to help guide to master problem towards feasible solutions. This requires the addition of another set of binary variables, $w_j$ that track whether vertex model $\phi(y_j)$ has a size greater than one. Then, $\forall y_j \in \mathbf{V}(Y)$:

$$\sum_{i:x_i \in \mathbf{V}(X)} \alpha_{ij} - n \cdot w_j \leq 1, \tag{16a}$$

$$n(1 - \alpha_{ij}) + \sum_{\ell:(x_i,x_\ell) \in \mathbf{E}(X)} \alpha_{\ell j} + \sum_{\ell:(x_\ell,x_i) \in \mathbf{E}(X)} \alpha_{\ell j} \geq w_j \quad \forall x_i \in \mathbf{V}(X). \tag{16b}$$

This constraints ensure that the variable $w_j$ is one if the node $y_j$ is mapped to more than one node $x_i$. Equations (13)–(16), together with the cuts generated by the subproblems, define the master problem.

## 4.2    Subproblems

The subproblem validates if there exist vertices in the embedding belonging to the same vertex model $\phi(y_j)$ which are not connected in the target graph. If this is the case, it returns a constraint that either: i) encourages connectivity in future iterations, or ii) removes occurrences of the disconnected vertex models from the graph. For each vertex model with more than one vertex on the embedding, it checks at each vertex on the target graph that belongs to that vertex model. If that vertex does not contain an edge that connects it to another vertex of that vertex model, then the checking procedure returns disconnected.

## 4.3    Cuts

If a particular vertex model is found to be disconnected in the solution, we add a constraint to remove the current solution and prevent future solutions from having the same disconnectivity. Let the set of disconnected vertices in the source graph be denoted as $\hat{j} : y_{\hat{j}} \in \hat{Y}$. Let the set of vertices in the target graph that belong to this vertex model, $y_{\hat{j}}$, in the current incumbent solution, be represented as the vertex model $\phi(y_{\hat{j}}) \subseteq X$. Let the set of vertices that are adjacent to any vertex in $\phi(y_{\hat{j}})$, but are not assigned value $y_{\hat{j}}$, be denoted $\phi'(y_{\hat{j}})$. The constraint generated in the current iteration for disconnected qubit $y_{\hat{j}}$ is then given by:

$$|\phi(y_{\hat{j}})| - \sum_{i:x_i \in \phi(y_{\hat{j}})} \alpha_{i\hat{j}} + \sum_{i:x_i \in \phi'(y_{\hat{j}})} \alpha_{i\hat{j}} \geq 1. \tag{17}$$

This removes the current infeasible solution from the search space and requires the master problem to: i) include at least one fewer vertex with this vertex model (bracketed term), or ii) include at least one more vertex with this vertex model, among the set of vertices that could improve connectivity (non-bracketed term).

Notice that we reformulated the pullback condition from the Eq. (9) in terms of $\delta^{\parallel}$ and $\delta^{\perp}$ into the variables $z_{e_x e_y}$ and its corresponding constraints, while the fiber condition is relaxed with the subproblem and cut generation procedure. Following the intuition in [4], where the heuristic method tries to obtain embeddings with a small qubit footprint, the default objective function implemented in the master problem is to minimize size, as in Eq. (12). This objective leads the master problem to return compact assignments of variables. In the case that the feasibility objective is considered within this approach, the optimization procedure is stopped when the first feasible solution is found.

## 5 Results

The model in Sect. 3 was implemented using the Python package Pyomo [16], which interfaces with several open-source and commercial solvers. The decomposition approach presented in Sect. 4 is implemented in C++ and uses the CPLEX 12.9 solver [6]. Our approaches are compared with the D-Wave default heuristic minorminer, introduced in [4] (available at github.com/dwavesystems/minorminer). Unless otherwise stated, the monolithic IP method assumes a value of maximum fiber size $k = 3$, which is justified for the structured random graphs given their construction. This provides the monolithic method an advantage with respect to the decomposition method given that the infeasibility proofs are contingent on $k$. All experimental results were obtained using a laptop running Ubuntu 18.04 with an Intel Core i7-6820HQ CMU @ 2.7GHz with 8 threads and 16 GB of RAM.

### 5.1 Illustrative Example

Our illustrative example is taken from [9], where a $K_{4,4}$ bipartite graph is connected through a single edge to a structured graph with 4 nodes and is embedded in a $C_{4,1,2}$ *Chimera* graph, as seen in Fig. 2. This embedding is challenging for heuristic methods that search vertex models outside of the blocks [9]. The embedding with the minimal size is given when one of the nodes in the 4-node block is embedded in a chain of length 2, resulting in an embedding of length 13. The heuristic implemented in minorminer fails roughly 50% of the 1000 runs (i.e., it is not able to find a valid embedding in half of the experiments). We consider solving this problem using the CAG approach proposed in [9], by computing the Groebner basis of the polynomial ideal. When using the software Maple 2017 [20], which includes Faugère's algorithm [10,11], the Groebner basis computation is unable to find a solution after 5 h of computation, before running out of memory. We apply our IP approach with the open-source solvers GLPK 4.61 [24] and CBC 2.9.6 [12], as well as the commercial solvers Gurobi 8.1 [15] and CPLEX 12.9 [6]. We use a time limit of one minute per experiment.

The open-source solvers fail to provide feasible solutions within the time limit when there is a constant objective function. The CBC solver, however, can find a solution when minimizing embedding size, illustrating that including an objective function can be beneficial. In this case, although it finds the optimal solution, the CBC solver is unable to prove it is optimal (with a gap at the end of the runtime of 8.3%). The commercial solvers, on the other hand, can provide both feasible and optimal solutions in under a minute of computation. In particular, Gurobi takes 1.3 s to find an initial feasible solution and 31.2 s to find and prove the optimal solution, while CPLEX takes 3.5 s and 9.4 s for the same tasks, respectively. As expected, the time required to provide a feasible solution is less than that taken to give optimality guarantees.

Finally, the decomposition approach provides feasible solutions more efficiently, and with higher quality, than the other approaches. Specifically, it took 0.4 s for the decomposition approach to provide a feasible solution which was nearly optimal, with a 7% optimality gap. Finding a provably optimal solution required 46.5 s. These results suggest that the usage of commercial solvers is required for solving these challenging IP problems. The performance difference between commercial and open-source solvers for a given mathematical formulation is driven by a large number of factors; recent solver benchmarks have identified that commercial IP solvers, overall, significantly outperform their open-source counterparts [21]. For the problem presented here, the open-source solvers evaluate the root node relaxation less efficiently than the commercial approaches. Furthermore, the commercial solvers appear to have much more effective branching heuristics resulting in fewer nodes explored, overall. Our initial experimentation indicated that, out of the commercial solvers, CPLEX had superior performance to Gurobi, and so we only report CPLEX results for the remainder of the manuscript.

## 5.2   Random Graphs

*1. Random Structured Graphs.* Here we generalize the example above. We consider the bipartite graph $K_{4,4}(p_{inter})$ parameterized by $p_{inter}$, which is the probability of the existence of edges between the two partitions. We randomly choose $\zeta$ edges, which we contract into nodes (each edge into a single node). This graph is then connected to a complete $K_{4,4}$ bipartite graph by 4 edges chosen with a probability $p_{intra}$. By construction, the resulting graph is a subgraph of $C_{4,1,2}$, and its size is $m + \zeta$. It is the smallest minor of the corresponding graph without contraction. The example of Sect. 5.1 is obtained with $\zeta = 1$ (and $m = 12$). Fixing $p_{inter} = p_{intra} = 0.5$, for each value of contracted edges $\zeta \in \{0, \ldots, 3\}$, we generate 10 random graphs. These random graphs were embedded in a $C_{4,1,2}$ graph with a time limit of 300 s. Figure 3a gives the runtimes for the monolithic IP and the decomposition methods solved using CPLEX. This figure also shows the boxplots for the 1000 runs of `minorminer`. For this case, given the way the random structured graphs are constructed, we see that the longest fiber will be at most of size 3, which we encode for the monolithic IP approach using the parameter $k = 3$. Notice that this observation biases the results in favor of the monolithic IP approach with respect to the decomposition approach. For finding a feasible solution, the decomposition approach is more efficient than the monolithic IP approach. When $\zeta = 0$, where finding a feasible embedding is practically finding the minimally sized embedding, there is no difference in time performance between the cases of embedding size minimization and finding a feasible solution. For $\zeta > 0$, the embedding size minimization becomes more expensive, in particular for the decomposition approach. The monolithic IP and the decomposition approaches were able to find smaller or equally sized embeddings for 33 and 30 cases out of the 40 experiments, respectively. When the objective function is the size minimization, this number increased for all instances in all cases and is strictly better in 22 cases for the monolithic IP and 21 cases for the decomposition approach. We note that the monolithic IP approach was able to find an embedding for one instance which was smaller than any of the 1000 runs of the heuristic method.

Larger instances of random structured graphs can be generated by combining two graphs like the ones described above, and including the edges appearing in a $C_{4,2,2}$ graph between the cells with probability $p_{intra}$. As before, we generated 10 random instances with values of $\zeta$ ranging from 0 to 3. The time performance of the different methods is shown in Fig. 3a. Out of the 40 instances, the monolithic IP and the decomposition method are still able to find embeddings as compact as the median heuristic behavior in 17 and 23 instances when trying to find a feasible solution, and in 20 and 17 instances when trying to minimize the size of the embedding, respectively. Similar to the previous case, the monolithic IP approach is able to find smaller embeddings than any of the 1000 runs of the heuristic method for two problem instances.

Figure 3b shows a comparison of the embedding median sizes obtained by the heuristic method versus the ones obtained for the IP methods. The size of the markers represents the heuristic failure rate fraction, computed from the 1000 runs of the heuristic method for each instance. Both the monolithic and the decomposition approaches (different colors) and the feasibility and size minimization objectives (different marker shapes) are represented in this figure. In total, out of the 80 structured instances, the heuristic failed more than 50% for 47 instances, more than 80% for 12 instances, and more than 90% for 3 instances. These instances appear on the right side of Fig. 3b. For those instances, the IP approaches were able to find a feasible solution in less than 20 s, and only for 19, 7, and 2 instances, respectively, the optimal solution could not be guaranteed within the time limit. For high failure rate problems (>50% failure rate) for the heuristic, our methods find a feasible solution in under 20 s and prove optimality in more than half of the instances in less than 5 min. Notice that most of the runs corresponding to finding a feasible embedding (circles) are above the diagonal line, indicating a larger embedding size for the IP methods compared to the heuristic, while size minimization runs (triangles) lie on the diagonal or below it.

2. *Erdös-Rényi Graphs.* These graphs are parametrized by the number of nodes $\nu$ and the probability of an edge existing between each pair of nodes $p$. We consider a set of 10 random instances for each combination of $\nu \in \{5, 6, \ldots, 16\}$ and $p \in \{0.3, 0.5, 0.7\}$. Each of these graphs is embedded in different sizes of *Chimera*, $C_{4,1,1}, C_{4,2,1}, C_{4,3,1}$, and $C_{4,2,2}$ . We set the time limit to 60 s. In this experiment, we considered 2160 instances. In 1100 of them, the heuristic method

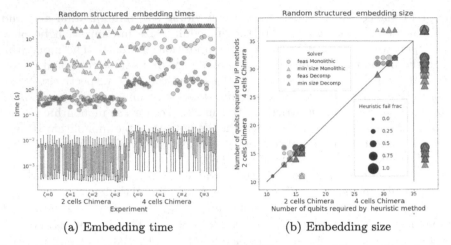

(a) Embedding time                    (b) Embedding size

**Fig. 3.** Embedding time and size comparison for different embedding methods for structured random graphs in $\mathcal{C}_{4,1,2}$ and $\mathcal{C}_{4,2,2}$ with respect to median behavior of `minorminer`. Values beyond the red lines represent embeddings where the heuristic median performance (right) or the IP methods (top) failed to return an embedding. (Color figure online)

could not find any feasible embedding after 1000 runs. In 94% of these cases, at least one of the monolithic IP methods does not time out, meaning that the methods could prove the infeasibility of the embedding or find a feasible embedding. This proves that the methods proposed in this work are valid for providing guarantees of the embeddability of graph minors in cases where the current heuristics are unable to answer this satisfiability question. In the trivially infeasible case where $\nu > n$ our methods could almost immediately identify the infeasibility, contrary to the `minorminer` heuristic. The conclusion is that the runtime for the monolithic IP methods increases with the size of the target and source graphs, the density of the source graph given by $p$, and when the objective function is to minimize the embedding sizes.

We complete our benchmark of random graphs by embedding larger problems. In this case we, consider 5 random instances for each combination of $\nu \in \{10, 15, \ldots, 35\}$ and $p \in \{0.1, 0.3, 0.5, 0.7\}$ embedded into $\mathcal{C}_{4,4,4}$, where the longest fiber size was increased to $k = 5$. We observe that, for these instances, the only IP solver that does not run out time is CPLEX implementing either the monolithic IP or the decomposition approach with the feasible solution objective.

Figure 4 presents the embedding size and time comparison for the small random graph experiments. For this test case, in 60% of the instances the decomposition approach yielded embeddings with sizes equal to, or smaller than, the median of the ones returned by the heuristic, when looking for a feasible solution, and in 90% of the instances when minimizing the embedding size. The monolithic IP approach was more efficient to declare infeasibility in non-trivial cases ($\nu \leq n$) than the decomposition approach, Figs. 4a and 4b, as the values

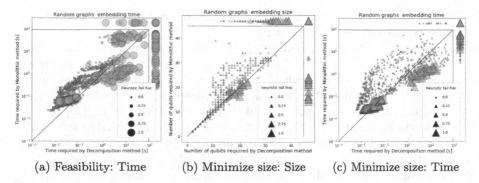

(a) Feasibility: Time    (b) Minimize size: Size    (c) Minimize size: Time

**Fig. 4.** Embedding size and time comparison for Erdös-Rényi graphs ($\nu \in \{5, 6, \ldots, 16\}$, $p \in \{0.3, 0.5, 0.7\}$) given different objectives. Values beyond the red lines represent embedding where the decomposition (right) or the monolithic IP methods (top) failed to return an embedding or went over the time limit.

are below the diagonal with large heuristic failure fractions and longer runtimes. When compared to the minimal size found after the 1000 runs of the heuristic method, the monolithic IP methods are still able to find smaller embeddings for around 5% of the cases. The comparison in Fig. 4 highlights that the sizes of the embeddings found by the decomposition approach are in most cases as small or smaller than the monolithic IP approach. In terms of the computational time, we observe that those instances that were challenging for the heuristic (large markers) are more easily solved by the decomposition approach, especially when minimizing the size of the embedding. The remaining instances appear to be solved more efficiently using the monolithic IP approach. The larger and more challenging instances lead to different results. Out of the 120 instances solved, the decomposition approach outperforms the monolithic IP approaches, obtaining equally good or better embeddings than the median heuristic behavior in 30% of the cases, compared to around 10% for the monolithic IP approaches. The solution requires larger fibers, which affected directly the formulation size of the monolithic case making it more challenging to solve. Only for one instance, a smaller embedding than any of the observed heuristic solutions is obtained, in this case by the decomposition approach.

## 5.3    Applications

*1. Gadgets.* It has been shown previously [8] that all of the cubic gadgets can be embedded in a single *Chimera* cell, but three of the quartic gadgets require more than a single *Chimera* cell. The three gadgets were $K_6 - e$, *Double $K_4$*, and $K_6$.

We find more efficient embeddings for several of the gadgets, namely $K_5$ (with 2 and 1 auxiliary variables), $K_6$, and $K_6 - e$ compared to [8]. The embeddings found can be guaranteed to be the minimal size within a few seconds of computation. For the case of the quartic gadgets, all but one (double $K_4$) could be embedded in a single *Chimera* cell, in which case our method could provide infeasibility guarantees in less than 10 s.

*2. Spanning Tree.* An example of an application is the communication of vehicles/agents with a central control station that can be disrupted in a particular area and can be routed through $\Delta$ agents/vehicles. Finding the communication routing of the vehicles that minimizes the distance, is equivalent to finding the minimum spanning tree with bounded degree $\Delta$. Rieffel et al. [26] propose three different formulations of this problem that can be embedded in a quantum annealer. Given the graph defined by the agents/vehicles, the distances among them might change but not the graph itself. At the same time, the degree of the spanning tree $\Delta$ is fixed by the communication equipment. We generate 80 instances with graphs of 4 vertices and between 3 and 5 edges. When reformulating the problems as a QUBO, we obtain instances where the target graph ranges in size between 19 and 29 nodes and with 35 to 70 edges. The resulting QUBOs were embedded using the decomposition approach and compared to the heuristic in `minorminer`. We obtain smaller (or equally sized) embeddings than the median length of the heuristic in 12.5% (15%) of the instances in 5 min of computational time (compared to 1000 runs of the heuristic).

*3. Protein Folding.* Perdomo-Ortiz et al. [25] encode the different configuration of the amino acids in a protein in terms of a QUBO representing the overall energy of the system. Minimizing this QUBO with respect to the different number of bonds between amino-acids would yield the protein's least-energy configuration. [25] not only encodes the problem as a QUBO, but also provides a custom algorithm to embed it in the D-Wave One chip, which has hardware described by a faulty $C_{4,4,4}$ graph. The largest instance solved in this paper involved embedding a QUBO of 19 variables in a target graph of 127 qubits. The resulting embedding involved 81 qubits with the largest vertex model of length 5, being at the time the largest problem embedded and solved with D-Wave's quantum annealers. We highlight two qualities of our approach: 1) we are not making any assumption about the source or target graphs, allowing us to work with faulty *Chimera* graphs as targets; and 2) we can exploit the fact of having an existing embedding to initialize our procedures, allowing us to solve our IP problems more efficiently. Initializing with the embedding provided by Perdomo-Ortiz et al. [25] while restricting the $k = 5$ in the monolithic IP approach, we find an embedding of length 77 (4.9% qubit footprint reduction) within an hour of computation. Allowing longer fibers, we find an embedding of size 74 (8.6% qubit footprint reduction) with a fiber of size 6. These embeddings are not guaranteed to be optimal, but in both cases improve those previously found.

## 6   Conclusions

Integer programming (IP) approaches are proposed to solve the graph minor-embedding problem. Specifically, we develop a monolithic IP derived from the polynomial equations presented in Dridi et al. [9], and a decomposition approach, both of which are capable of identifying infeasible instances and providing bounds on solution quality. These approaches are also agnostic to the source and target graphs. Both approaches were implemented and tested using a range of different source graphs with various sizes, densities, and structures. The target graphs used follow the architecture of the chips in D-Wave's current and future quantum annealers. Although slower overall than the currently-employed heuristic method [4], the proposed methods prove to be a viable solution approach for highly structured source graphs, where the heuristic fails with a higher probability.

The results presented highlight more general approaches to minor-embedding using IP. The proposed formulations and results are a baseline for future methods that can work at larger scales. Applications to gadget embeddings, spanning tree problems, and protein folding demonstrate the advantages of our approaches. Promising future directions include using symmetries and the invariant formulation as suggested in [9], or the design of chip-specific cuts/techniques. Another direction is to reduce the search space by imposing certain limitations on the embedding, e.g., by allowing only certain topologies for the vertex models or by fixing certain embedding characteristics, such as maximum fiber size. Initial attempts to include these approximations show a promising decrease in the computation time with an acceptable trade-off in quality.

**Acknowledgements.** We thank Prof. Ignacio Grossmann and Dr. Eleanor Rieffel for the constructive discussions during the preparation of this work. DB, KB, and DV are supported/partially supported by NASA NAMS (NNA16BD14C), AFRL NYSTEC Contract (FA8750-19-3-6101). DB is also supported by the USRA Feynman Quantum Academy and the Center for Advanced Process Decision Making (CAPD) at CMU. NASA QuAIL acknowledges support from the Office of the Director of National Intelligence (ODNI) and the Intelligence Advanced Research Projects Activity (IARPA), via IAA 145483. The views and conclusions contained herein are those of the authors and should not be interpreted as necessarily representing the official policies or endorsements, either expressed or implied, of ODNI, IARPA, AFRL, or the U.S. Government.

# References

1. Bian, Z., Chudak, F., Israel, R., Lackey, B., Macready, W.G., Roy, A.: Discrete optimization using quantum annealing on sparse Ising models. Front. Phys. **2**, 56 (2014)
2. Bixby, R.E.: A brief history of linear and mixed-integer programming computation. Documenta Mathematica · Extra (2012)
3. Bunyk, P.I., et al.: Architectural considerations in the design of a superconducting quantum annealing processor. IEEE Trans. Appl. Superconductivity **24**(4), 1–10 (2014)
4. Cai, J., Macready, W.G., Roy, A.: A practical heuristic for finding graph minors. arXiv:1406.2741 (2014)
5. Choi, V.: Minor-embedding in adiabatic quantum computation: II. Minor-universal graph design. Quant. Inf. Process. **10**(3), 343–353 (2011)
6. Cplex: 12.9 user's manual (2019)
7. Date, P., Patton, R., Schuman, C., Potok, T.: Efficiently embedding qubo problems on adiabatic quantum computers. Quant. Inf. Process. **18**(4), 117 (2019)

8. Dattani, N., Chancellor, N.: Embedding quadratization gadgets on Chimera and Pegasus graphs. arXiv:1901.07676 (2019)

9. Dridi, R., Alghassi, H., Tayur, S.: A novel algebraic geometry compiling framework for adiabatic quantum computations. arXiv:1810.01440 (2018)

10. Faugère, J.C.: A new efficient algorithm for computing Gröbner bases (F4). J. Pure Appl. Algebra **139**(13), 61–88 (1999)

11. Faugère, J.C.: A new efficient algorithm for computing Gröbner bases without reduction to zero (F5). In: Proceedings of the 2002 International Symposium on Symbolic and Algebraic Computation, ISSAC 2002, pp. 75–83. ACM, New York (2002)

12. Forrest, J., Lougee-Heimer, R.: CBC user guide. In: Emerging Theory, Methods, and Applications (2005)

13. Geoffrion, A.M.: Generalized benders decomposition. J. Optim. Theory Appl. **10**(4), 237–260 (1972)

14. Goodrich, T.D., Sullivan, B.D., Humble, T.S.: Optimizing adiabatic quantum program compilation using a graph-theoretic framework. Quant. Inf. Process. **17**(5), 1–26 (2018). https://doi.org/10.1007/s11128-018-1863-4

15. Gurobi Optimization, L.: Gurobi optimizer reference manual (2019)

16. Hart, W.E., et al.: Pyomo-Optimization Modeling in Python, vol. 67. Springer, Heidelberg (2017). https://doi.org/10.1007/978-3-319-58821-6

17. Hooker, J.N., Ottosson, G.: Logic-based Benders decomposition. Math. Program. **96**(1), 33–60 (2003)

18. Johnson, M.W., et al.: Quantum annealing with manufactured spins. Nature **473**(7346), 194–198 (2011)

19. Kaminsky, W.M., Lloyd, S.: Scalable architecture for adiabatic quantum computing of NP-hard problems. In: Leggett, A.J., Ruggiero, B., Silvestrini, P. (eds.) Quantum Computing and Quantum Bits in Mesoscopic Systems, pp. 229–236. Springer, Heidelberg (2004). https://doi.org/10.1007/978-1-4419-9092-1_25

20. Maplesoft: Algorithms for Groebner basis, Maple 2017 (2019)

21. Mittelmann, H.D.: Benchmarking optimization software - a (hi)story. SN Oper. Res. Forum **1**(1), 2 (2020). ISSN 2662–2556, https://doi.org/10.1007/s43069-020-0002-0

22. Neven, H., Denchev, V.S., Drew-Brook, M., Zhang, J., Macready, W.G., Rose, G.: NIPS 2009 demonstration: binary classification using hardware implementation of quantum annealing (2009)

23. Okada, S., Ohzeki, M., Terabe, M., Taguchi, S.: Improving solutions by embedding larger subproblems in a D-wave quantum annealer. Sci. Rep. **9**(1), 2098 (2019)

24. Oki, E.: GLPK (GNU Linear Programming Kit). In: Linear Programming and Algorithms for Communication Networks (2012)

25. Perdomo-Ortiz, A., Dickson, N., Drew-Brook, M., Rose, G., Aspuru-Guzik, A.: Finding low-energy conformations of lattice protein models by quantum annealing. Sci. Rep. (2012)

26. Rieffel, E.G., et al.: From Ansätze to Z-gates: a NASA View of Quantum Computing. arXiv:1905.02860 (2019)

27. Roshanaei, V., Booth, K.E.C., Aleman, D.M., Urbach, D.R., Beck, J.C.: Branch-and-check methods for multi-level operating room planning and scheduling. Int. J. Prod. Econ. (2019)

28. Serra, T., Huang, T., Raghunathan, A., Bergman, D.: Template-based Minor Embedding for Adiabatic Quantum Optimization. arXiv:1910.02179 (2019)

29. Sugie, Y., et al.: Graph minors from simulated annealing for annealing machines with sparse connectivity. In: Fagan, D., Martín-Vide, C., O'Neill, M., Vega-Rodríguez, M.A. (eds.) TPNC 2018. LNCS, vol. 11324, pp. 111–123. Springer, Cham (2018). https://doi.org/10.1007/978-3-030-04070-3_9
30. Tichy, W.: Is quantum computing for real? An interview with Catherine McGeoch of D-wave systems. Ubiquity **2017**(July), 1–20 (2017)
31. Venegas-Andraca, S.E., Cruz-Santos, W., McGeoch, C., Lanzagorta, M.: A cross-disciplinary introduction to quantum annealing-based algorithms. Contemp. Phys. **59**(2), 174–197 (2018)
32. Yang, Z., Dinneen, M.J.: Graph minor embeddings for D-wave computer architecture. Technical report, Department of Computer Science, The University of Auckland, New Zealand (2016)

# An Ising Framework for Constrained Clustering on Special Purpose Hardware

Eldan Cohen[1]([✉]), Arik Senderovich[2], and J. Christopher Beck[1]

[1] Department of Mechanical & Industrial Engineering, University of Toronto,
Toronto, Canada
{ecohen,jcb}@mie.utoronto.ca
[2] Faculty of Information, University of Toronto, Toronto, Canada
sariks@mie.utoronto.ca

**Abstract.** The recent emergence of novel hardware platforms, such as quantum computers and Digital/CMOS annealers, capable of solving combinatorial optimization problems has spurred interest in formulating key problems as Ising models, a mathematical abstraction shared by a number of these platforms. In this work, we focus on constrained clustering, a semi-supervised learning task that involves using limited amounts of labelled data, formulated as constraints, to improve clustering accuracy. We present an Ising modeling framework that is flexible enough to support various types of constraints and we instantiate the framework with two common types of constraints: pairwise instance-level and partition-level. We study the proposed framework, both theoretically and empirically, and demonstrate how constrained clustering problems can be solved on a specialized CMOS annealer. Empirical evaluation across eight benchmark sets shows that our framework outperforms the state-of-the-art heuristic algorithms and that, unlike those algorithms, it can solve problems that involve combinations of constraint types. We also show that our framework provides high quality solutions orders of magnitudes more quickly than a recent constraint programming approach, making it suitable for mainstream data mining tasks.

## 1 Introduction

Recent years have seen the emergence of novel computational platforms, including adiabatic and gate-based quantum computers, Digital/CMOS annealers, and neuromorphic computers (for a review see [8]). These machines represent a challenge and opportunity to AI and OR researchers: how can specialized models of computation as embodied by the new hardware be harnessed to better solve AI/OR problems. Several new hardware platforms have adopted Ising models [19] as their mathematical formulation and, consequently, a number of existing problems have been formulated as Ising models, including clustering [22], community detection [34], and partitioning, covering, and satisfiability [26].

Constrained clustering is a semi-supervised learning task that exploits small amounts of labelled data, provided in the form of constraints, to improve clustering performance [35]. In the past two decades, this topic has received significant

© Springer Nature Switzerland AG 2020
E. Hebrard and N. Musliu (Eds.): CPAIOR 2020, LNCS 12296, pp. 130–147, 2020.
https://doi.org/10.1007/978-3-030-58942-4_9

attention and algorithms that support different types of constraints have been proposed [6,24,29]. As finding an optimal solution to the (semi-supervised) clustering problem is an NP-hard problem [27], the commonly used algorithms rely on heuristic methods that quickly converge to a local optimum.

In a recent work, Kumar et al. [22] presented an Ising model for unsupervised clustering and observed mixed results using a quantum annealer. However, formulating constrained clustering problems as Ising models and solving them in hardware has not been studied. In this work, we introduce and analyze a novel Ising modeling framework for semi-supervised clustering that supports the combination of different types of constraints and we instantiate it with pairwise instance-level and partition-level constraints. We demonstrate the performance on the Fujitsu Digital Annealer [28,33], and discuss several hardware-related considerations when embedding our framework on this hardware.

Our main contributions are summarized as follows:

- We introduce an Ising framework for constrained clustering with pairwise and partition-level constraints that can be solved on a variety of novel hardware platforms.
- We demonstrate the performance of our framework on a specialized CMOS annealer and show that it outperforms the state-of-the-art heuristic methods for constrained clustering and produces approximately equal or better solutions compared to a constraint programming model in a small fraction of the runtime (i.e., a two orders of magnitude speed-up).
- We show that the framework can seamlessly solve semi-supervised clustering problems with both pairwise and partition constraints, problems that cannot be solved by the existing heuristic techniques.
- We discuss some of the challenges in embedding Ising models onto quantum and quantum-inspired hardware.

## 2    Background

Let $X = \{x_i\}_{i=1}^n$ be the set of $n$ data points with $x_i$ being a finite-sized feature vector and $K$ be the number of clusters ($K<n$). Combinatorial clustering algorithms attempt to find a partition of $X$ into $K$ disjoint subsets, $S = S_1 \cup \cdots \cup S_K$, that minimizes a chosen objective function, typically the total within-cluster scatter [32] based on pairwise dissimilarities, $d(x_i, x_j)$. When the dissimilarity is represented by the squared Euclidean distance the objective is:

$$\min \sum_{k=1}^{K} \sum_{\substack{i<j: \\ x_i, x_j \in S_k}} d(x_i, x_j) = \sum_{k=1}^{K} \sum_{\substack{i<j: \\ x_i, x_j \in S_k}} \|x_i - x_j\|^2. \tag{1}$$

In the Euclidean case, another commonly used objective function is the sum of squared errors [18],

$$\min \sum_{k=1}^{K} \sum_{x_i \in S_k} \|x_i - \mu_k\|^2, \tag{2}$$

where $\mu_k$ is the mean vector of the points in cluster $k$.

## 2.1   Constrained Clustering

In a semi-supervised setting, we assume some amount of labelled data in the form of constraints. Constrained clustering is the problem of finding a partition that satisfies the provided constraints [35]. First, we consider two pairwise constraints: must-link (ML) and cannot-link (CL) [4]. ML constraints are defined by a set, $\mathcal{M}$, of pairs of points that must be assigned to the same cluster, $(x_i, x_j) \in \mathcal{M} \Rightarrow s(x_i) = s(x_j)$, where $s(x_i)$ denotes the cluster that $x_i$ is assigned to, $s(x_i) = k \iff x_i \in S_k$. CL constraints are defined by a set, $\mathcal{C}$, of pairs of points that must be assigned to different clusters, $(x_i, x_j) \in \mathcal{C} \Rightarrow s(x_i) \neq s(x_j)$.

Bilenko et al. [6] proposed the *Pairwise Constrained K-Means (PCK-Means)* problem that incorporates the constraints in the objective function:

$$\min \sum_{k=1}^{K} \sum_{x_i \in S_k} \|x_i - \mu_j\|^2 + \sum_{(x_i, x_j) \in \mathcal{M}} w_{i,j} \mathbb{1}[s(x_i) \neq s(x_j)]$$
$$+ \sum_{(x_i, x_j) \in \mathcal{C}} \overline{w}_{i,j} \mathbb{1}[s(x_i) = s(x_j)] \tag{3}$$

where $\mathbb{1}[true] = 1$ and $\mathbb{1}[false] = 0$. PCK-Means is solved using a greedy iterative algorithm, adapted from the K-Means algorithm [25]. Note that Eq. (3) allows violation of ML and CL constraints depending on the weights $w_{i,j}$ and $\overline{w}_{i,j}$ that correspond to the confidence in the external information [23]. *Metric PCK-Means (MPCK-Means)* [6] is a combination of PCK-Means with distance-metric learning [36] that outperforms PCK-Means [9].

Other well-known approaches include *Constrained Vector Quantization Error (CVQE)* [13] that augments the clustering objective to account for constraint violations, but uses the distances between the centroids to compute the violation costs, and *linear-time CVQE (LCVQE)* [29] that computes the violation costs based on the distances between objects and centroids. LCVQE was found to be competitive in terms of accuracy with CVQE while violating fewer constraints [9].

We also consider partition-level (PL) constraints, where some points have predefined cluster labels. Formally, assuming an arbitrary labeling of clusters $k$, $X_k \subseteq X$ denotes the set of points that must be assigned to cluster $k$. For example, in clustering of patients into two cancer risk categories, $X_1$ ($X_2$) is the set of patients known to have low (high) risk of having cancer.

To handle PL constraints, Liu et al. [24] proposed the *Partition-Level Constrained Clustering (PLCC)* problem that uses the following objective:

$$\min \sum_{k=1}^{K} \sum_{x_i \in S_k} \|d_i^{(1)} - m_k^{(1)}\|^2 + \Lambda \mathbb{1}[d_i \in P]\|d_i^{(2)} - m_k^{(2)}\|^2 \tag{4}$$

where the first term is the squared distance from centroid and the second term is the constraint violation weighted by $\Lambda$. PLCC is solved using a K-Means-like algorithm.

Several works have applied model-based exact techniques to constrained clustering, including constraint programming [10–12] and integer linear programming [2]. In a recent work, Dao et al. [12] proposed a constraint programming (CP) approach for constrained clustering that minimizes within-cluster pairwise dissimilarity (Eq. (1)) using a dedicated global constraint. In an earlier work, they showed that a similar CP approach for minimizing sum of squared errors outperforms integer programming [11]. Although exact techniques are able to find and prove optimal solutions, they are often several orders of magnitude slower than heuristic techniques and for large problems can be intractable. Furthermore, they do not return a solution in case of contradictory constraints.

## 2.2 Ising Models

Ising models are graphical models that comprise a set of nodes $\mathcal{N}$ representing *spin* variables, $\sigma_i \in \{-1, 1\}, i \in \mathcal{N}$ and a set of edges $\mathcal{E}$ representing *interactions* between spin variables, $(i, j) \in \mathcal{E}$. The problem is parameterized by the *biases* $h_i$ and the *couplers* $J_{i,j}$. The objective is to minimize the *energy* of the model given by the Hamiltonian:

$$E(\sigma) = \sum_{(i,j)\in\mathcal{E}} J_{i,j}\sigma_i\sigma_j + \sum_{i\in\mathcal{N}} h_i\sigma_i. \tag{5}$$

*Quadratic unconstrained binary optimization (QUBO)* models are equivalent representations used to model problems with *binary* decision variables. Specifically, a QUBO model has $n$ decision variables, $q_i \in \{0, 1\}, i \in [1..n]$, with corresponding *biases*, $c_i$, and *couplers*, $c_{i,j}$. The objective of the QUBO is to minimize the following quadratic function:

$$E(q) = \sum_{i=1}^{n} c_i q_i + \sum_{i<j} c_{i,j} q_i q_j. \tag{6}$$

QUBO models can be converted to Ising models by setting $\sigma_i = 2q_i - 1$ [5] and thus we refer to them as Ising models.

## 2.3 Unsupervised Clustering with Ising Models

Kumar et al. [22] presented a QUBO model for unsupervised clustering,

$$E(q) = \sum_{i<j} c_{i,j} \sum_{k=1}^{K} q_k^i q_k^j + \sum_{i=1}^{n} \lambda_i \phi_i. \tag{7}$$

The first term in the objective is the within-cluster all-pairs dissimilarity. The cluster assignment for each data point is represented using one-hot encoding, i.e., $K$ binary variables $q_k^i$ such that $q_k^i{=}1 \iff x_i{\in}S_k$. Since each point is assigned to exactly one cluster, the QUBO model includes a quadratic penalty term to ensure the one-hot encoding holds:

$$\phi_i = \left( \sum_{k=1}^{K} q_k^i - 1 \right)^2. \tag{8}$$

If $x_i$ is assigned to exactly one cluster $\phi_i = 0$, otherwise $\phi_i \geq 1$ and the objective is penalized by $\lambda_i \phi_i$.

Kumar et al. [22] could only fit very small instances on a quantum annealer (up to 40 points) and used classical solver for larger instances. Their results were, at best, competitive with the K-Means heuristic in terms of solution quality.

## 3    A Framework for Constrained Clustering

We start by formulating the semi-supervised constrained clustering problem as a constrained optimization problem (COP). Given a problem instance defined by $\langle X, K, \mathcal{M}, \mathcal{C}, \{X_k\}_1^K \rangle$, we wish to find a partition, $S = S_1 \cup \cdots \cup S_K$, that minimizes the objective in Eq. (1) while satisfying the constraints:

$$\min_{S} \sum_{k=1}^{K} \sum_{\substack{i<j: \\ x_i, x_j \in S_k}} \|x_i - x_j\|^2$$

$$\text{s.t.} \quad \begin{aligned} s(x_i) &= s(x_j), & \forall (x_i, x_j) \in \mathcal{M} \\ s(x_i) &\neq s(x_j), & \forall (x_i, x_j) \in \mathcal{C} \\ s(x_i) &= k, & \forall k \in K, \forall x_i \in X_k. \end{aligned} \tag{9}$$

### 3.1    A QUBO Model for Constrained Clustering

We modify the unsupervised clustering model (Eq. (7)) to include clustering constraints. Specifically, we introduce the pairwise and partition-level constraints as quadratic penalty terms in the energy function:

$$E(q) = \sum_{i<j} c_{i,j} \sum_{k=1}^{K} q_k^i q_k^j + \sum_{i=1}^{n} \lambda_i \phi_i \sum_{\substack{i<j: \\ (x_i, x_j) \in \mathcal{M}}} w_{i,j}^{\mathcal{M}} \psi_{(i,j)}^{M}$$

$$+ \sum_{i<j:(x_i, x_j) \in \mathcal{C}} w_{i,j}^{\mathcal{C}} \psi_{(i,j)}^{C} + \sum_{k=1}^{K} \sum_{i:x_i \in X_k} w_{i,k}^{P} \psi_{(i,k)}^{P}. \tag{10}$$

The cost function is $c_{i,j} = \|x_i - x_j\|^2$ and the terms $\lambda_i \phi_i$ enforce the one-hot encoding (Eq. (8)). The terms $w_{i,j}^{\mathcal{M}} \psi_{(i,j)}^{M}$ enforce must-link constraints by penalizing the energy function if $x_i$ and $x_j$ are assigned to different clusters,

$$\psi_{(i,j)}^{M} = \sum_{k=1}^{K} (q_k^i - q_k^j)^2, \tag{11}$$

with $(q_k^i - q_k^j)^2$ being quadratic terms equal to one if $q_k^i \neq q_k^j$ and zero if $q_k^i = q_k^j$.[1]

The terms $w_{i,j}^C \psi_{(i,j)}^C$ enforce the cannot-link constraints by penalizing the energy function if $x_i$ and $x_j$ are in the same cluster, i.e., there exists $k$ such that $q_k^i = 1$ and $q_k^j = 1$,

$$\psi_{(i,j)}^C = \sum_{k=1}^{K} q_k^i q_k^j. \tag{12}$$

The terms $w_{i,k}^P \psi_{(i,k)}^P$ enforce the partition-level constraints by penalizing the energy function for assigning a data point $x_i \in X_k$ in a cluster $m \neq k$,

$$\psi_{(i,k)}^P = \sum_{\substack{m=1, \\ m \neq k}}^{K} q_m^i. \tag{13}$$

Once we obtain a solution to the QUBO in Eq. (10), each point $x_i$ is represented by $K$ bits $q_k^i, k \in [1..K]$ where $q_k^i = 1$ if and only if $x_i$ is in cluster $k$. We can extract the cluster for each point using the following function:

$$z_i(q) = \underset{k \in [1..K]}{\arg\max}\, q_k^i. \tag{14}$$

If the one-hot encoding constraint is satisfied, $z_i$ is bijective and therefore the partition can be obtained as follows:

$$x_i \in S_k \iff z_i(q) = k. \tag{15}$$

## 3.2   Choosing the Weights

Given Eq. (10), we must choose weights for the penalty terms to control the constraint violation. In most practical cases, the one-hot encoding is a hard constraint that we do not want violated. However, depending on the confidence we have in each of the constraints, we may be willing to violate some of these constraints in favor of satisfying others.

We consider the case in which our constraints come from a trusted source and we wish to find a partition that satisfies all constraints. Setting the weights for all penalty terms to be $n\tilde{d}$, where $\tilde{d} = \max c_{i,j}$, guarantees that the optimal solution to the QUBO model in Eq. (10) is an optimal solution for the COP in Eq. (9).

**Theorem 1** *Consider a constrained clustering problem defined by $\langle X, K, \mathcal{M}, \mathcal{C}, \{X_k\}_1^K \rangle$, such that the COP in Eq. (9) is satisfiable. Let $E(q)$ be the energy function in our QUBO model (Eq. (10)), with the following weights for the penalty terms $\lambda_i = w_{i,j}^M = w_{i,j}^C = w_{i,k}^P = n\tilde{d}$. Let $\bar{q}$ be an optimal solution to our QUBO model. Then the corresponding partition $\bar{S}$, $x_i \in \bar{S}_k \iff z_i(q) = k$, is an optimal solution to the COP in Eq. (9).*[2]

---

[1] If the one-hot encoding constraint is satisfied, violating a must-link constraint will apply two penalty terms, one for each of the two clusters of the data points.

[2] All proofs appear in tidel.mie.utoronto.ca/pubs/constrained-clustering-proofs.pdf.

### 3.3    An Efficient Encoding for $K = 2$

In the special case of $K = 2$, we can use an encoding that only requires $n$ variables, rather than $Kn$ variables:[3]

$$E_B(p) = \sum_{i<j} c_{i,j}(p^i + p^j - 1)^2 + \sum_{i<j:(x_i,x_j)\in\mathcal{M}} \hat{w}_{i,j}^M \sigma_{(i,j)}^M$$
$$+ \sum_{i<j:(x_i,x_j)\in\mathcal{C}} \hat{w}_{i,j}^C \sigma_{(i,j)}^C + \sum_{k=1}^{K} \sum_{i:x_i\in X_k} \hat{w}_{i,k}^P \sigma_{(i,k)}^P. \tag{16}$$

The variables $p^i$ represent the partition: $x_i$ is in the first cluster if $p^i = 0$ and in the second cluster otherwise. The terms $\sigma_{(i,j)}^M = (p^i - p^j)^2$ enforce the must-link constraints, the terms $\sigma_{(i,j)}^C = (p^i + p^j - 1)^2$ enforce the cannot-link constraints, and the terms $\sigma_{(i,k)}^P = [p^i - (k-1)]^2$ enforce the partition-level constraints.

Theorem 2 shows that the equivalence between the efficient encoding and the general model for $K = 2$. The bound in Theorem 1 is therefore applicable for this model.

**Theorem 2** *Consider a constrained clustering problem defined by $\langle X, K, \mathcal{M}, \mathcal{C}, \{X_k\}_1^K \rangle$ such that $K = 2$. Let $q_k^i$ be an assignment of variable for the K-clustering model in Eq. (10). We set $\hat{w}_{i,j}^M = 2w_{i,j}^M$, $\hat{w}_{i,j}^C = w_{i,j}^C$ and $\hat{w}_{i,j}^P = w_{i,j}^P$. If the one-hot encoding constraint is satisfied (i.e., $\phi_i = 0$ in Eq. (10)), then $E(q) = E_B(p)$ where $p^i$ is equal to zero if $q_1^i = 1$ and equal to one if $q_2^i = 1$.*

## 4    Constrained Clustering on the Fujitsu Digital Annealer

The Fujitsu Digital Annealer (DA) is recent CMOS hardware designed for Ising optimization problems formulated as a QUBO [28,33]. We use the first generation of the DA that is capable of representing problems with up to 1024 variables with 16-bit precision for the couplers and 26-bit precision for the biases.

The DA algorithm is based on simulated annealing [21], however it takes advantage of the massive parallelization provided by the custom CMOS hardware [1]. Furthermore, it has several key differences compared to simulated annealing:

- It starts every run from the same arbitrary state to reduce computational effort.
- It uses a *parallel-trial* scheme in which each Monte Carlo step considers all possible one-bit flips, in parallel. If more than one flip is accepted, one of accepted flips is chosen uniformly at random.
- It uses *dynamic offset* to increase the energy of a state in order to escape local minima.

---

[3] Kumar et al. [22] presented a model for unsupervised clustering with $n$ variables for $K = 2$. Their model uses spin-glass variables and does not optimize the energy function in Eq. (10).

## 4.1 Embedding Problems on the DA

When solving constrained clustering problems on the DA we have to make some practical representation and configuration choices. Due to the precision limit, we need to embed the couplers and biases on a scale with limited granularity. We therefore make the following implementation choices:

1. The distances $d(x_i, x_j)$ are normalized in the discrete range of $[0, 150]$.
2. The chosen weights cannot be arbitrarily high and the bound in Theorem 1 cannot be met. Instead we use the highest supported value for $\lambda$, the weight that enforces the one-hot encoding.
3. The bound in Theorem 1 guarantees that all constraints are satisfied if the problem is solved to optimality. In practice, the DA does not necessarily solve problems to optimality and instead terminates after a specified time limit. To avoid cases where the DA violates a one-hot encoding constraint in favor of satisfying a clustering constraint, we empirically find that it is better to use a lower weight for the penalty terms of the clustering constraints. In our experiments, we used a ratio of 1:4, $w^{\mathcal{M}} = w^{\mathcal{C}} = w^{\mathcal{P}} = \frac{1}{4}\lambda$.

The optimization parameters that represent the temperature schedule are tuned once per data set based solely on the obtained objective value (we do not use the true labels).

Unlike K-Means-based algorithms that run until convergence, our method runs for a given time limit and returns the best solution encountered. We therefore need to define a time limit to use in the evaluation of our approach. Considering the run time of heuristic techniques can vary significantly (for example, Liu and Fu [23] found LCVQE average run time varies between 0.01 to 76.73 s across different data sets) and the needs of practical applications, we arbitrarily choose 5 s as a time limit for each execution of our model (see Sect. 5.6 for further discussion).

## 5 Empirical Evaluation

We perform an empirical evaluation of our method across eight benchmark data sets. As the commonly used methods only support one type of constraint (pairwise or partition-level), we first compare performance on problems with one constraint type. Then, we evaluate our method on problems that involve *both* pairwise and partition-level constraints. To demonstrate the advantages of using special purpose hardware for combinatorial optimization, we compare our method to constraint programming [12] and two CPU solvers for Ising models.

### 5.1 Data Sets

We run experiments on eight data sets: *Breast Cancer, Ionosphere, Pima, Sonar, Seeds, Optdigits, Letters* [15], and *Protein* [36]. *Optdigits-389* is a randomly sampled subset of the UCI handwritten digits data set containing only the digits

**Table 1.** Description of data sets

| Data set | Instances | Features | Classes | CV |
|----------|-----------|----------|---------|-------|
| Breast cancer | 683 | 9 | 2 | 0.424 |
| Ionosphere | 351 | 34 | 2 | 0.399 |
| Pima[†] | 768 | 8 | 2 | 0.427 |
| Sonar | 208 | 60 | 2 | 0.095 |
| Seeds | 210 | 7 | 3 | 0.000 |
| Protein | 116 | 20 | 6 | 0.330 |
| Optdigits-389 | 283 | 64 | 3 | 0.032 |
| Letters-IJLT | 250 | 16 | 4 | 0.168 |

[†]Data is normalized using the standard deviation.

$\{3, 8, 9\}$, generated by sampling each instance with a probability of 0.15. *Letters-IJLT* is a randomly sampled subset of 250 instances from the letter recognition data set containing only the letters $\{I, J, L, T\}$.

Table 1 reports the number of instances, features, and classes. The coefficient of variation (CV) [14] describes the degree of class imbalance: zero indicates perfectly balanced classes, while higher values indicate higher class imbalance.

### 5.2 Algorithms

For problems with pairwise constraints, we compare our model to MPCK-Means[4] and LCVQE.[5] For problems with partition-level constraints, we compare our model to PLCC.[6] For MPCK-Means and PLCC we used the weights proposed in the original papers. Increasing the weights did not lead to a significant change in results.

If $K = 2$, we use the efficient QUBO encoding (Eq. (16)). Otherwise, we use the general QUBO model (Eq. (10)).

### 5.3 Evaluation Measures

Since labels are available for the data sets, we use the following measures to evaluate and compare the different methods.

**Adjusted Rand Index (ARI).** Rand Index [30] measures agreement between two partitions of the same data, $P_1$ and $P_2$. Each partition represents $\binom{n}{2}$ decisions over all pairs, assigning them to the same or different clusters. Let $a$ be

---

[4] Obtained from www.cs.utexas.edu/users/ml/risc/code.

[5] Obtained from github.com/danyaljj/constrained_clustering.

[6] As the code is not available, we implemented PLCC in Python.

the number of pairs assigned to the same cluster in both $P_1$ and $P_2$. Let $b$ be the number of pairs assigned to different clusters. Rand Index is defined as follows:

$$RI(P_1, P_2) = \frac{a + b}{\binom{n}{2}},$$

while the Adjusted Rand Index (ARI) [17] is a correction for RI, based on its expected value:

$$ARI = \frac{RI - \mathbb{E}(RI)}{Max(RI) - \mathbb{E}(RI)}.$$

An ARI of zero indicates the partition is not better than a random assignment, while one indicates a perfect match.

**Normalized Mutual Information (NMI).** Mutual information quantifies the statistical information shared between two distributions [31]. We use $MI(P_1, P_2)$ to denote the mutual information between partitions $P_1$ and $P_2$, and $H(P_i)$ to denote the entropy of partition $P_i$. Normalized mutual information (NMI) [31] is normalized using a generalized mean (e.g., arithmetic or geometric) of $H(P_1)$ and $H(P_2)$:

$$NMI(P_1, P_2) = \frac{MI(P_1, P_2)}{Mean(H(P_1), H(P_2))}$$

Values close to zero indicate independent partitions, while values close to one indicate a significant agreement between $P1$ and $P2$. We use NMI based on arithmetic mean.

**Fraction of Violated Constraints.** We compute the mean fraction of constraints that were violated in the partition.

### 5.4 Empirical Results

**Instance-Level Pairwise Constraints.** We compare our framework with MPCK-Means and LCVQE, on clustering with different numbers of randomly generated pairwise constraints. Following Covões et al. [9], each constraint is generated by randomly selecting two different instances in the data set and adding an ML constraint if they are in the same cluster and a CL constraint otherwise.

Figure 1 shows the performance for a varying number of pairwise constraints, measured by ARI. Each point in the plot is the average of 50 runs with different, randomly generated, sets of constraints. The bands represent the 95% confidence interval obtained using bootstrapping with 1000 replications. Note that the graphs *do not* share the same y-axis to increase readability (each graph presents data for a different data set and we do not compare across data sets). Results for NMI exhibited similar patterns and are omitted due to space.

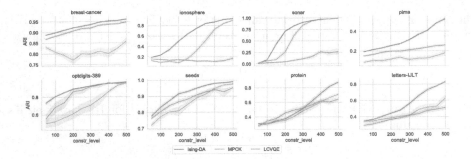

**Fig. 1.** Comparison of ARI scores for clustering with pairwise constraints.

In all cases but one, our framework outperforms the other methods. In *Breast Cancer, Ionosphere, Sonar, Pima, Optdigits-389, Seeds, Letters-IJLT* our framework is at least as good, and usually significantly better, across all numbers of constraints. In *Protein* there is no dominating algorithm and our framework is the best performing one for problems with large number of constraints (approximately 300 or more) while LCVQE is the best performing algorithm for problems with smaller number of constraints (less than 300).

Interestingly, there is no clear winner between LCVQE and MPCK-Means. In three data sets LCVQE outperforms MPCK-Means, in two data sets MPCK-Means outperforms LCVQE, and in the rest they are comparable. In contrast, our framework clearly outperforms the other methods.

Figure 2 shows the average fraction of violated must-link constraints and the 95% confidence interval for four data sets. In all data sets but *Breast Cancer*, we find that our method violates fewer constraints than the other methods, and in most cases does not violate any of the constraints. On *Breast Cancer*, our method and LCVQE outperform MPCK-Means, but do not dominate each other. Analysis of violated CL constraints is omitted due to space. As with ML constraints, our method is as good or better than the other methods in all cases except for *Breast Cancer*.

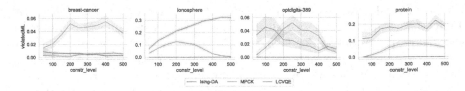

**Fig. 2.** Average fraction of violated must-link constraints.

**Partition-Level Constraints.** We compare our framework with PLCC on clustering with different numbers of randomly generated partition-level constraints, taken from the true labels. To be consistent with previous work [23, 24], we present the number of constraints as the fraction of the labeled data points. Figure 3 shows the performance of PLCC and our algorithm, measured by ARI. Results for NMI exhibit similar patterns and are omitted due to space.

Our method is consistently at least as good as PLCC, and in most cases better. Interestingly, for PL constraints, the improvement observed for general clustering problems is larger than the one observed for problems with $K = 2$.

Next, we analyze the fraction of violated partition-level constraints. When $K = 2$, we found that both algorithms satisfy approximately 100% of the constraints, with no significant differences. For the data sets with $K > 2$, PLCC violates a significant portion of the partition-level constraints while our method continues to satisfy all of them (see Fig. 4). This may account for the larger difference in performance between the two algorithms for data sets with $K > 2$.

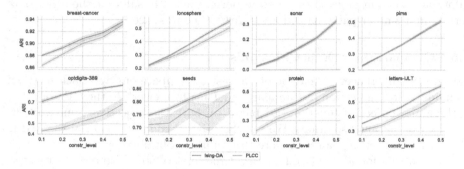

**Fig. 3.** ARI scores for clustering with partition-level constraints.

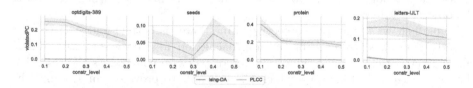

**Fig. 4.** Average fraction of violated partition-level constraints for $K > 2$.

**Mixed Constraint Types.** One of the advantages of our method, based on a mathematical model solved using a general optimization technique, is the ability to easily combine different types of constraints without the need to create a specialized algorithm.

To demonstrate this ability, we present results for problems that involve both pairwise and partition-level constraints. As far as we are aware, such problems cannot be solved by any existing heuristic techniques. Figure 5 reports

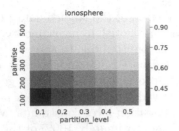 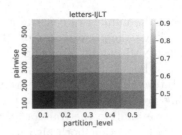

**Fig. 5.** ARI for mixed constraints.

the ARI for *Ionosphere* and *Letters-IJLT* for different combinations of pairwise and partition-level constraints. We can see that fusing different types of side-information can improve the clustering performance. Results for the other data sets exhibit similar patterns and are omitted due to space.

### 5.5   Comparison to Exact Methods

Despite the differences, it may be of interest to compare our approach to exact techniques. In this section, we compare our Ising framework to the CP approach with similar objective function [12] based on both the objective value and the accuracy of obtained solutions. We use the original code that is implemented in the Gecode solver [16][7] and compare the solutions obtained by the DA after 5 s to the solutions found by Gecode with a time limit of 500 s. Note that objective value is only comparable in case both methods satisfy the same set of constraints. For CP, solutions that satisfy all pairwise constraints were found for all instances. For DA, solutions that satisfy all pairwise constraints were found for 595 out of the 600 instances. In each of the other five instances only a single pairwise constraint was not satisfied, however we remove these instances when comparing the objective values.

Table 2 shows the average objective value (lower is better) and ARI (higher is better) obtained by each of the approaches on four data sets with different characteristics and a varying number of pairwise constraints. We also list the percentage of instances for which CP was able to prove optimality and the average per-instance objective ratio between the two methods (DA/CP). In the majority of cases Gecode was not able to prove optimality within the time limit. Furthermore, solutions found by the DA within 5 s are approximately equal or better for all configurations. In terms of clustering accuracy (measured by ARI), our approach outperforms CP for Ionosphere and Protein while in Sonar and Optdigits the methods are comparable.

### 5.6   Comparison to CPU Baselines

Our interest in Ising models is motivated by their ability to be efficiently solved by a variety of specialized hardware platforms. To demonstrate the benefit of

---

[7] Obtained from cp4clustering.github.io.

**Table 2.** Comparison between our Ising approach and Constraint Programming.

| Data set | Num. constr. | Ising-DA (5 s) | | CP (500 s limit) | | | DA/CP |
|---|---|---|---|---|---|---|---|
| | | Obj. | ARI | Obj. | ARI | % Opt | Obj. |
| Sonar ($K = 2$) | 50 | 31764.7 | 0.02 | 32992.2 | 0.04 | 0% | 0.9632 |
| | 150 | 35501.9 | 0.43 | 35760.6 | 0.41 | 0% | 0.9927 |
| | 350 | 36588.7 | 0.94 | 36588.5 | 0.94 | 100% | 1.0000 |
| Ionosphere ($K = 2$) | 50 | 433763.2 | 0.20 | 464745.6 | 0.13 | 0% | 0.9344 |
| | 150 | 478007.9 | 0.34 | 500041.8 | 0.30 | 0% | 0.9566 |
| | 350 | 514919.0 | 0.84 | 514922.3 | 0.84 | 94% | 1.0000 |
| Optdigits ($K = 3$) | 50 | 18790486.7 | 0.73 | 18840370.7 | 0.73 | 0% | 0.9974 |
| | 150 | 18862921.9 | 0.85 | 18947962.6 | 0.87 | 0% | 0.9955 |
| | 350 | 18955340.4 | 0.94 | 18957524.6 | 0.94 | 0% | 0.9999 |
| Protein ($K = 6$) | 50 | 226791.6 | 0.31 | 260764.2 | 0.23 | 0% | 0.8701 |
| | 150 | 245273.6 | 0.35 | 270906.9 | 0.29 | 0% | 0.9070 |
| | 350 | 259862.0 | 0.62 | 269950.0 | 0.56 | 6% | 0.9643 |

specialized hardware, we compare the results of the DA, a CMOS annealer, to two CPU baselines for Ising models: `neal`, a simulated annealer for Ising models, and `qbsolv`, a decomposing solver that splits QUBO problems into sub-problems solved by a tabu search (both are part of D-Wave's Ocean software package).[8]

We compare the quality of solutions obtained by these tools after 10 s and after 30 s to the solutions obtained by the DA after one and 5 s. Table 3 reports the mean ARI for four selected data sets for different numbers of pairwise constraints. Solutions that violate the one-hot encoding are considered to have an ARI of zero. As solutions obtained by the CPU solvers often do not satisfy all constraints, we do not compare the methods based on objective value.

We can see that the DA achieves better performance compared to the CPU baseline, even when we allow the CPU baselines longer time limits. In all but one configuration, DA with 5 s outperforms `neal` and `qbsolv` with 30 s. Interestingly, even when given only one second the DA performs well and in most configurations obtain solutions that are equal or better than those found by the CPU baselines in 30 s.

---

[8] Both tools obtained from github.com/dwavesystems.

**Table 3.** Mean ARI for DA vs. CPU solvers.

| Num const. | DA 1s | 5s | neal 10s | 30s | qbsolv 10s | 30s |
|---|---|---|---|---|---|---|
| 50 | .02 | .02 | .02 | .02 | .02 | .02 |
| 150 | .41 | .43 | .38 | .40 | .39 | .40 |
| 350 | .94 | .94 | .94 | .94 | .94 | .94 |

(a) Sonar ($K=2$)

| Num const. | DA 1s | 5s | neal 10s | 30s | qbsolv 10s | 30s |
|---|---|---|---|---|---|---|
| 50 | .19 | .20 | .18 | .19 | .16 | .16 |
| 150 | .33 | .34 | .26 | .31 | .32 | .33 |
| 350 | .84 | .84 | .82 | .83 | .80 | .82 |

(b) Ionosphere ($K=2$)

| Num const. | DA 1s | 5s | neal 10s | 30s | qbsolv 10s | 30s |
|---|---|---|---|---|---|---|
| 50 | .71 | .73 | .35 | .48 | .27 | .29 |
| 150 | .85 | .85 | .57 | .72 | .32 | 0.35 |
| 350 | .94 | .94 | .89 | .91 | .82 | 0.83 |

(c) Optdigits ($K=3$)

| Num const. | DA 1s | 5s | neal 10s | 30s | qbsolv 10s | 30s |
|---|---|---|---|---|---|---|
| 50 | .29 | .31 | .02 | .02 | .26 | .27 |
| 150 | .30 | .35 | .06 | .05 | .28 | .31 |
| 350 | .59 | .62 | .32 | .37 | .61 | .63 |

(d) Protein ($K=6$)

# 6   Discussion and Limitations

Our empirical evaluation shows that our method, based on an Ising model and specialized hardware, outperforms state-of-the-art K-Means-like methods. In unsupervised clustering, Kumar et al. [22] found that using Ising models for clustering achieves, at best, equal performance to K-means. Our results suggest that in the semi-supervised setting, where the problems include a set of constraints, using specialized hardware is a promising direction. The comparison to CP and CPU baselines shows that our approach can provide high quality solutions fast, making it an attractive solution for modern data mining tasks.

Our framework can be extended to other scenarios: representing new types of constraints (e.g., cluster-size constraints [7]), tuning the weights of the constraints if they are not fully-trusted, and evaluating our model with constraints arising from active learning [3] are all potential extensions of our work. While our models can incorporate any constraint that can be represented as a quadratic equality or inequality over the binary variables, some constraints may require additional auxiliary or slack variables. Investigating ways to efficiently encode other types of constraints is also an interesting direction for future work.

Our method is sensitive to hardware-related limitations. For example, the number of data points is limited by the number of variables supported by the hardware and our ability to represent the objective is limited by the precision. However, new hardware allows for larger problems and increased precision (e.g., [1,37]) and improved optimization schemes can reduce the need to tune the temperature schedule and potentially yield superior performance [20].

Our model can be solved on any platform that supports Ising models. As a large number of novel computational platforms (including quantum computers) have chosen Ising as their main abstraction [8], experimenting with new and different hardware platforms is an important direction of future work.

## 7 Conclusion

We address the problem of semi-supervised clustering on specialized hardware and present an Ising formulation that can be solved on a variety of novel hardware platforms. Our empirical analysis shows that our method outperforms the state-of-the-art heuristic methods for semi-supervised clustering and, unlike those algorithms, can support combinations of constraint types. The use of a mathematical model means that our framework is easily extended to support other types of constraints and hardware platforms.

**Acknowledgements.** This research was supported by the Natural Sciences and Engineering Research Council of Canada, Fujitsu Co-Creation Research Laboratory at the University of Toronto, and the Lyon Sachs Fellowship.

## References

1. Aramon, M., Rosenberg, G., Valiante, E., Miyazawa, T., Tamura, H., Katzgraber, H.G.: Physics-inspired optimization for quadratic unconstrained problems using a digital annealer. Front. Phys. **7**, 48 (2019)
2. Babaki, B., Guns, T., Nijssen, S.: Constrained clustering using column generation. In: Simonis, H. (ed.) CPAIOR 2014. LNCS, vol. 8451, pp. 438–454. Springer, Cham (2014). https://doi.org/10.1007/978-3-319-07046-9_31
3. Basu, S., Banerjee, A., Mooney, R.J.: Active semi-supervision for pairwise constrained clustering. In: SDM, pp. 333–344. SIAM (2004)
4. Basu, S., Davidson, I., Wagstaff, K.: Constrained Clustering: Advances in Algorithms, Theory, and Applications. CRC Press (2008)
5. Bian, Z., Chudak, F., Macready, W.G., Rose, G.: The ising model: teaching an old problem new tricks. D-wave Syst. **2** (2010)
6. Bilenko, M., Basu, S., Mooney, R.J.: Integrating constraints and metric learning in semi-supervised clustering. In: ICML, pp. 81–88 (2004)
7. Bradley, P., Bennett, K., Demiriz, A.: Constrained k-means clustering. Microsoft Res. Redmond **20** (2000)
8. Coffrin, C., Nagarajan, H., Bent, R.: Evaluating ising processing units with integer programming. In: CPAIOR, pp. 163–181 (2019)
9. Covões, T.F., Hruschka, E.R., Ghosh, J.: A study of k-means-based algorithms for constrained clustering. Intel. Data Anal. **17**(3), 485–505 (2013)
10. Dao, T.-B.-H., Duong, K.-C., Vrain, C.: A declarative framework for constrained clustering. In: Blockeel, H., Kersting, K., Nijssen, S., Železný, F. (eds.) ECML PKDD 2013. LNCS (LNAI), vol. 8190, pp. 419–434. Springer, Heidelberg (2013). https://doi.org/10.1007/978-3-642-40994-3_27
11. Dao, T.-B.-H., Duong, K.-C., Vrain, C.: Constrained minimum sum of squares clustering by constraint programming. In: Pesant, G. (ed.) CP 2015. LNCS, vol. 9255, pp. 557–573. Springer, Cham (2015). https://doi.org/10.1007/978-3-319-23219-5_39

12. Dao, T.B.H., Duong, K.C., Vrain, C.: Constrained clustering by constraint programming. Artif. Intell. **244**, 70–94 (2017)
13. Davidson, I., Ravi, S.: Clustering with constraints: feasibility issues and the k-means algorithm. In: SDM, pp. 138–149 (2005)
14. DeGroot, M.H., Schervish, M.J.: Probability and Statistics. Pearson Education (2012)
15. Dua, D., Graff, C.: UCI machine learning repository (2017). http://archive.ics.uci.edu/ml
16. Gecode Team: http://www.gecode.org
17. Hubert, L., Arabie, P.: Comparing partitions. J. Classif. **2**(1), 193–218 (1985)
18. James, G., Witten, D., Hastie, T., Tibshirani, R.: An Introduction to Statistical Learning. Springer, NewYork (2013). https://doi.org/10.1007/978-1-4614-7138-7
19. Johnson, M.W., et al.: Quantum annealing with manufactured spins. Nature **473**(7346), 194 (2011)
20. Katzgraber, H.G., Trebst, S., Huse, D.A., Troyer, M.: Feedback-optimized parallel tempering Monte Carlo. J. Stat. Mech: Theory Exp. **2006**(03), P03018 (2006)
21. Kirkpatrick, S., Gelatt, C.D., Vecchi, M.P.: Optimization by simulated annealing. Science **220**(4598), 671–680 (1983)
22. Kumar, V., Bass, G., Tomlin, C., Dulny, J.: Quantum annealing for combinatorial clustering. Quantum Inf. Process. **17**(2), 1–14 (2018). https://doi.org/10.1007/s11128-017-1809-2
23. Liu, H., Fu, Y.: Clustering with partition level side information. In: IEEE ICDM, pp. 877–882 (2015)
24. Liu, H., Tao, Z., Fu, Y.: Partition level constrained clustering. IEEE TPAMI **40**(10), 2469–2483 (2018)
25. Lloyd, S.: Least squares quantization in PCM. IEEE Trans. Inf. Theory **28**(2), 129–137 (1982)
26. Lucas, A.: Ising formulations of many np problems. Front. Phys. **2**, 5 (2014)
27. Mahajan, M., Nimbhorkar, P., Varadarajan, K.: The planar k-means problem is NP-hard. In: Das, S., Uehara, R. (eds.) WALCOM 2009. LNCS, vol. 5431, pp. 274–285. Springer, Heidelberg (2009). https://doi.org/10.1007/978-3-642-00202-1_24
28. Matsubara, S., et al.: Ising-model optimizer with parallel-trial bit-sieve engine. In: Barolli, L., Terzo, O. (eds.) CISIS 2017. AISC, vol. 611, pp. 432–438. Springer, Cham (2018). https://doi.org/10.1007/978-3-319-61566-0_39
29. Pelleg, D., Baras, D.: $K$-means with large and noisy constraint sets. In: Kok, J.N., Koronacki, J., Mantaras, R.L., Matwin, S., Mladenič, D., Skowron, A. (eds.) ECML 2007. LNCS (LNAI), vol. 4701, pp. 674–682. Springer, Heidelberg (2007). https://doi.org/10.1007/978-3-540-74958-5_67
30. Rand, W.M.: Objective criteria for the evaluation of clustering methods. J. Am. Stat. Assoc. **66**(336), 846–850 (1971)
31. Strehl, A., Ghosh, J.: Cluster ensembles–a knowledge reuse framework for combining multiple partitions. J. Mach. Learn. Res. **3**, 583–617 (2002)
32. Trevor, H., Robert, T., Friedman, J.: The Elements of Statistical Learning: Data Mining, Inference, and Prediction. Springer, Heidelberg (2009). https://doi.org/10.1007/978-0-387-84858-7
33. Tsukamoto, S., Takatsu, M., Matsubara, S., Tamura, H.: An accelerator architecture for combinatorial optimization problems. Fujitsu Sci. Tech. J **53**(5), 8–13 (2017)

34. Ushijima-Mwesigwa, H., Negre, C.F., Mniszewski, S.M.: Graph partitioning using quantum annealing on the d-wave system. In: Proceedings of the Second International Workshop on Post Moores Era Supercomputing, pp. 22–29. ACM (2017)
35. Wagstaff, K., Cardie, C., Rogers, S., Schrödl, S.: Constrained k-means clustering with background knowledge. In: ICML, pp. 577–584 (2001)
36. Xing, E.P., Jordan, M.I., Russell, S.J., Ng, A.Y.: Distance metric learning with application to clustering with side-information. In: NIPS, pp. 521–528 (2003)
37. Yamaoka, M., Yoshimura, C., Hayashi, M., Okuyama, T., Aoki, H., Mizuno, H.: A 20k-spin ising chip to solve combinatorial optimization problems with CMOS annealing. IEEE J. Solid-State Circuits 51, 303–309 (2016)

# From MiniZinc to Optimization Modulo Theories, and Back

Francesco Contaldo[✉], Patrick Trentin[✉], and Roberto Sebastiani[✉]

DISI, University of Trento, Trento, Italy
francesco.contaldo@alumni.unitn.it,
{patrick.trentin,roberto.sebastiani}@unitn.it

**Abstract.** Optimization Modulo Theories (OMT) is an extension of SMT that allows for finding models that optimize objective functions. In this paper we aim at bridging the gap between Constraint Programming (CP) and OMT, *in both directions*. First, we have extended the OMT solver OPTIMATHSAT with a FLATZINC interface – which can also be used as FLATZINC-to-OMT encoder for other OMT solvers. This allows OMT tools to be used in combination with MZN2FZN on the large amount of CP problems coming from the MiniZinc community. Second, we have introduced a tool for translating SMT and OMT problems on the linear arithmetic and bit-vector theories into MiniZinc. This allows MiniZinc solvers to be used on a large amount of SMT/OMT problems.

We have discussed the main issues we had to cope with in either directions. We have performed an extensive empirical evaluation comparing three state-of-the-art OMT-based tools with many state-of-the-art CP tools on (i) CP problems coming from the MiniZinc challenge, and (ii) OMT problems coming mostly from formal verification. This analysis also allowed us to identify some criticalities, in terms of efficiency and correctness, one has to cope with when addressing CP problems with OMT tools, and vice versa.

## 1 Introduction

The last two decades have witnessed the rise of Satisfiability Modulo Theories (SMT) [12] as efficient tool for dealing with several applications of industrial interest, in particular in the contexts of Formal Verification (FV). SMT is the problem of finding value assignments satisfying some formula in first-order logic wrt. some background theory. Optimization Modulo Theories (OMT) [15,33,34,36,39,42,53,56] is a more-recent extension of SMT searching for the *optimal* value assignment(s) w.r.t. some objective function(s), by means of a combination of SMT and optimization procedures. (Since OMT extends SMT, hereafter we often simply say "OMT" for both SMT and OMT.)

Several distinctive traits of OMT solvers –like, e.g., the efficient combination of Boolean and arithmetical reasoning, incrementality, the availability of decision procedures for infinite-precision arithmetic and the ability to produce conflict explanations– are a direct consequence of their tight relationship with the FV domain and its practical needs. On the whole, it appears that OMT can be a potentially interesting and efficient technology for dealing with Constraint Programming (CP) problems as well. At the

© Springer Nature Switzerland AG 2020
E. Hebrard and N. Musliu (Eds.): CPAIOR 2020, LNCS 12296, pp. 148–166, 2020.
https://doi.org/10.1007/978-3-030-58942-4_10

same time, modeling CP problems for OMT solvers requires a higher-level of expertise, because the same CP instance can have many possible alternative formulations, s.t. the performance of SMT solvers on each encoding are hardly predictable [26,28,29].

On the other hand, the availability, efficiency and expressiveness of CP tools makes them of potential interest as backend engines also for FV applications (e.g., [22,23,31]), in particular with SW verification, where currently SMT is the dominating backend technology, s.t. a large amount of SMT-encoded FV problems are available [11].

In this paper we aim at bridging the gap between CP and OMT, *in both directions*.

In the CP-to-OMT direction, we have extended the state-of-the-art OMT solver OPTIMATHSAT [58] with a FLATZINC interface (namely "FZN2OMT"). In combination with the standard MZN2FZN encoder [38], this new interface can be used to either (i) solve CP models with OPTIMATHSAT directly or (ii) generate OMT formulas encoded in the SMT-LIB [25] format with optimization extensions, to be fed to other OMT solvers, such as BCLT [16] and Z3 [15]. This allows state-of-the-art OMT technology to be used on MINIZINC problems coming from the CP community.

In the OMT-to-CP direction, we have introduced a tool for translating SMT and OMT problems on the theories of linear arithmetic over the integers and rational ($\mathcal{LIRA}$) and bit-vector ($\mathcal{BV}$) into MINIZINC models (hereafter "OMT2MZN"). This allows MINIZINC solvers to be used on OMT problems, giving them access to a large amount of OMT problems, mostly coming from formal verification.

With both directions, we first present and discuss the challenges we encountered and the solutions we adopted to address the differences between the two formalisms. Then we present an extensive empirical evaluation comparing three OMT tools with many state-of-the-art CP tools on (i) CP problems coming from the MINIZINC challenge, and (ii) OMT problems coming mostly from formal verification. This analysis allowed us to identify some criticalities, in terms of efficiency and correctness, one has to cope with when addressing CP problems with OMT tools, and vice versa.

Overall, our new compilers FZN2OMT and OMT2MZN in combination with the standard compiler MZN2FZN [38] provide a framework for translating problems encoded in the SMT-LIB or the MINIZINC format in either direction. This framework enables also for a comparison between OMT solvers and CP tools on problems that do not belong to their original application domain. To the best of our knowledge, this is the first time that such a framework has been proposed, and that the OMT and CP technologies have been extensively compared on problems coming from both fields.

*Related Work.* The tight connection between SMT and Constraint Programming (CP) has been known for a relatively long period of time [43] and it has previously been subject to investigation. Some works considered a direct encoding of CP [28,29] and weighted CP [8] into SMT and MAXSMT, or an automatic framework for translating MINIZINC –a standard CP modeling language [40]– into SMT-LIB –the standard SMT format– [17,18]. Other works explored the integration of typical SAT and SMT techniques within CP solvers [27,45]. Nowadays, several MINIZINC solvers –like, e.g., HAIFACSP [61] and PICAT [62]– are at least partially based on SAT technology.

To this extent, our first contribution FZN2OMT also obviates the loss, due to obsolescence, of the FZN2SMT compiler proposed by Bofill et al. in [17,18]. FZN2SMT is not compatible with the changes that have been introduced to the MINIZINC and FLATZ-INC standards starting from version 2.0 of the MINIZINC distribution. Since some of these changes are not backward compatible, it is also not possible to use FZN2SMT in conjunction with an older version of the MZN2FZN compiler when dealing with recent MINIZINC models. Furthermore, FZN2SMT translates satisfaction problems into the Version 1 of the SMT-LIB standard and produces no SMT-LIB output in the case of optimization problems, that are solved directly. However, the optimization interface of modern OMT solvers is based on the Version 2 of the SMT-LIB standard. This makes it difficult to use it together with OMT solvers. Unfortunately, the FZN2SMT compiler is closed source, with only the binaries being freely distributed, and seemingly no longer maintained. This made it necessary to provide a new alternative solution to FZN2SMT. To this extent, our new FLATZINC interface of OPTIMATHSAT, FZN2OMT, translates both satisfaction and optimization problems in the Version 2 of the SMT-LIB standard enriched with the optimization extensions for OMT described in [58].

*Content.* The rest of the paper is organized as follows. In Sect. 2 we provide some background on OMT, MINIZINC and FLATZINC. In Sect. 3 we describe the process from MINIZINC to OMT. In Sect. 4 we describe the process from OMT to MINIZINC. In Sect. 5 we describe an empirical evaluation comparing a OMT-based tool with many state-of-the-art CP tools. Finally, in Sect. 6 we conclude and point out some further research directions.

A longer and more detailed version of this paper is publicly available as [24].

## 2    Background

Satisfiability Modulo Theories (SMT) is the problem of deciding the satisfiability of a first-order formula $\varphi$ with respect to a combination of decidable first-order theories. Typical theories of SMT interest are (the theory of) linear arithmetic over the rationals ($\mathcal{LRA}$), the integers ($\mathcal{LIA}$) or their combination ($\mathcal{LIRA}$), non-linear arithmetic over the rationals ($\mathcal{NLRA}$) or the integers ($\mathcal{NLIA}$), arrays ($\mathcal{AR}$), bit-vectors ($\mathcal{BV}$), floating-point arithmetic ($\mathcal{FP}$), and their combinations thereof. (See [12,44,52] for an overview.). The last two decades have witnessed the development of very efficient SMT solvers based on the so-called lazy-SMT schema [12,52]. This has brought previously-intractable problems to the reach of state-of-the-art SMT solvers.

Optimization Modulo Theories (OMT), [15,34,36,39,42,53,56,60], is an extension to SMT that allows for finding a model of a first-order formula $\varphi$ that is *optimal* with respect to some objective function expressed in some background theory, by means of a combination of SMT and optimization procedures. State-of-the art OMT tools allow optimization in a variety of theories, including linear arithmetic over the rationals (OMT($\mathcal{LRA}$)) [53] and the integers (OMT($\mathcal{LIA}$)) [15,56], bit-vectors (OMT($\mathcal{BV}$)) [39] and floating-point numbers (OMT($\mathcal{FP}$)) [60].

A relevant strict subcase of OMT($\mathcal{LRA}$) is OMT with Pseudo-Boolean objective functions (OMT($\mathcal{PB}$)) in the form $\sum_i w_i A_i$ s.t. $w_i$ are rational values and $A_i$ are Boolean variables whose values are interpreted as $\{0, 1\}$. Notice that OMT($\mathcal{PB}$) is also equivalent to (partial weighted) MAXSMT, the SMT extension of MAXSAT, and that

OMT($\mathcal{PB}$) and MAXSMT can be encoded into OMT($\mathcal{LRA}$) but not vice versa [54]. Encoding OMT($\mathcal{PB}$)/MAXSMT into OMT($\mathcal{LRA}$), however, is not the most efficient way to solve them, so that modern OMT solvers such as BCLT [16], OPTIMATHSAT [58] and Z3 [15] implement specialized OMT($\mathcal{PB}$)/MAXSMT procedures which are much more efficient than general-purpose OMT($\mathcal{LRA}$) ones [15,57,58].

We stress the fact that —unlike with purely-combinatorial problems, which are encoded into SAT or MAXSAT and are thus solved by purely-Boolean search– typically OMT problems involve the interleaving of *both* Boolean and arithmetical search: search not only for the best truth-value assignment to the atomic subformulae, but also for the best values to the numerical variables compatible with such truth-value assignment [54].

To this date, few OMT solvers exist, namely BCLT [16], CEGIO [9], HAZEL [39], OPTIMATHSAT [58], PULI [33], SYMBA [36] and Z3 [15]. To this aim, we observe that (i) some of these solvers are quite recent, (ii) most of these solvers focus on different, partially overlapping, niche subsets of Optimization Modulo Theories, and (iii) the lack of an official Input/Output interface for OMT makes it hard to compare some of these tools with one another. OMT finds applications in the context of static analysis [19,32], formal verification and model checking [37,48], scheduling and planning with resources [33,35,46,50], software security and requirements engineering [41], workflow analysis [13], machine learning [59], and quantum computing [14].

A distinctive trait of SMT (and OMT) solvers is the trade-off of speed against the ability to certify the correctness of the result of any computation, which is particularly important in the contexts of Formal Verification (FV) and Model Checking (MC). When dealing with linear arithmetic in particular, SMT solvers employ *infinite-precision arithmetic* software libraries to avoid numerical errors and overflows.

SMT-LIB [25] is the standard input format by SMT solvers, it provides a standardized definition of the most prominent theories supported by SMT solvers and the corresponding language primitives to use these features. At present, there is no standard input format for modeling optimization problems targeting OMT solvers, although there exist only minor syntactical differences between the major OMT solvers. The tools presented in this paper conform to the *extended* SMT-LIB format for OMT presented in [58], that includes language primitives for modeling objectives.

OPTIMATHSAT [53–58] is a state-of-the-art OMT solver based on the MATHSAT5 SMT solver [3,21]. OPTIMATHSAT features both single- and multi-objective optimization over arbitrary sets of $\mathcal{LRA}$, $\mathcal{LIA}$, $\mathcal{LIRA}$, $\mathcal{BV}$, $\mathcal{FP}$, Pseudo-Boolean ($\mathcal{PB}$) and MAXSMT cost functions. Multiple objective functions can be combined with one another into a Lexicographic or a Pareto optimization problem, or independently solved in a single run (for the best efficiency).

MINIZINC [38,40] is a widely adopted high-level declarative language for modeling Constraint Satisfaction Problems (CSP) and Constraint Optimization Problems (COP). The MINIZINC format defines three scalar types (bool, int and float) and two compound types (sets and fixed-size arrays of some scalar type). The standard provides an extensive list of predefined *global constraints*, a class of high-level language primitives that allows one to encode complex constraints in a compact way.

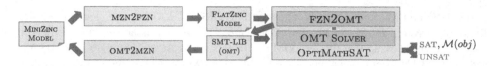

**Fig. 1.** Circular translation schema from MINIZINC to SMT-LIB and back, resulting from the composition of MZN2FZN, OPTIMATHSAT and OMT2MZN. In this picture, OPTIMATHSAT acts both as a FLATZINC/OMT solver, and also as a FLATZINC to SMT-LIB compiler.

FLATZINC is a lower-level language whose purpose is to bridge the gap between the high-level modeling in MINIZINC and the need for a fixed, and easy-to-parse, format that can simplify the implementation of the input interface of MINIZINC solvers. A MINIZINC model is typically flattened into a FLATZINC instance using the MZN2FZN compiler [38], and then solved with some MINIZINC tool.

## 3   From MINIZINC to OMT

We consider the problem of translating MINIZINC models into OMT problems first. Similarly to other MINIZINC solvers, we assume that the MINIZINC model is first translated into FLATZINC using the MZN2FZN standard compiler, as depicted in Fig. 1. We describe the main aspects of FZN2OMT, focusing on the challenges we have encountered and on the solutions we have adopted.

FLATZINC *data-types.* The first challenge is to find a suitable representation of the data-types supported by FLATZINC in SMT-LIB.

One possible choice for modeling the three basic scalar types of FLATZINC –namely bool, int and float– with SMT-LIB are the Boolean, bit-vector and floating-point theories respectively. However, the decision procedures for the bit-vector and floating-point numbers theories can be significantly more resource demanding than the decision procedure for the linear arithmetic theory ($\mathcal{LIRA}$), in particular when dealing with a substantial amount of arithmetic computations. For this reason, we have opted to model FLATZINC int and float data-types with the SMT-LIB integer and rational types respectively, by default. For the case in which no substantial linear arithmetic computation is performed, we also optionally allow for encoding the FLATZINC int data-type as a SMT-LIB bit-vector.

For what concerns the two compound types of FLATZINC, that is the set and array data-types, we have chosen to proceed as follows. Given that OPTIMATHSAT lacks a decision procedure for the theory of finite sets [7], we model a set using the Boolean and integer theories, similarly to what has been done in [17]. The basic idea is to introduce a fresh Boolean variable for each element in the domain of a set, and use such variable as a placeholder for the membership of an integer element to the set instance. Differently from [17], we make an extensive use of cardinality networks [10] to encode constraints over the sets because they are handled more efficiently, for their nice arc-consistency properties. No action is required to encode a FLATZINC array into SMT-LIB, because it is used only as a container for other variables.

*Floating-Point Precision.* A consequence of encoding the FLATZINC `int` and `float` data-types with the linear arithmetic theory is that all of our computation is performed with infinite-precision arithmetic. This can result in a performance disadvantage wrt. other MINIZINC solvers using finite-precision arithmetic, due to the increased cost of each operation, but it has the benefit of guaranteeing the correctness of the final result of the computation.

Currently, the MINIZINC language does not allow one to express a certain quantity as an infinite-precision fraction between two constant numbers. Instead, the MZN2FZN compiler computes on-the-fly the result of any division operation between two constant integers or floating-point numbers applying the rules corresponding to the type of the operands. However, there are some instances in which we really need to be able to both express quantities and perform computation with infinite-precision arithmetic. One of such situations is to double-check the correctness of the MINIZINC models generated by the OMT2MZN compiler described in Sect. 4 (we have done this for the experimental evaluations in Sect. 5.2). In order to get around this limitation we developed a simple wrapper around the MZN2FZN compiler, called EMZN2FZN [5], that replaces any fraction among two constant `floats` with a fresh variable and, after the basic MZN2FZN compiler generated the FLATZINC model, the EMZN2FZN wrapper restores the original fractional values using the FLATZINC constraint `float_div()`.

FLATZINC *Constraints.* The SMT-LIB encoding of the majority of FLATZINC constraints in OPTIMATHSAT follows their definition in the FLATZINC Standard, with the exception of Pseudo-Boolean constraints, which we examine in detail later on. Several global constraints are also supported in the same way, because the OMT-solver currently lacks ad hoc and efficient decision procedures for dealing with them. Constant values and alias variables (e.g. those arising from the definition of some arrays) are propagated through the formula, so as to keep the set of problem variables as compact as possible. Those constraints requiring non-linear arithmetic –like, e.g., trigonometric, logarithmic and exponential functions– are currently not supported; this situation may change soon due to the recent extension of MATHSAT5 with a procedure for it [20].

*Pseudo-Boolean Constraints.* When dealing with Pseudo-Boolean sums of the form $\sum_{i=1}^{i=N} A_i \cdot w_i$, where $A_i$ is a Boolean variable and $w_i$ is a numerical weight, the MZN2FZN compiler associates a fresh 0/1 integer variable $x_i$ to each $A_i$, and encodes the sum as $\sum_{i=1}^{i=N} x_i \cdot w_i$. Notice that the original $A_i$s may not be eliminated from the FLATZINC model, because they typically occur elsewhere in the problem, i.e. as part of a Boolean formula. From our own experience, this situation arises frequently, because Pseudo-Boolean sums are typically used to express cardinality constraints that have a variety of uses. As described in [57], one limitation of this naive approach is that SMT and OMT solvers do not typically handle this encoding efficiently. The main reason is that the pruning power of the conflict clause resulting from a conflicting assignment is typically limited to one specific Boolean assignment at a time, meaning that a large number of conflict clauses (possibly exponential) has to be generated along the search. As shown in [57], SMT and OMT solvers can benefit from encoding Pseudo-Boolean constraints with cardinality networks.

FZN2OMT goes through some effort in order to recognize Pseudo-Boolean sums over the integers, and replace the naive encoding with one based on cardinality networks. We note that using this technique generally results in a trade-off between solving time and the overhead of generating cardinality networks prior to starting the search, especially when dealing with a large number of variables.

*Multi-objective Optimization.* FZN2OMT allows for multiple optimization goals, of heterogeneous type, being defined within the same FLATZINC model. This is a non-standard extension to the FLATZINC format. Multiple objectives can be solved independently from one another, or combined into a Lexicographic or Pareto optimization goal. We refer the reader to [58] for details on the input encoding and the solver configuration.

*Functionality.* Given a satisfiability or optimization problem encoded in the FLATZINC format, OPTIMATHSAT can be used in the following ways (Fig. 1):

- to directly solve the problem, optionally enumerating any sub-optimal solution found during the search or all possible solutions with the same optimal value;
- to produce an OMTproblem encoded with the extended SMT-LIB format described in [58]. This problem can be directly solved with OPTIMATHSAT or, with minor transformations[1], fed as input to other OMT solvers such as BCLT and Z3.

## 4   From OMT to MINIZINC

In this section, we consider the problem of translating OMT formulas, encoded in the optimization-extended SMT-LIB format of [58], into MINIZINC models. Hereafter, we describe the main challenges we have faced and the solutions we have adopted. Further details about this conversion are available in [4].

*General Translation Approach.* The main challenge is to design an encoding from OMT to MINIZINC that is *correct* (i.e., it preserves in full the semantics of the input OMT problems), *effective* (i.e., it produces as output MINIZINC models which are as compact and easy-to-solve as possible), and *efficient* (i.e. it does it with the least consumption of time and memory). To this extent, one critical design choice is the way in which the internal representation of the input OMT formula is organized and converted in terms of MINIZINC primitives. After a preliminar experimental evaluation we determined that the sweet-spot, in terms of compactness and easiness to solve of the resulting MINIZINC model, is to adopt what we call "$\geq$2-father DAG-ification": a Directed-Acyclic-Graph (DAG) internal representation of the formula where a fresh label is associated to all and only DAG nodes with at least two fathers, inlining all other nodes (see [24] for details).

---

[1] To make this step as easy as possible, we collected our scripts into a public repository [1].

*Theories Restriction.* The SMT-LIB standard describes a wide number of SMTtheories, most of which have no direct counterpart in MINIZINC due to the few data-types supported (see Sect. 2)). Hence, of linear rational and integer arithmetic, and their combination. On this regard, we note that even though OMT2MZN can also handle the theory of bit-vectors, we do not cover it here because it is not used in the experimental evaluation in Sect. 5 (We cover it in the long version of this paper [24]). We leave the handling of other SMT theories to future work.

*Linear Arithmetic Theory.* On the surface, encoding linear arithmetic constraints coming from OMT in MINIZINC, using the *int* and *float* data-types, looks like a trivial task. In reality, this poses several challenges and it is subject to several limitations, due to a couple of facts.

First, in SMT-LIB the linear arithmetic theory requires the capability to perform *infinite-precision* computations. Unfortunately, to the best of our knowledge, no MINIZINC solver provides infinite-precision arithmetic reasoning, and the MZN2FZN compiler itself prevents representing arbitrarily-large and arbitrarily-precise quantities (e.g. the fine-grained decimal weights of the machine learning application in [59]).

Second, in OMT linear arithmetic variables are not required to be bounded and have quite often no explicit domain (i.e. they lack a lower-bound, an upper-bound or both), because it is not necessary for the problem at hand or it is implied by other constraints. This is in contrast with MINIZINC, whereby linear arithmetic variables are expected to have a finite domain and, when they lack one, their domain appears to be capped with some solver-dependent pair of values.

These restrictions are currently part of the MINIZINC language and the target application domain, and we do not see any obvious work-around solving them. We note that although there exist methods for bounding all variables in a given LP problem (e.g. [47]), these have been deemed too impractical at this stage of our investigation. Nonetheless, we have chosen to translate SMT-LIB linear arithmetic constraints with a corresponding MINIZINC encoding based on the *int* and *float* data-types. Although the encoding is not always applicable, it does still allow one to correctly translate a number of interesting OMT problems into MINIZINC, as witnessed by our experimental evaluation in Sect. 5.2. More in detail, the translation is done as follows. We declare each integer variable as unbounded, and then extend the MINIZINC model with the appropriate constraints bounding its domain when the input OMT formula contains any such information. Our empirical observation is that MINIZINC models generated in this way are correctly handed by all MINIZINC solvers which we have tried, with the exception of GUROBI, which returns an "unsupported" message. Floating-Point variables, instead, are always declared with a user-defined domain. This is because all of the MINIZINC solvers we have tried, among those that can handle floating-point constraints, require such information.

*Other* OMT *Functionalities.* Several problems of OMT interest require the capability of dealing with soft-constraints (i.e. Weighted MAXSMT) and also with multiple objectives, that are either considered independent goals or combined in a Lexicographic or Pareto-like fashion. To the best of our knowledge, the MINIZINC standard does not allow for an explicit encoding of soft-constraints, nor to deal with more than one objective function at a time.

We encode (weighted) MAXSMT problems using a standard Pseudo-Boolean encoding, such as the one used in [53]. When dealing with OMT problems that contain $N$ goals $obj_1, ..., obj_N$, for $N > 1$, we use the following approach. If these objectives are independent targets, we generate $N$ MINIZINC models, each with a different goal $obj_i$, and separately solve each model. If instead the multiple objectives belong to a Lexicographic OMT problem, then we generate a unique MINIZINC model that leverages the lexicographic-optimization functionality provided by MINISEARCH [49]. (In all other cases, MINISEARCH is not used). We do not have any encoding for dealing with Pareto-optimization, yet.

## 5 Experimental Evaluations

In this section we present an extensive empirical evaluation comparing OMT tools with many state-of-the-art CP tools on CP problems coming from the MINIZINC challenge (Sect. 5.1), and on OMT problems coming mostly from formal verification (Sect. 5.2).

The OMT solvers under evaluation are BCLT, OPTIMATHSAT (v. 1.6.0) and Z3 (v. 4.8.5). These are compared with some of the top-scoring solvers that participated at recent editions of the MINIZINC challenge, including CHOCO (v. 4.0.4), CHUFFED, G12(FD) (v. 1.6.0), GECODE (v. 6.0.1), GUROBI (v. 8.0.1), HAIFACSP (v. 1.3.0), JACOP (v. 4.5.0), IZPLUS (v. 3.5.0), OR-TOOLS (v. 6.7.4981) and PICAT (v. 2.4).

*Remark 1.* We could not include FZN2SMT [17, 18] in our experimental evaluation because it is not compatible with the features of MINIZINC that have been added since version 2.0.

We run all these experimental evaluations on two identical *8-core 2.20 Ghz Xeon* machines with 64 GB of RAM and running *Ubuntu Linux*. All the benchmark-sets, the tools and the scripts used to run these experiments, and some of the plots for the results in Tables 1, 2 and 3 which could not fit into this paper, can be downloaded from [2].

We stress the fact that the goal of these experiments is not to establish a winner among OMT and MINIZINC tools; rather, it is to assess the correctness, effectiveness and efficiency of our OMT-to-CP and CP-to-OMT encoders and, more generally, to investigate the feasibility of solving MINIZINC problems with OMT tools and vice versa, and to identify the criticalities in terms of efficiency and correctness in these processes.

### 5.1 Evaluation on MINIZINC Benchmark Sets

We consider the benchmark-sets used at the MINIZINC Challenge of 2016 (MC16) and 2019 (MC19), each comprised by 100 instances. For compatibility reasons, the version of MZN2FZN used to convert the problems to the FLATZINC format differs between the two benchmark-sets. We use version 2.2.1 and 2.3.2 (with patches) for the problems in MC16 and MC19 respectively. Due to recent changes in the FLATZINC format that affect the benchmarks in MC19, the version of some MINIZINC tools differs from what described in Sect. 5 (see Table 1). In some cases, we had to download and compile the latest source available for the tool, i.e. the "nightly" version.

We run each MINIZINC solver with the corresponding directory of global constraints, and we run each MINIZINC and OMT tool with the default options. We consider two OMT encodings of the original FLATZINC problems, LA and BV. The first encodes the FLATZINC *int* type with the theory of *linear integer arithmetic*, whereas the second is based on the theory of *bit-vectors*. We evaluate each OMT solver on both SMT-LIB encodings, except for BCLT that has no support for bit-vector optimization. For uniformity reasons with the other OMT solvers, we evaluate OPTIMATHSAT using its SMT-LIB interface only, using thus its FZN2OMT interface as an external tool, like with the other OMTsolvers. We note that the solving time for all OMT solvers includes the time required for translating the formula from the FLATZINC to the SMT-LIB format. Each solver, either OMT or MINIZINC, is given up to $1200s$. to solve each problem, not including the time taken by MZN2FZN to flatten it.

We verify the correctness of the results by automatically checking that all terminating solvers agree on the (possibly optimal) solution and, when this is not the case, we manually investigate the inconsistency.

*Experiment Results.* The results of this experiment are shown in Table 1, with separate numbers for satisfiability (**s**) and optimization (**o**) instances in each benchmark-set. Using the experimental data, we separately computed the virtual best configuration among all MINIZINC solvers (i.e. VIRTUAL BEST(MINIZINC)), all OMT solvers (i.e. VIRTUAL BEST(OMT)), and also the virtual best among all tools considered in the experiment (i.e. VIRTUAL BEST(ALL)). The last two columns in the table list the number of problems solved by the given configuration in the same amount of time as the VIRTUAL BEST() of each group (col. *BT1*) and as the VIRTUAL BEST(all) (col. *BT2*).

We start by looking at the MINIZINC solvers in Table 1. The performance ladder is dominated by OR-TOOLS(SAT) and PICAT(SAT), closely followed by GUROBI, HAIFACSP and CHUFFED (in MC19). By looking at column *BT1*, we observe that the top-performing MINIZINC solvers tend to dominate over all the others. Looking at the results of the MC19 experiment, we notice a significant increase in the number of errors with respect to the benchmark-set of the MC16 edition, as well as a handful of problems solved incorrectly. In the case of GUROBI and PICAT(SAT), the MZN2FZN compiler encountered an error over a few instances. As a consequence, the total number of problems is smaller than 100 for both tools. After taking a closer look, we ascribe this phenomenon to the recent changes in the MINIZINC/FLATZINC format, that has created some minor issues with some tools that have not been adequately updated.

Looking at the OMT tools only, we observe that Z3 has leading performance over the other solvers. When compared to the MINIZINC solvers, the OMT solvers place themselves in the middle of the rank on both benchmark-sets. Given the fact that none of the OMT solvers has specialized procedures or encodings for dealing with global constraints, we consider this an interesting result.

## 5.2 Evaluation on OMT Benchmark Sets

In this experimental evaluation we use OMT formulas taken from well-known, publicly available, repositories. We characterize these benchmark-sets as follows:

**Table 1.** MINIZINC Challenge formulas. The columns list the total number of instances (inst.), of timeouts (timeout), of run-time errors (error), of unsupported problems (unsup.), of incorrectly solved instances (incor.), of correctly solved instances (correct), the total solving time for all solved instances (time), the number of instances solved in the shortest time within the same category (BT1) and those solved in the shortest time considering all tools (BT2).

| tool, configuration & encoding | inst. s | o | timeout s | o | error s | o | unsup. s | o | incor. s | o | correct s | o | time (s.) s | o | BT1 s | o | BT2 s | o |
|---|---|---|---|---|---|---|---|---|---|---|---|---|---|---|---|---|---|---|
| **MINIZINC Challenge 2016** | | | | | | | | | | | | | | | | | | |
| PICAT(CP) | 15 | 85 | 9 | 70 | 0 | 0 | 0 | 0 | 0 | 0 | 6 | 15 | 2281 | 6043 | 0 | 0 | 0 | 0 |
| G12(FD) | 15 | 85 | 4 | 71 | 1 | 3 | 0 | 0 | 0 | 0 | 10 | 11 | 4436 | 4220 | 0 | 0 | 0 | 0 |
| CHOCO() | 15 | 85 | 3 | 50 | 0 | 0 | 0 | 0 | 0 | 0 | 12 | 35 | 4256 | 11423 | 1 | 0 | 1 | 0 |
| IZPLUS() | 15 | 85 | 6 | 44 | 0 | 0 | 0 | 0 | 0 | 0 | 9 | 41 | 999 | 5492 | 3 | 4 | 3 | 4 |
| CHUFFED() | 15 | 85 | 2 | 40 | 0 | 0 | 5 | 0 | 0 | 0 | 8 | 45 | 635 | 4187 | 0 | 5 | 0 | 5 |
| JACOP() | 15 | 85 | 3 | 39 | 0 | 0 | 0 | 0 | 0 | 0 | 12 | 46 | 3411 | 12825 | 0 | 0 | 0 | 0 |
| GUROBI() | 15 | 85 | 6 | 22 | 0 | 0 | 0 | 0 | 0 | 0 | 9 | 63 | 2346 | 3037 | 0 | 15 | 0 | 15 |
| HAIFACSP() | 15 | 85 | 4 | 23 | 0 | 0 | 0 | 0 | 0 | 0 | 11 | 62 | 591 | 4444 | 0 | 11 | 0 | 11 |
| PICAT(SAT) | 15 | 85 | 1 | 26 | 0 | 0 | 0 | 0 | 0 | 0 | **14** | 59 | 151 | 7293 | 10 | 1 | 10 | 1 |
| OR-TOOLS(SAT) | 15 | 85 | 1 | 15 | 0 | 0 | 0 | 0 | 0 | 0 | **14** | 70 | 555 | 1338 | 1 | 45 | 1 | 45 |
| VIRTUAL BEST(MINIZINC) | 15 | 85 | 0 | 7 | 0 | 0 | 0 | 0 | 0 | 0 | 15 | 78 | 146 | 3514 | - | - | - | - |
| OPTIMATHSAT(INT) | 15 | 85 | 10 | 38 | 0 | 0 | 0 | 0 | 0 | 0 | 5 | 47 | 604 | 4856 | 1 | 20 | 0 | 0 |
| OPTIMATHSAT(BV) | 15 | 85 | 2 | 42 | 0 | 0 | 0 | 0 | 0 | 0 | **13** | 43 | 3664 | 8561 | 11 | 2 | 0 | 0 |
| BCLT(INT) | 15 | 85 | 10 | 33 | 0 | 0 | 0 | 0 | 0 | 0 | 5 | 52 | 1117 | 5998 | 0 | 15 | 0 | 2 |
| Z3(INT) | 15 | 85 | 10 | 32 | 0 | 0 | 0 | 0 | 0 | 0 | 5 | 53 | 676 | 10424 | 0 | 11 | 0 | 0 |
| Z3(BV) | 15 | 85 | 5 | 28 | 0 | 0 | 0 | 0 | 0 | 0 | 10 | **57** | 2938 | 11113 | 2 | 19 | 0 | 0 |
| VIRTUAL BEST(OMT) | 15 | 85 | 1 | 21 | 0 | 0 | 0 | 0 | 0 | 0 | 14 | 64 | 3842 | 6432 | - | - | - | - |
| VIRTUAL BEST(ALL) | 15 | 85 | 0 | 7 | 0 | 0 | 0 | 0 | 0 | 0 | 15 | 78 | 146 | 3514 | - | - | - | - |
| **MINIZINC Challenge 2019** | | | | | | | | | | | | | | | | | | |
| PICAT(CP) [2.7B12] | 10 | 90 | 8 | 67 | 0 | 11 | 0 | 5 | 0 | 0 | 2 | 7 | 54 | 1440 | 1 | 0 | 1 | 0 |
| IZPLUS() | 10 | 90 | 5 | 71 | 0 | 4 | 0 | 0 | 0 | 0 | 5 | 15 | 14 | 3077 | 1 | 3 | 1 | 3 |
| G12(FD) | 10 | 90 | 5 | 64 | 0 | 10 | 0 | 0 | 0 | 0 | 5 | 16 | 323 | 4010 | 0 | 0 | 0 | 0 |
| CHOCO(STD) | 10 | 90 | 4 | 63 | 0 | 5 | 0 | 0 | 0 | 0 | 6 | 22 | 415 | 4312 | 0 | 0 | 0 | 0 |
| GECODE() [6.2.0] | 10 | 90 | 4 | 63 | 0 | 0 | 0 | 0 | 0 | 0 | 6 | 27 | 420 | 5094 | 0 | 6 | 0 | 6 |
| JACOP() [4.8] | 10 | 90 | 4 | 55 | 0 | 6 | 0 | 0 | 0 | 0 | 6 | 29 | 260 | 5467 | 0 | 1 | 0 | 1 |
| HAIFACSP() | 10 | 90 | 0 | 47 | 0 | 10 | 0 | 0 | 0 | 2 | 8 | 31 | 2 | 6408 | 4 | 7 | 4 | 4 |
| CHUFFED() [NIGHTLY] | 10 | 90 | 0 | 43 | 0 | 0 | 5 | 10 | 0 | 0 | 5 | 37 | 1 | 4886 | 3 | 19 | 3 | 19 |
| GUROBI() [8.1.1] | 10 | 80 | 0 | 48 | 0 | 0 | 0 | 0 | 0 | 0 | **10** | 32 | 705 | 2895 | 2 | 6 | 2 | 4 |
| PICAT(SAT) [2.7B12] | 10 | 90 | 0 | 45 | 0 | 5 | 0 | 0 | 0 | 1 | **10** | 39 | 275 | 9894 | 0 | 7 | 0 | 5 |
| OR-TOOLS(SAT) [NIGHTLY] | 10 | 90 | 5 | 42 | 0 | 3 | 0 | 0 | 0 | 0 | 5 | **45** | 8 | 7239 | 0 | 13 | 0 | 11 |
| VIRTUAL BEST(MINIZINC) | 10 | 90 | 0 | 29 | 0 | 0 | 0 | 0 | 0 | 0 | 10 | 61 | 9 | 5247 | - | - | - | - |
| OPTIMATHSAT(INT) [1.6.4.1] | 10 | 90 | 5 | 62 | 0 | 0 | 0 | 0 | 0 | 0 | 5 | 28 | 4 | 3650 | 2 | 10 | 0 | 0 |
| OPTIMATHSAT(BV) [1.6.4.1] | 10 | 90 | 4 | 59 | 0 | 5 | 0 | 0 | 0 | 0 | 6 | 26 | 484 | 7271 | 0 | 1 | 0 | 0 |
| BCLT(INT) | 10 | 90 | 5 | 60 | 0 | 0 | 0 | 0 | 0 | 0 | 5 | 30 | 6 | 3369 | 0 | 6 | 0 | 5 |
| Z3(INT) | 10 | 90 | 5 | 64 | 0 | 0 | 0 | 0 | 0 | 0 | 5 | 26 | 4 | 5358 | 3 | 6 | 0 | 1 |
| Z3(BV) | 10 | 90 | 0 | 55 | 0 | 2 | 0 | 0 | 0 | 0 | **10** | 33 | 1629 | 7550 | 5 | 17 | 0 | 3 |
| VIRTUAL BEST(OMT) | 10 | 90 | 0 | 48 | 0 | 2 | 0 | 0 | 0 | 0 | 10 | 40 | 1624 | 5179 | - | - | - | - |
| VIRTUAL BEST(ALL) | 10 | 90 | 0 | 29 | 0 | 0 | 0 | 0 | 0 | 0 | 10 | 61 | 9 | 4919 | - | - | - | - |

- *SAL* [integers]: 66 SMT-based Bounded Model Checking and K-Induction parametric problems created with the SAL model checker [6];
- *SAL* [rationals]: as above, with problems on the rationals;
- *Symba* [rationals]: 2632 *bounded*[2] software verification instances derived from a set of C programs used in the Software Verification Competition of 2013 [36];
- *Jobshop and Strip Packing* [rationals]: 190 problems taken from [51,53];
- *Machine Learning* [rationals]: 510 OMTinstances generated with the PYLMT tool based on Machine Learning Modulo Theories [59].

---

[2] We discarded any *unbounded* instance in the original benchmark-set in [36].

The first benchmark-set is on the integers, whereas the other four are on the rationals. We stress the fact that all formulas contained in all benchmark-sets are *satisfiable*.

*Remark 2.* Although there exists a repository of multi-objective OMT formulas (e.g. [36,56]), we have chosen to not include these in our experimental evaluation. The reason for this is twofold. First, such comparison would likely be unfair wrt. CP tools because that the workaround for dealing with multi-independent OMT formulas described in Sect. 4 is not competitive with the integrated optimization schema provided by OMT solvers [36,56]. In fact, the experimental evidence in [36,56] collected on a group of OMT solvers indicates that the latter approach can be an order of magnitude faster than the former one. Second, the workaround for dealing with lexicographic-optimization is limited by the fact that MINISEARCH is not fully compatible with recent versions of MINIZINC, and it only works with a restricted set of tools.

We have used the OMT2MZN tool described in Sect. 4 to translate each OMT formula to the MINIZINC format. OMT2MZN is written in Python and it is built on top of PYSMT [30], a general-purpose Python library for solving SMT problems, and it is available at [4]. During this step, it has been necessary to impose a finite domain to any unconstrained SMT-LIB rational variable, because otherwise none of the MINIZINC solvers would have been able to deal with them. We have experimented with two different domains: the largest feasible domain for floating-point variables of 32 bits (i.e. $\pm 3.402823e + 38$) for the first two benchmark-sets, and the largest feasible domain for integer variables (i.e. $\pm 2^{31}$) for the last two.

We consider two OPTIMATHSAT configurations: OPTIMATHSAT(SMT), solving the original OMTformulas, and OPTIMATHSAT(FZN), executed on the generated MINIZINC instances. The benefits of this choice is two-fold. First, we can double-check the correctness of such encoding, by comparing the optimum models generated in the two cases. Second, we can verify whether there is any performance loss caused by the encoding of the formula.

Only four of the MINIZINC solvers listed in Sect. 5 support floating-point reasoning. This limited the number of tools that could be used with some OMT benchmark-sets. The running-time of each MINIZINC solver reported in these experiments (including OPTIMATHSAT(FZN)) is comprehensive of the time taken by the MZN2FZN compiler, because the latter can sometime solve the input formulas on its own. The overall timeout is set to 600 s.

Notice that the optimal solutions found by OPTIMATHSAT(SMT) have been previously independently verified with a third-party SMT tool as reported in previous publications [55–57].[3] Therefore, we verify the correctness of the results found by any other configuration by comparing them with those found by OPTIMATHSAT(SMT), and otherwise mark the result as "unverified".

*Experimental Results over the Integers.* In this experiment, we evaluate the *SAL (over integers)* benchmark-set. The results are collected in Table 2.

---

[3] For every OMT problem $\langle \varphi, \text{obj} \rangle$ s.t. OPTIMATHSAT(SMT) returns a minimum value $min$ for obj on the formula $\varphi$, we say $min$ is correct iff $\varphi \wedge (\text{obj} = min)$ is satisfiable and $\varphi \wedge (\text{obj} < min)$ is unsatisfiable. (Dual for maximization.).

**Table 2.** SAL over integers. A SAT result is marked as *correct* when the objective value matches the reference solution provided by OPTIMATHSAT(SMT) (when run without a timeout), as *incorrect* otherwise.

| tool & configuration | inst. | timeout | tool-er. | unsupp. | incor. | correct | terminated tot. time (s.) | avg. time (s.) | med. time (s.) |
|---|---|---|---|---|---|---|---|---|---|
| GUROBI() | 66 | 0 | 0 | 66 | 0 | 0 | 0 | 0.00 | 0.00 |
| G12(FD) | 66 | 0 | 66 | 0 | 0 | 0 | 0 | 0.00 | 0.00 |
| IZPLUS() | 66 | 0 | 66 | 0 | 0 | 0 | 0 | 0.00 | 0.00 |
| JACOP() | 66 | 0 | 66 | 0 | 0 | 0 | 0 | 0.00 | 0.00 |
| CHUFFED() | 66 | 19 | 47 | 0 | 0 | 0 | 0 | 0.00 | 0.00 |
| OR-TOOLS(SAT) | 66 | 57 | 9 | 0 | 0 | 0 | 0 | 0.00 | 0.00 |
| CHOCO() | 66 | 66 | 0 | 0 | 0 | 0 | 0 | 0.00 | 0.00 |
| HAIFACSP() | 66 | 66 | 0 | 0 | 0 | 0 | 0 | 0.00 | 0.00 |
| PICAT(CP) | 66 | 66 | 0 | 0 | 0 | 0 | 0 | 0.00 | 0.00 |
| GECODE() | 66 | 66 | 0 | 0 | 0 | 0 | 0 | 0.00 | 0.00 |
| GUROBI(L) | 66 | 63 | 0 | 0 | 0 | 3 | 166 | 55.49 | 52.44 |
| PICAT(SAT) | 66 | 62 | 0 | 0 | 0 | 4 | 1667 | 416.85 | 467.09 |
| VIRTUAL BEST(MINIZINC) | 66 | 62 | 0 | 0 | 0 | 4 | 718 | 179.51 | 78.54 |
| OPTIMATHSAT(FZN) | 66 | 18 | 0 | 0 | 0 | 48 | 7113 | 148.20 | 70.52 |
| VIRTUAL BEST(FZN) | 66 | 18 | 0 | 0 | 0 | 48 | 7113 | 148.20 | 70.52 |
| OPTIMATHSAT(SMT) | 66 | 22 | 0 | 0 | 0 | 44 | 2657 | 60.41 | 18.72 |
| VIRTUAL BEST(ALL) | 66 | 16 | 0 | 0 | 0 | 50 | 5037 | 100.75 | 25.13 |

**Table 3.** OMT Problems defined over the rationals. A SAT result is marked as *correct* when the objective value matches the reference solution provided by OPTIMATHSAT(SMT) with an absolute error $\Delta < 10^{-6}$. A result is marked as *unverified* when we have no reference solution and *incorrect* if neither of the previous two conditions apply.

| tool & configuration | instances | timeout | tool-errors | terminated incorrect | verified | unverified | tot. time (s.) | avg. time (s.) | med. time (s.) | incorrect results unsat | $\Delta \leq 10^{-6}$ | $\Delta \leq 10^{-3}$ | $\Delta \leq 10^{-1}$ | $\Delta \leq 10^{0}$ | $\Delta \leq 10^{1}$ |
|---|---|---|---|---|---|---|---|---|---|---|---|---|---|---|---|
| *SAL, Symba, Jobshop and Strippacking* | | | | | | | | | | | | | | | |
| GECODE() | 2888 | 2733 | 0 | 0 | 155 | 0 | 10800 | 69.68 | 18.63 | 0 | 0 | 0 | 0 | 0 | 0 |
| G12(MIP) | 2888 | 10 | 0 | 2855 | 0 | 23 | 317 | 13.79 | 11.21 | 2765 | 90 | 90 | 86 | 39 | 0 |
| GUROBI() | 2888 | 48 | 0 | 2728 | 104 | 8 | 3961 | 35.37 | 2.14 | 2684 | 44 | 32 | 32 | 1 | 0 |
| VIRTUAL BEST(MINIZINC) | 2888 | 0 | 0 | 2628 | 237 | 23 | 13801 | 53.08 | 6.97 | - | - | - | - | - | - |
| OPTIMATHSAT(FZN) | 2888 | 31 | 0 | 0 | 2854 | 3 | 22320 | 7.81 | 0.40 | 0 | 0 | 0 | 0 | 0 | 0 |
| VIRTUAL BEST(FZN) | 2888 | 0 | 0 | 11 | 2854 | 23 | 20674 | 7.19 | 0.40 | - | - | - | - | - | - |
| OPTIMATHSAT(SMT) | 2888 | 23 | 0 | 0 | 2865 | 0 | 15676 | 5.47 | 0.08 | 0 | - | - | - | - | - |
| VIRTUAL BEST(ALL) | 2888 | 0 | 0 | 0 | 2865 | 23 | 15183 | 5.26 | 0.08 | - | - | - | - | - | - |
| *Machine Learning* | | | | | | | | | | | | | | | |
| GECODE() | 510 | 322 | 0 | 164 | 24 | 0 | 11 | 0.44 | 0.43 | 147 | 17 | 17 | 2 | 0 | 0 |
| G12(MIP) | 510 | 108 | 0 | 400 | 2 | 0 | 225 | 112.47 | 112.47 | 400 | 0 | 0 | 0 | 0 | 0 |
| GUROBI() | 510 | 9 | 0 | 472 | 28 | 1 | 201 | 6.92 | 3.17 | 468 | 4 | 4 | 2 | 0 | 0 |
| VIRTUAL BEST(MINIZINC) | 510 | 9 | 0 | 464 | 36 | 1 | 383 | 10.34 | 0.46 | - | - | - | - | - | - |
| OPTIMATHSAT(FZN) | 510 | 7 | 0 | 237 | 263 | 3 | 2797 | 10.52 | 2.21 | 177 | 60 | 59 | 0 | 0 | 0 |
| OPTIMATHSAT(FZN+E) | 510 | 92 | 0 | 0 | 415 | 3 | 1197 | 2.86 | 2.03 | 0 | 0 | 0 | 0 | 0 | 0 |
| VIRTUAL BEST(FZN) | 510 | 7 | 0 | 83 | 417 | 3 | 1366 | 3.25 | 2.03 | - | - | - | - | - | - |
| OPTIMATHSAT(SMT) | 510 | 10 | 0 | 0 | 500 | 0 | 5766 | 11.53 | 12.15 | 0 | - | - | - | - | - |
| VIRTUAL BEST(ALL) | 510 | 7 | 0 | 0 | 500 | 3 | 2290 | 4.55 | 2.05 | - | - | - | - | - | - |

We notice first that OPTIMATHSAT(FZN) always produces correct results and it shows comparable performances in terms on number of problems solved wrt. the baseline OPTIMATHSAT(SMT), solving even 4 problems more. (We conjecture that the latter fact should be attributed to the limited, but effective, deduction capabilities of the MZN2FZN compiler, that may have helped OPTIMATHSAT in solving the input formulas.) This suggests that, at least on problems on the integers, OMT2MZN is efficient and effective and does not affect correctness.

In general, MINIZINC solvers do not seem to deal efficiently with this benchmark-set. Some tools have experienced some internal error (e.g. dumped-core, segmentation fault), some others have been killed to a high memory consumption (over 32GB), whereas the majority of the remaining tools had a timeout.

We explain this behavior with the fact that the given benchmark set is characterized by the presence of a heavy Boolean structure combined with arithmetical constraints, which requires the efficient combination of strong Boolean-reasoning capabilities (e.g., efficiently handling chains of unit propagations) with strong arithmetical-solving&optimization capabilities, which is a typical feature of OMT solvers.

None of the input formulas was initially supported by GUROBI. After restricting the bound of every integer variable to $\pm 10^6$, GUROBI(L) was able to solve 3 instances within the timeout. Among the MINIZINC solvers, the best result is obtained by PICAT(SAT), that solved 4 problems out of 66.

*Experimental Results Over the Rationals.* We consider first the first three benchmark-sets over the rationals: SAL over rationals, Symba, JobShop&Strip-Packing. (Separate tables for the four benchmarks are reported in the extended version of this paper [24].) Of all MINIZINC solvers we have tried, only three are able to deal with floating-point constraints. The results are shown in Table 3. Since each of the input formulas is satisfiable, we consider a result incorrect either when it is equal to UNSAT, or when the relative error $\Delta$ exceeds $10^{-6}$, s.t.: $\Delta \stackrel{\text{def}}{=} \frac{|o_{smt} - o_{fzn}|}{|o_{smt}|}$, $o_{smt}$ and $o_{fzn}$ being the optimal value found by OPTIMATHSAT(SMT) and the optimal value found by the MINIZINC solver under test respectively. (Recall that the former was previously checked to be correct.)

Similarly to the previous experiment on the integers, OPTIMATHSAT(FZN) always produces correct results, and display comparable performance wrt. OPTIMATH-SAT(OMT·) in terms of number of instances being solved, solving somewhat fewer problems. This is not the case of the other three MINIZINC solvers. Among these, GECODE experienced a timeout on the majority of the formulas being considered, G12(MIP) returned mostly incorrect answers, whereas GUROBI seems to have the best performance, in particular on the third benchmark-set.

We attribute the large number of incorrect results returned by all three MINIZINC solvers to the fact that these tools use finite-precision floating-point arithmetic internally. The incorrect behavior of some of these solvers (e.g. GUROBI) can also be partially explained with the large domain of floating-point variables in these problems. However, given the nature of these input instances, it was not possible for us to assign a smaller domain to each variable in the problem *a priori*.

We analyze separately the results for the last benchmark-set reported in Table 3. The peculiar aspect of the Machine Learning benchmark-set [59] is that it is characterized by Pseudo-Boolean sums over rational weights, and by very fine-grained rational values[4].

---

[4] For example, $\frac{1799972218749879}{2251799813685248}$ is a sample weight value from problems in [59].

Unfortunately, these fine-grained rational values are rounded by the standard MZN2FZN compiler, which causes the incorrect results even of OPTIMATHSAT(FZN) in Table 3, despite the fact that OPTIMATHSAT uses infinite-precision arithmetic.

In order to overcome this issue, we leverage the EMZN2FZN compiler described in Sect. 3 so that the original fractional values are preserved in the resulting FLATZ-INCmodel, and show that with this approach OPTIMATHSAT does not produce incorrect results any longer (configuration OPTIMATHSAT(FZN+E) in Table 3), solving correctly 152 problems more than OPTIMATHSAT(FZN).

Overall, since there are at least 237 formulas affected by the above issue with the MZN2FZN compiler, we avoid an in-depth discussion of the results obtained by the other MINIZINC solvers. However, at a first glance the situation does not seem to differ from the other benchmark sets over the rationals.

### 5.3   Discussion

On the whole, from our experiments, OMT tools appear to be still at some disadvantage when dealing with MINIZINC problems wrt. specific tools, and vice versa.

On the one hand, OMT solvers seem to be penalized by their lack of efficient ad hoc decision procedures for dealing with global constraints. Moreover, the approach taken by the MZN2FZN compiler, that creates lots of alias Boolean, integer and floating-point variables for dealing with Pseudo-Boolean constraints, is particularly challenging to deal with efficiently by an OMTsolver.

On the other hand, MINIZINC solvers seem to suffer with problems needing an arithmetic-reasoning component combined with heavy Boolean-reasoning component. Even more importantly, the lack of infinite-precision linear arithmetic procedures causes a number of incorrect results when dealing with OMT problems over the rationals. Both of these points need to be addressed in order to deal with the vast number of Formal Verification and Model Checking applications in the SMT/OMT domain.

## 6   Conclusions and Future Work

In this paper we have taken a first step forward towards bridging the MINIZINC and the OMT communities. The ultimate goal is to obtain a correct, effective and efficient fully-automated system for translating problems from one community to the other, so as to extend the application domain of both communities. With our experimental evaluation, we have identified some criticalities that need to be addressed by each community in order to solidify this union.

We plan to push this investigation forward as follows. In the short term, we plan to address the inefficient handling of Pseudo-Boolean constraints over the rationals revealed by the experimental evaluation in Sect. 5.2. In order to deal with those FLATZ-INC constraints that require non-linear arithmetic, we envisage an opportunity to either extend OPTIMATHSAT with proper handling of the non-linear arithmetic theory [20] or to experiment with an encoding based on the floating-point theory [60]. This objective goes hand in hand with the extension of OMT2MZN to deal with other SMT theories. In the long term, OMT solving may also benefit from adopting efficient ad hoc decision procedures for frequently used global constraints. Finally, we plan to broaden the scope of our investigation and include other OMT solvers in our study.

# References

1. FZN2OMT. https://github.com/PatrickTrentin88/fzn2omt
2. Benchmarks, Tools and Data. http://disi.unitn.it/trentin/resources/cpaior2020.tar.xz
3. MathSAT 5. http://mathsat.fbk.eu/
4. OMT2MZN. https://github.com/cespio/omt2mzn
5. EMZN2FZN Repository. https://github.com/PatrickTrentin88/emzn2fzn
6. Sal, symbolic analysis laboratory. http://sal.csl.sri.com
7. SMT-LIB Format for Finite Lists, Sets and Maps. https://www.cprover.org/SMT-LIB-LSM/
8. Ansótegui, C., Bofill, M., Palahí, M., Suy, J., Villaret, M.: Solving weighted CSPs with meta-constraints by reformulation into satisfiability modulo theories. Constraints 18(2), 236–268 (2013). https://doi.org/10.1007/s10601-012-9131-1
9. Araújo, R., Bessa, I., Cordeiro, L.C., Filho, J.E.C.: SMT-based verification applied to non-convex optimization problems. In: 2016 VI Brazilian Symposium on Computing Systems Engineering (SBESC), November 2016
10. Asín, R., Nieuwenhuis, R., Oliveras, A., Rodríguez-Carbonell, E.: Cardinality networks: a theoretical and empirical study. Constraints 16(2), 195–221 (2011). https://doi.org/10.1007/s10601-010-9105-0
11. Barrett, C., Ranise, S., Stump, A., Tinelli, C.: The satisfiability modulo theories library (SMT-LIB) (2010). http://www.smtlib.org
12. Barrett, C., Sebastiani, R., Seshia, S.A., Tinelli, C.: Satisfiability Modulo Theories, vol. 185, chap. 26, pp. 825–885. IOS Press, February 2009
13. Bertolissi, C., dos Santos, D.R., Ranise, S.: Solving multi-objective workflow satisfiability problems with optimization modulo theories techniques. In: SACMAT. ACM (2018)
14. Bian, Z., Chudak, F., Macready, W., Roy, A., Sebastiani, R., Varotti, S.: Solving SAT and MaxSAT with a quantum annealer: foundations and a preliminary report. In: Dixon, C., Finger, M. (eds.) FroCoS 2017. LNCS (LNAI), vol. 10483, pp. 153–171. Springer, Cham (2017). https://doi.org/10.1007/978-3-319-66167-4_9
15. Bjørner, N., Phan, A.-D., Fleckenstein, L.: vZ - an optimizing SMT solver. In: Baier, C., Tinelli, C. (eds.) TACAS 2015. LNCS, vol. 9035, pp. 194–199. Springer, Heidelberg (2015). https://doi.org/10.1007/978-3-662-46681-0_14
16. Bofill, M., Nieuwenhuis, R., Oliveras, A., Rodríguez-Carbonell, E., Rubio, A.: The barcelogic SMT solver. In: Gupta, A., Malik, S. (eds.) CAV 2008. LNCS, vol. 5123, pp. 294–298. Springer, Heidelberg (2008). https://doi.org/10.1007/978-3-540-70545-1_27
17. Bofill, M., Palahí, M., Suy, J., Villaret, M.: Solving constraint satisfaction problems with SAT modulo theories. Constraints 17(3), 273–303 (2012). https://doi.org/10.1007/s10601-012-9123-1
18. Bofill, M., Suy, J., Villaret, M.: A system for solving constraint satisfaction problems with SMT. In: Strichman, O., Szeider, S. (eds.) SAT 2010. LNCS, vol. 6175, pp. 300–305. Springer, Heidelberg (2010). https://doi.org/10.1007/978-3-642-14186-7_25
19. Candeago, L., Larraz, D., Oliveras, A., Rodríguez-Carbonell, E., Rubio, A.: Speeding up the constraint-based method in difference logic. In: Creignou, N., Le Berre, D. (eds.) SAT 2016. LNCS, vol. 9710, pp. 284–301. Springer, Cham (2016). https://doi.org/10.1007/978-3-319-40970-2_18
20. Cimatti, A., Griggio, A., Irfan, A., Roveri, M., Sebastiani, R.: Incremental linearization for satisfiability and verification modulo nonlinear arithmetic and transcendental functions. ACM Trans. Comput. Logic 19(3), 1–52 (2018)

21. Cimatti, A., Griggio, A., Schaafsma, B.J., Sebastiani, R.: The MathSAT5 SMT solver. In: Piterman, N., Smolka, S.A. (eds.) TACAS 2013. LNCS, vol. 7795, pp. 93–107. Springer, Heidelberg (2013). https://doi.org/10.1007/978-3-642-36742-7_7

22. Collavizza, H., Rueher, M.: Exploration of the capabilities of constraint programming for software verification. In: Hermanns, H., Palsberg, J. (eds.) TACAS 2006. LNCS, vol. 3920, pp. 182–196. Springer, Heidelberg (2006). https://doi.org/10.1007/11691372_12

23. Collavizza, H., Rueher, M., Van Hentenryck, P.: CPBPV: a constraint-programming framework for bounded program verification. Constraints 15(2), 238–264 (2010). https://doi.org/10.1007/s10601-009-9089-9

24. Contaldo, F., Trentin, P., Sebastiani, R.: From minizinc to optimization modulo theories, and back (extended version) (2019). https://arxiv.org/abs/1912.01476

25. SMTURL: SmtLibav2. www.smtlib.cs.uiowa.edu/

26. Elgabou, H.Λ., Frisch, A.M.: Encoding the lexicographic ordering constraint in SAT modulo theories. In: Proceedings of Thirteenth International Workshop on Constraint Modelling and Reformulation, September 2014

27. Feydy, T., Stuckey, P.J.: Lazy clause generation reengineered. In: Gent, I.P. (ed.) CP 2009. LNCS, vol. 5732, pp. 352–366. Springer, Heidelberg (2009). https://doi.org/10.1007/978-3-642-04244-7_29

28. Frisch, A.M., Giannaros, P.A.: SAT Encodings of the At-Most-k Constraint Some Old, Some New, Some Fast, Some Slow (2010)

29. Frisch, A.M., Palahí, M.: Anomalies in SMT solving: difficulties in modelling combinatorial problems. In: Proceedings of Thirteenth International Workshop on Constraint Modelling and Reformulation, September 2014

30. Gario, M., Micheli, A.: PySMT: a solver-agnostic library for fast prototyping of SMT-based algorithms. In: SMT Workshop 2015, 13th International Workshop on Satisfiability Modulo Theories (2015)

31. Grinchtein, O., Carlsson, M., Pearson, J.: A constraint optimisation model for analysis of telecommunication protocol logs. In: Blanchette, J.C., Kosmatov, N. (eds.) TAP 2015. LNCS, vol. 9154, pp. 137–154. Springer, Cham (2015). https://doi.org/10.1007/978-3-319-21215-9_9

32. Karpenkov, G.E.: Finding inductive invariants using satisfiability modulo theories and convex optimization. Theses, Université Grenoble Alpes (2017)

33. Kovásznai, G., Erdélyi, B., Biró, C.: Investigations of graph properties in terms of wireless sensor network optimization. In: 2018 IEEE International Conference on Future IoT Technologies (Future IoT), January 2018

34. Larraz, D., Oliveras, A., Rodríguez-Carbonell, E., Rubio, A.: Minimal-model-guided approaches to solving polynomial constraints and extensions. In: Sinz, C., Egly, U. (eds.) SAT 2014. LNCS, vol. 8561, pp. 333–350. Springer, Cham (2014). https://doi.org/10.1007/978-3-319-09284-3_25

35. Leofante, F., Abraham, E., Niemueller, T., Lakemeyer, G., Tacchella, A.: Integrated synthesis and execution of optimal plans for multi-robot systems in logistics. Inf. Syst. Front. 21(1), 87–107 (2018)

36. Li, Y., Albarghouthi, A., Kincad, Z., Gurfinkel, A., Chechik, M.: Symbolic optimization with SMT solvers. In: POPL (2014)

37. Liu, T., Tyszberowicz, S.S., Beckert, B., Taghdiri, M.: Computing exact loop bounds for bounded program verification. In: Larsen, K.G., Sokolsky, O., Wang, J. (eds.) SETTA 2017. LNCS, vol. 10606, pp. 147–163. Springer, Cham (2017). https://doi.org/10.1007/978-3-319-69483-2_9

38. MiniZinc. www.minizinc.org

39. Nadel, A., Ryvchin, V.: Bit-vector optimization. In: Chechik, M., Raskin, J.-F. (eds.) TACAS 2016. LNCS, vol. 9636, pp. 851–867. Springer, Heidelberg (2016). https://doi.org/10.1007/978-3-662-49674-9_53

40. Nethercote, N., Stuckey, P.J., Becket, R., Brand, S., Duck, G.J., Tack, G.: MiniZinc: towards a standard CP modelling language. In: Bessière, C. (ed.) CP 2007. LNCS, vol. 4741, pp. 529–543. Springer, Heidelberg (2007). https://doi.org/10.1007/978-3-540-74970-7_38

41. Nguyen, C.M., Sebastiani, R., Giorgini, P., Mylopoulos, J.: Multi-objective reasoning with constrained goal models. Requirements Eng. **23**(2), 189–225 (2016). https://doi.org/10.1007/s00766-016-0263-5

42. Nieuwenhuis, R., Oliveras, A.: On SAT modulo theories and optimization problems. In: Biere, A., Gomes, C.P. (eds.) SAT 2006. LNCS, vol. 4121, pp. 156–169. Springer, Heidelberg (2006). https://doi.org/10.1007/11814948_18

43. Nieuwenhuis, R., Oliveras, A., Rodríguez-Carbonell, E., Rubio, A.: Challenges in satisfiability modulo theories. In: Baader, F. (ed.) RTA 2007. LNCS, vol. 4533, pp. 2–18. Springer, Heidelberg (2007). https://doi.org/10.1007/978-3-540-73449-9_2

44. Nieuwenhuis, R., Oliveras, A., Tinelli, C.: Solving SAT and SAT modulo theories: from an abstract Davis-Putnam-Logemann-Loveland procedure to DPLL(T). J. ACM **53**(6), 937–977 (2006)

45. Ohrimenko, O., Stuckey, P.J., Codish, M.: Propagation via lazy clause generation. Constraints **14**(3), 357–391 (2009). https://doi.org/10.1007/s10601-008-9064-x

46. Oliver, R.S., Craciunas, S.S., Steiner, W.: IEEE 802.1Qbv gate control list synthesis using array theory encoding. In: 2018 IEEE Real-Time and Embedded Technology and Applications Symposium (RTAS), April 2018

47. Papadimitriou, C.H.: On the complexity of integer programming. J. ACM **28**(4), 765–768 (1981)

48. Ratschan, S.: Simulation based computation of certificates for safety of dynamical systems. In: Abate, A., Geeraerts, G. (eds.) FORMATS 2017. LNCS, vol. 10419, pp. 303–317. Springer, Cham (2017). https://doi.org/10.1007/978-3-319-65765-3_17

49. Rendl, A., Guns, T., Stuckey, P.J., Tack, G.: MiniSearch: a solver-independent meta-search language for MiniZinc. In: Pesant, G. (ed.) CP 2015. LNCS, vol. 9255, pp. 376–392. Springer, Cham (2015). https://doi.org/10.1007/978-3-319-23219-5_27

50. Roselli, S.F., Bengtsson, K., Åkesson, K.: SMT solvers for job-shop scheduling problems: models comparison and performance evaluation. In: 2018 IEEE 14th International Conference on Automation Science and Engineering (CASE), August 2018

51. Sawaya, N.W., Grossmann, I.E.: A cutting plane method for solving linear generalized disjunctive programming problems. Comput. Chem. Eng. **29**(9), 1891–1913 (2005)

52. Sebastiani, R.: Lazy satisfiability modulo theories. J. Satisf. Boolean Model. Comput. JSAT **3**(3–4), 141–224 (2007)

53. Sebastiani, R., Tomasi, S.: Optimization in SMT with $\mathcal{L}A(\mathbb{Q})$ cost functions. In: Gramlich, B., Miller, D., Sattler, U. (eds.) IJCAR 2012. LNCS (LNAI), vol. 7364, pp. 484–498. Springer, Heidelberg (2012). https://doi.org/10.1007/978-3-642-31365-3_38

54. Sebastiani, R., Tomasi, S.: Optimization modulo theories with linear rational costs. ACM Trans. Comput. Logics **16**(2), 1–43 (2015)

55. Sebastiani, R., Trentin, P.: OptiMathSAT: a tool for optimization modulo theories. In: Kroening, D., Păsăreanu, C.S. (eds.) CAV 2015. LNCS, vol. 9206, pp. 447–454. Springer, Cham (2015). https://doi.org/10.1007/978-3-319-21690-4_27

56. Sebastiani, R., Trentin, P.: Pushing the envelope of optimization modulo theories with linear-arithmetic cost functions. In: Baier, C., Tinelli, C. (eds.) TACAS 2015. LNCS, vol. 9035, pp. 335–349. Springer, Heidelberg (2015). https://doi.org/10.1007/978-3-662-46681-0_27

57. Sebastiani, R., Trentin, P.: On optimization modulo theories, MaxSMT and sorting networks. In: Legay, A., Margaria, T. (eds.) TACAS 2017. LNCS, vol. 10206, pp. 231–248. Springer, Heidelberg (2017). https://doi.org/10.1007/978-3-662-54580-5_14

58. Sebastiani, R., Trentin, P.: OptiMathSAT: a tool for optimization modulo theories. J. Autom. Reason. **64**(3), 423–460 (2018)

59. Teso, S., Sebastiani, R., Passerini, A.: Structured learning modulo theories. Artif. Intell. **244**, 166–187 (2017)

60. Trentin, P., Sebastiani, R.: Optimization modulo the theory of floating-point numbers. In: Fontaine, P. (ed.) CADE 2019. LNCS (LNAI), vol. 11716, pp. 550–567. Springer, Cham (2019). https://doi.org/10.1007/978-3-030-29436-6_33

61. Veksler, M., Strichman, O.: Learning general constraints in CSP. Artif. Intell. **238**, 135–153 (2016)

62. Zhou, N.-F., Kjellerstrand, H.: Optimizing SAT encodings for arithmetic constraints. In: Beck, J.C. (ed.) CP 2017. LNCS, vol. 10416, pp. 671–686. Springer, Cham (2017). https://doi.org/10.1007/978-3-319-66158-2_43

# Transfer-Expanded Graphs
# for On-Demand Multimodal Transit
# Systems

Kevin Dalmeijer[(✉)] [ID] and Pascal Van Hentenryck [ID]

Georgia Institute of Technology, Atlanta, GA 30332, USA
dalmeijer@gatech.edu, pvh@isye.gatech.edu

**Abstract.** This paper considers a generalization of the network design problem for On-Demand Multimodal Transit Systems (ODMTS). An ODMTS consists of a selection of hubs served by high frequency buses, and passengers are connected to the hubs by on-demand shuttles which serve the first and last miles. This paper generalizes prior work by including three additional elements that are critical in practice. First, different frequencies are allowed throughout the network. Second, additional modes of transit (e.g., rail) are included. Third, a limit on the number of transfers per passenger is introduced. Adding a constraint to limit the number of transfers has a significant negative impact on existing Benders decomposition approaches as it introduces non-convexity in the subproblem. Instead, this paper enforces the limit through transfer-expanded graphs, i.e., layered graphs in which each layer corresponds to a certain number of transfers. A real-world case study is presented for which the generalized ODMTS design problem is solved for the city of Atlanta. The results demonstrate that exploiting the problem structure with transfer-expanded graphs results in significant computational improvements.

**Keywords:** Combinatorial optimization · Multimodal transportation · Benders decomposition · Transfer-expanded graphs

## 1 Introduction

This paper is motivated by the design and implementation of an On-Demand Multimodal Transit System (ODMTS) for the city of Atlanta. The share of public transit in Atlanta (about 2–3%) is very low compared to other American cities (e.g., about 15% in Boston) and Atlanta is also the 8th most congested city in the world. There is thus a strong need for a modern transit systems that leverages the train and bus infrastructure of the city and complements it with innovative mobility concepts.

This paper considers the design of an ODMTS for Atlanta that combines a network of trains and buses with on-demand multimodal shuttles that act as feeders to/from the bus/rail network and serve local demand. ODMTS address

© Springer Nature Switzerland AG 2020
E. Hebrard and N. Musliu (Eds.): CPAIOR 2020, LNCS 12296, pp. 167–175, 2020.
https://doi.org/10.1007/978-3-030-58942-4_11

the first/last mile problem that plagues transit systems, while mitigating congestion on high-density corridors and leveraging economy of scale. ODMTS and their design challenge was introduced in [11], which also presents an overview of related work. The main contribution of this paper is to generalize prior work by including three additional elements that are critical for ODMTS in large cities such as Atlanta. First, different frequencies are allowed throughout the network. Second, additional modes of transit (e.g., rail) are included. Third, a limit on the number of transfers per passenger is introduced. Adding a constraint to limit the number of transfers has a significant negative impact on existing Benders decomposition approaches as it introduces non-convexity in the subproblem. Instead, this paper enforces the limit through transfer-expanded graphs, i.e., layered graphs in which each layer corresponds to a certain number of transfers. A real-world case study is presented for which the generalized ODMTS design problem is solved for the city of Atlanta, which has the 8th largest transit system in the US by ridership. The results demonstrate that exploiting the problem structure through transfer-expanded graphs results in significant computational improvements.

## 2   The Generalized ODMTS Design Problem

This section presents the generalized ODMTS design problem that enhances the model from [11] along several dimensions: The choice of bus frequencies, additional transportation modes and, most importantly, a constraint on the number of transfers. The Benders decomposition approach in [11] exploits a natural decomposition of the ODMTS design problem. The network design is determined by the *master problem*, while the routing of the passengers for a given design is determined by the *subproblem*. A major benefit of this decomposition is that the subproblem can be solved for each trip independently. The same decomposition is used in this paper.

### 2.1   The Master Problem for Network Design

Consider a directed multigraph $G = (V, A)$, with vertices $V = \{1, \ldots, n\}$ and arc set $A$. Let $F$ be the set of possible frequencies, i.e., the total number of vehicles during the time horizon, let $M$ be the set of possible transportation modes, which may include shuttles, and let $K$ be the total number of arcs that each passenger may travel. By definition, $K$ is equal to the maximum number of transfers plus one. In the multigraph $G$, each arc $a \in A$ is uniquely defined by the quadruple $a = (i, j, m, f) \in V \times V \times M \times F$, $i \neq j$. Using arc $a$ means traveling from $i$ to $j$ with mode $m$, which departs with frequency $f$. For a given arc $a \in A$, these elements are referred to as $i(a)$, $j(a)$, $m(a)$, and $f(a)$, respectively.

Designing a generalized ODMTS amounts to deciding which arcs $a \in A$ are made available to passengers. Let the binary variable $z_a \in \mathbb{B}$ be equal to one if arc $a$ is made available, and zero otherwise. The cost of enabling arc $a$ is given by the parameter $\beta_a$. It is assumed that $\beta_a \geq 0$ for all $a \in A$.

For a given design, a cost is incurred due to passengers traveling trough the network. This cost $\Phi(z)$ is a function of the values of the $z$-variables that define the design. The value of $\Phi(z)$ can be found by solving the subproblem, which is discussed in Sect. 2.2. If the subproblem is not feasible, then $\Phi(z) = \infty$.

A formulation for the master problem is presented in Fig. 1. For convenience, $\delta^+(i)$ is defined as the set of all arcs going out of $i \in V$. Similarly, the set $\delta^+(i, m)$ is defined as the set of all arcs with mode $m \in M$ going out of $i \in V$. The sets $\delta^-(i)$ and $\delta^-(i, m)$ are defined analogously for the incoming arcs.

$$
\min \quad \sum_{a \in A} \beta_a z_a + \Phi(z), \tag{1a}
$$

$$
\text{s.t.} \quad \sum_{a \in \delta^+(i,m)} f(a) z_a - \sum_{a \in \delta^-(i,m)} f(a) z_a = 0 \quad \forall i \in V, m \in M, \tag{1b}
$$

$$
\sum_{f \in F | (i,j,m,f) \in A} z_{(i,j,m,f)} \leq 1 \quad \forall i \in V, j \in V, m \in M, \tag{1c}
$$

$$
z_a \in \mathbb{B} \quad \forall a \in A. \tag{1d}
$$

**Fig. 1.** Formulation for the generalized ODMTS design problem.

Objective (1a) minimizes the cost of the design plus the cost of routing the passengers through the network. Constraints (1b) ensure that the frequencies for each mode are balanced at each vertex. For example, if three buses arrive during the time horizon, then three buses should also depart. Constraints (1c) enforce that only one frequency can be selected for a given connection and a given mode. Equations (1d) state the integrality requirements.

## 2.2   The Subproblem: Routing Passengers Through the Network

For a given design, the passenger trips are routed through the network at minimum cost. Let $T$ be the set of all passenger trips, and let each trip $r \in T$ be defined by an origin $o(r)$, a destination $d(r)$, and a number of passengers $p(r)$. If trip $r \in T$ is routed through arc $a \in A$, then a cost of $\gamma_a^r$ is incurred. The total cost of routing all passenger trips, $\Phi(z)$, is the sum over the costs per trip. It is assumed that $\gamma_a^r > 0$ for every arc $a \in A$ and trip $r \in T$, such that the optimal route is a simple path from $o(r)$ to $d(r)$.

Solving the subproblem amounts to solving a shortest path problem from $o(r)$ to $d(r)$ for each trip $r \in T$, with the additional restriction that the number of arcs in the path is at most $K$. This problem is known as the cardinality-constrained shortest path problem (CSP) [6]. Note that the cardinality constraint follows from the limit on the number of transfers. Without this limit, the subproblem is an (unconstrained) shortest path problem (SP), as is the case in [11].

It is well-known that SP possesses total unimodularity and can be solved by linear programming (LP). Adding an additional constraint, however, typically destroys this structure [1]. This is indeed the case when a cardinality constraint is added to the subproblem formulation in [11]. As a result, the cost function $\Phi(z)$ would change from convex to non-convex, which negatively impacts Benders decomposition approaches (see Sect. 3).

*To remedy this limitation, this paper presents a new formulation for the subproblem that enforces the transfer limit without destroying total unimodularity.* This formulation uses transfer-expanded graphs, i.e., layered graphs for which each layer corresponds to a number of transfers. Transfer-expanded graphs encode the transit constraints directly, making it possible to use shortest-path algorithms.

## 2.3  Transfer-Expanded Graphs

Transfer-expanded graphs share some similarities with time-expanded networks, where each vertex has multiple copies for different periods of time. This is the case, for example, for modern algorithms for evacuation planning and scheduling [9,13,14]. Reference [12] also uses a layered network to solve the dynamic generalized assignment problem. As a result, some of the side-constraints do not need to be handled explicitly. See [3] for a recent literature review on time-expanded graphs. Reference [8] discusses more general layered graph approaches.

Let $\bar{G}^r = (\bar{V}^r, \bar{A}^r)$ be the transfer-expanded graph for a given trip $r \in T$. This graph contains multiple copies of the original arcs and vertices, organized in $K+1$ layers. It is assumed that $K \geq 2$, as the subproblem is trivial for $K = 1$. A vertex $\bar{v} = (i, k) \in \bar{V}^r$ in the transfer-expanded graph is defined by a vertex $i \in V$ in the original graph and by a layer $k \in \{1, \ldots, K+1\}$. Similarly, the definition of an arc is extended to $\bar{a} = (a, k, l)$, in which $a \in A$ is the original arc, $k \in \{1, \ldots, K\}$ is the layer of the starting vertex of $\bar{a}$ and $l \in \{2, \ldots, K+1\}$ is the layer of the ending vertex.

The transfer-expanded graph is constructed as follows. For convenience, Fig. 2 provides an example for $K = 3$. First, the vertex set $\bar{V}^r$ is defined. For the origin and the destination of the trip, introduce the vertices $(o(r), 1)$ and $(d(r), K+1)$. For the other vertices $i \in V \backslash \{o(r), d(r)\}$ of the original graph, add the copies $(i, k)$ for $k \in \{2, \ldots, K\}$ to the transfer-expanded graph. The arc set $\bar{A}^r$ is constructed based on the arcs of the original graph, as follows:

1. For each arc starting in the origin, i.e., $a \in \delta^+(o(r))$, add the arc $(a, 1, 2)$ if $j(a) \neq d(r)$, or the arc $(a, 1, K+1)$ if $j(a) = d(r)$.
2. For each arc not adjacent to the origin or the destination, i.e., $a \in A$ and $i(a), j(a) \notin \{o(r), d(r)\}$, add the arcs $(a, k, k+1)$ for all $k \in \{2, \ldots, K-1\}$.
3. For each arc ending in the destination that does not start in the origin, i.e., $a \in \delta^-(d(r))$, $i(a) \neq o(r)$, add the arcs $(a, k, K+1)$ for all $k \in \{2, \ldots, K\}$.

By construction, it follows that solving CSP on the original graph is equivalent to solving SP on the transfer-expanded graph. Figure 3 formulates the subproblem as a collection of SPs on transfer-expanded graphs. Let $y_{\bar{a}}^r \in \mathbb{B}$ be the flow on arc $\bar{a} \in \bar{A}^r$ of trip $r \in T$. For convenience, define $\bar{\delta}_r^+(\bar{v})$ to be the set of all arcs in $\bar{A}^r$ coming out of $\bar{v} \in \bar{V}^r$. Similarly, let $\bar{\delta}_r^-(\bar{v})$ be the set of incoming arcs.

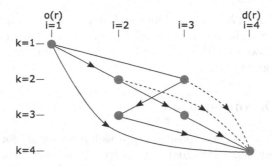

**Fig. 2.** A transfer-expanded graph for $K = 3$, $|M| = 1$, $|F| = 1$, for a complete graph as the original graph. The dotted arcs can potentially be removed (discussed below).

$$\Phi(z) = \min \sum_{r \in T} \sum_{\bar{a}=(a,k,l) \in \bar{A}^r} \gamma_{\bar{a}}^r y_{\bar{a}}^r, \tag{2a}$$

$$\text{s.t.} \quad y_{\bar{a}}^r \leq z_a \qquad \forall r \in T, \bar{a} = (a,k,l) \in \bar{A}^r, \tag{2b}$$

$$\sum_{\bar{a} \in \bar{\delta}_r^+(\bar{v})} y_{\bar{a}}^r - \sum_{\bar{a} \in \bar{\delta}_r^-(\bar{v})} y_{\bar{a}}^r = \begin{cases} 1 & \text{if } \bar{v} = (o(r), 1) \\ -1 & \text{if } \bar{v} = (d(r), K+1) \quad \forall r \in T, \bar{v} \in \bar{V}^r, \\ 0 & \text{else} \end{cases} \tag{2c}$$

$$y_{\bar{a}}^r \geq 0 \qquad \forall r \in T, \bar{a} \in \bar{A}^r. \tag{2d}$$

**Fig. 3.** Formulation for the subproblem on transfer-expanded graphs.

Objective (2a) minimizes the cost of all trips. Constraints (2b) state that passengers can only use arcs available in the design. Constraints (2c) enforce flow conservation, and Equations (2d) define the variables. Due to total unimodularity of the SPs, no integrality conditions are required.

The main advantage of using tranfer-expanded graphs is that the limit on the number of transfers can be enforced without destroying total unimodularity. A potential downside is that the number of variables and constraints in the subproblem increases linearly with $K$. In public transit, however, the number of transfers that passengers are willing to take, and therefore the value of $K$, is typically very low. Furthermore, a larger subproblem does not necessarily mean that the subproblem is more difficult to solve, as algorithms may benefit from the fact that the transfer-expanded graph is acyclic. When the $z$-variables are integers, for example, the acyclic subproblem for each trip can be solved in linear time through topological sorting [5].

Finally, it is worth pointing out that if $o(r)$ and $d(r)$ are only served by shuttles, and shuttles satisfy the triangle inequality, and a direct shuttle trip is possible, then some arcs may be removed from the transfer-expanded graph without sacrificing optimality. Specifically, using a shuttle on the path $(o(r), 1) \rightarrow (i, 2) \rightarrow (d(r), K+1)$ for $i \in V$ is always dominated by using a direct shuttle from $(o(r), 1)$ to $(d(r), K + 1)$. It follows that the shuttle arcs between $(i, 2)$ and $(d(r), K + 1)$ may be removed for all $i \in V$, as also indicated in Fig. 2. For $K \leq 3$, it then follows that the transfer-expanded graph does not require more edges than the original graph.

## 3    Benders Decomposition

Following [11], a Benders decomposition approach is presented for the generalized ODMTS design problem. The goal is to solve the master problem (1), which is complicated by the fact that $\Phi(z)$ is defined implicitly. To apply Benders decomposition, replace $\Phi(z)$ in Objective (1a) by a new variable $\theta \in \mathbb{R}$, and add the constraint $\theta \geq \Phi(z)$. Note that this does not change the problem, as $\theta = \Phi(z)$ in any optimal solution. In Benders decomposition, the constraint $\theta \geq \Phi(z)$ is enforced through *Benders cuts*. For subproblem (2), these cuts are

$$\theta \geq \Phi(\bar{z}) + \sum_{r \in T} \sum_{a \in A} \sum_{k=1}^{K} \lambda_a^{rk}(\bar{z})(z_a - \bar{z}_a), \tag{3}$$

with $\lambda_a^{rk}(z)$ the dual values of Constraints (2b) and $\bar{z}$ any feasible solution to the LP relaxation of the master problem [2]. For the case study in this paper, the subproblem is always feasible. If this assumption is not satisfied, *Benders feasibility cuts*, which are similar to (3), may also be included [2].

The Benders decomposition approach is implemented in C++ and Gurobi 8.1.1. The master problem is the main model, and the Benders cuts (3) are separated in both the MIP solution callback (in case the $z$-variables are integer) and in the MIP node callback (in case the $z$-variables are fractional). The subproblem for each trip is also solved with Gurobi, and dual simplex is used to ensure that the basis remains feasible when the subproblem is solved for different values of $z$. To prevent excessive calls to the subproblem, feasibility heuristics are disabled. The number of cut separation rounds in the root node is set to the maximum value to get the best possible bound. Finally, the $2\epsilon$-trick is used to stabilize the master problem [7]. This stabilization uses $\epsilon = 0.00001$ and the trivial core point obtained by assigning $z_a = \frac{1}{4}$ to every bus arc.

Without transfer-expanded graphs, the subproblem is not totally unimodular and $\Phi(z)$ is not convex (see Sect. 2.2). In that case, Benders decomposition cannot be applied directly. Instead, $\theta \geq \Phi(z)$ may be enforced by adding *combinatorial Benders cuts* in the MIP solution callback and Benders cuts for the LP relaxation of the subproblem in both callbacks [4,10]. It follows from these references that combinatorial Benders cuts for the ODMTS design problem are given by $\theta \geq \Phi(\bar{z}) \left(1 - \sum_{r \in T} \sum_{a \in A} (\bar{z}_a(1 - z_a) + (1 - \bar{z}_a)z_a)\right)$.

## 4    Atlanta as a Case Study

The generalized ODMTS design problem was solved for the city of Atlanta. In Atlanta, the Metropolitan Atlanta Rapid Transit Authority (MARTA) operates two modes: bus and rail. The case study adds on-demand shuttles and the bus system is redesigned accordingly. More precisely, define the three modes $M = \{S, B, R\}$ for shuttle, bus, and rail respectively. Shuttle arcs are introduced to connect from origins to hubs and from hubs to destinations, as well as to serve the local demand. The corresponding $z_a$ variables are fixed to one, as shuttles are always available. Following [11], the cost of using a shuttle is a weighted sum of cost and convenience, controlled by the parameter $\alpha \in [0, 1]$. Let $d_a$ and $t_a$ be the travel distance and the travel time of arc $a \in A$, respectively. The parameter $c^S$ is the cost of using a shuttle per person per unit of distance. The cost of traversing arc $a \in A$ for trip $r \in T$ is then defined as $\gamma_a^r = p(r) \left( (1 - \alpha) c^S d_a + \alpha t_a \right)$. Note that the frequency does not affect the cost, and can be set to an arbitrary value.

Bus arcs are defined between the potential hub locations and between each hub and the three nearest rail stations. The cost of enabling bus arc $a \in A$ is given by $\beta_a = (1 - a) c^B f(a) d_a$. That is, the distance is multiplied by the cost per unit distance and the number of buses over the time horizon. The cost of traversing a bus arc is given by $\gamma_a^r = \alpha \left( t_a + L + \frac{H}{2f(a)} \right)$. Here $L$ is the fixed time required for a transfer, $H$ is the time horizon, and $\frac{H}{2f(a)}$ is the expected waiting time before the next bus arrives, which depends on the frequency. Rail arcs are defined between all rail stations that are connected by the same rail line. The costs of traversing an arc is defined in the same way as for the buses. For each rail arc $a \in A$, the variable $z_a$ is fixed to one, which makes the cost of enabling an arc irrelevant.

The case study is based on passenger trip data provided by MARTA for March 16, 2018, between 6am and 10am. Connecting trips have been chained together to obtain origin and destination pairs. This resulted in 2588 unique trips, with 7167 passengers in total. There are 5563 bus stops and rail stations in total, and their locations were also provided by MARTA. Eleven hubs were selected manually on the map. The data shows that demand is very stable and predictable over time, with about 90% of the trips being recurrent.

For the distances $d_a$, great-circle distances are used. To estimate travel times $t_a$, the distances are divided by a constant speed of 30 mph. The cost parameters are set to $c^S = 5$ and $c^B = 1$. The fixed transfer time is chosen to be five minutes, i.e., $L = 5$ min, and the time horizon is set to four hours, i.e., $H = 240$ min. To balance cost and convenience, $\alpha = 0.5$ is used. The rail frequency is assumed to be fixed to six per hour, i.e., $f(a) = 6 \times 4 = 24$, and bus frequencies are determined by the model to be either three per hour or six per hour. At most two transfers are allowed, i.e., $K = 3$. This is feasible, as shuttles can always be used to decrease the number of transfers (at a cost).

Figure 4a presents the result of solving the generalized ODMTS problem using transfer-expanded graphs. In total, it took 122 s to obtain the optimal solution and prove optimality, with a objective value of 131,905. Without transfer-expanded graphs, i.e., when adding combinatorial Benders cuts, it was not possible to obtain an optimal solution in reasonable time. Instead, the evaluation considered a relaxation in which the combinatorial Benders cuts are ignored: Only the Benders cuts for the LP relaxation of the subproblem were added.

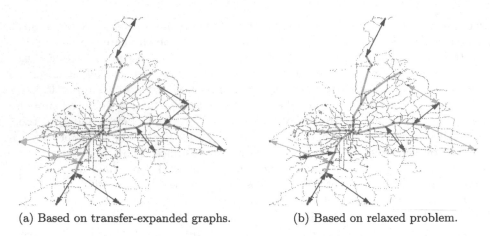

(a) Based on transfer-expanded graphs.          (b) Based on relaxed problem.

**Fig. 4.** Network designs for Atlanta showing rail (lines), low frequency buses (orange/light), and high frequency buses (purple/dark). (Color figure online)

Solving this relaxation to optimality took 3.8 h. Keep in mind that this relaxation explores routes that may require many transfers. To evaluate the quality of the design obtained by the relaxation, the passengers were routed through the transfer-expanded formulation with the $z$-variables fixed to their values found in the relaxation. The result is presented in Fig. 4b and has an objective value of 131,965. The higher cost is due to the fact that the transfer limit was not enforced when designing the network, i.e., passenger willingness to transfer was overestimated, which results in a network with less public transit.

In summary, the main benefit of the transfer-expanded formulation is the significant computational benefits it provides in capturing the transfer limit. Without transfer-expanded paths, it can be optimal to fractionally select long paths that do not adhere to this constraints. These longer fractional paths likely play a role in the difference of computational performance.

## 5    Conclusion

This paper presented a generalization of the ODMTS design problem that introduces three elements that are critical in practice: different frequencies, additional transit modes, and a limit on the number of transfers. Transfer-expanded graphs are introduced to handle the transfer limit without negatively impacting existing Benders decomposition approaches. The Atlanta case study demonstrates that this approach is very effective, as transfer-expanded graphs significantly improve computational performance. As the case study involves a real world instance and a significant transit network (the 8th largest in the US by ridership), this is a good indication that the same approach may be effective for other real-world instances. Exploiting the problem structure through transfer-expanded graphs opens the door to designing increasingly realistic networks in the future. One possible extension is to incorporate the capacity of the on-demand shuttles. As capacity of these shuttles is typically small, expanded networks could also be used to model capacity efficiently.

**Acknowledgements.** This research is partly supported by NSF Leap HI proposal NSF-1854684.

# References

1. Aneja, Y.P., Nair, K.P.K.: The constrained shortest path problem. Naval Res. Logistics Q. **25**(3), 549–555 (1978). https://doi.org/10.1002/nav.3800250314
2. Benders, J.F.: Partitioning procedures for solving mixed-variables programming problems. Numer. Math. **4**(1), 238–252 (1962)
3. Boland, N., Hewitt, M., Marshall, L., Savelsbergh, M.: The continuous-time service network design problem. Oper. Res. **65**(5), 1303–1321 (2017). https://doi.org/10.1287/opre.2017.1624
4. Codato, G., Fischetti, M.: Combinatorial Benders' cuts for mixed-integer linear programming. Oper. Res. **54**(4), 756–766 (2006). https://doi.org/10.1287/opre.1060.0286
5. Cormen, T.H., Leiserson, C.E., Rivest, R.L., Stein, C.: Introduction to Algorithms. The MIT Press (2009)
6. Dahl, G., Realfsen, B.: The cardinality-constrained shortest path problem in 2-graphs. Networks **36**(1), 1–8 (2000)
7. Fischetti, M., Ljubić, I., Sinnl, M.: Redesigning Benders decomposition for large-scale facility location. Manag. Sci. **63**(7), 2146–2162 (2017). https://doi.org/10.1287/mnsc.2016.2461
8. Gouveia, L., Leitner, M., Ruthmair, M.: Layered graph approaches for combinatorial optimization problems. Comput. Oper. Res. **102**, 22–38 (2019). https://doi.org/10.1016/j.cor.2018.09.007
9. Hasan, M.H., Van Hentenryck, P.: A column-generation algorithm for evacuation planning with elementary paths. In: Beck, J.C. (ed.) CP 2017. LNCS, vol. 10416, pp. 549–564. Springer, Cham (2017). https://doi.org/10.1007/978-3-319-66158-2_35
10. Laporte, G., Louveaux, F.V., Van Hamme, L.: An integer L-shaped algorithm for the capacitated vehicle routing problem with stochastic demands. Oper. Res. **50**(3), 415–423 (2002). https://doi.org/10.1287/opre.50.3.415.7751
11. Mahéo, A., Kilby, P., Van Hentenryck, P.: Benders decomposition for the design of a hub and shuttle public transit system. Transp. Sci. **53**(1), 77–88 (2019). https://doi.org/10.1287/trsc.2017.0756
12. Moccia, L., Cordeau, J.F., Monaco, M.F., Sammarra, M.: A column generation heuristic for a dynamic generalized assignment problem. Comput. Oper. Res. **36**(9), 2670–2681 (2009). https://doi.org/10.1016/j.cor.2008.11.022
13. Pillac, V., Cebrian, M., Van Hentenryck, P.: A column-generation approach for joint mobilization and evacuation planning. Constraints **20**(3), 285–303 (2015). https://doi.org/10.1007/s10601-015-9189-7
14. Pillac, V., Van Hentenryck, P., Even, C.: A conflict-based path-generation heuristic for evacuation planning. Transp. Res. Part B: Methodol. **83**, 136–150 (2016). https://doi.org/10.1016/j.trb.2015.09.008

# Reinforcement Learning for Variable Selection in a Branch and Bound Algorithm

Marc Etheve[1,3]([envelope])[ID], Zacharie Alès[2,3][ID], Côme Bissuel[1][ID], Olivier Juan[1][ID], and Safia Kedad-Sidhoum[3][ID]

[1] EDF R&D, Paris, France
{marc.etheve,come.bissuel,olivier.juan}@edf.fr
[2] ENSTA Paris, Institut Polytechnique de Paris, Paris, France
zacharie.ales@ensta-paris.fr
[3] CNAM Paris, CEDRIC, Paris, France
safia.kedad_sidhoum@cnam.fr

**Abstract.** Mixed integer linear programs are commonly solved by Branch and Bound algorithms. A key factor of the efficiency of the most successful commercial solvers is their fine-tuned heuristics. In this paper, we leverage patterns in real-world instances to learn from scratch a new branching strategy optimised for a given problem and compare it with a commercial solver. We propose FMSTS, a novel Reinforcement Learning approach specifically designed for this task. The strength of our method lies in the consistency between a local value function and a global metric of interest. In addition, we provide insights for adapting known RL techniques to the Branch and Bound setting, and present a new neural network architecture inspired from the literature. To our knowledge, it is the first time Reinforcement Learning has been used to fully optimise the branching strategy. Computational experiments show that our method is appropriate and able to generalise well to new instances.

**Keywords:** Reinforcement learning · Mixed integer linear programming · Neural network · Branch and bound · Branching strategy

## 1 Introduction

Mixed Integer Linear Programming (MILP) is an active field of research due to its tremendous usefulness in real-world applications. The most common method designed to solve MILP problems is the Branch and Bound (B&B) algorithm (see [1] for an exhaustive introduction). B&B is a general purpose procedure dedicated to solve any MILP instance, based on a divide and conquer strategy and driven by generic heuristics and bounding procedures.

Recently, a lot of attention has been paid to the interactions between MILP and machine learning. As pointed out in [2], learning methods may compensate

© Springer Nature Switzerland AG 2020
E. Hebrard and N. Musliu (Eds.): CPAIOR 2020, LNCS 12296, pp. 176–185, 2020.
https://doi.org/10.1007/978-3-030-58942-4_12

for the lack of mathematical understanding of the B&B method and its variants [3,4]. The plethora of different approaches in this young field of research gives evidence of the variety of ways in which learning can be leveraged. For instance, a natural idea is to bypass the whole B&B procedure to directly learn solutions of MILP instances [5]. If one wants to preserve the optimality guarantee provided by the B&B algorithm, a solution could be to rather learn the output of a computationally expensive heuristic used in a B&B scheme [6–8]. Alternatively, [9] suggests learning to select the best cut among a set of available cuts at each node of the B&B tree. Whether it is by Imitation Learning or by Reinforcement Learning (RL), these solutions are often limited by their scope: they seek to take decisions according to a local criterion.

In the present work, we propose FMSTS (*Fitting for Minimising the SubTree Size*), a novel approach based on Reinforcement Learning aiming at optimising a global criterion at the scale of the whole B&B tree. We learn a branching strategy from scratch, independent of any heuristic.

The paper is structured as follows. First, we define the general setting of our study. Using RL to minimise a global criterion, we then demonstrate that, under certain assumptions, a specific kind of value functions enforces the optimality of such criterion. Next, we propose to adapt known generic learning methods and neural network architectures to the Branch and Bound setting. We illustrate our proposed method on industrial problems and discuss it before concluding.

## 2   General Setting

In real-world applications, companies often optimise systems on a regular basis given fluctuating data. This case has been studied in the literature for different purposes, such as learning an approximate solution [5] or imitating heuristics [8]. However, to our knowledge, no concrete contribution has been made regarding the use of Reinforcement Learning for variable selection (branching) in this setting. The present work fills this gap.

Throughout this paper, we are interested in the following setting. For a given problem $\mathcal{P}$, the instances are perceived as randomly distributed according to an unknown distribution $\mathcal{D}$. This distribution, emanating from real-world systems, governs the fluctuating data $(A, b, c)$ across instances, written as

$$p \in \mathcal{P} : \begin{cases} \min_{x \in \mathbb{R}^n} c^\top x \\ \text{s.t.} \quad \begin{aligned} Ax &\leq b \\ x_{\mathcal{J}} &\in \{0,1\}^{|\mathcal{J}|}, \ x_{-\mathcal{J}} \in \mathbb{R}^{n-|\mathcal{J}|} \end{aligned} \end{cases} \tag{1}$$

with $A \in \mathbb{R}^{m \times n}$, $b \in \mathbb{R}^m$ and $c \in \mathbb{R}^n$. In practice, as the instances come from a single problem, they share the same structure. Especially, the set of null coefficients, the number of constraints $m$, of variables $n$ and the set of binary variables $\mathcal{J}$ are the same for every instance of a given problem.

In this setting, we seek to learn and optimise a branching strategy to solve to optimality any instance of a given problem. As pointed out in [10], the problem of optimising the decisions along a B&B tree is naturally formulated as a control problem on a sequential decision-making process. More specifically, it is equivalent to solving a finite-horizon deterministic Markov Decision Process (MDP) and may thus be tackled by Reinforcement Learning (see [11] for an introduction).

## 3    Fitting for Minimising the SubTree Size (FMSTS)

In a finite-horizon setting, Reinforcement Learning aims at optimising an agent to produce sequences of actions that achieve a global objective. The agent is guided by local costs associated with the actions it takes. Starting from an initial state in a possibly stochastic environment, it must learn to take the best sequence of actions and thus transitioning from state to state to minimise the overall costs. Exact Q-learning solves such problem by updating a table mapping (Q-function) from state/action pairs to discounted future costs (Q-values).

In the following, we define the RL problem of interest and, as exact Q-learning is not practicable, the approximate framework considered. Next, we propose a specific informative Q-function allowing us to use this framework in practice.

### 3.1    Approximate Q-Learning with Observable Q-Values

Let us denote by $\mathcal{S}$ the set of every reachable state for a given problem $\mathcal{P}$, a state being defined as all the information available when taking a branching decision in a B&B tree. Under perfect information, a state associated to a B&B node is the whole B&B tree as it has been expanded at the time the branching decision is taken. We write $\mathcal{A}$ the set of actions, *i.e.* the set of available branching decisions on a specific problem (the set of binary variables $\mathcal{A} = \mathcal{J}$ in our case) and $\pi$ a policy mapping any state to a branching decision:

$$\pi \; : \; \begin{cases} \mathcal{S} \to \mathcal{A} \\ s \mapsto \pi(s) = a \end{cases}.$$

The transitions between states are governed by the B&B solver, and the MDP is regarded as deterministic: given an instance and a state, performing an action will always lead to the same next state. In practice, such assumption is met as soon as the solver's decisions (apart from branching) are non stochastic.

We call agent any generator $\Pi$ of branching sequences following policy $\pi$ and denote $\Pi(p)$ the sequence of decisions that maps an instance $p$ to a complete B&B tree. Let $\mu\left(\Pi(p)\right)$ be any metric of interest on the tree generated by $\Pi$ for an instance $p$, and assume this metric is to be minimised. For instance, we can think of $\mu$ as the size of the generated tree, the number of simplex iterations, etc. In this setting, we are looking for the $\mu-$optimal agent $\Pi^*$ such that

$$\Pi^* \in \arg\min_{\Pi} \mathbb{E}_{p \sim \mathcal{D}} \left[\mu\left(\Pi\left(p\right)\right)\right]. \tag{2}$$

Note here that the expectation is only on the MILP instances, as the MDP is deterministic.

Let us assume that one can define a Q-function $Q^\pi$ which is consistent with $\mu$, in the sense that $\mu$ is minimised if $\pi(s) = \arg\min_{a \in \mathcal{A}} Q^\pi(s, a)$. Even in this case, exact Q-learning cannot be used to minimise $\mu$. First, maintaining an exact table for the Q-function is not tractable due to the size of $\mathcal{S}$, including for small real-world problems. Second, the transition from a state to the next is too complex to be modeled since it partly results from a linear optimisation.

To bypass these problems, we approximate the Q-function (see [11] for an introduction to Approximate Q-learning) by a neural network $\hat{Q}$ parametrised by $\theta$ and optimised by a dedicated gradient method as in the DQN (Deep Q-Network) approach [12]. We define the policy $\pi_\theta$ resulting from the Q-network as $\pi_\theta(s) = \arg\min_{a \in \mathcal{A}} \hat{Q}(s, a; \theta)$.

When facing a deterministic MDP, the exact Q-value $Q^{\pi_\theta}(s, a)$ of an action $a$ given a state $s$ and a policy $\pi_\theta$ is not stochastic and thus may be observable. In that case, the classic Temporal Difference loss used in [12] for training the Q-Network comes down to the simple expression

$$L(\theta) = \mathbb{E}_{s,a\sim\rho(.)} \left[ \left( Q^{\pi_\theta}(s, a) - \hat{Q}(s, a; \theta) \right)^2 \right] \tag{3}$$

where $\rho$ is the behaviour distribution of our agent, as stated in [12]. Note that, in Eq. (3), the observed Q-values are naturally influenced by parameter $\theta$ through the policy. Such loss is actually intuitive: if $Q^{\pi_\theta}$ is consistent with $\mu$, if each action has non-zero probability to be taken in any encountered state and if $L(\theta) = 0$, then each B&B tree built by agent $\Pi_\theta$ (following $\pi_\theta$) is optimal with respect to $\mu$ with probability 1.

## 3.2   Using the Subtree Size as Value Function

As highlighted in [8], an important difficulty when applying Reinforcement Learning to B&B algorithms is the credit assignment problem [13]: in order to determine the actions that lead to a specific outcome, one may define non-sparse informative local costs (negative rewards) consistent with the global objective.

This is not mandatory in a RL setting but may facilitate the learning task.

We choose the number of nodes in the generated tree as the global metric $\mu$. This metric is often used to compare B&B methods (see for instance [6–8]), as it is a proxy for computational efficiency and independent of hardware considerations.

One of the main contributions of this paper is to propose a local Q-function $Q^\pi$ which is consistent with the chosen global metric $\mu$. We take advantage of the deterministic aspect of the environment and define $Q^\pi(s, a)$ as the size of the subtree rooted in the B&B node corresponding to $s$ generated by action $a$ and policy $\pi$. As stated in Proposition 1, this particular Q-function is not consistent in general with our choice of global criterion $\mu$. Nonetheless, Proposition 2 asserts its optimality when using Depth First Search as node selection strategy.

**Proposition 1.** *In general, minimising the size of the subtree under any node in a B&B tree is not optimal with respect to the tree size.*

The proof is omitted for the sake of conciseness, but one can prove that minimising the subtree size can be sub-optimal when using Breadth First Search as the node selection strategy.

**Proposition 2.** *When using Depth First Search (DFS) as the node selection strategy, minimising the whole B&B tree size is achieved when any subtree is of minimal size.*

*Proof.* Let us call $\mathcal{O}$ the set of open nodes at a given iteration of the B&B process for a specific instance. The set of closed nodes (either by pruning or branching) is denoted $\mathcal{C}$.

We write $V^\pi(s \,|\, \zeta, \eta)$ the size of the subtree under $s$, entirely determined by the policy $\pi$ followed in this subtree, a set of primal bounds $\zeta$ found in other subtrees and a node selection strategy $\eta$. When using DFS, the subtrees under each open node are expanded and fully solved sequentially, thus we can assume with no loss of generality that $\mathcal{O}$ is equal to $\{1, ..., k\}$ and is sorted according to the planned visiting order. In that case, the size $V$ of the whole B&B tree can be expressed as

$$V = |\mathcal{C}| + \sum_{i=1}^{k} V^{\pi_i}\left(s_i \,\left|\, \{z_0\} \cup \left(\bigcup_{j<i} \zeta_j\right)\right., \eta = DFS\right)$$

with $\zeta_i$ the set of all bounds to be found in the subtree rooted in $s_i$ and $z_0$ the best bound obtained in $\mathcal{C}$.

It remains to prove that $\pi_1$ is optimal only if it leads to the minimal subtree under $s_1$. As two separate subtrees can only affect each other through their best primal bound under DFS, we have

$$V = |\mathcal{C}| + V^{\pi_1}(s_1 \,|\, \{z_0\}, \eta = DFS) + \sum_{i=2}^{k} V^{\pi_i}(s_i \,|\, \{z_{i-1}\}, \eta = DFS)$$

with $z_{i-1} = \min\left\{\{z_0\} \cup \left(\bigcup_{j<i} \zeta_j\right)\right\}$.

Since the B&B procedure (with a gap set to zero) guarantees that we find the best primal bound of any expanded subtree, $(z_{ii=1})^k$ are completely independent of the branching policies, which gives, for any $\pi_j$, $j \in \{2, ..., k\}$:

$$\arg\min_{\pi_1 \in \Pi_1} V = \arg\min_{\pi_1 \in \Pi_1} V^{\pi_1}(s_1 \,|\, \{z_0\}, \eta = DFS)$$

with $\Pi_1$ the set of all valid branching policies under $s_1$. Therefore, choosing any other policy than $\pi_1 \in \arg\min_{\pi_1 \in \Pi_1} V^\pi(s_1 \,|\, \{z_0\}, \eta = DFS)$ is sub-optimal with respect to the tree size. □

In the remaining, we use DFS as the node selection strategy according to Proposition 2. We now focus on optimising the branching strategy (variable selection) to minimise at each node the size of the underlying subtree. If we write $D_0^{\pi(s)}(s)$ and $D_1^{\pi(s)}(s)$ the child nodes of $s$ following policy $\pi$, such value function satisfies the Bellman Equation (4). The relationship between the value and the Q-function is trivially defined by $Q^\pi(s,a) = 1 + V^\pi(D_0^a(s)) + V^\pi(D_1^a(s))$.

$$V^\pi(s) = 1 + V^\pi\left(D_0^{\pi(s)}(s)\right) + V^\pi\left(D_1^{\pi(s)}(s)\right) \tag{4}$$

This value function has two advantages. First, it is observable as assumed earlier: we only need to count the number of inheriting nodes once the B&B tree is fully expanded. Second, it is a local objective which guarantees the optimality of a global criterion, hence allowing us to perform RL without designing a suboptimal reward using any domain knowledge.

### 3.3    Algorithm

Using Approximate Q-learning and the subtree size as value function leads us to propose the FMSTS algorithm (Algorithm 1). Using Experience Replay and $\varepsilon$-greedy exploration as in [12], the algorithm essentially boils down to consecutively solving a MILP instance following the current policy or random choices with probability $\varepsilon$, fitting the observed values sampled from an experience replay buffer and iterating with the updated policy.

---

**Algorithm 1. FMSTS**

---

   **for** t = 0,...,N-1 **do**

      Draw randomly an instance $p$.

      Solve $p$ following $\pi_{\theta_t}$ with $\varepsilon$-greedy exploration.

      Collect experiences along the generated tree $\left(s_i, a_i, Q^{\pi_{\theta_t}}(s,a), \hat{Q}(s,a;\theta_t)\right)$ and store them into an experience replay buffer $\mathcal{B}$.

      Update to $\theta_{t+1}$ using loss (3) on experiences drawn from $\mathcal{B}$.

   **end for**

---

## 4    Adapting Learning to the Branch and Bound Setting

To ensure the success of the FMSTS method (Algorithm 1) with respect to the objective (2), we need to adapt some components to the Branch and Bound setting. First, we adapt the loss guiding the neural network's training. Next, we use Prioritized Experience Replay while normalising probabilities. Last, we propose a new neural network architecture inspired from the literature.

## 4.1    Minimising an Expectation on the Instance Distribution

The loss defined by Eq. (3) does not seem to correspond to our objective (2). Indeed, it naturally gives more importance to the biggest trees, which can be heavily instance dependent. To neutralise this effect, we weight the loss by the inverse of the size of the corresponding B&B tree generated by the agent:

$$L(\theta) = \mathbb{E}_{s,a \sim \rho(.)} \left[ \frac{1}{V^{\pi_\theta}(r(s))} \left( Q^{\pi_\theta}(s,a) - \hat{Q}(s,a;\theta) \right)^2 \right] \tag{5}$$

with $r(s)$ the root node of the tree containing $s$, such that $V^{\pi_\theta}(r(s))$ corresponds to the size of this tree. Then, any instance has equal weight in loss (5).

## 4.2    Performing Prioritized Experience Replay

Prioritized Experience Replay [14] biases the uniform replay sampling of Experience Replay [12] towards experiences with high Temporal Difference errors, *i.e.* when the predicted Q-values are far from their target. In FMSTS, an experience is a 4-tuple $\left( s_j, a_j, Q^{\pi_{\theta_j}}(s_j, a_j), \hat{Q}(s_j, a_j; \theta_j) \right)$ and the target $Q^{\pi_{\theta_j}}(s_j, a_j)$ is observed, which reduces the error to the simple expression $|Q^{\pi_{\theta_j}}(s_j, a_j) - \hat{Q}(s_j, a_j; \theta_j)|$.

In the context of sampling experiences in a B&B tree, one should take into account that the scale of the target $Q^{\pi_{\theta_j}}$ may vary exponentially both along the tree and across instances. As the scale of the error may likely vary with that of the target, we normalise this error by the target to get the sampling probability in the experience replay buffer

$$p_j \propto \frac{|Q^{\pi_{\theta_j}}(s_j, a_j) - \hat{Q}(s_j, a_j; \theta_j)|}{Q^{\pi_{\theta_j}}(s_j, a_j)}. \tag{6}$$

## 4.3    Designing a Regressor for the Q-Function

As in [6], we use both static and dynamic features to represent a state. Although many features may be relevant for the states' encoding, we opted to keep them limited in the present work. For static information, we perform a dimension reduction by PCA [15]: each instance is represented as the concatenation of its data $(A, b, c)$ and PCA is applied on the resulting vectors. Our representation also includes the following dynamic features: the node's depth, the distance of the current primal solution to the bounds and the branching state. Concretely, the branching state $B$ is one-hot encoded in a $3|\mathcal{J}|$ vector. Let us call $\mathcal{B}_0$ and $\mathcal{B}_1$ the set of variables that have been respectively set to 0 and 1 in the ascendant nodes of the current state. With no loss of generality, let us assume that $\mathcal{J} = \{1, ..., J\}$. Then we have $B_j = \mathbb{1}_{x_j \in \mathcal{B}_0}$, $B_{j+J} = \mathbb{1}_{x_j \in \mathcal{B}_1}$ and $B_{j+2J} = 1 - B_j - B_{j+J}$ for any $j \in \mathcal{J}$.

The chosen Q-function is essentially multiplicative, in the sense that the ratio between the targets in two consecutive states may be of magnitude 2 due to the binary tree structure. In addition, the scale of these targets may strongly vary between instances. A basic feedforward neural network, based on summations, may struggle to handle such phenomena. To compensate for these effects and adapt to the B&B setting, we take inspiration from the Dueling architecture of [16] and propose the Multiplicative Dueling Architecture (MDA). As shown in Fig. 1, MDA implements the product between the 1-D output of a block of fully-connected layers fed with static features and the $|\mathcal{J}|$-D output of a block fed with both static and dynamic features. A linear activation on the 1-D output allows our agent to capture the variability of the chosen Q-function.

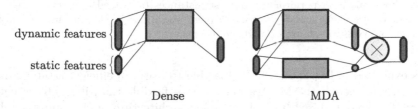

dynamic features

static features

Dense                     MDA

**Fig. 1.** Dense and Multiplicative Dueling architectures for the Q-network. The rectangles represent consecutive dense layers, the lightblue block being fed with all the features whereas the lightgreen one is only fed with static features (darkgreen). The output of the MDA is the product between a single unit and a $|\mathcal{J}|$-unit dense layer. (Colour figure online)

## 5 Experiments and Discussion

We test our algorithms on two sets of instances provided by Electricité de France (EDF), a french electric utility company. They are drawn from two different problems, one is related to energy management in a microgrid ($\mathcal{P}_1$) whereas the other one comes from a hydroelectric valley ($\mathcal{P}_2$). The problems have respectively 186 and 282 constraints, 120 and 207 variables, and 54 and 96 binary variables.

We compare our algorithms to the default branching strategy of CPLEX (denoted CPLEX in the following) and full Strong Branching (denoted SB). We use CPLEX 12.7.1 [17] under DFS while turning off all presolving and cutting.

To avoid any dependency of our results to the train or the test set, we present cross-validated results. Algorithm 1 is run 100 times independently on randomly partitioned train and test sets. Each time, 200 instances are used for training while 500 unseen instances are used for testing. Figure 2 shows the averaged number of nodes in the complete B&B trees on the test sets during the learning process: test instances are solved using the strategy learned on train instances at the current iteration of Algorithm 1.

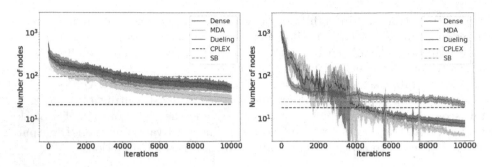

**Fig. 2.** Cross-validated performance on test instances (averaged number of nodes in log scale) for $\mathcal{P}_1$ (left) and $\mathcal{P}_2$ (right) through iterations of Algorithm 1. Gaussian confidence intervals are shown around the means.

As exhibited in Fig. 2, our method is able to learn an efficient strategy from scratch. As expected, the MDA agent is more flexible than its additive counterpart (Dueling) inspired from the Dueling architecture of [16] and the fully-connected agent (Dense). It outperforms systematically the Strong Branching policy, and finds comparable or better strategies than CPLEX, depending on the problem. Results on training data are not displayed for the sake of conciseness, but it is worth mentioning that our agents do not overfit and are able to generalise well. In addition, the computation time is negligible compared with full Strong Branching as an action comes only at the price of a forward pass in our neural network.

Despite these good performances, some limits have to be pointed out at this stage. First, our framework requires DFS as the node selection strategy, which can be far from optimal for certain problems. Note that using another strategy may be complicated to handle due to more complex dependencies, but may also turn out to be effective as targetting small subtrees makes sense in general. Second, we only showed promising results on easy problems. With more difficult problems, the training becomes computationally prohibitive as a randomly initialised agent produces exponential trees. To tackle these limitations, we encompass different solutions such as fine-tuning the features and network architecture or using supervision to decrease the size of the generated trees during the first episodes. To reduce the cost of exploration, one could apply the same methodology with a set of branching heuristics as action set, similarly to what is proposed in [18].

## 6   Conclusion

In this paper, we presented a novel Reinforcement Learning framework designed to learn from scratch the branching strategy in a B&B algorithm. In addition to the specific metrics used in our FMSTS method, we introduced a new neural network architecture designed to tackle the multiplicative nature of the value function. Besides, we adapted some known RL techniques to the B&B setting.

We ran experiments on real-world problems to validate our method and showed better or comparable performances with existing strategies.

It is worthwhile to highlight that our method is generic enough to be applied to other metrics than the tree size, e.g. the number of simplex iterations or even the computation time. If one is not interested in proving optimality, many other value functions may be encompassed. Furthermore, it may be interesting to enlarge the scope of the method, especially to include Branch and Cut algorithms as they usually are more efficient.

# References

1. Wolsey, L.A.: Integer programming. Wiley, Hoboken (1998)
2. Bengio, Y., Lodi, A., Prouvost, A.: Machine learning for combinatorial optimization: a methodological tour d'horizon. arXiv preprint arXiv:1811.06128 (2018)
3. Barnhart, C., Johnson, E.L., Nemhauser, G.L., Savelsbergh, M.W., Vance, P.H.: Branch-and-price: column generation for solving huge integer programs. Oper. Res. **46**(3), 316–329 (1998)
4. Mitchell, J.E.: Branch-and-cut algorithms for combinatorial optimization problems. In: Handbook of Applied Optimization, vol. 1, pp. 65–77 (2002)
5. Rachelson, E., Abbes, A.B., Diemer, S.: Combining mixed integer programming and supervised learning for fast re-planning. In: 2010 22nd IEEE International Conference on Tools with Artificial Intelligence, vol. 2, pp. 63–70. IEEE (2010)
6. Khalil, E.B., Le Bodic, P., Song, L., Nemhauser, G., Dilkina, B.: Learning to branch in mixed integer programming. In: Thirtieth AAAI Conference on Artificial Intelligence (2016)
7. Balcan, M.-F., Dick, T., Sandholm, T., Vitercik, E.: Learning to branch. arXiv preprint arXiv:1803.10150 (2018)
8. Gasse, M., Chételat, D., Ferroni, N., Charlin, L., Lodi, A.: Exact combinatorial optimization with graph convolutional neural networks. arXiv preprint arXiv:1906.01629 (2019)
9. Tang, Y., Agrawal, S., Faenza, Y.: Reinforcement learning for integer programming: learning to cut. arXiv preprint arXiv:1906.04859 (2019)
10. He, H., Daume III, H., Eisner, J.M.: Learning to search in branch and bound algorithms. In: Ghahramani, Z., Welling, M., Cortes, C., Lawrence, N.D., Weinberger, K.Q. (eds.) Advances in Neural Information Processing Systems, vol. 27, pp. 3293–3301. Curran Associates Inc., New York (2014)
11. Sutton, R.S., Barto, A.G.: Reinforcement Learning: An Introduction. MIT Press, Cambridge (2018)
12. Mnih, V., et al.: Human-level control through deep reinforcement learning. Nature **518**(7540), 529 (2015)
13. Minsky, M.: Steps toward artificial intelligence. Proc. IRE **49**(1), 8–30 (1961)
14. Schaul, T., Quan, J., Antonoglou, I., Silver, D.: Prioritized experience replay. arXiv preprint arXiv:1511.05952 (2015)
15. Pearson, K.: LIII. On lines and planes of closest fit to systems of points in space. London Edinburgh Dublin Philos. Mag. J. Sci. **2**(11), 559–572 (1901)
16. Wang, Z., Schaul, T., Hessel, M., Van Hasselt, H., Lanctot, M., De Freitas, N.: Dueling network architectures for deep reinforcement learning. arXiv preprint arXiv:1511.06581 (2015)
17. Manual, C.U.: IBM ILOG CPLEX Optimization Studio (1987)
18. Di Liberto, G., Kadioglu, S., Leo, K., Malitsky, Y.: Dash: dynamic approach for switching heuristics. Eur. J. Oper. Res. **248**(3), 943–953 (2016)

# Duplex Encoding of Staircase At-Most-One Constraints for the Antibandwidth Problem

Katalin Fazekas[1]([✉])(iD), Markus Sinnl[2](iD), Armin Biere[1](iD), and Sophie Parragh[2](iD)

[1] Institute for Formal Models and Verification, Johannes Kepler University, Linz, Austria
{katalin.fazekas,armin.biere}@jku.at
[2] Institute of Production and Logistics Management, Johannes Kepler University, Linz, Austria
{markus.sinnl,sophie.parragh}@jku.at

**Abstract.** Decision and optimization problems can be tackled with different techniques, such as Mixed Integer Programming, Constraint Programming or SAT solving. An important ingredient in the success of each of these approaches is the exploitation of common constraint structures with specialized (re-)formulations, encodings or other techniques. In this paper we present a new linear SAT encoding using binary decision diagrams over multiple variable orders as intermediate representation of a special form of constraints denoted as staircase at-most-one-constraints. The use of these constraints is motivated by recent work on the antibandwidth problem, where an iterative solution procedure using feasibility-mixed integer programs based on such constraints was most effective. In a computational study we compare the effectiveness of our new encoding against traditional SAT-encodings for staircase at-most-one-constraints. Additionally we compare against previous exact solution methods for the antibandwidth problem, such as a constraint programming approach and the one based on feasibility-mixed integer programs.

## 1 Introduction

An important ingredient in the success of computational approaches, such as Mixed Integer Programming (MIP), Constraint Programming (CP) or propositional satisfiability solving (SAT), for solving optimization and decision problems is the exploitation of common constraint structures with specialized encodings, (re-)formulations or other techniques (see e.g. [1–3]).

In this paper we present a new and specialized SAT encoding of problems where an at-most-one constraint slides over a sequence of Boolean variables. We denote this special case of sliding sequence constraints [4–7] as *staircase at-most-one constraint* (SCAMO) and illustrate the reason for this name with the following example.

© Springer Nature Switzerland AG 2020
E. Hebrard and N. Musliu (Eds.): CPAIOR 2020, LNCS 12296, pp. 186–204, 2020.
https://doi.org/10.1007/978-3-030-58942-4_13

*Example 1.* Given a sequence of variables $X = \langle x_1 \, x_2 \cdots x_{10} \rangle$, the staircase at-most-one constraint set of width 4 is the following formula:

$$
\begin{aligned}
x_1 + x_2 + x_3 + x_4 & \qquad\qquad\qquad\qquad\qquad\quad \leq 1 \wedge \\
x_2 + x_3 + x_4 + x_5 & \qquad\qquad\qquad\qquad\quad \leq 1 \wedge \\
x_3 + x_4 + x_5 + x_6 & \qquad\qquad\qquad\quad \leq 1 \wedge \\
x_4 + x_5 + x_6 + x_7 & \qquad\qquad\quad \leq 1 \wedge \\
x_5 + x_6 + x_7 + x_8 & \qquad\quad \leq 1 \wedge \\
x_6 + x_7 + x_8 + x_9 & \quad \leq 1 \wedge \\
x_7 + x_8 + x_9 + x_{10} & \leq 1.
\end{aligned}
$$

This research is motivated by recent work [8] of the second author on the *antibandwidth problem* (ABP). The ABP is a *graph labeling problem* (see e.g. [9] for more on such problems) where the goal is to maximize the smallest difference between labels of neighbouring nodes. It has various applications, such as scheduling [10], obnoxious facility location [11], radio frequency assignment [12] and map-coloring [13]. It has been studied from a theoretical point of view (see e.g. [14–19]), and several heuristics and metaheuristics (e.g. [20–23]) have been designed for it. In [21], aside from a metaheuristic, also a MIP approach was presented to solve the ABP exactly.

In [8] new MIP formulations were presented, and based on one of them, an iterative solution procedure, which repeatedly solved feasibility-MIPs, was designed. For a given number $k$, these MIPs encode the question whether there exists a solution with antibandwidth greater than $k$. This iterative procedure actually proved to be the most effective one in the computational study of [8].

Our proposed encoding can be used for more difficult problem structures than the one given in Example 1. In the ABP, for example, the difference of labels of neighbouring nodes is restricted by combining two SCAMO constraints on two sequences of variables. Aside from the ABP (and other labeling problems), the SCAMO constraints can potentially be used in many further application contexts, such as scheduling problems (see e.g. [24–26]) or in staff rostering [27, 28] and car sequencing problems [29,30], when at most one variable is allowed to take a given value in every sequence of variables.

As at-most-one constraints are ubiquitous in applications of SAT they are featured prominently in the literature, see e.g. [31–36]. They are forming a special case of cardinality constraints [37–39], which in turn are instances of Pseudo-Boolean constraints [40–43] and thus 0/1 integer linear programs. Encoding constraints (for an overview see [36]) instead of handling them natively (as in [38]) allows to make full use of the power of SAT solving. For some applications mixed strategies [44] are better though. In practice, size is the most important criteria to evaluate such encodings, while at least in theory also propagation strength is considered. See [45] for a discussion of these trade-offs. In particular, the path based encoding of binary decision diagrams introduced in [45] has the goal to improve propagation. However, as the authors point out, it can not be used for encoding shared constraints, which is the main reason of the efficiency in our encoding. Thus we also provide a new set of benchmarks for which such sharing occurs naturally.

## 2  Preliminaries

A propositional formula in conjunctive normal form (CNF) consists of a set of clauses, where each clause $C$ is a disjunction of literals, which are Boolean (also called 0/1) variables (e.g. $x$) or their negation ($\neg x$ or $1 - x$). A truth assignment $T$ maps truth values (0/1 values) to Boolean variables and can be represented by a set of consistent literals; it satisfies a literal $\ell$ (i.e. assigns value 1 to $\ell$) if $\ell \in T$, and falsifies it (assigns value 0 to $\ell$) if $\neg \ell \in T$, where $\neg \ell = \neg x$ if $\ell = x$ and $\neg \ell = x$ if $\ell = \neg x$. The satisfiability problem (SAT) for a formula in CNF asks whether there is a truth assignment such that all clauses contain at least one satisfied literal. A truth assignment satisfying a formula is also called a model.

An *at-most-one* (AMO) constraint is an expression of the form $\sum_{i=1}^n x_i \leq 1$, where $x_1, x_2, \ldots, x_n$ are Boolean variables. Similarly, we can formulate *at-most-zero* (AMZ) constraints (as $\sum_{i=1}^n x_i \leq 0$), which actually states that each variable must be false (i.e. assigned value 0). Further, an exactly-one (EO) constraint is an expression of the form $\sum_{i=1}^n x_i = 1$. Notice that we define and use these constraints over Boolean variables, but they are trivially extensible to literals.

A binary decision diagram (BDD, see e.g. [46,47]) is a rooted, directed, acyclic graph with at most two leafs, labeled with $\bot$ (false or 0) and $\top$ (true or 1). Every non-leaf (also called nonterminal) node of a BDD is labeled with a Boolean variable and has exactly two outgoing edges (called *low* and *high* in [46]). In this paper we use BDDs to represent AMO and AMZ constraints. Figure 1a depicts an example BDD of an AMO constraint over variables $x_1, x_2$ and $x_3$. Each path from the root of the BDD that ends in the true leaf ($\top$) is a model of $x_1 + x_2 + x_3 \leq 1$. Whenever the low or high child (marked with dashed resp. solid line in Fig. 1) of a node labeled with variable $x$ is taken, it means that $x$ is assigned to be false (true respectively) on that path. Since all our BDDs represent AMO or AMZ constraints, we will depict them rather in an expanded form where each node contains the whole Boolean expression represented by the sub-graph starting from it, as it can be seen on Fig. 1b. To emphasize the decision variables of the nodes, we mark them explicitly on the edges. Further, beyond the non-terminal (i.e. non-leaf) nodes we distinguish non-unit nodes that are representing a constraint over more than one variable. For example, the BDD of Fig. 1b contains two leaf nodes ($\top$ and $\bot$), two unit nodes (over $x_3$) and three non-unit non-leaf nodes. The ordering of the variables appearing in BDDs is fixed (e.g. $x_1 < x_2 < x_3$ in Fig. 1), i.e. we use ordered BDDs (OBDD in short). Even though we merge isomorphic subtrees in our BDDs, they are not reduced because nodes with identical children are kept (see e.g. $x_3$ in Fig. 1). Thus we use partially reduced ordered BDDs (ROBDD) over multiple variable orders.

Given a graph $G = (V, E)$, a feasible solution to the antibandwidth problem consists of assigning each node $v \in V$ a unique label from the range $1, \ldots, |V|$.

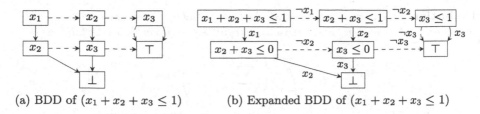

(a) BDD of $(x_1 + x_2 + x_3 \leq 1)$    (b) Expanded BDD of $(x_1 + x_2 + x_3 \leq 1)$

**Fig. 1.** Different BDD representations of AMO constraint $(x_1 + x_2 + x_3 \leq 1)$.

Given such a labeling $f$, the antibandwidth $AB_f(v)$ of a node $v$ is defined as $\min\{|f(v) - f(v')| : \{v, v'\} \in E\}$, and the antibandwidth $AB_f(G)$ is defined as $\min\{AB_f(v) : v \in V\}$. The goal of the ABP is to find a labeling $f^*$, such that $f^* = \arg\max_{f \in \mathcal{F}(G)} AB_f(G)$, where $\mathcal{F}(G)$ denotes the set of all labelings of $G$.

We briefly discuss previous work [8] on which our new SAT solution is based. Let binary variables $x_i^\ell = 1$ if and only if vertex $i$ is assigned label $\ell$ (i.e. $f_i = \ell$). For a given $k$, the question, whether there exists a solution with $AB(G) \geq k+1$, can be formulated as MIP as follows. We will denote this formulation as $F_e(k)$.

$$\max 0$$

$$\sum_{i \in V} x_i^\ell = 1 \qquad\qquad \forall \ell \in \{1, \ldots, |V|\} \qquad \text{(LABELS)}$$

$$\sum_{\ell \in \{1,\ldots,|V|\}} x_i^\ell = 1 \qquad\qquad \forall i \in V \qquad \text{(VERTICES)}$$

$$\sum_{\lambda \leq \ell \leq \lambda+k} (x_i^\ell + x_{i'}^\ell) \leq 1 \qquad \forall \{i, i'\} \in E, 1 \leq \lambda \leq |V| - k \qquad \text{(OBJ}_k\text{)}$$

$$x_i^\ell \in \{0, 1\} \qquad \forall i \in V, \forall \ell \in \{1, \ldots, |V|\}$$

Constraints (LABELS) make sure that each label is used only once and constraints (VERTICES) ensure that each node $i \in V$ gets assigned one label. Thus, the solution encoded by these constraints corresponds to a labeling. Constraints (OBJ$_k$) describe that for each edge $\{i, i'\}$, the labels $f_i$, $f_{i'}$ are not allowed to be within a range of $k$. Thus, any solution of the above constraints corresponds to a labeling with antibandwidth at least $k + 1$. The iterative algorithm of [8] starts with a value of $k$ obtained by a heuristic, which constructs a feasible labeling, and then iteratively solves $F_e(k)$ and increases $k$ by one, until either $F_e(k)$ becomes infeasible (proving optimality of $k$) or a time limit is reached.

## 3    Staircase At-Most-One Constraint Sets

As a first step we define and illustrate the main concept of our paper, the so-called staircase AMO constraint set (SCAMO). Following that, in the next section we demonstrate step-by-step our proposed SAT encoding of these constraints.

**Definition 1.** *Given a sequence of Boolean variables* $X = \langle x_1 \, x_2 \cdots x_n \rangle$ *and a width $w$ s.t. $1 < w \leq n$, a staircase constraint set is formulated as follows:*

$$\text{SCAMO}(X, w) = \bigwedge_{i=0}^{(n-w)} \left( \sum_{j=i+1}^{(i+w)} x_j \leq 1 \right) \quad \text{where } n = |X|.$$

Notice that this constraint is a special sub-case of SEQUENCE constraints (see e.g. [4–7]) and so could be formulated as SEQUENCE$(0, 1, w, X, \{1\})$.

In Example 1 we saw, that there is an ordering of the constraints in that problem such that each constraint differs only slightly from the previous one. For instance, in Example 1 the 1st and 2nd constraints both include the sum of $x_2, x_3$ and $x_4$ while the 2nd and 3rd both contain the sub-expression $x_3 + x_4 + x_5$. Since addition is associative, the sum of the variables can be calculated regardless of the grouping of the variables. However, if we would like to reuse previous calculations, it is more beneficial to evaluate the first AMO constraint for example as $x_1 + (x_2 + x_3 + x_4)$ instead of considering any other variable grouping (e.g. $(x_1 + x_2) + (x_3 + x_4)$). Doing so, the second constraint can just simply consider the result of $(x_2 + x_3 + x_4)$ together with $x_5$. Continuing the evaluation with the next constraint, we could reuse $(x_3 + x_4)$ from $(x_2 + x_3 + x_4)$, in case we calculated it as $x_2 + (x_3 + x_4)$, to decide $x_3 + x_4 + x_5 + x_6 \leq 1$ by combining it with $(x_5 + x_6)$. In general, each constraint shares a sub-sum over $w - 1$ variables with the previous and at the same time with the next constraint.

Evaluating the very first constraint in this example in a right associative way allows us to reuse (at least once) all its sub-expression in the following three (i.e. $w - 1$) constraints. However, in order to reuse these sub-expressions we need a left associative grouping of variables in the constraint $x_5 + x_6 + x_7 + x_8 \leq 1$, since in the second constraint we need $x_5$, then $(x_5 + x_6)$ and then $(x_5 + x_6 + x_7)$ to complement the reused sub-sums of $x_1 + x_2 + x_3 + x_4$.

All in all, considering only the first $w$ constraints, we see that we need a right associative evaluation of the first constraint and a left associative grouping of the $(w + 1)$'th constraint. Figure 2 depicts how these variable groupings can be "bonded" together to reconstruct the original constraints of Example 1. Extending this pattern to the whole set of constraints, we can see that each $w$ consecutive constraints need to be considered once left associative to combine with the previous $w$ constraints' sub-expressions and once right associative, to combine with the next $w$ constraints. Thus, in Fig. 2 the sum over variables $x_5, x_6, x_7$ and $x_8$ is actually considered twice, once with a left and once with a right associative variable ordering. This duplicate view of constraints is the main concept behind our proposed duplex encoding.

## 4   Duplex Encoding of Staircase Constraint Sets

Our goal is to exploit sharing of sub-expressions between constraints to obtain a compact encoding. Again, the main idea of our approach can be seen in Fig. 2 where we identified common sub-sums. In our concrete encoding we have to

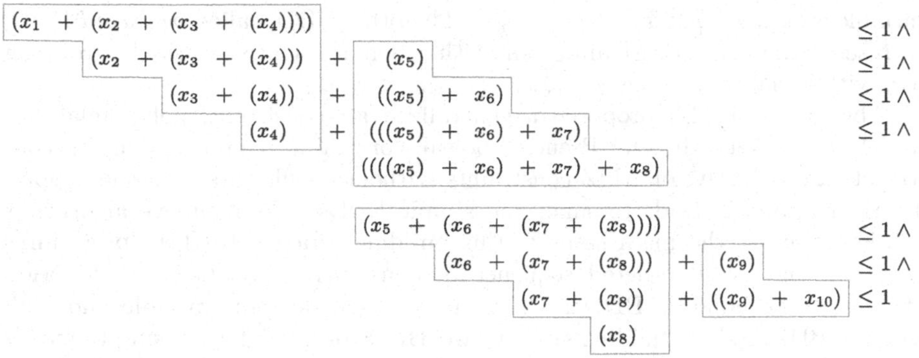

$$
\begin{array}{llll}
\boxed{(x_1 + (x_2 + (x_3 + (x_4))))} & & & \leq 1 \wedge \\
\boxed{(x_2 + (x_3 + (x_4)))} & + & \boxed{(x_5)} & \leq 1 \wedge \\
\boxed{(x_3 + (x_4))} & + & \boxed{((x_5) + x_6)} & \leq 1 \wedge \\
\boxed{(x_4)} & + & \boxed{(((x_5) + x_6) + x_7)} & \leq 1 \wedge \\
& & \boxed{((((x_5) + x_6) + x_7) + x_8)} & \\
\end{array}
$$

$$
\begin{array}{llll}
\boxed{(x_5 + (x_6 + (x_7 + (x_8))))} & & & \leq 1 \wedge \\
\boxed{(x_6 + (x_7 + (x_8)))} & + & \boxed{(x_9)} & \leq 1 \wedge \\
\boxed{(x_7 + (x_8))} & + & \boxed{((x_9) + x_{10})} & \leq 1 \\
\boxed{(x_8)} & & & \\
\end{array}
$$

**Fig. 2.** Decomposition of the staircase AMO constraint set of Example 1.

go one step further though and actually have to share sub-constraints. This is achieved by decomposing longer AMO constraints into two smaller ones using the following proposition. While the original longer constraints may be used only once, smaller constraints potentially can be shared and reused multiple times.

**Proposition 1.** *A constraint* $x_1 + \cdots + x_n \leq 1$ *holds iff for all* $1 \leq i < n$

$$(x_1 + \ldots + x_i \leq 1) \wedge (x_{i+1} + \ldots + x_n \leq 1) \wedge (x_1 + \ldots + x_i \leq 0 \vee x_{i+1} + \ldots + x_n \leq 0).$$

### 4.1  Sub-constraint Construction

As a first step, given a sequence of variables $X = \langle x_1 \cdots x_n \rangle$ and width $w$, we partition the variables into $M = \lceil \frac{n}{w} \rceil$ consecutive *windows* $\omega_1, \omega_2, \ldots, \omega_M$, where $\omega_1$ contains variables $x_1, \ldots, x_w$, $\omega_2$ contains $x_{w+1}, \ldots, x_{2w}$ etc. Note that unless $(n \bmod w) = 0$, the very last window contains fewer than $w$ variables.

*Example 2.* Continuing the previous example, our width $w = 4$ splits $X$ into three windows: $\omega_1 = \{x_1, x_2, x_3, x_4\}$, $\omega_2 = \{x_5, x_6, x_7, x_8\}$ and $\omega_3 = \{x_9, x_{10}\}$.

To encode a SCAMO set of constraints as compositions of smaller constraints, we build two BDDs for each window with two different variable orderings (hence the name "duplex"). Notice that any SAT encoding technique of AMO constraints could be employed instead of BDDs (as long as we do duplex encoding by considering both directions). However, beyond the smaller AMO constraints, we further need AMZ constraints in order to connect the parts together (see the binary clause in Proposition 1). One benefit of BDDs is that we get these constraints automatically already by encoding the AMO constraints. Thus in this paper we will focus only on this BDD based approach.

Given window $\omega_i$ over variables $X_i = \{x_{i_1}, \ldots x_{i_w}\}$, we construct two two-rooted BDDs, both representing the same two constraints $x_{i_1} + \cdots + x_{i_w} \leq 1$ and $x_{i_1} + \cdots + x_{i_w} \leq 0$. The first BDD, which we call *forward* BDD, considers the AMO and AMZ constraints with a right associative variable grouping (i.e. with variable ordering $x_{i_1} < x_{i_2} < \ldots < x_{i_w}$). The other BDD, called *backward* BDD, represents the same constraints but with a left associative variable grouping (i.e. with variable ordering $x_{i_w} < x_{i_{w-1}} < \ldots < x_{i_1}$).

Abío et al. in [42] proposed a generalized arc-consistent, polynomial size ROBDD-based encoding for Pseudo-Boolean constraints. In our setting the constraints are all AMO or AMZ constraints without coefficients, and thus applying their approach leads to small and simple BDDs. The recursive algorithms in Fig. 3 present the main steps of this building process. In these procedures $\langle x_i \cdots x_j \rangle$ means an ordered sequence of consecutive variables and function if-then-else builds a BDD node with the given decision variable and high and low BDD nodes. Building the forward BDDs of a window $\omega_i$ simply means to call BDD-AMO and BDD-AMZ with $\langle x_{i_1} \cdots x_{i_w} \rangle$ as parameter. To build the backward BDDs, we need to call the methods with $\langle x_{i_w} \cdots x_{i_1} \rangle$ as argument. The result in both cases (see Example 3) will be a two-rooted BDD with height of at most $(w + 1)$.

---

**BDD-AMO** (consecutive variables $\langle x_i \cdots x_j \rangle$)

1   $\mathcal{B} := \text{Search-AMO}(\langle x_i \cdots x_j \rangle)$
2   **if** $\mathcal{B} = \emptyset$ **then**
3     **if** $|\langle x_i \cdots x_j \rangle| = 1$ **then**
4       $\mathcal{B}_T, \mathcal{B}_F := \top, \top$
5     **else**
6       $\mathcal{B}_T := \text{BDD-AMZ}(\langle x_{i+1} \cdots x_j \rangle)$
7       $\mathcal{B}_F := \text{BDD-AMO}(\langle x_{i+1} \cdots x_j \rangle)$
8     $\mathcal{B} := \text{if-then-else}(x_i, \mathcal{B}_T, \mathcal{B}_F)$
9   **return** $\mathcal{B}$

**BDD-AMZ** (consecutive variables $\langle x_i \cdots x_j \rangle$)

1   $\mathcal{B} := \text{Search-AMZ}(\langle x_i \cdots x_j \rangle)$
2   **if** $\mathcal{B} = \emptyset$ **then**
3     **if** $|\langle x_i \cdots x_j \rangle| = 1$ **then**
4       $\mathcal{B}_T, \mathcal{B}_F := \bot, \top$
5     **else**
6       $\mathcal{B}_T := \bot$
7       $\mathcal{B}_F := \text{BDD-AMZ}(\langle x_{i+1} \cdots x_j \rangle)$
8     $\mathcal{B} := \text{if-then-else}(x_i, \mathcal{B}_T, \mathcal{B}_F)$
9   **return** $\mathcal{B}$

**Fig. 3.** Algorithms BDD-AMO and BDD-AMZ to construct binary decision diagrams for constraints over a given sequence of consecutive Boolean variables.

---

Consider the following *layers* of these constructed BDDs. A non-leaf layer $l_j$ (where $1 \leq j \leq w$) of a forward BDD (backward BDD) consists of two nodes, one capturing the AMO and another node representing the AMZ constraint over variables $\langle x_{i_j} \cdots x_{i_w} \rangle$ (respectively $\langle x_{i_{w-(j-1)}} \cdots x_{i_1} \rangle$ for the backward BDD).

*Example 3.* The upper part of Fig. 4 shows what the forward BDD of $\omega_1$ in Example 2 looks like. The BDD is the result of calling BDD-AMO($\langle x_1 \, x_2 \, x_3 \, x_4 \rangle$) and BDD-AMZ($\langle x_1 \, x_2 \, x_3 \, x_4 \rangle$). Notice that due to the search for already existing BDDs at the beginning of each method (Search-AMO and Search-AMZ), the two calls result in a single shared structure (i.e. we have a partially reduced ordered BDD). Further notice that though node $x_4 \leq 1$ could be reduced simply to $\top$, we kept this node in the representation. In this BDD we can distinguish four layers $(l_1 - l_4)$ that refer to four sub-constraints of the root expressions.

The lower part of the figure depicts the backward BDD of $\omega_2$ in Example 2, resulting from calls BDD-AMO($\langle x_8\,x_7\,x_6\,x_5\rangle$) and BDD-AMZ($\langle x_8\,x_7\,x_6\,x_5\rangle$). The variable ordering here is $x_8 < x_7 < x_6 < x_5$. Notice that the structure of the two BDDs are identical, they just talk about different variables in different orders.

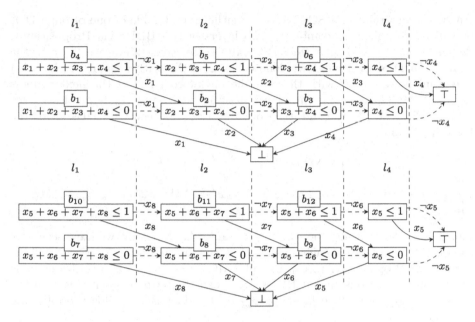

**Fig. 4.** Forward BDD of $\omega_1$ with variable ordering $x_1 < x_2 < x_3 < x_4$ and backward BDD of $\omega_2$ with ordering $x_8 < x_7 < x_6 < x_5$. Two-rooted partially reduced OBDDs to represent constraints $x_1 + x_2 + x_3 + x_4 \leq K$ with right and $x_5 + x_6 + x_7 + x_8 \leq K$ with left associative variable groupings, where $K \in \{0, 1\}$.

## 4.2  CNF Encoding of BDDs

During BDD construction (e.g. after Line 5 in both algorithms of Fig. 3), or later in an independent traversal, we can assign new Boolean variables to each non-unit non-leaf node. Notice that top nodes of the forward and backward BDDs over the same variables can use the same Boolean variable.

Now, given a node with auxiliary Boolean variable $b$, that decides on variable $x_i$ and has a true child node with variable $t$ and a false child node with variable $f$, we introduce clauses to encode $x_i \rightarrow (b \leftrightarrow t)$ and $\neg x_i \rightarrow (b \leftrightarrow f)$. However, there are several simplification possibilities due to the structure of our BDDs and our problem. For instance, all AMZ nodes have $\bot$ as a true child (see Fig. 4) and all AMO nodes are assumed as unit clauses (due to using them with Proposition 1). Nodes of a constraint $x_i \leq 1$ are simply encoded as $\top$, while nodes of constraints $x_i \leq 0$ are encoded as $\neg x_i$ in the clausal representation of the parent nodes.

*Example 4.* On Fig. 4 the introduced new Boolean variables are represented together with their nodes. For example, variable $b_6$ belongs to the node of constraint $x_3 + x_4 \leq 1$. The introduced clause regarding this node is $(\neg x_3 \vee \neg x_4)$.

### 4.3  Bonding Stairs

An AMO constraint of a SCAMO set is either a root node of one of our BDDs or can be described by combining two layers of two BDDs via Proposition 1. As last step of encoding a whole SCAMO set of constraints, we traverse the forward BDD of each window (denoted as $\omega_i^f$-BDD with $i \in \{1, \ldots, M - 1\}$) and combine its nodes with those of the backward BDD of the next window ($\omega_{i+1}^b$-BDD). Thus, we combine layer $l_j$ of $\omega_i^f$ with layer $l_{(w\ j)+2}$ of $\omega_{i+1}^b$ for each $j = 2, \ldots, w$. At the end, the bonding of two consecutive BDDs yields the following formula:

$$\mathrm{BOND}(\omega_i^f, \omega_{i+1}^b) = \omega_i^f\text{-}l_1\text{-AMO} \wedge$$

$$\bigwedge_{j=2}^{w} (\omega_i^f\text{-}l_j\text{-AMO} \wedge \omega_{i+1}^b\text{-}l_{(w-j)+2}\text{-AMO} \wedge (\omega_i^f\text{-}l_j\text{-AMZ} \vee \omega_{i+1}^b\text{-}l_{(w-j)+2}\text{-AMZ})).$$

*Example 5.* We continue the running example. At this point we have seen how to construct a BDD for each small stair structure in Fig. 2. Next we combine them using Proposition 1 to capture all AMO constraints. Figure 5 depicts how the layers of the constructed BDDs are meant to be paired with each other. Applying Proposition 1 on layers of $\omega_1^f$-BDD and $\omega_2^b$-BDD yields the following formula:

$$(x_1 + x_2 + x_3 + x_4 \leq 1) \wedge$$
$$(x_2 + x_3 + x_4 \leq 1) \wedge (x_5 \leq 1) \wedge (x_2 + x_3 + x_4 \leq 0 \vee x_5 \leq 0) \wedge$$
$$(x_3 + x_4 \leq 1) \wedge (x_5 + x_6 \leq 1) \wedge ((x_3 + x_4 \leq 0) \vee (x_5 + x_6 \leq 0)) \wedge$$
$$(x_4 \leq 1) \wedge (x_5 + x_6 + x_7 \leq 1) \wedge ((x_4 \leq 0) \vee (x_5 + x_6 + x_7 \leq 0)),$$

that translates to the clauses $b_4 \wedge b_5 \wedge \top \wedge (b_2 \vee \neg x_5) \wedge b_6 \wedge b_{12} \wedge (b_3 \vee b_9) \wedge \top \wedge b_{11} \wedge (\neg x_4 \vee b_8)$. Notice that with this set of clauses, together with the BDD clauses, we encoded the first four AMO constraints of our SCAMO problem.

### 4.4  Arc Consistency of Duplex Encoding

Notice that AMO, AMZ and SCAMO constraints are all monotonic decreasing Boolean functions, i.e. setting any of the variables to false does not restrict any other variables. Thus setting a variable to true affects only those variables that share at least one AMO constraint with it. Note that decomposing each AMO constraint of a SCAMO set based on Proposition 1 results in an equivalent problem. Although our constructed BDDs for this decomposition share most of their nodes with each other (due to the chosen variable orders), our method is still a BDD-based translation of each AMO and AMZ constraint into clauses.

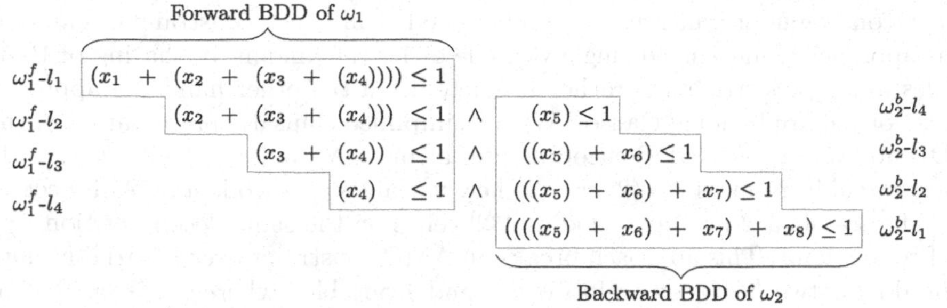

Forward BDD of $\omega_1$

| | | | |
|---|---|---|---|
| $\omega_1^f\text{-}l_1$ | $(x_1 + (x_2 + (x_3 + (x_4)))) \leq 1$ | | |
| $\omega_1^f\text{-}l_2$ | $(x_2 + (x_3 + (x_4))) \leq 1$ $\wedge$ | $(x_5) \leq 1$ | $\omega_2^b\text{-}l_4$ |
| $\omega_1^f\text{-}l_3$ | $(x_3 + (x_4)) \leq 1$ $\wedge$ | $((x_5) + x_6) \leq 1$ | $\omega_2^b\text{-}l_3$ |
| $\omega_1^f\text{-}l_4$ | $(x_4) \leq 1$ $\wedge$ | $(((x_5) + x_6) + x_7) \leq 1$ | $\omega_2^b\text{-}l_2$ |
| | | $((((x_5) + x_6) + x_7) + x_8) \leq 1$ | $\omega_2^b\text{-}l_1$ |

Backward BDD of $\omega_2$

**Fig. 5.** Combining forward and backward BDDs to encode SCAMO constraints.

Thus, applying an arc consistent encoding [48,49] on each BDD node (e.g. the one in Minisat+ [41]) makes our encoding arc consistent as well.

In fact, notice that our bonding clauses contain a unit clause for each AMO constraint in order to enforce the output of the corresponding (sub-)BDD to be true. Beyond that, it is not hard to see that setting an input variable to true falsifies the output variable of each AMZ-BDD containing it. Thus the binary clauses of the bonding clauses enforce the root-node of each respective AMZ constraint to be true, and in turn unit propagation, the main inference rule of SAT solvers, falsifies all the variables in them.

## 5    Comparing Encodings of Staircase Constraints

In this section we discuss commonly used existing SAT encodings of AMO constraints and possible SEQUENCE encodings of SCAMO constraints. We compare them to our proposed duplex encoding in the context of SCAMOs.

Let $N = (n-w)+1$ be the number of AMO constraints in a staircase problem set over $n$ variables and width $w$. A *naive* (also called *pair-wise* or *binomial*) encoding of a $w$-long AMO constraint is $\bigwedge_{i=1}^{(w-1)} \bigwedge_{j=i+1}^{(w)} (\neg x_i \vee \neg x_j)$. Although this approach does not require any additional Boolean variable, the number of clauses constructed with that encoding over $N$ $w$-long AMO constraints is $N \cdot ((w-1) + (w-2) + \ldots + (w-(w-1))) = N \cdot \frac{(w-1) \cdot w}{2}$.

Using the naive encoding on the SCAMO constraint set would produce more than once many of the binary clauses. Eliminating duplicated clauses yields the *reduced naive* encoding with $\frac{(w-1) \cdot w}{2} + (N-1) \cdot (w-1)$ unique clauses.

Sinz introduced in [37] a *sequential* counter encoding for Boolean cardinality constraints. Applying it to an AMO constraint over $w$ variables produces $3 \cdot w - 5$ binary clauses and introduces $w-2$ auxiliary variables. With $N$ AMO constraints this gives $N \cdot (3 \cdot w - 5)$ clauses and $N \cdot (w-2)$ new variables.

The BDD-*based* encoding for Pseudo-Boolean constraints [41,42] applied to AMO constraints is comparable to the sequential counter encoding. However, for a fixed variable order, the BDD built for each $w$-long AMO constraint of a SCAMO set, will always either contain a variable that does not occur in any other constraint or will miss a variable needed in other constraints. Thus for this approach using a fixed single variable order the amount of sharing of BDD nodes among constraints is rather restricted. On the other hand the approach does not require bonding clauses. With a simplified clausal representation of the BDD nodes, the naive BDD encoding uses at most $N \cdot (3 \cdot (w-2) + 2 \cdot (w-1) - 1)$ clauses and introduces $N \cdot (2 \cdot w - 3)$ new variables to encode a SCAMO set.

The so-called *2-product* encoding [32] relies on the same decomposition rule as Proposition 1. This approach breaks an AMO constraint over $w$ variables into a product of two AMO constraints over $p$ and $q$ variables, where $p*q \geq w$. To simplify the calculation we use $p = \lceil \sqrt{w} \rceil$ and $q = \lceil w/p \rceil$ as recommended in [32] and assume recursive 2-product encoding of the resulting smaller constraints. Even though this approach can efficiently encode a single AMO constraint, making use of shared sub-expressions is not straightforward. Thus, based on the estimations given in [32], the number of clauses is $N \cdot (2 \cdot w + 4 \cdot \sqrt{w} + O(\sqrt[4]{w}))$. Further, the number of newly introduced variables is $N \cdot (2 \cdot \sqrt{w} + O(\sqrt[4]{w}))$ again following [32].

Instead of focusing on specialized AMO encodings, it is also possible to encode a complete SCAMO set with more generic approaches, like the ones in [6]. For example, encoding SCAMO as a REGULAR constraint yields similar results as a naive BDD-based approach with a single variable order (i.e. $O(n \cdot w)$ size).

Another encoding (also from [6]) based on cumulative sums or difference constraints requires an internal representation which is at least quadratic size in the worst case. Similarly, partial sums (again see [6]) would consider every possible sub-sums which also yields $O(n \cdot w^2)$ constraints.

The size-wise most competitive sequence encoding from [6] is the log-based approach where a SCAMO set could be represented as $O(n \cdot log\, w)$ constraints.

## 5.1 Duplex Encoding

For a given constraint set over $n$ variables of width $w$ we construct two BDDs of the same size (each having $2 \cdot (w+1)$ nodes) for $M = \lceil \frac{n}{w} \rceil$ windows. To simplify the calculation, we will assume that each BDD has the same size (even though the last window is most of the time way smaller) and that we encode the first and last windows in both directions. Thus, we provide here just an upper bound on the actual values. With these assumptions we have $2 \cdot M$ BDDs. For each BDD we construct three clauses for the non-unit non-leaf AMZ nodes and at most two clauses for the non-unit non-leaf AMO nodes. Beyond these clauses, we need to bond together each layer of the neighbouring forward and backward BDDs, resulting in $M - 1$ bond-clause sets, each consisting of two unit and a binary clause. All in all, the final number of clauses in the encoding is as follows:

$$\#\text{BDD-clauses} \leq 2 \cdot M \cdot (3 \cdot (w-1) + 2 \cdot (w-1) - 1) = 10 \cdot M \cdot w - 12 \cdot M$$
$$\#\text{BOND-clauses} \leq (M-1) \cdot (3 \cdot (w-1) + 1) = 3 \cdot M \cdot w - 2 \cdot M - 3 \cdot w + 2$$
$$\#\text{BDD} + \#\text{BOND-clauses} \leq 13 \cdot M \cdot w - 14 \cdot M - 3 \cdot w + 2$$

The number of new variables at the very end of the encoding is at most $4 \cdot M \cdot (w-1)$ introducing one for each non-leaf non-unit node of our BDDs.

## 5.2    Comparison Summary

Table 1 summarizes the sizes of different SAT encodings expressed as functions over the number $n$ of all variables in a SCAMO constraint set and the width $w$ of the individual AMO constraints, combined into $N = (n - w) + 1$ (the number of AMO constraints) and $M = \lceil \frac{n}{w} \rceil$ (the number of windows in duplex encoding). The columns capture the number of auxiliary variables and number of clauses of the encodings. Notice that $M$ is significantly smaller than $N$. The last column gives the worst case of each approach, assuming $w = n/2$, where $N$ is approximately $n/2$ too. In this scenario existing encodings are quadratic or even cubic. However, in our duplex encoding we have $M = 2$ in that case and thus it remains linear.

Figure 6 visualizes the difference between SAT encodings for the fixed number of variables $n = 500$. The horizontal axis ranges over all possible widths $w$. Note that the naive encoding is only partially shown here, and further, that in our application $n/2$ is an upper bound on the width $w$, and thus only the left part of Fig. 6 is interesting up to the middle $w = n/2 = 250$.

The asymptotic behavior of the last column of Table 1 can be observed in Fig. 6 too. Again, the largest difference between the encodings occurs for $w = n/2$. According to Fig. 6 the reduced naive encoding turns out to be the best SAT-based alternative to our approach in terms of number of clauses. Though Fig. 6 focuses only on SAT encodings, note that the smallest sequence-based alternative (in [6]) would have size $\mathcal{O}(n \cdot log\, n)$ when $w = n/2$, that is smaller than most SAT encodings but larger than our proposed linear encoding.

**Table 1.** Comparison of size of SAT encodings of $w$-long SCAMO sets over $n$ variables. Columns #NewVars and #Clauses show the number of additional variables and clauses of each approach, where $N = (n - w) + 1$ and $M = \lceil \frac{n}{w} \rceil$.

| Encoding | #NewVars | #Clauses | WorstCase |
|---|---|---|---|
| Naive | 0 | $N \cdot \frac{(w-1)\cdot w}{2}$ | $\mathcal{O}(n^3)$ |
| Reduced | 0 | $\frac{(w-1)\cdot w}{2} + (N-1)\cdot(w-1)$ | $\mathcal{O}(n^2)$ |
| Sequential | $N \cdot (w - 2)$ | $N \cdot (3 \cdot (w-2) + 1)$ | $\mathcal{O}(n^2)$ |
| BDD | $N \cdot (2 \cdot w - 3)$ | $N \cdot (3 \cdot (w-2) + 2 \cdot (w-1) - 1)$ | $\mathcal{O}(n^2)$ |
| 2-Product | $N \cdot (2 \cdot \sqrt{w} + O(\sqrt[4]{w}))$ | $N \cdot (2 \cdot w + 4 \cdot \sqrt{w} + O(\sqrt[4]{w}))$ | $\mathcal{O}(n^2)$ |
| Duplex | $4 \cdot M \cdot (w - 1)$ | $13 \cdot M \cdot w - 14 \cdot M - 3 \cdot w + 2$ | $\mathcal{O}(n)$ |

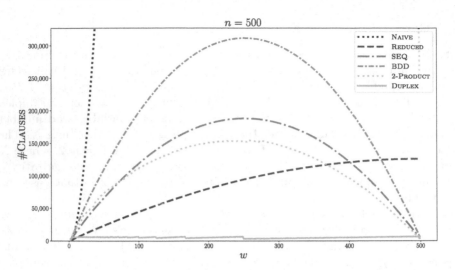

**Fig. 6.** Comparison of number of clauses for different encodings of a single SCAMO constraint set on $n = 500$ variables and width $w$ between 2 and 500.

## 6 Experimental Evaluation

Formulating the antibandwidth problem iteratively, as it was proposed in [8] (see Sect. 2), asks whether there exists a labelling for a graph $G = (V, E)$ s.t. $AB(G) \geq k + 1$. The question has $2 \cdot |V|$ pieces of $|V|$-long exactly-one constraints (as (LABELS) and (VERTICES)) and for each edge of the graph (i.e. $|E|$ times) a $(|V| - k)$ big set of AMO constraints, each over $2 \cdot k$ variables (as (OBJ$_k$)).

An off-the-shelf SAT solution could encode each of the AMO and exactly-one constraints one-by-one (e.g. as in Sect. 5). However, for a given edge between nodes $i, i'$ (i.e. $\{i, i'\} \in E$) constraint (OBJ$_k$) can be reformulated as

$$\bigwedge_{\lambda=1}^{(|V|-k)} \left( \sum_{\ell=\lambda}^{(\lambda+k)} x_i^\ell + x_{i'}^\ell \leq 1 \right) \overset{Prop.\ 1}{\equiv}$$

$$\bigwedge_{\lambda=1}^{(|V|-k)} \left( \sum_{\ell=\lambda}^{(\lambda+k)} x_i^\ell \leq 1 \wedge \sum_{\ell=\lambda}^{(\lambda+k)} x_{i'}^\ell \leq 1 \wedge \left( \sum_{\ell=\lambda}^{(\lambda+k)} x_i^\ell \leq 0 \vee \sum_{\ell=\lambda}^{(\lambda+k)} x_{i'}^\ell \leq 0 \right) \right).$$

In that form we have exactly two SCAMO sets of width $k + 1$, one over the variables of node $i$ and another over variables of $i'$. The third component of the decomposition takes the disjunction of AMZ constraints that can be constructed easily by combining our smaller AMZ nodes corresponding to the SCAMO sets.

The staircase structure in (OBJ$_k$) allows to apply our new duplex encoding by simply encoding a SCAMO set of width $k + 1$ for each node of the graph (i.e. $|V|$ times) and combining the corresponding AMZ constraints (with less than $4 \cdot (|V| - k)$ binary clauses for each edge). This encodes all AMO constraints of the problem. Also note that we can reuse the Boolean variables representing the root nodes of the constructed AMO BDDs to encode the (VERTICES) constraints.

**Experimental Results**

We implemented a framework to compare off-the-shelf SAT encodings in practice to our proposed SCAMO based duplex encoding on the antibandwidth problem (as formulated in Sect. 2). Beyond SAT encodings, we also compared our approach against alternative exact methods to solve the problem, like Constraint Programming or the iterative method presented in [8] based on feasibility-MIPs.

The experiments considered 24 matrices of the Harwell-Boeing Sparse Matrix Collection [50], containing 12 relatively small and 12 rather large graphs (as in [8]). For each graph lower bounds (by a construction heuristic) and theoretical upper bounds of the antibandwidth were provided in [8]. These values were reused in our experiment as starting and ending points for the iterative methods and as lower bounds in the CP approaches. All reported results were experimented on our cluster with Intel Xeon E5-2620 v4 @ 2.10 GHz CPUs.

Table 2 summarizes our results.[1] For each graph it shows the number of nodes and edges, the starting width or lower bound and last width to check of the solving methods (columns $|V|, |E|$ and LB, UB). Then for each solving technique we report the best found solution together with the time (in seconds) and memory consumption (in MB). Each approach was limited to 1800 s and 120 GB memory. This rather high main memory limit is due to trying to solve the alternative SAT encodings with a large number of clauses as well, while the other methods never exceeded 4 GB.

We compare the 2-product [32] and reduced naive AMO encodings to our proposed duplex SCAMO encoding as the first three techniques in Table 2. All three techniques are implemented in the same framework and follow the same method: encode (considering LB as width of SCAMO or as $k$ of the AMO constraints) and solve the SAT representation of the problem with a SAT solver (we used CaDiCaL 1.2.1 [51]). If it is satisfiable, increase the width and start again to encode and solve the new problem. If it is unsatisfiable or the width is UB, it means that the optimal solution of ABP was found and the process ends. At the moment when the 1800 s or 120 GB is exceeded, the method stops (with TO or MO respectively). The reported solutions are the highest widths with what the formula was still successfully constructed and solved. In case even the first formula was too hard to solve, it is marked with "–".

While the 2-product encoding of the largest instance had a memory out during solving the first formula (after a successful encoding), the reduced naive approach required less memory and even solved a few of the larger problems with more than one width in 1800 s. The duplex encoding required significantly less memory and was faster in encoding and solving the problems compared to the other SAT approaches. It performed well also compared to further techniques.

The next two approaches, $F_e(k)$ and CP-CPLEX, are taken from [8] as is, and were executed on our cluster for comparison. Note that while CP-CPLEX

---

[1] Source code, data and benchmarks are available at http://fmv.jku.at/duplex/.

**Table 2.** Results of different approaches to solve the antibandwidth problem (TO = 1800 s and MO = 120 GB).

| Instance | \|V\| | \|E\| | LB | UB | 2-Product | | | Reduced Naive | | | Duplex | | | $F_e(k)$ [8] | | | CP-CPLEX [8] | | | CP-MZ-Chuffed | | |
|---|---|---|---|---|---|---|---|---|---|---|---|---|---|---|---|---|---|---|---|---|---|---|
| | | | | | Obj. | Time | MB | Obj. | Time | MB | Obj. | Time | MB | Obj. | Time | MB | Obj. | Time | MB | Obj. | Time | MB |
| A-pores_1 | 30 | 103 | 6 | 8 | 6 | 206.85 | 80 | 6 | 166.48 | 68 | 6 | 185.52 | 52 | 6 | 23.71 | 29 | 6–8 | TO | 57 | 6 | 5.97 | 11 |
| B-ibm32 | 32 | 90 | 9 | 9 | 9 | 14.06 | 51 | 9 | 46.03 | 47 | 9 | 1.30 | 11 | 9 | 28.57 | 29 | 9 | 7.35 | 20 | 9 | 17.4 | 11 |
| C-bcspwr01 | 39 | 46 | 16 | 17 | 17 | 83.12 | 69 | 17 | 56.02 | 59 | 17 | 3.85 | 13 | 17 | 6.64 | 28 | 17 | 18.78 | 21 | 17 | TO | 11 |
| D-bcsstk01 | 48 | 176 | 8 | 9 | 9 | 14.41 | 139 | 9 | 8.59 | 47 | 9 | 0.25 | 14 | 9 | 62.28 | 36 | 9 | 20.15 | 21 | 9 | 6.35 | 12 |
| E-bcspwr02 | 49 | 59 | 21 | 22 | 21 | 36.17 | 76 | 21 | 53.01 | 80 | 21 | 3.37 | 13 | 21 | 774.02 | 205 | 21 | 22.84 | 19 | 21 | 673.44 | 11 |
| F-curtis54 | 54 | 124 | 12 | 13 | 13 | 20.89 | 139 | 13 | 1.02 | 41 | 13 | 1.33 | 18 | 13 | 12.56 | 32 | 13 | 34.66 | 21 | 13 | 2.14 | 11 |
| G-will57 | 57 | 127 | 12 | 14 | 13 | 108.79 | 164 | 13 | 26.8 | 79 | 13 | 0.57 | 19 | 13 | 15.4 | 33 | 13 | 44.75 | 21 | 13 | 2.69 | 11 |
| H-impcol_b | 59 | 281 | 8 | 8 | 8 | 5.51 | 173 | 8 | 0.47 | 52 | 8 | 0.54 | 22 | 8 | 0.47 | 24 | 8–22 | TO | 63 | 8 | 23.3 | 12 |
| I-ash85 | 85 | 219 | 19 | 27 | 21 | TO | 794 | 21 | TO | 658 | 23 | TO | 331 | 20 | TO | 133 | 22–31 | TO | 37 | 21 | TO | 12 |
| J-nos4 | 100 | 247 | 32 | 40 | 32 | TO | 1037 | 32 | TO | 911 | 35 | 585.33 | 190 | | TO | 106 | 34–47 | TO | 31 | | TO | 12 |
| K-dwt_234 | 117 | 162 | 46 | 58 | 47 | TO | 924 | 47 | TO | 957 | 49 | TO | 477 | 48 | TO | 26 | 51–57 | TO | 33 | | TO | 11 |
| L-bcspwr03 | 118 | 179 | 39 | 39 | 39 | 22.82 | 662 | 39 | 6.92 | 436 | 39 | 0.99 | 58 | 39 | 0.52 | 21 | 39 | 110.92 | 22 | 39 | 26.42 | 12 |
| M-bcsstk06 | 420 | 3720 | 28 | 72 | | TO | 53392 | 29 | TO | 22076 | 34 | TO | 1621 | 33 | TO | 625 | | TO | 20 | | TO | 35 |
| N-bcsstk07 | 420 | 3720 | 28 | 72 | | TO | 53392 | 29 | TO | 22097 | 34 | TO | 1621 | 33 | TO | 634 | | TO | 20 | | TO | 35 |
| O-impcol_d | 425 | 1267 | 91 | 173 | | TO | 30306 | 92 | TO | 22285 | 99 | TO | 1043 | 95 | TO | 691 | | TO | 18 | | TO | 24 |
| P-can_445 | 445 | 1682 | 78 | 120 | | TO | 41572 | | TO | 27030 | | TO | 1581 | | TO | 644 | | TO | 18 | | TO | 24 |
| Q-494_bus | 494 | 586 | 219 | 246 | | TO | 25944 | | TO | 29640 | | TO | 1167 | 220 | TO | 905 | | TO | 18 | | TO | 21 |
| R-dwt_503 | 503 | 2762 | 46 | 71 | | TO | 56611 | 47 | TO | 35227 | 62 | TO | 1680 | 52 | TO | 911 | | TO | 19 | | TO | 31 |
| S-sherman4 | 546 | 1341 | 256 | 272 | | TO | 73031 | | TO | 59860 | | TO | 1129 | | TO | 1033 | | TO | 19 | | TO | 24 |
| T-dwt_592 | 592 | 2256 | 103 | 150 | | TO | 85816 | | TO | 62936 | | TO | 2253 | | TO | 1068 | | TO | 20 | | TO | 37 |
| U-662_bus | 662 | 906 | 219 | 220 | | TO | 63844 | | TO | 68402 | 220 | 319.73 | 1564 | | TO | 1320 | | TO | 19 | | TO | 28 |
| V-nos6 | 675 | 1290 | 326 | 337 | | TO | 101724 | | TO | 90129 | | TO | 1571 | | TO | 1434 | | TO | 20 | | TO | 28 |
| W-685_bus | 685 | 1282 | 136 | 136 | | TO | 76110 | | TO | 72839 | 136 | 14.33 | 1428 | 136 | 9.24 | 37 | | TO | 20 | | TO | 29 |
| X-can_715 | 715 | 2975 | 112 | 142 | | 686.23 | MO | | TO | 106462 | | TO | 3312 | | TO | 1468 | | TO | 21 | | TO | 39 |

knows LB, $F_e(k)$ constructs it internally. The last reported approach is based on Chuffed [44,52] via the MiniZinc language [53]. This hybrid solver employs lazy clause generation and combines the strengths of SAT and finite domain solving techniques. Note that both CP approaches encode the ABP naively as a labeling problem to maximize smallest neighbour-distances, using state-of-the-art solvers off-the-shelf. All in all we can see that the SCAMO based duplex encoding of the ABP is comparable and most of the time even better than other approaches.

# 7 Conclusion and Outlook

In this paper we have proposed a new SAT encoding for at-most-one constraints with a staircase structure, i.e. where consecutive constraints share sequences of sub-expressions in a structured way. This structure is exploited in an encoding which relies on binary decision diagrams using two variable orderings. Compared to alternative encodings for the ABP, our encoding outperforms the existing ones.

In the future we plan to integrate and interleave the MIP based approach of [8] and the SAT approach proposed here. Further, we want to apply the proposed method to other problems featuring at-most-one constraints with a staircase structure. Another intriguing direction for future work is to explore how symbolic optimization techniques using decision diagrams [54] can take advantage of multiple variable orders simultaneously, which is essential to keep our encoding compact.

**Acknowledgments.** This research has been supported by the Austrian Science Fund (FWF) under projects W1255-N23, S11408-N23 and by the LIT AI Lab funded by the State of Upper Austria. The authors would like to thank the reviewers for their useful suggestions and helpful comments.

# References

1. Achterberg, T.: SCIP: solving constraint integer programs. Math. Program. Comput. **1**(1), 1–41 (2009). https://doi.org/10.1007/s12532-008-0001-1
2. Vielma, J.P.: Mixed integer linear programming formulation techniques. SIAM Rev. **57**(1), 3–57 (2015)
3. Bofill, M., Coll, J., Suy, J., Villaret, M.: SAT encodings of pseudo-Boolean constraints with at-most-one relations. In: Rousseau, L.-M., Stergiou, K. (eds.) CPAIOR 2019. LNCS, vol. 11494, pp. 112–128. Springer, Cham (2019). https://doi.org/10.1007/978-3-030-19212-9_8
4. Beldiceanu, N., Contejean, E.: Introducing global constraints in CHIP. Math. Comput. Model. **20**, 97–123 (1996)
5. Bessiere, C., Hebrard, E., Hnich, B., Kiziltan, Z., Walsh, T.: SLIDE: a useful special case of the CARDPATH constraint. In: Ghallab, M., Spyropoulos, C.D., Fakotakis, N., Avouris, N.M. (eds.) ECAI 2008–Proceedings of the 18th European Conference on Artificial Intelligence, Patras, Greece, 21–25 July 2008, Volume 178 of Frontiers in Artificial Intelligence and Applications, pp. 475–479. IOS Press (2008)

6. Brand, S., Narodytska, N., Quimper, C.-G., Stuckey, P., Walsh, T.: Encodings of the SEQUENCE constraint. In: Bessière, C. (ed.) CP 2007. LNCS, vol. 4741, pp. 210–224. Springer, Heidelberg (2007). https://doi.org/10.1007/978-3-540-74970-7_17

7. van Hoeve, W.-J., Pesant, G., Rousseau, L.-M., Sabharwal, A.: Revisiting the sequence constraint. In: Benhamou, F. (ed.) CP 2006. LNCS, vol. 4204, pp. 620–634. Springer, Heidelberg (2006). https://doi.org/10.1007/11889205_44

8. Sinnl, M.: A note on computational approaches for the antibandwidth problem. Cent. Eur. J. Oper. Res. (2020). https://doi.org/10.1007/s10100-020-00688-4

9. Gallian, J.A.: A dynamic survey of graph labeling. Electron. J. Comb. 16(6), 1–219 (2009)

10. Leung, J.Y., Vornberger, O., Witthoff, J.D.: On some variants of the bandwidth minimization problem. SIAM J. Comput. 13(3), 650–667 (1984)

11. Cappanera, P.: A survey on obnoxious facility location problems (1999)

12. Hale, W.K.: Frequency assignment: theory and applications. Proc. IEEE 68(12), 1497–1514 (1980)

13. Gansner, E.R., Hu, Y., Kobourov, S.: GMap: visualizing graphs and clusters as maps. In: 2010 IEEE Pacific Visualization Symposium (PacificVis), pp. 201–208. IEEE (2010)

14. Miller, Z., Pritikin, D.: On the separation number of a graph. Networks 19(6), 651–666 (1989)

15. Liu, Y., Yuan, J.: The dual bandwidth problem for graphs. J. Zhengzhou Univ. 35(1), 1–5 (2003)

16. Raspaud, A., Schröder, H., Sỳkora, O., Torok, L., Vrt'o, I.: Antibandwidth and cyclic antibandwidth of meshes and hypercubes. Discrete Math. 309(11), 3541–3552 (2009)

17. Wang, X., Wu, X., Dumitrescu, S.: On explicit formulas for bandwidth and antibandwidth of hypercubes. Discrete Appl. Math. 157(8), 1947–1952 (2009)

18. Dobrev, S., Královič, R., Pardubská, D., Török, L., Vrt'o, I.: Antibandwidth and cyclic antibandwidth of hamming graphs. Discrete Appl. Math. 161(10–11), 1402–1408 (2013)

19. Bekos, M.A., Kaufmann, M., Kobourov, S., Veeramoni, S.: A note on maximum differential coloring of planar graphs. J. Discrete Algorithms 29, 1–7 (2014)

20. Bansal, R., Srivastava, K.: Memetic algorithm for the antibandwidth maximization problem. J. Heuristics 17(1), 39–60 (2011). https://doi.org/10.1007/s10732-010-9124-4

21. Duarte, A., Martí, R., Resende, M.G., Silva, R.M.: GRASP with path relinking heuristics for the antibandwidth problem. Networks 58(3), 171–189 (2011)

22. Lozano, M., Duarte, A., Gortázar, F., Martí, R.: Variable neighborhood search with ejection chains for the antibandwidth problem. J. Heuristics 18(6), 919–938 (2012). https://doi.org/10.1007/s10732-012-9213-7

23. Scott, J., Hu, Y.: Level-based heuristics and hill climbing for the antibandwidth maximization problem. Numer. Linear Algebra Appl. 21(1), 51–67 (2014)

24. van den Akker, J.: LP-based solution methods for single-machine scheduling problems. Ph.D. thesis, Technische Universiteit Eindhoven - Department of Mathematics and Computer Science (1994)

25. Boland, N., Kalinowski, T., Waterer, H., Zheng, L.: Mixed integer programming based maintenance scheduling for the hunter valley coal chain. J. Sched. 16(6), 649–659 (2013). https://doi.org/10.1007/s10951-012-0284-y

26. Maravelias, C.T.: On the combinatorial structure of discrete-time MIP formulations for chemical production scheduling. Comput. Chem. Eng. 38, 204–212 (2012)

27. Burke, E.K., Causmaecker, P.D., Berghe, G.V., Landeghem, H.V.: The state of the art of nurse rostering. J. Sched. **7**(6), 441–499 (2004). https://doi.org/10.1023/B: JOSH.0000046076.75950.0b
28. Ernst, A.T., Jiang, H., Krishnamoorthy, M., Sier, D.: Staff scheduling and rostering: a review of applications, methods and models. Eur. J. Oper. Res. **153**(1), 3–27 (2004)
29. Dincbas, M., Simonis, H., Hentenryck, P.V.: Solving the car-sequencing problem in constraint logic programming. In: Kodratoff, Y. (ed.) Proceedings of the 8th European Conference on Artificial Intelligence, ECAI 1988, Munich, Germany, 1–5 August 1988, pp. 290–295. Pitmann Publishing, London (1988)
30. Solnon, C., Cung, V., Nguyen, A., Artigues, C.: The car sequencing problem: overview of state-of-the-art methods and industrial case-study of the ROADEF'2005 challenge problem. Eur. J. Oper. Res. **191**(3), 912–927 (2008)
31. Prestwich, S.D.: CNF encodings. In: Biere, A., Heule, M., van Maaren, H., Walsh, T. (eds.) Handbook of Satisfiability. Volume 185 of Frontiers in Artificial Intelligence and Applications, pp. 75–97. IOS Press (2009)
32. Chen, J.: A new sat encoding of the at-most-one constraint. In: Proceedings of the Constraint Modelling and Reformulation (2010)
33. Manthey, N., Heule, M.J.H., Biere, A.: Automated reencoding of Boolean formulas. In: Biere, A., Nahir, A., Vos, T. (eds.) HVC 2012. LNCS, vol. 7857, pp. 102–117. Springer, Heidelberg (2013). https://doi.org/10.1007/978-3-642-39611-3_14
34. Hölldobler, S., Nguyen, V.H.: On SAT-encodings of the at-most-one constraint. In: Katsirelos, G., Quimper, C.G. (eds.) Proceedings of the Twelfth International Workshop on Constraint Modelling and Reformulation, Uppsala, Sweden, 16–20 September 2013, pp. 1–17 (2013)
35. Knuth, D.E.: The Art of Computer Programming, Volume 4B, Fascicle 6: Satisfiability. Addison-Wesley, Boston (2015)
36. Nguyen, V.: SAT encodings of finite-CSP domains: a survey. In: Proceedings of the Eighth International Symposium on Information and Communication Technology, Nha Trang City, Viet Nam, 7–8 December 2017, pp. 84–91. ACM (2017)
37. Sinz, C.: Towards an optimal CNF encoding of Boolean cardinality constraints. In: van Beek, P. (ed.) CP 2005. LNCS, vol. 3709, pp. 827–831. Springer, Heidelberg (2005). https://doi.org/10.1007/11564751_73
38. Liffiton, M.H., Maglalang, J.C.: A cardinality solver: more expressive constraints for free. In: Cimatti, A., Sebastiani, R. (eds.) SAT 2012. LNCS, vol. 7317, pp. 485–486. Springer, Heidelberg (2012). https://doi.org/10.1007/978-3-642-31612-8_47
39. Biere, A., Le Berre, D., Lonca, E., Manthey, N.: Detecting cardinality constraints in CNF. In: Sinz, C., Egly, U. (eds.) SAT 2014. LNCS, vol. 8561, pp. 285–301. Springer, Cham (2014). https://doi.org/10.1007/978-3-319-09284-3_22
40. Roussel, O., Manquinho, V.M.: Pseudo-Boolean and cardinality constraints. In: Biere, A., Heule, M., van Maaren, H., Walsh, T. (eds.) Handbook of Satisfiability. Volume 185 of Frontiers in Artificial Intelligence and Applications, pp. 695–733. IOS Press (2009)
41. Eén, N., Sörensson, N.: Translating pseudo-Boolean constraints into SAT. JSAT **2**(1–4), 1–26 (2006)
42. Abío, I., Nieuwenhuis, R., Oliveras, A., Rodríguez-Carbonell, E., Mayer-Eichberger, V.: A new look at BDDs for pseudo-Boolean constraints. J. Artif. Intell. Res. **45**, 443–480 (2012)

43. Philipp, T., Steinke, P.: PBLib – a library for encoding pseudo-Boolean constraints into CNF. In: Heule, M., Weaver, S. (eds.) SAT 2015. LNCS, vol. 9340, pp. 9–16. Springer, Cham (2015). https://doi.org/10.1007/978-3-319-24318-4_2

44. Feydy, T., Stuckey, P.J.: Lazy clause generation reengineered. In: Gent, I.P. (ed.) CP 2009. LNCS, vol. 5732, pp. 352–366. Springer, Heidelberg (2009). https://doi.org/10.1007/978-3-642-04244-7_29

45. Abío, I., Gange, G., Mayer-Eichberger, V., Stuckey, P.J.: On CNF encodings of decision diagrams. In: Quimper, C.-G. (ed.) CPAIOR 2016. LNCS, vol. 9676, pp. 1–17. Springer, Cham (2016). https://doi.org/10.1007/978-3-319-33954-2_1

46. Bryant, R.E.: Graph-based algorithms for Boolean function manipulation. IEEE Trans. Comput. 35(8), 677–691 (1986)

47. Bryant, R.E.: Binary decision diagrams. Handbook of Model Checking, pp. 191–217. Springer, Cham (2018). https://doi.org/10.1007/978-3-319-10575-8_7

48. Gent, I.P.: Arc consistency in SAT. In: van Harmelen, F. (ed.) Proceedings of the 15th European Conference on Artificial Intelligence, ECAI 2002, Lyon, France, 2002 July 2002, pp. 121–125. IOS Press (2002)

49. Bacchus, F.: GAC via unit propagation. In: Bessière, C. (ed.) CP 2007. LNCS, vol. 4741, pp. 133–147. Springer, Heidelberg (2007). https://doi.org/10.1007/978-3-540-74970-7_12

50. Rodriguez-Tello, E., Romero-Monsivais, H., Ramírez-Torres, J., Lardeux, F.: Harwell-boeing graphs for the CB problem (2015). https://www.researchgate.net/publication/272022702_Harwell-Boeing_graphs_for_the_CB_problem

51. Biere, A.: CaDiCaL at the SAT Race 2019. In: Heule, M., Järvisalo, M., Suda, M. (eds.) Proceedings of SAT Race 2019 - Solver and Benchmark Descriptions. Volume B-2019-1 of Department of Computer Science Series of Publications B, pp. 8–9. University of Helsinki (2019)

52. Stuckey, P.J.: Lazy clause generation: combining the power of SAT and CP (and MIP?) solving. In: Lodi, A., Milano, M., Toth, P. (eds.) CPAIOR 2010. LNCS, vol. 6140, pp. 5–9. Springer, Heidelberg (2010). https://doi.org/10.1007/978-3-642-13520-0_3

53. Nethercote, N., Stuckey, P.J., Becket, R., Brand, S., Duck, G.J., Tack, G.: MiniZinc: towards a standard CP modelling language. In: Bessière, C. (ed.) CP 2007. LNCS, vol. 4741, pp. 529–543. Springer, Heidelberg (2007). https://doi.org/10.1007/978-3-540-74970-7_38

54. Bergman, D., Ciré, A.A., van Hoeve, W., Hooker, J.N.: Decision Diagrams for Optimization. Artificial Intelligence: Foundations, Theory, and Algorithms. Springer, Cham (2016). https://doi.org/10.1007/978-3-319-42849-9

# Core-Guided and Core-Boosted Search for CP

Graeme Gange[1(✉)], Jeremias Berg[2], Emir Demirović[3], and Peter J. Stuckey[1]

[1] Monash University, Melbourne, Australia
{graeme.gange,peter.stuckey}@monash.edu
[2] HIIT, Department of Computer Science, University of Helsinki, Helsinki, Finland
jeremias.berg@helsinki.fi
[3] University of Melbourne, Melbourne, Australia
emir.demirovic@unimelb.edu.au

**Abstract.** Core-guided search has proven to be the state-of-the-art in finding optimal solutions for maximum Boolean satisfiability and these techniques have recently been successfully imported in constraint programming. While effective on a wide range of problems, the methods are direct translations of their propositional logic counterparts. We propose two reformulation techniques that take advantage of the rich formalism offered by constraint programming rather than relying on propositional logic strategies, and generalise two existing techniques to improve core-extraction and the overall performance. Our experiments demonstrate the effectiveness of our approaches over the conventional (core-guided) CP methods, both in terms of proving optimality and quickly computing high-quality solutions.

## 1 Introduction

Discrete optimisation problems are ubiquitous: they include scheduling, rostering, production planning, and many other important questions. Optimal or good solutions to these problems result in more efficient use of scarce resources, saving time, money and the environment. Because of their importance, there are many paradigms to solve optimisation problems, including Mixed Integer Programming (MIP), Constraint Programming (CP), Maximum Satisfiability (MaxSAT) and local search. In this work, we focus on constraint programming, and in particular on improving core-guided search for CP.

There are two main approaches to optimization in constraint programming: 1) *branch-and-bound*, that iteratively improves a best known solution during search, and 2) *core-guided search*, where the algorithm assumes all constraints can be satisfied, and upon detecting infeasibility, relaxes the assumptions and reiterates. Branch-and-bound and core-guided search can be seen as upper and lower bounding methods, respectively. Branch and bound is by far the most used approach in CP.

Core-guided search originates from the MaxSAT community, where problems are specified as propositional logic formulae. It is one of the central approaches

© Springer Nature Switzerland AG 2020
E. Hebrard and N. Musliu (Eds.): CPAIOR 2020, LNCS 12296, pp. 205–221, 2020.
https://doi.org/10.1007/978-3-030-58942-4_14

to *complete* MaxSAT solving, as a large portion complete solvers in the annual MaxSAT Evaluation use the core-guiding methodology. In contrast, in CP, core-guided approaches have only recently been developed. In particular, the solver LCG-Glucose-UC, based on core-guiding, achieved the third highest score in the MiniZinc Challenge 2016, and OR-tools has introduced core-guided search in 2018 as part of their parallel solver. While these techniques have seen success in CP, they do not, in fact, use the expressivity offered by the constraint programming framework. Indeed, the methods are direct translations of the MaxSAT approaches, which were originally developed for the low-level language of propositional logic.

In this work, we advance the state-of-the-art for core-guided search in CP by exploiting the high-level language constructs offered by constraint programming. We provide two novel reformulation techniques for CP which are unique to constraint programming. Moreover, we generalise two existing techniques for CP, namely assumption probing to improve core-extraction and core-boosting [7] to increase the overall performance. We also discuss an issue with using explanation lifting with the conventional CP core-guided approach and note a number of techniques adapted from the MaxSAT community which play an important role in obtain high quality results. Our experiments on benchmarks from the MiniZinc Challenge show improvements over the conventional (core-guided) CP approaches, both in terms of the number of instances solved to optimality and the ability to quickly produce high-quality solutions.

The rest of the paper is organised as follows. In the next section, we introduce basic notations; constraint programming solvers with explanations, and core-guided MaxSAT methods along with their conventional translation for CP. Our main contributions are given in Sect. 3, together with additonal techniques that improve empirical performance. A report on the experimental evaluation is given in Sect. 4. We conclude in Sect. 5.

## 2    Preliminaries

**Notation:** A Boolean variable can take values *true* (1) or *false* (0). A literal is a Boolean variable $b$ or its negation $\neg b$. A clause is a disjunction (set of) of literals. The set $\text{VAR}(\mathcal{F})$ and $\text{LIT}(\mathcal{F})$ contain all variables (resp. literals) of the formula $\mathcal{F}$. The binary variable $\langle x \square k \rangle$ is an indicator for the condition $x \square k$ being true, e.g., $\langle x > k \rangle$ is true if the integer variable $x$ is assigned a value greater than $k$. For two formulas $R$ and $L$, $R \Rightarrow L$ denotes logical entailment, i.e., all models $R$ are also models of $L$. We use the notation $unsat(L)$ to indicate that formula $L$ is unsatisfiable (i.e. has no models). We will use $[\![x]\!]_l^u$ to denote the function returning the *excess* of $x$ above $l$, up to $u$. That is, $[\![x]\!]_l^u = \max(0, \min(u, x) - l)$. For convenience, we write $[\![x]\!]_l^\infty$ as $[\![x]\!]_l$.

CP solvers typically (implicitly) reason about an integer variables $x$ taking values in $[l...u]$ by using the *atomic constraints*: $x \geqslant d$, $x \leqslant d$, $x = d$, $x \neq d$ and *false* for $d \in [l..u]$. Given a set of constraints $\mathcal{F}$, the current domain $D$, seen as a formula containing conjunction of atomic constraints, represents all possible values that each variable $x \in \text{VAR}(\mathcal{F})$ can take. We denote the upper and lower bound (in the current domain) of an integer variable $x$ by $\mathsf{ub}(x)$ and $\mathsf{lb}(x)$, respectively. A propagator $f_c$ for a constraint $c \in \mathcal{F}$ takes the current domain $D$ and returns a set $f_c(D)$ of atomic constraints such that each $r \in f_c(D)$ is entailed by $D \wedge c$ but not $D$ alone, i.e. $D \not\Rightarrow r$ and $D \wedge c \Rightarrow r$. If $false \in f_c(D)$, the propagator has detected unsatisfiability, i.e. that the current domain is inconsistent with the constraint $c$. For example, if $c$ is a clause, then $f_c(D) = \{false\}$ if the current domain sets all literals in $c$ to 0. If instead, $D$ sets all but one literal $l \in c$ to false, then $f_c(D) = \{l = 1\}$.

We consider the problem of minimising a linear objective function $\sum_i w_i x_i$, subject to a set of constraints $\mathcal{F}$. Here each $x_i$ is an integer variable taking values in some domain $[l..u]$ assigned a weight $w_i$. Whenever all variables are binary and all constraints in $\mathcal{F}$ are clauses, we talk about a (weighted partial) MaxSAT problem. We say that a literal $b_i \in \mathrm{LIT}(\mathcal{F})$ for which $w_i > 0$ is an *objective* variable and denote the set of all objective variables by $S(\mathcal{F})$. A model $\tau$ of $\mathcal{F}$ is a solution and has cost $\mathrm{COST}(\mathcal{F}, \tau) = \sum_{b_i \in S(\mathcal{F})} w_i \tau(b_i)$. A solution $\tau$ is optimal if $\mathrm{COST}(\mathcal{F}, \tau) \leqslant \mathrm{COST}(\mathcal{F}, \tau')$ holds for all solutions $\tau'$ to $\mathcal{F}$. Note that the traditional description of MaxSAT is somewhat different to the above, but this is closer to the mathematical view of optimisation problems, and more closely reflects how MaxSAT solvers (including core-guided solvers) work internally [9].

An important concept in this work is that of an (unsatisfiable) *core*. Given a set $\mathcal{F}$ of constraints and conjunction of *assumption* atomic constraints $A$, a core $\kappa$ is a set $\kappa$ of atomic constraints s.t. $\mathcal{F} \Rightarrow \kappa$ and $\kappa \Rightarrow \neg A$. In other words, a core is a nogood made up of negated assumptions. A key observation for core-guided search methods is that the existence of a core $\kappa$ that only contains objective variables implies a lower bound $w^\kappa = \min\{w_i \mid b_i \in \kappa\}$ on the objective function $\sum_{i=1}^n w_i b_i$. Core-guided search methods make use of this fact by reformulating the instance during search. More specifically, given a lower bound $\mathsf{lb}(x_i)$ on each (integer) objective variable $x_i$, the objective function can be rewritten as $\sum_{i=1}^n w_i x_i = \sum_{i=1}^n w_i \mathsf{lb}(x_i) + \sum_{i=1}^n w_i \llbracket x_i \rrbracket_{\mathsf{lb}(x_i)} = C_{lb} + \sum_{i=1}^n w_i \llbracket x_i \rrbracket_{\mathsf{lb}(x_i)}$ where $C_{lb}$ is constant. We also note that, for an integer variable $x$, the following holds: $\llbracket x \rrbracket_k = \langle x > k \rangle + \llbracket x \rrbracket_{k+1}$, and $\langle x > k \rangle = \llbracket x \rrbracket_k^{k+1}$. For an integer variable $x$ with initial domain $[l..u]$ then $x = l + \llbracket x \rrbracket_l^u$.

**Lifting Explanations in CP Solving:** Similarly to how core-guided MaxSAT solvers make extensive use of conflict driven clause learning (CDCL) SAT solving under assumptions [13], core-guided CP solvers make extensive use of *lazy clause generation solving* (LCG) under assumptions [23]. Given a set $F$ of propagators (representing a set of constraints), a current domain $D_{orig}$ and a set $A$ of assumptions (in the form of atomic constraints) over integer variables, an LCG solver $\mathsf{LCG}(F, D, A)$ determines if there exists an assignment $\theta$ to the variables that entails: (i) all constraints, (ii) all assumptions and (iii) the original domain $D_{orig}$. If so, the solver returns $\mathrm{SAT}(\theta)$. Otherwise the solver returns $\mathrm{UNSAT}(\kappa)$ where $\kappa$ a core of the instance.

As we use LCG solvers in a black-box manner, we will not go into detail on how such solvers operate (see e.g. [14] for more details). A central concept for applying LCG solvers in core-guided CP is that of *explanations for propagations*. Each propagator $f_c$ in a LCG solver must be able to *explain* its propagation of an atomic constraint $r$ in the form of a clause, i.e., compute an explanation clause $\mathsf{expl}(c, D, r) \equiv (E \to r)$ where $E$ is a conjunction of atomic constraints and $D \Rightarrow E$, as well as $c \Rightarrow E \to r$. During conflict analysis, a learnt clause is derived by starting from the conflict $C$, and repeatedly replacing some atom $r \in C$ with its reason, i.e. $E$. Analogously to how learned clauses over assumptions represent cores in core guided MaxSAT, the explanations for failure over assumptions represent cores in core-guided CP. We note that for a single propagation there can be (and often are) are many different explanations. Instead any explanation that is correct in the current domain suffices. The following examples demonstrate that some of the explanations are better for core guided CP than others.

*Example 1.* Consider the propagation of the linear inequality $2x + 3y + 4z \leqslant 27$ with the current domain $x \in [5..7]$, $y \in [4..9]$, $z \in [5..8]$. The propagator detects unsatisfiability, a simple explanation is: $\langle x \geqslant 5 \rangle \wedge \langle y \geqslant 4 \rangle \wedge \langle z \geqslant 5 \rangle \to false$ i.e. the clause $C_1 = (\langle x < 5 \rangle \vee \langle y < 4 \rangle \vee \langle z < 5 \rangle)$. This is not however, the only explanation. By relaxing bounds we obtain the *lifted* explanation $\langle x \geqslant -2 \rangle \wedge \langle y \geqslant 4 \rangle \wedge \langle z \geqslant 5 \rangle \to false$ i.e. the clause $C_2 = (\langle x < -2 \rangle \vee \langle y < 4 \rangle \vee \langle z < 5 \rangle)$, for the same propagation. Observe that the lifted explanation is stronger than the simple one as $C_2 \Rightarrow C_1$. Some lifted explanations can be particularly attractive, for example if the original domain sets $x \geqslant 0$, then the lifted explanation is equivalent to $\langle y \geqslant 4 \rangle \wedge \langle z \geqslant 5 \rangle \to false$, containing one less literal.     □

One way of obtaining stronger explanations is through *lifting*, informally speaking a lifted explanation can be computed by making use of the original propagator in order to compute an explanation for an atom in the (partial) learned clause, instead of the explanation graph. The reason computed by the propagator will frequently be *weaker* than the atom that was originally inferred, and allow the construction of a more general explanation.

*Example 2.* Consider a propagator $f_c$ for the linear inequality $c = 7y + 4t \geqslant 34$, and a current domain $y \in [0..10]$ and $t \in [0..2]$. The propagator can propagate $\langle y \geqslant 4 \rangle$ with explanation $\langle t \leqslant 2 \rangle \to \langle y \geqslant 4 \rangle$. Now suppose a partial learned clause of form $\neg \langle y \geqslant 1 \rangle \vee Q$, is encountered during conflict analysis. Since $\langle y \geqslant 4 \rangle \Rightarrow \langle y \geqslant 1 \rangle$ the original explanation can be used to obtain $C_1 = \neg \langle t \leqslant 2 \rangle \vee Q$. However, the propagator $f_c$ can return the lifted explanation $\langle t \leqslant 8 \rangle \to \langle y \geqslant 1 \rangle$ which allows deriving the learned clause $C_2 = \neg \langle t \leqslant 8 \rangle \vee Q$. We observe that $C_2 \Rightarrow C_1$, i.e. the learned clause obtained with the lifted explanation is stronger than the original one.     □

**Objective Probing:** Branch-and-bound (B&B) CP solvers iteratively search for better solutions by constraining that the objective value must be better than in the previous found solution. A common issue that arises is slow convergence: after finding a solution, B&B solvers typically generate many incrementally better solutions before reaching the true optimal solution. One strategy for improving convergence rate is *optimistic partitioning* [19]: after finding a solution with objective value $\hat{z}$ given lower bound $z_{lb}$, optimistic partitioning speculatively posts a constraint $\langle z < \frac{\hat{z}+z_{lb}}{2}\rangle$ (instead of $\langle z < \hat{z}\rangle$). If this succeeds, we have a much better solution; otherwise, the lower bound is greatly increased.

**Core-Guided MaxSAT:** that originated with the Fu-Malik algorithm [15] is today one of three central approaches to complete industrial MaxSAT solving, together with the implicit hitting set approach [10,24] and the model improving (corresponding to branch-and-bound in CP) approach [12,18,20,22]. Core-Guided search is a lower bounding approach based on first assuming that all objective literals can be set to false and relaxing the assumption whenever new cores (i.e. sources of unsatisfiability) are detected.

In more detail, when minimising an objective $\sum_i w_i x_i$ subject to a set $\mathcal{F}$ of clauses, modern core-guided MaxSAT solvers maintain a working instance $\mathcal{F}^w$, initialised to $\mathcal{F}$. During each iteration, a SAT solver is used to determine if there exists a solution of cost 0 to $\mathcal{F}^w$ by querying it on the clauses of $\mathcal{F}^w$ while assuming $A = \{\neg x | x \in S(\mathcal{F}^w)\}$, that is that all objective variables are false. If the result is satisfiable, a solution satisfying $\mathcal{F}^w$ and $A$ will be an optimal solution to $\mathcal{F}$. Otherwise, the solver returns a set of variables $\kappa \subset S(\mathcal{F}^w)$ that represents a core of $\mathcal{F}^w$. Next $\mathcal{F}^w$ is relaxed (reformulated) based on $\kappa$. First, the weight of each objective variable in $\kappa$ is lowered by $w^\kappa = \min\{w(x) \mid x \in \kappa\}$ (this is known as *weight-splitting*). Second, new objective variables and clauses REFORM($\kappa$) that rule out $\kappa$ as a source of unsatisfiability are added to $\mathcal{F}^w$.

Most core-guided solvers differ mainly in the instantiation of REFORM($\kappa$). We detail the OLL algorithm [1,21] as it been shown to be the most effective in the MaxSAT evaluations and can be naturally extended to CP. Assuming $|\kappa| = n$ OLL adds new objective variables $\langle o_\kappa > 1\rangle, ..., \langle o_\kappa > n-1\rangle$, each of weight $w^\kappa$ and clauses corresponding to AS-CNF($\sum_{l\in\kappa}(l) > k \rightarrow \langle o_\kappa > k\rangle$) for each $k \in [1..n-1]$ (the CNF encoding of the constraint). Assuming one of the commonly used encodings [4,6], the clauses enforce that setting $k > 1$ literals of $\kappa$ to true propagates the literals $\langle o_\kappa > 1\rangle, ..., \langle o_\kappa > k-1\rangle$ to true, incurring $(k-1)w^\kappa$ additional cost. Informally, the new clauses allow setting one objective variable in $\kappa$ to true for free while incurring more cost for any additional ones.

*Example 3.* Consider the following problem:

$$\min z = 3x_1 + 2x_2 + 2x_3 + 4x_4$$
$$s.t. \quad \max(x_1, x_2) \geqslant 2, \quad \max(x_2, x_3) \geqslant 2, \quad \max(x_3, x_4) \geqslant 2.$$

where each $x_i$ is an integer variable with domain $[0..3]$. To solve this problem with the MaxSAT OLL algorithm, we consider the equivalent problem:

$$\min z = \sum_{k=1}^{3} 3\langle x_1 \geqslant k\rangle + \sum_{k=1}^{3} 2\langle x_2 \geqslant k\rangle + \sum_{k=1}^{3} 2\langle x_3 \geqslant k\rangle + \sum_{k=1}^{3} 4\langle x_4 \geqslant k\rangle$$
$$s.t \quad \langle x_i \geqslant k\rangle \rightarrow \langle x_i \geqslant k-1\rangle \text{ for } i \in [1..4], k \in [2..3]$$
$$\langle x_1 \geqslant 2\rangle \vee \langle x_2 \geqslant 2\rangle, \quad \langle x_2 \geqslant 2\rangle \vee \langle x_3 \geqslant 2\rangle, \quad \langle x_3 \geqslant 2\rangle \vee \langle x_4 \geqslant 2\rangle.$$

We sketch an execution of the MaxSAT OLL algorithm. The initial solver call is made assuming $\langle x_i \geqslant k \rangle$ to false for all $i \in [1..4]$, and $k \in [1..3]$. Let $\kappa_1 = \{\langle x_2 \geqslant 2 \rangle, \langle x_3 \geqslant 2 \rangle\}$ be the first core. First the weights of both variables in the core are lowered by $\min\{w(\langle x_2 \geqslant 2 \rangle), w(\langle x_3 \geqslant 2 \rangle)\} = 2$. Then a new variable $\langle o_1 > 1 \rangle$ (with weight 2) defined with clauses corresponding to $\left(\neg \langle o_1 > 1 \rangle \rightarrow \sum_{l \in \kappa_1}(l) \leqslant 1\right)$ is introduced to the instance before the solver reiterates. In the next iteration, the objective variables of the instance are $\langle o_1 > 1 \rangle$, $\langle x_i \geqslant k \rangle$ for $k \in [1..3]$ and $i \in \{1, 4\}$ as well as $\langle x_j \geqslant 1 \rangle$, $\langle x_j \geqslant 3 \rangle$ for $j \in \{2, 3\}$. Let $\kappa_2 = \{\langle x_1 \geqslant 2 \rangle, \langle o_1 > 1 \rangle, \langle x_4 \geqslant 2 \rangle\}$ be the next core. As in the first iteration, the weight of each objective variable in the core is lowered by 2 and new objective variables $\langle o_2 > 1 \rangle$ and $\langle o_2 > 2 \rangle$ (weight 2) are introduced and defined with constraints AS-CNF $\left(\neg \langle o_2 > k \rangle \rightarrow \sum_{l \in \kappa_2}(l) \leqslant k\right)$.

Next the algorithm extracts two unit cores arising due to the order-encoding of integers before terminating with the satisfying assignment that sets $\langle x_2 \geqslant 1 \rangle$, $\langle x_3 \geqslant 1 \rangle$, $\langle x_2 \geqslant 2 \rangle$, $\langle x_3 \geqslant 2 \rangle$ and $\langle o_1 > 1 \rangle$ to true and all other variables to false. This assignment has cost 8 and corresponds to $x_1 = x_4 = 0$ and $x_2 = x_3 = 2$, an optimal assignment to the original problem.                                    □

**Core-Boosting** is a recently proposed [7] search strategy for MaxSAT that combines core-guided search with an anytime approach (originally a B&B type search for MaxSAT). The intuition underlying core-boosting is that core-guided search is mostly an "all-or-nothing" strategy. In its most basic form, core-guided search only finds one feasible solution during search, an optimal one. Furthermore, core-guided search tends to be somewhat bimodal [7], either proving optimality fairly quickly or not terminating within a reasonable time. Core-boosted search is designed to take advantage of the fact that core-guided search may rule out a significant number of cores from the instance that would cause trouble for approaches like B&B.

More specifically, given a total resource budget, core-boosting spends a small fraction of its budget running in a core-guided mode. If this budget is exhausted and optimality has not yet been proven it rebuilds the objective based on the cores found so far, and then spends its remaining time optimizing the *reformulated* objective in a branch-and-bound mode.

## 3   Advancing Core-Guided Search for CP

In this section, we overview core-guided search for CP and discuss our contributions toward advancing its performance. We begin with what we call *slice-based* reformulation, i.e. the conventional translation of the OLL algorithm for MaxSAT to CP. We also discuss potential issues when applying explanation lifting to the slice based formulation. Motivated by these we then detail our main contributions: *coefficient elimination* and *variable-based reformulation*, two novel core-guided reformulations specific for CP. Finally, we also discuss improvements and generalisations of existing search heuristics from MaxSAT and CP; assumption probing and core-boosting.

## 3.1   Slice-Based Reformulation

The following restatement of Example 3 for CP provides intuition for the slice based reformulation.

*Example 4.* In contrast to MaxSAT, the OLL algorithm with slice based reformulation works directly on the original problem of Example 3. For each variable $x_i$, we track a *threshold*: initially $\mathsf{lb}(x_i)$, the threshold is the amount of $x_i$ which is already accounted for elsewhere in the objective. Initially, the LGC solver is called while assuming all variables to their initial threshold, i.e. $\langle x_i \leqslant 0 \rangle$ for $i \in [1..4]$. Let $\kappa_1 = \{\neg \langle x_2 \leqslant 0 \rangle, \neg \langle x_3 \leqslant 0 \rangle\}$, i.e. $\{\langle x_2 \geqslant 1 \rangle, \langle x_3 \geqslant 1 \rangle\}$ be the first core obtained. The algorithm now introduces a new *integer* variable $o_1 \geqslant \langle x_2 \geqslant 1 \rangle + \langle x_3 \geqslant 1 \rangle$ with an initial domain $o_1 = \{1,2\}$ (notice that $o_1 \neq 0$ as at least one of the literals in the core has to be false). For an alternative view, $o_1$ could be seen as the variable $o_1 = \langle x_2 \geqslant 1 \rangle + \langle x_3 \geqslant 1 \rangle$ that has its upper bounds enforced by assumptions. Next the objective is reformulated using $o_1$ to $z = 2 + 2\llbracket o_1 \rrbracket_1 + 3x_1 + 2\llbracket x_2 \rrbracket_1 + 2\llbracket x_3 \rrbracket_1 + 4x_4$. For some intuition, notice for example that the term $2\llbracket x_2 \rrbracket_1$ corresponds to the term $2\sum_{k=2}^{3} \langle x_2 \geqslant k \rangle$ in the MaxSAT objective and that the term $2\llbracket o_1 \rrbracket_1 + 2$ can be seen as $2\langle x_2 \geqslant 1 \rangle + 2\langle x_3 \geqslant 1 \rangle$ plus the lower bound implied by $\kappa_1$.

In the next iteration, all variables are again assumed to their current threshold i.e. $\langle x_1 \leqslant 0 \rangle$, $\langle x_2 \leqslant 1 \rangle$, $\langle x_3 \leqslant 1 \rangle$, $\langle x_4 \leqslant 0 \rangle$, and $\langle o_1 \leqslant 1 \rangle$. Notice how the threshold for $x_2$ and $x_3$ is 1, conceptually, the (potential) weight for $x_2 = x_3 = 0$ is accounted for by the variable $o_1$. Assume the next core obtained is $\kappa_2 = \{\langle x_2 \geqslant 2 \rangle, \langle x_3 \geqslant 2 \rangle\}$. Similarly to before, a new variable $o_2$ and constraint $o_2 \geqslant \langle x_2 \geqslant 2 \rangle + \langle x_3 \geqslant 2 \rangle$ is introduced, and the objective reformulated to $z = 4 + 2\llbracket o_1 \rrbracket_1 + 2\llbracket o_2 \rrbracket_1 + 3x_1 + 2\llbracket x_2 \rrbracket_2 + 2\llbracket x_3 \rrbracket_2 + 4x_4$ .

In the next iteration, the thresholds for the variables are $\langle x_1 \leqslant 0 \rangle$, $\langle x_2 \leqslant 2 \rangle$, $\langle x_3 \leqslant 2 \rangle$, $\langle x_4 \leqslant 0 \rangle$, $\langle o_1 \leqslant 1 \rangle$, and $\langle o_2 \leqslant 1 \rangle$. Assume that the next core extracted is $\kappa_3 = \{\langle x_1 \geqslant 1 \rangle, \langle o_2 \geqslant 2 \rangle, \langle x_4 \geqslant 1 \rangle\}$ after which the constraint $o_3 \geqslant \langle x_1 \geqslant 1 \rangle + \langle o_2 \geqslant 2 \rangle + \langle x_4 \geqslant 1 \rangle$ is introduced and the objective reformulated to $z = 6 + 2\llbracket o_1 \rrbracket_1 + \llbracket x_1 \rrbracket_0^1 + 3\llbracket x_1 \rrbracket_1 + 2\llbracket x_2 \rrbracket_2 + 2\llbracket x_3 \rrbracket_2 + 2\llbracket x_4 \rrbracket_0^1 + 4\llbracket x_4 \rrbracket_1$.

Finally, the solver still extracts the core $\kappa_4 = \{\langle o_1 \geqslant 2 \rangle\}$ and reformulates the instance one last time before terminating with the solution $x_1 = x_4 = 0$, $x_2 = x_3 = 2$ and $o_1 = o_2 = 2$, $o_3 = o_4 = 1$.    □

Example 4 gives some intuition for the term *slice based* reformulation. In each iteration the algorithm *slices off* the current threshold value of all variables appearing in a core $\kappa$, packaging the removed values into a new penalty term $o_\kappa$.

Algorithms 1 and 2 detail OLL with slice based reformulation for CP. Given a set $\mathcal{F}$ of constraints and an objective $z$ to minimize, the algorithm initially checks the feasibility of the problem by calling the LCG solver on the constraints without assumptions. If the problem has feasible solutions, the algorithm enters its main search loop. On each iteration, the LCG solver is invoked on the instance while assuming all objective terms to their current thresholds. These thresholds are maintained in a mapping $E$ that maps each variable $x_i$ to a tuple $(t_i, u_i, w_i)$ containing its threshold $(t_i)$, its residual weight $u_i$ and its full weight $w_i$. If the

---

**Algorithm 1:** OLL for constraint programming using slice-based reformulation

---

**Data:** Constraints $\mathcal{F}$, an original domain $D_o$ and objective
$\quad\quad z = w_1 x_1 + \ldots + w_m x_m$.
**Result:** Optimal solution $\theta^*$.
$z_{lb} \leftarrow \sum_{i=1}^{n} w_i \mathsf{lb}(x_i)$
**switch** $LCG(F, D_o, \varnothing)$ **do**
$\quad$ **case** $\mathrm{UNSAT}(\varnothing)$
$\quad\quad$ | **return** UNSAT
$\quad$ **case** $\mathrm{SAT}(\theta)$
$\quad\quad$ $E \leftarrow \{x_i \mapsto (\mathsf{lb}(x_i), w_i, w_i) \mid i \in 1 \ldots n\}$
$\quad\quad$ **while true do**
$\quad\quad\quad$ **switch** $LCG(F, D_o, \{\langle x_i \leqslant t_i \rangle \mid E[x_i] = (t_i, u_i, w_i)\})$ **do**
$\quad\quad\quad\quad$ **case** $\mathrm{SAT}(D)$
$\quad\quad\quad\quad\quad$ | **return** $D, z_{lb}$
$\quad\quad\quad\quad$ **case** $\mathrm{UNSAT}(\kappa)$
$\quad\quad\quad\quad\quad$ | $\mathcal{F}, z_{lb}, E \leftarrow \text{REFORMULATE-SLICE}(\mathcal{F}, z_{lb}, E, \kappa)$

---

**Algorithm 2:** REFORMULATE-SLICE($\mathcal{F}, z_{lb}, E, \kappa$)

---

$w^\kappa \leftarrow \min\{u_i \mid \langle x_i \geqslant k_i \rangle \in \kappa, E[x_i] = (t_i, u_i, w_i)\}$
$o_\kappa \leftarrow \text{NEW-VAR}(\mathcal{F}, [1, |\kappa|])$
$R \leftarrow 0$
**for** $\langle x_i \geqslant k_i \rangle \in \kappa$ **do**
$\quad$ $(t_i, u_i, w_i) \leftarrow E[x_i]$
$\quad$ $R \leftarrow R + \langle x_i \geqslant t_i + 1 \rangle$
$\quad$ **if** $u_i = w^\kappa$ **then**
$\quad\quad$ | $E[x_i] \leftarrow (t_i + 1, w_i, w_i)$
$\quad$ **else**
$\quad\quad$ | $E[x_i] \leftarrow (t_i, u_i - w^\kappa, w_i)$
$E[o_\kappa] \leftarrow (1, w^\kappa, w^\kappa)$
$\mathcal{F} \leftarrow \mathcal{F} \wedge (o_\kappa \geq R)$
**return** $\mathcal{F}, z_{lb} + w^\kappa, E$

---

solver returns $\mathrm{SAT}(D)$ the obtained domain will be an optimal solution to the problem so the algorithm terminates. Otherwise, the solver returns a core $\kappa$. The algorithm then reformulates the instance using Algorithm 2. Analoguously to MaxSAT, slice based reformulation of the instance means: (i) computing $w^\kappa$, the minimum residual weight of all literals in the core, (ii) lowering the (residual) weight of each literal in the core by $w^\kappa$ and (iii) introducing a new variable $o_\kappa$ with lower bound (and threshold) 1 and full weight $w^\kappa$ as well as new constraints $o_\kappa \geqslant \sum_{l \in \kappa}(l)$. Any variable whose residual weight gets lowered to 0 during step (ii) gets its threshold by one and residual weight reset to its full weight

*Example 5.* Consider core $\kappa_3 = \{\langle x_1 \geqslant 1 \rangle, \langle o_2 \geqslant 2 \rangle, \langle x_4 \geqslant 1 \rangle\}$ from Example 4. The current state of the objective is $E[o_1] = (1,2,2)$, $E[o_2] = (1,2,2)$, $E[x_1] = (0,3,3)$, $E[x_2] = (2,2,2)$, $E[x_3] = (2,2,2)$, $E[x_4] = (0,4,4)$, with $z_{lb} = 4$. We determine $w^\kappa = 2$ as the minimum of the $u_i$ coefficients of $\{x_1, o_2, x_4\}$. We create new variable $o_3$ and collect the expression $R = \langle x_1 \geqslant 1 \rangle + \langle o_2 \geqslant 2 \rangle + \langle x_4 \geqslant 1 \rangle$, updating the $E$ entries to $E[x_1] = (0,1,3)$, $E[o_2] = (3,2,2)$, $E[x_4] = (0,2,4)$ We set $E[o_3] = (1,2,2)$ and add the constraint $o_3 \geqslant R$. We return $z_{lb} = 6$. The resulting objective is exactly as shown in Example 4.

The following example demonstrates a potential weakness of slice-based reformulation, motivating the novel reformulation strategies we propose in the next section.

*Example 6.* Consider the problem defined in Example 3 and the initial LCG call made by OLL for CP with the assumptions $\langle x_i \leqslant 0 \rangle$ for $i \in [1..4]$. Assume now that we obtain the lifted core $\kappa_1' = \{\langle x_2 \geqslant 2 \rangle, \langle x_3 \geqslant 2 \rangle\}$ and introduce a single new variable $o_\kappa = \langle x_2 \geqslant 1 \rangle + \langle x_2 \geqslant 2 \rangle + \langle x_3 \geqslant 1 \rangle + \langle x_3 \geqslant 2 \rangle = [\![x_2]\!]_0^2 + [\![x_3]\!]_0^2$ If the solver later derives $\langle x_2 \geq 1 \rangle$ and $\langle x_3 \geq 1 \rangle$, the lower bound on $o_\kappa$ is set to 2. However, with the reformulations performed in Example 4 the algorithm has already derived $\langle x_2 \geqslant 2 \rangle \vee \langle x_3 \geqslant 2 \rangle$, implying a lower bound of 3 on $o_\kappa$.  $\square$

In other words, slice based reformulation makes using lifted cores difficult. Hence, instead of the approach presented in Example 6 we instead perform reformulation similarly to Example 4 instead. We do however add the lifted core to the model, thus allowing the algorithm to extract it later without search.

## 3.2   Novel Core-Guided Reformulations for CP

Next we detail the main contribution of this work, two novel reformulation strategies for the CP OLL algorithm: 1) *Coefficient Elimination* and 2) *Variable-based reformulation*. Coefficient elimination seeks to increase the number of variables whose lower bounds are increased during reformulation steps., thus increasing the rate at which the lower bounds of the variables increase. Variable-based reformulation attempts to make better use of the information provided by lifted cores in order to increase the lower bound on the objective faster.

**Coefficient Elimination.** Let $\kappa$ be a set of literals corresponding to a core obtained during an iteration of OLL for CP and $w^\kappa$ the smallest (residual) weight of the literals in the core. Consider now the weighted sum of the literals in the core, i.e the variable $o_\kappa = \sum_{x_i \in \kappa} w_i \langle x_i \geqslant t_i + 1 \rangle$. Since $\kappa$ corresponds to a core, the lower bound of $o_\kappa$ is $w^\kappa$ and the objective could be reformulated using $\sum_{x_i \in \kappa} w_i [\![x_i]\!]_{t_i} = w^\kappa + [\![o_\kappa]\!]_{w^\kappa} + \sum_{x_i \in \kappa} w_i [\![x_i]\!]_{t_i + 1}$ . Notice how, in contrast to the strategy described in Sect. 3.1 and Example 4, coefficient elimination in this form results in the lower bound of *all* variables in $\kappa$ being increased by one. The drawback is instead the (potential) increase in complexity of the subsequent LCG solver calls.

*Example 7.* Consider the following problem:

$$\textbf{minimize } 1000p + \sum_{i=1}^{n} x_i + y_i \quad \textbf{s.t.} \qquad \neg p \rightarrow x_i + y_i \geq 1, \ \forall i \in 1 \ldots n$$

With weight splitting, the OLL algorithm for CP generates $\max(n, 1000)$ cores of form $\{\langle p \geqslant 1 \rangle, \langle x_i \geqslant 1 \rangle, \langle y_i \geqslant 1 \rangle\}$ $i \in [1..]$, each time decreasing the coefficient of $p$ by 1. Since $p$ is never removed from the objective, we expect extracting each core to require approximately similar amounts of computational effort (see also independent core extraction detailed later in this section).

With coefficient elimination, the algorithm instead introduces the variable $o_\kappa = 1000p + x_1 + y_1$ with a lower bound of 1 when reformulating $\kappa = \{\langle p \geqslant 1 \rangle, \langle x_1 \geqslant 1 \rangle \langle y_1 \geqslant 1 \rangle\}$. In the next iteration, the variable $p$ is no longer directly in the objective. Instead the next core extracted will be $\{\langle o_\kappa \geqslant 2 \rangle, \langle x_2 \geqslant 1 \rangle, \langle y_2 \geqslant 1 \rangle\}$ instead. Informally speaking, all of the subsequent cores will depend on the reformulation variables introduced in previous iterations thus making extracting cores require increasing amounts of computational effort. [5]                                                                      □

The version of coefficient elimination that we consider is a hybrid strategy designed to balance the number of variables whose lower bounds can be increased during each reformulation with the potential of extracting independent cores during subsequent iterations. More specifically, when reformulating on a set $\kappa$ of literals having minimum weight $w^\kappa$, we fully reformulate all literals in $\kappa$ that have weight less than $Bw^\kappa$ where $B$ is a boundary parameter, and slice the rest. More formally, coefficient elimination introduces a variable $o_\kappa = \sum_{x_i \in \kappa} c_i x_i$ where $c_i = \max(w_i, Bw^\kappa)$ and reformulates the objective using $\sum_{x_i \in \kappa} w_i \llbracket x_i \rrbracket_{t_i} = w^\kappa + \llbracket o_\kappa \rrbracket_{w^\kappa} + \sum_{x_i \in \kappa} \max(0, w_i - Bw^\kappa) \llbracket x_i \rrbracket_{t_i} + \sum_{x_i \in \kappa} w_i \llbracket x_i \rrbracket_{t_i + 1}$ .

**Variable-Based Reformulation** attempts to overcome the difficulties that slice based reformulation has with exploiting the full potential of lifted cores, i.e. that slice-based reformulation can only ever increase the lb of variables by 1 and thus the objective by the minimum (residual) weight of the variables in the core.

Recall for example the lifted core $\{\langle x_2 \geqslant 2 \rangle, \langle x_3 \geqslant 2 \rangle\}$ discussed in Example 6. When reformulating with variable-based reformulation, the OLL algorithm for CP introduces the variable $o_\kappa \geqslant x_2 + x_3$ with an initial domain of $[2 \ldots)$. In more general terms, variable based reformulation merges all integer variables appearing in a core into a single new variable, and assigns the new variable an initial lower bound equal to the sum of $\text{lb}(x_i)$ plus the smallest gap between the some $\text{lb}(x_i)$ and the corresponding value in the core (we do not maintain a separate threshold). Notice that the potential benefits of variable based elimination are directly related to the size of the domains of the involved variables. In this particular case, the approach lifts the lower bound by 2 but it is easy to create examples where the increase is higher, which we observed to also occur frequently in practice.

In addition to more effectively exploiting information in lifted cores, we often observed that variable based reformulation resulted in unit cores being extracted in subsequent iterations. Unit cores are particularly attractive for the OLL algorithm for CP as no new variables nor constraints need to be introduced. Instead it suffices to increase the lower bound of the variable in the core.

### 3.3 Generalisations of Existing Techniques

Before reporting on an experimental evaluation of the new reformulation strategies, we briefly describe new generalisations and improvements to existing heuristics in both CP and MaxSAT solving that we make use of in this work.

**Core-Boosting for CP.** We extended core-boosted search from MaxSAT to CP. A key difference when applying core boosting in CP compared to MaxSAT is the need to explicitly encode the objective function (which is only implicitly defined during the core-guided phase) before switching to branch-and-bound search.

Explicitly encoding the objective function when using variable based reformulation is fairly straightforward. During each reformulation step, a set $x_1, \ldots, x_k$ of variables in the objective are replaced with a variable $o$ representing their sum. When switching to B&B search, the same procedure is used to remove all remaining terms and merging them into a single new variable. In contrast, combining core-boosted search with slice-based reformulation is more intricate. Consider a possible (implicit) objective:

$$
\begin{aligned}
z = \quad & c_1 \llbracket x_1 \rrbracket_{d_1} \quad + c_2 \llbracket x_2 \rrbracket_{d_2} + \ldots \quad + c_k \llbracket x_m \rrbracket_{d_m} \\
+ \ & b_1 \langle x_1 \geqslant d_1 \rangle + b_2 \langle x_2 \geqslant d_2 \rangle + \ldots + b_k \langle x_m \geqslant d_m \rangle + z_{lb}
\end{aligned}
$$

obtained after several iterations of core-guided search with sliced based reformulation. A simple approach to making $z$ explicit is to introduce fresh variables for each sub-term, i.e. let $x'_i = \max(0, x_i - d_i)$, $x''_i = \langle x_i \geq d_i \rangle$ and $z = z_{lb} + \sum_{i=1}^{m} c_i x'_i + \sum_{i=1}^{m} b_i x''_i$.

A more efficient method makes use of the monotonicity of $\llbracket x_i \rrbracket_{d_i}$ which in turn implies that any atomic constraint $\langle \llbracket x_i \rrbracket_{d_i} \geqslant c \rangle$ can be expressed as an equivalent atom $\langle x_i \geq c' \rangle$:

$$
\langle \llbracket x_i \rrbracket_{l_i}^{u_i} \geqslant c \rangle =
\begin{cases}
true & \text{if } c \leqslant 0 \\
\langle x_i \geqslant l_i + c \rangle & \text{if } 0 < c \leqslant u_i - l_i \\
false & \text{if } c > u_i - l_i
\end{cases}
$$

Hence we can use a form of variable view [25] to encode the expressions $\llbracket x_i \rrbracket_{d_i}$ and $\langle x_i \geqslant d_i \rangle \equiv \llbracket x_i \rrbracket_{d_i}^{d_i+1}$, thus avoiding the need to introduce new variables.

**Progressive Probing.** Recall that when given an incumbent solution with cost $\hat{z}$ and a lower bound $z_{lb}$ on the objective, objective (optimistic) probing attempts to improve the solution and find a solution of cost $(\hat{z} + z_{lb})/2$. In practice, we observed that objective probing is a risky strategy since the jump from $\hat{z}$ to $(\hat{z} + z_{lb})/2$ is quite aggressive, and thus can result in difficult unsatisfiable subproblems. Instead we consider a more conservative strategy that we call *progressive probing*, an idea resembling to the use of progression in MaxSAT solving [16]. For geometrically increasing values of $\delta$ (more precisely, in iteration $i$ $\delta_i = 2^i stepsize$ where *stepsize* is a parameter) the solver is queried for a solution of cost $\hat{z} - \delta_i$. The procedure reiterates until the solver either returns UNSAT or runs out of the resources allocated for probing.

In addition to B&B search, we also make use of probing during core-guided search, inspired by techniques from the MaxSAT community [3,17]. Anytime a singleton core $\langle u > k \rangle$ is extracted, the bound $k$ is probed by repeated invocations of $\mathsf{LCG}(F, D, \{\langle u \leqslant k + \delta_i \rangle\})$ for the geometrically increasing values of $\delta_i$ defined above. Core-probing like this is particularly effective in combination with variable-based reformulation: if $\mathsf{lb}(x) + \mathsf{lb}(y)$ is much smaller than the true lower bound of $u = x + y$, core probing will quickly push up the bound of $z$, skipping many of the intermediate steps. Recall also that variable based reformulation often results in unit cores being extracted in subsequent iterations.

## 3.4   Additional Techniques

Finally, we also considered a number of fairly direct translations of MaxSAT techniques to CP.

*Independent core extraction* [8] is a strategy for obtaining *simpler* cores. Given some core $\kappa$ of instance, the reformulation (i.e. introduction of new variables and constraints) is delayed and instead only the assumptions in the core are relaxed (i.e. their weight is lowered by $w^\kappa$ or $Tw^\kappa$ in the case of variable based reformulations). Since at least one of the weights will be lowered to zero, the solver can be invoked to extract another core without needing to reformulate. Note that reformulating the instance makes it more complicated, and thus delaying is beneficial. The process continues until no more cores can be found, at which point all found cores are reformulated.

*Stratification* [2] starts by posting assumptions using only literals with high weights, and throughout the search introduces the remaining literals. This allows high-weighted core to be extracted early in the search. As a side effect, feasible solutions can be generated in the process.

*Hardening* [2] can be used to enforce satisfaction for certain literals. Given an upper $z_{ub}$ and lower bound $z_{lb}$ on the optimal cost, *hardening* will set false any Booleans with weight $w > z_{ub} - z_{lb}$. The same rule can be generalised to integer variables $x_i$ by setting $\mathsf{ub}(x_i) = t_i + \lfloor \frac{z_{ub} - z_{lb}}{c_i} \rfloor$ where $c_i$ is the coefficient and $t_i$ the threshold of $x_i$.

*Solution-guided search* [11,12] is a value-selection heuristic that assigns a branching-variable the value it takes in the current best solution if possible, and otherwise resorts to the default value-selection strategy. This focuses the search around the best solution, quickly finding local improvements.

## 4    Experiments

In order to experimentally evaluate the improvements to core-guided search we integrated the described core-guided optimisation methods into **geas** (https:// bitbucket.org/gkgange/geas), a lazy-clause generation solver. The core engine of **geas** is written in C++, with a **FlatZinc** frontend written in OCaml. The core-guided optimisation techniques are entirely implemented in the OCaml frontend, using the engine's assumption interface to handle the cores and reformulation. Propagators in **geas** implement lazy explanation with lifting, so it can extract lifted cores.

As benchmarks, we took the set of models and instances from the MiniZinc Challenge [26] for years 2015–2018, and selected all optimization models with a linear objective. This resulted in 48 models, and 249 instances. We then ran **geas** on this data-set, comparing its branch-and-bound configuration (bb), with all combinations of the following core-based configurations:

- core-guided (core), or core-boosted (boost), using a 10% of time limit (i.e. 60 s) core-guided phase before switching to branch and bound search
- slice-based (slice), or variable-based (var) reformulations.
- weight splitting (split) or coefficient elimination with boundary $B = 2$ (elim).

All core-based methods were run with stratification, independent core extraction and hardening. All methods were run using *free search* (alternating programmed- and activity-driven search) and a geometric restart sequence. Each instance was run with a 600 s time-limit, reporting the time to prove optimality as well as the best objective value found.

Figure 1 compares the overall performance of each set of parameters across the dataset. We observe that branch and bound performs slightly better than "the basic version" of core guided search, i.e. core-slice-split and core-slice-elim. Variable based reformulation improves over slice based reformulation, obtaining performance superior to B&B. The best overall performance is obtained by boost-var-split making use of core-boosted search, variable based reformulation and weight splitting, although the difference between coefficient elimination and weight splitting is minor.

Figure 2 gives a per-instance breakdown of the results, comparing core-guided search with branch-and-bound as well as the reformulation strategies. We observe that B&B and core-guide search are fairly orthogonal in the sense that there are many instances on which B&B search finished quickly while core-guided search times out and vice versa. This observation provides a possible explanation for the good overall performance of core-boosted search, notice that most of the instances where core-guided outperforms branch-and-bound are clustered in the bottom-right of the figure. The other side of the figure also clearly demonstrates the superior performance of variable based reformulation compared to slice based reformulation.

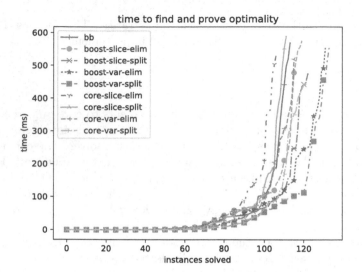

**Fig. 1.** Time to prove optimality for all methods.

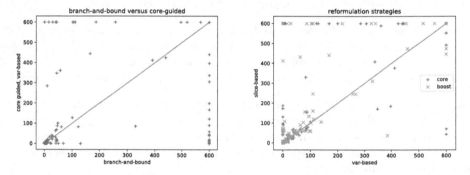

**Fig. 2.** Comparing time-to-optimality. Left: branch-and-bound versus core guided using a variable-based reformulation. Right: slice-based versus variable-based reformulation.

In addition to proving optimality we investigate the anytime behaviour of the methods i.e. how good are the solutions obtained when optimality is not proven? Tables 1 and 2 compare the quality of solutions found by each method across all (Table 1) and a representative (Table 2) set of benchmarks. The tables again demonstrates the orthogonality of the methods we consider, no individual method dominates all others. However, pure core-guided methods were much less competitive as anytime methods supporting the intuition that core-guided methods typically either prove optimality quickly, or fail to produce solutions of reasonable quality. And for anytime search, using variable elimination rather than splitting is worthwhile: variable elimination paired with core-boosting most reliably produced the best feasible solution.

**Table 1.** How many times did each method (row) report a *strictly better* objective value than each other method (column) and the best objective value found overall (column BEST).

| | | | BB | BOOST | | | | CORE | | | | |
|---|---|---|---|---|---|---|---|---|---|---|---|---|
| | | | | VAR | | SLICE | | VAR | | SLICE | | |
| | | | | ELIM | SPLIT | ELIM | SPLIT | ELIM | SPLIT | ELIM | SPLIT | BEST |
| | | BB | 0 | 25 | 22 | 33 | 38 | 100 | 93 | 89 | 87 | 196 |
| BOOST | VAR | ELIM | 39 | 0 | 27 | 36 | 36 | 102 | 95 | 93 | 91 | 211 |
| | | SPLIT | 39 | 23 | 0 | 33 | 36 | 103 | 93 | 94 | 91 | 205 |
| | SLICE | ELIM | 37 | 18 | 20 | 0 | 25 | 101 | 92 | 91 | 88 | 196 |
| | | SPLIT | 33 | 22 | 20 | 28 | 0 | 101 | 93 | 92 | 87 | 197 |
| CORE | VAR | ELIM | 20 | 4 | 4 | 10 | 8 | 0 | 22 | 39 | 34 | 144 |
| | | SPLIT | 19 | 4 | 5 | 9 | 8 | 24 | 0 | 37 | 31 | 152 |
| | SLICE | ELIM | 22 | 8 | 8 | 8 | 8 | 50 | 47 | 0 | 13 | 151 |
| | | SPLIT | 22 | 7 | 8 | 9 | 8 | 57 | 54 | 25 | 0 | 154 |

**Table 2.** Quality scores for selected models. Quality of a solution with objective value $z$ is defined as $\frac{z_{vbs}-z_{lb}+1}{z-z_{lb}+1}$, where $z_{lb}$ is the initial lower bound, and $z_{vbs}$ the best solution found by any solver.

| MODEL | BB | BOOST | | | | CORE | | | |
|---|---|---|---|---|---|---|---|---|---|
| | | VAR | | SLICE | | VAR | | SLICE | |
| | | ELIM | SPLIT | ELIM | SPLIT | ELIM | SPLIT | ELIM | SPLIT |
| cargo_coarsePiles | 0.99 | **1.00** | 0.98 | 0.90 | 0.89 | 0.81 | 0.81 | 0.79 | 0.79 |
| celar | 0.59 | 0.97 | 0.98 | **0.99** | **0.99** | 0.68 | 0.68 | 0.94 | 0.91 |
| oc-roster | 0.93 | **1.00** | 0.97 | **1.00** | 0.96 | 0.66 | 0.51 | 0.55 | 0.52 |
| seat-moving | 0.84 | 0.95 | 0.95 | 0.95 | 0.95 | **1.00** | **1.00** | **1.00** | **1.00** |
| vrplc_service | **1.00** | 0.99 | 0.99 | 0.99 | **1.00** | 0.80 | 0.80 | 0.74 | 0.74 |

# 5 Conclusion

In this paper, we revisit the use of unsatisfiable core approaches for CP – both standalone, and as part of a hybrid (core-boosted) strategy. We exploit the extra expressiveness of lazy clause generation solvers to build more compact OLL-style reformulations, and to opportunistically tighten lower and upper bounds. We experimentally evaluated the new methods and draw the following conclusions 1) Core-boosting is generally worthwhile, both for anytime performance *and* proving optimality. 2) Variable-based reformulations are typically better for proving optimality, but this is model-dependent. 3) If using core-boosting, variable-based reformulations also produce better solutions. 4) Surprisingly, slice-based reformulations yield better solutions for core-guided; but still not as good as those for core-boosted. 5) Coefficient elimination finds the best solution slightly more frequently in combination with variable-based core-boosting (but is slightly worse at proving optimality). In other configurations, it is worse than weight splitting.

# References

1. Andres, B., Kaufmann, B., Matheis, O., Schaub, T.: Unsatisfiability-based optimization in clasp. In: Proceedings of the ICLP Technical Communications. LIPIcs, vol. 17, pp. 211–221. Schloss Dagstuhl - Leibniz-Zentrum fuer Informatik (2012)
2. Ansótegui, C., Bonet, M.L., Gabàs, J., Levy, J.: Improving SAT-based weighted MaxSAT solvers. In: Milano, M. (ed.) CP 2012. LNCS, pp. 86–101. Springer, Heidelberg (2012). https://doi.org/10.1007/978-3-642-33558-7_9
3. Ansótegui, C., Gabàs, J.: WPM3: An (in)complete algorithm for weighted partial MaxSAT. Artif. Intell. **250**, 37–57 (2017)
4. Asín, R., Nieuwenhuis, R., Oliveras, A., Rodríguez-Carbonell, E.: Cardinality networks and their applications. In: Kullmann, O. (ed.) SAT 2009. LNCS, vol. 5584, pp. 167–180. Springer, Heidelberg (2009). https://doi.org/10.1007/978-3-642-02777-2_18
5. Bacchus, F., Narodytska, N.: Cores in core based MaxSat algorithms: an analysis. In: Sinz, C., Egly, U. (eds.) SAT 2014. LNCS, vol. 8561, pp. 7–15. Springer, Cham (2014). https://doi.org/10.1007/978-3-319-09284-3_2
6. Bailleux, O., Boufkhad, Y.: Efficient CNF encoding of boolean cardinality constraints. In: Rossi, F. (ed.) CP 2003. LNCS, vol. 2833, pp. 108–122. Springer, Heidelberg (2003). https://doi.org/10.1007/978-3-540-45193-8_8
7. Berg, J., Demirović, E., Stuckey, P.J.: Core-boosted linear search for incomplete MaxSAT. In: Rousseau, L.-M., Stergiou, K. (eds.) CPAIOR 2019. LNCS, vol. 11494, pp. 39–56. Springer, Cham (2019). https://doi.org/10.1007/978-3-030-19212-9_3
8. Berg, J., Järvisalo, M.: Weight-aware core extraction in SAT-based MaxSAT solving. In: Beck, J.C. (ed.) CP 2017. LNCS, vol. 10416, pp. 652–670. Springer, Cham (2017). https://doi.org/10.1007/978-3-319-66158-2_42
9. Berg, J., Järvisalo, M.: Unifying reasoning and core-guided search for maximum satisfiability. In: Calimeri, F., Leone, N., Manna, M. (eds.) JELIA 2019. LNCS (LNAI), vol. 11468, pp. 287–303. Springer, Cham (2019). https://doi.org/10.1007/978-3-030-19570-0_19
10. Davies, J., Bacchus, F.: Exploiting the power of MIP solvers in MAXSAT. In: Järvisalo, M., Van Gelder, A. (eds.) SAT 2013. LNCS, vol. 7962, pp. 166–181. Springer, Heidelberg (2013). https://doi.org/10.1007/978-3-642-39071-5_13
11. Demirović, E., Stuckey, P.J.: Local-style search in the linear MaxSAT algorithm: a computational study of solution-based phase saving. In: Pragmatics if SAT Workshop (2018)
12. Demirović, E., Stuckey, P.J.: Techniques inspired by local search for incomplete MaxSAT and the linear algorithm: varying resolution and solution-guided search. In: Schiex, T., de Givry, S. (eds.) CP 2019. LNCS, vol. 11802, pp. 177–194. Springer, Cham (2019). https://doi.org/10.1007/978-3-030-30048-7_11
13. Eén, N., Sörensson, N.: Temporal induction by incremental SAT solving. Electr. Notes Theor. Comput. Sci. **89**(4), 543–560 (2003)
14. Feydy, T., Stuckey, P.J.: Lazy clause generation reengineered. In: Gent, I.P. (ed.) CP 2009. LNCS, vol. 5732, pp. 352–366. Springer, Heidelberg (2009). https://doi.org/10.1007/978-3-642-04244-7_29

15. Fu, Z., Malik, S.: On solving the partial MAX-SAT problem. In: Biere, A., Gomes, C.P. (eds.) SAT 2006. LNCS, vol. 4121, pp. 252–265. Springer, Heidelberg (2006). https://doi.org/10.1007/11814948_25
16. Ignatiev, A., Morgado, A., Manquinho, V.M., Lynce, I., Marques-Silva, J.: Progression in maximum satisfiability. In: Proceedings of the 21st European Conference on Artificial Intelligence, pp. 453–458 (2014)
17. Ignatiev, A., Morgado, A., Marques-Silva, J.: RC2: a python-based MaxSAT solver. In: MaxSAT Evaluation 2018, p. 22 (2018)
18. Koshimura, M., Zhang, T., Fujita, H., Hasegawa, R.: QMaxSAT: a partial max-sat solver. J. Satisf. Boolean Model. Comput. 8, 95–100 (2012)
19. Marriott, K., Stuckey, P.: Programming with Constraints: An Introduction. MIT-Press, Cambridge (1998)
20. Martins, R., Manquinho, V.M., Lynce, I.: Improving linear search algorithms with model-based approaches for MaxSAT solving. J. Exp. Theor. Artif. Intell. 27(5), 673–701 (2015)
21. Morgado, A., Dodaro, C., Marques-Silva, J.: Core-guided MaxSAT with soft cardinality constraints. In: O'Sullivan, B. (ed.) CP 2014. LNCS, vol. 8656, pp. 564–573. Springer, Cham (2014). https://doi.org/10.1007/978-3-319-10428-7_41
22. Morgado, A., Heras, F., Marques-Silva, J.: Model-guided approaches for MaxSAT solving. In: Proceedings of the ICTAI, pp. 931–938. IEEE Computer Society (2013)
23. Ohrimenko, O., Stuckey, P., Codish, M.: Propagation via lazy clause generation. Constraints 14(3), 357–391 (2009)
24. Saikko, P., Berg, J., Järvisalo, M.: LMHS: a SAT-IP hybrid MaxSAT solver. In: Creignou, N., Le Berre, D. (eds.) SAT 2016. LNCS, vol. 9710, pp. 539–546. Springer, Cham (2016). https://doi.org/10.1007/978-3-319-40970-2_34
25. Schulte, C., Tack, G.: View-based propagator derivation. Constraints 18(1), 75–107 (2013)
26. Stuckey, P.J., Feydy, T., Schutt, A., Tack, G., Fischer, J.: The MiniZinc challenge 2008–2013. AI Mag. 35(2), 55–60 (2014)

# Robust Resource Planning for Aircraft Ground Operations

Yagmur S. Gök[1]([⊠]), Daniel Guimarans[2], Peter J. Stuckey[2],
Maurizio Tomasella[1], and Cemalettin Ozturk[3]

[1] Business School, The University of Edinburgh,
29 Buccleuch Place, Edinburgh EH8 9JS, UK
{Yagmur.Gok,Maurizio.Tomasella}@ed.ac.uk
[2] Faculty of Information Technology, Monash University,
900 Dandenong Road, Melbourne, VIC 3145, Australia
{daniel.guimarans,peter.stuckey}@monash.edu
[3] United Technologies Research Center,
4th Floor, Penrose Business Centre, Penrose Wharf, Cork T23 XN53, Ireland
OzturkC@utrc.utc.com

**Abstract.** Aircraft turnaround scheduling and airport ground services team/equipment planning directly concern both the airport operator and service providers. We first ensure airport-wide global optimality by solving a resource-constrained project scheduling problem (RCPSP) for minimal overall delays. We then support decentralized allocation of teams/vehicles to flights, independently by each service provider. Either a multiple traveling salesman problem with time-windows (mTSPTW), or a vehicle routing problem with time-windows (VRPTW) are solved for this purpose, by taking advantage of both constraint programming (CP) and mixed integer programming (MIP) solvers. We also exploit these models in a matheuristic approach based on large neighborhood search used to reach good solutions in reasonable time for real-world instances. Unlike the classical VRP objective of minimizing traveling time, we maximize the total slack time between team visits, and show that doing this fosters robustness of the generated plans. We assess the robustness of solutions through a discrete-event simulation model, and conclude by validating our approach with data provided by a major ground handling company for a day of operations at Barcelona El Prat Airport.

**Keywords:** Optimization · Scheduling · Routing · Aviation · Airport operations

## 1 Introduction

Effective planning and scheduling is crucial in many areas of airport operations, where decisions are interconnected with each other and the potential for flight delays due to knock-on effects is rather high. Careful planning for the

© Springer Nature Switzerland AG 2020
E. Hebrard and N. Musliu (Eds.): CPAIOR 2020, LNCS 12296, pp. 222–238, 2020.
https://doi.org/10.1007/978-3-030-58942-4_15

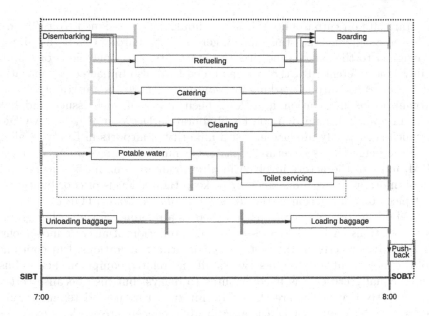

**Fig. 1.** An example of aircraft turnaround operations and precedence relationships between them within a timeline

day of operations is essential. With many factors out of the control of any airport decision maker—weather conditions, aircraft technical faults, delays, late passengers on boarding, etc.—, plans that are in some form *robust* should be sought. This is particularly true for aircraft turnaround and airport ground services. When an aircraft lands, it proceeds to a parking stand. Here, it undergoes a sequence of operations to get ready for the following take-off. A mix of operations may be needed: passenger disembarking/boarding, baggage unloading/loading, refueling, cabin cleaning, catering, toilet and potable water servicing, and aircraft push-back. Precedence relations do apply—e.g. refueling often cannot start before passengers have disembarked due to safety regulations. Figure 1 shows an example of a single aircraft turnaround. Conjunctive arcs represent the precedence relations. There exist specific time windows through which each activity should be performed. Each time window shortens/stretches out depending on what happens to other related operations. The turnaround should ideally start as soon as an aircraft arrives at its stand, ideally at the time for which it was scheduled to arrive there—or start of *in-block time* (*SIBT*)—and should be completed by the time it was scheduled to be pushed-back into the taxiway on its way to the runway—also called its start of *off-block time* (*SOBT*).

Turnaround operations are often handled by different service providers (SPs). A great deal of coordination is required to make sure they do not delay any aircraft, and for delays not to propagate to other aircraft. Cross-turnaround delay propagation may happen if teams from SPs are scheduled back-to-back between subsequent aircraft: team A finishing late on aircraft turnaround X will surely start late on turnaround Y if it was immediately assigned to it. Team A will also be occupying the stand for longer, with turnaround Z of the aircraft next assigned to the same stand likely to start late, and operations of team B (possibly of a different SP), that was assigned to Z also impacted. Propagation of delays across the airport parking lot—*apron* for short—is a certainty.

Organizations in aviation have long been aware of such issues and have devised an approach named Airport Collaborative Decision Making (A-CDM) ([5]), which is currently in place in just a minority of airports in Europe—albeit some of the biggest ones in terms of passenger numbers per year. According to A-CDM, most of the actors involved in aircraft operations must share certain pieces of information with one another to keep tighter levels of coordination. A central piece of information is the *Target Off-Blocks Time* (TOBT), which is calculated on the day of operations and used as a reference for all other ground service operations. A-CDM focuses on the coordination of aircraft movements, seen, probably correctly, as the center-piece of airport operations. But coordination of the movements of ground service staff and related equipment across busy aprons is equally critical, as it contributes to delays, but also because certain pieces of costly equipment are shared by multiple turnaround teams working for separate SPs. While A-CDM is an appealing concept, to work as expected, a certain degree of information sharing among the airport and the various SPs need to be in place, which happens typically only if enforced. Busier hub airports represent the typical example where the enabling conditions are met. In non-A-CDM airports, though, coordination is also needed, but without data sharing mechanisms it is virtually impossible to achieve. The remainder of this paper proposes a feasible and robust way to support this more general scenario.

## 2    Related Literature

The Operations Research (OR) community has, in the past three decades, worked on the modeling and solution of problems related to coordinating the movements, usage, and sharing of turnaround teams and equipment, but only partially.

The body of work by Norin and colleagues [18,19] shares quite a few elements with our study. As we, and others do (e.g., [1] and [21]), the problem is modeled as a form of Vehicle Routing Problem with Time Windows (VRPTW) [22]. Similarly to us, they work in an A-CDM setting, albeit with a different objective function: they minimize the weighted sum of flight delays and traveling distance. Due to the computational complexity of the underlying problem, they also focus on finding reasonable solutions in short times adopting a form of Greedy Randomized Adaptive Search Procedure (GRASP). They also make use of simulation to test the robustness of their heuristic plans under a range of uncertain conditions. Differently from us, the authors focus on one turnaround operation only: de-icing, with one type of vehicle/related staff to operate them. Stockholm Arlanda provides for their data set and motivating example.

In another closely related study [12], the authors did not see an option to adopt a fully centralized solution process, as the ground service providers are effectively separate legal entities making independent decisions. They also did not see the point of adopting a complex negotiation mechanism between different service providers, essentially ruling out in their assumptions any form of information sharing. This is a very relevant paper based on the same problem as ours, but examined through a substantially different perspective.

Padron et al. [21] probably represents the closest work to our study. They also consider scheduling all the ground handling vehicles at an airport, and they combine CP with heuristic search techniques such as Large Neighborhood Search (LNS) and Variable Neighborhood Descent (VND). Unlike us, their turnarounds have a fixed sequence of tasks. In a subsequent paper [20], as we do here, they integrate simulation at the bottom-end of their methodology to investigate the robustness of the generated heuristic solutions.

Among the remaining studies, less related to ours, [7] represents a more recent example of VRPTW formulation, but focuses on individual workers and their synchronization with vehicles, and has no view on collaborative mechanisms. Works such as [1,10], and [23] provide different heuristic approaches to solve the vehicle routing problems implied by aircraft turnaround operations at airports. Finally, many older studies had approached apron resource allocation problems from a multi-agent distributed planning perspective. Among these, [11,14,16] and [13] modeled their problems—as we also do in one of our steps—after the Resource-Constrained Project Scheduling Problem (RCPSP) [2], while [6] modeled the problem as a form of job shop scheduling.

# 3 Planning Process and Related Models

There are two levels to the turnaround planning problem studied in this paper—*long-term* and *short-term* (Fig. 2). All decisions are of tactical nature, as they take place ahead of the day of operation. Real-time management of apron operations and related resources is out of scope for the present paper.

In the longer term, flight schedules for the next few months are known to both the airport operator (AO) and the SPs. The AO needs this information to coordinate the scheduling of all the turnaround operations for each of the days in the planning horizon, aiming at minimum delays. The main decision maker in the longer-term is the AO, wanting to fix time windows for all operations whilst keeping overall resource requirements for the airport reasonably contained. After these decisions are made centrally, each SP can start thinking about their own resource requirements for each day of operation, and run their own staff rostering processes.

In the shorter term, approximately a week before the day of operations, the updated flight schedule is shared with all SPs and the same tactical reasoning can be rerun, with time windows revisited by the AO and rosters updated by the SPs.

In the even shorter term, closer to the start of the day of operations, time windows remain fixed and turnaround teams are known to SPs with a high degree of certainty. Then, each SP makes sure they can optimally route their

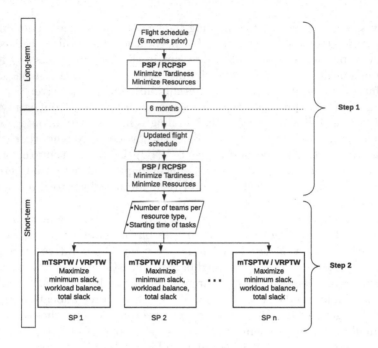

**Fig. 2.** Proposed turnaround planning process

staff through all turnaround tasks, by keeping some slack time to compensate for unforeseeable delays. At this stage, the objectives and constraints can differ by SP, but all SPs still have to stick to the time windows set by the AO.

Although our approach would ultimately help to plan for all kinds of resources in apron operations, we focus on planning for 'teams' of handling agents of each SP, i.e. we focus on human resources. Each of the turnaround operations is normally executed by small teams of employees, of size known to each SP, and planning for the sequence of turnarounds to be visited and serviced by each team during the day of operation is of utmost importance. In the following, we assume that all needed equipment is either carried over by the teams as they move from one turnaround to the next, or sourced across the apron area as they move through their jobs for the day, and as such is not modeled directly.

From an OR perspective, two classes of problems are involved in the description given above: *project scheduling* provides a convenient framework to the AO's problem of fixing time windows and keeping resource requirements under control, while *vehicle routing* comes to the rescue of SPs for optimal routing of their turnaround teams.

We consider (Subsect. 3.1) both versions of project scheduling problems (PSPs), with and without constraints in the number of resources (e.g., teams for unloading/loading baggage, cleaning, catering service, refueling, etc.), the latter class of problems taking the name of resource-constrained project scheduling problems (RCPSP). The first objective we seek is to minimize the overall tardiness for the airport. After achieving a minimal-delay *ideal* schedule, the total resource requirement for the airport is minimized by enforcing any optimal solution found at this stage to maintain at least the same level of tardiness coming from the PSP.

Moving to optimal routing of teams for each SP (Subsect. 3.2), the frameworks of reference are two: multi-traveling salesmen problem with time windows (mTSPTW) [4] and vehicle routing problem with time windows (VRPTW). The choice depends on whether the vehicles that are operated by teams are *capacitated* or not. Most resources/teams use a vehicle with limited capacity in order to service an aircraft, such as catering trucks, where the catering team is responsible for loading a certain number of galley trolleys into the aircraft. Other resources are uncapacitated and have no limitations on the number of aircraft that can be serviced before returning to the depot (e.g., push-back vehicles). Unlike the typical objective of both mTSPTW and VRPTW, which is minimizing the traveling distance, we first maximize the slack between tasks, in order to give enough time to absorb any small disruptions. This is meant to enforce a certain degree of robustness to the plan. After that, we try and maximize the workload balance among teams of the same SP, as much as possible, to foster fairness of the plan. Finally, we maximize the total slack time, when the minimum slack time cannot be improved further, with the effect of increasing all slack times between tasks except the minimum among all.

Details of our models are provided in the next two sub-sections. Many powerful global constraints from the Constraint Programming (CP) community are available for PSP/RCPSP and mTSPTW/VRPTW, hence our CP formulations. By employing such global constraints and taking advantage of the strength of different solvers, one would expect better computational performance with respect to, say, Mixed-Integer Programming (MIP) formulations and solvers. All models were developed in MiniZinc [17], a solver-independent modeling framework that allows the model to be run on many different solvers. This feature was crucial to enable our solution approach (Sect. 4).

## 3.1   Project Scheduling Models

For a given day of operations, we set the start time $start_i \in [0, t_{max}]$ of all tasks $i \in I = \{1 \ldots \phi\}$ that cover all aircraft turnarounds expected at the given airport ($t_{max}$ is the length of the day of operation). We do so in two steps. In the first (PSP, Eqs. (1a) and (2)–(6)), we aim at minimum costs resulting from tardy turnarounds, assuming unlimited resources. In the second (RCPSP, Eqs. (1b), (2)–(6), (7b) and (8b)), we aim for minimal resource needs whilst maintaining tardiness performance established in the first step.

Each task has an expected processing time $duration_i$. Based on the known flight timetables, both the Scheduled Time of Arrival (STA) and Scheduled Time of Departure (STD) of each aircraft are known. As a result of this and of the precedence relations among all tasks (Fig. 1), earliest start times $sta_i$ and earliest end times $std_i$ of all tasks are also known in advance. The set of all tasks $j \in I$ which can only start after a given task $i \in I$ is completed is denoted as $S_i$. This set will be empty for push-back tasks, which represent the natural conclusion of the related turnarounds. In between any two tasks, a fixed setup time $setup_i$ will ensure resources can effectively be gathered and moved from one location to another across the apron. This parameter can be estimated as a function of the maximum distance among all stands.

Each task represents a specific activity $A = \{1 \dots \alpha\}$, e.g. baggage loading. By $AT_a$ we denote the set of all tasks of type $a \in A$. Sets $SO$ and $SI$ represent, respectively, activities that are only allowed to start a certain time before STD, or after STA (mostly due to process specifications).

Certain pairs of tasks cannot be performed simultaneously, e.g. potable water and toilet servicing of the same turnaround. $P = \{1 \dots N\}$ is the set of such forbidden pairings, and $D_p$ is the set of (two) tasks for each $p \in P$.

Each task requires, uninterruptedly from start to end, a given amount $rr_{i_k}$ of a given type of resource $k \in K = \{1 \dots \kappa\}$. Resource types effectively represent teams of handling agents providing services of different nature. Resource capacity per resource type $rc_k$ also needs to be decided.

In the joint CP formulation of the PSP/RCPSP steps that follows, we also denote (Objective $Z_1$, see (1a)) the cost of tardy turnarounds per unit of time as $costtardy$, while parameter $sobt_a$—see constraint (4)—states that certain activities need to be completed within given bounds from the planned departure time. Finally, we employ two global constraints: global constraint (6) ensures non permitted task pairs are scheduled separately, while global constraint (8b) ensures resource levels are not exceeded at any time.

## CP Formulation

$$Z_1^* = \min \; Z_1 = \min \sum_{\substack{i \in I \\ \text{where} \\ S_i = \{\}}} costtardy \times max\{0, start_i + duration_i - std_i\}$$

$$\text{(1a)}$$

$$Z_2^* = \min \; Z_2 = \min \sum_{k \in K} rc_k$$

$$\text{(1b)}$$

subject to

$$start_i \geq sta_i \;\; \forall i \in I \tag{2}$$

$$start_j \geq start_i + duration_i \;\; \forall i \in I, \forall j \in S_i \tag{3}$$

$$start_i + duration_i \geq std_i - sobt_a \;\; \forall a \in SO, \forall i \in AT_a \tag{4}$$

$$start_i = sta_i \;\; \forall a \in SI, \forall i \in AT_a \tag{5}$$

$$disjunctive([start_i | i \in D_p], [duration_i | i \in D_p]) \;\; \forall p \in P \tag{6}$$

$$Z_1 = Z_1^* \tag{7b}$$

$$cumulative([start_i \mid i \in I], [duration_i + setup_i \mid i \in I], [rr_{k,i} \mid i \in I], rc_k)$$

$$\forall k \in K \text{ where } rc_k < \sum_{i \in I} rr_{k,i} \tag{8b}$$

## 3.2 Routing Models

After solving the above project scheduling problems, both the AO and all SPs know all time windows in which tasks should be executed on the given day of operation, as well as the number of teams of all types that are likely required to do the job. Each SP then takes this information to optimally schedule for their own teams to cover all turnaround tasks they are contracted to service. In the following we will see how this single-SP decision can be supported.

As in Subsect. 3.1, we provide a joint formulation following a lexicographic approach, with three objectives/sub-problems in this case. The sequence of three is then repeatedly solved as many times as the resource types managed by the given SP. For some resource types (teams), the vehicles used, e.g. re-fueling trucks, have finite capacity, hence they may need to be replenished before visiting the next turnaround. For these resource types, capacity constraints (20)–(22) will need to be included in the three models. The sub-problems take then the form of a VRPTW, irrespective of the objective/step in the sequence. For other resource types (teams), e.g. push-back trucks, capacity is not an issue, constraints (20)–(22) are excluded, and the sub-problems take the form of an mTSPTW.

The objective of utmost importance, and the one to pursue first (Eqs. (9) and (12)–(23)), is to maximize the minimum slack time between any two tasks, in an attempt to absorb short delays and prevent minor knock-on effects. The second step in the sequence (Eqs. (10), (12)–(23) and (24b)) looks at maximizing the workload balance among teams, to enforce some form of fairness in the plan, something which would be required in highly-unionized settings. Workload equity and its calculation is a subject of interest in the literature [15] on routing problems. The general suggestion points at minimizing the maximum distance in order to achieve a balanced workload while still ensuring the minimization of traveling distances. However, we are not as concerned with traveling times in between tasks as we are with the much higher processing times for each task. The last step (Eqs. (11), (12)–(23), (24b) and (24c)) then seeks to maximize the total slack time in the plan, in a way to increase its robustness.

On the given day of operation, we focus on a given working shift $S = [startshift, endshift]$ for which a staff roster of the given SP is available. Within that, we know the number of teams $t \in T_{SP,k} = \{1 \dots t_{SP,k}\}$ of resource type $k$, who need to cover, overall, a known number of tasks $i \in I_{SP,k} = \{1 \dots \phi_{SP,k}\} \subset I$, by moving across a given number of parking stands $h \in H = \{1 \dots \eta\}$, where tasks are performed. The SP wants to set, for each task $i$:

- the start time of the task, or $stime_i \in S$;
- the team $rt_i \in T_{SP,k}$ assigned to $i$;

– the task $s_i$ immediately following $i$;
– whether replenishment is needed prior to moving to $s_i$.

For each available team, a specific route needs to be set up for the given shift (hence the letters 'rt' in $rt_i$), where the first task is a dummy task the label of which is a function of the label/index of the team in question, while all other tasks are 'genuine' tasks from set $I_{SP,k}$. Task labels then, whether genuine or dummy, take on value in set $N = I_{SP,k} \cup \{\phi_{SP,k} + 1, \ldots, \phi_{SP,k} + t_{SP,k}\}$. As a result, any task that is not the first task in each route is a task $s_i \in I_{SP,k}$. Constraint (14) ensures all the dummy nodes represent the start of individual routes. Constraint (15) makes sure both $i$ and $s_i$ belong to the same route.

## CP Formulation

$$z_1^* = \max \ z_1 = \max \ \min_{i \in I_{SP,k}} slack_i \tag{9}$$

$$z_2^* = \max \ z_2 = \max \ (\min_{t \in T_{SP,k}} workload_t - \max_{t \in T_{SP,k}} workload_t) \tag{10}$$

$$z_3^* = \max \ z_3 = \max \ \sum_{i \in I_{SP,k}} slack_i \tag{11}$$

subject to

$$circuit\,([s_i \mid i \in N]) \tag{12}$$

$$alldifferent\,([s_i \mid i \in N]) \tag{13}$$

$$rt_{\phi+t} = t \ \forall t \in T_{SP,k} \tag{14}$$

$$rt_{s_i} = rt_i \ \forall i \in I_{SP,k} \tag{15}$$

$$stime_i = start_i \ \forall i \in I_{SP,k} \tag{16}$$

$$busy_i = \begin{cases} stime_i + duration_i + \\ traveltime_{h_i,h_j} + x_i \times replenish, & \forall i \in I_{SP,k} \mid s_i \in I_{SP,k} \\ endshift, & otherwise \end{cases} \tag{17}$$

$$stime_{s_i} \geq busy_i \ \forall i \in I_{SP,k} \tag{18}$$

$$slack_i = stime_{s_i} - busy_i \ \forall i \in I_{SP,k} \tag{19}$$

$$q_i = cap \ \forall i \in N \setminus I_{SP,k} \tag{20}$$

$$q_{s_i} = q_i - demand_{s_i} \ \forall i \in N \setminus I_{SP,k} \tag{21}$$

$$q_{s_i} = \begin{cases} cap - demand_{s_i}, & if \ x_i = 1 \\ q_i - demand_{s_i}, & otherwise \end{cases} \ \forall i \in I_{SP,k} \tag{22}$$

$$workload_t = \sum_{\substack{i \in I_{SP,k} \\ \text{where } rt_i = t}} duration_i \ \forall t \in T_{SP,k} \tag{23}$$

$$z_1 = z_1^* \tag{24b}$$
$$z_2 \geq z_2^* \tag{24c}$$

Tasks should start at a time $stime_i$ that is no earlier than the $start_i$ that was assigned at the PSP/RCPSP stage. Later starts could be advisable/needed, e.g. because of resource limitations. Hence, although potentially contributing to causes of delays, at this routing stage decisions on task start times could be reconsidered, in principle at least. In our approach we simplify this aspect by fixing $start_i$ exactly as in constraint (16), as this is more akin to maximizing the form of slack in the system that we have in two out of three objectives.

Tasks should also not start until their immediate predecessor has been completed, as in constraints (18) and (17). Any time available in between tasks $i$ and $s_i$ is defined as slack—see constraint (19). The replenishment decision is enacted through binary decision variable $x_i$, which takes on value 1 if replenishment needs to happen between task $i$ and task $s_i$, 0 otherwise. Each replenishment takes $replenish$ time. Moving between two consecutive tasks requires $traveltime_{h_i,h_j}$, with $h_i, h_j \in H$, $h_i \neq h_j$. Task duration is again denoted as $duration_i$. The sum of the duration of all tasks assigned to a given team contribute to defining the total workload $workload_t$ for the team, as in constraint (23).

For capacity constrained turnaround services, initial capacity $q_i$ for the team and related vehicle is set to $cap$—constraint (20); capacity then depletes as required by the demand from each subsequent task $s_i$ but is topped-up any time a replenishment decision is made—see constraints (21) and (22).

Global constraint (12) builds a single overall sequence for all tasks of all routes, with dummy tasks signposting the start of each team's own route. Global constraint (13) is redundant and added to help propagation.

## 4 Solution Approach

From the above discussion, we know that our overall approach develops as in Fig. 2, with perspectives from both the AO and all the SPs supported by the models presented, respectively, in Subsects. 3.1 and 3.2.

Table 1 shows further details around solvers and search strategies we used, as well as additional parameters around any time limits adopted as stopping criterion, or whether we made use of a warm-start. Numbers in the first column to the left correspond to component steps of Steps 1 and 2 from Fig. 2. Variable names in the table refer to the formulations from Sect. 3.

All models were implemented in MiniZinc, which enabled us to test the performance of different solvers for each model, and ultimately select the most suitable for use in each case. In the case of steps 1.1 and 1.2, Chuffed [3] clearly outperformed all other available solvers and we were able to prove optimality for

**Table 1.** Solution steps

| Step | Model | Obj. | Additional input | Solver | Search strategy* | Time limit(s) | Warm start |
|------|-------|------|------------------|--------|------------------|---------------|------------|
| 1.1 | PSP | $Z_1$ | – | Chuffed | smallest, indomain_ min | – | – |
| 1.2 | RCPSP | $Z_2$ | $Z_1^*$ | Chuffed | smallest, indomain_min | – | – |
| 2.1.1 | mTSPTW/ VRPTW | $z_1$ | $t_{SP,k}, start$ | Gecode | first_fail/smallest, indomain_max/min | 10 | – |
| 2.1.2 | mTSPTW/ VRPTW | $z_1$ | $t_{SP,k}, start,$ $s, rt$ | Gurobi | | – | Yes |
| 2.2 | mTSPTW/ VRPTW | $z_2$ | $t_{SP,k}, start,$ $z_1^*$ | Gecode | first_fail/smallest, indomain_max/min | 20 | – |
| 2.3 | mTSPTW/ VRPTW | $z_3$ | $t_{SP,k}, start,$ $z_1^*, z_2^*$ | Gecode | first_fail/smallest, indomain_max/min | 90 | – |

all instances. The first sub-problem of Step 2 though, that is the maximization of minimum slack time, proved slightly different, and had to be broken down into two further component steps. We first used CP with a specialized search strategy (step 2.1.1) which allowed us to reach the maximum as quickly as possible with Gecode [8]. However, it took very long for CP to prove optimality in almost all cases. Thus, by taking advantage of the warm-start possibility, we used the same model and the solution provided by Gecode as a warm start for a MIP solver (Gurobi [9]) (2.1.2), thanks to which we managed to prove optimality for all instances in a very short time.

Choosing the right search strategy also proved decisive in terms of solving times wherever we adopted a CP approach. In steps 1.1 and 1.2, we chose the $start_i$ variables to lead the search. The variable selection strategy *smallest* means a variable is chosen with the smallest value in its domain, and the assignment of the value to that variable is done using *indomain_min*, meaning that it will get assigned the minimum value in its domain. On the other hand, for the mTSPTW/VRPTW component, we noticed the model was unable to solve quickly without specifying any search strategies. We then noticed that the two forms of the problem claim for different choices of search strategy. We observed that *first_fail*, where the variable with the smallest domain is chosen, outperforms other strategies for all mTSPTWs, while *smallest* performed better for all VRPTWs. On the variable assignment for mTSPTW, *indomain_max* ruled out the rest, meaning the assignment was made with the maximum value in its domain. For VRPTW, on the other hand, *indomain_min* performed better.

---

**Algorithm 1:** Large Neighborhood Search

---

**Input:** $rt, s, z_3$ from CP model (step 2.3 )
$rt^b \leftarrow rt$
$s^b \leftarrow s$
$z_3^b \leftarrow z_3$
**while** $iter < maxIter$ **do**
    $k1 \leftarrow$ a random number from $\{1 \ldots t_{SP,k}\}$
    $k2 \leftarrow$ a random number from $\{1 \ldots t_{SP,k}\} \setminus \{k1\}$
    **for** $i \leftarrow 1$ **to** $\phi$ **do**
        **while** $rt_i \in [k1,k2]$ **do**
            | Destroy $rt_i$ and $s_i$
        **end**
    **end**
    Repair $rt, s, z_3$ with CP model(step 2.3)
    $rt^t \leftarrow rt$
    $s^t \leftarrow s$
    $z_3^t \leftarrow z_3$
    **if** $z_3^t < z_3^b$ **then**
        $z_3 \leftarrow z_3^t$
        $rt \leftarrow rt^t$
        $s \leftarrow s^t$
    **else**
        $z_3^b \leftarrow z_3^t$
        $rt^b \leftarrow rt^t$
        $s^b \leftarrow s^t$
    **end**
**end**

---

The very last step of our approach involves adopting a Large Neighborhood Search (LNS) schema (Algorithm 1) to further improve the solutions obtained from each of the routing sub-problems composing step 2.2. In our implementation of LNS, we take the solution from maximizing $z_3$ as a starting point, then 'destroy' two routes, chosen at random from the given solution, and finally use again the same model from step 2.3 to 'repair' it. If the new solution is better than the incumbent, we update the record of the best solution, and repeat the process for up to 200 iterations.

## 5   Experiments

In this study we used real data coming from Europe's 6th busiest commercial airport, Barcelona - El Prat (BCN). Our data relate to one given day of operation and include seven resource types, with each type handled by a different SP, and ten different turnaround activity types, for a total of 914 tasks to be scheduled at the PSP/RCPSP stage. At the mTSPTW/VRPTW stage, we considered the two shifts per day as currently adopted at the given airport, irrespective of resource type/SP. There are approximately 50 turnarounds in each shift, amounting to approximately the same number of tasks to be assigned per resource type and shift. In some cases from out data set, two teams are required to perform a task— e.g., baggage loading/unloading for wide-body aircraft. In these cases, the tasks are duplicated to ensure two teams, not one, perform the same task.

We ran all models on a personal laptop (1.6 GHz Intel Core i5) running mac-OS High Sierra. The overall integration was achieved in Python 3.7 using MiniZinc Python (MiniZinc version 2.3.2), which allowed us to solve the models in an incremental way, whilst providing a platform for easy integration of our LNS implementation. Computation times are not of primary concern when dealing with our problem, as clear from Fig. 2, hence the average 45 min taken to run the whole solution approach end to end does not represent a problem, to start with. In reality though, the AO will only run the PSP/RCPSP stage, which will take only approximately 7 min. Each SP will instead run the mTSPTW/VRPTW separately, on its own, for two shifts, which will take up to around 6 min for each SP, LNS step included. The RCPSP is proven to be optimal for minimizing tardiness, as well as for minimizing total resources/teams per SP. Moreover, in the second stage, we prove optimality for the maximum minimum slack time for the given tardiness and resource levels. Work balance is between 0 and 66 min, and 15 min on average, within the allowed time limit. Ensuring optimal workload balance proved too challenging, hence we limited the time available for this stage.

The most time-consuming part of the whole approach was to find the optimal solution for total slack time for each VRPTW. This is due to the workload balance objective and constraints. When we relaxed this and maximized it for $z_1$ and $z_3$, we could get an optimal solution using Gurobi for almost all resource types and shifts, with only a few instances not proven in 5 hours of solving time. We compared the results of the objective bounds with our lexicographic approach (excluding workload balance objective and constraint). The average gap was 0.68% for 14 instances of a mix of mTSPTW and VRPTW (each for one shift). The maximum gap was 2.72%. The average solution time for our approach was 1.78 min, while it required 60 min for Gurobi to prove optimality, if reached.

Padron and Guimarans [20] tackled an extremely similar ground-handling problem on the exact same data set. Compared to them, our approach was able to reduce the number of resources used per resource type when tested in the same instance for BCN. This was largely because in the RCPSP stage we give flexibility to the ordering of tasks for an individual turnaround, rather than using a preset plan for each turnaround as they do.

To test our optimization results under the uncertainty that normally permeates real airport settings, we developed, validated and used a discrete event simulator. Uncertain factors in our problem include: aircraft arrival time, task duration, traveling time between stands, and replenishment time. A bounded exponential probability distribution was used for the traveling time, and triangular distributions for the rest.

In addition to the real case from BCN, we generated several instances with different mix of aircraft and frequencies of arrivals and departures. Instances are presented as $ta[\theta]\_t[tmax]$, where $\theta$ is the number of turnarounds and $tmax$ is the planning horizon in minutes. We study these scenarios, together with the one from Barcelona Airport, by setting different levels of variability (normal and high variability). We use this experimental setting to evaluate our approach using two different objectives: maximizing slack time and minimizing traveling time, as in typical VRPTWs. Ten independent replications were produced for each scenario. Table 2 provides a summary of relevant indicators. $\Sigma$ is the total delay time, $N$ is the number of delays, $N_\%$ is the percentage of delays and $\Sigma_{>15}$ shows the sum of delays that are in excess of 15 min. Superscripts $s$ and $t$ refer to the two approaches: maximizing slack and minimizing traveling distance, respectively. Computation time, including solving the RCPSP plus the average solving time (in seconds) of all VRPs and excluding simulation for a deterministic bound, are indicated by $\tau^s$ and $\tau^t$ for the respective approaches. Times for the simulated instances with _normal and _high variability are not provided in the table since the deterministic instances' computation times (_det) are the main indicators of the solution approach and simulation is not part of the solution but only used as a tool for evaluation.

**Table 2.** Simulation Key Performance Indicators (KPIs) for slack time maximization and travel time minimization approaches

| Instance | $N_\%^s$ | $N_\%^t$ | $\Sigma^s$ | $\Sigma^t$ | $\Sigma_{>15}^s$ | $\Sigma_{>15}^t$ | $\Delta\Sigma_{>15}$ | $\tau^s$ | $\tau^t$ |
|---|---|---|---|---|---|---|---|---|---|
| ta24_t120_normal | 62.50 | 67.50 | 87.67 | 106.89 | 4.61 | 6.58 | 1.97 | – | – |
| ta48_t240_normal | 66.04 | 76.25 | 262.98 | 362.66 | 15.67 | 36.18 | 20.51 | – | – |
| ta72_t360_normal | 60.28 | 69.44 | 376.97 | 512.09 | 21.65 | 55.86 | 34.21 | – | – |
| ta96_t480_normal | 69.69 | 78.23 | 570.39 | 792.53 | 25.79 | 88.98 | 63.19 | – | – |
| bcn_normal | 46.67 | 51.51 | 200.36 | 242.1 | 0.08 | 0.82 | 0.74 | – | – |
| ta24_t120_high | 86.67 | 92.50 | 290.36 | 279.45 | 77.69 | 48.21 | −29.48 | – | – |
| ta48_t240_high | 83.54 | 89.38 | 462.05 | 670.61 | 71.76 | 165.66 | 93.90 | – | – |
| ta72_t360_high | 84.17 | 91.67 | 708.97 | 1128.36 | 131.77 | 337.09 | 205.32 | – | – |
| ta96_t480_high | 85.52 | 92.81 | 1015.62 | 1620.01 | 154.56 | 509.48 | 354.92 | – | – |
| bcn_high | 61.72 | 69.35 | 423.13 | 541.47 | 9.65 | 21.67 | 12.02 | – | – |
| ta24_t120_det | 25.00 | 25.00 | 32.00 | 32.00 | 1.00 | 1.00 | 0.00 | 21 | 33 |
| ta48_t240_det | 41.67 | 41.67 | 164.00 | 164.00 | 2.00 | 2.00 | 0.00 | 558 | 120 |
| ta72_t360_det | 40.28 | 40.28 | 252.00 | 252.00 | 2.00 | 2.00 | 0.00 | 961 | 415 |
| ta96_t480_det | 53.13 | 53.13 | 370.00 | 370.00 | 3.00 | 3.00 | 0.00 | 1171 | 322 |
| bcn_det | 22.58 | 22.58 | 71.00 | 72.00 | 0.00 | 0.00 | 0.00 | 327 | 270 |

In the deterministic case, the KPIs are the same since there are enough teams to perform the given tasks on time, no matter what the objective is. In the normal and high variability cases, except one, we observe that our approach maximizing slack outperforms the typical minimization of traveling time. In the real case of BCN, we only have partial information provided by a ground handling company, and the instance does not correspond to the whole operation at the airport—i.e., it only includes the flights corresponding to the airlines currently having a contract with the ground handler. This implies a lower arrival frequency in the instance, causing long idle times between the majority of tasks. In these cases, our approach is not significantly better than simply minimizing traveling time, as long as variability remains low. However, as aircraft arrival frequency and variability increase, we observe a significant difference between the two objectives. Figure 3 shows how our approach is able to outperform travel time minimization, reducing the total delay across all turnarounds. This figure also shows that our approach provides more predictable delays, with a more contained spread across simulations for all instances.

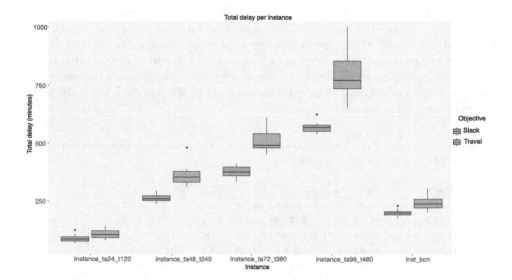

**Fig. 3.** Total delay per instance among 10 simulation replications

Table 2 also includes other important key performance indicators (KPIs), such as percentage of delayed turnarounds and total minutes of delay exceeding the *on time* threshold of 15 min. The latter is a big concern for SPs, since failing to meet this target carries penalties and potential further delays due to air traffic management. Our approach clearly reduces the total delay over 15 min, for up to 70% over the day of operation, except for one instance under high variability. Considering each minute of delay incurs a cost, deploying our approach in real-life scenarios could potentially result in significant cost savings.

## 6    Conclusion

In this work, we proposed a novel two-step solution approach to the airport ground service scheduling and team planning problem. With respect to earlier approaches, the RCPSP step allows service providers to operate their busy schedules with potentially fewer human resources. Our focus on maximizing minimum slack in the second step ensures they can do so efficiently. Our simulation proves the robustness of our approach. Still, tighter links between simulation and the optimization components could help, in the future, to enhance the performance of our LNS-based approach, e.g., by generating cuts from the simulation results for the benefit of the heuristic search component or using simulation within the CP search strategy.

# References

1. Andreatta, G., Capanna, L., De Giovanni, L., Monaci, M., Righi, L.: Efficiency and robustness in a support platform for intelligent airport ground handling. J. Intell. Transp. Syst.: Technol. Plan. Oper. 18(1), 121–130 (2014). https://doi.org/10.1080/15472450.2013.802160
2. Blazewicz, J., Lenstraand, J., Rinnooy Kan, A.: Scheduling subject to resource constraints: classification and complexity. Disc. Appl. Math. 5, 11–24 (1983)
3. Chu, G.: Improving combinatorial optimization. Ph.D. thesis, The University of Melbourne (2011). http://hdl.handle.net/11343/36679
4. Desrosiers, J., Dumas, Y., Solomon, M.M., Soumis, F.: Chapter 2 time constrained routing and scheduling. In: Network Routing, Handbooks in Operations Research and Management Science, vol. 8, pp. 35–139. Elsevier (1995). https://doi.org/10.1016/S0927-0507(05)80106-9
5. Eurocontrol: Airport Collaborative Decision Making (A-CDM) (2018). http://www.eurocontrol.int/articles/airport-collaborative-decision-making-cdm
6. Fan, W., Xue, F.: Optimize cooperative agents with organization in distributed scheduling system. In: Huang, D.-S., Li, K., Irwin, G.W. (eds.) ICIC 2006. LNCS (LNAI), vol. 4114, pp. 502–509. Springer, Heidelberg (2006). https://doi.org/10.1007/978-3-540-37275-2_61
7. Fink, M., Desaulniers, G., Frey, M., Kiermaier, F., Kolisch, R., Soumis, F.: Column generation for vehicle routing problems with multiple synchronization constraints. Eur. J. Oper. Res. 272(2), 699–711 (2019). https://doi.org/10.1016/j.ejor.2018.06.046
8. Gecode Team: Gecode: generic constraint development environment (2017). http://www.gecode.org
9. Gurobi: Gurobi software. http://www.gurobi.com/
10. Ip, W.H., Wang, D., Cho, V.: Aircraft ground service scheduling problems and their genetic algorithm with hybrid assignment and sequence encoding scheme. IEEE Syst. J. 7(4), 649–657 (2013). https://doi.org/10.1109/JSYST.2012.2196229
11. Kuster, J., Jannach, D.: Handling airport ground processes based on resource-constrained project scheduling. In: Advances in Applied Artifical Intelligence, pp. 166–176 (2006). https://doi.org/10.1007/11779568_20
12. van Leeuwen, P., Witteveen, C.: Temporal decoupling and determining resource needs of autonomous agents in the airport turnaround process. In: 2009 IEEE/WIC/ACM International Joint Conference on Web Intelligence and Intelligent Agent Technology. vol. 2, pp. 185–192 (2009). https://doi.org/10.1109/wi-iat.2009.149
13. Mao, X., Roos, N., Salden, A.: Distribute the selfish ambitions. In: Belgian/Netherlands Artificial Intelligence Conference, pp. 137–144 (2008)

14. Mao, X., Ter Mors, A., Roos, N., Witteveen, C.: Agent-based scheduling for aircraft deicing. In: Proceedings of the 18th Belgium-Netherlands Conference on Artificial Intelligence, BNVKI, pp. 229–236 (2006)
15. Matl, P., Hartl, R., Vidal, T.: Workload equity in vehicle routing problems: a survey and analysis. Transp. Sci. **52**(2), 239–260 (2018). https://doi.org/10.1287/trsc.2017.0744
16. Neiman, D.E., Hildum, D.W., Lesser, V.R., Sandholm, T.W.: Exploiting meta-level information in a distributed scheduling system. In: Proceedings of the National Conference on Artificial Intelligence, vol. 1, pp. 394–400 (1994)
17. Nethercote, N., Stuckey, P.J., Becket, R., Brand, S., Duck, G.J., Tack, G.: MiniZinc: towards a standard CP modelling language. In: Bessière, C. (ed.) CP 2007. LNCS, vol. 4741, pp. 529 543. Springer, Heidelberg (2007). https://doi.org/10.1007/978-3-540-74970-7_38
18. Norin, A., Yuan, D., Granberg, T.A., Värbrand, P.: Scheduling de-icing vehicles within airport logistics: a heuristic algorithm and performance evaluation. J. Oper. Res. Soc. **63**(8), 1116–1125 (2012). https://doi.org/10.1057/jors.2011.100
19. Norin, A., Granberg, T.A., Värbrand, P., Yuan, D.: Integrating optimization and simulation to gain more efficient airport logistics. In: Eighth USA/Europe Air Traffic Management Research and Development Seminar (2009)
20. Padron, S., Guimarans, D.: Using simulation for evaluating ground handling solutions reliability under stochastic conditions. In: 2018 ROADEF Lorient, France, pp. 1–6 (2018)
21. Padron, S., Guimarans, D., Ramos, J.J., Fitouri-Trabelsi, S.: A bi-objective approach for scheduling ground-handling vehicles in airports. Comput. Oper. Res. **71**, 34–53 (2016). https://doi.org/10.1016/j.cor.2015.12.010
22. Solomon, M.M., Desrosiers, J.: Survey paper–time window constrained routing and scheduling problems. Transp. Sci. **22**(1), 1–13 (1988). https://doi.org/10.1287/trsc.22.1.1
23. Trabelsi, S.F., Mora-Camino, F., Padron, S.: A decentralized approach for ground handling fleet management at airports. In: 2013 International Conference on Advanced Logistics and Transport, ICALT 2013, pp. 302–307 (2013). https://doi.org/10.1109/ICAdLT.2013.6568476

# Primal Heuristics for Wasserstein Barycenters

Pierre-Yves Bouchet[1,2] , Stefano Gualandi[1(✉)] ,
and Louis-Martin Rousseau[3]

[1] Dipartimento di Matematica "F. Casorati",

Università degli Studi di Pavia, Pavia, Italy
stefano.gualandi@unipv.it
[2] Dept. de Mathématiques et Génie Industriel,
GERAD-Polytechnique Montréal, Montreal, Canada
[3] Dept. de Mathématiques et Génie Industriel,
CIRRELT-Polytechnique Montréal, Montreal, Canada

**Abstract.** This paper presents primal heuristics for the computation of Wasserstein Barycenters of a given set of discrete probability measures. The computation of a Wasserstein Barycenter is formulated as an optimization problem over the space of discrete probability measures. In practice, the barycenter is a discrete probability measure which minimizes the sum of the pairwise Wasserstein distances between the barycenter itself and each input measure. While this problem can be formulated using Linear Programming techniques, it remains a challenging problem due to the size of real-life instances. In this paper, we propose simple but efficient primal heuristics, which exploit the properties of the optimal plan obtained while computing the Wasserstein Distance between a pair of probability measures. In order to evaluate the proposed primal heuristics, we have performed extensive computational tests using random Gaussian distributions, the MNIST handwritten digit dataset, and the Fashion MNIST dataset introduced by Zalando. We also used Translated MNIST, a modification of MNIST which contains original images, rescaled randomly and translated into a larger image. We compare the barycenters computed by our heuristics with the exact solutions obtained with a commercial Linear Programming solver, and with a state-of-the-art algorithm based on Gaussian convolutions. Our results show that the proposed heuristics yield in very short run time and an average optimality gap significantly smaller than 1%.

**Keywords:** Wasserstein Barycenter · Kantorovich-Wasserstein distance · Linear programming · Constrained optimization

## 1 Introduction

The theory of Optimal Transport has recently received a renewed interest from the Machine Learning community as a mathematical tool to compare probability

© Springer Nature Switzerland AG 2020
E. Hebrard and N. Musliu (Eds.): CPAIOR 2020, LNCS 12296, pp. 239–255, 2020.
https://doi.org/10.1007/978-3-030-58942-4_16

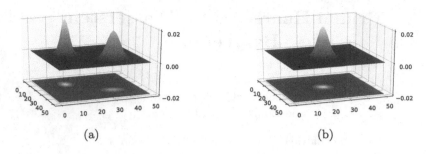

**Fig. 1.** Two Gaussian probability distributions and their Wasserstein Barycenter.

distributions by using the Kantorovich-Wasserstein distance [18,21,28,31,32]. The Wasserstein distance is also known as the *Earth Mover Distance* [27], because it can be seen as an analogy of the transportation cost of a given "mass of earth" distributed among the first distribution to the "required quantities of earth" distributed among the second distribution. Since the late nineties, the Wasserstein distance is used by the Computer Vision community as a tool to compare feature histograms of images [19,23], to implement adaptive color transfer [26], or to perform point set registration [8]. The same distance, but re-branded as *Word Mover Distance*, has proved to be extremely efficient for text classification, outperforming previous state-of-the-art algorithms [16]. The Wasserstein distance is used also in deep learning for solving optimization problems in the framework of Generative Adversarial Network (GAN) [4]. In general, the main idea for using Wasserstein distances consists of interpreting the problem data as (discrete) probability measures, and then, to compare such measures by solving an optimal transport problem. For further application domains, we refer the reader to [24].

In a more recent trend, the theory of Optimal Transport has been used to perform statistical inference on the space of probability distributions [1,22]. The main interest, in this case, is the possibility to take as input a given number of probability distributions, and then to compute a single distribution that represents them all. Essentially, the idea is to extend the notion of Fréchet mean to the space of probability functions using the Wasserstein distance [1]. We recall that the Fréchet mean generalizes the notion of centroid: for a given number of points in a metric space, it looks for a single point that minimizes the sum of distances to all other points given in input. For instance, on the field of the real numbers endowed with the Euclidean distance, the Fréchet mean gives the arithmetic means; on the field of the real numbers endowed with the hyperbolic distance, the Fréchet mean yields the geometric mean.

If we consider the space of (discrete) probability distributions endowed with a Wasserstein distance, and we look for their Fréchet mean, we get what is a called a Wasserstein Barycenter. For instance, Fig. 1(a) shows two Gaussian probability density functions defined on $\mathbb{R}^2$, while Fig. 1(b) shows their corresponding Wasserstein Barycenter. When we deal with discrete probability measures, that are probability measures whose support points are defined on a finite and discrete set of support points, we can compute their Discrete Wasserstein Barycenter, which is the main subject of this paper. Fig. 2 shows two translated digits 9 (interpreted as discrete measures) along with their Euclidean mean, a Convolutional Wasserstein Barycenter (heuristic), and the optimal Wasserstein Barycenter.

| Measure 1 | Measure 2 | Euclidean | Conv. [29] | Optimal |

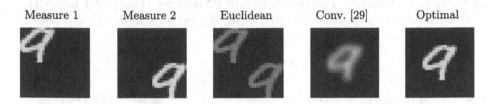

**Fig. 2.** Comparison among the Euclidean mean, the Convolutional Wasserstein Barycenters [29], and the optimal Wasserstein Barycenter.

*Related Works.* The Wasserstein Barycenter problem is introduced in [1], where the authors give the conditions of existence of a unique barycenter, and propose a fixed point iterative algorithm to compute the barycenter of a given set of Gaussian distributions. In the discrete setting, a fast algorithm to compute Wasserstein Barycenters, based on the Sinkhorn's algorithm [10], is proposed in [11]. When the discrete measures are supported on a regular grid (e.g., 2D images), the Wasserstein Barycenter can be computed by using Gaussian convolutions [29]. Other algorithms are reported in [12,20,25,30]. Using a Linear Programming perspective, most of these algorithms can be interpreted as dual algorithms, which search for a barycenter without guarantying primal feasibility (i.e., one or more problem constraints might be violated).

*Main Contribution.* The main contribution of this paper is to propose primal heuristics to compute efficiently feasible approximations of discrete Wasserstein Barycenters of order 2. Our primal heuristics first decompose the problem into smaller subproblems, and then iteratively interpolates the Wasserstein Barycenter between pairs of discrete measures. The quality of the primal solutions is evaluated by measuring the optimality gap, using as a baseline the optimal solutions obtained by solving the Wasserstein Barycenter problem with a commercial Linear Programming solver. Extensive computational tests with different datasets confirms that the proposed primal heuristics achieves a very good tradeoff between average optimality gap and run time.

*Outline.* This paper is organized as follows. Section 2 reviews the main concepts of the theory of Optimal Transport used in this paper. The definition of the Wasserstein Barycenter problem is given in Sect. 3. Section 4 presents our novel primal heuristics, that are the main contribution of this paper. Section 5 reports our extensive computational tests.

## 2    Background on Optimal Transport

In this section, we review the main notions of the theory of optimal transport used in this paper, with a specific focus on discrete probability measures. For an introduction on the theory of optimal transport, we refer the interested reader to the textbooks [24, 28, 32].

**Definition 1 (Discrete Finite Probability Measure (Chap. 2.1 in [24])).** *A discrete measure $\mu$ with weights $\mu_1, \ldots, \mu_n$ defined on a finite set of $n$ points $x_1, \ldots, x_n \in X$ is defined as*

$$\mu = \sum_{i=1}^{n} \mu_i \delta_{x_i},$$

*where $\delta_x$ is the Dirac at position $x$, intuitively a unit of mass concentrated at location $x$. If the weights are non-negative and $\sum_{i=1}^{n} \mu_i = 1$, that is, the vector of weights $\mu_i$ belongs to the simplex $S^n$, then $\mu$ is a discrete finite probability measure.*

**Definition 2 (Kantorovich-Rubinstein functional).** *Given two discrete finite probability measures $\mu = \sum_{i=1}^{n} \mu_i \delta_{x_i}$ and $\nu = \sum_{j=1}^{n} \nu_j \delta_{y_j}$ with $x_i, y_j \in X$, a cost function $c$ defined on $X \times X$, the Kantorovich-Rubinstein functional is equivalent to the following Linear Programming problem:*

$$\mathcal{W}_c(\mu, \nu) := \min \quad \sum_{i=1}^{n} \sum_{j=1}^{n} c(x_i, y_j) \pi(x_i, y_j) \tag{1}$$

$$s.t. \quad \sum_{j=1}^{n} \pi(x_i, y_j) = \mu_i \qquad\qquad i = 1, \ldots, n \tag{2}$$

$$\sum_{i=1}^{n} \pi(x_i, y_j) = \nu_j, \qquad\qquad j = 1, \ldots, n \tag{3}$$

$$\pi(x_i, y_j) \geq 0 \qquad\qquad i = 1, \ldots, n, \quad j = 1, \ldots, n. \tag{4}$$

*The optimal values of the variables $\pi(x_i, y_j)$ yields an optimal transportation plan, herein denoted by $\pi_{ij}^*$.*

Note that (1)–(4) is a standard Koopmans-Hitchcock transportation problem [15], where the decision variables $\pi(x_i, y_j)$ indicates the amount of mass moving from the support point $x_i$ to $y_j$. The problem can be formulated and solved with Linear Programming, or it can be formulated as an uncapacitated minimum cost flow problem on a bipartite graph, as in [2,5,7,14].

**Definition 3 (Wasserstein distance of order $p$).** *When the cost in the Kantorovich-Rubinstein functional is the $p^{th}$ power of a distance defined over $X$, that is, $c(x_i, y_j) = d(x_i, y_j)^p$, the Wasserstein distance of order $p$ is defined as*

$$W_p(\mu, \nu) := \mathcal{W}_{d^p}(\mu, \nu)^{\min\{1, \frac{1}{p}\}}. \tag{5}$$

The Wasserstein distance $W_p$ is a distance function defined on probability functions over a space $X$, that is, it can be shown that $W_p$ satisfies the axioms of a distance: (i) non-negativity, (ii) symmetry, and (iii) the triangle inequality (for details, see, e.g., [32]). As a consequence, the Wasserstein distance $W_p$ is commonly used to compare probability distribution functions.

In the remaining of this paper, we focus on the Wasserstein distance of order 2 and we restrict to the Euclidean distance $d^p(x_i, y_j) = ||x_i - y_j||_2^2$.

**Fig. 3.** Weighted Wasserstein Barycenter of two measures with $(\lambda_1, \lambda_2) = (1 - t, t)$.

## 3    Wasserstein Barycenters

Suppose that we are interested in computing the average $\hat{\rho}$ between two Dirac $\delta$-measures $\mu = \delta_x, \nu = \delta_y$. A possibility is to average over the two weight vectors, obtaining a new measure supported in $x$ and $y$, a sum of two $\delta$-measures with weights equal to $\frac{1}{2}$:

$$\hat{\rho} = \frac{1}{2}\delta_x + \frac{1}{2}\delta_y.$$

However, since we are averaging over probability distributions defined on a space $X \subseteq \mathbb{R}^k$, we can, alternatively, define the "average" as a new $\delta$-measure having a single support point located at the *mean* location of the two points $x$ and $y$. That is, we get a new measure

$$\rho = \delta_{\frac{x+y}{2}}.$$

Moreover, if we want to compute a weighted average, with weights $\lambda_1, \lambda_2 \geq 0$ satisfying $\lambda_1 + \lambda_2 = 1$, we can compute the measure $\rho = \delta_{(\lambda_1 x + \lambda_2 y)}$. If we extend this basic example to two discrete probability measures defined on a larger set of support points, as in Fig. 3, we can use the following lemma.

**Lemma 1 (Interpolation between two measures (Chap. 7, in [24])).**
*Given (i) two weights $(\lambda_1, \lambda_2) \in \mathbb{R}_+$ satisfying $\lambda_1 + \lambda_2 = 1$, (ii) two discrete measures $\mu$ and $\nu$ defined on $X$:*

$$\mu = \sum_{i=1}^{n} \mu_i \delta_{x_i} \quad and \quad \nu = \sum_{j=1}^{n} \nu_j \delta_{y_j},$$

*and (iii) an optimal transportation plan $\pi^*$ minimizing the functional $W_2(\mu, \nu)$, that is, an optimal solution of Problem (1)–(4) with $c(x_i, y_j) = ||x_i - y_j||_2^2$, the interpolated average measure $\rho$ between $\mu$ and $\nu$ is*

$$\rho = f(\mu, \nu, \lambda_1, \lambda_2) := \sum_{i=1}^{n} \sum_{j=1}^{n} \pi_{ij}^* \delta_{(\lambda_1 x_i + \lambda_2 y_j)}. \tag{6}$$

While interpolating between two discrete probability measures defined over a discrete set (e.g. $X$ is a regular grid), it might happen that one (or more) interpolated support points do not belong to $X$, that is, $(\lambda_1 x_i + \lambda_2 y_j) \notin X$. In this case, in post processing, we have to select a point $z \in X$ that minimizes the discretization error $\epsilon = ||(\lambda_1 x_i + \lambda_2 y_j) - z||_2$.

The concept of Wasserstein Barycenter is introduced when dealing with a set of $m$ discrete finite probability measures $\mu_1, \ldots, \mu_m$, with $m \geq 2$, assumed (without loss of generality) to be defined over the same finite set $X$.

**Definition 4 (Wasserstein Barycenter (Chap. 5 in [28])).** *The Weighted Wasserstein Barycenter of $m$ measures $\boldsymbol{\mu} = (\mu_1, \ldots, \mu_m)$, with given weights $\boldsymbol{\lambda} = (\lambda_1, \ldots, \lambda_m)$, such that $\lambda_i \geq 0$ and $\sum_{i=1}^{m} \lambda_i = 1$, is defined as*

$$\rho^*(\boldsymbol{\mu}, \boldsymbol{\lambda}) := \arg\min_{\rho \in \mathcal{S}^n} \sum_{k=1}^{m} \lambda_k (W_2(\mu_k, \rho))^2. \tag{7}$$

If $\mu_k$ are discrete probability measures and $\mu_{ik}$ is the $i$-th element of the measure $\mu_k$, if we use the cost $c(x_i, y_j) = ||x_i - y_j||_2^2$ in the objective function, and if we fix a set of possible locations $y_j$ for the support points of the barycenter $\rho$, then problem (7) is equivalent to the following Linear Program [3]:

$$\mathcal{B}(\boldsymbol{\mu}, \boldsymbol{\lambda}) = \min \sum_{k=1}^{m} \lambda_k \left( \sum_{i=1}^{n} \sum_{j=1}^{n} ||x_i - y_j||_2^2 \, \pi_{ijk} \right) \tag{8}$$

$$\text{s.t.} \quad \sum_{j=1}^{n} \pi_{ijk} = \mu_{ik} \qquad i = 1, \ldots, n, k = 1, \ldots, m \tag{9}$$

$$\sum_{i=1}^{n} \pi_{ijk} = \rho_j \qquad j = 1, \ldots, n, k = 1, \ldots, m \tag{10}$$

$$\sum_{j=1}^{n} \rho_j = 1 \tag{11}$$

$$\pi_{ij_k} \geq 0, \rho_j \geq 0, \ i, j = 1, \ldots, n, k = 1, \ldots, m \tag{12}$$

Constraints (9) and (10) replicates the constraints of Problem (1)–(4) for computing the distance between the barycenter measure $\rho$ and each input measure $\mu_k$. The constraint (11) and the non-negative constraints in (12) force $\rho$ to belong to the simplex $\mathcal{S}^n$. Whenever the support points of the measures $\rho$ and $\mu_k$ are fixed, we can solve Problem (8)–(12) with any Linear Programming solver.

Herein, we denote by $\rho^*$ the values of the optimal decision variables $\rho_j$ which corresponds to an optimal solution of Problem (8)–(12). We remark that by solving Problem (6) and by discretizing in post-processing the support points of $\rho$, we can recover a nearly optimal solution of Problem (8)–(12) with $m = 2$ (it is only "nearly optimal" because of possible discretization errors). Indeed, by solving (6) we obtain the Wasserstein Barycenter between two discrete measures on a continuous space.

The LP model (8)–(12) is valid also when the distance $W_2(\mu_k, \rho)$ in (7) is replaced by the Wasserstein distance of order 1 $W_1(\mu_k, \rho)$. In the latter case, we get a different variant of the Wasserstein Barycenter which corresponds to a Fréchet *median*, instead of a Fréchet *mean*. Efficient LP models for the Fréchet median, based on a network flow formulations of the problem, are studied in [6].

## 4   Primal Heuristics

In this section, we present two types of heuristics to compute primal solutions for Problem (7). The only constraint that a feasible solution $\bar{\rho}$ of (7) must satisfies is to belong to the simplex $\mathcal{S}^n$. The simplest method to obtain a feasible solution for (7) is to compute the *Euclidean mean* of the input measures. If we denote by $z_1, \ldots, z_n$ the union of all the support points of the input measures $\mu_1, \ldots, \mu_m$, then we can define the Euclidean mean as a discrete measure as follows:

$$\xi = \sum_{i=1}^{n} \xi_i \delta_{z_i} = \sum_{i=1}^{n} \left( \frac{1}{n} \sum_{k=1}^{m} \tilde{\mu}_{ik} \right) \delta_{z_i}, \tag{13}$$

where $\tilde{\mu}_{ik} = \mu_{ik}$ if $z_i$ is a support point of the measure $\mu_k$, and, $\tilde{\mu}_{ik} = 0$ otherwise. Figure 2 shows that $\xi$ is far from the optimal Wasserstein Barycenter.

### 4.1   Sequential Heuristics

*The Iterative Heuristic (IH).* The first heuristic we propose approximates $\rho^*$ by a discrete probability measure $\bar{\rho}$ obtained by iteratively computing the barycenters between pairs of measures. At each iteration, the barycenter between two measures is computed using the interpolating function (6).

Given the input measures with a fixed order $\mu_1, \ldots, \mu_m$, our heuristic computes $\bar{\rho}^{IH}$ by solving the following recursion:

$$\theta^{(k)} = \begin{cases} \mu_1 & \text{if } k = 1, \\ f\left(\mu_k, \theta^{(k-1)}, \frac{1}{k}, \frac{k-1}{k}\right) & \text{if } k > 1, \end{cases} \tag{14}$$

$$\bar{\rho}^{IH} = \theta^{(m)}. \tag{15}$$

We call this heuristic the *Iterative Heuristic*. When $k = 2$, it computes the barycenter $\theta^{(2)}$ of the first two input measures $\mu_1$ and $\mu_2$, using the same weights $\lambda_1 = \lambda_2 = \frac{1}{2}$. When $k = 3$, the barycenter $\theta^{(3)}$ is computed using the third measure $\mu_3$ with weight $\lambda_1 = \frac{1}{3}$, and the barycenter $\theta^{(2)}$ with weight $\lambda_2 = \frac{2}{3}$. At the very last iteration, the heuristic computes the barycenter between $\mu_m$ with weight $\frac{1}{m}$, and the accumulated barycenter $\theta^{(m-1)}$ with weight $\frac{m-1}{m}$.

Since the order of the input measures has an impact on the final measure $\bar{\rho}^{IH}$ in the iterative heuristic, we have investigated different criteria for sorting the input sequence. Let us denote by $J_1 = \{1, \ldots, m\}$ and by $J_k = J_{k-1} \setminus \{i_{k-1}\}$ where the sequence $(i_k)$ is a permutation of $\{1, \ldots, m\}$ representing the sorted sequence of inputs.

*The Iterative Closest Heuristic (CH).* The first idea is, when $k = 1$, to begin with the measure that is the closest to the Euclidean barycenter $\xi$. Then, for every $k > 1$, we select, among the measures indexed by $J_k$, the measure $\mu_{i_k}$ that is the closest to $\theta^{(k-1)}$. Here, "closest" have to be understood in the sense of the two measures which minimises a function $d$ defined over two finite discrete probability measures over $X$ (which is not necessarily a distance). The iterative heuristic is:

$$\theta^{(1)} = \mu_{i_1}, \qquad\qquad \text{where } i_1 \in \arg\min_{j \in J_1} d(\mu_j, \xi), \qquad (16)$$

$$\theta^{(k)} = f\left(\mu_{i_k}, \theta^{(k-1)}, \frac{1}{k}, \frac{k-1}{k}\right), \quad \text{where } i_k \in \arg\min_{j \in J_k} d(\mu_j, \theta^{(k-1)}), \quad (17)$$

$$\bar\rho^{CH} = \theta^{(m)}. \qquad (18)$$

This recursion differs from (14)–(15) because the sequence $\mu_1, \ldots, \mu_m$ is dynamically reordered into $\mu_{i_1}, \ldots, \mu_{i_m}$, by selecting at each iteration $k$ the "closest" measure to $\theta^{(k-1)}$, while ignoring the measures that were already selected at the previous iterations. Regarding the function $d$ appearing in (16) and (17), we could use the Wasserstein distance $W_2$. However, since we have to trade solution accuracy for run time, and since computing the Wasserstein distance is a computationally demanding task with complexity $O(n^3 \log(n))$, we decided to use in our tests the Euclidean distance computed on the weight vectors of the two discrete distributions, that is, $d(\mu, \nu) = \sqrt{\sum_{i=1}^{n} (\mu_i - \nu_i)^2}$.

*The Iterative Farthest Heuristic (FH).* If we replace in Eqs. (16) and (17) the arg min operator with an arg max, we get our third heuristic, which is called the *Iterative Farthest Heuristic.* The farthest heuristic is defined as:

$$\theta^{(1)} = \mu_{i_1}, \qquad\qquad \text{where } i_1 \in \arg\max_{j \in J_1} d(\mu_j, \xi), \qquad (19)$$

$$\theta^{(k)} = f\left(\mu_{i_k}, \theta^{(k-1)}, \frac{1}{k}, \frac{k-1}{k}\right), \quad \text{where } i_k \in \arg\max_{j \in J_k} d(\mu_j, \theta^{(k-1)}), \quad (20)$$

$$\bar\rho^{FH} = \theta^{(m)}. \qquad (21)$$

*Remarks.* There exists an ordering of the input sequence $\mu_1, \ldots, \mu_m$ such that the iterative heuristic will generate $\bar\rho^{CH}$ (or $\bar\rho^{FH}$). However, the worst-case complexity of the closest and farthest heuristic is higher than the iterative heuristic, since at each iteration we have to perform a linear scan over $O(m)$ elements.

## 4.2   Pairwise Heuristic

We present in this section another class of heuristics that can solve in parallel the computation of pairwise barycenters. Let us first suppose that the number of input measures $\mu_1, \ldots, \mu_m$ is a power of two, that is, $m = 2^h$. Later, we discuss in detail the case when $2^h < m < 2^{h+1}$.

*The Pairwise Heuristic.* The main idea of this heuristic is to iteratively divide the size of the input sequence by two, while computing pairwise barycenters of consecutive measures. After $h$ iterations, we will get a single discrete probability measure which is used to approximate the barycenter. We begin with a vector $\Theta^{(0)}$ that is equal to the input sequence of measures. Then, at each iteration $k$, we compute the barycenter of every consecutive pair of measures $\Theta_{2i-1}^{(k-1)}$ and $\Theta_{2i}^{(k-1)}$ with weights equal to $\frac{1}{2}$, and we get a new vector $\Theta^{(k)}$ of size $q_k = \frac{m}{2^k}$. More formally, the pairwise heuristic is defined by the following procedure:

$$\Theta^{(0)} = \{\mu_1, \ldots, \mu_m\}, \lambda^{(0)} = \{1, \ldots, 1\} \tag{22}$$

$$\Theta^{(k)} = \left\{ f\left( \Theta_{2i-1}^{(k-1)}, \Theta_{2i}^{(k-1)}, \frac{\lambda_{2i-1}^{(k-1)}}{\lambda_{2i-1}^{(k-1)} + \lambda_{2i}^{(k-1)}}, \frac{\lambda_{2i}^{(k-1)}}{\lambda_{2i-1}^{(k-1)} + \lambda_{2i}^{(k-1)}} \right) \right\},$$
$$i = 1, \ldots, q_k \tag{23}$$

$$\lambda^{(k)} = \{\lambda_{2i-1}^{(k-1)} + \lambda_{2i}^{(k-1)}\}, \quad i = 1, \ldots, q_k \tag{24}$$

$$\bar{\rho}^{PR} = \Theta^{(h)} \tag{25}$$

In (23), each element of the vector $\Theta^{(k)}$ can be computed in parallel, since the barycenter for each consecutive pair is independent from the others.

*Dealing with a number of measures $2^h < m < 2^{h+1}$.* When the number of measures $m$ is not a power of 2, we have to distinguish two cases: either $q_k$ is odd or $q_k$ is even, where $q_k$ is the number of elements left in in $\Theta^{(k)}$. If $q_k$ is even, we can compute the barycenters between consecutive measures, while dividing the number of elements in $\Theta^{(k)}$ by 2. If $q_k$ is odd, the last measure is unpaired, and we have to leave it out. Hence, in order to avoid to underweight the last measures, at the begging of each iteration $k$, the heuristic randomly shuffles the order of the measures stored in $\Theta^{(k)}$. For this reason, we call the heuristic (22)–(25) with the addition of the shuffling effect, the *Pairwise Random Heuristic (PR)*. The number of iterations to obtain the primal solution $\bar{\rho}^{PR}$ is $O(\log_2 m)$.

*The Pairwise Farthest Heuristic (PF).* The pairwise heuristic (22)–(25) relies on the initial order of the input sequence. Similarly to the closest and the farthest iterative heuristic, we can change the order of the sequence $\Theta^{(k)}$ at run time. We describe next the idea for the farthest heuristic. Given the vector $\Theta_1^{(k)}, \ldots, \Theta_q^{(k)}$, we want to reorder it into $\Theta_{i_1}^{(k)}, \ldots, \Theta_{i_q}^{(k)}$ in such a way that $(\Theta_{i_1}^{(k)}, \Theta_{i_2}^{(k)})$ is the farthest pair of measures, $(\Theta_{i_3}^{(k)}, \Theta_{i_4}^{(k)})$ is the pair of farthest measures among the set $\Theta^{(k)} \setminus \{\Theta_{i_1}^{(k)}, \Theta_{i_2}^{(k)}\}$ of remaining measures, and so on. More formally, defining $J_0^k = \{1, \ldots, q_k\}$, the sequence $i_1, \ldots, i_{q_k}$ is constructed in the following way:

$$\text{for } j = 1, \ldots, q_k : \begin{cases} (h^*, \ell^*) \in \underset{(h,\ell) \in (J_0^k \setminus J_{j-1}^k)^2, h \neq \ell}{\arg\max} \ d\left(\Theta_h^{(k)}, \Theta_\ell^{(k)}\right), \\ (i_{2j-1}, i_{2j}) = (h^*, \ell^*), \\ J_j^k = J_{j-1}^k \setminus \{h^*, \ell^*\}. \end{cases} \tag{26}$$

Note that we are reordering the input sequence at each iteration by iteratively computing the pairwise farthest measures. When $j = 1$, we are taking the two farthest measures; when $j = 2$, we are taking the second pair of pairwise farthest measures among the remaining, and so on. Indeed, this heuristic has a higher computational cost per iteration, since we have to sort the measures in $\Theta^{(k)}$. However, the number of iterations $k$ remains in $O(\log_2 m)$.

### 4.3  Improved LP Model for Regular Grids

The objective of our computational tests is to measure the gap between the optimal solution $\rho^*$ of (7), and the solution obtained with any of the previous primal heuristics. Unfortunately, if we try to solve directly the LP (8)–(12) for standard benchmark, we run out of memory already for small values of $m$. Hence, we have extended the tripartite model introduced in [5] for computing Wasserstein distance of order 2 on regular grids, to the computation of a Wasserstein Barycenter defined as in (7). Although the improved LP model is not a major contribution of this paper, it was essential in order to measure the quality of our primal heuristics in terms of optimality percentage gap with the larger instances.

## 5  Computational Results

We run extensive computational tests to evaluate the tradeoff between the solution quality and the run time of the primal heuristics presented in the previous section, which are herein denoted by Iterative (IH), Farthest (FH), PairRnd (PR), and PairFar (PF).

The primary objective of our tests is to evaluate the quality of the heuristic solutions with respect to the optimal Wasserstein Barycenter $\rho^*$ obtained via the LP model (8)–(12). As a measure of quality, we use the percentage gap computed as $\frac{\bar{\rho} - \rho^*}{\rho^*}\%$, where $\bar{\rho}$ is the obtained with any of our primal heuristics. The secondary objective of our test is to measure how the run time scales as a function of the number of input measures $m$. Finally, we visualize the barycenters obtained with the methods, in order to show a qualitative measure of accuracy. In addition, we compare our primal heuristics with a state-of-the-art algorithm for computing the Wasserstein Barycenter of 2D images (discrete measures), namely, the Convolutional Wasserstein Barycenter presented in [29], and implemented in the Python Optimal Transport (POT) library [13].

**Table 1.** Solution values of the Wasserstein Barycenter problem for $m$ random Gaussian distributions, with $m = 2$ and $m = 10$.

| $m$ | $\rho^*(\boldsymbol{\mu}, \boldsymbol{\lambda})$ | $\lfloor \rho^*(\boldsymbol{\mu}, \boldsymbol{\lambda}) \rceil$ | $\mathcal{B}(\boldsymbol{\mu}, \boldsymbol{\lambda})$ | $\bar{\rho}^{PR}$ |
|---|---|---|---|---|
| 2 | 52.2 | 43.3 | 43.3 | 43.3 |
| 10 | 801.2 | 654.0 | 653.7 | 656.7 |

*Datasets.* We use four different benchmarks. First, we randomly generate a number of Gaussian probability distributions in $\mathbb{R}^2$. For this type of measures, we can compute the optimal barycenters, and, hence, we can estimate the discretization error. Second, we used the MNIST handwritten digit dataset [17], and the Fashion MNIST dataset [33]. These two datasets are composed of a large number of grey scale images of resolution $28 \times 28$ pixels, divided into 10 classes. Finally, we use a rescaled and translated set of images from the MNIST dataset, for the sake of comparison with the method proposed in [29].

*Implementation Details.* We have implemented all our algorithms in Python 3.7. In order to compute the optimal transport plan between a pair of measures, we use the emd algorithm of the POT library [13], which implements the network simplex algorithm proposed in [9]. The LP models are solved using the commercial solver Gurobi v8.1. All the tests are run using a single thread on a Linux CentOS workstation equipped with an Intel Xeon Gold 6130 CPU, working with a base frequency of 2.1 Ghz.

### 5.1   Barycenter of Gaussian Distributions

First, we evaluate the pairwise farthest heuristic using the barycenter of Gaussian distributions. The Wasserstein Barycenter of Gaussian probability distributions can be computed with a fix-point iterative algorithm (see Chap. 9 in [24]), and hence, we can evaluate the impact of discretization of continuous distributions.

Table 1 reports the objective function values of the Wasserstein Barycenter problems for $m$ random Gaussian distributions. We consider two main cases: first, the Gaussian distributions are considered as continuous probability distributions, and, second, the Gaussian are discretized and converted into discrete probability measures with support points located on a regular grid of dimension $28 \times 28$. Table 1 gives for each $m$, (i) the optimal solution value $\rho^*(\boldsymbol{\mu}, \boldsymbol{\lambda})$ of problem (7) when the Gaussians and the barycenter are continuous distributions; (ii) the optimal value $\lfloor \rho^*(\boldsymbol{\mu}, \boldsymbol{\lambda}) \rceil$ when the Gaussians are discretized and the support points of $\rho^*$ are selected among the grid points; (iii) the optimal solution value $\mathcal{B}(\boldsymbol{\mu}, \boldsymbol{\lambda})$ of problem (8)–(12) for the discretized Gaussian distributions; and (iv) the objective function value of the pairwise farthest heuristic attained by $\bar{\rho}^{PF}$.

The results of Table 1 shows that the discretization of the input measures and the restrictions for the support points has a strong effect on the overall solution value. Even though our primal heuristics are sub-optimal, their optimality gaps are noticeably lower than the error introduced by the discretization.

### 5.2   MNIST and Fashion MNIST

The MNIST dataset [17] is an entry level dataset for classification and clustering algorithms, and, recently, it was used as a benchmark for measuring the scalability of algorithms that compute Wasserstein Barycenters. The dataset contains 60 000 grey scale images of resolution $28 \times 28$ pixels, with approximately 6 000 images representing a given "class" (a digit). The digits are rescaled and centred in such a way that the center of mass is in the middle of the square grid. Indeed, we consider every single image as a discrete probability function, by (i) normalizing each pixel intensity so that the overall sum of pixel intensities is equal to one, and (ii) by considering that the support points are located on a square regular grid in $\mathbb{R}^2$ of dimension $28 \times 28$. The Fashion-MNIST dataset [33] is an harder variant of MNIST. It is also constituted of $28 \times 28$ pixels grey scale images. Figure 4 shows some of the data in each dataset. In the following tests, the algorithms are run on a given number of inputs images belonging to the same class.

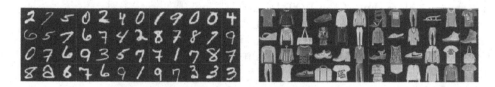

**Fig. 4.** MNIST and Fashion MNIST dataset.

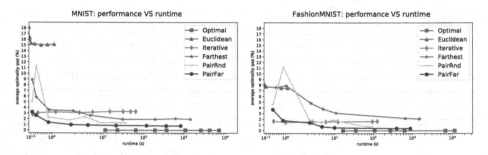

**Fig. 5.** Comparison of runtime versus average percentage gap. For each method, there is a marker for each value of $m \in \{10, 20, 50, 100, 200, 500, 1\,000, 1\,500, 2\,000\}$.

*Runtime vs. Gap.* Figure 5 shows the aggregate results for the MNIST and the Fashion MNIST datasets. The plot reports, for each method, on the $x$-axis the average run time (averaged over each classes), and on the $y$-axis the average percentage gap for computing the barycenter. Each dot in the plot represents the pair *(runtime, gap)* averaged over the 10 classes, for a fixed number $m$ of input measures, with $m \in \{10, 20, 50, 100, 200, 500, 1\,000, 1\,500, 2\,000\}$. The runtime for computing the Wasserstein Barycenter for $2\,000$ inputs reaches the two hours (last dot of the red line), while the Pairwise Farthest heuristic achieves a solution with a gap smaller than 1% (last circle marker) in a few hundreds seconds. The Pairwise Random heuristic (cross marker), is three orders of magnitude faster,

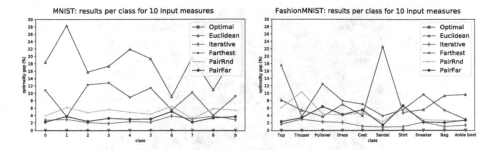

**Fig. 6.** Average percentage gaps for the MNIST and Fashion MNIST datasets.

while achieving a percentage gap smaller than 2%. The Euclidean mean $\xi$ is as fast as inaccurate.

*Gap vs. Image Class.* Figure 6 shows the percentage gap as a function of image class, for the two datasets. Clearly, the computation of the barycenter is harder for digits 1, 4, and 9, and for the class of products 1, 5, and 9, which correspond respectively to Trousers, Sandals, and Ankle boots. While the Fashion MNIST dataset was designed to be more difficult than the MNIST dataset, for the computation of the barycenter, and with respect to the average percentage gap, the Fashion MNIST looks easier than the MNIST, since for most of the methods (Euclidean mean included) the average gaps are smaller.

*Visual Qualitative Impact.* Figure 7 shows the Wasserstein Barycenters of each classes computed by the different methods. Indeed, the Euclidean mean gives a fuzzy mean of all the images, while the optimal solutions obtained with Gurobi are the sharpest. The Euclidean means are fuzzy images because the input digits does not share the exact same support points, and hence the resulting measure spread the overall mass on the union of all the support points of the input measures (e.g., digit 1). Some barycenters in the Fashion MNIST dataset are not relevant (e.g., the dresses), in the sense that it does not represent the cloth. This shows that when the image classes are highly dissimilar, then their barycenter is not very representative. However, even though the barycenter is not visually relevant, it remains the optimal solution of the LP problem (7). Moreover, for all of our primal heuristics, the approximated barycenter is very similar to the optimal solution.

## 5.3  Rescaled and Translated MNIST Images

A modified MNIST dataset, herein called the *Translated MNIST*, is used in the literature for stress testing algorithms that compute Wasserstein Barycenters [11]. In this dataset, the input images are randomly rescaled and translated into a larger grid of dimension $h \times w = 56 \times 56$ pixels [11]. Figure 8 shows a small sample of translated and rescaled images of the digits 9. We used the Translated

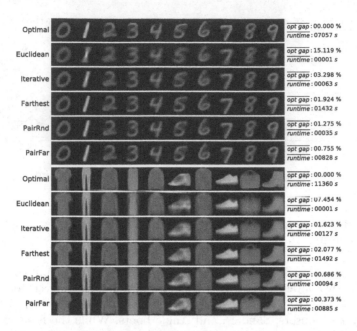

**Fig. 7.** Wasserstein Barycenters of 2000 inputs per class for MNIST and Fashion MNIST. The sharpest images are consistent with the lowest average optimality gaps.

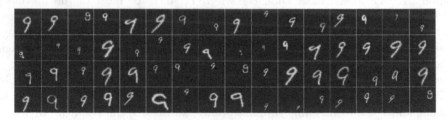

**Fig. 8.** Random sample of images from Translated MNIST.

MNIST benchmark in order to compare our algorithm with the Convolutional Wasserstein Barycenters computed using the approach introduced in [29], and implemented in the POT library [13]. This algorithm depends on a regulariza-tion term $\gamma$ which has to be tuned. In our preliminary test, we tried several different values of this parameter, but unfortunately the method proved to be numerically unstable. In the following, we report the results obtained by fixing the value of the parameter to $\gamma(m) = 2m\frac{h}{28}\frac{w}{28}10^{-3}$, since this formula recovers closely the best values of $\gamma$ we obtained for a given $m$, on all datasets. All other parameters of the convolutional algorithm are left at their default values.

Figures 9 and 10 show that the Translated MNIST dataset is challenging. The Convolutional Wasserstein Barycenter algorithm is faster than the LP solver that computes the optimal solution, but it is not competitive with our best primal heuristics. Notably, the high sensitivity to the regularisation parameter

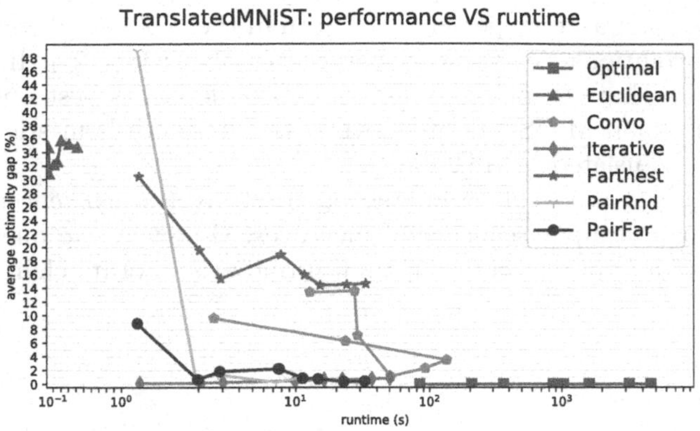

**Fig. 9.** Translated MNIST: comparison of runtime versus average percentage gap. For each method, there is a marker for each value of $m \in \{10, 15, 20, 50, 75, 100, 150, 200\}$.

$\gamma$, made it to fail on high number of inputs. On the contrary, the Iterative Heuristic performs very well both in terms of average percentage gap and run time, since it runs in less than 60 s for the large instance, while achieving an average percentage gap smaller than 0.7%.

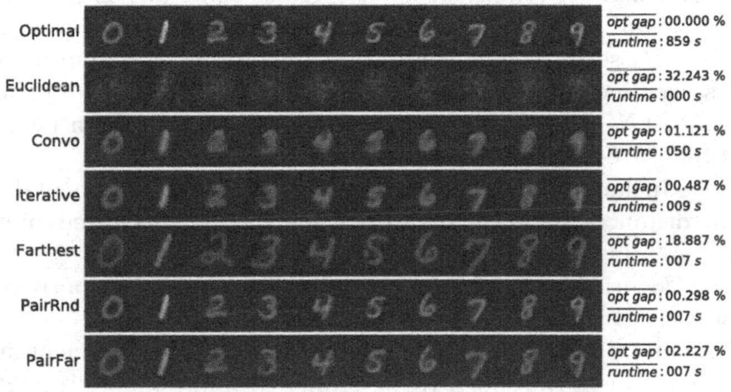

**Fig. 10.** Translated MNIST: Wasserstein barycenters of 50 inputs per class.

## 6 Conclusions

In this paper, we have introduced two type of primal heuristics to compute the Wasserstein barycenter of a given number of probability measures: the iterative heuristics and the pairwise heuristics. Although simple in spirit, our primal heuristics reach near-optimal solutions in very short time, and generate better solutions in terms of average percentage gap than those obtained with the state-of-the-art Convolutional Wasserstein barycenter algorithm [29]. In particular, our results show that the pairwise farthest heuristic is the best option for the MNIST and Fashion MNIST dataset, while the iterative heuristic is the best option for the Translated MNIST dataset.

As future work, it would be interesting to study the existence of an optimal ordering of the input sequence of measures. However, the question about the existence of an optimal ordering, and, if one exists, of its characterization, remains open.

# References

1. Agueh, M., Carlier, G.: Barycenters in the Wasserstein space. SIAM J. Math. Anal. **43**(2), 904–924 (2011)
2. Ahuja, R.K., Magnanti, T.L., Orlin, J.B.: Network flows: Theory, Algorithms, and Applications. Cambridge, Mass.: Alfred P. Sloan School of Management, Massachusetts Institute of Technology (1988)
3. Anderes, E., Borgwardt, S., Miller, J.: Discrete Wasserstein Barycenters: optimal transport for discrete data. Math. Methods Oper. Res. **84**(2), 389–409 (2016)
4. Arjovsky, M., Chintala, S., Bottou, L.: Wasserstein GAN. arXiv preprint arXiv:1701.07875 (2017)
5. Auricchio, G., Bassetti, F., Gualandi, S., Veneroni, M.: Computing kantorovich-wasserstein distances on $d$-dimensional histograms using $(d+1)$-partite graphs. In: Advances in Neural Information Processing Systems, pp. 5793–5803 (2018)
6. Auricchio, G., Bassetti, F., Gualandi, S., Veneroni, M.: Computing Wasserstein Barycenters via linear programming. In: Rousseau, L.-M., Stergiou, K. (eds.) CPAIOR 2019. LNCS, vol. 11494, pp. 355–363. Springer, Cham (2019). https://doi.org/10.1007/978-3-030-19212-9_23
7. Bassetti, F., Gualandi, S., Veneroni, M.: On the computation of Kantorovich-Wasserstein distances between 2D-histograms by uncapacitated minimum cost flows. arXiv preprint arXiv:1804.00445 (2018)
8. Bonneel, N., Coeurjolly, D.: Spot: sliced partial optimal transport. ACM Trans. Graph. **38**(4), 1–13 (2019)
9. Bonneel, N., Van De Panne, M., Paris, S., Heidrich, W.: Displacement interpolation using Lagrangian mass transport. ACM Trans. Graph. **30**, 158–160 (2011)
10. Cuturi, M.: Sinkhorn distances: lightspeed computation of optimal transport. In: Advances in Neural Information Processing Systems, pp. 2292–2300 (2013)
11. Cuturi, M., Doucet, A.: Fast computation of Wasserstein Barycenters. In: International Conference on Machine Learning, pp. 685–693 (2014)
12. Dvurechenskii, P., Dvinskikh, D., Gasnikov, A., Uribe, C., Nedich, A.: Decentralize and randomize: faster algorithm for Wasserstein Barycenters. In: Advances in Neural Information Processing Systems, pp. 10760–10770 (2018)
13. Flamary, R., Courty, N.: POT: Python Optimal Transport library (2017). https://github.com/rflamary/POT
14. Goldberg, A.V., Tardos, É., Tarjan, R.: Network flow algorithms. Cornell University Operations Research and Industrial Engineering, Technical report (1989)
15. Koopmans, T.C.: Optimum utilization of the transportation system. Econom. J. Econom. Soc. **17**, 136–146 (1949)

16. Kusner, M., Sun, Y., Kolkin, N., Weinberger, K.: From word embeddings to document distances. In: International Conference on Machine Learning, pp. 957–966 (2015)
17. LeCun, Y., Cortes, C., Burges, C.J.: MNIST dataset. http://yann.lecun.com/exdb/mnist/. Accessed 12 Mar 2019
18. Levina, E., Bickel, P.: The Earth mover's distance is the mallows distance: some insights from statistics. In: IEEE International Conference on Computer Vision, vol. 2, pp. 251–256 (2001)
19. Ling, H., Okada, K.: An efficient earth mover's distance algorithm for robust histogram comparison. IEEE Trans. Pattern Anal. Mach. Intell. **29**(5), 840–853 (2007)
20. Luise, G., Salzo, S., Pontil, M., Ciliberto, C.: Sinkhorn Barycenters with free support via Frank-Wolfe algorithm. arXiv preprint arXiv:1905.13194 (2019)
21. Monge, G.: Mémoire sur la théorie des déblais et des remblais. Histoire de l'Académie Royale des Sciences de Paris (1781)
22. Panaretos, V.M., Zemel, Y.: Statistical aspects of Wasserstein distances. Ann. Rev. Stat. Appl. **6**, 405–431 (2019)
23. Pele, O., Werman, M.: Fast and robust earth mover's distances. In: IEEE International Conference on Computer vision, pp. 460–467 (2009)
24. Peyré, G., Cuturi, M., et al.: Computational optimal transport. Found. Trends Mach. Learn. **11**(5–6), 355–607 (2019)
25. Qian, Y., Pan, S.: A proximal ALM method for computing Wasserstein Barycenter in d2-clustering of discrete distributions. arXiv preprint arXiv:1809.05990 (2018)
26. Rabin, J., Ferradans, S., Papadakis, N.: Adaptive color transfer with relaxed optimal transport. In: 2014 IEEE International Conference on Image Processing, pp. 4852–4856 (2014)
27. Rubner, Y., Tomasi, C., Guibas, L.J.: The earth mover's distance as a metric for image retrieval. Int. J. Comput. Vision **40**(2), 99–121 (2000)
28. Santambrogio, F.: Optimal Transport for Applied Mathematicians, pp. 99–102. Birkäuser (2015)
29. Solomon, J., et al.: Convolutional Wasserstein distances: efficient optimal transportation on geometric domains. ACM Trans. Graph. **34**(4), 66 (2015)
30. Staib, M., Claici, S., Solomon, J.M., Jegelka, S.: Parallel streaming Wasserstein Barycenters. In: Advances in Neural Information Processing Systems, pp. 2647–2658 (2017)
31. Vershik, A.M.: Long history of the Monge-Kantorovich transportation problem. Math. Intell. **35**(4), 1–9 (2013)
32. Villani, C.: Optimal Transport: Old and New, vol. 338. Springer, Heidelberg (2008)
33. Xiao, H., Rasul, K., Vollgraf, R.: Fashion-MNIST: a novel image dataset for benchmarking machine learning algorithms. arXiv preprint arXiv:1708.07747 (2017)

# An Exact CP Approach for the Cardinality-Constrained Euclidean Minimum Sum-of-Squares Clustering Problem

Mohammed Najib Haouas[✉], Daniel Aloise[✉], and Gilles Pesant[✉]

Polytechnique Montréal, Montreal, Canada
{mohammed-najib.haouas,daniel.aloise,gilles.pesant}@polymtl.ca

**Abstract.** Clustering consists in finding hidden groups from unlabeled data which are as homogeneous and well-separated as possible. Some contexts impose constraints on the clustering solutions such as restrictions on the size of each cluster, known as cardinality-constrained clustering. In this work we present an exact approach to solve the Cardinality-Constrained Euclidean Minimum Sum-of-Squares Clustering Problem. We take advantage of the structure of the problem to improve several aspects of previous constraint programming approaches: lower bounds, domain filtering, and branching. Computational experiments on benchmark instances taken from the literature confirm that our approach improves our solving capability over previously-proposed exact methods for this problem.

## 1 Introduction

Data analysis has become an important field of study in an age dominated by substantial and indiscriminate data collection. One of the most direct ways to extract information from a set of data observations takes the form of a clustering procedure wherein data is grouped in homogeneous and/or well separated bundles based on some measure of similarity/dissimilarity. The partitioned data offers a more tractable presentation of the unlabeled observations. Depending on the criterion on which the partitioning is based, different clusters may be achieved.

**Definition 1.** *Let $O = \{o_1, o_2, ..., o_n\}$ be a set of $n$ data observations in some space and $d : O^2 \mapsto \mathbb{R}^+$ a dissimilarity measure (not necessarily a distance). A $k$-partition $(k < n)$ $\Delta \in \mathcal{A}$ of $O$ into a set of classes $\mathcal{C} = \{C_c\}_{1 \leq c \leq k}$ (with $\mathcal{A}$ the set of all possible $k$-partitions) is such that :*

$$C_c \neq \varnothing \ \forall \, 1 \leq c \leq k, \quad \cup_{1 \leq c \leq k} C_c = O, \quad C_c \cap C_{c'} = \varnothing \ \forall \, 1 \leq c < c' \leq k$$

*Let $\gamma_d : \mathcal{A} \mapsto \mathbb{R}^+$ be a partitioning criterion based on $d$. $\Delta^*$ is an* optimal partition *if $\Delta^* = \operatorname{argmin}_{\Delta \in \mathcal{A}} \gamma_d(\Delta)$.*

© Springer Nature Switzerland AG 2020
E. Hebrard and N. Musliu (Eds.): CPAIOR 2020, LNCS 12296, pp. 256–272, 2020.
https://doi.org/10.1007/978-3-030-58942-4_17

One popular partitioning criterion is the Euclidean Minimum Sum-of-Squares Clustering (MSSC) which is widely used to produce high quality, homogeneous, and well-separated clusters [2]. At its core, it minimizes intra-cluster variance.

**Definition 2.** *Consider observations in $\mathbb{R}^s$. MSSC aims to find the cluster centers $c_j \in \mathbb{R}^s$, as well as cluster assignments $w_{ij}$ that solve the following program [2]*

$$\text{minimize} \quad \sum_{i=1}^{n} \sum_{j=1}^{k} w_{ij} \|o_i - c_j\|^2$$

$$s.t. \quad \sum_{j=1}^{k} w_{ij} = 1 \qquad \forall 1 \leq i \leq n$$

$$w_{ij} \in \{0, 1\} \qquad \forall 1 \leq i \leq n, \forall 1 \leq j \leq k,$$

*where $w_{ij} = 1$ represents the assignment of observation $o_i$ to cluster $C_j$.*

MSSC is NP-hard in general dimension [1].

Often, prior information is known about the data and can be introduced to the clustering process in order to increase performance as well as solution quality. This is possible through expression of custom constraints on the observations or the resulting clusters [6, 26]. In this paper we propose an exact approach to solve a specific variant of constrained MSSC: one that involves cardinality constraints on the resulting clusters (ccMSSC). Strict cardinality constraints in clustering can be encountered in various fields such as image segmentation [17], distributed clustering [4], category management in business [5], document clustering [5], and workgroup composition [15]. Cardinality constraints can also be used to reinforce the clustering procedure against the presence of outliers as well as groups that are either too large or too small [23, 25]. The ccMSSC is already NP-hard in one dimension for $k \geq 2$ [9]. In principle, existing Constraint Programming (CP) approaches for MSSC [13, 14, 16] may be extended in order to handle such a variant by adding a global cardinality constraint. Our contribution shows that we can achieve better performance with specialized global constraints with targeted filtering algorithms as well as an adapted search heuristic, both designed to take advantage of the special structure of the problem in order to quickly reduce the search space. Furthermore, in using CP we ensure easy extension of this work to include independent user-defined constraints.

In the rest of the paper, Sect. 2 defines the CP model used to solve MSSC, to which constraints can be added for the special case of ccMSSC, among others. Section 3 is devoted to a review of the literature surrounding MSSC as well as constrained MSSC. Sections 4 and 5 present our contributions: two filtering algorithms dedicated to ccMSSC resolution as well as an updated and more robust version of an existing search heuristic for MSSC. Section 6 summarizes experimental results as well as comparisons to existing methods. Finally, Sect. 7 provides a brief summary of our work and discusses future research avenues.

## 2    Basic CP Model

The problem stated in Definition 2 was modeled in CP by Dao et al [14]:

*Variables.* Each observation $o_i$ is represented as an integer variable $x_i$, $D(x_i) = \{1, \ldots, k\}$ representing the index of the class to which the corresponding observation belongs. A variable $n_c$, $D(n_c) = \{0, 1, \ldots, n\}$ is introduced for each cluster to represent its cardinality.

*Objective.* Recall MSSC involves finding an optimal set of cluster centers that minimize the intra-cluster variance, per Definition 2. However there is an equivalent formulation [13] of the objective which circumvents these centers, enabling us to solve the problem without making them explicit:

$$\text{minimize} \quad \sum_{c=1}^{k} \frac{1}{2} \frac{1}{|C_c|} \sum_{o,o' \in C_c} \|o - o'\|^2. \tag{1}$$

Using reified constraints, the objective can be further simplified and rewritten as follows:

$$\text{minimize} \quad \sum_{c=1}^{k} \frac{1}{n_c} \sum_{i=1}^{n-1} \sum_{j=i+1}^{n} (x_i = c \wedge x_j = c) \cdot \|o_i - o_j\|^2 \tag{2}$$

The objective expression in Eq. 2 can be constrained to be equal to a real variable $Z$, $D(Z) = [0, \infty[$ known as the Within Cluster Sum of Squares (WCSS), from which the new objective is:

$$\text{minimize} \quad Z \tag{3}$$

*Constraints.* A Global Cardinality Constraint (GCC) [21] constrains variables $n_c$ to take on the cardinality of their corresponding cluster:

$$\texttt{GCC} \left( \{n_c\}_{1 \leq c \leq k}, \{1, 2, \ldots, k\}, \{x_i\}_{1 \leq i \leq n} \right) \tag{4}$$

This model contains a value symmetry which can hinder performance (cluster indices are interchangeable). One way to overcome this is to maintain pairwise integer value precedence on the branching variables as follows [27]:

$$\texttt{intValPrecedence} \left( \{x_i\}_{1 \leq i \leq n}, c - 1, c \right) \qquad \forall \, 1 < c \leq k \tag{5}$$

In essence, each instance of the above constraint ensures that if $x_i = c$ then $\exists \, j < i$ such that $x_j = c - 1$. A higher level of propagation can theoretically be achieved by considering each possible pair (as opposed to only adjacent pairs) of values. However, this comes at a price for virtually no benefit to domain reductions in practice [20].

## 3   Related Work

MSSC is a very well-studied problem and one that is often tackled through heuristics due to its extremely hard nature. K-means is perhaps the most important and widely-used algorithm to solve the unconstrained MSSC problem [28]. It performs a local search to find a partition with minimal within-cluster variance, iteratively relocating cluster centers and stopping at a local optimum. Among exact methods, *CP Clustering* (CPC) presented in [13] is a first successful attempt at using CP for MSSC. Improving on the model presented in Sect. 2, the authors suggest a simple search heuristic as well as a global constraint to efficiently navigate the search space looking for a globally optimal solution to the problem. The authors leverage calculation of lower bounds to filter the objective variable as well as perform cost-based filtering on the branching variables. *CP Repetitive Branch and Bound* (CP RBBA) presented in [16] is a second attempt at leveraging CP to solve MSSC. Its operation is inspired from Repetitive Branch and Bound (RBBA) in [10] where MSSC is divided into subproblems, each treated as an independent CP model in CP RBBA. This enables the use of a range of user constraints (which RBBA doesn't support) as well as the computation of tighter bounds, leading to substantially better performance for many instances.

Turning now to constrained variants of MSSC, the K-means heuristic approach has been extended to support various constraints [8,26]. A special case of ccMSSC, the balanced MSSC, is approached in [12] using a simple Variable Neighborhood Search. Through constant-time reevaluations of the objective after each reassignment as well as carefully selected local search neighborhoods, the authors are able to find the best known values of several large instances. More relevant to us, the authors of [23] suggest a method for solving the ccMSSC using convex relaxations of the problem, whose solutions can be "rounded" to a valid one for the main problem. Their approach distinguishes itself from the others by providing *a posteriori* guarantees on the sub-optimality of the solutions obtained. In fact, based on these guarantees, the authors are able to declare several of the solutions they found as being globally optimal. An exact Column Generation framework for solving constrained MSSC was proposed in [3], supporting anti-monotone constraints which can be used to restrict the maximum cardinality of the clusters. Of course the CP methods previously described for MSSC, CPC and CP RBBA, can solve the ccMSSC by simply adding a GCC but a contribution of our work is to show that, for such a constraint, a more integrated approach is much more productive.

## 4   Filtering Based on Cardinality-Constrained Clustering

In this section we present two filtering algorithms for a global constraint [13] aimed at accelerating resolution of the model in Sect. 2 for the case of ccMSSC.

## 4.1   Basic Filtering Derived from CPC

This first filtering algorithm represents a specialization of CPC for the ccMSSC. We both accelerate and tighten its bound computation by exploiting the fact that cluster sizes $\{n_c\}_{1 \le c \le k}$ are fixed.

The global constraint in CPC evaluates, at each search tree node, the minimum contribution $\underline{Z}(C_c, m)$ to the objective $Z$ for each cluster $C_c$ whenever any $m$ free observations are assigned to it:

$$\underline{Z}(C_c, m) = \frac{Z(C_c) \cdot |C_c| + \sum_{i=1}^{m} R_i(c)}{|C_c| + m} \tag{6}$$

where $Z(C_c)$ represents the WCSS of the partially filled $C_c$ and $(R_i(c))_{i-1}^q$ is a non-decreasing sequence where each term represents the lowest individual contribution of the $i$-th free observation to $C_c$ (among $q$ which are unassigned at the current node) such that:

$$R_i(c) = r_2(i, c) + \sum_{j=1}^{m} r_{3,j}(i) \tag{7}$$

where $r_2(i, c)$ is the contribution of the $i$-th free observation due to the observations already in $C_c$ and $(r_{3,j}(i))_{j=1}^q$ is a non-decreasing sequence where each term represents half the distance between that same $i$-th free observation and a nearby element in $U$, the set of free observations (itself included, i.e. $r_{3,1}(i) = 0 \, \forall i$). Refer to Fig. 1 for an illustration.

The authors of CPC make use of dynamic programming in conjunction with Eq. 6 to compute lower bounds for the general MSSC problem as well as to perform the necessary filtering on variables [13].

*Global Lower Bound for ccMSSC.* We observe that at each node of the ccMSSC resolution, one knows exactly how many observations are to be assigned to each cluster $C_c$ to complete it to its target cardinality $n_c$. As such, given a partial assignment, a lower bound on the cost of a full solution can be more simply computed as follows without resorting to dynamic programming to compute terms for different values of $m$:

$$\underline{Z}(\mathcal{C}) = \sum_{c=1}^{k} \underline{Z}(C_c, n_c - |C_c|) = \sum_{c=1}^{k} \underline{Z}(C_c, m_c) = \sum_{c=1}^{k} \underline{Z}_0(C_c) \tag{8}$$

where we denote as $\underline{Z}_0(C_c)$ the minimum individual contribution of $C_c$ when it is completed to its target cardinality $n_c$, using $m_c$ observations ($m_c := n_c - |C_c|$). Equation 8 filters the objective $Z$ by tightening its lower bound. It also prunes branches that cannot result in a solution better than the incumbent.

*Cost-based Filtering on Cluster Assignment Variables.* It is possible to recycle computations in order to reevaluate a global lower bound to the problem for each value-variable assignment in order to enable effective cost-based filtering.

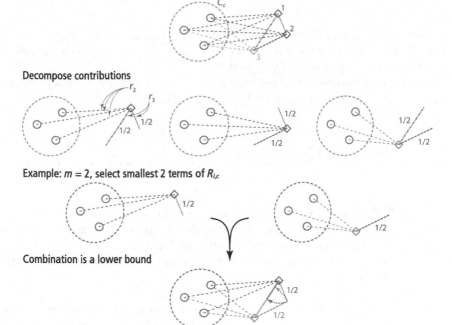

**Partial cluster & 3 free observations**

**Decompose contributions**

**Example: $m = 2$, select smallest 2 terms of $R_{i,c}$**

**Combination is a lower bound**

**Fig. 1.** Illustration of the computation of $\underline{Z}(C_c, m)$

Consider assigning the $\ell$-th free observation $o'$ (w.r.t. the order of the sequence $(R_i(c))_{i=1}^q$) to cluster $C_c$. Let $\mathcal{C}' = \{C_1, \ldots, C_c', \ldots, C_k\}$ denote the set of partially filled clusters identical to $\mathcal{C}$ except for $C_c'$ which also contains $o'$ ($C_c' = C_c \cup \{o'\}$). It is then possible to write the following:

$$\underline{Z}(\mathcal{C}') = \underline{Z}(\mathcal{C}) - \underline{Z}_0(C_c) + \underline{Z}_0(C_c') \tag{9}$$

All the terms in Eq. 9 are available except the last one. Therefore we devise a simple way to get a lower bound on it:

$$
\begin{aligned}
\underline{Z}_0(C_c') &= \frac{\underline{Z}(C_c, m_c - 1) \cdot (|C_c| + m_c - 1) + \ell\text{-th observation's contribution}}{|C_c| + m_c} \\
&\geq \frac{Z(C_c) \cdot |C_c| + \sum_{i=1}^{m_c-1} R_i(c) + \sum_{i=1}^{m_c-1} r_{3,m_c}(i) + r_{2,c}(\ell) + \sum_{j=1}^{m_c} r_{3,j}(\ell)}{|C_c| + m_c} \\
&= \frac{Z(C_c) \cdot |C_c| + \sum_{i=1}^{m_c-1} \left[ r_2(i,c) + \sum_{j=1}^{m_c} r_{3,j}(i) \right] + r_{2,c}(\ell) + \sum_{j=1}^{m_c} r_{3,j}(\ell)}{|C_c| + m_c} \\
&= \frac{(|C_c| + m_c - 1) \cdot \underline{Z}_1(C_c) + r_{2,c}(\ell) + \sum_{j=1}^{m_c} r_{3,j}(\ell)}{|C_c| + m_c} \tag{10}
\end{aligned}
$$

The $\ell$-th observation's contribution represents the sum of the following quantities:

- the sum of dissimilarities between it and $C_c$'s components: $r_{2,c}(\ell)$;
- half dissimilarities between it and $m_c - 1$ other free observations, which is greater than or equal to $\sum_{j=1}^{m_c} r_{3,j}(\ell)$;
- the other half dissimilarities between it and $m_c - 1$ other free observations, which is greater than or equal to $\sum_{i=1}^{m_c-1} r_{3,m_c}(i)$.

with

$$\underline{Z}_1(C_c) = \frac{Z(C_c) \cdot |C_c| + \sum_{i=1}^{m_c-1} R_i(c)}{|C_c| + m_c - 1} \qquad \text{where } R_i(c) = r_2(i,c) + \sum_{j=1}^{m_c} r_{3,j}(i)$$

(11)

which is similar to $\underline{Z}_0(C_c)$ with the difference being we only select $m_c - 1$ terms of $(R_i(c))_{i=1}^q$ in $\underline{Z}_1(C_c)$ instead of $m_c$. This enables the sequential computation of both values with the same complexity. Comparing Eq. 9 against the upper bound of $Z$ for each assignment considered enables filtering of values that cannot result in a solution better than the incumbent.

*Summary.* Propagation algorithms are called whenever the domain of some variable $x_i$ changes or bounds on $Z$ are tightened. Algorithm 4.1 summarizes the results of this section.

---

**Algorithm 4.1. propagate method: basic filtering**

(Computation of $r_2$ and $r_3$ not shown for brevity and are identical to [13])
1: **for** $c \leftarrow 1..k$ **do**
2:     **for** $v \leftarrow 0..1$ **do**
3:         **for** $i \leftarrow 1..n$ where $|D(x_i)| > 1$ **do**   ▷ there are $q$ unassigned observations
4:             **if** $m_c - v > 0$ **then**
5:                 $R[i] \leftarrow r_2[c,i] + r_3[i, m_c - 1]$▷ $r_3$ represents the sum in Eq. 7 directly
6:         **sort** $(R[i \in 1..n : |D(x_i)| > 1])$ ▷ sort the contribs of the $q$ free observations
7:         $\underline{Z}_v(C_c) \leftarrow \frac{Z(C_c) \cdot |C_c| + \sum_{i=1}^{m_c-v} R[i]}{m_c + |C_c| - v}$        ▷ $\underline{Z}_0(C_c)$ and $\underline{Z}_1(C_c)$
8: LB$(Z) \leftarrow \sum_{c=1}^k \underline{Z}_0(C_c)$          ▷ filter objective, Eq. 8
9: **for** $c \leftarrow 1..k$ **do**          ▷ cost-based filtering
10:     LB$_E \leftarrow$ LB$(Z) - \underline{Z}_0(C_c)$
11:     **for** $\ell \leftarrow 1..n$ where $|D(x_\ell)| > 1$ **do**
12:         **if** $c \in D(x_\ell)$ **then**
13:             LB$_P \leftarrow \frac{(|C_c|+m_c-1) \cdot \underline{Z}_1(C_c) + r_2[c,\ell] + r_3[\ell, m_c-1]}{|C_c|+m_c}$     ▷ Eq. 10
14:             **if** (LB$_E +$ LB$_P \geq$ UB$(Z)$) **then**
15:                 $D(x_\ell) \leftarrow D(x_\ell) \setminus \{c\}$     ▷ filter if incumbent cost exceeded

---

The modified CPC filtering algorithm specialized for ccMSSC has a time complexity in $\mathcal{O}(qn + q^2 \log q + kq \log q + k + kq) = \mathcal{O}(qn + q^2 \log q)$ (down from $\mathcal{O}(qn + kq^2 \log q)$ [13]) and a space complexity in $\mathcal{O}(n^2)$. The reduction in asymptotic complexity is less important than the tighter bounds produced which enable more aggressive domain reduction for the case of ccMSSC compared to the original version of the constraint. Computing individual $r_2$ and $r_3$ contributions incrementally has a detrimental effect in practice due to the overhead involved in pinpointing the changes that have occurred since the last node.

However, this filtering algorithm is limited by each cluster's individual minimum contribution being computed at a local level, regardless of that of other clusters. This means that it is possible for a given observation to be considered for the minimum contribution of two distinct clusters, hindering lower bound quality. We propose a way to correct this in the next section.

## 4.2 Improved Filtering

*A Tighter Global Lower Bound for ccMSSC.* In computing the smallest cost of the solution extended from a partial assignment (i.e., the global lower bound at a certain node of the search tree), it helps to consider all clusters as a whole rather than each of them separately while distributing free observations between them. This eliminates the issue identified with the basic filtering discussed above and can be achieved by solving a minimum-cost flow (MCF) problem. At each node of the search tree, where $|U| = q$ observations are unassigned, a network can be built as follows:

1. start from a bipartite assignment graph where the first set of vertices represents the $q$ free observations and the second set of vertices represents the $k'$ ($k' \leq k$) incomplete clusters;
2. supply each of the vertices representing the observations with one unit of flow using a common source;
3. connect each of the vertices representing the partially filled clusters with arcs of capacity $m_c$ to a common sink;
4. all other arcs have a capacity equal to 1;
5. only arcs connecting observations to clusters bear a cost, equal to $R_i(c)/n_c$. Such arcs only exist if the assignment is possible.

The MCF solution is integral because the constraints matrix for the corresponding linear program is Totally Unimodular and all other coefficients are integers. Arcs selected by the MCF represent the optimal division of the free observations between the incomplete clusters. The corresponding cost incurred by this completion, based on the minimum individual contributions of each free observation, necessarily leads to a lower bound on the cost of the solution derived from the current partial assignment:

$$\underline{Z}(\mathcal{C}) = \sum_{c=1}^{k} \frac{Z(C_c) \cdot |C_c|}{n_c} + \mathrm{MCF}^*_{\mathrm{cost}} \tag{12}$$

This lower bound is greater or equal to the one given by Eq. 8.

On the surface, the method being discussed here resembles a GCC with costs [22]. However it is inapplicable here due to changing costs at each node of the search tree. Indeed, the cost incurred by the assignment of an individual observation is not known *a priori* as it changes every time a cluster is modified (which happens repeatedly in the search tree). Moreover, this continuously varying nature of the problem prevents us from taking advantage of most incremental computations involved in maintaining arc consistency in GCC with costs.

A new MCF instance must be solved each time an impactful change occurs in the search tree. We define such a change as one where an assignment variable has been fixed or one where a value has been filtered from the domain of an assignment variable such that it eliminates a flow-carrying arc in the current MCF solution. Otherwise the latter solution is still valid. If a new MCF solution must be computed, we use Network Simplex due to its speed and the fact that implementations of it are readily available.

*A More Thorough Cost-Based Filtering.* The same way adopting a global view of the problem facilitates generation of tighter bounds on $Z$, it is possible to leverage the flow formulation discussed above to perform a more powerful filtering of the decision variables. This is done through forcing flow on an arc using augmenting constraints in the current MCF problem to mandate a particular assignment. If a bound calculated using an augmenting constraint is higher than the cost of the incumbent solution, the value corresponding to the assumption made is filtered.

For the sake of efficiency, instead of recomputing a solution to the MCF problem for every possible augmenting constraint, we start by modifying the one that has been computed for the global lower bound. Such a modification will result in an infeasible solution (Fig. 2, left) because one cluster will be overfilled by one unit (red arc in violation) while another will be missing one unit (transparent bold arc). The task shifts to reestablishing a feasible and optimal solution from the situation depicted.

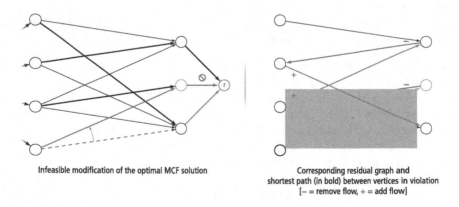

Infeasible modification of the optimal MCF solution

Corresponding residual graph and
shortest path (in bold) between vertices in violation
[− = remove flow, + = add flow]

**Fig. 2.** MCF solution update for cost-based filtering

One way to fix this is to proceed, in an alternating fashion, to the removal and the addition of flow in arcs of the network designed to send the excess unit of flow from the vertex corresponding to the overfilled cluster to the one corresponding to the underfilled cluster. For the solution to be optimal, this alternating sequence of removals and additions should result in the lowest possible added cost.

A more straightforward way to look at this operation is through a residual graph derived from the optimal MCF solution (Fig. 2, right). A flow-carrying arc in the current solution is flipped in the residual graph and given the opposite cost. Restoring the optimal solution becomes a shortest path problem between the vertices in violation. This again is reminiscent of GCC with costs [22]. However our case is a more targeted one where the resulting residual graph is a simple bipartite digraph. We use the Bellman-Ford algorithm [7] (due to presence of negative-cost arcs) to solve the shortest path problem for each assumption. If a path cannot be found, then the augmented MCF problem is inconsistent and the value corresponding to the assumption made must be filtered. The cost increase of the MCF solution after introduction of the augmenting constraint is equal to the variation due to relocating the free observation between clusters plus the weight of the shortest path.

*Summary.* Below is an algorithmic summary of the advanced filtering propagation. Changes with respect to the basic filtering algorithm are shown in green.

---

**Algorithm 4.2.** propagate method: advanced filtering

(Computation of $R$, $r_2$, and $r_3$ not shown for brevity; identical to [13] and Alg. 4.1)
1: **if** impactfulChangeHasOccurred() **then**
2:     makeMCFModel()
3:     solution ← solveMCFModel()                                    ▷ Network Simplex
4:     LB(Z) ← partial WCSS + solution.cost()                        ▷ filter objective, Eq. 12
5: **for** $c \leftarrow 1..k$ **do**                                  ▷ cost-based filtering
6:     **for** $\ell \leftarrow 1..n$ where $|D(x_\ell)| > 1$ **do**
7:         ▷ Only consider non-redundant assumptions, hence the test below
8:         **if** $c \in D(x_\ell) \wedge$ solution.hasFlow($x_\ell$, $c$) **then**
9:             $\delta \leftarrow$ shortestDistUpdate(solution, $x_\ell$, $c$)
10:            **if** $\delta = \infty \vee LB(Z) + \delta \geq UB(Z)$ **then**
11:                $D(x_\ell) \leftarrow D(x_\ell) \setminus \{c\}$   ▷ filtrer if bound exceeded or pb inconsistent

---

The time and space complexities of Algorithm 4.2 are dominated by Network Simplex when called. Depending on the implementation [19], these vary and can be linked to arc costs. In practice, complexity analysis around Network Simplex rarely represents a faithful depiction of real world performance. The function impactfulChangeHasOccurred() runs in $\mathcal{O}(n)$ time. Solving the shortest path problem using the Bellman-Ford algorithm is done in time $\mathcal{O}(qk(k+q))$ due to the graph comprising $\mathcal{O}(qk)$ arcs and $\mathcal{O}(q+k)$ vertices [7]. Overall time complexity for the cost-based filtering of assignment variables is thus $\mathcal{O}(q^2k^2(k+q))$.

# 5    Search Strategy

The search heuristic discussed in this section is inspired from the one proposed for CPC [13] with two key improvements.

## 5.1    Bootstrapping from a Heuristic Solution

To solve MSSC, CPC starts from a feasible, heuristically generated solution whose cost is used for the first domain reductions (recall that CPC makes use of a cost-based filtering mechanism). A superior left branch in our search tree, leading to an initial solution, helps to reduce its size by acting on two separate aspects of the problem: it provides a tighter initial upper bound and it moves potentially unsuccessful alternate, future branches near the top of the search tree to avoid revisiting them repeatedly.

*Initial Solution Generation.* Instead of starting from a greedy assignment whose results depend on the order of the observations in $O$ [13] and which produces poor results when user constraints are present, suppose a feasible good-enough initial solution $\tau_0$ is known in advance (which may or may not be globally optimal). We can use $\tau_0$ as a guide for the first $n$ branching assignments of the CP search to ensure the first solution found is equal to $\tau_0$, thus also ensuring the initial upper bound is equal to the cost of $\tau_0$. Generating this solution can be done using any number of existing heuristic methods (some of which have been discussed in Sect. 3) to solve a constrained MSSC.

The cost of $\tau_0$ is only part of what helps the CP resolution. The order in which individual assignments appear has its importance.

*Order of Initial Assignments.* Authors of [11] demonstrate the substantial impact of initial data sequencing on branch and bound searches. They show that solution times can spread over several orders of magnitude for randomly sampled sequences. It is possible to prune off large sub-trees by ordering the data of a heuristically generated solution in an careful manner.

Based on this, given an initial solution $\tau_0$ with initial cluster centers, we suggest two variable orderings in the left branch of the search tree which showed competitive results in our empirical testing:

- *Decreasing distances to own cluster's center*: this method orders assignments in $\tau_0$ from the one whose corresponding observation is farthest from the center of its cluster to the one that is closest. It tries to place potentially disruptive, hard to assign observations near the root of the search tree, where we have greater flexibility to recover from a poor choice.
- *Decreasing minimal distances to other clusters' centers*: this method is similar to the previous one with the difference being that the ordering is based on the minimum distance between each observation and centers of clusters which are not its own. Therefore, it maximizes the likelihood alternate branches will fail the closer they are to the top, eliminating bigger sub-trees.

## 5.2   Dynamic Tie-Breaking

Once the left branch has been generated, another branching strategy takes over. The heuristic in CPC is adequate but displays a major weakness. For each unassigned $x_i$ and $c \in D(x_i)$, CPC computes $t_{i,c}$, the WCSS increase on $Z$ if $x_i = c$. It then branches on the variable given by $\text{argmax}_{i\,:\,|D(x_i)|>1} \min_{c \in D(x_i)} t_{i,c}$ [13]. However, ties may occur whenever a cluster becomes empty upon backtracking: if every unassigned variable has this cluster in its domain, the minimum will be zero for all. In that case, the heuristic essentially falls back to a lexicographical one. To correct this, we design a dynamic tie-breaking strategy.

When presented with a tie as a result of a cluster becoming empty, we branch in a way that assigns to the empty cluster the observation whose sum-of-squares between it and other unassigned observations is the highest:

$$o^* = \underset{o \in U}{\text{argmax}} \sum_{o' \in U} \|o - o'\|^2 \tag{13}$$

Since bound computations in CPC directly involve the sum of squared distances between observations of each cluster, the choice depicted in Eq. 13 is akin to a fail-first strategy: we initiate a cluster with the observation which is most likely to produce worse solutions through elevation of its cluster's contribution.

# 6   Experiments

We compare our CP-centered approach to solving ccMSSC to the works discussed in Sect. 3. In it, we cited CPC in [13] and CP RBBA in [16] as CP frameworks for solving MSSC. These two approaches can easily be extended to solve ccMSSC through the introduction of adequate cardinality constraints to their CP models. We also cited a numerical method for solving ccMSSC with guarantees on the sub-optimality of the solutions [23] as well as a column generation framework to solve constrained MSSC [3]. However the latter's current implementation does not support solving ccMSSC and would require a significant amount of work to add the necessary constraints.[1]

To carry out our experiments we select 19 instances, summarized in Table 1. All of them are available in the UCI Machine Learning Repository[2] except for HA [18] and RU [24]. Instances from exact methods presented in Sect. 3 all appear in Table 1 and have been completed with randomly sampled datasets from the UCI repository with 200 data points or less and with numerical attributes.

Table 1 legend: (1) Instance with balanced classes; (2) Number of classes and/or target cardinalities decided randomly by us; (3) Multiple versions available, version *Small* used here; (4) Multiple versions available, version with 215 observations used here.

---

[1] Personal communication from one of the authors.
[2] https://archive.ics.uci.edu.

**Table 1.** Description of selected instances

| Code | Name | n | s | k | Targeted cluster cardinalities | Notes |
|------|------|---|---|---|-------------------------------|-------|
| AI3 | Acute Inflammations | 120 | 6 | 3 | 40 40 40 | 1, 2 |
| AI4 | Acute Inflammations | 120 | 6 | 4 | 30 30 30 30 | 1, 2 |
| BC | Breast Cancer Coimbra | 116 | 9 | 3 | 38 39 39 | 1, 2 |
| BT | Breast Tissue | 106 | 9 | 6 | 18 17 17 18 18 18 | 1, 2 |
| CB | Connectionist Bench | 208 | 60 | 2 | 111 97 | |
| CS | Concrete Slump Test | 103 | 7 | 3 | 34 34 35 | 1, 2 |
| GI | Glass Identification | 214 | 9 | 6 | 70 17 9 29 76 13 | |
| HA | Hatco | 100 | 14 | 3 | 34 33 33 | 1, 2 |
| FI2 | Fisher's Iris | 150 | 4 | 2 | 60 90 | 2 |
| FI3 | Fisher's Iris | 150 | 4 | 3 | 50 50 50 | 1 |
| PA | Parkinson's | 195 | 22 | 2 | 147 48 | |
| PR | Planning Relax | 182 | 12 | 2 | 130 52 | |
| RU | Ruspini | 75 | 2 | 4 | 18 19 19 19 | 1, 2 |
| SE | Seeds | 210 | 7 | 3 | 70 70 70 | 1 |
| SY | Soybean | 47 | 35 | 4 | 11 12 12 12 | 1, 2, 3 |
| TH | Thyroid Disease | 215 | 5 | 3 | 72 71 72 | 1, 2, 4 |
| UL | Urban Land Cover | 168 | 147 | 9 | 14 16 14 25 15 23 17 15 29 | |
| WN0 | Wine | 178 | 13 | 3 | 60 40 78 | 2 |
| WNB | Wine | 178 | 13 | 3 | 59 60 59 | 1, 2 |

New algorithms[3] were implemented in C++ using IBM ILOG CPLEX Optimization Studio 12.9.0. CPC was reimplemented using this same software framework. CP RBBA's C++ GECODE implementation is publicly available[4] and was used with slight alterations (to introduce the necessary constraints for ccMSSC).

All algorithms were compiled using Intel C++ Compiler 19.0.3.199 and run on stock Intel Xeon Gold 6148 processors. Each process was allocated 1 GB of memory and one core. Maximum runtime was set to 86400 seconds (i.e., 1 day).

## 6.1    Impact of Dynamic Tie-Breaking

We present in Table 2 the impact of our dynamic tie-breaking strategy on the performance of CPC. To isolate the effect of the tie-breaking, we do not introduce any cardinality constraints to the model for this specific test. Dashes represent a run that has timed out. Due to space constraints, we show results for 6 instances that faithfully convey the general trend.

**Table 2.** Impact of dynamic tie-breaking on the performance of CPC

| Inst. | No tie-breaking | | | Dynamic sum-of-squares tie-breaking | | |
|-------|----------|--------|----------|----------|--------|----------|
|       | Time [s] | Fails  | Branches | Time [s] | Fails  | Branches |
| BC    | 886.2    | 79.08k | 158.21k  | 56.6     | 4.03k  | 8.06k    |
| BT    | —        | —      | —        | 184      | 7.73k  | 15.45k   |
| HA    | 338.4    | 40.63k | 81.27k   | 289.8    | 33.56k | 67.14k   |
| FI3   | 1337.2   | 56.59k | 113.21k  | 911.5    | 41.08k | 82.19k   |
| RU    | 0.4      | 83     | 170      | 0.4      | 82     | 168      |
| SY    | 3.2      | 1.69k  | 3.38k    | 2.2      | 1.15k  | 2.31k    |

Table 2 shows a clear and generalized improvement brought on by the tie-breaking strategy, both with respect to search space size as well as run time.

---

[3] Source-code can be retrieved from: https://github.com/mnhaouas/card-const-MSSC.

[4] https://cp4clustering.github.io/.

Notably, *Breast Tissue* can only be solved by applying the dynamic tie-breaking to the search process.

## 6.2 Resolution of ccMSSC

We suggest in Sect. 5.1 starting the CP search from known good solutions to the problem and arranging them in the search tree to improve performance. To this end, we make use of two heuristic approaches to start ccMSSC resolution: LIMA-VNS [12] (discussed in Sect. 3) for the balanced instances and Constrained K-Means Clustering [8] for the others. We use a third-party public implementation[5] for the latter and an executable supplied by the authors for the former.

Both heuristic algorithms are seeded using `/dev/random` and run 10 times for each instance. The median of the 10 runs is picked as a starting point for the CP search.

Each instance is solved 8 times: twice through CP RBBA using two distinct observation orderings recommended by its authors (FF for Farthest First and NN for Nearest Neighbor) and twice through each of CPC, the basic approach in

**Table 3.** ccMSSC resolution statistics for all algorithms

| Inst. | Ord. | CP RBBA & GCC | | CPC & GCC | | | Basic filtering | | | Advanced filtering | | | |
|---|---|---|---|---|---|---|---|---|---|---|---|---|---|
| | | Time [s] | Fails | Ord. | Time [s] | Fails | Gap [%] | Time [s] | Fails | Gap [%] | Time [s] | Fails | Gap [%] |
| AI3 | NN | − | − | OC | − | − | 53.2 | − | − | 53.2 | − | − | 38.0 |
| | FF | − | − | RC | − | − | 55.4 | − | − | 55.4 | − | − | 37.7 |
| AI4 | NN | − | − | OC | − | − | 57.9 | − | − | 57.8 | − | − | 30.0 |
| | FF | − | − | RC | − | − | 55.7 | − | − | 55.7 | − | − | 29.6 |
| BC | NN | − | − | OC | − | − | 81.6 | 42.6 | 29.22k | − | 34.7 | 6.82k | − |
| | FF | − | − | RC | − | − | 81.6 | 12.2 | 5.52k | − | **5.4** | **787** | − |
| BT | NN | − | − | OC | − | − | 99.9 | 23.0 | 12.98k | − | **9.1** | **886** | − |
| | FF | − | − | RC | − | − | 99.9 | − | − | 4.0 | − | − | 3.7 |
| CB | NN | − | − | OC | − | − | 36.2 | − | − | 35.9 | − | − | 17.4 |
| | FF | − | − | RC | − | − | 35.9 | − | − | 35.6 | − | − | 17.3 |
| CS | NN | − | − | OC | − | − | 45.7 | − | − | 45.7 | − | − | 28.0 |
| | FF | − | − | RC | − | − | 45.5 | − | − | 45.5 | − | − | 28.1 |
| GI | NN | − | − | OC | − | − | 96.5 | − | − | 86.4 | − | − | 55.2 |
| | FF | − | − | RC | − | − | 96.5 | − | − | 90.3 | − | − | 58.9 |
| HA | NN | 877.1 | 195.78k | OC | 60.1 | 14.46k | − | 6.0 | 2.58k | − | 4.0 | 559 | − |
| | FF | 2.0 | 161.82k | RC | 46.0 | 11.31k | − | 3.7 | 1.72k | − | **1.4** | **127** | − |
| FI2 | NN | 24.7 | 19.11k | OC | 380.3 | 43.51k | − | 10.8 | 2.53k | − | 9.4 | 1.02k | − |
| | FF | 2.0 | 91.56k | RC | 23.9 | 6.17k | − | 3.2 | 706 | − | **1.5** | **142** | − |
| FI3 | NN | 18.9 | 1.30M | OC | 2498.7 | 207.59k | − | 321.6 | 58.07k | − | 147.6 | 9.94k | − |
| | FF | − | − | RC | 350.6 | 33.49k | − | 54.3 | 11.93k | − | **7.8** | **530** | − |
| PA | NN | − | − | OC | − | − | 79.6 | − | − | 68.0 | 27022.1 | 1.74M | − |
| | FF | − | − | RC | − | − | 79.6 | − | − | 68.0 | 24867.4 | 1.66M | − |
| PR | NN | − | − | OC | − | − | 49.3 | − | − | 40.4 | − | − | 25.2 |
| | FF | − | − | RC | − | − | 49.3 | − | − | 40.4 | − | − | 25.2 |
| RU | NN | 1056.9 | 103.27M | OC | 5290.6 | 3.52M | − | 157.5 | 167.28k | − | 107.1 | 16.58k | − |
| | FF | 5192.1 | 586.64M | RC | 3341.4 | 2.21M | − | 38.8 | 50.73k | − | **9.3** | **1.74k** | − |
| SE | NN | − | − | OC | − | − | 47.7 | − | − | 46.8 | − | − | 24.2 |
| | FF | − | − | RC | − | − | 47.7 | − | − | 46.8 | 37197.0 | 1.69M | − |
| SY | NN | 13543.4 | 2.51G | OC | 442.3 | 766.06k | − | 6.0 | 16.43k | − | 13.1 | 3.54k | − |
| | FF | 58074.8 | 11.67G | RC | 99.7 | 283.89k | − | **0.9** | **2.84k** | − | 1.8 | 643 | − |
| TH | NN | − | − | OC | − | − | 87.4 | − | − | 83.3 | − | − | 37.8 |
| | FF | − | − | RC | − | − | 87.4 | − | − | 83.3 | − | − | 37.8 |
| UL | NN | − | − | OC | − | − | 92.2 | − | − | 60.8 | − | − | 30.3 |
| | FF | − | − | RC | − | − | 92.2 | − | − | 60.8 | − | − | 30.3 |
| WN0 | NN | − | − | OC | − | − | 70.6 | 32239.3 | 6.60M | − | 6420.8 | 567.35k | − |
| | FF | − | − | RC | − | − | 70.6 | 58373.0 | 9.28M | − | 11335.3 | 755.86k | − |
| WNB | NN | − | − | OC | − | − | 69.4 | 497.7 | 154.16k | − | 240.5 | 31.71k | − |
| | FF | − | − | RC | − | − | 69.4 | 95.4 | 15.69k | − | **13.7** | **853** | − |

---

[5] https://github.com/Behrouz-Babaki/MinSizeKmeans.

Sect. 4.1, and the advanced approach in Sect. 4.2. For each, we make use of the two observation orderings suggested in Sect. 5.1 (OC for *Decreasing distances to own cluster's center* and RC for *Decreasing minimal distances to remote clusters' centers*). We also make use of our tie-breaking strategy to accelerate all methods except CP RBBA due to its fundamentally different nature.

Table 3 shows a clear advantage for both variants of the filtering algorithms proposed and particularly for the advanced, flow-based approach. Eight of the 19 instances could not be solved to optimality. However, on the flip side, the advanced approach is capable of solving two instances none of the other methods could solve in the allotted time. The basic approach, while fast, produces lower quality bounds and loses its advantage to substantially bigger search trees.

Overall, ordering RC yields the best results. However, it is not superior for all instances shown. For example, OC is best for the non balanced version of *Wine* as well as *Breast Tissue*.

CPC is also able to take advantage of our improved search strategy to show competitive results compared to CP RBBA, tighter bounds computed by the latter notwithstanding. Without our improved search, CPC with GCC is only able to solve HA, RU and SY in 12398, 15462 and 6006 s respectively. This confirms the important role of a reinforced search strategy for ccMSSC.

### 6.3   Comparison with IBM CP Optimizer Default Search

In order to frame the performance of our search strategy in a recognizable reference, we compared it to the default strategy shipped with IBM CP Optimizer. Solving the 6 instances in Table 2 as ccMSSC problems using the advanced filtering method yielded, on average, search trees 27 times bigger for CP Optimizer search and run times were increased by a factor of 20.

### 6.4   Comparison with the Conic Optimization Approach

The semidefinite programming lower bound and the rounding heuristic of [23] were able to prove the optimality of FI3, SE, PR, and PA in 584, 3823, 2637, and 2000 s, respectively, thus surpassing our best advanced filtering approach except for FI3 where we show vastly improved results. They are also able to guarantee a solution to CB and UL with gaps of 0.001% and 3% respectively while our approach's best gaps are equal to 17% and 30% respectively for these instances.

However this numerical method does not allow easy expression of user constraints (ours can leverage the flexibility of CP to quickly and easily introduce any extra constraints). Besides, our CP method is designed as a global optimization method which ends its execution only when all possibilities in the search space have been exhausted. The method of [23] is not conceived towards obtaining the global optimum of the problem shall the upper bound produced by the rounding method not coincide with the lower bound obtained via the semidefinite programming relaxation.

# 7    Conclusion

We presented in this paper a CP approach for exact resolution of the cardinality-constrained MSSC problem. We suggest both a bolstered search strategy as well as a global constraint with two distinct filtering schemes: a basic one and a more advanced one. Experiments on widely used data sets confirm our approach outperforms previously available exact methods for solving ccMSSC.

Our work can be improved upon by identifying ways that can extend our global constraint developed for ccMSSC to support soft cardinality constraints where deviations from target cardinalities could be allowed if it meant obtaining a lower cost solution. Moreover, as seen previously, performance is heavily dependent on bound quality. Therefore, looking for more innovative ways to fully exploit the structure of ccMSSC for even tighter bounds could be another avenue for future research.

**Acknowledgements.** Financial support from a Natural Sciences and Engineering Research Council of Canada (NSERC) graduate scholarship is gratefully acknowledged.

# References

1. Aloise, D., Deshpande, A., Hansen, P., Popat, P.: NP-hardness of euclidean sum-of-squares clustering. Mach. Learn. **75**(2), 245–248 (2009)
2. Aloise, D., Hansen, P.: Evaluating a branch-and-bound rlt-based algorithm for minimum sum-of-squares clustering. J. Global Optim. **49**(3), 449–465 (2011)
3. Babaki, B., Guns, T., Nijssen, S.: Constrained clustering using column generation. In: Simonis, H. (ed.) CPAIOR 2014. LNCS, vol. 8451, pp. 438–454. Springer, Cham (2014). https://doi.org/10.1007/978-3-319-07046-9_31
4. Balcan, M.F., Ehrlich, S., Liang, Y.: Distributed k-means and k-median clustering on general topologies. In: Proceedings of the 26th International Conference on Neural Information Processing Systems - Volume 2, NIPS 2013, pp. 1995–2003, USA. Curran Associates Inc. (2013)
5. Banerjee, A., Ghosh, J.: Scalable clustering algorithms with balancing constraints. Data Min. Knowl. Disc. **13**(3), 365–395 (2006)
6. Basu, S., Davidson, I., Wagstaff, K.: Constrained Clustering: Advances in Algorithms, Theory, and Applications, 1st edn. Chapman & Hall/CRC (2008)
7. Bellman, R.: On a routing problem. Q. Appl. Math. **16**(1), 87–90 (1958)
8. Bennett, K.P., Bradley, P.S., Demiriz, A.: Constrained k-means clustering. Technical report MSR-TR-2000-65, Microsoft Research, May 2000
9. Bertoni, A., Goldwurm, M., Lin, J., Saccà, F.: Size constrained distance clustering: separation properties and some complexity results. Fundamenta Informaticae **115**, 125–139 (2012)
10. Brusco, M.J.: A repetitive branch-and-bound procedure for minimum within-cluster sums of squares partitioning. Psychometrika **71**(2), 347–363 (2006)
11. Carbonneau, R.A., Caporossi, G., Hansen, P.: Extensions to the repetitive branch and bound algorithm for globally optimal clusterwise regression. Comput. Opera. Res. **39**(11), 2748–2762 (2012)
12. Costa, L.R., Aloise, D., Mladenović, N.: Less is more: basic variable neighborhood search heuristic for balanced minimum sum-of-squares clustering. Inf. Sci. **415–416**, 247–253 (2017)

13. Dao, T.-B.-H., Duong, K.-C., Vrain, C.: Constrained minimum sum of squares clustering by constraint programming. In: Pesant, G. (ed.) CP 2015. LNCS, vol. 9255, pp. 557–573. Springer, Cham (2015). https://doi.org/10.1007/978-3-319-23219-5_39

14. Dao, T.-B.-H., Duong, K.-C., Vrain, C.: Constrained clustering by constraint programming. Artif. Intell. **244**, 70–94 (2017). Combining Constraint Solving with Mining and Learning

15. Desrosiers, J., Mladenović, N., Villeneuve, D.: Design of balanced mba student teams. J. Oper. Res. Soc. **56**(1), 60–66 (2005)

16. Guns, T., Dao, T.-B.-H., Vrain, C., Duong, K.-C.: Repetitive branch-and-bound using constraint programming for constrained minimum sum-of-squares clustering. In: Proceedings of the Twenty-Second European Conference on Artificial Intelligence, ECAI 2016, pp. 462–470. IOS Press, Amsterdam (2016)

17. Hagen, L., Kahng, A.B.: New spectral methods for ratio cut partitioning and clustering. IEEE Trans. Comput. Aided Des. Integr. Circuits Syst. **11**(9), 1074–1085 (1992)

18. Hair, J.F., Tatham, R.L., Anderson, R.E., Black, W.: Multivariate Data Analysis, 5th edn. Pearson, New York (1998)

19. Jungnickel, D.: The network simplex algorithm. In: Graphs, Networks and Algorithms. Algorithms and Computation in Mathematics, pp. 321–339. Springer, Heidelberg (2005)

20. Law, Y.C., Lee, J.H.M.: Global constraints for integer and set value precedence. In: Wallace, M. (ed.) CP 2004. LNCS, vol. 3258, pp. 362–376. Springer, Heidelberg (2004). https://doi.org/10.1007/978-3-540-30201-8_28

21. Quimper, C.-G., López-Ortiz, A., van Beek, P., Golynski, A.: Improved algorithms for the global cardinality constraint. In: Wallace, M. (ed.) CP 2004. LNCS, vol. 3258, pp. 542–556. Springer, Heidelberg (2004). https://doi.org/10.1007/978-3-540-30201-8_40

22. Régin, J.-C.: Arc consistency for global cardinality constraints with costs. In: Jaffar, J. (ed.) CP 1999. LNCS, vol. 1713, pp. 390–404. Springer, Heidelberg (1999). https://doi.org/10.1007/978-3-540-48085-3_28

23. Rujeerapaiboon, N., Schindler, K., Kuhn, D., Wiesemann, W.: Size matters: cardinality-constrained clustering and outlier detection via conic optimization. SIAM J. Optim. **29**(2), 1211–1239 (2019)

24. Ruspini, E.H.: Numerical methods for fuzzy clustering. Inf. Sci. **2**(3), 319–350 (1970)

25. Tang, W., Yang, Y., Zeng, L., Zhan, Y.: Size constrained clustering with milp formulation. IEEE Access **8**, 1587–1599 (2020)

26. Wagstaff, K., Cardie, C., Rogers, S., Schrödl, S.: Constrained k-means clustering with background knowledge. In: Proceedings of the Eighteenth International Conference on Machine Learning, ICML 2001, pp. 577–584. Morgan Kaufmann Publishers Inc., San Francisco (2001)

27. Walsh, T.: Symmetry breaking constraints: Recent results. In: AAAI Conference on Artificial Intelligence (2012)

28. Wu, X., et al.: Top 10 algorithms in data mining. Knowl. Inf. Syst. **14**(1), 1–37 (2008)

# Minimum Cycle Partition with Length Requirements

Kai Hoppmann[1,2]([✉])(iD), Gioni Mexi[1](iD), Oleg Burdakov[3](iD),
Carl Johan Casselgren[3](iD), and Thorsten Koch[1,2](iD)

[1] Zuse Institute Berlin, Takustr. 7, 14195 Berlin, Germany
{kai.hoppmann,mexi,koch}@zib.de
[2] Chair of Software and Algorithms for Discrete Optimization, TU Berlin,
Str. des 17. Juni 135, 10623 Berlin, Germany
[3] Department of Mathematics, Linköping University, 58183 Linköping, Sweden
{carl.johan.casselgren,oleg.burdakov}@liu.se

**Abstract.** In this article we introduce a Minimum Cycle Partition Problem with Length Requirements (CPLR). This generalization of the Travelling Salesman Problem (TSP) originates from routing Unmanned Aerial Vehicles (UAVs). Apart from nonnegative edge weights, CPLR has an individual critical weight value associated with each vertex. A cycle partition, i.e., a vertex disjoint cycle cover, is regarded as a feasible solution if the length of each cycle, which is the sum of the weights of its edges, is not greater than the critical weight of each of its vertices. The goal is to find a feasible partition, which minimizes the number of cycles. In this article, a heuristic algorithm is presented together with a Mixed Integer Programming (MIP) formulation of CPLR. We furthermore introduce a conflict graph, whose cliques yield valid constraints for the MIP model. Finally, we report on computational experiments conducted on TSPLIB-based test instances.

**Keywords:** Travelling salesman problem · Combinatorial optimization · Mixed integer linear programming · Conflict graph · Unmanned Aerial Vehicles

## 1 Motivation

UAVs are widely used to execute surveillance tasks, since they can gather information about areas from long distance and high altitude. In particular, they are able to visit areas that are not accessible in any other way. Applications include monitoring critical infrastructure such as gas pipelines [11], fighting forest-fires [15], and analyzing widespread animal populations [4]. The Minimum Cycle Partition Problem with Length Requirements (CPLR) originates from a routing problem regarding these UAVs.

Given a set of areas $V = \{v_1, \ldots, v_n\}$, the goal is to determine the minimum number of UAVs necessary to visit all areas, while their individual flying routes have to fulfill three conditions. First, the UAVs must fly tours, which means that

© Springer Nature Switzerland AG 2020
E. Hebrard and N. Musliu (Eds.): CPAIOR 2020, LNCS 12296, pp. 273–282, 2020.
https://doi.org/10.1007/978-3-030-58942-4_18

a UAV starts and ends its tour at the same area and visits all other areas assigned to it exactly once. We assume that a UAV continues on the same tour after finishing it without any delay. Second, each area is visited by exactly one UAV and therefore contained in exactly one tour. This is because possible interferences resulting from the intersection of tours shall be avoided. Third, each area $v_i \in V$ is associated with a critical weight $T_i \in \mathbb{R}_{\geq 0}$, which is an upper bound on the duration for which it can be left unattended, and a scanning time $S_i \in \mathbb{R}_{\geq 0}$, which is the amount of time a UAV needs to scan it. Thus, after scanning $v_i$ for $S_i$ time units, the UAV has to return and rescan it within $T_i$ time units.

## 2    Problem Formulation

For CPLR we are given a complete graph $G = (V, E)$, where $V = \{v_1, \ldots, v_n\}$ denotes the set of vertices and $E = \{\{v_i, v_j\} \in V \times V \mid 1 \leq i < j \leq n\}$ the set of edges. For each vertex $v_i \in V$ we are given a critical weight $T_i \in \mathbb{R}_{\geq 0}$ and a scanning time $S_i \in \mathbb{R}_{\geq 0}$ with $S_i \leq T_i$. The weight of edge $e_{ij} := \{v_i, v_j\} \in E$ is given by $\hat{L}_{ij} \in \mathbb{R}_{\geq 0}$ and the edge weights respect the triangle inequality, i.e., $\hat{L}_{ij} \leq \hat{L}_{ik} + \hat{L}_{kj}$ holds for all $v_i, v_j, v_k \in V$.

In the following, we call a nonempty cycle $C_k$ proper if $|C_k| \geq 2$ and singleton if $|C_k| = 1$. Further, $C_k$ is called feasible if $\tau_k \leq T_i$ holds for each $v_i \in C_k$, where

$$\tau_k := \sum_{e_{ij} \in C_k} \hat{L}_{ij} + \sum_{v_i \in C_k} S_i$$

denotes the length of $C_k$. In other words, the length of the cycle is not allowed to be greater than the critical weight of each of its vertices. A solution for CPLR is a cycle partition $\mathcal{C} = \{C_1, \ldots, C_m\}$ of $V$, i.e., a set of cycles such that each vertex is contained in exactly one of them. It is feasible if each of its cycles is feasible. The goal of CPLR is to determine a feasible cycle partition $\mathcal{C}$ of minimum size w.r.t. the cardinality $|\mathcal{C}|$.

In the following, we assume w.l.o.g. that $S_i = 0$ for all $v_i \in V$, since the scanning times can be included in the edge weights: Let $L_{ij} := \hat{L}_{ij} + \frac{S_i + S_j}{2}$ for each $e_{ij} \in E$ and consider some cycle $C_k$. If it is proper, all vertices have degree two and we have

$$\tau_k = \sum_{e_{ij} \in C_k} \hat{L}_{ij} + \sum_{v_i \in C_k} S_i = \sum_{e_{ij} \in C_k} (\hat{L}_{ij} + \frac{S_i + S_j}{2}) = \sum_{e_{ij} \in C_k} L_{ij}.$$

Note that the new edge weights still respect the triangle inequality. Thus, in the following we denote an instance of CPLR as a four-tuple $(V, E, T, L)$ with $T \in \mathbb{R}_{\geq 0}^{|V|}$ and $L \in \mathbb{R}_{\geq 0}^{|E|}$.

**Lemma 1.** *CPLR is NP-hard.*

*Proof.* Consider an instance $(V, E, L)$ of the Travelling Salesman Problem (TSP) and some $B \geq 0$. Setting $T_i := B$ as critical weight for each $v_i \in V$ induces an instance of CPLR. An optimal solution for the CPLR instance consists of exactly one tour if and only if there exists a Hamiltonian cycle with length not greater than $B$. Thus, since the decision variant of the TSP is known to be NP-complete [9], it follows that CPLR is NP-hard.

# 3    Related Work

Optimization problems in which the vertices of a graph have to be visited under various timing or length constraints are a field of active research. However, to the best of our knowledge there exists no previous work regarding CPLR.

Drucker et al. [6] consider the Cyclic Routing of UAVs (CR-UAV) problem, which is a generalization of CPLR. Here, cyclic routes have to be determined, which have to start and end at the same vertex and can visit vertices multiple times. Additionally, waiting is possible and the routes are allowed to intersect. The goal is to determine the minimum number of UAVs that is necessary to jointly satisfy the critical weight requirements of all vertices. Ho and Ouaknine [12] showed that the corresponding decision problem is PSPACE-complete even in the case of a single UAV. A solution approach based on solving satisfiability problems is presented in [8]. However, the proposed method does not guarantee an optimal solution. In [7] on the other hand, a reduction to model-checking is suggested and an algorithm, which runs in parallel a bounded model checker to detect feasible solutions and an explicit-state search attempting to prove their absence, is presented. Further, Asghar et al. [2] introduce a factor $\mathcal{O}(\log \rho)$ approximation algorithm, where $\rho$ is the ratio of maximum and minimum critical weight. It is based on solving Minimum Cycle Cover Problems (MCCPs) on a partition of the vertices.

Given a graph $G = (V, E)$ and $\lambda \in \mathbb{R}_{\geq 0}$, MCCP is to determine the minimum number of cycles covering the vertex set, such that the length of each cycle is not greater than $\lambda$. In contrast to CPLR, the cycles do not have to be disjoint. Yu et al. [20] present a $\frac{32}{7}$-approximation algorithm for MCCP.

Besides TSP, there are several other well-known combinatorial optimization problems, which are closely related to CPLR. One is the Vehicle Routing Problem with Time Windows (VRPTW), see Solomon and Desrosiers [18] or Desrochers et al. [5] for surveys on the problem. The goal is to determine a collection of routes for a fleet of homogeneous vehicles. The routes have to start and end at a common depot $v_0$ and to jointly visit a given set of customers $\{v_1, \ldots, v_n\}$. Each customer $v_i \in V$ has some service requirement $q_i$ which has to be satisfied within a time window $[l_i, u_i]$ by exactly one of the vehicles. The goal is to minimize the number of necessary vehicles while the accumulated requirements of the customers are not allowed to exceed the capacity of their assigned vehicle.

# 4    Conflict Graph for CPLR

In this section, we determine vertex pairs, that cannot be contained in a common feasible cycle. Therefore, we introduce the notion of a conflict graph for CPLR and show that its cliques give rise to a set of valid constraints. Conflict graphs are used in many research areas including for example the conflict analysis in MIP, which has its origin in solving satisfiability problems [1,13,19,21].

**Definition 1.** *Let $(V, E, T, L)$ be an instance of CPLR with underlying graph $G = (V, E)$. Its conflict graph $G^c = (V, E^c)$ consists of the vertex set $V$ and the* *edge set $E^c$, where there is an edge $e_{ij}^c \in E^c$ between $v_i \in V$ and $v_j \in V$ if no feasible cycle containing both vertices exists.*

Due to the triangle inequality, the edge set of the conflict graph is equal to the set of edges in $G$, that cannot be contained in any feasible cycle.

**Lemma 2.** *Let $(V, E, T, L)$ be an instance of CPLR and let $e_{ij} \in E$. There exists a feasible cycle containing $e_{ij}$ if and only if $2 \cdot L_{ij} \le \min\{T_i, T_j\}$.*

*Proof.* Let $C_k$ be a feasible cycle containing $e_{ij} \in E$. This implies that $C_k$ contains $v_i \in V$ and $v_j \in V$, too. Further, $C_k$ can be split into $e_{ij}$ and a simple path $p_{ij}$ between these two vertices. Thus, since $C_k$ is feasible and because the edges respect the triangle inequality, we have

$$2 \cdot L_{ij} \le \ell(p_{ij}) + L_{ij} = \sum_{e_{ij} \in C_k} L_{ij} = \tau_k \le \min_{v_k \in C_k} T_k \le \min\{T_i, T_j\}.$$

Conversely, if $2 \cdot L_{ij} \le \min\{T_i, T_j\}$ the cycle induced by $v_i$, $v_j$ and $e_{ij}$ is feasible.

**Corollary 1.** *Let $(V, E, T, L)$ be an instance of CPLR. The edge set of the conflict graph is given by $E^c = \{e_{ij}^c \in V \times V \mid 2 \cdot L_{ij} > \min\{T_i, T_j\}\}$.*

Using $G^c$, we can derive a lower bound on the size of feasible cycle partitions.

**Lemma 3.** *Let $(V, E, T, L)$ be a CPLR instance and let $U \subseteq V$ be a clique of size $|U| = m$ in $G^c$. For each feasible cycle partition $C$ we have $|C| \ge m$.*

*Proof.* Let $C$ be a cycle partition with $|C| < m$. By the pigeonhole principle there exists a cycle containing at least two vertices from $U$. Thus, this cycle cannot be feasible and therefore $C$ is not feasible.

**Corollary 2.** *Let $(V, E, T, L)$ be a CPLR instance and let $U \subseteq V$ be a maximum clique in $G^c$. Then $|U|$ is a lower bound on the optimal solution value of CPLR.*

## 5     Cheapest Insertion Heuristic for CPLR

Next, we present a Cheapest Insertion Heuristic for CPLR, which is inspired by the corresponding heuristic for TSP [17].

In each iteration $k$ a feasible cycle $C_k$ is determined. First, some vertex $v_x = \operatorname{argmin}_{v_i \in V} T_i$ with minimum critical weight is selected. If $v_x$ is the only vertex left in $G$ or the length of a shortest proper cycle containing it, which consists of $v_x$ and a vertex $v_y$ being closest to it w.r.t. the edge weights, exceeds $T_x$, $v_x$ is removed as singleton $C_k$ from $G$ and we continue with the construction of the next cycle. Otherwise, we proceed with a shortest feasible cycle as described above. Next, we determine two vertices $v_a$ and $v_b$, which are adjacent in $C_k$, and a vertex $v_c \in V \setminus C_k$ such that the insertion of $v_c$ between $v_a$ and $v_b$ yields a minimum increase in the cycle length. If the augmented cycle length does not exceed $T_x$, we insert $v_c$ into $C_k$ and search for more suitable triples of vertices. Otherwise, we remove $C_k$ from $G$ and continue with the construction of the next cycle. The algorithm terminates when all vertices have been inserted into a cycle.

**Lemma 4.** *Cheapest Insertion for CPLR has no constant approximation ratio.*

*Proof.* Consider the CPLR instance based on the complete graph $G = (V, E)$ with $|V| = n = 2k^2$ vertices. For each $v_i \in \{v_1, \ldots, v_k\} =: V_1$ let $T_i = 2k$ and for each $v_i \in V \setminus V_1 =: V_2$ let $T_i = 2k^2 - k$. Additionally, let $L_{ij} = 2$ for each edge in $e_{ij} \in V_1 \times V_1$ and let $L_{ij} = 1$ otherwise. An optimal solution consists of two cycles: One cycle containing all nodes in $V_1$ and the other cycle containing all nodes in $V_2$. However, the heuristic produces a solution with $k$ cycles, each featuring one node from $V_1$ and $2k - 1$ nodes from $V_2$. Hence, Cheapest Insertion for CPLR does not admit a constant approximation ratio.

In contrast, the corresponding TSP heuristic has an approximation factor of 2.

# 6    A Mixed Integer Programming Model for CPLR

In this section we present a mixed integer programming (MIP) model for CPLR. In the following, we consider the induced directed graph $G = (V, A)$, whose arc set features two directed arcs $a_{ij}, a_{ji} \in A$ for each edge $e_{ij} \in E$. Further, both arcs are assigned $L_{ij}$ as their weights.

For each potential proper feasible cycle $C_k$ with $k \in \{1, \ldots, \lfloor \frac{n}{2} \rfloor\} := \mathcal{K}$ we introduce a binary variable $u_k$ indicating whether it contains any vertices or not and a nonnegative continuous variable $\tau_k$ representing its length. Additionally, for each vertex $v_i \in V$ we introduce a binary variable $y_i$ indicating whether the vertex forms a singleton or not. Further, for each vertex $v_i \in V$ and each potential proper cycle $C_k$ we introduce a binary variable $z_i^k$ indicating whether $v_i \in C_k$ or not and analogously for each arc $a_{ij} \in A$ a binary variable $x_{ij}^k$ indicating whether $a_{ij} \in C_k$ or not. The model states as

$$\min \sum_{v_i \in V} y_i + \sum_{k \in \mathcal{K}} u_k \tag{1}$$

$$\text{s.t.} \qquad y_i + \sum_{k \in \mathcal{K}} z_i^k = 1 \qquad \forall v_i \in V \tag{2}$$

$$z_i^k \leq u_k \qquad \forall v_i \in V, \forall k \in \mathcal{K} \tag{3}$$

$$\sum_{a_{ij} \in \delta^+(v_i)} x_{ij}^k = z_i^k \qquad \forall v_i \in V, \forall k \in \mathcal{K} \tag{4}$$

$$\sum_{a_{ji} \in \delta^-(v_i)} x_{ji}^k = z_i^k \qquad \forall v_i \in V, \forall k \in \mathcal{K} \tag{5}$$

$$\sum_{a_{ij} \in A} L_{ij} x_{ij}^k = \tau_k \qquad \forall k \in \mathcal{K} \tag{6}$$

$$T_i + (M_k - T_i)(1 - z_i^k) \geq \tau_k \qquad \forall v_i \in V, \forall k \in \mathcal{K} \tag{7}$$

$$u_k, y_i, z_i^k, x_{ij}^k \in \{0, 1\}$$

$$\tau_k \in \mathbb{R}_{\geq 0}$$

The objective function (1) aims at minimizing the number of cycles. Constraints (2) ensure that each vertex either forms a singleton or is assigned to a proper cycle. If vertex $v_i$ is assigned to the proper cycle $C_k$, then (3) indicates that it is nonempty and $v_i$ has to have an outgoing and an ingoing arc due to constraints (4) and (5), respectively. Next, constraints (6) keep track of the cycle lengths, while (7) ensures that the critical weights of the vertices are respected. Here, $M_k$ denotes the $k$-th biggest critical weight among all vertices.

The formulation ensures that each vertex is contained in exactly one cycle and that all critical weight requirements are satisfied. However, it can happen that the vertices assigned to some proper cycle $C_k$ form subtours. Hence, we extend the model adapting the idea of Miller, Tucker and Zemlin [14] (MTZ) to prohibit subtours. Here, each vertex is assigned a positive weight while the starting vertex has value zero. For each pair of consecutive vertices in a tour the weights must increase except for the last and the starting vertex. A straightforward use for CPLR is not possible, since we cannot fix starting vertices for the proper cycles in advance. Thus, for each $k \in \mathcal{K}$ and each vertex $v_i \in V$ we introduce additional binary variables $s_i^k \in \{0, 1\}$ indicating whether $v_i$ is the starting vertex of cycle $C_k$ or not. Weight variables $w_i^k \in \mathbb{Z}_{\geq 0}$ together with constraints

$$\sum_{v_j \in V} s_j^k = u_k \qquad \forall k \in \mathcal{K} \qquad (8)$$

$$s_i^k \leq z_i^k \qquad \forall v_i \in V, \forall k \in \mathcal{K} \qquad (9)$$

$$\sum_{v_j \in V} z_j^k - u_k \geq w_i^k \qquad \forall v_i \in V, \forall k \in \mathcal{K} \qquad (10)$$

$$w_i^k - w_j^k + |V| \cdot (x_{ij}^k - s_j^k) \leq |V| - 1 \qquad \forall a_{ij} \in A, \forall k \in \mathcal{K}$$

$$s_i^k \in \{0, 1\} \qquad \forall v_i \in V, \forall k \in \mathcal{K}$$

$$w_i^k \in \mathbb{Z}_{\geq 0} \qquad \forall v_i \in V, \forall k \in \mathcal{K}. \qquad (11)$$

model the MTZ idea for CPLR. Constraints (8) determine a starting vertex for each cycle, which has to be part of it due to constraint (9). Furthermore, the weights necessary for each proper cycle are bounded by (10). Eventually, constraints (11) are the Miller-Tucker-Zemlin constraints as explained above. Thus, the MIP formulation consisting of (1)–(7) and (8)–(11) models CPLR.

## 6.1   Symmetry Breaking Constraints

The solution space of the MIP model can be highly symmetric. Given a feasible solution, all permutations of the proper cycle indices respecting constraints (7) are feasible. Assume w.l.o.g. that the vertices are ordered non-increasingly by their critical weights. Then constraints

$$z_i^k \leq \sum_{j=1}^{i-1} z_j^{k-1} \quad \forall v_i \in V, \forall k \in \mathcal{K} \setminus \{1\} \qquad (12)$$

ensure that only the permutation with the proper cycles sorted by the minimum index of their vertices is feasible.

## 6.2 Conflict Graph Clique Constraints

Let $(V, E, T, L)$ be a CPLR instance and let $G^c = (V, E^c)$ be the corresponding conflict graph. Further, let $\mathcal{U}$ denote the set of all cliques in $G^c$ and let $U \in \mathcal{U}$. From the proof of Lemma 3 we derive that no two vertices from $U$ can be contained in a common feasible cycle. Hence,

$$\sum_{v_i \in U} z_i^k \leq 1 \quad \forall U \in \mathcal{U}, \forall k \in \mathcal{K} \tag{13}$$

are valid constraints for CPLR. In addition, by Corollary 2 the cardinality of a maximum clique is a lower bound on the number of singletons and proper cycles contained in an optimal solution. Hence, another valid constraint is

$$\sum_{v_i \in V} y_i + \sum_{k \in \mathcal{K}} u_k \geq \max_{U \in \mathcal{U}} |U|. \tag{14}$$

# 7 Computational Experiments

For our computational experiments, we generated two sets of test instances based on the graphs of the 28 instances from the TSPLIB [16], which have 100 or less vertices. Let $\tau^*$ denote the length of an optimal tour for the corresponding TSP instance. For the CPLR instances, we assigned each vertex a random integer from the interval $[\frac{\tau^*}{6}, \frac{\tau^*}{2}]$ and from the interval $[\frac{\tau^*}{8}, \frac{\tau^*}{4}]$ as critical weight in the first and second test set, respectively. All instances can be downloaded from https://cloud.zib.de/s/CPLR_data/download.

In our MIP formulations, we excluded all edges, which cannot be contained in feasible cycles by Lemma 2. Beside the symmetry breaking constraints (12), we computed all maximal cliques of the conflict graph using the algorithm of Cazals and Karande [3] and added the corresponding constraints (13) as well as the lower bound (14). Additionally, we used the solution computed by the Cheapest Insertion Heuristic as initial incumbent and derived a tighter bound on the necessary size of $\mathcal{K}$ in all cases. All mentioned calculations were done in less than 0.1 seconds for each instance.

We ran our experiments on a cluster of machines composed of two Intel Xeon Gold 5122 running at 3.60 GHz, which provide 8 cores and 96 GB of RAM in total. All algorithms were implemented in Python and we used the corresponding interface of Gurobi v9.0 [10] to solve our MIP models with a time limit of 24 h.

The computational results can be found in Table 1. Here, the number of removed edges for each instance is given. Furthermore, the value of the solution found by the Cheapest Insertion Heuristic is shown as UB (heur) and the lower bound from the size of a maximum clique in the conflict graph is LB (max-c). Finally, the upper and the lower bound at the end of the solving process, which was either reached when the problem was solved or due to the time limit (indicated by TL), are shown.

**Table 1.** Computational results for the instances of the two test sets.

| Instance | Critical weights from: $[\frac{\tau^*}{6}, \frac{\tau^*}{2}]$ | | | | | | Critical weights from: $[\frac{\tau^*}{8}, \frac{\tau^*}{4}]$ | | | | | |
|---|---|---|---|---|---|---|---|---|---|---|---|---|
| | Removed edges | UB (heur) | LB (max-c) | UB | LB | Time (sec) | Removed edges | UB (heur) | LB (max-c) | UB | LB | Time (sec) |
| burma14 | 40 | 5 | 4 | 5 | 5 | 3 | 67 | 8 | 6 | 6 | 6 | 1 |
| ulysses16 | 39 | 5 | 4 | 4 | 4 | 1 | 65 | 6 | 5 | 6 | 6 | 1 |
| gr17 | 61 | 5 | 4 | 5 | 5 | 1 | 96 | 8 | 7 | 8 | 8 | 1 |
| gr21 | 86 | 6 | 3 | 5 | 5 | 9 | 141 | 9 | 6 | 8 | 8 | 11 |
| ulysses22 | 66 | 5 | 4 | 5 | 5 | 5 | 127 | 7 | 6 | 7 | 7 | 9 |
| gr24 | 53 | 6 | 4 | 5 | 5 | 89 | 172 | 8 | 7 | 7 | 7 | 1 |
| fri26 | 105 | 6 | 5 | 6 | 6 | 291 | 188 | 8 | 6 | 8 | 8 | 21 |
| bayg29 | 103 | 6 | 4 | 5 | 5 | 5758 | 238 | 10 | 7 | 8 | 8 | 178 |
| bays29 | 111 | 7 | 5 | 6 | 6 | 2661 | 235 | 10 | 6 | 8 | 8 | 206 |
| dantzig42 | 243 | 7 | 4 | 7 | 5 | TL | 478 | 11 | 7 | 9 | 8 | TL |
| swiss42 | 160 | 8 | 4 | 7 | 5 | TL | 451 | 10 | 7 | 9 | 8 | TL |
| att48 | 299 | 8 | 5 | 6 | 5 | TL | 605 | 10 | 6 | 9 | 7 | TL |
| gr48 | 197 | 8 | 4 | 7 | 5 | TL | 536 | 10 | 5 | 10 | 7 | TL |
| hk48 | 282 | 7 | 4 | 7 | 5 | TL | 548 | 10 | 6 | 10 | 7 | TL |
| eil51 | 127 | 7 | 3 | 7 | 5 | TL | 521 | 9 | 5 | 9 | 7 | TL |
| berlin52 | 264 | 8 | 4 | 8 | 5 | TL | 515 | 12 | 6 | 10 | 8 | TL |
| brazil58 | 227 | 6 | 3 | 6 | 4 | TL | 649 | 10 | 5 | 8 | 7 | TL |
| st70 | 217 | 7 | 3 | 7 | 4 | TL | 1073 | 10 | 5 | 10 | 6 | TL |
| eil76 | 184 | 7 | 3 | 7 | 5 | TL | 775 | 12 | 5 | 12 | 6 | TL |
| pr76 | 208 | 8 | 3 | 8 | 4 | TL | 925 | 11 | 6 | 10 | 6 | TL |
| gr96 | 406 | 8 | 3 | 8 | 4 | TL | 1587 | 10 | 5 | 10 | 6 | TL |
| rat99 | 586 | 8 | 3 | 8 | 5 | TL | 1673 | 11 | 5 | 11 | 6 | TL |
| kroA100 | 803 | 8 | 3 | 8 | 4 | TL | 2141 | 11 | 5 | 11 | 6 | TL |
| kroB100 | 643 | 8 | 3 | 8 | 4 | TL | 1979 | 10 | 5 | 10 | 6 | TL |
| kroC100 | 856 | 7 | 3 | 7 | 4 | TL | 2183 | 12 | 6 | 11 | 7 | TL |
| kroD100 | 635 | 8 | 3 | 8 | 5 | TL | 1962 | 11 | 6 | 11 | 6 | TL |
| kroE100 | 678 | 8 | 3 | 8 | 4 | TL | 2092 | 12 | 5 | 12 | 6 | TL |
| rd100 | 445 | 8 | 4 | 8 | 4 | TL | 1817 | 11 | 5 | 11 | 6 | TL |

The results show that each of the 18 instances with up to 29 vertices was solved in at most 96 min. All other instances with 42 and more nodes could not be solved within 24 h. Furthermore, the Cheapest Insertion Heuristic produced solutions with at most two extra cycles w.r.t. the upper bounds in UB. Additionally, the first test set seems to be harder than the second one considering the solved instances. One reason may be that due to the larger critical weights there are more degrees of freedom for determining feasible cycles.

## 8  Conclusion and Outlook

In this article we introduced CPLR, developed a heuristic, and formulated a MIP model, which features clique constraints derived from a conflict graph. We were able to solve test instances of small and medium size.

There are several directions of future research that seem worth to be investigated. From a theoretical point of view, it remains an open question whether CPLR is contained in APX or not. Additionally, we are currently developing a

MIP model based on a variant of subtour elimination constraints. Furthermore, extending the conflict graph concept to hypergraphs seems promising. Finally, it appears natural to study the generalization of CPLR where the triangle inequality condition on the edge weights is dropped.

# References

1. Achterberg, T.: Conflict analysis in mixed integer programming. Disc. Optim. 4(1), 4–20 (2007)
2. Asghar, A.B., Smith, S.L., Sundaram, S.: Multi-Robot Routing for Persistent Monitoring with Latency Constraints. arXiv preprint arXiv:1903.06105 (2019)
3. Cazals, F., Karande, C.: A note on the problem of reporting maximal cliques. Theor. Comput. Sci. 407(1–3), 564–568 (2008)
4. Chamoso, P., Raveane, W., Parra, V., González, A.: UAVs applied to the counting and monitoring of animals. In: Ramos, C., Novais, P., Nihan, C.E., Corchado Rodríguez, J.M. (eds.) Ambient Intelligence - Software and Applications. AISC, vol. 291, pp. 71–80. Springer, Cham (2014). https://doi.org/10.1007/978-3-319-07596-9_8
5. Desroches, M., Lenstra, J., Savelbergh, M., Soumis, F.: Vehicle routing with time windows: optimization and approximation. In: Golden B.L., Assad, A.A. (eds.) Vehicle routing: Methods and Studies, North-Holland, Amsterdam, pp. 65–84 (1988)
6. Drucker, N., Penn, M., Strichman, O.: Cyclic routing of unmanned air vehicles. Information Systems Engineering Technical Reports. IE/IS-2014-02 (2014)
7. Drucker, N., Ho, H.M., Ouaknine, J., Penn, M., Strichman, O.: Cyclic-routing of unmanned aerial vehicles. J. Comput. Syst. Sci. 103, 18–45 (2019)
8. Drucker, N., Penn, M., Strichman, O.: Cyclic routing of unmanned aerial vehicles. In: Quimper, C.-G. (ed.) CPAIOR 2016. LNCS, vol. 9676, pp. 125–141. Springer, Cham (2016). https://doi.org/10.1007/978-3-319-33954-2_10
9. Garey, M.R., Johnson, D.S.: Computers and Intractability, vol. 29. W.H, Freeman New York (2002)
10. Gurobi Optimization, L.: Gurobi Optimizer Reference Manual, Version 9.0.0 (2019). http://www.gurobi.com
11. Hausamann, D., Zirnig, W., Schreier, G.: Monitoring of gas transmission pipelines - a customer driven civil UAV application. In: ODAS Conference (2003)
12. Ho, H.-M., Ouaknine, J.: The cyclic-routing UAV problem is PSPACE-complete. In: Pitts, A. (ed.) FoSSaCS 2015. LNCS, vol. 9034, pp. 328–342. Springer, Heidelberg (2015). https://doi.org/10.1007/978-3-662-46678-0_21
13. Marques-Silva, J.P., Sakallah, K.A.: GRASP: a search algorithm for propositional satisfiability. IEEE Trans. Comput. 48(5), 506–521 (1999)
14. Miller, C.E., Tucker, A.W., Zemlin, R.A.: Integer programming formulation of traveling salesman problems. J. ACM (JACM) 7(4), 326–329 (1960)
15. Ollero, A., Martínez de Dios, J.R., Merino, L.: Unmanned aerial vehicles as tools for forest-fire fighting. Forest Ecol. Manage. 234(1), S263 (2006)
16. Reinelt, G.: TSPLIB - A Traveling Salesman Problem Library. ORSA Journal on Computing 3(4), 267–384 (1991)
17. Rosenkrantz, D.J., Stearns, R.E., Lewis II, P.M.: An analysis of several heuristics for the traveling salesman problem. SIAM J. Comput. 6(3), 563–581 (1977)

18. Solomon, M.M., Desrosiers, J.: Survey paper - time Window constrained routing and scheduling problems. Transp. Sci. **22**(1), 1–13 (1988)
19. Witzig, J., Berthold, T., Heinz, S.: Experiments with conflict analysis in mixed integer programming. In: International Conference on AI and OR Techniques in Constraint Programming for Combinatorial Optimization Problems, pp. 211–220. Springer (2017). https://doi.org/10.1007/978-3-319-59776-8_17
20. Yu, W., Liu, Z., Bao, X.: New approximation algorithms for the minimum cycle cover problem. Theor. Comput. Sci. **793**, 44–58 (2019)
21. Zhang, L., Madigan, C.F., Moskewicz, M.H., Malik, S.: Efficient conflict driven learning in a boolean satisfiability solver. In: IEEE/ACM International Conference on Computer Aided Design. ICCAD 2001. IEEE/ACM Digest of Technical Papers (Cat. No. 01CH37281), pp. 279–285. IEEE (2001)

# Optimizing Student Course Preferences in School Timetabling

Richard Hoshino$^{(\boxtimes)}$ and Irene Fabris

Quest University Canada, Squamish, BC, Canada
richard.hoshino@gmail.com, irene.fabris@questu.ca

**Abstract.** School timetabling is a complex problem in combinatorial optimization, requiring the best possible assignment of course sections to teachers, timeslots, and classrooms. There exist standard techniques for generating a school timetable, especially in cohort-based programs where students take the same set of required courses, along with several electives. However, in small interdisciplinary institutions where there are only one or two sections of each course, and there is much diversity in course preferences among individual students, it is very difficult to create an optimal timetable that enables each student to take their desired set of courses while satisfying all of the required constraints.

In this paper, we present a two-part school timetabling algorithm that was applied to generate the optimal Master Timetable for a Canadian all-girls high school, enrolling students in 100% of their core courses and 94% of their most desired electives. We conclude the paper by explaining how this algorithm, combining graph coloring with integer linear programming, can benefit other institutions that need to consider student course preferences in their timetabling.

**Keywords:** School timetabling · Post Enrollment Course Timetabling Problem · Integer Programming · Graph coloring · Optimization

## 1 Introduction

Every educational institution needs to produce a Master Timetable, listing the complete set of offered courses, along with the timeslot and classroom for each section of that course. This timetable allows teachers to know what courses they are teaching, and enables students to enroll in a subset of these courses.

As many school administrators know, creating a timetable is incredibly difficult, requiring the careful balance of numerous *requirements* (hard constraints) and *preferences* (soft constraints). When timetables are constructed by hand, the process is often 10% mathematics and 90% politics [4], leading to errors, inefficiencies, and resentment among teachers and students.

To address these concerns, scholars in Operations Research have analyzed the School Timetabling Problem (STP) ever since the 1960s [10]. Various heuristics have been applied to create timetables for schools in Argentina, Brazil, Denmark, Germany, Greece, Italy, Netherlands, South Africa, and Vietnam [23].

© Springer Nature Switzerland AG 2020
E. Hebrard and N. Musliu (Eds.): CPAIOR 2020, LNCS 12296, pp. 283–299, 2020.
https://doi.org/10.1007/978-3-030-58942-4_19

In the most basic version of the STP, the objective is to assign courses to teachers, timeslots, and classrooms, subject to the following constraints: a teacher cannot teach two courses in the same timeslot, no classroom can be used by two courses simultaneously, and each teacher has a set of unavailable teaching timeslots. This problem is NP-complete [7].

Real-life timetabling problems involve additional constraints that must be satisfied [24], further increasing the complexity of the STP. These variations include *event* constraints (e.g. Course X must be scheduled before Course Y), and *resource* constraints (e.g. scheduling only one lab-based course in any timeslot). At large universities, there are additional constraints that must be considered, such as taking into account the time students need to walk from one end of the campus to the other.

Over the past five decades, numerous algorithms have been applied to generate optimal (or nearly-optimal) timetables for STP benchmark instances. These techniques include constraint programming [6], evolutionary algorithms [26], simulated annealing [18], and tabu search [20]. The complete list of methods appears in a comprehensive survey paper published earlier this decade [23].

Given how hard the STP is, a common practice is to focus only on teacher requirements and preferences, ignoring the wishes of the students (i.e., the individuals most affected by the timetable). This assumption is made because many school programs are *cohort-based* [15], where students are divided into fixed groups and take the same sequence of courses to complete their education. At many high schools and universities, the timetabling is done via *homogeneous sectioning* [4], where students are grouped according to their interests or majors: for example, students in the Arts stream versus students in the Sciences stream.

There are obvious deficiencies to this practice, most notably in small interdisciplinary institutions where cohorts do not exist, and each student takes a unique set of courses from all departments. Many such institutions are private schools, where their revenue comes exclusively from student tuition. If students cannot enroll in their desired courses, they (or their parents) will go to a different school that will accommodate their preferences. Thus, these schools are under tremendous pressure to create a timetable that satisfies teachers *and* students. This is the motivation for the *Post-Enrollment Course Timetabling Problem (PECTP)*, an active area of research in the field of automated timetabling.

This paper proceeds as follows. In Sect. 2 and 3, we define the PECTP and provide a brief literature review on related work that incorporates student course preferences in timetabling. In Sect. 4 and 5, we describe our solution to the PECTP, which is a two-part algorithm that generates an optimal coloring of a weighted conflict graph for single-section courses, after which an integer linear program is solved to generate the final timetable. In Sect. 6, we generate the optimal Master Timetable for an interdisciplinary all-girls high school in Canada, and demonstrate the speed and quality of our two-part algorithm. And in Sects. 7 and 8, we explore the strengths and limitations of our timetabling algorithm to large universities, and conclude the paper with some ideas and directions for future research.

## 2    Problem Definition

The standard School Timetabling Problem (STP) is an example of a constraint satisfaction problem, which asks whether there exists a feasible assignment of course sections to teachers and timeslots. To avoid confusion, we will rename timeslots as *blocks*, so that $T$ will denote the set of teachers and $B$ will denote the set of blocks during which the courses will take place.

The more general version of the STP is a combinatorial optimization problem, which asks for the best assignment satisfying all of the hard constraints while maximizing the preferences of the teachers being assigned their desired courses in specific blocks.

Both versions of the STP can be set up as a 0–1 integer linear program (ILP), in which each unknown variable $X_{t,c,b}$ represents whether teacher $t$ is assigned to a section of course $c$ in block $b$. The total number of variables is $n = |T||C||B|$, where $|T|$ is the number of teachers, $|C|$ is the number of offered courses, and $|B|$ is the number of blocks.

Let $D_{t,c,b}$ be the desirability of teacher $t$ assigned to course $c$ in block $b$. This coefficient will be a function of teacher $t$'s ability and willingness to teach course $c$, combined with their preference for teaching that course in block $b$.

Then, subject to all of the hard constraints, we want to maximize

$$\sum_{t \in T} \sum_{c \in C} \sum_{b \in B} D_{t,c,b} \cdot X_{t,c,b}.$$

The Post-Enrollment Course Timetabling Problem (PECTP) was introduced just over a decade ago [17], as part of the second International Timetabling Competition. In the PECTP, points are awarded for enrolling students in any section of a desired course. For example, if there are ten different sections of Calculus 101, a student wishing to take Calculus 101 needs to be assigned to exactly one of these ten sections.

In addition to all of the constraints in the STP (e.g. no teacher can be assigned to two courses in overlapping blocks), the PECTP involves additional student-related hard constraints, such as ensuring that no student is enrolled in multiple sections of the same course.

Let $Y_{s,c,b}$ be the binary decision variable representing whether student $s$ is assigned to a section of course $c$ in block $b$, and let $P_{s,c,b}$ be the preference of student $s$ being enrolled in course $c$ in block $b$.

Then, subject to all of the hard constraints, we want to maximize

$$\sum_{t \in T} \sum_{c \in C} \sum_{b \in B} D_{t,c,b} \cdot X_{t,c,b} + \sum_{s \in S} \sum_{c \in C} \sum_{b \in B} P_{s,c,b} \cdot Y_{s,c,b}.$$

This is the most basic formulation of the STP and PECTP. There are extensions that we will not consider in this paper, such as adding a penalty function whenever student $s$ has a class in the last block of the day or has a class in three consecutive blocks. For a full discussion and treatment of these PECTP extensions, we refer the reader to [19].

# 3  Related Work

The PECTP was introduced in 2007 as one of the tracks of the International Timetabling Competition (ITC). Over the past twelve years, different teams of Operations Research scholars have developed algorithms to tackle hard instances of the PECTP. The majority of these approaches rely on multi-stage heuristics.

Fonseca et al. [8] propose a three-stage hybrid solver involving graph algorithms, metaheuristics, and "matheuristics". Nothegger et al. [22] present an iterative three-step ant colony optimization algorithm. One of the finalists [5] for ITC 2007 employs a multiphased heuristic solver based on a stochastic local search, whereas the winning team [3] applies a two-stage local search approach combining tabu search and simulated annealing.

Heuristics are advantageous for they easily compute within the strict time limit imposed by the ITC, yet they cannot guarantee the optimality of the output solution. The latter is a particular weakness of local search approaches, which lack the flexibility of moving in the space of feasible solutions and get stuck in local minima despite the large size of the search neighborhood [3].

Recent papers have made much progress. Cambazard et al. [3] find provably-optimal solutions to three of the PECTP benchmark instances, by augmenting simulated annealing with a large neighborhood search. Kristiansen et al. [15] embed exact repair methods within an adaptive large neighborhood search (ALNS) to find a feasible solution first, which they optimize by solving a mixed integer program (MIP). Their ALNS finds timetables within 1% of the optimal solution, outperforming Gurobi, a state-of-the-art MIP solver, on large instances.

Parallel to the noteworthy progress in heuristic approaches, the last decade was marked by a significant advance in general-purpose MIP solvers. An obvious advantage of Integer Programming is its ability to issue certificates of optimality [14]. Since it is NP-complete to solve a 0–1 Integer Program [13], it has become common practice to decompose IP models into smaller sub-problems [27]. Van Den Broek et al. [29] use the lexicographical optimization of four ILP sub-problems to solve a real-world instance of PECTP at a Dutch university, and Kristiansen et al. [14] devise a two-step MIP algorithm for Danish high schools.

The problem considered in this paper is most similar to the formulation of the generalized PECTP presented by Mendez-Diaz et al. [19] and Carter [4], which were inspired by university timetabling problems in Argentina and Canada, respectively. In these two papers, the researchers first assign course sections to blocks, and then assign students to course sections. On the next page, we present our mathematical model that explains how we can perform both assignments simultaneously.

We make two contributions in this paper. First, we present a complete algorithm that guarantees fast optimal PECTP solutions for small educational institutions. Secondly, we provide a graph-theoretic framework to demonstrate how courses can be "bundled" together and treated as a single super-course, which significantly reduces the time required to generate a nearly-optimal timetable. This makes our algorithm scalable for larger schools and universities.

## 4    Mathematical Model

Let $T$ be the set of teachers, $S$ be the set of students, $C$ be the set of courses, and $B$ be the set of blocks.

For each $t \in T, c \in C, b \in B$, let $X_{t,c,b}$ be the binary variable that equals 1 if teacher $t$ is assigned to course $c$ in block $b$, and is 0 otherwise. Similarly, for each $s \in S, c \in C, b \in B$, let $Y_{s,c,b}$ be the binary variable that equals 1 if student $s$ is enrolled in course $c$ in block $b$, and is 0 otherwise.

Earlier we defined the desirability coefficient $D_{t,c,b}$ and the preference coefficient $P_{s,c,b}$. Our Integer Linear Program (ILP) has the following objective function:

$$\sum_{t \in T} \sum_{c \in C} \sum_{b \in B} D_{t,c,b} \cdot X_{t,c,b} + \sum_{s \in S} \sum_{c \in C} \sum_{b \in B} P_{s,c,b} \cdot Y_{s,c,b}.$$

We now present our hard constraints.

No teacher can be assigned to two different classes in the same block, and at most one section of any course is offered in any given block.

$$\sum_{c \in C} X_{t,c,b} \leq 1 \qquad \forall\, t \in T,\, b \in B \tag{1}$$

$$\sum_{t \in T} X_{t,c,b} \leq 1 \qquad \forall\, c \in C,\, b \in B \tag{2}$$

Define $O_c$ to be the number of offered sections of course $c$.

$$\sum_{b \in B} \sum_{t \in T} X_{t,c,b} = O_c \qquad \forall\, c \in C \tag{3}$$

No student can be enrolled in more than one course in the same block, nor can any student be enrolled in two sections of the same course.

$$\sum_{c \in C} Y_{s,c,b} \leq 1 \qquad \forall\, s \in S,\, b \in B \tag{4}$$

$$\sum_{b \in B} Y_{s,c,b} \leq 1 \qquad \forall\, s \in S,\, c \in C \tag{5}$$

No student can be enrolled in a course during a block in which that course is not offered by any teacher.

$$Y_{s,c,b} \leq \sum_{t \in T} X_{t,c,b} \qquad \forall\, s \in S,\, c \in C,\, b \in B \tag{6}$$

Let $R$ be the number of available rooms in the school.

$$\sum_{t \in T} \sum_{c \in C} X_{t,c,b} \leq R \qquad \forall\, b \in B \tag{7}$$

Let $M$ be the maximum size of a class.

$$\sum_{s \in S} Y_{s,c,b} \leq M \qquad \forall\, c \in C,\, b \in B \tag{8}$$

Our ILP maximizes the objective function subject to these eight constraints.

This model has a total of $(|T| + |S|) \cdot |C||B|$ binary decision variables. In practice, the large majority of these variables $X_{t,c,b}$ and $Y_{s,c,b}$ will be pre-set to 0, since teachers are qualified to only teach a small subset of the offered courses, and likewise, students will only want to be enrolled in a small subset of these courses. By fixing these zero variables, we can solve the PECTP using the above ILP, guaranteeing an optimal timetable whenever $|T|$, $|S|$, and $|C|$ are of reasonable size. But when these values are large, like at most universities, simplifications are required to ensure tractability.

There are two natural ways to simplify the problem: assume there are *cliques of teachers* or assume there are *cohorts of students*. In the former, the Master Timetable is generated one clique at a time: first assign course sections to the math teachers and fix those assignments, then do the same with the science teachers, and so on. In the latter, the students are pre-divided into fixed groups, and each group is assigned to the same set of course sections.

Unfortunately, these two approaches fail when there are many teachers who teach different subjects (e.g. Ms. X teaches Grade 12 Math and Grade 7 French), and when cohorts do not exist and students wish to take a unique combination of courses from two different faculties (e.g. an undergraduate attempting a double-major in Chemistry and Sociology).

Our approach is not to bundle teachers or bundle students, but rather to *bundle courses*. We now present a graph-theoretic approach that efficiently partitions one-section courses into discrete bundles that enable us to significantly reduce the running time of the ILP.

## 5    Bundling One-Section Courses

$C$ is the set of courses. Some of these courses will be sought by many students, and so multiple sections of the course must be offered in the timetable. The rest are specialized courses that will attract only a small number of students, and so only a single section is required. Let $C = C_M \cup C_O$, where $C_M$ is the set of multiple-section courses and $C_O$ is the set of one-section courses.

While there is much flexibility to timetabling courses in $C_M$, courses in $C_O$ can only be assigned to a single block, and so we must ensure that the courses in $C_O$ avoid any type of scheduling conflict: by teacher, by room, or by student.

Define $G$ to be the *weighted conflict graph*, where $C_O$ is the set of vertices. For each pair $x, y \in C_O$, we calculate the edge weight $w(x, y)$ as follows:

(a) Add a weight of $w_t$ if the same teacher is required to teach both $x$ and $y$.
(b) Add a weight of $w_r$ if the same room must be used for both $x$ and $y$.
(c) Add a weight of $w_s$ for *each* student who wishes to take both $x$ and $y$.

The weights $w_t, w_r, w_s$ can vary, though in practice it is most logical to set high values of $w_t$ and $w_r$ and low values for $w_s$ (e.g. $w_t = 100, w_r = 100, w_s = 1$).

For each integer $i \geq 0$, define $G_i$ to be the graph with vertex set $C_O$ whose edge set only consists of edges with weight greater than $i$. By definition, there exists a sufficiently large integer $i$ for which $G_i$ is an empty graph with no edges.

For each $G_i$, the *chromatic number* $\chi(G_i)$ is the fewest number of colors needed to color the vertices of $G_i$ so that no two vertices joined by an edge share the same color.

If $\chi(G_0)$ is at most $|B|$, the number of blocks in the timetable, then all of the one-section courses assigned the same color can be "bundled" together in the same block. This guarantees that every student will be able to take all of their desired one-section courses, since no pair will be offered at the same time. These bundles can be thought of as the "supernodes" of the conflict graph [2].

If $\chi(G_0) > |B|$, then by definition, it is impossible to create a timetable that enables every student to get into all of their desired courses. In this case, we find the smallest index $t$ for which $\chi(G_t) \leq |B|$, and once again, the color classes correspond to our bundles for the one-section courses.

Although it is NP-complete to determine the chromatic number of a general graph [13], for many large graphs we can compute $\chi(G)$ using state-of-the-art algorithms based on local search [12]. We can also compute $\chi(G)$ by solving the corresponding 0–1 ILP, and adding the constraint that no color class can contain more than $R$ courses, where $R$ is the number of rooms available for teaching.

This motivates our solution to the PECTP, where we use graph coloring to reduce the number of variables in our ILP.

(i) Construct the weighted conflict graph $G$, where the vertex set is $C_O$.
(ii) Starting with $i = 0$, calculate $\chi(G_i)$. If $\chi(G_i) \leq |B|$, then stop. Otherwise increment $i$ by 1 until we find some index $i = t$ for which $\chi(G_i) \leq |B|$.
(iii) Find a $|B|$-coloring of $\chi(G_i)$ where the number of one-section courses in each color class is at most the number of available rooms. Let $X_j$ be the set of courses in $C_O$ assigned to color $j$.
(iv) Redefine $C$ to equal $C_M \cup X_1 \cup X_2 \cup \ldots \cup X_{|B|}$, where there are $|C_M|$ courses that have multiple sections, and $|B|$ bundles, each of which is a one-section "super-course" with multiple teachers that can be assigned to any number of students. We then solve the previously-defined ILP, using this new set $C$.

For example, suppose that there are $|C| = 120$ courses to be timetabled into $|B| = 10$ blocks, where $|C_M| = 20$ and $|C_O| = 100$. The above algorithm bundles the 100 one-section courses into $|B| = 10$ bundles. Thus, instead of considering $|C| = 120$ courses in our ILP, we now only need to consider $|C_M| + |B| = 30$ courses. By reducing the number of variables by a factor of four, we create a massive reduction in the total running time while only sacrificing a small percentage in quality, as measured by the value of our objective function.

Our approach is particularly useful in small interdisciplinary institutions that offer numerous one-section courses desired by different sets of students. We now provide an example of such an educational institution, and apply our algorithm to create the optimal Master Timetable for this all-girls independent school.

# 6   Application

St. Margaret's School (SMS) is located in Victoria, the capital city of the Canadian province of British Columbia. Since 1908, educators at SMS have dedicated themselves to inspiring girls who want to change the world and helping them become women who do change the world. The school has an enrollment of approximately 375 students, starting from Junior Kindergarten (age 3 and 4).

As mentioned in their online handbook [28], SMS prides itself on their small-scale learning environment, which provides teachers with the flexibility to personalize learning for each student and challenge each girl to realize her own potential. In order to achieve this goal, the school spends several months each year constructing the Master Timetable, by hand, with several dozen iterations.

The biggest challenge is timetabling the courses for the Grade 11 and 12 students, i.e., the juniors and seniors at the high school. Unlike students in the lower grades who take mostly required (core) courses, there are numerous elective courses in the final two years, and each student wants to enroll in a different combination of courses from the *eighty* offerings that are available, including Advanced Placement Calculus, Law Studies, Studio Art, and Creative Writing.

The $|S| = 58$ students going into Grade 11 and 12 completed a survey indicating their course choices for the following year. The administrators used these responses to decide to offer $|C| = 39$ of the 80 possible courses. To ensure a maximum class size of 18, the administrators assigned two sections to $|C_M| = 9$ courses requested by more than 18 students, and one section for the remaining $|C_O| = 30$ courses that were requested by at most 18 students.

There are five periods in each day, and a total of $|B| = 9$ blocks. The nine blocks are fixed in the schedule, as follows:

| Monday | Tuesday | Wednesday | Thursday | Friday |
|---|---|---|---|---|
| Block 1 | Block 2 | Assembly | Block 1 | Block 2 |
| Block 7 | Block 8 | Block 8 | Block 9 | Block 3 |
| Block 4 | Block 3 | Block 6 | Block 5 | Block 7 |
| Block 5 | Block 9 | Block 4 | Block 6 | Block 8 |
| Block 6 | Block 7 | Block 5 | Block 4 | Block 9 |

Of the $|C| = 39$ courses, 18 are "short" courses offered in blocks 1/2/3 with two weekly classes, and the other 21 are "long" courses in blocks 4/5/6/7/8/9 with three weekly classes. Each student is required to take a set of core courses, with the rest being freely-chosen electives. Most (but not all) of the core courses are long, and most (but not all) of the elective courses are short.

Each student $s$ updated their survey with their most desired courses for 2019–2020, listing up to 3 short courses and up to 6 long courses. This selection included the core courses of English and Career/Leadership, as well as an additional English Language Learner course for non-native English speakers. Finally,

Grade 11 students were required to take a Physical Education course. Thus, all Grade 11s had at most 4 core courses, while Grade 12s students had at most 3.

The $|S| = 58$ students requested a total of 447 courses, which is fewer than the maximum total of $|S| \times |B| = 58 \times 9 = 522$. This occurred because some students had no preference for certain electives (e.g. they viewed every Social Science course as interchangeable), and also because many of the students qualified to take eight courses with their ninth one being a "self-study period".

For each block $b$, we set the preference coefficient $P_{s,c,b}$ as follows:

(i)   10 points if $c$ is a core course
(ii)  3 points if $c$ is an elective course and $s$ is a Grade 12 student
(iii) 1 point if $c$ is an elective course and $s$ is a Grade 11 student

The SMS leadership team pre-assigned each of the $9 + 9 + 30 = 48$ course sections to one of the $|T| = 19$ teachers in the Senior School, based on extensive consultations with each teacher. With teacher preferences pre-assigned, this reduces our ILP's objective function to maximizing student preferences. Mathematically this is equivalent to setting the desirability coefficient $D_{t,c,b}$ to equal 0 if teacher $t$ could be assigned to course $c$ in some block $b$, and $-1000$ otherwise.

In addition to the eight constraints we mentioned previously, we also included extra constraints requested by the school. For example, some of the teachers work part-time, and are only available to teach on Mondays, Wednesdays, and Thursdays. This forced all of their teaching blocks to be 1, 4, 5, or 6. Two courses are "two-block combination courses" (e.g. Pre-Calculus 11 and Pre-Calculus 12), and these double courses must be offered in Block 1 and Block 8.

Our optimization program, written in Python, requires two Excel sheets as input: one called "Student Data" and one called "Course Data". These two documents encapsulate all of the information described above. For the actual optimization, we use COIN-OR Branch and Cut (CBC), an open-source MIP solver, with the Google OR-Tools linear solver wrapper [9]. The final model has a total of $(|T| + |S|) \cdot |C||B| = 77 \times 39 \times 9 = 27027$ binary decision variables.

Our ILP generates the optimal timetable in 201.6 s on a stand-alone laptop, specifically a 8 GB Lenovo running Windows 10 with a 2.1 GHz processor. The student author has created a repository containing all of the Python code used in this paper, as well as the input files of the student course choices. The repository can be found at https://github.com/ifabrisarabellapark/Timetabling.

Here are the summary statistics.

|                            | Grade 12 | Grade 11 |
|----------------------------|----------|----------|
| Total students             | 25       | 33       |
| Requested core courses     | 64       | 103      |
| Enrolled core courses      | 64       | 103      |
| Requested elective courses | 129      | 151      |
| Enrolled elective courses  | 122      | 141      |

Every teacher is assigned to their desired set of courses, with at most 18 students in any class. Our timetable enrolls students in 167 out of 167 core courses (100%) and 263 out of 280 elective courses (94%), corresponding to an objective value of $167 \times 10 + 122 \times 3 + 141 \times 1 = 2177$.

Of the $|S| = 58$ students, 41 receive all of their desired courses, while the remaining 17 receive all courses except for one elective. This provably-optimal Master Timetable, presented below, was accepted by St. Margaret's School for the 2019–2020 academic year.

| Optimal timetable for St. Margaret's School | | |
|---|---|---|
| Block 1 | Block 2 | Block 3 |
| Culinary Arts 11A (DR) | Culinary Arts 11B (DR) | Comp. Prog. 11 (WF) |
| Core French Intro 11 (AS) | Life Education 11A (DH) | Life Education 11B (DH) |
| Philosophy 12 (JP) | Entrepreneurship 11 (CJ) | Spoken Language 11 (MC) |
| Economics 12 (SW) | Drama 11/12 (NC) | Japanese Intro 11 (MH) |
| Life Connections 12A (KD) | Life Connections 12B (KD) | Composition 12 (NP) |
| EarthSci 11/12 Combo (CJ) | Law Studies 12 (SW) | Spanish 12 (BP) |
| Pre-Calc 11/12 Combo (CT) | Social Justice 12 (JP) | Comp. Cultures 12 (SW) |
| Block 4 | Block 5 | Block 6 |
| Chemistry 11A (SB) | Art Studio 11 (LH) | Physics 11 (CT) |
| Active Living 11 (JS) | Fitness 11 (JS) | Life Sciences 11 (DR) |
| AP Studio Art 12 (LH) | Japanese 12 (MH) | Pre-Calculus 11A (WF) |
| English Studies 12A (NP) | Pre-Calculus 12A (CT) | English Studies 12B (NP) |
| AP Calculus 12 (CT) | Chemistry 12A (SB) | |
| Block 7 | Block 8 | Block 9 |
| Media 11 (NP) | Pre-Calculus 11B (WF) | Chemistry 11B (SB) |
| Creative Writing 11A (CN) | Creative Writing 11B (NP) | Core French 12 (AS) |
| Anatomy/Physio 12 (SB) | Chemistry 12B (SB) | Human Geog 12 (LZ) |
| Pre-Calculus 12B (CT) | EarthSci 11/12 Combo (CJ) | Univ/Grad Prep 12 (KD) |
| Physics 12 (WF) | Pre-Calc 11/12 Combo (CT) | . |

In the above timetable, the teacher's initials appear in parentheses, and multiple-section courses are indicated - e.g. Chemistry 11A and Chemistry 11B.

Our optimal Master Timetable allocates seven Grade 11 and 12 courses in blocks 1/2/3, and four or five Grade 11 and 12 courses in blocks 4/5/6/7/8/9, resulting in a symmetric well-balanced schedule. Once these course assignments were confirmed, it was easy to manually schedule the Grade 9 and 10 courses since these students take the same set of required courses, taught by the same set of teachers.

The final Senior School timetable was delivered in May 2019, months before the start of the 2019–2020 academic year. The early deliverable enabled the school guidance counsellor to have one-on-one meetings with each of the students before they left for the summer. The seventeen Grade 11 and 12 students who were not given one of their most-desired elective courses worked with the guidance counsellor to select an alternative course in the same subject area.

Given that our ILP solved to optimality in just over three minutes, there was no need to apply the time-reducing "course-bundling algorithm" we developed prior to receiving the final data sets from the school. However, we now provide this information to illustrate the effectiveness of this approach, especially when we have more than $|T| = 19$ teachers, $|S| = 58$ students, and $|C| = 39$ courses.

We create our conflict graph $G$ on our $|C_O| = 30$ one-section courses. For each pair of courses $c_1$ and $c_2$, we assign a weight of 100 if these two courses must be taught by the same teacher or if they must take place in the same classroom. For each student $s$ desiring both $c_1$ and $c_2$, we assign a weight of $\min(P_{s,c_1,b}, P_{s,c_2,b})$. This conflict graph $G$ has 30 vertices and 94 edges.

$G$ is a two-component graph, since the set of "short" courses that must be scheduled in blocks $1/2/3$ is disjoint from the set of "long" courses that must be scheduled in blocks $4/5/6/7/8/9$. We determine that $\chi(G) = 5 + 7 = 12$ using GrinPy, an open source program that quickly calculates graph invariants [11].

Let $G_1$ be the same graph as $G$, except we only include edges with weight more than 1. Then $G_1$ becomes a graph with 30 vertices and 69 edges, with $\chi(G_1) = 3 + 6 = 9$. Following the algorithm described in Sect. 5, we find a 3-coloring of the short courses and a 6-coloring of the long courses.

We then take each pair of one-section courses in the same bundle (e.g. $c_1$ and $c_2$), determine the assigned teachers for those two courses (e.g. $t_1$ and $t_2$) and add the following constraint to our ILP:

$$X_{t_1,c_1,b} = X_{t_2,c_2,b} \qquad \forall\, b \in B \qquad (9)$$

This constraint ensures that courses $c_1$ and $c_2$ are assigned to the same block, i.e., bundled together. Our modified ILP has only $|C_M| + 3 + 6 = 18$ courses instead of 39, since have cleverly combined one-section courses to virtually eliminate student conflicts.

This ILP is rapidly solved, requiring only 5.3 s of computation time. The generated timetable enrolls students in 167 out of 167 core courses (100%) and 246 out of 280 elective courses (88%), corresponding to an objective value of $167 \times 10 + 118 \times 3 + 128 \times 1 = 2152$.

Of course, the quality of the final timetable is dependent on the initial 3-coloring of the short courses and 6-coloring of the long courses. Thus, we run a simulation of 1000 trials, using the NetworkX package for graph coloring [21]. We use the built-in *greedy_color* method to randomly order the vertices and assign the first available color to each vertex. (If the greedy coloring of $G_1$ requires more than $\chi(G_1)$ colors, then we re-run the method until a valid coloring is attained.)

The average objective value of our 1000 trials is 2155.03, with a minimum of 2141 and a maximum of 2170. The average running time is 4.17 s, with a minimum of 1.4 s and a maximum of 8.9 s. Thus, for this data set, our graph-theoretic bundling algorithm reduces the total running time by 98% (from 201.6 s to an average of 4.17 s) while reducing the solution quality by just 1% (from an objective value of 2177 to an average of 2155.03).

In all 1000 trials, the generated timetable enrolls students in 167 out of 167 core courses (100%), and on average, in 249 out of 280 elective courses (89%). The latter result is lower than our rate of 94% in our optimal ILP, yet it is only moderately lower given the substantial time improvement.

Our bundling technique reduced the total running time by 98% at virtually no cost to the objective function. However, as we will see in the following section, our graph-theoretic bundling technique does not guarantee nearly-optimal solutions in all cases. We now explore the general applicability of our approach, and discuss the strengths and limitations of our methods for much more complex instances of School Timetabling.

# 7  Strengths and Limitations

At many large university campuses, each faculty is its own distinct entity, with their own buildings, classrooms, courses, professors, and registered students. Because there is little to no overlap between faculties (e.g. between the Business School and the Medical School), we can think of the university-wide timetabling process as solving hundreds of discrete timetabling problems, and combining these non-overlapping solutions to produce a single Master Timetable.

For example, the largest university in our province has over 50,000 undergraduate students, with nearly 20% in the Faculty of Science. Within the Faculty of Science are twelve different Departments, including botany, mathematics, and zoology. Each department chair is responsible for their own timetabling, and no effort is made to coordinate timetabling to serve the small minority of students who have multiple specializations, e.g. a mathematics and zoology double major.

This is why our two-step timetabling approach holds much promise, even for large institutions. Each department chair is responsible for timetabling a number of third-year and fourth-year courses, many of which will be single-section courses appealing only to the students pursuing a major offered by that department. If the department has thousands of registered students, generating an optimal timetable might be computationally intractable, but a nearly-optimal timetable could be generated using our graph-theoretic course bundling approach.

At St. Margaret's School, most of the teachers have their own dedicated classroom, which meant that we could ignore constraints such as "there are only 3 science labs available, and so we cannot schedule 4 lab-based courses in the same block". Fortunately, there exist ways we can ensure feasible room assignments in situations like this, by analyzing a specific partial transversal polytope [16], to complement our two-step timetabling algorithm.

We can also consider the effect of relaxing or eliminating the pre-allocation of teachers to courses. If each teacher has a ranked preference list of courses they wish to teach, and each teacher is qualified to teach dozens of courses, then this will significantly increase the running time of our algorithm. For St. Margaret's School, all courses were pre-assigned to a teacher. For a high school whose timetable we are creating now, only student preferences are to be considered in the optimization. Thus, all teacher assignments are made *after* the courses are assigned to blocks, and we will optimally assign teachers to courses to ensure everyone has a feasible schedule maximizing the total "preference score".

Despite the promising results we have found thus far, there are limitations of course bundling. As an extreme example, consider the following scenario where we wish to schedule 5 one-section courses $\{a, b, c, d, e\}$ and 2 two-section courses $\{x, y\}$ into a 3-block timetable. Suppose that the students wish to take the following set of courses, in any order:

| Student | Course 1 | Course 2 | Course 3 |
|---|---|---|---|
| $S_1$ | $a$ | $b$ | $e$ |
| $S_2$ | $c$ | $d$ | $e$ |
| $S_3$ | $a$ | $b$ | $x$ |
| $S_4$ | $c$ | $d$ | $x$ |
| $S_5$ | $a$ | $b$ | $y$ |
| $S_6$ | $c$ | $d$ | $y$ |
| $S_7$ | $a$ | $y$ | $e$ |
| $S_8$ | $d$ | $y$ | $e$ |
| $S_9$ | $b$ | $x$ | $e$ |
| $S_{10}$ | $c$ | $x$ | $e$ |

It is straightforward to show there is only one optimal timetable that enables all ten students to enroll in a desired course in each of the three blocks.

$$\text{Block } 1 = \{a, d, x\}$$
$$\text{Block } 2 = \{b, c, y\}$$
$$\text{Block } 3 = \{x, y, e\}$$

For example, student $S_{10}$ can register for $c$, $x$, $e$ in Blocks 2, 1, 3, respectively.

Let $C_O = \{a, b, c, d, e\}$ be our single-section courses. We construct the conflict graph $G$ by adding a weight of 1 to each pair of courses desired by a student. Below is the graph $G$ with all edges of non-zero weight (Fig. 1).

Clearly, $\chi(G) = 3$, and there are two possible colorings up to isomorphism: the color classes are either $[\{a, d\}, \{b, c\}, \{e\}]$ or $[\{a, c\}, \{b, d\}, \{e\}]$. In the former, we bundle $a$ and $d$ into Block 1 and bundle $b$ and $c$ into Block 2. This immediately generates the optimal solution shown on the previous page.

In the latter, we bundle $a$ and $c$ into Block 1 and bundle $b$ and $d$ into Block 2. In this case, it is impossible to create a 3-block timetable that allows every student to get into all of their desired courses. The best timetable, shown below, necessitates a conflict for students $S_8$ and $S_{10}$.

$$\text{Block } 1 = \{a, c, x\}$$
$$\text{Block } 2 = \{b, d, y\}$$
$$\text{Block } 3 = \{x, y, e\}$$

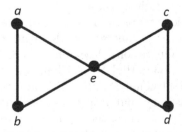

**Fig. 1.** The conflict graph of the single-section courses

Thus, our course bundling algorithm has a 50% chance of creating a timetable that reduces by 20% the number of students getting into all of their desired courses. While this is an extreme example, this result highlights the importance of not relying on a single coloring, since an unlucky assignment of colors to one-section courses may generate a sub-optimal Master Timetable.

However, if we run our course bundling algorithm $n$ times on this instance, and accept the best result of all $n$ iterations, then we have a probability of $1 - \frac{1}{2^n}$ of generating the optimal timetable.

In general, if there are $|B|$ blocks in the timetable, at least one $|B|$-coloring of the conflict graph must be identical to the optimal solution. By running our course bundling algorithm many times and selecting the best result, we increase the probability of producing the best possible timetable.

For St. Margaret's School, course bundling reduced the computation time by an average of 98% while having a minimal reduction on the quality of the generated timetable. We are optimistic that this technique can be applied to larger timetabling instances, to produce solutions to previously-intractable timetabling instances that are close to optimal.

# 8    Conclusion

In this paper, we presented an ILP-based model to optimally solve a real-life instance of the Post-Enrollment Course Timetabling Problem. Our Master Timetable for St. Margaret's School (SMS) enrolled Grade 11 and 12 students into 100% of their core courses and 94% of their most desired elective courses.

We also developed a "pre-processing phase" that partitioned one-section courses into discrete bundles using graph coloring. On the SMS data set, this approach decreased the total running time by 98%, with only a 1% reduction in the value of our objective function.

The collaboration with St. Margaret's School was a tremendous success, and our Master Timetable has been well-received by the school leadership, and more importantly, from their teachers and students.

During the past month, we have been approached by three different independent high schools in British Columbia, who have learned about the new "happiness-maximizing timetable" at St. Margaret's School and have requested our services to design their 2020–2021 Master Timetable. We are looking forward to these collaborations and learning ways to further improve the speed and quality of our timetabling algorithm on larger data sets.

There are many directions for future research. One natural direction is to test the performance of our two-step timetabling algorithm on benchmark instances maintained by scholars in Operations Research. Specifically, we propose testing our algorithm on XHSTT-2011, a High School Timetabling Archive containing 21 real-life instances from eight countries [25]. This archive includes an evaluator that checks syntax consistency and returns a cost value proportional to the number of violated constraints. Thus, the online evaluator allows us to compare the quality of our solution against multiple published algorithms.

To generate the best possible results, we will need to apply more sophisticated ways to color the nodes of the conflict graph beyond our approach of randomly ordering the vertices and assigning the first available color to each vertex. There exist systematic and deterministic approaches on how to color the nodes, including ordering the vertices by decreasing degree. The Python NetworkX package contains various algorithms, including one based on "degree saturation" [1], where the vertex order is determined dynamically, based on the number of colors that cannot be used because of conflicts with previously colored vertices.

Another direction is to commercialize a cloud-based application to allow school administrators to run our timetabling algorithm on their own machines. The current product requires the client to email us two Excel documents from which our Python program generates the optimal Master Timetable in the form of another Excel document. We are excited by the prospect of developing a cloud-based solution that can be deployed by the administration of high schools and universities from around the world, especially those institutions who wish to create timetables that optimize for student course preferences.

**Acknowledgments.** The authors thank the reviewers for their insightful comments that significantly improved the final version of this paper. We are grateful to Darlene DeMerchant and Megan Hedderick at St. Margaret's School for making this collaboration possible. Finally, we acknowledge that the student author's research was sponsored by a Student Project Grant awarded by the Research and Scholarly Works Committee at Quest University Canada.

# References

1. Brélaz, D.: New methods to color the vertices of a graph. Commun. ACM **22**(4), 251–6 (1979)
2. Burke, E.K., Mareček, J., Parkes, A., Rudová, H.: A supernodal formulation of vertex colouring with applications in course timetabling. Ann. Oper. Res. **179**(1), 105–130 (2010). https://doi.org/10.1007/s10479-010-0716-z
3. Cambazard, H., Hébrard, E., O'Sullivan, B., Papadopoulos, A.: Local search and constraint programming for the post enrolment-based course timetabling problem. Ann. Oper. Res. **194**(1), 111–135 (2012). https://doi.org/10.1007/s10479-010-0737-7
4. Carter, M.W.: A comprehensive course timetabling and student scheduling system at the University of Waterloo. In: Burke, E., Erben, W. (eds.) PATAT 2000. LNCS, vol. 2079, pp. 64–82. Springer, Heidelberg (2001). https://doi.org/10.1007/3-540-44629-X_5
5. Chiarandini, M., Fawcett, C., Hoos, H.H.: A modular multiphase heuristic solver for post enrolment course timetabling. In: PATAT (2008)
6. Demirović, E., Stuckey, P.J.: Constraint programming for high school timetabling: a scheduling-based model with hot starts. In: van Hoeve, W.-J. (ed.) CPAIOR 2018. LNCS, vol. 10848, pp. 135–152. Springer, Cham (2018). https://doi.org/10.1007/978-3-319-93031-2_10
7. Even, A.S.S., Itai, A., Shamir, A.: On the complexity of timetable and multicommodity flow problems. SIAM J. Comput. 5(4), 691–703 (1976)

8. Fonseca, G.H., Santos, H.G., Carrano, E.G., Stidsen, T.J.: Modelling and solving university course timetabling problems through XHSTT. In: PATAT, vol. 16, pp. 127–138 (2016)
9. Google OR-Tools: Fast and portable software for combinatorial optimization. https://developers.google.com/optimization. Accessed 17 Feb 2020
10. Gotlieb, C.C.: The construction of class-teacher timetables. In: IFIP Congress, Amsterdam, vol. 62, pp. 73–77 (1963)
11. GrinPy: A NetworkX extension for calculating graph invariants. https://pypi.org/project/grinpy/. Accessed 17 Feb 2020
12. Hébrard, E., Katsirelos, G.: A hybrid approach for exact coloring of massive graphs. In: Rousseau, L.-M., Stergiou, K. (eds.) CPAIOR 2019. LNCS, vol. 11494, pp. 374–390. Springer, Cham (2019). https://doi.org/10.1007/978-3-030-19212-9_25
13. Karp, R.M.: Reducibility among combinatorial problems. In: Miller, R.E., Thatcher, J.W., Bohlinger, J.D. (eds.) Complexity of Computer Computations. IRSS, pp. 85–103. Springer, Boston (1972). https://doi.org/10.1007/978-1-4684-2001-2_9
14. Kristiansen, S., Sørensen, M., Stidsen, T.R.: Integer programming for the generalized high school timetabling problem. J. Sched. 18(4), 377–392 (2014). https://doi.org/10.1007/s10951-014-0405-x
15. Kristiansen, S., Stidsen, T.R.: Elective course student sectioning at Danish high schools. Ann. Oper. Res. 239(1), 99–117 (2014). https://doi.org/10.1007/s10479-014-1593-7
16. Lach, G., Lübbecke, M.E.: Optimal university course timetables and the partial transversal polytope. In: McGeoch, C.C. (ed.) WEA 2008. LNCS, vol. 5038, pp. 235–248. Springer, Heidelberg (2008). https://doi.org/10.1007/978-3-540-68552-4_18
17. Lewis, R., Paechter, R., McCollum, B.: Post enrolment based course timetabling: a description of the problem model used for track two of the second international timetabling competition (2007)
18. Liu, Y., Zhang, D., Leung, S.C.: A simulated annealing algorithm with a new neighborhood structure for the timetabling problem. In: Proceedings of the First ACM/SIGEVO Summit on Genetic and Evolutionary Computation, pp. 381–386 (2009)
19. Méndez-Díaz, I., Zabala, P., Miranda-Bront, J.J.: An ILP based heuristic for a generalization of the post-enrollment course timetabling problem. Comput. Oper. Res. 76, 195–207 (2016)
20. Minh, K.N.T.T., Thanh, N.D.T., Trang, K.T., Hue, N.T.T.: Using Tabu search for solving a high school timetabling problem. In: Nguyen, N.T., Katarzyniak, R., Chen, S.M. (eds.) Advances in Intelligent Information and Database Systems. SCI, vol. 283, pp. 305–313. Springer, Heidelberg (2010). https://doi.org/10.1007/978-3-642-12090-9_26
21. NetworkX: Software for complex networks. https://networkx.github.io/. Accessed 17 Feb 2020
22. Nothegger, C., Mayer, A., Chwatal, A., Raidl, G.R.: Solving the post enrolment course timetabling problem by ant colony optimization. Ann. Oper. Res. 194(1), 325–339 (2012). https://doi.org/10.1007/s10479-012-1078-5
23. Pillay, N.: A survey of school timetabling research. Ann. Oper. Res. 218(1), 261–293 (2014). https://doi.org/10.1007/s10479-013-1321-8
24. Post, G., et al.: An XML format for benchmarks in high school timetabling. Ann. Oper. Res. 194(1), 385–397 (2012). https://doi.org/10.1007/s10479-010-0699-9

25. Post, G., et al.: XHSTT: an XML archive for high school timetabling problems in different countries. Ann. Oper. Res. **218**(1), 295–301 (2011). https://doi.org/10.1007/s10479-011-1012-2
26. Raghavjee, R., Pillay, N.: An application of genetic algorithms to the school timetabling problem. In: Proceedings of the 2008 Annual Research Conference of the South African Institute of Computer Scientists and Information Technologists on IT Research in Developing Countries: Riding the Wave of Technology, pp. 193–199 (2008)
27. Sørensen, M., Dahms, F.H.W.: A two-stage decomposition of high school timetabling applied to cases in Denmark. Comput. Oper. Res. **1**(43), 36–49 (2014)
28. St. Margaret's Boarding School Viewbook. https://www.stmarg.ca/wp-content/uploads/SMS-Boarding-Viewbook.pdf. Accessed 17 Feb 2020
29. Van Den Broek, J., Hurkens, C., Woeginger, G.: Timetabling problems at the TU Eindhoven. Eur. J. Oper. Res. **196**(3), 877–885 (2009)

# Adaptive CP-Based Lagrangian Relaxation for TSP Solving

Nicolas Isoart[(✉)] and Jean-Charles Régin

Université Côte d'Azur, CNRS, I3S, Nice, France
nicolas.isoart@gmail.com, jcregin@gmail.com

**Abstract.** M. Sellmann showed that CP-based Lagrangian relaxation gave good results but the interactions between the filtering algorithms and the Lagrangian multipliers were quite difficult to understand. In other words, it is difficult to determine when filtering algorithms should be triggered. There are two main reasons for this: the best multipliers do not lead to the best filtering and each filtering disrupts the solving of the Lagrangian multiplier problem. In this paper, we study these interactions for the Traveling Salesman Problem (TSP) because the resolution of the TSP in CP is mainly based on a Lagrangian relaxation. We propose to match the calls to the filtering algorithms with the strong variations of the objective value. In addition, we try to avoid oscillations of the objective function. We introduce Scope Sizing Subgradient algorithm, denoted by SSSA, which is an adaptive algorithm, that implements this idea. We experimentally show the advantage of this approach by considering different search strategies or additional constraints. A gain of a factor of two in time is observed compared to the state of the art.

## 1 Introduction

The Traveling Salesman Problem (TSP) consists in finding a circuit of minimum total weight that visits all vertices of a graph. It is a very common problem in the industry, directly as in routing problems or indirectly as in scheduling problems where cities correspond to tasks that should be performed and arcs to transition times between tasks. It is often found associated with other constraints such as time windows that specify the time period during which a node can be visited, or precedence constraints between nodes. When side constraints are involved the famous TSP solver Concorde [1] can no longer be applied and constraint programming (CP) becomes a competitive approach [11,17].

The TSP is expressed in CP by the weighted circuit constraint which ensures that a given graph contains a set of arcs that form a circuit and whose sum of weights (*i.e.* the weight of the circuit) is less than a given value. The performance of the filtering algorithms associated with this constraint is therefore very important because they will immediately have an impact on the solving of many industrial problems.

Several filtering algorithms (FAs) can be applied to the weighted circuit constraint [6,10,21]. However, according to recent papers [8,9], the best filtering is

© Springer Nature Switzerland AG 2020
E. Hebrard and N. Musliu (Eds.): CPAIOR 2020, LNCS 12296, pp. 300–316, 2020.
https://doi.org/10.1007/978-3-030-58942-4_20

the one proposed by Benchimol et al. [3]. It is based on the 1-tree relaxation of Held and Karp, which is a common Lagrangian Relaxation of the TSP.

Lagrangian relaxation (LR) is a relaxation method which approximates $P$, a difficult problem of constrained optimization, by a simpler problem [2]. It consists in removing difficult constraints by integrating them into the objective function. It is therefore appropriate for solving problems where the constraints can be partitioned into two parts: a set of constraints that can be easily solved and a set that contains the other constraints. The constraints of the second group are moved to the objective, so only constraints that are easy to solve remain. The satisfaction of difficult constraints is achieved by penalizing them in the objective by introducing a cost for each constraint that measures the distance to satisfaction and by multiplying this cost by a multiplier. For each set of multipliers the optimal solution of the LR of $P$ is a lower bound of the optimal solution of $P$. These lower bounds are often used in conjunction with a branch-and-bound algorithm to accelerate the search for an optimal solution of $P$.

The LR of the TSP [14] can be defined as follows. The TSP can be seen as the search for a 1-tree (*i.e.* a node associated with two arcs joining a spanning tree) of minimum weight such that each node of the 1-tree has a degree two. The search for a minimum weight 1-tree is equivalent to the search for a minimum spanning tree, and so can be solved in polynomial time. However, we do not know how to deal at the same time with the constraint on degrees, since the decision version of the TSP is NP-Complete. The Lagrangian relaxation transfers these degree constraints into the objective. Thus for each node $v$, the expression $\mu_i(degree(v) - 2)$ with $\mu \geq 0$ is added to the objective, where the degree of $v$ is expressed as the number of the arcs in the 1-tree with $v$ as an endpoint.

Since for any set of multipliers $\mu$ the optimal value of the LR of $P$ is a lower bound of the optimal value of $P$, it can be used to remove some values of the variables. Consider $UB$, an upper bound of the optimal solution of $P$ (for example any solution of $P$, therefore not necessarily optimal), and $x = a$ an assignment, if for $x = a$ the optimal value of the LR of $P$ is greater than $UB$ then we can remove $a$ from $D(x)$ since we know that $x = a$ does not belong to the optimal solution. From this idea, the CP-based Lagrangian relaxation has been introduced [25] and successfully used to solve many problems [4,5,7,12,16,19]. It consists in modeling the problem so that one or more cost based FAs can be used on the easy part of the problem. Difficult constraints are moved to the objective function and the cost based FAs are used when looking for good multipliers.

Sellmann made two important observations about the relationship between the LR and FAs [24]:

- Suboptimal multipliers can be more efficient for filtering than the optimal multipliers for the original problems.
- It is not clear whether FAs should actually take place during the optimization of the Lagrangian multipliers, because the standard approach for the optimization of the multipliers are not guaranteed to be robust enough to

enable a change (*i.e.* the removal of a value) of the underlying subproblem during the optimization.

These observations show the complexity of the interactions between FAs and multipliers, which have important consequences, such as losing the monotony so dear to CP[1].

In addition, it is important to note that the CP-based LR is usually associated with a branch-and-bound algorithm. We therefore have no reason to seek to converge the LR towards the optimum. The lower bounds it provides are sufficient for our purpose (*i.e.* having the most effective FAs) and we will obtain the optimal solution thanks to the search algorithm.

From these considerations, we can formulate the problem of the interaction between LR and FAs by the following question: for which set of multipliers should FAs be called and how do we get them?

Most of the articles in the literature using CP-based LR do not address this issue and it is by reading the source code of the programs that we discover precisely when FAs are called. Some authors (L-M Rousseau and X. Lorca) have confirmed that the call conditions were determined experimentally after numerous tests.

In this article, we study some of the interactions between FAs and LR and propose a method to determine when FAs should be called.

First of all, we propose to use a subgradient optimization algorithm, because it gives us access to suboptimal multipliers that can be quickly computed. The problem of the slow convergence of this type of algorithm does not arise in our case since we also use a search procedure.

By doing so, the problem that needs to be solved becomes: when are the FAs called in the subgradient algorithm?

The subgradient algorithms used in CP-based LR are most often variants of Beasley's [2]. Conceptually, they proceed by successive iterations for different calculation accuracies, called agility. At each iteration the agility is divided by a power of 2. For a given agility value, the subgradient algorithm iteratively adjust the Lagrangian multipliers to find values that improve the lower bound. For a given agility value, the number of internal iterations, which we call scope, is the unknown we are looking for.

Thus, the problem becomes: for which scope values FAs should be called?

To answer this question, we propose to study the variation in the value of the LR objective as a function of the scope. We have observed that this value often stagnates or oscillates and we have experimentally measured that these variations do not bring anything in terms of filtering. Thus, we recommend to detect stagnations and oscillations and immediately stop iterations when they occur, then call the FAs after these iterations. Stopping multiplier computations is not a problem, because convergence towards optimality is done using a search procedure and not only with the LR. However, it is important to note that

---

[1] Normally, in CP, when $F_2$, a FA, is added to $F_1$, another FA, all values eliminated by $F_1$ are also eliminated by the combination of $F_1$ and $F_2$.

stopping multiplier computations prematurely can lead to a weaker bound that leads to a larger search tree. Thus, we need to find a good trade-off. The Scope sizing subgradient algorithm (SSSA), which implements this approach is detailed in this paper. The experimental results show an improvement in the resolution of the state-of-the-art instances of the TSP about a factor of two.

The article is organized as follows. We recall some definitions. Then we study the behavior of the optimal value of the LR for some multipliers corresponding to scope values and introduce SSSA, a scope sizing subgradient algorithm implementing the scope determination we propose. Before concluding, we give some experimental results.

## 2 Preliminaries

**Lagrangian Relaxation.** The Lagrangian relaxation procedure uses the idea of relaxing some difficult constraints by bringing them into the objective function with associated Lagrangian multipliers $\mu \geq 0$. The application of LR to a mixed integer program can be defined as follows.

$$Z = \min c \cdot x \qquad\qquad Z_{LR}(\mu) = \min c \cdot x + \mu(A_1 \cdot x - b_1)$$
$$\text{s.t.} \begin{cases} A_1 \cdot x \leq b_1 \\ A_2 \cdot x \leq b_2 \\ x \in X \end{cases} \longrightarrow \text{s.t.} \begin{cases} A_2 \cdot x \leq b_2 \\ x \in X \end{cases}$$

We will denote by $LR(P)$ the Lagrangian relaxation of the problem $P$ and by $LR(P, \mu)$ the LR of $P$ associated with the multiplier set $\mu$.

Assume that the constraint $A_1 \cdot x \leq b_1$ is difficult to solve whereas constraint $A_2 \cdot x \leq b_2$ is easy. LR moves the first one into the objective. If $A_1 \cdot x \leq b_1$ is violated then $A_1 \cdot x > b_1$ and so $d = A_1 \cdot x - b_1 > 0$. This value $d$ measures the distance to the satisfaction of this constraint. Intuitively, the larger $d$ is, the more the constraint should be penalized and the smaller $d$ is, the less the constraint should be penalized. This result is obtained by adding the value $(A_1 x - b_1)$ in the objective. Lagrangian relaxation proposes to use a non-negative multiplier $\mu$ for each constraint introduced in the objective. The interest of the multipliers is shown by the following property:

**Property 1.** *For any vector $\mu$, the value of $Z_{LR}(\mu)$ is a lower bound of $Z$.*

The Lagrangian multiplier problem consists of searching for the best multipliers. The two most popular types of methods for solving it are the subgradient and the bundle methods [13]. This second type of method converges faster than the first one. Since we only need to use suboptimal multipliers to filter we will focus our attention on the first type.

**Subgradient Algorithm.** Subgradient algorithms work in steps and reoptimize locally the multipliers according to a certain precision, called agility. Beasley's algorithm [2] is one of the most widely used. Its structure is depicted in Algorithm 1. It calls Function SOLVELR which computes the optimal value of the LR for a given set of multipliers and define new multipliers. The number of agility values is defined by #*agility*, which is close to 6 most of the time. Usually, and as mentioned in Algorithm 1, the agility starts at 2 and is divided by 2 at each iteration. For a given agility value, #*scope* is the maximal number of internal iterations of the LR, called scope. For each value of scope, the optimal value of the LR is computed and multipliers are updated accordingly. We have also introduced Function STOPCONDITION(...), which takes some parameters and tests if some stopping conditions of the current loop are met. For instance, the program can be stopped when there is no more progression of the objective function value according to the current $UB$.

The subgradient algorithm (FLR) used by Fages et al. [9] in their experiments corresponds to the values of the parameters #*agility* = 5 and #*scope* = 30 of Algorithm 1 and makes the agility slightly differently since it uses the following update formula: $\pi \leftarrow \pi/\beta$; $\beta \leftarrow \beta/2$ with $\beta = 1/2$ at initialization. It should also be noted that Fages et al. repeat the call to the algorithm as long as the lower bound of the 1-tree is increased.

---

**Algorithm 1:** SUBGRADIENTSOLVE algorithm of Beasley

---

SUBGRADIENTBEASLEY($P,Z_{ub},\mu$): returns ($\mu^{k+1}, x^k$)

    $\pi \leftarrow 2$ // Subgradient agility ;
    $k \leftarrow 0$ ;
    $\mu^0 \leftarrow \mu$ // We start with the current values of multipliers ;
    **for each** *iterAgility* = 1..#*agility* **do**
        $scope \leftarrow 0$ ;
        **while** $scope <$ #*scope* **do**
            ($\mu^{k+1}, x^k, Z^k$) $\leftarrow$ SOLVELR($P,Z_{ub}, \pi, \mu^k, k$) ;
            **if** $Z^k = Z_{ub}$ **then return** ($\mu^{k+1}, x^k$)  // optimal sol. ;
            $k \leftarrow k + 1$ ;
            $scope \leftarrow scope + 1$ ;
            **if** STOPCONDITION(...) **then break** ;
        $\pi \leftarrow \pi/2$ ;
    **return** ($\mu^{k+1}, x^k$) ;

SOLVELR($P,Z_{ub}, \pi, \mu^k, k$): returns ($\mu^{k+1}, x^k, Z^k$)

    $x^k \leftarrow$ solve $LR(P, \mu^k)$ to optimality ;
    // the optimal value of $LR(P)$ is computed ;
    $R \leftarrow |\mu|$ // Number of relaxed constraints ;
    $Z^k \leftarrow obj(x^k) + \sum_{1 \leq r \leq R} \mu_r^k obj_r(x^k)$ ;
    $\Delta^k \leftarrow \frac{\pi(Z_{ub} - Z^k)}{\sum_{1 \leq r \leq R}(obj_r(x^k))^2}$ // step ;
    // Multipliers are updated ;
    $\forall 1 \leq r \leq R : \mu_r^{k+1} \leftarrow \max(0, \mu_r^k + \Delta^k obj_r(x^k))$ ;
    **return** ($\mu^{k+1}, x^k, Z^k$) ;

---

**CP-Based Lagrangian Relaxation.** According to Sellmann [24], CP-based LR consists in the following procedure: Assuming we are given a linear optimization problem that consists in the conjunction of two constraint families $A$ and $B$ for which an efficient filtering algorithm $prop(B)$ is known, we try to optimize Lagrangian multipliers for $A$ and use $prop(B)$ for filtering in each Lagrangian subproblem $LR(P, \mu)$.

It is not necessary for constraints $A$ or $B$ to be linear (something that is not imposed in CP). We need to ensure that the relaxation we calculate for any multiplier set is a relaxation of $P$. So, we just need to make sure that $prop(B)$ remains valid when the objective becomes that of the LR.

Sellmann defined a particular consistency based on the continuous relaxation of $P$, but it does not matter in this paper. He also defined the following property:

**Property 2.** *Suboptimal multipliers can be more efficient for filtering than the optimal multipliers for the original problems.*

This property is explained by the fact that a value $x = a$ can be removed when the optimal value of $P \wedge (x = a)$ is greater than $UB$, a given upper bound. By considering the Lagrangian relaxation we consider the problem $LR(P)$ and not $LR(P \wedge (x = a))$ and there is no reason why the best multipliers for $LR(P)$ should also be the best for $LR(P \wedge (x = a))$.

In CP, it is also possible to express the violation of the constraint in different ways, we can also decide not to measure the distance to the violation [12]. Since we will relax only equality constraints we will not detail it here.

**TSP Model.** The TSP consists in searching for an Hamiltonian cycle whose sum of the cost of its edges is minimum. We model it with the weighted circuit constraint (WCC) [3], which is based on the Held and Karp LR of the TSP [14]:

A 1-tree of a graph $G$ is formed by a node $x$, two edges having $x$ as an extremity and a spanning tree of $G - x$ (the graph $G$ in which $x$ has been removed). Held and Karp proposed to represent the TSP as the search for a 1-tree where each vertex has a degree two and whose sum of the costs of the edges it contains is minimum. Searching for a minimum 1-tree is an easy task because it is related to the search for a minimum spanning tree. However, the constraints on the degree modify the complexity of the problem. Held and Karp proposed to use the Lagrangian relaxation on these constraints (the degree of a node $x$ is expressed as the number of arcs taken with $x$ as an endpoint).

Different CP models have been tested [8] and the conclusion is that the WCC give the best results for the TSP. Recently, the resolution of TSP has been improved by using the k-cutset [15] constraint in conjunction with the WCC. This constraint is mainly based on the structure of the graphs, more precisely the edge cuts of G. Different search strategies have also been tested [9] for the CP models and the conclusion is that three search strategies associated with the LCFirst policy (*i.e.* keep one of the endpoints of the last branching edge and selects the edges from the neighborhood of the kept node) give similar results that are better than the others: maxCost (*i.e.* selects the edge with the maximum cost), minReplacementCost (*i.e.* selects the edge whose removal involves the smallest increase in the relaxation value), minDeltaDegree (*i.e.* selects the edge for which the sum of the endpoint degrees in the upper bound minus the sum of the endpoint degrees in the set variable lower bound is minimal). Thus, it can be considered that the WCC and k-cutset used in conjunction with one of the previous search strategies is the state of the art of TSP modeling in CP.

# 3  Scope Sizing Subgradient Algorithm

As mentioned previously, our problem can be summarized by the question: for which scope values FAs should be called?

In order to answer this question, we propose to observe the LR optimal values computed by Beasley's algorithm according to some scope values. We set $\#agility = 6$.

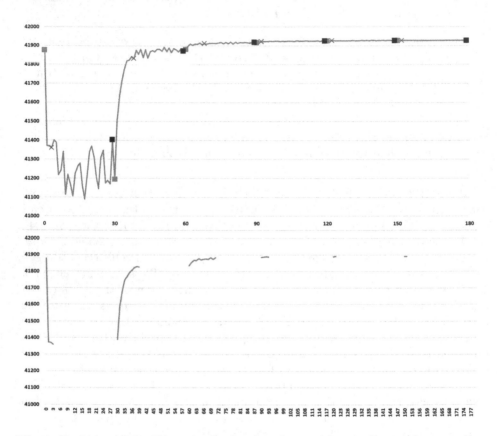

**Fig. 1.** Evolution of the LR optimal value (on the y-axis) according to the *scope* (on the x-axis). (top graph) Beasley's algorithm with $\#scope = 30$. A segment between green and dark corresponds to one agility value. Computations after red crosses are identified as useless. (bottom graph) Scope sizing subgradient algorithm. Computations are stopped at red crosses of the top graph. (Color figure online)

Figure 1 (top) is quite representative of what is frequently observed, namely:

- a strong variation followed by oscillations (this is the case for the two first agility values)
- a weak variation followed by oscillations (third agility value)
- almost no variation (the last three agility values)

We have also observed weak or strong variations without oscillations.

For FAs to be effective, successive FAs must be called with relatively large variations, otherwise it is unlikely that successive filtering will remove more values than the previous. The orientation of the variations should not play a role since we are mainly looking for various multipliers. It is therefore certainly interesting to trigger the FAs after a strong variation in the objective value.

What about oscillations? They do not really provide any information. The multipliers are changed but very slightly between two oscillations. Nor do they provide much in terms of boundaries. We therefore propose to avoid them as much as possible.

There is still the case of the absence of variations or very small variations (*i.e.* stagnations). It is in our interest to stop the calculations as soon as possible because the multipliers change very little.

We therefore propose to proceed as follows:

**Rule 1.** As soon as we no longer measure any variation in the objective value, we stop changing the scope value and trigger the FAs.

Figure 1 represents our choice using red crosses. Indeed, the red crosses mark the end of the search for new multipliers. If we implement our proposal then we get the results of the bottom graphic of Fig. 1 that we made coincide with the top graphic to show the difference. We can see that there is little difference in the calculated objective values in the end. There is a slight decrease for many of the avoided calculations. This can be due to the stop of the computations as well as the call of the FAs that modify the problem to be solved and therefore impacts the LR.

One could objectively ask whether it is appropriate to trigger the FAs when almost no variation is measured. According to our tests, it doesn't change much to trigger the FAs in this case, for the sake of simplicity of the algorithm we decided to trigger the FAs when the multipliers are no longer searched. Note that FAs are called only once per agility value.

The scope sizing subgradient algorithm (see Algorithm 2) is a possible implementation of our approach. It is a direct adaptation of Beasley's algorithm. Note that the number of agility values that are considered is 6. We tested different values but we did not observe enough changes to justify the introduction of an additional parameter. The maximum value of scope is 12 because we almost always observe oscillations or stagnation for larger values. The stop conditions are not tested for all scope values but only one time out of two (internal loop of the $q$ variable). This allows us to detect a large part of the oscillations. We measure for two iterations the variations and if the sum of these two variations does not deviate enough from the value at the beginning of the two iterations then we no longer change the scope value. In this case, either an absence of variations or two variations of the same amplitude in opposite directions (*i.e.* an oscillation) will be detected. We define the minimum deviation as 1% of the difference between the current upper bound and the current objective value.

**Algorithm 2:** SSSA: Scope Sizing Subgradient Algorithm

SSSA($P$,$Z_{ub}$,$\mu$): returns ($\mu^{k+1}$, $x^k$) ;
  $\pi \leftarrow 2$ // subgradient agility ;
  $k \leftarrow 0$ ;
  $\mu^0 \leftarrow \mu$ // We start with the current values of multipliers ;
  **for each** $iterAgility = 1..6$ **do**
    $scope \leftarrow 0$ ;
    **while** $scope < 12$ **do**
      $mean \leftarrow 0$ ;
      $d \leftarrow Z_{ub} - Z^k$    // distance to UB ;
      **for each** $q = 1..2$ **do**
        $prevBound \leftarrow Z^k$ ;
        ($\mu^{k+1}$, $x^k$, $Z^k$) $\leftarrow$ SOLVELR($P$,$Z_{ub}$, $\pi$, $\mu^k$, $k$) ;
        **if** $Z^k = Z_{ub}$ **then** return ($\mu^{k+1}$, $x^k$) // optimal sol. ;
        $mean \leftarrow mean + Z^k - prevBound$ ;
        $k \leftarrow k + 1$ ;
        $scope \leftarrow scope + 1$ ;
      **if** $|mean/2| \leq 0.01d$ **then** break ;
    // trigger the filtering algorithms ;
    RUNPROPAGATION($P$, $x^{k-1}$, $Z_{ub}$, $\mu^k$) ;
    **if** *solver failed* **then** return nil ;
    $\pi \leftarrow \pi/2$ ;
  return ($\mu^{k+1}$, $x^k$) ;

## 4 Experiments

The algorithms have been implemented in Java 11 in a locally developed CP solver. The experiments were performed on a Windows 10 machine using an Intel Core i7-3930K CPU @ 3.20 GHz and 64 GB of RAM. The reference instances are from the TSPLib [23], a library of reference graphs for the TSP and the set of instances is the same as in [9] that can be seen as state-of-the-art instances. All instances considered are symmetrical graphs. The name of each instance is suffixed by its number of nodes. The TSP is modeled by the WCC using CP-based LR in conjunction with the k-cutset constraint [15].

### 4.1 Scope Impacts

The best scope for an instance is the scope value that gives the smallest solving time for the given instance. It should be noted that a predefined scope is always the same for all agility values. For a set of instances, we define the best global scope as the scope value that obtains the best geometric average of the solving times of the instances.

First of all, we observe the impact of different global scope values for the three best search strategies. The results (See Table 1) show that it is necessary to find a trade-off in the mean time and number of backtracks. Filtering with refined boundaries (*i.e.* with the largest scope) reduces the number of backtracks but this reduction is not worthwhile. It is also important to note that there is no monotony: for example, a scope equal to 8 for LCFirstMinRepCost has less backtracks than 4 and 10.

**Table 1.** Comparison of solving times (in s) and backtracks (in 1000 s) between predefined scope values for Algorithm 1.

| scope | LCFirstMaxCost | | | | LCFirstMinRepCost | | | | LCFirstMinDeltaDeg | | | |
|---|---|---|---|---|---|---|---|---|---|---|---|---|
| | mean | | geo mean | | mean | | geo mean | | mean | | geo mean | |
| | time | #*bk* | time | #bk | time | #bk | time | #bk | time | #bk | time | #bk |
| 2 | 221.4 | 132.5 | 13.3 | 9.1 | 81.4 | 50.4 | 6.9 | 4.2 | 106.9 | 39.0 | 8.6 | 3.6 |
| 4 | 185.3 | 101.0 | 10.0 | 5.9 | 41.8 | 20.9 | 5.0 | 2.5 | 66.3 | 22.3 | 6.2 | 2.1 |
| 6 | 168.7 | 86.9 | 10.5 | 5.8 | 33.5 | 15.5 | 4.4 | 2.0 | 33.9 | 10.0 | 4.8 | 1.5 |
| 8 | 157.6 | 71.2 | 10.2 | 5.1 | 39.3 | 16.9 | 4.4 | 1.7 | 71.8 | 19.6 | 5.4 | 1.6 |
| 10 | 164.1 | 73.7 | 10.8 | 4.9 | 36.2 | 14.1 | 4.6 | 1.8 | 53.1 | 13.0 | 6.0 | 1.8 |
| 12 | 163.1 | 59.8 | 10.1 | 4.2 | 37.8 | 13.3 | 4.4 | 1.5 | 71.4 | 16.5 | 4.9 | 1.2 |

**Table 2.** Comparison of solving times (in s) between best and worst predefined scope values for Algorithm 1 with LCFirstMinDeltaDeg strategy.

| Instance | LCFirstMinDeltaDeg | | | | | |
|---|---|---|---|---|---|---|
| | Best time | | Worst time | | ratio | scope = 6 |
| | scope | time | scope | time | w/b | time |
| a280 | 12 | 12.1 | 4 | 35.7 | 3.0 | 13.6 |
| bier127 | 10 | 0.2 | 10 | 0.4 | 1.9 | 0.3 |
| brg180 | 10 | 0.3 | 2 | 22.9 | 72.6 | 0.4 |
| ch130 | 8 | 1.9 | 12 | 4.6 | 2.4 | 2.7 |
| ch150 | 12 | 1.8 | 4 | 4.3 | 2.4 | 2.4 |
| d198 | 12 | 16.6 | 2 | 23.1 | 1.4 | 17.8 |
| gr120 | 8 | 0.6 | 10 | 0.9 | 1.5 | 0.7 |
| gr137 | 12 | 2.6 | 2 | 4.1 | 1.6 | 3.0 |
| gr202 | 6 | 2.2 | 2 | 3.1 | 1.4 | 2.2 |
| gr96 | 8 | 0.5 | 4 | 1.6 | 3.4 | 0.8 |
| kroA100 | 4 | 1.6 | 2 | 2.8 | 1.7 | 2.1 |
| kroA150 | 10 | 4.6 | 2 | 13.3 | 2.9 | 8.2 |
| kroB100 | 4 | 1.5 | 8 | 5.9 | 3.9 | 3.5 |
| kroB150 | 12 | 291.4 | 2 | 1163.2 | 4.0 | 358.6 |
| kroB200 | 10 | 186.3 | 2 | 475.6 | 2.6 | 230.0 |
| kroC100 | 2 | 0.9 | 6 | 1.7 | 2.0 | 1.7 |
| kroD100 | 12 | 0.3 | 4 | 0.6 | 1.7 | 0.4 |
| kroE100 | 4 | 2.4 | 2 | 4.6 | 1.9 | 2.9 |
| lin318 | 4 | 27.4 | 2 | 64.3 | 2.3 | 47.6 |
| pr124 | 10 | 3.4 | 12 | 6.7 | 2.0 | 3.5 |
| pr136 | 8 | 32.9 | 2 | 50.2 | 1.5 | 36.7 |
| pr144 | 6 | 1.5 | 2 | 3.7 | 2.5 | 1.5 |
| pr226 | 8 | 1.5 | 2 | 2.5 | 1.6 | 2.0 |
| pr264 | 12 | 5.3 | 2 | 10.1 | 1.9 | 6.0 |
| rat195 | 10 | 48.1 | 2 | 65.0 | 1.4 | 51.3 |
| rat99 | 8 | 0.1 | 2 | 0.2 | 1.7 | 0.2 |
| tsp225 | 6 | 135.7 | 2 | 328.2 | 2.4 | 135.7 |
| u159 | 12 | 0.8 | 2 | 1.2 | 1.5 | 1.0 |

The difference between the best global scope and the best scope for each instance is shown by Table 2. Only the LCFirstMinDeltaDeg strategy is considered (for reasons of space). Similar results are obtained for any other search strategy. We observe that the best scope per instance changes quite often, as well as rather important differences between instances. The best overall scope is also clearly dependent on the set of instances considered. For all instances, the best scope is 6, but it is easy to build an instance set whose best overall scope value is not 6. The instances bier127, kroA150, kroB200, pr124 form such a set since scope = 10 is the best scope for each of them. This table also shows that selecting the best scope for each instance significantly improves the average resolution time (the last column gives the time obtained with the best global scope (scope = 6)). Unfortunately, this information is not available without solving the same problem several times. For each instance, we observe very different resolution times between the selected scope values. The w/b column indicates the time ratio between the worst and best scope. We are very often beyond the factor 2. All these observations show that it is not easy to determine a priori a good scope value and that an error can have a strong impact on solving times.

## 4.2    Best Searches

Table 3 compares our approach with the three best search strategies. The "best" scope column indicates that the best scope is considered for each instance, otherwise the best mean scope is used. It is very clear that SSSA is competitive with the best scope per instance. We remind you that it is impossible to know a priori the best scope value per instance and that we can only speculate on the best scope value for a given search strategy after having done tests. We observe almost systematically a gain against the best global scope value When SSSA does not improve the best global scope, which is rare, the loss is low compared to the worst possible scope choice. We are almost certain of a gain and this should lead to the adoption of SSSA in the future.

## 4.3    Other Search

We also tested SSSA's behaviour with other search strategies that do not use the LCFirst policy. Table 4 presents the results obtained for the MinRepCost strategy, which are quite representative of the other tests we have conducted for other search strategies. The gain is higher (1.68) than with search strategies using the LCFirst policy. This policy seems to better manage the structure of the graph and in particular allows to strongly reduce the number of backtracks, also it seems to reduce a little the impact of SSSA for problems that can be solved quickly. In the given results, SSSA never loses significantly.

**Table 3.** Comparison of solving times (in s) between predefined scope values for Algorithm 1 and SSSA for the recommended searches. The *best* scope is defined for each instance and not globally.

| Instance | LCFirstMaxCost | | | LCFirstMinRepCost | | | LCFirstMinDeltaDeg | | |
|---|---|---|---|---|---|---|---|---|---|
| | scope | | | scope | | | scope | | |
| | 8 | best | SSSA | 6 | best | SSSA | 6 | best | SSSA |
| a280 | 66.6 | 44.4 | 19.3 | 13.6 | 12.1 | 18.6 | 38.2 | 35.1 | 34.3 |
| bier127 | 0.3 | 0.3 | 0.3 | 0.3 | 0.2 | 0.4 | 2.6 | 0.7 | 0.5 |
| brg180 | 6.7 | 5.9 | 7.0 | 0.4 | 0.3 | 0.4 | 0.4 | 0.3 | 0.4 |
| ch130 | 6.7 | 4.0 | 4.4 | 2.7 | 1.9 | 2.2 | 1.6 | 1.6 | 1.6 |
| ch150 | 4.8 | 2.7 | 5.2 | 2.4 | 1.8 | 2.6 | 2.0 | 2.0 | 1.2 |
| d198 | 107.9 | 87.1 | 65.6 | 17.8 | 16.6 | 12.4 | 60.9 | 41.4 | 30.8 |
| gr120 | 1.3 | 0.9 | 0.7 | 0.7 | 0.6 | 0.6 | 0.8 | 0.8 | 0.7 |
| gr137 | 5.8 | 5.8 | 4.6 | 3.0 | 2.6 | 1.6 | 3.0 | 2.4 | 2.8 |
| gr202 | 4.0 | 3.5 | 2.8 | 2.2 | 2.2 | 2.2 | 2.3 | 2.1 | 3.9 |
| gr96 | 1.9 | 1.5 | 1.5 | 0.8 | 0.5 | 0.6 | 1.0 | 0.5 | 0.7 |
| kroA100 | 5.7 | 5.1 | 5.2 | 2.1 | 1.6 | 1.9 | 2.5 | 2.5 | 1.3 |
| kroA150 | 77.2 | 62.0 | 51.4 | 8.2 | 4.6 | 6.5 | 5.6 | 5.6 | 8.4 |
| kroB100 | 10.7 | 8.9 | 11.5 | 3.5 | 1.5 | 3.6 | 2.1 | 1.5 | 1.2 |
| kroB150 | 531.0 | 531.0 | 525.3 | 358.6 | 291.4 | 238.3 | 189.1 | 189.1 | 441.1 |
| kroB200 | 855.6 | 855.6 | 752.0 | 230.0 | 186.3 | 197.1 | 144.0 | 87.9 | 88.4 |
| kroC100 | 2.0 | 1.7 | 3.0 | 1.7 | 0.9 | 1.3 | 0.6 | 0.6 | 0.4 |
| kroD100 | 0.6 | 0.5 | 0.5 | 0.4 | 0.3 | 0.7 | 0.4 | 0.2 | 0.2 |
| kroE100 | 6.0 | 3.6 | 3.3 | 2.9 | 2.4 | 2.2 | 6.4 | 3.0 | 2.5 |
| lin318 | 43.9 | 42.7 | 57.9 | 47.6 | 27.4 | 22.4 | 32.7 | 31.3 | 14.6 |
| pr124 | 1.3 | 1.1 | 1.0 | 3.5 | 3.4 | 2.7 | 1.6 | 1.1 | 1.0 |
| pr136 | 250.7 | 213.2 | 175.0 | 36.7 | 32.9 | 28.4 | 58.1 | 53.4 | 33.0 |
| pr144 | 1.1 | 1.0 | 1.0 | 1.5 | 1.5 | 1.6 | 1.6 | 0.8 | 0.6 |
| pr226 | 5.3 | 2.7 | 2.0 | 2.0 | 1.5 | 1.3 | 3.6 | 1.5 | 2.2 |
| pr264 | 9.9 | 8.6 | 6.9 | 6.0 | 5.3 | 4.3 | 7.8 | 7.1 | 5.1 |
| rat195 | 605.6 | 562.4 | 485.0 | 51.3 | 48.1 | 45.9 | 76.8 | 76.8 | 96.9 |
| rat99 | 0.2 | 0.1 | 0.1 | 0.2 | 0.1 | 0.2 | 0.1 | 0.1 | 0.1 |
| tsp225 | 1800.0 | 1800.0 | 1800.0 | 135.7 | 135.7 | 214.2 | 303.0 | 303.0 | 479.8 |
| u159 | 0.3 | 0.3 | 0.3 | 1.0 | 0.8 | 0.9 | 0.6 | 0.3 | 0.8 |
| mean | 157.6 | 152.0 | 142.6 | 33.5 | 28.0 | 29.1 | 33.9 | 30.4 | 44.8 |
| geo mean | 10.2 | 8.3 | 7.9 | 4.4 | 3.4 | 3.8 | 4.8 | 3.5 | 3.5 |

**Table 4.** Comparison of solving times (in s) between predefined scope values for Algorithm 1 and SSSA for the MinRepCost Strategy.

| | MinRepCost | | |
|---|---|---|---|
| | scope | | |
| | 10 | best | SSSA |
| a280 | 1800.0 | 1800.0 | 1800.0 |
| bier127 | 2.7 | 2.1 | 1.5 |
| brg180 | 0.3 | 0.3 | 0.4 |
| ch130 | 2.0 | 1.5 | 1.4 |
| ch150 | 3.8 | 3.1 | 2.9 |
| d198 | 1161.9 | 814.2 | 1057.6 |
| gr120 | 1.1 | 0.9 | 0.8 |
| gr137 | 1.7 | 1.7 | 1.5 |
| gr202 | 4.5 | 2.6 | 1.4 |
| gr96 | 1.0 | 0.8 | 1.4 |
| kroA100 | 2.4 | 2.1 | 1.6 |
| kroA150 | 18.5 | 12.3 | 8.6 |
| kroB100 | 1.2 | 1.2 | 1.3 |
| kroB150 | 736.9 | 668.3 | 840.3 |
| kroB200 | 272.0 | 149.3 | 79.3 |
| kroC100 | 0.7 | 0.5 | 0.4 |
| kroD100 | 0.5 | 0.2 | 0.3 |
| kroE100 | 5.3 | 5.3 | 4.5 |
| lin318 | 41.8 | 27.7 | 15.8 |
| pr124 | 1.1 | 0.8 | 0.8 |
| pr136 | 110.7 | 88.4 | 55.9 |
| pr144 | 0.8 | 0.8 | 0.6 |
| pr226 | 57.1 | 11.2 | 7.0 |
| pr264 | 7.2 | 6.3 | 4.5 |
| rat195 | 212.3 | 127.1 | 103.7 |
| rat99 | 0.1 | 0.1 | 0.1 |
| tsp225 | 1557.3 | 1333.0 | 437.3 |
| u159 | 1.0 | 0.5 | 0.8 |
| mean | 214.5 | 180.8 | 158.3 |
| geo mean | 8.7 | 6.2 | 5.4 |

## 4.4   Overall Improvement

We can now present the comparison between our approach and the FLR approach [9] which is the best method known to date. Table 5 shows that SSSA saves

**Table 5.** Comparison of solving times (in s) between FLR and SSSA for the three recommended searches.

| | LCFirst MaxCost | | | LCFirst MinRepCost | | | LCFirst MinDeltaDeg | | |
|---|---|---|---|---|---|---|---|---|---|
| | FLR | SSSA | ratio | FLR | SSSA | ratio | FLR | SSSA | ratio |
| a280 | 175.0 | 19.3 | 9.1 | 15.6 | 18.6 | 0.8 | 111.4 | 34.3 | 3.3 |
| bier127 | 0.5 | 0.3 | 1.4 | 0.3 | 0.4 | 0.8 | 3.4 | 0.5 | 6.9 |
| brg180 | 46.9 | 7.0 | 6.7 | 0.1 | 0.4 | 0.4 | 0.5 | 0.4 | 1.3 |
| ch130 | 8.9 | 4.4 | 2.0 | 3.8 | 2.2 | 1.7 | 1.8 | 1.6 | 1.1 |
| ch150 | 10.7 | 5.2 | 2.1 | 3.2 | 2.6 | 1.2 | 7.9 | 1.2 | 6.5 |
| d198 | 157.4 | 65.6 | 2.4 | 28.9 | 12.4 | 2.3 | 45.7 | 30.8 | 1.5 |
| gr120 | 1.9 | 0.7 | 2.9 | 0.6 | 0.6 | 1.1 | 1.0 | 0.7 | 1.4 |
| gr137 | 9.3 | 4.6 | 2.0 | 5.4 | 1.6 | 3.3 | 3.6 | 2.8 | 1.3 |
| gr202 | 8.4 | 2.8 | 3.0 | 4.1 | 2.2 | 1.9 | 3.7 | 3.9 | 1.0 |
| gr96 | 4.3 | 1.5 | 2.8 | 1.0 | 0.6 | 1.7 | 2.4 | 0.7 | 3.2 |
| kroA100 | 14.3 | 5.2 | 2.8 | 4.8 | 1.9 | 2.5 | 7.7 | 1.3 | 6.1 |
| kroA150 | 109.9 | 51.4 | 2.1 | 25.8 | 6.5 | 4.0 | 19.7 | 8.4 | 2.4 |
| kroB100 | 33.5 | 11.5 | 2.9 | 8.5 | 3.6 | 2.3 | 5.9 | 1.2 | 4.9 |
| kroB150 | 1503.1 | 525.3 | 2.9 | 789.1 | 238.3 | 3.3 | 695.1 | 441.1 | 1.6 |
| kroB200 | 1800.0 | 752.0 | 2.4 | 307.5 | 197.1 | 1.6 | 219.0 | 88.4 | 2.5 |
| kroC100 | 4.5 | 3.0 | 1.5 | 1.9 | 1.3 | 1.4 | 0.5 | 0.4 | 1.3 |
| kroD100 | 0.8 | 0.5 | 1.5 | 0.5 | 0.7 | 0.7 | 0.3 | 0.2 | 1.6 |
| kroE100 | 9.4 | 3.3 | 2.9 | 5.9 | 2.2 | 2.7 | 5.8 | 2.5 | 2.3 |
| lin318 | 109.0 | 57.9 | 1.9 | 87.9 | 22.4 | 3.9 | 41.3 | 14.6 | 2.8 |
| pr124 | 3.4 | 1.0 | 3.5 | 6.2 | 2.7 | 2.3 | 2.7 | 1.0 | 2.6 |
| pr136 | 253.8 | 175.0 | 1.5 | 59.3 | 28.4 | 2.1 | 92.4 | 33.0 | 2.8 |
| pr144 | 2.1 | 1.0 | 2.1 | 4.2 | 1.6 | 2.6 | 1.2 | 0.6 | 1.9 |
| pr226 | 4.0 | 2.0 | 2.0 | 1.6 | 1.3 | 1.3 | 3.4 | 2.2 | 1.5 |
| pr264 | 3.6 | 6.9 | 0.5 | 3.0 | 4.3 | 0.7 | 2.3 | 5.1 | 0.5 |
| rat195 | 815.6 | 485.0 | 1.7 | 121.7 | 45.9 | 2.6 | 207.3 | 96.9 | 2.1 |
| rat99 | 0.2 | 0.1 | 2.4 | 0.2 | 0.2 | 1.3 | 0.2 | 0.1 | 2.1 |
| tsp225 | 1800.0 | 1800.0 | 1.0 | 250.8 | 214.2 | 1.2 | 771.9 | 479.8 | 1.6 |
| u159 | 0.5 | 0.3 | 1.5 | 1.5 | 0.9 | 1.6 | 1.4 | 0.8 | 1.7 |
| mean | 246.1 | 142.6 | | 62.3 | 29.1 | | 80.7 | 44.8 | |
| geo mean | 17.5 | 7.9 | 2.2 | 6.3 | 3.8 | 1.7 | 7.4 | 3.5 | 2.1 |

about a factor of 2 in time for each search strategy. There are greater gains for all difficult problems. These results clearly show the improvement achieved by our method.

## 5   Related Work

The best scope value of the subgradient algorithm could also be determined with a sampling method similar to Parallel Search Strategy [20] which aims to determine a priori the best search strategy. This method decomposes the initial problem into a large number of subproblems consistent with the propagation, as does the Embarrassingly Parallel Search (EPS) method [18, 22]. Then, it proceeds by sampling and solving: it randomly draws a set of subproblems and solves them in parallel by setting a timeout corresponding to twice the time of the best method in order to limit the time spent with "bad" search strategies. A Wilcoxon test is finally applied to eliminate the statistically worse search strategies. All remaining search strategies being equivalent, one is chosen that will be used to solve the other subproblems.

This approach is difficult to implement in our case because of the large number of subproblems it requires. Consider we have $k$ methods to compare and we set a factor of 2 as timeout. With a confidence level of 95% and sample size equal to 30, which is not a good value in general but could be fine for our purpose, and if you accept to spend t % of the solving time in the selection of the best method then it means that the minimum number of elements in the population should be: $pop = \frac{2 \times 30 \times k}{t}$. For $t = 3\%$ and $k = 36$ (6 values of scope for each agility) we have $pop = 72,000$. Unfortunately, it requires a lot of time to decompose some TSP instances into 72,000 subproblems. For instance, the decomposition of kroB150 in more than 30,000 subproblems requires more than 100s, whereas the solving time is around 150 s. This prevent us from using this method in practice for a lot of instances or a new way to decompose the instances should be found.

## 6   Conclusion

We have shown that the relationship between filtering algorithms and Lagrangian relaxation can be seen as the determination of the trigger time of these algorithms when calculating Lagrangian multipliers by a subgradient algorithm. We have introduced SSSA, a scope sizing subgradient algorithm, which proposes to stop local multiplier optimization when the objective value no longer varies or oscillates, and to call, at that time, the filtering algorithms. The experimental results we presented show the interest of our approach. The performance of the best CP model known so far for solving TSP is improved almost systematically by a factor of 2. The tests show permanent improvements in all the considered scenarios.

We believe that SSSA is taking a first step towards a better understanding of the interactions between FAs and LRs and that others will follow.

We hope that it will also allow similar results to be obtained for problems other than TSP that may lead to a general improvement of the CP-based LR.

# References

1. Applegate, D.L., Bixby, R.E., Chvatal, V., Cook, W.J.: The Traveling Salesman Problem: A Computational Study. Princeton University Press, Princeton (2006)
2. Beasley, J.E.: Lagrangian Relaxation, Chap. 6, pp. 243–303. Wiley, New York (1993)
3. Benchimol, P., van Hoeve, W.J., Régin, J., Rousseau, L., Rueher, M.: Improved filtering for weighted circuit constraints. Constraints **17**(3), 205–233 (2012). https://doi.org/10.1007/s10601-012-9119-x
4. Bergman, D., Cire, A.A., van Hoeve, W.-J.: Improved constraint propagation via Lagrangian decomposition. In: Pesant, G. (ed.) CP 2015. LNCS, vol. 9255, pp. 30–38. Springer, Cham (2015). https://doi.org/10.1007/978-3-319-23219-5_3
5. Cambazard, H., Fages, J.-G.: New filtering for ATMOSTNVALUE and its weighted variant: a Lagrangian approach. Constraints **20**(3), 362–380 (2015). https://doi.org/10.1007/s10601-015-9191-0
6. Caseau, Y., Laburthe, F.: Solving small TSPs with constraints. In: ICLP, vol. 97, p. 104 (1997)
7. Demassey, S.: Compositions and hybridizations for applied combinatorial optimization. Habilitation à Diriger des Recherches (2017)
8. Ducomman, S., Cambazard, H., Penz, B.: Alternative filtering for the weighted circuit constraint: comparing lower bounds for the TSP and solving TSPTW. In: Proceedings of the Thirtieth AAAI Conference on Artificial Intelligence, Phoenix, Arizona, USA, 12–17 February 2016, pp. 3390–3396 (2016)
9. Fages, J.-G., Lorca, X., Rousseau, L.-M.: The salesman and the tree: the importance of search in CP. Constraints **21**(2), 145–162 (2014). https://doi.org/10.1007/s10601-014-9178-2
10. Focacci, F., Lodi, A., Milano, M.: Embedding relaxations in global constraints for solving TSP and TSPTW. Ann. Math. Artif. Intell. **34**(4), 291–311 (2002). https://doi.org/10.1023/A:1014492408220
11. Focacci, F., Lodi, A., Milano, M.: A hybrid exact algorithm for the TSPTW. INFORMS J. Comput. **14**(4), 403–417 (2002)
12. Fontaine, D., Michel, L., Van Hentenryck, P.: Constraint-based Lagrangian relaxation. In: O'Sullivan, B. (ed.) CP 2014. LNCS, vol. 8656, pp. 324–339. Springer, Cham (2014). https://doi.org/10.1007/978-3-319-10428-7_25
13. Frangioni, A.: Generalized bundle methods. SIAM J. Optim. **13**(1), 117–156 (2002)
14. Held, M., Karp, R.M.: The traveling-salesman problem and minimum spanning trees: Part II. Math. Program. **1**(1), 6–25 (1971). https://doi.org/10.1007/BF01584070
15. Isoart, N., Régin, J.C.: Integration of structural constraints into TSP models. In: Schiex, T., de Givry, S. (eds.) CP 2019. LNCS, pp. 284–299. Springer International Publishing, Cham (2019). https://doi.org/10.1007/978-3-030-30048-7_17
16. Khemmoudj, M.O.I., Bennaceur, H., Nagih, A.: Combining arc-consistency and dual Lagrangean relaxation for filtering CSPs. In: Barták, R., Milano, M. (eds.) CPAIOR 2005. LNCS, vol. 3524, pp. 258–272. Springer, Heidelberg (2005). https://doi.org/10.1007/11493853_20
17. Kilby, P., Shaw, P.: Vehicle routing. In: Foundations of Artificial Intelligence, vol. 2, pp. 801–836. Elsevier (2006)
18. Malapert, A., Régin, J., Rezgui, M.: Embarrassingly parallel search in constraint programming. J. Artif. Intell. Res. (JAIR) **57**, 421–464 (2016)

19. Menana, J.: Automates et programmation par contraintes pour la planification de personnel. Ph.D. thesis, Université de Nantes (2011)
20. Palmieri, A., Régin, J.-C., Schaus, P.: Parallel strategies selection. In: Rueher, M. (ed.) CP 2016. LNCS, vol. 9892, pp. 388–404. Springer, Cham (2016). https://doi.org/10.1007/978-3-319-44953-1_25
21. Pesant, G., Gendreau, M., Potvin, J.Y., Rousseau, J.M.: An exact constraint logic programming algorithm for the traveling salesman problem with time windows. Transp. Sci. **32**(1), 12–29 (1998)
22. Régin, J.-C., Rezgui, M., Malapert, A.: Embarrassingly parallel search. In: Schulte, C. (ed.) CP 2013. LNCS, vol. 8124, pp. 596–610. Springer, Heidelberg (2013). https://doi.org/10.1007/978-3-642-40627-0_45
23. Reinelt, G.: TSPLIB-a traveling salesman problem library. ORSA J. Comput. **3**(4), 376–384 (1991)
24. Sellmann, M.: Theoretical foundations of CP-based Lagrangian relaxation. In: Wallace, M. (ed.) CP 2004. LNCS, vol. 3258, pp. 634–647. Springer, Heidelberg (2004). https://doi.org/10.1007/978-3-540-30201-8_46
25. Sellmann, M., Fahle, T.: Constraint programming based Lagrangian relaxation for the automatic recording problem. Ann. Oper. Res. **118**(1–4), 17–33 (2003). https://doi.org/10.1023/A:1021845304798

# Minimal Perturbation in University Timetabling with Maximum Satisfiability

Alexandre Lemos$^{(\boxtimes)}$ , Pedro T. Monteiro , and Inês Lynce

Instituto Superior Técnico, Universidade de Lisboa, INESC-ID,
Rua Alves Redol 9, 1000-029 Lisbon, Portugal
{alexandre.lemos,pedro.tiago.monteiro,ines.lynce}@tecnico.ulisboa.pt

**Abstract.** Every new academic year, scheduling new timetables due to disruptions is a major problem for universities. However, computing a new timetable from scratch may be unnecessarily expensive. Furthermore, this process may produce a significantly different timetable which in many cases is undesirable for all parties involved. For this reason, we aim to find a new feasible timetable while minimizing the number of perturbations relative to the original disrupted timetable.

The contribution of this paper is a maximum satisfiability (MaxSAT) encoding to solve large and complex university timetabling problem instances which can be subject to disruptions. To validate the MaxSAT encoding, we evaluate university timetabling real-world instances from the International Timetabling Competition (ITC) 2019. We consider the originally found solutions as a starting point, to evaluate the capacity of the proposed MaxSAT encoding to find a new solution with minimal perturbation. Overall, our model is able to efficiently solve the disrupted instances.

**Keywords:** MaxSAT · University Course Timetabling · Minimal perturbation

## 1 Introduction

Many real-life problems can be encoded as constraint optimization problems, being university timetabling problems a concrete example. Solving optimization problems is by itself a hard and complex computational task. When solving these problems, unexpected disruptions may cause the original solution to be no longer feasible. Therefore, one needs to solve the problem again subject to these unexpected disruptions. Universities, and in particular their timetables, are dynamical systems. Hence, it is natural that one often needs to solve new

The authors would like to thank the reviewers for their helpful comments and suggestions that contributed to an improved manuscript. This work was supported by national funds through Fundação para a Ciência e a Tecnologia (FCT) with reference SFRH/BD/143212/2019 (PhD grant), DSAIPA/AI/0033/2019 (project LAIfeBlood) and UIDB/50021/2020 (INESC-ID multi-annual funding).

E. Hebrard and N. Musliu (Eds.): CPAIOR 2020, LNCS 12296, pp. 317–333, 2020.
https://doi.org/10.1007/978-3-030-58942-4_21

timetables subject to disruptions. These types of real-world scenarios are still a significant research line [1,2].

The contribution of this paper is a MaxSAT encoding to solve university course timetabling problems which can be subject to different disruptions. We showcase the application of the MaxSAT encoding with the large data sets from the ITC-2019 benchmark [3]. Furthermore, these instances are subject to the most common disruptions in the literature.

This paper is organized as follows. Section 2 provides a concise background on university timetabling and minimal perturbation problems. Section 3 formally describes the problem of minimal perturbation in university timetabling and the MaxSAT encoding. Section 4 discusses the main computational results. Finally, Sect. 5 concludes the paper and discusses possible future directions.

## 2   Background

In this section, we provide an overview of university timetabling, followed by the background on the minimum perturbation problem and the MaxSAT problem.

### 2.1   University Timetabling

University timetabling problems [1,2] can be categorized as follows: examination timetabling [4], course timetabling [5] and student sectioning [6]. These problems are known to be NP-complete [7].

Examination timetabling is the problem of assigning exams to rooms subject to a set of constraints. Course timetabling can be informally defined as the problem of finding a feasible assignment for all the classes of all courses to a time slot and a room, subject to a set of constraints. Student sectioning is the problem of sectioning students, subject to capacity and schedule constraints, to all the classes required by the courses they are enrolled in. In the context of this paper, we only consider course timetabling and student sectioning problems. A formal and detailed description of both problems is given in Sect. 3.

In recent years, a significant improvement in solving university timetabling problems has been achieved [1,2]. In the literature, one can find distinct approaches to solve university timetabling problems, most notably: Constraint Programming (CP) [8,9], Answer Set Programming (ASP) [10], Boolean Satisfiability (SAT) [11], Maximum Satisfiability (MaxSAT) [12], Integer Linear Programming (ILP) [13–15] and local search [13,16].

The availability of benchmark data sets from previous competitions [5], based on data from Udine University, motivated the development of the above mentioned methods. However, a gap between theory and practice [1,3] still persists, given that the benchmark does not express the whole complexity and size of the worldwide university timetabling problem. Recently, to further reduce this gap, a new benchmark was made available as part of ITC-2019 [3].

## 2.2   Minimal Perturbation Problem

Consider a given problem, subject to a set of constraints, for which $s$ is a feasible solution. A set of disruptions may imply a change in the set of constraints and/or a change in the set of variables of the problem. The disruptions cause the solution $s$ to be no longer feasible.

This optimization problem can be described as a Minimal Perturbation Problem (MPP) [8,10,14,15,17] where the goal is to minimize the number of perturbations caused to $s$ in order to find a new feasible solution. In this paper, we consider the MPP as a multi-objective optimization problem where we use the Hamming distance (HD) and the overall quality as the optimization criterion. This makes MPP cardinality minimal and so more restricted than subset minimal. The problem of finding similar/diverse solutions [18] has similarities to MPP. However, the task of finding similar/diverse solutions usually does not consider an infeasible solution as a starting point.

*Example 1.* Let us consider a course timetabling problem instance with two classes ($c_i$ and $c_j$) that can be assigned to five different time slots denoted as $t_1 \ldots t_5$. Time slots $t_3 \ldots t_5$ have a penalty associated with both classes. Classes $c_1$ and $c_2$ have a *no overlap* constraint, to ensure that they are assigned to different time slots. Also, let us assume that the original solution $s$ is optimal and consists in the assignment of $c_i$ to the time slot $t_1$ and $c_j$ to $t_2$. Now, if a disruption causes $t_1$ to be unavailable to class $c_i$, then solution $s$ becomes infeasible, and needs to be modified. If one solves the problem instance from scratch, the optimal solution is the assignment of $c_i$ to time slot $t_2$ and $c_j$ to $t_1$, corresponding to a different solution. The solution with the smallest number of perturbations only implies changing $c_i$ to time slot $t_2$ despite the fact that it causes a loss in the overall quality of the timetable.

The application of MPP to course timetabling has been studied in the literature [8,10,14,15]. The most common approach to measure the perturbations is to apply the HD [19].

Müller *et al.* [8] proposed the iterative forward search algorithm, a local search method that does not ensure completeness. Phillips *et al.* [15] proposed a neighborhood based integer programming algorithm to solve MPP in instances from the University of Auckland. In the worst case, the neighborhood will include the whole search space.

Recently, two different tools have been proposed to compute the Pareto front using ASP [10] and ILP [14]. The Pareto front is computed based on two objectives: (i) the minimization of the cost of unsatisfied soft constraints; and (ii) the minimization of the number of perturbations.

Another approach is to create a robust solution in order to resist predictable disruptions [16]. However, this approach will not be discussed in this paper.

## 2.3   MaxSAT

A literal $l$, is either a Boolean variable $x$ (positive literal) or its negation $\neg x$ (negative literal). A clause is a disjunction of literals. A propositional formula in Conjunctive Normal Form (CNF) is a conjunction of clauses. SAT is the problem of deciding whether a given formula has an assignment that satisfies all the clauses in the formula.

The MaxSAT problem is a generalization of SAT, where the objective is to find an assignment that maximizes the number of satisfied clauses. A weighted partial MaxSAT formula ($\varphi = \varphi_h \cup \varphi_s$) consists of hard clauses ($\varphi_h$), soft clauses ($\varphi_s$), and a function $w^\varphi : \varphi_s \to \mathbb{N}$ associating an integer cost to each soft clause. The goal in weighted partial MaxSAT is to find an assignment such that all hard clauses in $\varphi_h$ are satisfied, while maximizing the weight of the satisfied soft clauses in $\varphi_s$.

In this paper, we will assume that all propositional formulas are in CNF. However, to simplify the writing of some constraints, we will use the definition of pseudo-Boolean (PB) constraints. PB constraints are commonly applied in pseudo-Boolean optimization [20], a related problem to weighted partial MaxSAT. PB constraints are linear constraints over Boolean variables, and can be generally written as follows: $\sum q_i x_i$ OP $K$, where $K$ and all $q_i$ are integer constants, all $x_i$ are Boolean variables, and OP $\in \{<, \leq, =, \geq, >\}$. This type of constraints can be translated into SAT [21].

## 3   MaxSAT Encoding

In this section, we formally describe the university course timetabling problem [3] and its MaxSAT encoding. Consider a set of consecutive time slots of five minutes $T \in \{1, ..., 288\}$ corresponding to all possible time slots of a day and a set of sets of weekdays $\mathcal{D} \in \{0000000, ..., 1111111\}$. Each subset of days $Days \in \mathcal{D}$ has $|Days| = 7$. $Days^d$ corresponds to a particular weekday with $0 < d \leq |Days|$ ($Days^1$ corresponds to Monday, $Days^2$ to Tuesday, and so on). A set of sets of weeks of a semester is represented by $\mathcal{W}$. Each subset of weeks $Weeks \in \mathcal{W}$ has $|Weeks| = 16$. $Weeks^w$ corresponds to week $w$ with $0 < w \leq |Weeks|$. A time period $p$ is represented with a 4-tuple $(W_p, D_p, h_p, len_p)$: a set of weeks ($W_p \subseteq \mathcal{W}$); a set of days ($D_p \subseteq \mathcal{D}$); an hour ($h_p \in T$); and its duration ($len_p > 1$).

Consider a set of courses $Co$. A course ($co \in Co$) is composed by a set of classes $C_{co}$. These classes are characterized by configurations ($Config_{co}$) and organized in parts ($Parts_{config}$). A student must attend the classes from a single configuration. A student enrolled in the course $co$ and attending the configuration $config \in Config_{co}$ must attend *exactly-one* class from each part $Parts_{config}$. The set of classes belonging to $part \in Parts_{config}$ is represented by $C_{part}$.

The university has a set $R$ of rooms where the classes of a course can be scheduled. The travel time, in slots, between two rooms $r_1 \in R$ and $r_2 \in R$ is represented as $travel_{r_2}^{r_1}$. Each room $r \in R$ has a set of unavailable periods $P_r$.

All university classes $C$ (from different courses) must have a schedule assigned to them. Each class $c \in C$ has a set of possible periods ($P_c$) to be scheduled in. Each possible period $p \in P_c$ has an associated penalty. Furthermore, a class may need to be assigned to a room. A class has a hard limit on the number of students that can attend it ($lim_c$). A class may have a set of possible rooms

($R_c$). Each room $r \in R_c$ has *capacity* $\geq lim_c$ and an associated penalty. Each class may also have parent-child relation with another class. The parent of class $c$ is represented by $parent_c$.

The university has a set of students $S$. Each student $s \in S$ is enrolled in a set of courses $Co_s$. To reduce the number of similar variables and constraints, we create groups of students sharing the same curricular plan [22]. Furthermore, we limit the size of the group to the value of the greatest common divisor between the total number of students enrolled in a course and the smallest capacity limit of the classes of that course [23]. This process ensures that it is possible to find a feasible solution to a problem instance, since it is possible to combine all groups of students into classes. However, we may remove the optimal solution by not allowing the assignment of a single student to a given class. For this reason, we define *Cluster* as a set of clusters of students. The number of students merged in the $id \in Cluster$ is represented by $|id|$.

There are four optimization criteria: (i) the cost of assigning a class to a room; (ii) the cost of assigning a class to a time slot; (iii) the number of student conflicts and (iv) a set of soft constraints. Each criterion has its weights. We solve university course timetabling in two sequential MaxSAT runs. First, we solve the course timetabling problem and then we solve the student sectioning problem. The sequential runs may result in the loss of the global optimum (*i.e.* it may remove the optimal solution in terms of student conflicts). Nevertheless, it produces a solution within the Pareto front if given enough time and memory resources. Moreover, it reduces the size of the global problem. Furthermore, allows us to tackle the MPP using only the first MaxSAT model.

### 3.1   Course Timetabling

Our course timetabling encoding has four types of Boolean decision variables:

- $w_c^{Week_p}$ represents the assignment of class $c$ to the set of weeks $Week_p$, with $c \in C$, $Week_p \in \mathcal{W}$ and $p \in P_c$;
- $d_c^{Day_p}$ represents the assignment of class $c$ to the set of days $Day_p$, with $c \in C$, $Day_p \in \mathcal{D}$ and $p \in P_c$;
- $h_c^{hour_p}$ represents the assignment of class $c$ to the hour $hour_p$, with $c \in C$ and $p \in P_c$;
- $r_c^{room}$ represents the assignment of class $c$ to the room $room$, with $c \in C$ and $room \in R_c$.

The scheduling possibilities of a class are usually just a small part of the complete set of possible combinations of weeks, days and hours. Consequently, we only define these variables for acceptable values of the class domain reducing the size of the problem. Furthermore, using four variables instead of one provides a more flexible approach when writing the associated constraints, reducing the size of the encoding. For example, one can write the constraints using only related variables (*e.g. SameDay* constraint uses only variable $d$).

To simplify the writing of the *exactly-one* constraints ($\sum \cdot = 1$) we define the auxiliary variable $t$, where $t_c^{slot}$ represents the assignment of class $c$ to the allocation slot $slot \in [0, \ldots, |P_c|]$.

Our encoding has the following constraints. If a class $c$ takes place in the hour $hour$ then all allocation slots including $hour$ are assigned. If we consider that $n$ allocation slots have the same $hour$, then the following equivalence is needed:

$$h_c^{hour} \iff \bigwedge_n t_c^{slot_n}. \tag{1}$$

This equivalence can be easily converted to SAT. Similarly, the same type of equivalence has to be written between the week/day variables and the t variables.

A class can only be taught in exactly one allocation slot. For each class $c \in C$:

$$\sum_{slot \in [0, \ldots, |P_c|]} t_c^{slot} = 1. \tag{2}$$

A class with $R_c \neq \emptyset$ can only be taught in exactly one room. For each $c \in C$:

$$\sum_{room \in R_c} r_c^{room} = 1. \tag{3}$$

We define the auxiliary variable $sd_{c_j}^{c_i}$ to represent two classes taught in the same day (*i.e.* with at least one day overlap). For each two classes $c_i$, $c_j$ with $i \neq j$, where $Day_0$ to $Day_n$ belong to the domain of class $c_i$, $Day_{n+1}$ to $Day_m$ belong to the domain of class $c_j$, with $0 < n < m$, and they overlap we add:

$$sd_{c_j}^{c_i} \iff (d_{c_i}^{Day_0} \vee \ldots \vee d_{c_i}^{Day_n}) \wedge (d_{c_j}^{Day_{n+1}} \vee \ldots \vee d_{c_j}^{Day_m}). \tag{4}$$

Similarly, one can define an auxiliary variable $sw_{c_j}^{c_i}$ to represent two classes overlapping in at least one week.

A class $c$ with $R_c \neq \emptyset$ must be taught in a room not assigned to another class in the specific time slot. For each two classes $c_i$, $c_j$, where $room \in R_{c_i}$, $room \in R_{c_j}$, $hour_{p_i} + len_{p_i} > hour_{p_j}$ and $hour_{p_j} + len_{p_j} > hour_{p_i}$ with $p_i \in P_{c_i}$ and $p_j \in P_{c_j}$, we add clause:

$$\neg sd_{c_j}^{c_i} \vee \neg sw_{c_j}^{c_i} \vee \neg h_{c_i}^{hour_{p_i}} \vee \neg h_{c_j}^{hour_{p_j}} \vee \neg r_{c_i}^{room} \vee \neg r_{c_j}^{room}. \tag{5}$$

The clause above could have a smaller number of literals if we used the auxiliary variable $t$. However, it would require to generate more constraints. This trade-off was tested and fewer constraints proved to be more efficient.

The rooms may have unavailability time slots, where no class can be taught. To enforce this constraint we add the following clause for each class $c$, room $r$ and unavailable period $p$:

$$\neg r_c^r \vee \neg t_c^p. \tag{6}$$

The next set of constraints can be hard or soft. These constraints involve always a pair of classes. In case of being soft, the penalty associated with each constraint incurs for every pair of classes. Consider two classes $c_i$ and $c_j$ with $i \neq j$ and two time slots $p_i \in P_{c_i}$ and $p_j \in P_{c_j}$.

*SameStart*: The classes have to start at the same time. For each pair $hour_{p_i}$, $hour_{p_j}$ where $hour_{p_i} \neq hour_{p_j}$ we add a clause:

$$\neg h_{c_i}^{hour_{p_i}} \vee \neg h_{c_j}^{hour_{p_j}} . \tag{7}$$

*DifferentTime (SameTime)*: The classes must be taught at a (the) different (same) hour. For each pair $hour_{p_i}$, $hour_{p_j}$ with (no) overlap in time, we add (7).

*WorkDay*(V): There must not be more than $V$ time slots between the start time of the first class and the end time of the last class on any day. For each pair $hour_{p_i}$, $hour_{p_j}$ where $hour_{p_i} + len_{p_i} - hour_{p_j} \geq V$, we add clause (7).

*DifferentDays (SameDays)*: The classes must be taught in different days (the same subset of days). For each pair $Day_{p_i}$, $Day_{p_j}$ where $Day_{p_i} \wedge Day_{p_j} = \emptyset$ ($Day_{p_i} \subseteq Day_{p_j}$), we add a clause:

$$\neg d_{c_i}^{Day_{p_i}} \vee \neg d_{c_j}^{Day_{p_j}} . \tag{8}$$

*DifferentWeeks (SameWeeks)*: The classes must be taught in different weeks (the same subset of weeks). For each pair $Week_{p_i}$, $Week_{p_j}$ where $Week_{p_i} \wedge Week_{p_j} = \emptyset$ ($Week_{p_i} \subseteq Week_{p_j}$), we add a clause:

$$\neg w_{c_i}^{Week_{p_i}} \vee \neg w_{c_j}^{Week_{p_j}} . \tag{9}$$

*DifferentRoom (SameRoom)*: The classes must be taught in different rooms (the same room). For each pair $room_i \in R_{c_i}$, $room_j \in R_{c_j}$ where $room_i = room_j$ ($room_i \neq room_j$), we add a clause:

$$\neg r_{c_i}^{room_i} \vee \neg r_{c_j}^{room_j} . \tag{10}$$

*SameAttendees*: The classes cannot overlap in time, days and weeks. Furthermore, the attendees must have sufficient time to travel between the rooms corresponding to consecutive classes. For each pair of hours $hour_{p_i}$, $hour_{p_j}$ and rooms $room_i$, $room_j$ where and $hour_{p_i} + len_{p_i} + travel_{room_j}^{room_i} > hour_{p_j}$, we add:

$$\neg sd_{c_j}^{c_i} \vee \neg sw_{c_j}^{c_i} \vee \neg h_{c_i}^{hour_{p_i}} \vee \neg h_{c_j}^{hour_{p_j}} \vee \neg r_{c_i}^{room_i} \vee \neg r_{c_j}^{room_j} . \tag{11}$$

*Overlap (NotOverlap)*: The classes must (not) overlap in time, day and week. For each pair $p_i$, $p_j$ with (no) overlaps in time, we add a clause:

$$\neg t_{c_i}^{p_i} \vee \neg t_{c_j}^{p_j} . \tag{12}$$

*Precedence*: The first meeting of a class in a week must be before the first meeting of another class. For each pair $p_i$, $p_j$ where $p_j$ proceeds $p_i$, we add (12).

$MinGap$(V): The classes that are taught on the same day and on the same set of weeks must be at least $V$ slots apart. For each pair $p_i$, $p_j$ that is taught in the same week, day and $hour_{p_i} + len_{p_i} + V \geq hour_{p_j}$, we add (12).

The next set of constraints involve a set of classes. In these cases, the penalty depends on the distance between the solution and the unsatisfied constraint.

$MaxDays$(V): The classes cannot be taught in more than $V$ different days. When the constraint is soft, the penalty is multiplied by the number of days that exceed $V$. For this reason, we define an auxiliary variable $dayofweek_d^{const}$, where $const$ is the identifier of the constraint $MaxDays$ and $d \in \{1, \ldots, |Days|\}$. This variable corresponds to having at least one class, of this constraint, assigned to weekday $d$. Consider $Day_{p_1}, \ldots, Day_{p_n}, Day_{p_{n+1}}, \ldots, Day_{p_m}$ where $p_1, \ldots, p_n \in P_{c_i}$, $p_{n+1}, \ldots, p_m \in P_{c_j}$ and $Day_1^d = 1, \ldots, Day_m^d = 1$ we add:

$$dayofweek_d^{const} \iff d_{c_i}^{Day_{p_1}} \vee \ldots \vee d_{c_i}^{Day_{p_n}} \vee d_{c_j}^{Day_{p_{n+1}}} \vee \ldots \vee d_{c_j}^{Day_{p_m}}. \quad (13)$$

Now, we only need to ensure that:

$$\sum_{c \in C} \sum_{p \in P_c} \sum_{d \in [1, \ldots, |Day_p|]} dayofweek_d^{const} \leq V. \quad (14)$$

$MaxDayLoad$(V): The classes must (should, if the constraint is soft) be spread over the days in a way that there is no more than a given number of occupied $V$ time slots on each day. When the constraint is soft, the penalty is multiplied by the division of the sum of the number of slots that exceed $V$ for each day by the number of weeks. Hence, we only need to ensure that:

$$\sum_{d=1}^{7} \sum_{c \in C} \sum_{\substack{p \in P_c, \\ Day_p^d = 1}} d_c^{Day_p} \times len_p \leq V. \quad (15)$$

$MaxBreaks$($V_1$,$V_2$): There are at most $V_1$ breaks throughout a day between a set of classes in this constraint. A break between two classes is a gap larger than $V_2$ time slots. When the constraint is soft, the penalty is multiplied by the number of new breaks. For every class $c_1$ to $c_n$ assigned to a period ($p_1 \in P_{c_1}$ to $p_n \in P_{c_n}$) in such a way that it forms a block of classes that breaks this constraint, we add the clause:

$$\neg t_{c_1}^{p_1} \vee \ldots \vee \neg t_{c_n}^{p_n}. \quad (16)$$

$MaxBlock$($V_1$,$V_2$): There are at most $V_1$ consecutive slots throughout a day between a set of classes in this constraint. Two classes are considered to be consecutive if the gap between them is less than $V_2$ time slots. When the constraint is soft, the penalty is multiplied by the number of new blocks of classes. For every class $c_1$ to $c_n$ assigned to a period ($p_1 \in P_{c_1}$ to $p_n \in P_{c_n}$) in such a way that it forms a block of classes that breaks this constraint, we add (16).

## 3.2   Student Sectioning

To solve student sectioning our encoding is extended with one decision variable $s_{id}^c$, wheré $c \in C$ and $id \in [1, \ldots, |Cluster|]$. To ensure a student can only be sectioned to a single course configuration, we define an auxiliary variable for each pair configuration-cluster of students. The variable is denoted as $conf_{id}^{config}$, where $id \in [1, \ldots, |Cluster|]$, $config \in Config_{co}$ and $co \in Co$.

Each cluster of students $id$ must be enrolled in *exactly-one* configuration of each course, $co \in Co_s$, and thus we add the clause:

$$\sum_{config \in Config_{co}} conf_{id}^{config} = 1. \tag{17}$$

To ensure that the class capacity is not exceed, we add for each class $c$:

$$\sum_{id \in [1, \ldots, |Cluster|]} |id| \times s_{id}^c \leq lim_c. \tag{18}$$

A cluster of students $id$ enrolled in a class $c$ must be enrolled in class $parent_c$:

$$\neg s_{id}^c \vee s_{id}^{parent_c}. \tag{19}$$

Finally, we need to ensure that a cluster of students $id$ is enrolled in *exactly-one* class of each part of a single configuration of the course $co$. Therefore, for each cluster of student $id$ and for each pair of two classes $c_i$, $c_j$ in the same $c_i, c_j \in C_{part}$ where $part \in Parts_{config}$, we add:

$$\neg conf_{id}^{config} \vee \neg s_{id}^{c_i} \vee \neg s_{id}^{c_j}. \tag{20}$$

For each cluster of students and for each $part \in Parts_{config}$ we add:

$$\neg conf_{id}^{config} \vee s_{id}^{c_i} \vee \ldots \vee s_{id}^{c_{|C_{part}|}}. \tag{21}$$

The conflicting schedule of classes attended by the same cluster of students is represented by a set of weighted soft clauses. For each cluster of students $id$ enrolled in two classes $c_i$, $c_j$ with overlapping time:

$$\neg s_{id}^{c_i} \vee \neg s_{id}^{c_j} \vee \neg sw_{c_j}^{c_i} \vee \neg sd_{c_j}^{c_i} \vee \neg h_{c_i}^{hour_{c_i}} \vee \neg h_{c_j}^{hour_{c_j}}. \tag{22}$$

## 3.3   Disruptions

In this work we consider the following disruptions: *invalid time* and *invalid room*. These disruptions reduce the domain of a specific class $c$ in terms of available time slots or rooms. Disruptions in the students enrollments would only cause changes in the student sectioning part. The problem definition has the underlining assumption that all the rooms in the domain of class have enough capacity for the students attending. As our original solutions are sub-optimal we do not consider disruptions in the enrollments.

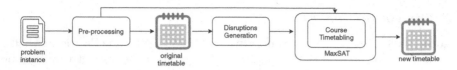

**Fig. 1.** Algorithm schema to solve university timetabling problems subject to disruptions.

*Invalid Time*: The time slot $t$ is no longer available for class $c$:

$$\neg t_c^t. \tag{23}$$

*Invalid Room*: The room $r$ is no longer available for class $c$:

$$\neg r_c^r. \tag{24}$$

When recovering from disruptions we apply lexicographic optimization with two criteria: (i) the HD and (ii) the overall quality of the solution (computed based on the four criteria defined above). This way we can take advantage of the disruption to improve the quality of the solution.

## 4    Experimental Evaluation

In this section, we discuss the main computational results obtained. First, we describe the setup used to validate our approach. Next, we discuss our results for both university timetabling problems and MPP.

### 4.1    Experimental Setup

The evaluation was performed using the *runsolver* tool [24] with a time out of **6,000** s. *Runsolver* was run on a computer with Fedora 14, with 32 CPUs 2.6 GHz and 126 Gb of RAM. To validate our approach, we used the benchmark obtained from ITC-2019 [3], which is divided into three groups (early, middle, late). The goal of the competition was to find the best solution for these instances with no time or memory limits. The organizers provided an validation tool[1], which we used to validate the correctness of our approach.

The proposed solution was implemented in C++, using the *TT-Open-WBO-Inc* [25, 26][2] MaxSAT solver. The solver was configured to use linear search with the clusters algorithm [27]. Moreover, a lexicographic optimization criterion [28] was used. *Exactly-one* constraints were encoded into CNF through the ladder encoding [29]. PB constraints were encoded to CNF using the adder encoding [30]. Our implementation is available at github.com/ADDALemos/MPPTimetables[3].

---

[1] https://www.itc2019.org/validator.

[2] *TT-Open-WBO-Inc* won the Weighted Incomplete category at *MaxSAT Evaluation 2019*. The results are available at https://maxsat-evaluations.github.io/2019.

[3] We use the *RAPIDXML* parser which is available at rapidxml.sourceforge.net/.

**Table 1.** Data sets per university (instances sorted by # of variables).

| | | $|C|$ | Avg. $|R_c|$ | Avg. $|P_c|$ | $|S|$ (k) | #MaxBreak | #MaxBlock | # Var. (k) |
|---|---|---|---|---|---|---|---|---|
| yach-fal17 | | 417 | 4 | 43 | 1 | 0 | 0 | 19 |
| nbi-spr18 | | 782 | 4 | 38 | 2 | 0 | 0 | 35 |
| tg* | Avg. | 693 | 11 | 24 | 0 | 0 | 0 | 42 |
| | Median | 693 | 11 | 24 | 0 | 0 | 0 | 42 |
| mun-f* | Avg. | 743 | 4 | 44 | 1 | 0 | 3 | 45 |
| | Med. | 700 | 4 | 30 | 1 | 0 | 2.5 | 38 |
| mary* | Avg. | 916 | 14 | 12 | 4 | 0 | 0 | 47 |
| | Med. | 916 | 14 | 12 | 4 | 0 | 0 | 47 |
| lums* | Avg. | 494 | 26 | 43 | 0 | 0 | 0 | 82 |
| | Med. | 494 | 26 | 43 | 0 | 0 | 0 | 82 |
| bet* | Avg. | 1,033 | 25 | 23 | 3 | 24 | 19 | 140 |
| | Med. | 1,033 | 25 | 23 | 3 | 24 | 19 | 140 |
| pu* | Avg. | 3,418 | 12 | 33 | 28 | 16 | 0 | 196 |
| | Med. | 1,929 | 12 | 30 | 31 | 17 | 0 | 125 |
| muni-pdf* | Avg. | 2,586 | 15 | 53 | 4 | 0 | 13 | 374 |
| | Med. | 2,526 | 17 | 56 | 3 | 0 | 10 | 373 |
| agh* | Avg. | 1,955 | 34 | 89 | 3 | 15 | 0 | 380 |
| | Med. | 1,239 | 10 | 75 | 2 | 14 | 0 | 340 |
| iku* | Avg. | 2,711 | 25 | 34 | 0 | 0 | 0 | 1,050 |
| | Med. | 2,711 | 25 | 34 | 0 | 0 | 0 | 1,050 |

Table 1 shows the different characteristics of the instances. One can see that the instances are distinct from each other. Instances from *iku** are the largest in terms of classes. However, they do not have students or *MaxBlock/MaxBreak*. They have one order of magnitude more variables than the next largest instance (despite not having students). The *muni-f** instances have a particular small search space in terms of possible rooms per class (only 4).

Figure 1 illustrates the process of solving the university timetabling problem subject to disruptions. The process starts with a problem instance and a timetable, and ends when a new feasible timetable is found. Each problem instance is pre-processed before generating the encoding. Our approach relies on two pre-processing methods: (i) identification of independent sub-sets in terms of courses; and (ii) merging students with exactly the same course enrollment plan. Method (i) divides the problem into self-contained sub-problems, while not removing any solution. Method (ii) was already discussed before (Section 3) and it may remove the optimal solution by not allowing the assignment of an individual student to a given class.

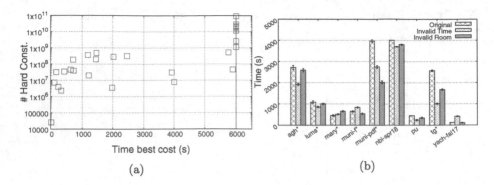

**Fig. 2.** (a) A comparison of the number of hard constraints versus CPU time, in seconds (log scale), for each instance. The dotted line represents the time limit. (b) A comparison of the CPU time per disruption scenario and university.

For each instance, we generated 50 different disruption instances. As our space is limited, we only show the results for the disruptions that are more likely to occur [14,15]. The disruptions were randomly generated following a uniform distribution with a probability of 21% and 25% for *invalid time* and *invalid room*, respectively. These percentages represent the probability of an assignment being invalid. These values were obtained by the academic office of our university, and applied to the ITC-2019 benchmark instances.

## 4.2   Computational Results

First, we discuss the results for the university course timetabling without subjecting it to disruptions. Next, we discuss the results for the MPP.

*University Course Timetabling.* The success of our approach is attested by having been ranked among the five finalists of the ITC-2019 competition. The creation of clusters has a significant impact on the number of variables. On average, one can reduce the number of variables relating to students up to 15%.

We are able to find a solution within limit in 20 out of 30 instances. However, the solver was not able to prove optimality within the time limit, on any of the instances considered in this paper. Note that, for most instances, the solver requires only a short amount of time to produce the best solution.

Figure 2(a) compares the number of hard constraints generated by the CNF encoding and the CPU time needed to find the best solution for each instance. One can see that most of these instances have a larger number of hard constraints. Most of them actually exceed in two orders of magnitude more constraints than the others (top right corner of Fig. 2(a)). Most of these constraints result from the *MaxBlock* and *MaxBreak* constraints.

A large contributing factor for these results is the size of $|P_c|$ and $|R_c|$ (see Table 1). In most cases, a larger size of these sets causes the instance to be harder

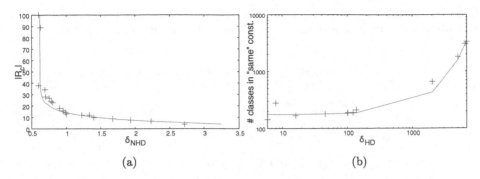

**Fig. 3.** (a) Room domain size ($R_c$) versus the normalized number of perturbations ($\delta_{NHD}$) for the room disruptions. (b) Number of classes involved on constraints of type *same* (log scale) versus the number of perturbations (HD) for the time disruptions (log scale). Data points represent the results and the line the best fit function.

to solve. In these cases, more options do not always mean more solutions. The *lums\** instances are an exception.

***Minimal Perturbation in University Timetabling.*** Our MaxSAT approach was compared with a modified integer programming approach based on [14]. The results showed that the integer programming approach is able to find the optimal solution for the MPP but only to a subset of instances compared to those solved by the MaxSAT approach. Furthermore, the MaxSAT approach is much faster.

Our approach is able to find feasible solutions to all disruptions tested. Moreover, the solver is able to find an optimal solution for all disrupted instances. Despite the fact that the disruptions only add new constraints, one can occasionally improve the cost of the solution. This can be explained by the fact that our original solutions are sub-optimal. Otherwise, the new solution could only, in the best case scenario, be as good as the original one.

The results for the disrupted instances with *invalid room* and *invalid time* are shown in Tables 2 and 3, respectively. The tables show the average and median required CPU time to find an optimal solution, as well as the distance between the two solutions ($\delta_{HD}$) and the change in the global cost ($\delta_{cost}$). It is important to take into consideration that the value of $\delta_{HD}$ is directly linked to the size of the instance due to the process of generating disrupted instances.

Figure 2(b) shows the CPU time, per university, for the instances with and without disruptions. In most cases, less time is needed to solve a problem instance subject to small disruptions than to solve the original problem instance. If the disruptions cause no perturbations in the original solution, then almost no time is needed (only parsing time). However, our disrupted instances were subject to significant disruptions. In most cases, the solver is able to find the optimal solution taking around the same time it took to find the best solution without disruptions. The time spent to find a solution increases with the number of perturbations required.

**Table 2.** Results for the *Invalid Room* disruption. $\delta_{HD}$ measures the number of perturbations and $\delta_{cost}$ measures the change in the global quality of the solution.

| | | Invalid Room | | | | | |
|---|---|---|---|---|---|---|---|
| | | Avg. Time (s) | Med. Time (s) | Avg. $\delta_{HD}$ | Med. $\delta_{HD}$ | Avg. $\delta_{cost}$ | Med. $\delta_{cost}$ |
| Early | agh-fis-spr17 | 1,460.4 | 1,612.7 | 22 | 29 | 42 | 39 |
| | agh-ggis-spr17 | 2,321.2 | 2,210.8 | 11 | 8 | 0 | 1 |
| | mary-spr17 | 231.5 | 253 | 25 | 29 | 54 | 55 |
| | muni-fi-spr16 | 2,133.2 | 2,317.9 | 15 | 18 | 4 | 6 |
| | muni-fsps-spr17 | 812 | 999.1 | 13 | 18 | 0 | 0 |
| | muni-pdf-spr16c | 4,114.8 | 4,101.2 | 42 | 38 | 26 | 21 |
| | pu-llr-spr17 | 142.5 | 143 | 35 | 36 | 6 | 6 |
| | tg-fal17 | 1,208.8 | 1,247 | 100 | 112 | 18 | 19 |
| Middle | agh-ggos-spr17 | 3,212.6 | 3,212.9 | 40 | 40 | 640 | 639 |
| | agh-h-spr17 | 679.9 | 699.9 | 19 | 20 | 57 | 60 |
| | lums-spr18 | 913.9 | 921.8 | 18 | 18 | 0 | 0 |
| | muni-fi-spr17 | 80.8 | 99 | 9 | 13 | 36 | 37 |
| | muni-fsps-spr17c | 888.4 | 977.3 | 39 | 44 | 20 | 20 |
| | muni-pdf-spr16 | 1,354.5 | 1,444.1 | 89 | 94 | 1,335 | 1,336 |
| | nbi-spr18 | 3,701.7 | 3,781 | 14 | 13 | 33 | 35 |
| | yach-fal17 | 415.56 | 420 | 56 | 66 | 84 | 86 |
| Late | lums-fal17 | 999.9 | 1,000.2 | 20 | 20 | 0 | 0 |
| | mary-fal18 | 788.9 | 812.1 | 20 | 24 | 40 | 42 |
| | tg-spr18 | 813.8 | 888 | 5 | 8 | 100 | 100 |
| | muni-fi-fal17 | 248.9 | 250.1 | 9 | 10 | 36 | 30 |

As one can see in Fig. 2(b), the *invalid room* disruptions are, in most cases, easier to sort out than *invalid time* disruptions. The CPU time is smaller since fewer perturbations are needed. The reduction in time can be also explained by the fact that a smaller number of hard constraints are, in fact, related to rooms. The solutions found are, usually, closer to the original one. This can be explained by the fact that most instances have fewer rooms than time slots available.

The *muni-f\** instances are, in most cases, the most difficult instances to solve after *invalid room* disruptions. This can be explained by the fact that these instances are very *tight* in terms of room space. On average, these instances only have 4 possible rooms by class versus an average of 14 in the other instances.

To evaluate the quality of the fittings the following metrics were defined. Root mean square error (RMSE) has a range from 0 to $\infty$, where the best fit model has a value closer to zero. Coefficient of determination (CD) has a range between 0 and 1, where the best fit model has a value closer to 1. To perform the fitting, we used the Microsoft Excel Solver [31].

Figure 3(a) shows the relation between the room domain size on $\delta_{HD}$. The RMSE of the fit function is 0.04. The CD is 0.95. Note that, for fairness we normalized the value of $\delta_{HD}$. The normalization simply takes into account the number of disruptions generated to the instance ($\delta_{NHD}$).

**Table 3.** Results for the *Invalid Time* disruption. $\delta_{HD}$ measures the number of perturbations and $\delta_{cost}$ measures the change in the global quality of the solution.

|         |                 | Invalid Time | | | | | |
|---------|-----------------|--------------|--------------|----------------|---------------|------------------|------------------|
|         |                 | Avg. Time (s) | Med. Time (s) | Avg. $\delta_{HD}$ | Med. $\delta_{HD}$ | Avg. $\delta_{cost}$ | Med. $\delta_{cost}$ |
| Early   | agh-fis-spr17   | 1596.22      | 1711.1       | 5001           | 5003          | 4                | 6                |
|         | agh-ggis-spr17  | 2358.2       | 2100.4       | 4              | 3             | 0                | 3                |
|         | mary-spr17      | 381.2        | 380.1        | 0              | 4             | 0                | 6                |
|         | muni-fi-spr16   | 1784.2       | 1794.2       | 16             | 18            | 0                | 0                |
|         | muni-fsps-spr17 | 212.4        | 218.4        | 45             | 46            | 0                | 0                |
|         | muni-pdf-spr16c | 2992.1       | 3001.2       | 6              | 6             | 4                | 4                |
|         | pu-llr-spr17    | 342.6        | 356          | 122            | 126           | 10               | 10               |
|         | tg-fal17        | 1408.7       | 1484         | 2021           | 2070          | 25               | 25               |
| Middle  | agh-ggos-spr17  | 5465.8       | 5466.1       | 92             | 93            | 276              | 139              |
|         | agh-h-spr17     | 919.1        | 920.9        | 97             | 98            | 290              | 289              |
|         | lums-spr18      | 961          | 978.8        | 6446           | 6436          | 0                | 0                |
|         | muni-fi-spr17   | 40.12        | 39           | 144            | 140           | 433              | 423              |
|         | muni-fsps-spr17c| 500.3        | 498.8        | 137            | 136           | 0                | 0                |
|         | muni-pdf-spr16  | 1035.3       | 1030         | 636            | 630           | 6363             | 6364             |
|         | nbi-spr18       | 3803.8       | 3991.1       | 164            | 186           | 3284             | 3289             |
|         | yach-fal17      | 112.56       | 111          | 100            | 100           | 0                | 4                |
| Late    | lums-fal17      | 1085.58      | 1100.1       | 6777           | 6787          | 0                | 0                |
|         | mary-fal18      | 800.12       | 812.1        | 269            | 270           | 807              | 900              |
|         | tg-spr18        | 933.2        | 934          | 568            | 559           | 1704             | 1705             |
|         | muni-fi-fal17   | 149.2        | 140.2        | 101            | 108           | 50               | 51               |

The *lums\** instances are the ones that have the largest $\delta_{HD}$ when tested subject to *invalid time* disruptions (see Table 3). This fact can be explained by the large number of constraints forcing the classes to be in the same allocation slot (*SameWeek*, *SameTime*, *SameDay* and *SameStart*). These constraints force a chain of perturbations for a single disruption. Figure 3(b) shows the relation between of the number of classes involved in constraints of type *Same* on the $\delta_{HD}$. The RMSE of the fit function is 131.8. The CD is 0.11.

## 5  Conclusion and Future Work

This paper discusses the real-world problem of solving university course timetabling problems which can be subject to disruptions. We propose a MaxSAT encoding to solve course timetabling and student sectioning problems. To validate our approach, we used the ITC-2019 benchmark. The approach is able to solve two thirds of the benchmark instances within the time limit of 6,000 s. Moreover, the proposed solution is able to efficiently solve them after the occurrence of the most common disruptions reported in the literature.

As future work, we recommend extending this work to explore the incremental nature of MPP. The application of an incremental algorithm would, in theory, reduce CPU time bypassing the repetition of decisions during the search for a feasible solution. Furthermore, one can study the performance of this implementation using different SAT solvers.

# References

1. McCollum, B.: University timetabling: bridging the gap between research and practice. In: 5th International Conference on the Practice and Theory of Automated Timetabling (PATAT), pp. 15–35(2006)
2. Oude Vrielink, R.A., Jansen, E.A., Hans, E.W., van Hillegersberg, J.: Practices in timetabling in higher education institutions: a systematic review. Ann. Oper. Res. **275**(1), 145–160 (2017). https://doi.org/10.1007/s10479-017-2688-8
3. Müller, T., Rudová, H., Müllerová, Z.: University course timetabling and international timetabling competition 2019. In: Proceedings of 12th International Conference on the Practice and Theory of Automated Timetabling (PATAT), p. 27 (2018)
4. Müller, T.: ITC-2007 solver description: a hybrid approach. Ann. Oper. Res. **172**(1), 429 (2009). https://doi.org/10.1007/s10479-009-0644-y
5. Di Gaspero, L., Schaerf, A., McCollum, B.: The second international timetabling competition (ITC-2007): Curriculum-based course timetabling (track 3). Technical report, Queen's University (2007)
6. Laporte, G., Desroches, S.: The problem of assigning students to course sections in a large engineering school. Comput. Oper. Res. **13**(4), 387–394 (1986)
7. Even, S., Itai, A., Shamir, A.: On the complexity of timetable and multicommodity flow problems. Soc. Ind. Appl. Math. J. Comput. **5**(4), 691–703 (1976). https://doi.org/10.1137/0205048
8. Müller, T., Rudová, H., Barták, R.: Minimal perturbation problem in course timetabling. In: Burke, E., Trick, M. (eds.) PATAT 2004. LNCS, vol. 3616, pp. 126–146. Springer, Heidelberg (2005). https://doi.org/10.1007/11593577_8
9. Atsuta, M., Nonobe, K., Ibaraki, T.: ITC-2007 track 2: an approach using a general CSP solver. In: 7th International Conference on the Practice and Theory of Automated Timetabling (PATAT), pp. 19–22 (2008)
10. Banbara, M., et al.: *teaspoon* : solving the curriculum-based course timetabling problems with answer set programming. Ann. Oper. Res. **275**(1), 3–37 (2019). https://doi.org/10.1007/s10479-018-2757-7
11. Bittner, P.M., Thüm, T., Schaefer, I.: SAT encodings of the At-Most-$k$ constraint - a case study on configuring university courses. In: Ölveczky, P.C., Salaün, G. (eds.) SEFM 2019. Lecture Notes in Computer Science, vol. 11724, pp. 127–144. Springer, Cham (2019). https://doi.org/10.1007/978-3-030-30446-1_7
12. Asín Achá, R., Nieuwenhuis, R.: Curriculum-based course timetabling with SAT and MaxSAT. Ann. Oper. Res. **218**(1), 71–91 (2012). https://doi.org/10.1007/s10479-012-1081-x
13. Lemos, A., Melo, F.S., Monteiro, P.T., Lynce, I.: Room usage optimization in timetabling: a case study at Universidade de Lisboa. Oper. Res. Perspect. **6**, 100092 (2019). https://doi.org/10.1016/j.orp.2018.100092
14. Lindahl, M., Stidsen, T., Sørensen, M.: Quality recovering of university timetables. Eur. J. Oper. Res. **276**(2), 422–435 (2019)
15. Phillips, A.E., Walker, C.G., Ehrgott, M., Ryan, D.M.: Integer programming for minimal perturbation problems in university course timetabling. Ann. Oper. Res. **252**(2), 283–304 (2016). https://doi.org/10.1007/s10479-015-2094-z
16. Gülcü, A., Akkan, C.: Robust university course timetabling problem subject to single and multiple disruptions. Eur. J. Oper. Res. **283**(2), 630–646 (2020)
17. Zivan, R., Grubshtein, A., Meisels, A.: Hybrid search for minimal perturbation in dynamic CSPs. Constraints **16**(3), 228–249 (2011). https://doi.org/10.1007/s10601-011-9108-5

18. Hebrard, E., Hnich, B., O'Sullivan, B., Walsh, T.: Finding diverse and similar solutions in constraint programming. In: 20th National Conference on Artificial Intelligence and 17th Innovative Applications of Artificial Intelligence, pp. 372–377 (2005)

19. Hamming, R.W.: Error detecting and error correcting codes. Bell Syst. Tech. J. **29**(2), 147–160 (1950). https://doi.org/10.1002/j.1538-7305.1950.tb00463.x

20. Biere, A., Heule, M., van Maaren, H.: Handbook of Satisfiability, vol. 185. IOS Press, Amsterdam (2009)

21. Eén, N., Sörensson, N.: Translating pseudo-Boolean constraints into SAT. J. Satisf. Boolean Model. Comput. **2**(1–4), 1–26 (2006)

22. Carter, M.W.: A comprehensive course timetabling and student scheduling system at the University of Waterloo. In: Burke, E., Erben, W. (eds.) PATAT 2000. LNCS, vol. 2079, pp. 64–82. Springer, Heidelberg (2001). https://doi.org/10.1007/3-540-44629-X_5

23. Schindl, D.: Optimal student sectioning on mandatory courses with various sections numbers. Ann. Oper. Res. **275**(1), 209–221 (2017). https://doi.org/10.1007/s10479-017-2621-1

24. Roussel, O.: Controlling a solver execution with the runsolver tool. J. Satisf. Boolean Model. Comput. **7**(4), 139–144 (2011)

25. Nadel, A.: Anytime weighted MaxSAT with improved polarity selection and bit-vector optimization. In: Proceedings of the 19th Conference on Formal Methods in Computer Aided Design (FMCAD) (2019)

26. Nadel, A.: TT-Open-WBO-Inc: tuning polarity and variable selection for anytime SAT-based optimization. In: Proceedings of the MaxSAT Evaluations (2019)

27. Joshi, S., Kumar, P., Martins, R., Rao, S.: Approximation strategies for incomplete MaxSAT. In: Hooker, J. (ed.) CP 2018. LNCS, vol. 11008, pp. 219–228. Springer, Cham (2018). https://doi.org/10.1007/978-3-319-98334-9_15

28. Marques-Silva, J., Argelich, J., Graça, A., Lynce, I.: Boolean lexicographic optimization: algorithms & applications. Ann. Math. Artif. Intell. **62**(3–4), 317–343 (2011). https://doi.org/10.1007/s10472-011-9233-2

29. Ansótegui, C., Manyà, F.: Mapping problems with finite-domain variables into problems with Boolean variables. In: Proceedings of the Seventh International Conference on Theory and Applications of Satisfiability Testing (SAT), vol. 3542, pp. 1–15 (2004)

30. Warners, J.P.: A linear-time transformation of linear inequalities into conjunctive normal form. Inf. Process. Lett. **68**(2), 63–69 (1998)

31. Fylstra, D.H., Lasdon, L.S., Watson, J., Waren, A.D.: Design and use of the microsoft excel solver. Interfaces **28**(5), 29–55 (1998). https://doi.org/10.1287/inte.28.5.29

# Leveraging Constraint Scheduling: A Case Study to the Textile Industry

Alexandre Mercier-Aubin, Jonathan Gaudreault, and Claude-Guy Quimper[✉]

Université Laval, Québec, QC G1V 0A6, Canada
claude-guy.quimper@ift.ulaval.ca

**Abstract.** Despite the significant progress made in scheduling in the past years, industrial problems with several hundred tasks remain intractable for some variants of the scheduling problems. We present techniques that can be used to leverage the power of constraint programming to solve an industrial problem with 800 non-preemptive tasks, 90 resources, and sequence-dependent setup times. Our method involves solving the traveling salesperson problem (TSP) as a simplification of the scheduling problem and using the simplified solution to guide the branching heuristics. We also explore large neighborhood search. Experiments conducted on a dataset provided by our partner from the textile industry show that we obtain non-optimal but satisfactory solutions.

**Keywords:** Multi-resource · Scheduling · Constraint programming · Traveling salesman problem

## 1 Introduction

Nowadays, most textiles are mass-produced using automated looms. In the context of the fourth industrial revolution [7,8], the textile industry seeks to expand the automation to planning and scheduling tasks. While recent progress made in constraint programming allows tackling many NP-Hard scheduling problems, the size of the scheduling instances that can be solved for some variants of the problem remain small compared to what the industry needs. It is common to observe industrial instances with 800 tasks and limited resources. In this context, constraint programming can still be used to obtain good, but not optimal, schedules. However, extra work on branching heuristics and the use of local search is often required to obtain satisfactory results.

We present a study case of a scheduling problem encountered by our industrial partner, a textile company. More than 800 tasks need to be scheduled over 90 automated looms. A team of technicians needs to set up the looms before starting new tasks. The duration of each setup depends on the tasks that precede and succeed the setup and no more setups than technicians should be simultaneously scheduled.

We explain how we succeed in obtaining good, but possibly non-optimal, solutions for large instances. We achieve this goal by solving a simplification of

© Springer Nature Switzerland AG 2020
E. Hebrard and N. Musliu (Eds.): CPAIOR 2020, LNCS 12296, pp. 334–346, 2020.
https://doi.org/10.1007/978-3-030-58942-4_22

the problem and using the simplification solution to find better solutions to the non-simplified problem. We also use a large neighborhood search to improve the solution. Note that similar problems were studied in the literature [4], but the approach to solve the problem relies on different optimization techniques such as Mixed Integer Programs.

The paper is divided as follows. Section 2 presents the preliminary concepts about constraint scheduling, the connection between scheduling with setup times and the traveling salesman problem (TSP), and the large neighborhood search. Section 3 formally introduces the industrial scheduling problem. Section 4 presents the mathematical models. Section 5 shows how to solve the models. Section 6 presents the experimental results. Finally, Sect. 7 concludes this work.

## 2  Background

### 2.1  Constraint Scheduling

We present the main components of common scheduling problems. The actual industrial scheduling problem we will solve is formally defined in Sect. 3.

A *scheduling problem* is composed of a set of tasks $\mathcal{I}$ (or activities) that need to be positioned on a time line. A task $i \in \mathcal{I}$ has for parameters its *earliest starting time* $\text{est}_i$, its *latest completion time* $\text{lct}_i$, a *processing time* $p_i$, a *due date* $d_i$, and a resource consumption rate $h_i$ also called *height*. We consider non-preemptive tasks, i.e. that when a task starts, it executes for exactly $p_i$ units of time without interruption. A task must start no earlier than its earliest starting time and complete no later than its latest completion time. A task that completes after its due date incurs a penalty that depends on the objective function.

A *cumulative resource* $r$ has capacity $C_r$. Multiple tasks can execute on a cumulative resource $r$ as long as the sum of their heights is no greater than $C_r$. By fixing $C_r = h_i$ for all tasks $i$, we obtain a *disjunctive resource* that can only execute one task at a time. On such a resource, it could happen that a setup needs to be performed between the execution of two tasks. The *setup time* $t_{i,j}$ is a minimum lapse of time that must occur between the completion of task $i$ and the starting time of task $j$, should task $i$ executes before task $j$. The setup times satisfy the triangle inequality $t_{i,j} + t_{j,k} \geq t_{i,k}$.

Constraint programming can be used to solve scheduling problems. One can define a *starting time* variable $S_i$ with domain $\text{dom}(S_i) = [\text{est}_i, \text{lct}_i - p_i]$. The constraints $\text{CUMULATIVE}([S_1, \ldots, S_n], [p_1, \ldots, p_n], [h_1, \ldots, h_n], C_r)$ or $\text{DISJUNCTIVE}([S_1, \ldots, S_n], [p_1, \ldots, p_n])$ ensures that the cumulative or disjunctive resource is not overloaded. Extra constraints can easily be added to the model such as a *precedence constraint* $S_i + p_i + t_{i,j} \leq S_j$ that forces a task $i$ to complete before a task $j$ can start.

Combining the concepts presented above results in a large variety of scheduling problems. For instance, the Resource-Constrained Project Scheduling Problem (RCPSP) contains cumulative resources and non-preemptive tasks subject to precedence constraints. Constraint solvers can find optimal solutions to instances of the RCPSP with 120 tasks [15]. However, with setup times, constraint solvers can only solve instances up to 120 tasks but never to optimality [1]. In order to improve the performances, search strategies and branching heuristics can be provided to the solver. It is also possible to use the constraint solver within a large neighborhood as explained in Sect. 2.2.

## 2.2  Local Search

A *local search* is a heuristic method that starts from a suboptimal and possibly unfeasible solution and tries to improve its feasibility and its objective value. At each iteration, an *operator* is applied, and the solution is modified. If the modified solution becomes more feasible or more optimal, it becomes the current solution. The operator modifies the values of a subset of the variables in the problem. When a large number of variables are modified, we say that the heuristic method is a *large neighborhood search*.

A constraint solver can be used as a large neighborhood search operator. One takes the current solution and forces a subset of variables to take the same values as the current solution. The remaining variables are let free. The result is a constraint satisfaction problem (CSP) with fewer variables to assign that is generally easier to solve. The solution of this CSP becomes the new current solution that can be further improved by selecting a different subset of variables.

There exists various strategies to select which variables to reassign in a scheduling problem. One could randomly select a fixed percentage of the tasks to reschedule [9]. It is also possible to select a time window from which all tasks whose execution is contained in this window are rescheduled [14]. Another option is to fix some precedences observed in the current solution between a subset of pairs of tasks and let the remaining pairs of tasks free from any precedence [13]. When minimizing the makespan in scheduling problems of 300 tasks, these techniques can significantly improve the objective function [5].

## 2.3  The Traveling Salesperson Problem

The traveling salesperson problem (TSP) is a classic optimization problem. In its directed variant, we have $n$ cities and a distance matrix $D$ such that $D_{i,j}$ is the distance to travel from $i$ to $j$. The matrix satisfies the triangle inequality $D_{i,j} + D_{j,k} \geq D_{i,k}$. The salesperson plans on finding the shortest circuit visiting each city exactly once.

There is a strong connection between the TSP and the scheduling problem with setup times. For a scheduling problem with $n$ tasks $\mathcal{I}$ and setup times $t_{i,j}$, one creates an instance of the TSP with $n+1$ cities and a distance matrix $D_{i,j} = t_{i,j}$ for $i, j \in \mathcal{I}$ and $D_{i,j} = 0$ otherwise. The extra city marks the beginning (and the end) of the schedule. The salesperson visits the cities (tasks) in order to minimize the sum of the distances (setup times). Figure 1 illustrates the reduction.

The solver Concorde [3] represents the state-of-the-art for solving the TSP. It can find and prove optimal solutions to instances with 85,000 cities. As it was designed solely for this type of problem, it cannot handle additional constraints.

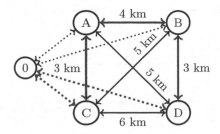

**Fig. 1.** Example of a scheduling problem with setup times reduced to a TSP problem. The node 0 is the dummy node. Dotted lines have null costs. The blue lines represent the optimal schedule: $C, A, B, D$. (Colour figure online)

For example, if the tasks have earliest starting times and latest completion times, the scheduling problem reduces to a TSP with time windows [11]. The solver Concorde cannot solve such problems.

Solving the TSP can help solving scheduling problems. For instance, Tran and Beck [17] use the TSP as the slave problem in a Benders decomposition to solve a problem with resource allocation and setup times.

## 3   Problem Description

We describe the industrial problem specified by our industrial partner. The parameters introduced in this section are summarized in Table 1. A task consists of weaving a textile on a loom. We therefore have a set of tasks $\mathcal{I}$ and a set of looms $L$. Each task $i \in \mathcal{I}$ is pre-assigned to a loom $l_i$ and has for processing time $p_i$. A loom $l \in L$ becomes available at time $a_l$. Prior to this time, the loom is busy terminating a task not in $\mathcal{I}$ that can neither be interrupted nor rescheduled. Each task $i \in \mathcal{I}$ has a style $z_i$, a due date $d_i$, and a priority $r_i$. We wish to minimize the total tardiness weighted by priority, i.e. $\sum_{i \in \mathcal{I}} r_i \cdot \max(0, S_i + p_i - d_i)$ where $S_i$ is the starting time of the task. The scheduling horizon spans from time 0 to time $H$. In practice, we have a horizon of 240 h with a time step of 15 min resulting in $H = 240 \times \frac{60}{15} = 960$.

*Major Setups:* Each loom $l \in L$ has an initial configuration $c_l^{\text{init}}$ and a final configuration $c_l^{\text{final}}$. If $c_l^{\text{init}} \neq c_l^{\text{final}}$ then there is a *major setup* of duration $p_l^{\text{major}}$ to change the configuration of the loom $l$. Only one major setup is possible during the scheduling horizon. A *specialized worker* is selected from a pool $W$ to achieve this major setup. A task $i \in \mathcal{I}$ needs to be executed on a loom $l_i$ when it has configuration $c_i \in \{c_{l_i}^{\text{init}}, c_{l_i}^{\text{final}}\}$. If $c_i = c_{l_i}^{\text{init}}$, the task needs to be executed prior to the major setup and after the major setup if $c_i = c_{l_i}^{\text{final}}$.

*Minor Setups:* A *minor setup* needs to be performed between two consecutive tasks $i, j \in \mathcal{I}$ on a loom. While a major setup entails a configuration change, a minor setup only gets the loom ready for its next job. This setup is decomposed

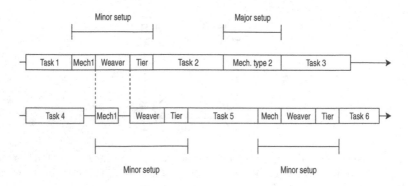

**Fig. 2.** Example of scheduling on two looms with one of each resource.

into several steps, each executed by a person of a different profession in $P$ which is disjoint from the set of workers $W$. The professions are sorted in order of execution and labeled with integers from 1 to $|P|$, i.e. that the first step of a minor setup is executed by profession 1, the second step is executed by profession 2, and so on. The order of execution is the same for all minor setups. The person of the profession $p \in P$ needs $t_{i,j,p}$ time to execute his/her part of the minor setup between task $i \in \mathcal{I}$ and $j \in \mathcal{I}$. There are $q_p$ people of the profession $p$. Consequently, no more than $q_p$ minor setup steps can be simultaneously executed by the people from the profession $p$.

Figure 2 shows a schedule on two looms with a worker of each category. We see conflicts delaying minor setups on the second loom.

**Table 1.** Parameters of the problem

| | |
|---|---|
| $\mathcal{I}$: | Set of tasks |
| $L$: | Set of looms |
| $P$: | Set of professions for minor setups |
| $W$: | Set of specialized workers for major setups |
| $r_i$: | Priority of task $i$ |
| $d_i$: | Due date of task $i$ |
| $z_i$: | Style of task $i$ |
| $l_i$: | Loom assigned to task $i$ |
| $c_l^{\text{init}}$: | Initial configuration of loom $l$ |
| $c_l^{\text{final}}$: | Final configuration of loom $l$ |
| $c_i$: | Required configuration for task $i$ |
| $p_i$: | Processing time of task $i$ |
| $p_l^{\text{major}}$: | Major setup time of loom $l$ |
| $t_{i,j,p}$: | Minor setup time between tasks $i$ and $j$ for the profession $p$ |
| $a_l$: | Earliest available time of loom $l$ |
| $q_p$: | Number of workers of the profession $p$ available for the minor setups |

**Table 2.** Variables and their domains

| Variable | Domain | Description |
|---|---|---|
| $S_l^{\text{major}}$ | $[a_l, H)$ | Start of the major setup on loom $i$ |
| $S_i$ | $[a_{l_i}, H)$ | Start of task $i$ |
| $S_{i,p}^{\text{minor}}$ | $[a_{l_i}, H)$ | Start of the minor setup for profession $p$ between task $i$ and its successor |
| $N_i$ | $\{j \in \mathcal{I} \mid l_j = l_i\} \cup \{\sigma_{l_i}\}$ | Next task after task $i$ |
| $F_l$ | $\{i \in \mathcal{I} \mid l_i = l\}$ | First task on loom $l$ |
| $T$ | $[0, \infty)$ | Total tardiness weighted by priority |

## 4   Models

We present an optimization model that is later submitted to a constraint solver. The variables and domains are summarized in Table 2. Constraints are stated from (1) to (11).

There are three types of events to schedule: tasks, major setups, and minor setups between a task and its successor. Let $S_i \in [a_l, H)$, $S_l^{\text{major}} \in [a_{l_i}, H)$, and $S_{i,p}^{\text{minor}} \in [a_{l_i}, H)$ be their starting time variables for tasks $i \in \mathcal{I}$, loom $l \in L$, and profession $p \in P$. Their domains prevent the events from starting before their respective loom $l$ becomes available.

The variable $F_l$ encodes the first task to execute on loom $l$. The variable $N_i$ encodes the task that succeeds task $i$ on the loom. If $i$ is the last task, its value is set to a sentinel. There is one sentinel per loom: $\sigma = \{\sigma_1, \ldots, \sigma_{|L|}\}$. The variable $N_i$ is also defined when $i \in \sigma$ is a sentinel. The next task of a sentinel is the first task on the next loom. Consequently, the vector $N$ is a permutation of $\mathcal{I} \cup \sigma$ with a single cycle.

The model contains the constraints (1) to (11). The objective function (1) minimizes the tardiness of the tasks, weighted by priority. Constraints (2) and (3) ensure that a task requiring its loom's initial configuration executes before the major setup or else waits after the major setup to execute. Constraint (4) ensures that the first step of the minor setup following task $i$ starts once task $i$ is completed. Constraint (5) is a precedence constraint over the different steps of a minor setup. Constraint (6) makes the task $N_i$ start immediately after the minor setup is completed. Indeed, once the loom is ready, there is no need to postpone the task. Constraints (7) and (8) ensure that the last task of a sentinel on a loom is the first task on the next loom. The loom that succeeds the last loom is the first loom. That creates a circuit visiting each task exactly once. This idea of a circuit is inspired from Focacci et al. [6] and led to the addition of constraint (9) to the model. This constraint [10] offers a strong filtering on the next variables $N$.

The model contains global constraints specialized for scheduling problems. We use the notation $[f(x) \mid x \in X]$ to represent the vector $[f(x_1), \ldots, f(x_n)]$ for $X = \{x_1, \ldots, x_n\}$. The constraint (10) limits to $q_p$ the number of simultaneous minor setups accomplished by a person of the profession $p$. The constraint (11) limits to $|W|$ the number of simultaneous major setups. The constraint (12) breaks a symmetry by forcing tasks producing the same product style on the same loom to execute in order of due dates.

Minimize $T$ subject to:

$$\text{Minimize} \qquad \sum_{i \in \mathcal{I}} r_i \cdot \max(0, S_i + p_i - d_i) \tag{1}$$

$$c_i = c_{l_i}^{\text{init}} \implies S_i + p_i \leq S_{A_i}^{\text{major}} \qquad\qquad \forall i \in \mathcal{I} \tag{2}$$

$$c_i \neq c_{l_i}^{\text{init}} \implies S_{A_i}^{\text{major}} + p_l^{\text{major}} \leq S_i \qquad\qquad \forall i \in \mathcal{I} \tag{3}$$

$$S_i + p_i \leq S_{i,1}^{\text{minor}} \qquad\qquad \forall i \in \mathcal{I} \tag{4}$$

$$S_{i,p+1}^{\text{minor}} \geq S_{i,p}^{\text{minor}} + t_{i,N_i,p} \qquad\qquad \forall i \in \mathcal{I}, \forall p \in P \setminus \{|P|\} \tag{5}$$

$$S_{N_i} = S_{i,|P|}^{\text{minor}} + t_{i,N_i,|P|} \qquad\qquad \forall i \in \mathcal{I} \tag{6}$$

$$N_{\sigma_l} = F_{l+1} \qquad\qquad \forall l \in [1, |L| - 1] \tag{7}$$

$$N_{\sigma_{|L|}} = F_1 \tag{8}$$

$$\text{Circuit}(N) \tag{9}$$

$$\text{Cumulative}([S_{i,p}^{\text{minor}} \mid i \in \mathcal{I}], [t_{i,N_i,p} \mid i \in \mathcal{I}], 1, q_p) \qquad \forall p \in P \tag{10}$$

$$\text{Cumulative}([S_l^{\text{major}} \mid l \in L], [p_l^{\text{major}} \mid l \in L], 1, |W|) \tag{11}$$

$$S_a \leq S_b \qquad\qquad \forall a, b \in \mathcal{I}, z_a = z_b \land l_a = l_b \land d_a \leq d_b \tag{12}$$

## 5 Resolution

We present four methods to solve the model from the previous section. Some are pure heuristics or are *rules of thumb* used in the industry. These methods are either used as a point of comparison or are integrated as a branching heuristics in the constraint solver.

### 5.1 The Greedy Method Based on Due Dates

The first method, denoted Greedy, consists of executing the tasks on a loom in non-decreasing order of due dates. Ties are arbitrarily broken. If a resource is unavailable to either execute the minor or major setup that precedes a task, it delays the execution of the task until the resource becomes available and has time to complete the setup.

While this method might return a sub-optimal solution, it is nevertheless a rule of thumb used by people in the industry to generate an initial schedule that can be improved later. It is also a point of comparison for other methods.

## 5.2   The CIRCUIT Method

The next approach denoted CIRCUIT focuses on the CIRCUIT constraint. We want to find the circuit that minimizes the sum of the setup times. While this objective function is not the weighted tardiness, it is correlated. Indeed, shorter setups lead to shorter idle times and therefore earlier completion times.

We solve the TSP instance induced by the CIRCUIT constraint using the solver Concorde [3]. Concorde is quick to solve TSP instances, especially for instances with fewer than 1000 cities. We sort the looms in non-decreasing amount of time available to execute the minor setups, i.e. for each loom $l \in L$, we compute $H - a_l - \sum_{i|l_i=l} p_i - p_l^{\text{major}}$. Loom by loom in the sorted order, we assign their tasks in the same order found by Concorde. We delay the minor and major setups until the resource is available. The resulting schedule minimizes the amount of time spent by the workers on the setups without considering tardiness.

## 5.3   The CP Method

The next approach denoted CP consists in coding the model with the MiniZinc [12] language and submitting the model to the constraint solver Chuffed [2].

As a branching heuristics, we generate a *template solution* before the search with either the method GREEDY or CIRCUIT. During the search, we choose a next variable $N_i$ that can be assigned to the same value as in the template solution. That was implemented by declaring a vector $B$ of Boolean variables connected to the model with the constraints $B_i = 1 \iff N_i = N_i^t$ where $N_i^t$ is the successor of task $i$ in the template solution. The heuristics branches on the vector $B$ by setting the variables to the value 1. This has for effect to set $N_i = N_i^t$. In case the value 0 is selected for $B_i$ (for instance, after a backtrack), that imposes the constraint $N_i! = N_i^t$ but this does not fix the variable $N_i$. Once the variables in $B$ are assigned, the solver chooses the starting variables of a task, a minor setup, or a major set up with the smallest value in its domain and assigns this variable to this smallest value. This has for effect to set all next variables that were not already assigned to a value.

## 5.4   The LNS Method

The large neighborhood search, denoted LNS, starts with an initial solution that could be, for instance, the solution obtained from the methods GREEDY and CIRCUIT. It iteratively improves this solution by randomly selecting looms. The CP method is then called to reschedule the tasks on these looms while leaving the tasks on unselected looms untouched.

A note about our implementation. For each iteration, we generate a MiniZinc data file that contains the execution time of the unselected tasks. The model has unary constraints of the form $S_i = v$ to fix the variables to a value. A constraint states that the objective value must improve over the best solution found so far. We solve for satisfaction, i.e. we stop the search once we found an improving solution. An iteration of the LNS is given a timeout (we use 5 minutes) after which, if no solution is found, we pass to the next iteration without changing the current solution but by reseting 10% fewer looms. Whenever the solver returns unsatisfiable, we reset 10% more looms until a solution is found or infeasibility is proven (which implies that the current solution is optimal).

Since the time for compiling the MiniZinc code is significant and could be avoided if the local search was directly implemented in C++ calling the Chuffed solver, we do not count the MiniZinc compilation time in the solving time. The search stops when the computation time, that excludes MiniZinc compilation time, reaches a timeout.

In a context of a local search, we do not use the branching heuristics described in Sect. 5.3 since this heuristics aims at finding a solution similar to a template solution. In a local search, one rather wants to find a solution that is different from the current one. We simply randomly assign the next variables $N$.

## 6    Experiments

### 6.1    Instances

Our industrial partner shared 4 instances with 571, 592, 756, and 841 tasks. From each instance, we create a dataset of 10 instances by randomly selecting 10%, 20%, ..., 100% of the tasks from the original instance. Once the tasks are selected, this selection is used for all the tests and all the solving methods. This allows to see how our algorithms scale with the number of tasks. The number of looms $|L| = 90$, the number of professions $|P| = 3$ with quantities $[q_1, q_2, q_3] = [5, 3, 2]$, and the number of workers $|W| = 1$ for the major setups remains constant for all instances of all datasets.

### 6.2    Experimental Setup

The CP model[1] was written in the MiniZinc language [12]. We use the solver Chuffed [2] with the free search parameter [16]. The LNS method is implemented in Python. We ran the experiments on a computer with the following configuration: Ubuntu 19.10, 16 GB ram, Processor Intel(R) Core(TM) i7-6700K CPU @ 4.00GHz, 4008 Mhz, 4 Cores, 8 Logical Processors.

### 6.3    Methodology

We solved all instances using the GREEDY and CIRCUIT method. For the CP method, we tried three different branching heuristics: CP+GREEDY assigns the next variables according to the solution of GREEDY, CP+CIRCUIT uses the solution from CIRCUIT, and CP+RANDOM randomly assigns the next variable and uses restarts with a Luby sequence with a scale of 250.

---

[1] The MiniZinc files are freely available on Claude-Guy Quimper's web site or directly at http://www2.ift.ulaval.ca/~quimper/publications/CPAIOR2020Submission.zip.

For the methods based on CP, we tried the three configurations with LNS and without LNS. At each iteration, 50% of the looms are rescheduled. A specific iteration is given a timeout of 5 min to improve the solution.

All methods are given a timeout of 15 min. As the search goes, the CP methods (with or without LNS) keeps improving their best solution. We keep track of the objective value and time of the solutions as we find them during the search.

### 6.4   Results

Figure 3 shows the performances of the methods GREEDY and CIRCUIT. Each point represents a solution obtained for an instance: on the $x$-axis is the number of tasks in the instance and on the $y$-axis is the objective value of the solution returned by the method. The color indicates after how much time (in seconds) the solution was found.

As expected, instances with a larger number of tasks get a larger weighted tardiness. Indeed, since the resources and the due dates remain the constant among all instances, it gets harder to deliver products on time if we increase the number of orders without increasing the resources or delaying the due dates.

Both the methods GREEDY and CIRCUIT solve every instance in less than a second. The quality of the solution is not competitive with any method using CP. However, we will see that GREEDY and CIRCUIT are nevertheless useful to guide the branching heuristics of the CP solver.

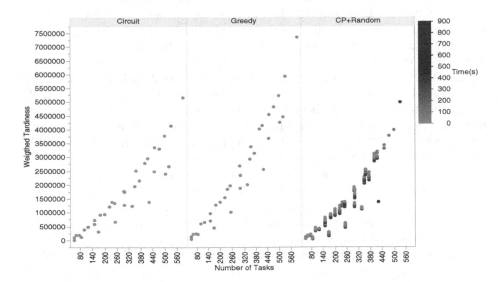

**Fig. 3.** Comparison between CIRCUIT, GREEDY, and one of the methods based on CP (Color figure online)

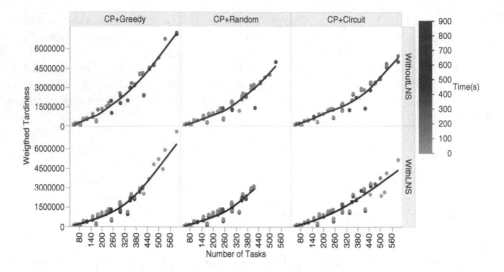

**Fig. 4.** Comparison of methods based on CP

Figure 4 presents a comparison of the six methods based on CP. The left, middle, and right graphs make the branching heuristics vary. It is either based on the CIRCUIT, GREEDY, or the random heuristics. The graphs on the top use the standard CP search while the graphs at the bottom use the LNS. For the LNS, the initial solutions are also based on CIRCUIT, GREEDY, and RANDOM. For CP+CIRCUIT+LNS and CP+GREEDY+LNS, the next variables $N_i$ are set in the initial solution according to the precedences in the solutions generated by CIRCUIT and GREEDY. For CP+RANDOM+LNS, the initial solution is completely random.

The first thing to analyze on Fig. 4 is how the methods behave on instances with more than 500 tasks. The methods CP+RANDOM CP+RANDOM+LNS are unable to solve all instances while other methods do. The CP+GREEDY method is more stable than CP+RANDOM while CP+RANDOM obtains better solutions. CP+CIRCUIT and CP+CIRCUIT+LNS outperform CP+RANDOM and CP+RANDOM+LNS in both stability and objective value.

Globally, CP+CIRCUIT and CP+CIRCUIT+LNS offer the solutions with the smallest weighted tardiness. This might look surprising at first sight because CIRCUIT aims at minimizing the amount of setups while GREEDY directly minimizes tardiness. However, the solution that CIRCUIT generates is optimal according to the amount of setup while the GREEDY algorithm only returns an approximation for the weighted tardiness. Guiding the search towards a solution that is optimal, even according to a different but correlated criteria, provides the best solution.

The third observation is that CP+CIRCUIT+LNS generally outputs better solutions than CP+CIRCUIT. The same goes for CP+GREEDY+LNS compared to CP+GREEDY. While other methods perform better with our homemade LNS, the random heuristic offers poor results.

Finally, we observe that using LNS is beneficial. The poor results of the LNS on the random heuristic is most likely due to a bad initial solution.

## 7  Conclusion

We presented a model to solve an industrial instance presented by our partner. We showed how a solution from a simplification (the TSP) can guide the search to obtain better solutions. Our model is now able to find solutions to the expected range of tasks. The integration of our program to the operations planning team is in progress.

## References

1. Chakrabortty, R.K., Sarker, R.A., Essam, D.L.: Resource constrained project scheduling with uncertain activity durations. Comput. Ind. Eng. **112**, 537–550 (2017). https://doi.org/10.1016/j.cie.2016.12.040. http://www.sciencedirect.com/science/article/pii/S0360835216305186
2. Chu, G., Stuckey, P.J., Schutt, A., Ehlers, T., Gange, G., Francis, K.: Chuffed, a lazy clause generation solver (2018)
3. Cook, W.: In Pursuit of the Traveling Salesman: Mathematics at the Limits of Computation. Princeton University Press, Princeton (2011)
4. Costa, A., Cappadonna, F.A., Fichera, S.: A hybrid genetic algorithm for job sequencing and worker allocation in parallel unrelated machines with sequence-dependent setup times. Int. J. Adv. Manuf. Technol. **69**, 2799–2817 (2013). https://doi.org/10.1007/s00170-013-5221-5
5. Danna, E., Perron, L.: Structured vs. unstructured large neighborhood search: a case study on job-shop scheduling problems with earliness and tardiness costs. In: Rossi, F. (ed.) CP 2003. LNCS, vol. 2833, pp. 817–821. Springer, Heidelberg (2003). https://doi.org/10.1007/978-3-540-45193-8_59
6. Focacci, F., Laborie, P., Nuijten, W.: Solving scheduling problems with setup times and alternative resources. In: Proceedings of the Fifth International Conference on Artificial Intelligence Planning Systems, pp. 92–101 (2000)
7. Hermann, M., Pentek, T., Otto, B.: Design principles for Industrie 4.0 scenarios. In: 49th Hawaii International Conference on System Sciences (HICSS), pp. 3928–3937 (2016)
8. Kagermann, H., Helbig, J., Hellinger, A., Wahlster, W.: Recommendations for implementing the strategic initiative Industrie 4.0: securing the future of German manufacturing industry; final report of the Industrie 4.0 working group. Technical report, Forschungsunion (2013)
9. Kajgård, E.: Route optimisation for winter road maintenance using constraint modelling (2015)
10. Lauriere, J.L.: A language and a program for stating and solving combinatorial problems. Artif. Intell. **10**(1), 29–127 (1978)
11. López-Ibáñez, M., Blum, C., Ohlmann, J.W., Thomas, B.W.: The travelling salesman problem with time windows: adapting algorithms from travel-time to makespan optimization. Appl. Soft Comput. **13**(9), 3806–3815 (2013)

12. Nethercote, N., Stuckey, P.J., Becket, R., Brand, S., Duck, G.J., Tack, G.: MiniZinc: towards a standard CP modelling language. In: Bessière, C. (ed.) CP 2007. LNCS, vol. 4741, pp. 529–543. Springer, Heidelberg (2007). https://doi.org/10.1007/978-3-540-74970-7_38

13. Psaraftis, H.N.: k-interchange procedures for local search in a precedence-constrained routing problem. Eur. J. Oper. Res. **13**(4), 391–402 (1983)

14. Savelsbergh, M.W.P.: Local search in routing problems with time windows. Ann. Oper. Res. **4**(1), 285–305 (1985). https://doi.org/10.1007/BF02022044

15. Schutt, A., Feydy, T., Stuckey, P.J., Wallace, M.G.: Why cumulative decomposition is not as bad as it sounds. In: Gent, I.P. (ed.) CP 2009. LNCS, vol. 5732, pp. 746–761. Springer, Heidelberg (2009). https://doi.org/10.1007/978-3-642-04244-7_58

16. Shishmarev, M., Mears, C., Tack, G., Garcia de la Banda, M.: Learning from learning solvers. In: Rueher, M. (ed.) CP 2016. LNCS, vol. 9892, pp. 455–472. Springer, Cham (2016). https://doi.org/10.1007/978-3-319-44953-1_29

17. Tran, T.T., Beck, J.C.: Logic-based benders decomposition for alternative resource scheduling with sequence dependent setups. In: Proceedings of the 20th European Conference on Artificial Intelligence ECAI 2012, pp. 774–779 (2012)

# Template Matching and Decision Diagrams for Multi-agent Path Finding

Jayanth Krishna Mogali[1]([⊠]) [iD], Willem-Jan van Hoeve[2][iD],
and Stephen F. Smith[1][iD]

[1] Robotics Institute, Carnegie Mellon University, Pittsburgh, USA
{jmogali,ssmith}@andrew.cmu.edu
[2] Tepper School of Business, Carnegie Mellon University, Pittsburgh, USA
vanhoeve@andrew.cmu.edu

**Abstract.** We propose a polyhedral cutting plane procedure for computing a lower bound on the optimal solution to multi-agent path finding (MAPF) problems. We obtain our cuts by projecting the polytope representing the solutions to MAPF to lower dimensions. A novel feature of our approach is that the projection polytopes we used to derive the cuts can be viewed as 'templates'. By translating these templates spatio-temporally, we obtain different projections, and so the cut generation scheme is reminiscent of the template matching technique from image processing. We use decision diagrams to compactly represent the templates and to perform the cut generation. To obtain the lower bound, we embed our cut generation procedure into a Lagrangian Relax-and-Cut scheme. We incorporate our lower bounds as a node evaluation function in a conflict-based search procedure, and experimentally evaluate its effectiveness.

**Keywords:** MAPF · Projection cuts · Template polytopes · Decision diagrams · Lagrangian relax and cut · Conflict based search

## 1 Introduction

Multi-agent path finding (MAPF) is the problem of finding paths for individual robots (agents), given a start and end vertex for each robot on some layout (graph), such that the paths are spatio-temporally conflict-free and an objective resembling travel costs is minimized. MAPF has found many applications in warehouse logistic systems [19] and robotics. MAPF is known to be NP-Hard to solve optimally on general graphs [21], nonetheless many techniques have been proposed and they come in different flavors.

Current approaches for MAPF can be broadly classified into search based methods [15,17], and solution methods that rely on polyhedral techniques such as the integer programming formulation of [20], and the branch-cut-price method of [12]. A significant challenge for either of these approaches is in developing strong lower bounding techniques. Such techniques are needed to prune search

© Springer Nature Switzerland AG 2020
E. Hebrard and N. Musliu (Eds.): CPAIOR 2020, LNCS 12296, pp. 347–363, 2020.
https://doi.org/10.1007/978-3-030-58942-4_23

regions that do not lead to an optimal solution. From the point of view of search based methods, this translates into developing strong admissible heuristics. For polyhedral techniques, strong lower bounds are typically obtained by developing cutting planes that tighten the linear description of the solution space.

In this paper, we propose a cut generation scheme that can be incorporated into search based methods for MAPF as well as polyhedral approaches. Incorporating cuts into techniques that use polyhedral approaches is common, but incorporating cuts into search based methods is rare and the focus of this paper.

**Contributions.** Main contributions of this work are 1) a new polyhedral approach for MAPF based on lower-dimensional 'templates' that can be translated spatio-temporally over the input graph, 2) the development of a cut generation scheme from these templates, which utilizes a decision diagram representation, 3) a Lagrangian Relax-and-Cut procedure to compute the lower bound, and 4) incorporating the resulting lower bound as a node evaluation function in a conflict-based search (CBS) procedure. Experimental evaluation shows that our lower bounds can be very effective when the MAPF problem is more constrained.

## 2 MAPF Problem Description

We consider the makespan-constrained version of the MAPF problem in this paper. We are given an undirected graph $G = (V, E)$, a set of $\mathbf{N}$ robots $\mathcal{R} = \{r_1, ..., r_\mathbf{N}\}$, and a makespan upper bound $\mathbf{T} \in \mathbb{Z}_+$, where $\mathbb{Z}_+$ represents the set of positive integers. Corresponding to each robot $r_i \in \mathcal{R}$, we are given a start vertex $s_i \in V$, and goal vertex $g_i \in V$. The task is to find a path for each robot, such that the robot paths do not conflict while minimizing the cumulative sum of path costs. A path $p$ can be viewed as a function $p : \{0, 1, ..., \mathbf{T}\} \to V$, where $p(t)$ returns a vertex in $V$ corresponding to time $t$. If $\mathcal{P} = \{p_1, ..., p_\mathbf{N}\}$ is a set of robot paths with 1 path for each robot, $\mathcal{P}$ is feasible to the MAPF problem iff:

1. $p_i(0) = s_i$ and $p_i(\mathbf{T}) = g_i$, $\forall i \in \{1, 2, ..., \mathbf{N}\}$.
2. For each robot $r_i \in \mathcal{R}$ and for all $t \in \{0, 1, ..., \mathbf{T} - 1\}$, we require $p_i(t) = p_i(t+1)$, or $(p_i(t), p_i(t+1)) \in E$. The robot either stays in its current vertex or moves to a neighbor.
3. To prevent vertex collisions, we require that $p_i(t) \neq p_j(t)$, for all pairwise distinct $i, j \in \{1, ..., \mathbf{N}\}$ and time $t \in \{0, 1, ..., \mathbf{T}\}$.
4. To prevent edge collisions, there should not exist a pair of robots $r_i, r_j$ and time $t \in \{0, 1, ..., \mathbf{T} - 1\}$ such that, $p_i(t) = p_j(t+1)$ and $p_i(t+1) = p_j(t)$.

We refer to any path $p_i$ satisfying 1 and 2 as a **start-end path** for robot $r_i$. The cost of start-end path $p_i$ is given by $c_i(p_i) = \sum_{t=0}^{\mathbf{T}-1} c_i(p_i(t), p_i(t+1))$, where

$$c_i(p_i(t), p_i(t+1)) = \begin{cases} 0, & \text{if } p_i(t) = p_i(t+1) = g_i \\ 1, & \text{otherwise} \end{cases} \quad (1)$$

Equation (1) assigns a cost of 0 if the robot is waiting at its goal vertex, else assigns a cost of 1. The goal of MAPF is to find a set of conflict-free robot paths

$p_1, ..., p_{\mathbf{N}}$ that minimizes the objective $\sum_{i=1}^{\mathbf{N}} c_i(p_i)$. The cost function in (1) slightly differs from the sum of completion times used in [9,13], where completion time is the earliest time the robot reaches its goal and remains stationary until time $\mathbf{T}$ at the goal vertex. We adopt the cost function shown in (1) to simplify the presentation of template construction presented in a later section.

## 3   Integer Programming Model for MAPF

We next provide a multi-commodity flow based Integer Programming (IP) model for the MAPF problem, similar to [20]. The IP model will be useful in deriving valid inequalities for the lower bounding procedure we propose in later sections. Below, for any $n \in \mathbb{Z}_+$ we will use the notation $[n]$ to denote the set $\{1, 2, ..., n\}$.

The IP model will make use of the so-called "time expanded graph". The time expanded graph is an arc-weighted directed acyclic graph defined for each robot, where the nodes can be partitioned into $\mathbf{T} + 1$ layers, and arcs into $\mathbf{T}$ layers. We shall denote the time expanded graph for robot $r_i$ by $F_i(N_i, A_i)$. Denote the nodes in layer $t \in \mathbb{Z}_+ \cup \{0\}$ by $N_i(t)$. Corresponding to each vertex $v \in V$, there exists a node in $N_i(t)$ if the shortest path from $s_i$ to $v$ in $G(V, E)$ passes through at most $t$ edges, and the shortest path from $v$ to $g_i$ passes through at most $\mathbf{T} - t$ edges. With a slight abuse of notation, we shall denote the node corresponding to vertex $v \in V$ in $N_i(t)$ by $v_i^t$. Throughout this paper, if a node from any of the graphs in the set $\{F_i\}_{i \in [\mathbf{N}]}$ is specified, we will assume that the vertex, time and robot associated with that node can be deduced from our notation. Arcs in $A_i$ connect nodes between adjacent layers, with the tail of the arc emanating from the node belonging to the lower indexed layer. Denote the arcs in level $t$ by $A_i(t)$. If $u_i^t \in N_i(t)$ and $v_i^{t+1} \in N_i(t+1)$, then there exists an arc $(u_i^t, v_i^{t+1}) \in A_i(t)$ iff $u = v$, or $(u, v) \in E$. We let $c_i(u_i^t, v_i^{t+1})$ denote the weight of arc $(u_i^t, v_i^{t+1})$, where $c_i(u_i^t, v_i^{t+1}) = 0$ if $u = v = g_i$, and 1 otherwise. There is a 1:1 correspondence between **start-end paths** for robot $r_i$ and $s_i^0 - g_i^{\mathbf{T}}$ paths in $F_i(N_i, A_i)$.

In describing our IP model, we use the following notation. For any node $v_i^t \in N_i(t)$, we denote $\delta_{F_i}^+(v_i^t)$ as the set of arcs in $A_i$ whose tail is the node $v_i^t$, and $\delta_{F_i}^-(v_i^t)$ as the set of arcs in $A_i$ whose head is the node $v_i^t$. For any vertex $u \in V$, we introduce the set $\overline{V}^t(u) = \{i \in [\mathbf{N}] | u_i^t \in N_i(t)\}$ for representing vertex collision constraints (CCs). For representing edge CCs, we define, for $(u, w) \in E$:

$$\overline{E}^t(u, w) = \{(i, j) \in [\mathbf{N}] \times [\mathbf{N}] | (u_i^t, w_i^{t+1}) \in A_i(t), (w_j^t, u_j^{t+1}) \in A_j(t), i \neq j\}$$

For each robot $r_i \in \mathcal{R}$, and each arc $a \in A_i$, we introduce a binary variable $x(a) \in \{0, 1\}$ to indicate whether robot $r_i$ traverses arc $a$ in a feasible solution to the MAPF problem. Let $|A| = \sum_{i \in [\mathbf{N}]} |A_i|$, where $|A_i|$ denotes cardinality of set $A_i$. The 0–1 IP formulation for the MAPF problem is:

$$\underset{x\in\{0,1\}^{|A|}}{\text{minimize}} \sum_{i=1}^{\mathbf{N}}\sum_{a\in A_i} c_i(a)x(a) \tag{2}$$

$$\text{s.t.} \sum_{a\in\delta_{F_i}^+(s_i^0)} x(a) = 1, \forall i \in [\mathbf{N}] \tag{3}$$

$$\sum_{a\in\delta_{F_i}^-(u_i^t)} x(a) = \sum_{a\in\delta_{F_i}^+(u_i^t)} x(a), \forall i \in [\mathbf{N}], \forall t \in [\mathbf{T}-1], \forall u_i^t \in N_i(t) \tag{4}$$

$$\sum_{i\in\overline{V}^t(u)}\sum_{a\in\delta_{F_i}^{-1}(u_i^t)} x(a) \leq 1, \forall u \in V, \forall t \in [\mathbf{T}] \tag{5}$$

$$x(u_i^t, w_i^{t+1}) + x(w_j^t, u_j^{t+1}) \leq 1, \forall(u,w) \in E, \forall t \in \{0,1,...,\mathbf{T}-1\}, \forall(i,j) \in \overline{E}^t(u,w) \tag{6}$$

Equations (3) and (4) (a.k.a flow balance constraints) ensure that a **start-end path** is chosen for every robot, while Eqs. (5) and (6) prevent vertex and edge collisions respectively.

## 4    Lower Bounds from Cut Generation

We provide lower bounds to the MAPF problem using a Lagrangian relax and cut (LRC) scheme [8] that makes use of a cut generation procedure. In this section we describe our cut generating procedure, which will later be incorporated into the Lagrangian relax and cut (LRC) scheme described in Sect. 5.

Let $\mathbf{P}$ denote the MAPF polytope as shown below:

$$\mathbf{P} = \mathbf{conv}(x \in \{0,1\}^{|A|}|x \text{ satisfies } (3) - (6)) \tag{7}$$

where $\mathbf{conv}$ denotes convex hull. Given a $\bar{x} \in \{0,1\}^{|A|}$ that violates some constraint in (5)–(6), we develop a cut generating procedure that outputs a cut $a^\mathsf{T}x \leq b$ that strictly separates $\bar{x}$ from $\mathbf{P}$ i.e. $\max_{x\in\mathbf{P}} a^\mathsf{T}x \leq b < a^\mathsf{T}\bar{x}$.

The cuts generated by our procedure are a form of projection cuts. The idea is to select a subset of arcs $S \subset A$, where assume $|S| = n$. We will construct a polytope $P(S) \subset \mathbb{R}^n$ such that $\text{Proj}_{x(S)}(\mathbf{P}) \subseteq P(S)$, where $x(S)$ denotes the variables corresponding to the arcs in $S$, and $\text{Proj}_{x(S)}(\mathbf{P})$ is the orthogonal projection of $\mathbf{P}$ onto the space spanned by the variables in $x(S)$:

$$\text{Proj}_{x(S)}(\mathbf{P}) = \{y \in \mathbb{R}^n : \exists w \in \mathbb{R}^{|A|-n}, \text{s.t. } (y,w) \in \mathbf{P}\} \tag{8}$$

In order to generate a cut that separates $\bar{x}$ from $\mathbf{P}$ we will output a face of $P(S)$, which separates $\text{Proj}_{x(S)}(\bar{x})$ and $P(S)$. Clearly, if $P(S)$ is a tight relaxation for $\text{Proj}_{x(S)}(\mathbf{P})$, then the cut obtained will also be deep.

Different choices for $S$ give rise to different $P(S)$, so different cuts separating $\bar{x}$ and $\mathbf{P}$ can be derived by varying $S$. From the perspective of projection cuts, the collision avoidance constraints in Eqs. (5), (6) contain variables belonging to a small spatio-temporal neighbourhood. For instance, if $\bar{x}$ violates Eq. (6), then Eqs. (6) can be viewed as a cut which separates $\mathrm{Proj}_{x(S)}(\mathbf{P})$ and $\mathrm{Proj}_{x(S)}(\bar{x})$, where $S = \{(u_i^t, w_i^{t+1}), (w_j^t, u_j^{t+1})\}$. In this work, we will typically choose larger spatio-temporal neighbourhoods for selecting the arcs in $S$. Consequently, the cuts that are generated by our approach tend to be deeper.

*On Selecting $S$ for an Infeasible $\bar{x}$:* The choice of $S$ for a given infeasible $\bar{x}$ will not be made arbitrarily. An arbitrary choice for $S$ can lead to poor cuts, and computing a good approximation to $\mathrm{Proj}_{x(S)}(\mathbf{P})$ is challenging. To specify $S$, we first parameterize $S$ by the sets $\mathcal{R}(S), T(S), L(S)$. $\mathcal{R}(S) \subseteq [\mathbf{N}]$ is a set of indices of robots, $T(S) = \{\mathbf{l}, \mathbf{l}+1, ..., \mathbf{u}\}$ is a discrete interval where $\mathbf{l}, \mathbf{u} \in \mathbb{Z}_+$ and $\mathbf{u} < \mathbf{T}$. For each $i \in \mathcal{R}(S)$ and each time $t \in T(S)$, let $L_i^t(S)$ denote a set of nodes in $N_i(t)$. Denote $L(S) = \underset{i \in \mathcal{R}(S), t \in T(S)}{\cup} L_i^t(S)$, $S$ is defined as the set of all incoming and outgoing arcs associated with nodes in $L(S)$

$$S = \underset{i \in \mathcal{R}(S), t \in T(S)}{\cup} \underset{v_i^t \in L_i^t(S)}{\cup} \left( \delta_{F_i}^+(v_i^t) \cup \delta_{F_i}^-(v_i^t) \right) \tag{9}$$

W.l.o.g let $\bar{x}$ contain a conflict between robots $r_1, r_2$ at time $t_c$, and let us denote the set of nodes from $N_1, N_2$ involved in the conflict by $Z_{cf}$. Note that if the conflict is a vertex conflict, then $Z_{cf}$ contains 2 nodes, while for an edge conflict $Z_{cf}$ contains 4 nodes. We say that $S$ is an appropriate selection for $\bar{x}$ iff $1, 2 \in \mathcal{R}(S)$, $t_c \in T(S)$ and $Z_{cf} \subseteq L(S)$ i.e. loosely speaking $S$ contains arcs relevant to the conflict present in $\bar{x}$. We provide an example for $S$ below.

Consider the set of robot paths for $r_1$ and $r_2$ shown in Fig. 1. Robot $r_1$ moves from location $(3,3)$ at time 4 to $(4,3)$ at time 5, while $r_2$ moves from $(4,3)$ at time 4 to $(3,3)$ at time 5, so we have an edge collision at time 4.

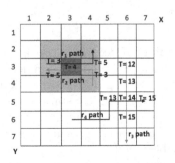

**Fig. 1.** One possible choice of $S$ for the edge conflict between $r_1$ and $r_2$ is: $\mathcal{R}(S) = \{1, 2\}$, and $T(S) = \{3, 4, 5\}$. For all $i \in [1, 2]$ and $\forall t \in T(S)$, $L_i^t(S)$ is set to nodes in $N_i(t)$ corresponding to all locations in the $3 \times 3$ grid centered at $(3, 3)$ (highlighted in yellow). $S$ for our example can be obtained from the parameters specified by applying Eq. (9). Clearly the arcs in the edge conflict are present in $S$. (Color figure online)

*Defining* $P(S)$: We retain as many relevant inequalities from the IP formulation in Sect. 3 for defining $P(S)$, thereby providing a tight relaxation for $\text{Proj}_{x(S)}(\mathbf{P})$.

$$P(S) = \mathbf{conv}\,(x(S) \in \{0,1\}^n | x \text{ satisfies } (11) - (14)) \tag{10}$$

$$\sum_{a \in \delta_{F_i}^-(v_i^t)} x(a) = \sum_{a \in \delta_{F_i}^+(v_i^t)} x(a), \, \forall i \in \mathcal{R}(S), \forall t \in T(S), \forall v_i^t \in L_i^t(S) \tag{11}$$

$$\sum_{j \in \overline{V}^t(v) \cap \mathcal{R}(S)} \sum_{a \in \delta_{F_j}^{-1}(v_j^t)} x(a) \leq 1, \, \forall t \in T(S), \forall i \in \mathcal{R}(S), \forall v_i^t \in L_i^t(S) \tag{12}$$

$$x(u_i^t, w_i^{t+1}) + x(w_j^t, u_j^{t+1}) \leq 1, \, {\scriptstyle \forall t \in \{l-1\} \cup T(S), \forall (i,j) \in \{(k,l) \in \mathcal{R}(S) \times \mathcal{R}(S) | k \neq l\}, \atop \forall (u,w) \in \{(p,q) \in E | (p_i^t, q_i^{t+1}), (q_j^t, p_j^{t+1}) \in S\}} \tag{13}$$

$$\sum_{a \in S \cap A_i(t)} x(a) \leq 1, \, \forall i \in \mathcal{R}(S), \forall t \in \{l-1\} \cup T(S) \tag{14}$$

Equation (11) ensures flow balance for all nodes in $L(S)$. Equation (12) prohibits vertex collisions on nodes in $L(S)$. Equation (13) prohibits edge collisions over arcs in $S$. Equation (14) ensures no two arcs present in $S$ and belonging to the same arc layer in the time expanded graph of a robot are both simultaneously selected. Clearly Eqs. (12) and (14) are implied from the MAPF IP formulation, while Eqs. (11) and (13) are present in the MAPF IP formulation, hence $\text{Proj}_{x(S)}(\mathbf{P}) \subseteq P(S)$.

*Separating* $\bar{x}$ *and* $P(S)$: Recall, we assumed that $\bar{x}$ violates some constraint from the set of Eqs. (5)–(6). Assuming we have selected an $S$ that is appropriate for $\bar{x}$, then we know that some constraint in the set of Eqs. (12)–(13) must be violated by $\bar{x}$. So by the strict hyper-plane separation theorem, we know that $\exists w \in \mathbb{R}^n$ such that $w^\mathsf{T} z < w^\mathsf{T} \text{Proj}_{x(S)}(\bar{x}), \forall z \in P(S)$. We can obtain such a $w$ by solving the optimization problem **CGLP** in Eq. (15). $Vert(P(S))$ shown in Eq. (16) refers to the vertices of polytope $P(S)$.

$$\mathbf{CGLP}: \max_{\|w\|_2 \leq 1} H(w), \text{ where } H(w) = w^\mathsf{T}(\text{Proj}_{x(S)}(\bar{x})) - h(w), \text{ and} \tag{15}$$

$$h(w) = \max_{y \in P(S)} \{w^\mathsf{T} y\} \iff \max_{v \in Vert(P(S))} \{w^\mathsf{T} v\} \tag{16}$$

The objective in **CGLP** is a piece-wise concave function, so **CGLP** can be solved using the well known projected sub-gradient ascent (PSGA) method, shown in Algorithm 1. For performing the maximization in line 3 of Algorithm 1, we require the vertices of $P(S)$. Drawing inspiration from the works of [2,7,16], in Sect. 7 we propose a compact representation for the vertices of $P(S)$ in terms of a decision diagram. The compact representation will enable us to perform maximization in a reasonable amount of time, at least when $|\mathcal{R}(S)|, |L(S)|$ are not too large. We provide details on how cuts are utilized for obtaining lower bounds to the MAPF problem in Sect. 5, and how the bound is integrated with CBS in Sect. 6.

---

**Algorithm 1.** PSGA for **CGLP**

1: **Initialize:** $k \leftarrow 1$, $w^{(k)} \leftarrow 0, \Delta \leftarrow 0, \mathbf{z} \leftarrow \text{Proj}_{x(S)}(\bar{x})$.
2: **while** Stopping criterion is not met **do**
3:    $v^{(k)} \leftarrow \arg \max_{v \in Vert(P(S))} v^\mathsf{T} w^{(k)}$
4:    $w^{(k+1)} \leftarrow \text{Proj}_{\|w\|_2 \leq 1}\left(w^{(k)} + \rho^{(k)}(\mathbf{z} - v^{(k)})\right)$, where $\rho^{(k)}$ is set to $\frac{1}{k}$
5:    **if** $H(w^{(k+1)}) > \max(0, \Delta)$ **then**
6:       $\Delta \leftarrow H(w^{(k+1)}), w^* \leftarrow w^{(k+1)}$
7:    $k \leftarrow k + 1$
8: **if** $\Delta > 0$ **then**
9:    **return** $w^*$

---

## 5   Lagrangian Relax-and-Cut

We will now describe a Non-Delayed Lagrangian Relax-and-Cut (LRC) procedure [14] to generate lower bounds to the MAPF problem. Consider the optimization problem shown in Eq. (17) obtained by omitting all CCs from the MAPF
IP formulation shown in Sect. 3.

$$\min_{x \in \{0,1\}^{|A|}} \{c^\mathsf{T} x | x \text{ satisfies Eqns (3)} - (4)\} \tag{17}$$

Notice, that the optimal solution to Eq. (17) (call it $\bar{x}$) consists of robot
start-end paths, which potentially may contain vertex and edge conflicts. We
can use our cut generation technique from the previous section, and generate
cuts (denote the set of inequalities generated by $Cx \leq d$) that separate $\bar{x}$ from
**P**. LRC incorporates the cuts generated by solving the Lagrangian dual problem:

$$\max_{\lambda \geq 0} \min_{x \in \{0,1\}^{|A|}} \{c^\mathsf{T} x + \lambda^\mathsf{T} (Cx - d) | x \text{ satisfies Eqns (3)} - (4)\} \tag{18}$$

Equation (18) is solved using the iterative PSGA procedure, where at each
iteration $\lambda$ is updated by using the solution to the inner minimization problem,
which is a min-cost flow problem. Denote the optimal solution to the min-cost
flow problem at iteration $k$ of PSGA by $\bar{x}_k$. Note that $\bar{x}_k$ represent start-end
paths for robots, and potentially contains conflicts. The key innovation of LRC
is that, if $\bar{x}_k$ contains conflicts, cuts are generated to separate $\bar{x}_k$ and **P**. Denote
the cuts generated by $E_k x \leq f_k$. $E_k x \leq f_k$ is incorporated into the optimization
problem by dualizing them with appropriate Lagrangian multipliers. In other
words, we can think of this operation as expanding $Cx \leq d$ at each iteration
by including $E_k x \leq f_k$. Adding cuts perturbs the current dual solution, and so
our approach is an attempt to dynamically strengthen the dual bound. An alternative motivation for adding cuts is through an interpretation of the procedure
from the primal side. By duality, the optimal objective values of Eq. (18) and
Eq. (19) are the same, see [10] for a proof. The feasible region in Eq. (19) is a
relaxation of **P**, since all constraints in Eq. (19) are valid for **P**. By introducing

an inequality that is violated by $\bar{x}_k$ at iteration $k$, we are dynamically strengthening this relaxation. As the objective in Eq. (19) is just the MAPF objective, the hope is that a tighter relaxation can lead to a better lower bound for the MAPF problem. If the cuts generated by our procedure are strong (deep), then we can expect to obtain a tighter relaxation. After a few iterations, no more cuts are added, at which point LRC becomes a standard dual ascent scheme.

$$\min_x \ \{c^\mathsf{T} x | \mathbf{conv}(x \text{ satisfies Eqns (3)} - (4), x \in \{0,1\}^{|A|}), Cx \le d\} \qquad (19)$$

A schematic overview of the LRC scheme described above is shown in Algorithm 2. Algorithm 2 takes as input a positive integer MAX_CUT_ITER, to decide when to stop adding cuts. In line 6, the $\lambda$ vector is updated with step size $\rho$. In our implementation, $\rho_k$ was set to $\mathcal{O}(\frac{1}{k})$ up-to MAX_CUT_ITER iterations, after which the step size was set according to the scheme in [1], to accelerate convergence. At iteration $k$, if $\bar{x}_k$ contains conflicts, we apply a primal repair procedure to try and convert $\bar{x}_k$ into a conflict free solution, for generating an upper bound.

Our primal repair procedure identifies a maximal independent set (MIS) of non conflicting robots from the paths provided in $\bar{x}_k$, and fixes the path of the robots in the MIS to as they are in $\bar{x}_k$. The paths for the remaining robots are computed sequentially, where the path assigned to a robot is the shortest start-end path that does not collide with any path fixed previously to other robots. The order in which the robots are chosen for their path to be computed is determined dynamically using the rule in [18]. If the procedure fails to compute a path for robot $r$, then the primal repair procedure is unsuccessful for the current iteration.

---

**Algorithm 2.** Lagrangian relax and cut algorithm

---

1: **Given:** MAX_CUT_ITER, $c$
2: **Output:** Inequalities $Cx \le d$, optimal Lagrangian multipliers $\lambda^*$, and upper bound(UB).
3: **Initialize:** $k \leftarrow 0$, $C \leftarrow \emptyset, d \leftarrow \emptyset$, $\lambda \leftarrow \emptyset$, UB $\leftarrow \infty$
4: **repeat**
5:     $\bar{x}_k \leftarrow \arg \min_{x \in \{0,1\}^{|A|}} \{c^\mathsf{T} x + \lambda^\mathsf{T}(Cx - d) | x \text{ satisfies Eqns (3)} - (4)\}$
6:     $\lambda \leftarrow (\lambda + \rho_k (C\bar{x}_k - d))_+$
7:     **if** $\bar{x}_k$ contains conflicts **then**
8:         **if** $k < $ MAX_CUT_ITER **then**
9:             Generate cuts $E_k x \le f_k$ separating $\bar{x}_k$ from **P**
10:            Append $E_k x \le f_k$ to $Cx \le d$. Introduce Lagrangian multipliers for $E_k x \le f_k$ initialized all to 1, and append it to the vector $\lambda$.
11:            Repair $\bar{x}_k$ to generate non-conflicting paths. If repair is successful and cost of repaired solution is less than UB, update UB.
12:        **else**
13:            **if** Cost$(\bar{x}_k) < $ UB **then**
14:                UB $\leftarrow$ Cost$(\bar{x}_k)$
15:     $k \leftarrow k + 1$
16: **until** Termination criterion is met

---

# 6 An LRC-Based Search Node Evaluation Function

In this section we will describe a new evaluation function for Conflict Based Search (CBS) that uses the output of the LRC procedure. We briefly describe only the relevant portions of CBS to our work. CBS performs a best first search on a search tree, where the most promising node among the previously unexplored nodes of the search tree is selected for exploration by applying an evaluation function. Each node in the search tree is characterized by a set of arcs that the robots are prohibited from using. An evaluation function takes as input any search tree node, and outputs a cost that does not overestimate the cost of the optimal solution to the MAPF problem with the added constraint that robots do not use any arcs prohibited in the search node. The node with the least evaluation cost is then selected. If the cost outputted by the evaluation function closely matches the true lower bound at every search tree node, then we should expect good search performance. Our goal is to improve existing evaluation functions.

Given a search node $sn$, let us denote the set of arcs that are prohibited in the node by $\bar{A}_{sn} \subset \cup_{i\in[N]}A_i$. We provide an evaluation function $\hat{f}_1(\cdot)$ based on Lagrangians. Before the search tree is created we apply Algorithm 2, and let $\hat{C}x \le \hat{d}$, $\hat{\lambda}$, and UB denote the outputs. $\hat{f}_1(sn)$ is computed as:

$$\hat{f}_1(sn) = \min_{\substack{x\in\{0,1\}^{|A|} \\ x(a)=0,\forall a\in\bar{A}_{sn}}} \{c^\mathsf{T}x + \hat{\lambda}^\mathsf{T}(\hat{C}x - \hat{d})|x \text{ satisfies Eqns } (3)-(4)\} \quad (20)$$

The validity of $\hat{f}_1(sn)$ as an evaluation function follows from the fact that $\hat{f}_1(sn)$ is a Lagrangian dual function where, $\hat{\lambda} \ge 0$ and the set of inequalities $\hat{C}x \le \hat{d}$ is valid for the feasible region of $sn$. Note that $\hat{f}_1(sn)$ can be computed using any shortest path algorithm on the time expanded graphs, but with arc costs reflecting the objective shown in Eq. (20). Our evaluation function is similar to obtaining Lagrangian lower bounds in Constraint Programming [3,4,11].

We can combine our proposed evaluation function with any other evaluation function $f$ previously proposed for the MAPF problem (see [9,13]), by taking the maximum i.e. $\max(f(sn), \hat{f}_1(sn))$ to yield a stronger evaluation function than either just $f$ or $\hat{f}_1$. If $\max(f(sn), \hat{f}_1(sn)) \ge UB$, we can omit descendants of $sn$ in the remainder of the search procedure i.e. we can prune the node $sn$.

We designed another evaluation function $\hat{f}_2(\cdot)$ inspired directly from the minimum vertex cover (MVC) heuristic of [9] and WDG heuristic of [13]. Observe that the minimization in Eq. (20) can be performed independently for the robots. Let us re-write Eq. (20) as $\hat{f}_1(sn) = -\hat{d}^\mathsf{T}\hat{\lambda} + \sum_{i\in[N]}\hat{f}_1(sn,i)$. Similar to the approach in WDG heuristic, for each pair of robots we can solve a 2-Agent MAPF problem to obtain lower bounds on the sum of pairwise costs. In this work, we use Lagrangian arc costs to obtain the pairwise bounds. The arc costs are modified to the Lagrangian objective shown in Eq. (20). Say for robots $r_i, r_j$, the optimal cost to the 2 agent MAPF problem with Lagrangian arc costs (i.e., the cost of each arc is set to the arc's co-efficient in $c^\mathsf{T}x + \hat{\lambda}^\mathsf{T}\hat{C}x$) is denoted by $l_{ij}(\hat{\lambda}, sn)$, we propose a node evaluation function $\hat{f}_2(sn)$ as:

$$\hat{f}_2(sn) = \min_{y \in \mathbb{R}^{\mathbf{N}}} -\vec{d}^{\mathsf{T}}\hat{\lambda} + \sum_{i \in [\mathbf{N}]} y_i$$

$$\text{s.t. } y_i + y_j \geq l_{ij}(\hat{\lambda}, sn), \forall i, j \in [\mathbf{N}] \times [\mathbf{N}] \text{ and } i < j$$

$$y_i \geq \hat{f}_1(sn, i), \forall i \in [\mathbf{N}] \tag{21}$$

## 7   Decision Diagram Representation of $P(S)$

We now address the problem of performing the maximization in line 3 of Algorithm 1 with the help of decision diagrams. We begin by first describing the decision diagram (DD) representation of $P(S)$. Borrowing notation from [7], we denote the DD for $P(S)$ by $\mathcal{D}(S) = (\mathcal{U}, \mathcal{A}, f)$, where $\mathcal{U}$ represents a set of nodes, $\mathcal{A}$ represents arcs in a top-down multi-graph, $f$ labels each an arc in $\mathcal{A}$ to some subset of arcs in $S$. $\mathcal{U}$ can be decomposed into $|T(S)| + 2$ layers $\mathcal{U}_0, \mathcal{U}_1, ..., \mathcal{U}_{|T(S)|+1}$, and $\mathcal{A}$ into $|T(S)| + 1$ layers $\mathcal{A}_0, \mathcal{A}_1, ..., \mathcal{A}_{|T(S)|}$. $\mathcal{U}_0$ contains a single node **sr** called source and $\mathcal{U}_{|T(S)|+1}$ contains a single node **sk** called sink. The tail of any arc in layer $j$ is connected to a node in $\mathcal{U}_j$ and its head to a node in $\mathcal{U}_{j+1}$.

To construct $\mathcal{D}(S)$ we will use the concept of a state transition diagram. The idea is to interpret each node in $\mathcal{U}$ as a state, a practice widely used for optimization using Decision Diagrams [5]. A state maps each $i \in \mathcal{R}(S)$ to a robot state. For $i \in \mathcal{R}(S)$, if robot $r_i$ occupies location $v \in V$ at time $t \in T(S)$, then the state of robot $r_i$ is defined as:

$$\begin{cases} v_i^t, & \text{if } v_i^t \in L_i^t(S) \\ \mathbf{o}, & \text{otherwise} \end{cases}$$

We motivate introducing **o** as a state for a robot with an example. Say robot $r_1$ traverses from location $a$ at time 1 to location $b$ at time 2 i.e $r_1$ traverses

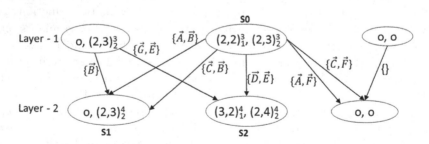

**Fig. 2.** A portion of the DD for the choice of $S$ in Fig. 1 is shown above, where $\vec{A} = ((2,2)_1^3, (2,1)_1^4)$, $\vec{B} = ((2,3)_2^3, (2,3)_2^4)$, $\vec{C} = ((2,2)_1^3, (1,2)_1^4)$, $\vec{D} = ((2,2)_1^3, (3,2)_1^4)$, $\vec{E} = ((2,3)_2^3, (2,4)_2^4)$, $\vec{F} = ((2,3)_2^3, (1,3)_2^4)$, $\vec{G} = ((3,1)_1^3, (3,2)_1^4)$.

the arc $(a_1^1, b_1^2) \in A_1(1)$ (refer $A_i(t)$ notation from Sect. 3), and then traverses the arc $(c_1^2, d_1^3) \in A_1(2)$, then first observe that such a path for $r_1$ is infeasible to the MAPF problem. However, when $b_1^2, c_1^2 \notin L_1^2(S)$ recall that $P(S)$ does not enforce flow balance at the nodes $b_1^2, c_1^2$, and so $P(S)$ may contain a vertex corresponding to the infeasible MAPF path, since after-all $P(S)$ is only a relaxation to $\mathrm{Proj}_{x(S)}(\mathbf{P})$. As we chose to adopt a state space representation for building $\mathcal{D}(S)$, introducing $\mathbf{o}$ allows us to interpret our example as: $r_1$ traverses state $a_1^1$ to state $\mathbf{o}$ using arc $(a_1^1, b_1^2)$, and then transitions from $\mathbf{o}$ to $d_1^3$ using arc $(c_1^2, d_1^3)$.

Observe that since $P(S)$ is parameterized with arc variables and any vertex of $P(S)$ corresponds to a series of arc traversals for the robots in $\mathcal{R}(S)$, we can interpret each vertex as a series of state transitions much the same way as we did earlier for the single robot. In constructing $\mathcal{D}(S)$, the overall aim is to ensure that there is a 1:1 correspondence between vertices in $P(S)$ and state transition paths in $\mathcal{D}(S)$. We next construct the nodes in $\mathcal{D}(S)$. Let us assume that $T(S) = \{\mathbf{l}, \mathbf{l}+1, ..., \mathbf{u}\}$. At time $t \in T(S)$, observe there are at most $\prod_{i \in \mathcal{R}(S)} (|L_i^t(S)| + 1)$ states for the robots. For each of those states at time $t \in T(S)$, we introduce a node in layer $t - \mathbf{l} + 1$ of $\mathcal{D}(S)$ i.e. $\mathcal{U}_{t-\mathbf{l}+1}$. For any node $u \in \mathcal{U}$, we shall use the notation $u[i]$ to denote the state of robot $r_i$ in $u$. The node $\mathbf{sr}$ in layer $\mathcal{U}_0$ and node $\mathbf{sk}$ in layer $\mathcal{U}_{T(S)+1}$, both correspond to a state where $\mathbf{sr}[i] = \mathbf{sk}[i] = \mathbf{o}, \forall i \in \mathcal{R}(S)$. Some of the nodes (states) populated in $\mathcal{U}$, may contain vertex collisions. A node $u \in \mathcal{U}$ is said to contain a vertex collision iff $\exists i, j \in \mathcal{R}(S)$ such that both $u[i]$ and $u[j]$ are different from $\mathbf{o}$, and $u[i], u[j]$ correspond to the same vertex in $V$. We remove all nodes that contain vertex collisions from $\mathcal{U}$, as such states can never be attained from any vertex of $P(S)$. Vertex collision checking with $\mathbf{o}$ was skipped because Eq. (12) has vertex CCs only for nodes in $L(S)$. The states for $\mathbf{sr}, \mathbf{sk}$ was defined that way because Eq. (12) is unconcerned with vertex collisions at times $\mathbf{l} - 1, \mathbf{u} + 1$.

Before describing how to populate arcs in $\mathcal{D}(S)$, we take a brief detour. For the choice of parameters of $S$ in Fig. 1, we show a portion of its corresponding DD in Fig. 2. Only a few states in layers $\mathcal{U}_1, \mathcal{U}_2$, and arcs in $\mathcal{A}_1$ with their labels are shown. For robot $r_1$ to transition from location $(2, 2)$ at time 3 (denoted by state $(2, 2)_1^3$) to $(3, 2)$ at time 4, $r_1$ needs to transition with arc $\overrightarrow{D}$ (refer Fig. 2 for definition). Likewise, for $r_2$ to transition from state $(2, 3)_2^3$ to $(2, 4)_2^4$, $r_2$ needs to transition with arc $\overrightarrow{E}$. Since the transition between those states for $r_1$ and $r_2$ does not lead to a collision, we connect states $\mathbf{S0}, \mathbf{S2}$ with an arc in $\mathcal{A}_1$ and label the arc with the set $\{\overrightarrow{D}, \overrightarrow{E}\}$, see Fig. 2. From the definition of $L_1^4(S)$ provided in the example in Fig. 1, the reader can easily verify that $r_1$ can transition to state $\mathbf{o}$ at time 4 from $(2, 2)_1^3$ using either $\overrightarrow{A}$ or $\overrightarrow{C}$. Consequently, note that $r_1, r_2$ can transition from $\mathbf{S0}$ to $\mathbf{S1}$ in 2 different ways as shown in Fig. 2.

Moving on from the example, we now provide a formal procedure to populate the arcs in $\mathcal{D}(S)$. Consider any node $v \in \mathcal{U}_k$ and any node $w \in \mathcal{U}_{k+1}$. To decide whether we should connect arcs from $v$ to $w$, reduces to first determining whether $w$ is a feasible state transition of $v$, and if yes, then determining all different ways in which the robots may transition from $v$ to $w$. By repeating this process for all pairs of nodes in $\mathcal{U}$ occurring between consecutive layers, we can populate $\mathcal{A}$.

To establish whether a state transition from $v$ to $w$ is feasible, for each $i \in \mathcal{R}(S)$ we will establish all the different ways in which robot $r_i$ can transition from $v[i]$ to $w[i]$ by traversing some arc in $S$, and store this information in the set $h(v, w, i) \subset S$. One of the cases below will be applicable for populating $h(v, w, i)$:

1. If $k = 0$ and $w[i] \in L_i^1(S)$, then first observe that $v[i] = \mathbf{o}$ since $v$ is the **sr** node. Robot $r_i$ can use any one of the arcs from the set $\delta_{F_i}^-(w[i])$ to transition to $w[i]$, and so $h(v, w, i) = \delta_{F_i}^-(w[i])$.
2. If $1 \leq k < |T(S)|$, we consider separately the following 3 cases:
   - If $v[i] \in L_i^{k+1-1}(S)$, $w[i] \in L_i^{k+1}(S)$ and $(v[i], w[i]) \in \delta_{F_i}^+(v[i])$, then clearly: $h(v, w, i) = \{(v[i], w[i])\}$.
   - If $v[i] = \mathbf{o}$ and $w[i] \in L_i^{k+1}(S)$, then :

$$h(v, w, i) = \{(p_i^{k+1-1}, w[i]) \in \delta_{F_i}^-(w[i]) | p_i^{k+1-1} \notin L_i^{k+1-1}(S)\}$$

   - If $v[i] \in L_i^{k+1-1}(S)$ and $w[i] = \mathbf{o}$, then:

$$h(v, w, i) = \{(v[i], p_i^{k+1}) \in \delta_{F_i}^+(v[i]) | p_i^{k+1} \notin L_i^{k+1}(S)\}$$

3. If $k = |T(S)|$ and $v[i] \in L_i^k(S)$, then $h(v, w, i) = \delta_{F_i}^+(v[i])$. Also note that $w$ is **sk** node, and so $w[i] = \mathbf{o}$.
4. If $v[i] = w[i] = \mathbf{o}$, then $h(v, w, i) = \emptyset$ since there are no arcs in $S$ that the robot can use to traverse between such a pair of states. However a state transition for the robot is still be feasible without using any arcs from $S$. For example, if at all times $t \in T(S)$ robot $r_i$ does not use any arc in $S$ in some start-end path which is feasible to the MAPF problem, then $r_i$ is in state $\mathbf{o}$ at all $t \in T(S)$.

If for some $i \in \mathcal{R}(S)$, $v[i], w[i]$ are not both $\mathbf{o}$, and no arc could be found for $h(v, w, i)$ by analyzing 1, 2 and 3, then robot $r_i$ cannot transition from $v[i]$ to $w[i]$. In such a case, we can safely conclude that the transition from $v$ to $w$ is not feasible, and so we do not need to insert any arc from $v$ to $w$ in $\mathcal{A}$. Assuming that this is not the case for $v, w$, we then proceed to check whether arcs can be added from $v$ to $w$. As the function $h(v, w, i)$ corresponds to the different ways in which the robot $r_i$ transitions from state $v[i]$ to $w[i]$, it is only natural that the elements of the set $H(v, w) = \prod_{i \in \mathcal{R}(S)} h(v, w, i)$ correspond to all the different ways in which robots can transition from $v$ to $w$. Note that $H(v, w)$ is a set product, and so each element of $H(v, w)$ is itself a subset of $S$ (includes the empty set). For any $i \in \mathcal{R}(S)$, note that each element of $H(v, w)$ contains at most one arc from $A_i$. Some elements of $H(v, w)$ may contain arcs from $S$, wherein robots transitioning using those arcs will result in an edge collision. We remove all those elements from $H(v, w)$ which will result in edge collisions. Corresponding to each element remaining in $H(v, w)$ after the previous edge collision filtration step, we add an arc $a$ from $v$ to $w$ in $\mathcal{A}_k$ and label (cf. $f$ function in definition of $\mathcal{D}(S)$) $a$ by the arcs from $S$ present in the element of $H(v, w)$.

There is a 1:1 correspondence between source-sink (**sr**−**sk**) paths in $\mathcal{D}(S)$ and vertices of $P(S)$, i.e. if $x_v$ is a vertex of $P(S)$ and let $Q = \{a \in S | x_v(a) = 1\}$, then there is a $\mathbf{sr} - \mathbf{sk}$ path in $\mathcal{D}(S)$ such that the labels occurring on the path coincide exactly with $Q$. Conversely for any $\mathbf{sr} - \mathbf{sk}$ path, if $\bar{S} \subset S$ are labels occurring on the path, then there is a vertex $x_v$ in $P(S)$ such that $x_v(\bar{S}) = 1, x_v(S \backslash \bar{S}) = 0$.

*Computing Line* 3 *in Algorithm* 1 *Using* $\mathcal{D}(S)$: To perform the maximization in line 3, we can make use of the correspondence between vertices of $P(S)$ and $\mathbf{sr} - \mathbf{sk}$ paths in $\mathcal{D}(S)$. We assign a cost to each arc in $\mathcal{A}$ depending on the labels on the arc. For instance, if arc $a \in \mathcal{A}$ is labelled with $b_1, b_2$, where $b_1, b_2 \in S$, then we simply assign a cost of $w^{(k)}(b_1) + w^{(k)}(b_2)$ to $a$, where $w^{(k)}(b_i)$ is the value corresponding to $b_i$ in vector $w^{(k)}$. If $a$ is not labeled with any arc from $S$, then we assign a cost of 0. After setting costs to all arcs in $\mathcal{A}$ in the manner just described, obtaining the arg max vertex in line 3 is equivalent to obtaining any longest $\mathbf{sr} - \mathbf{sk}$ path in $\mathcal{D}(S)$. The computational effort needed to obtain the longest path is $\mathcal{O}(|\mathcal{A}|)$, since $\mathcal{D}(S)$ is a directed acyclic graph.

# 8  Templates for Grids

As conflict locations and robots involved in conflicts vary, a different set of parameters for $S$ may need to be chosen in order to generate a cut for each conflict. Consequently, a different projection polytope ($P(S)$) needs to be built for each conflict, which is computationally expensive. When $G$ is a 4 or 8-connectivity grid, the neighborhood relative to any location on the grid is same across all locations on the grid, a property that allows us to build *Templates*.

Let us denote the polytope $P(S)$ described in Fig. 1 by $P_1$. Now consider the vertex conflict for robots $r_3$ and $r_4$ at time 14 shown in Fig. 1. For this conflict, we can create a polytope $P(S_2)$ with parameters: $\mathcal{R}(S_2) = \{3, 4\}$, and $T(S_2) = \{13, 14, 15\}$. For all $i \in [3, 4]$ and $\forall t \in T(S_2)$, $L_i^t(S_2)$ is set to nodes in $N_i(t)$ corresponding to all locations in the $3 \times 3$ grid centered at $(6, 5)$. Clearly $P(S_2)$ can also output a cut for the conflict between $r_3, r_4$. While polytopes $P_1, P(S_2)$ lie in different dimensions, the facial structure of both polytopes are identical. If we substitute $r_1$ for $r_3$, $r_2$ for $r_4$, advance the interval $T(S_2)$ by 10 time units, and translate all locations in $L(S_2)$ by 3 units along the negative X-axis and by 2 units along the negative Y-axis, we get back all the parameters for $S$ described in Fig. 1. Hence, we claim that both $P_1$ and $P(S_2)$ are manifestations of the same base template polytope.

While working with structured graphs such as grids, we can precompute a library of different templates, and use those templates to generate all cuts. By spatio-temporally shifting the parameters of the template about the conflict, multiple cuts can be generated using the same base template. While generating cuts with templates, some locations may be physically blocked on the grid, and so certain states in the DD representation of the template cannot be attained by the robots. In that case, we adjust the longest $\mathbf{sr} - \mathbf{sk}$ path computation procedure to avoid paths that pass through infeasible states.

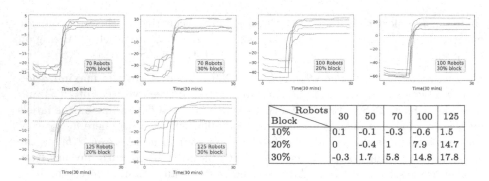

| Robots Block | 30 | 50 | 70 | 100 | 125 |
|---|---|---|---|---|---|
| 10% | 0.1 | -0.1 | -0.3 | -0.6 | 1.5 |
| 20% | 0 | -0.4 | 1 | 7.9 | 14.7 |
| 30% | -0.3 | 1.7 | 5.8 | 14.8 | 17.8 |

**Fig. 3.** On Y-axis, we plot the difference in lower bounds outputted by LR-DG* and DG* with time for 5 instances. For each problem scenario, the table reports the difference in bounds averaged over 10 instances after 30 mins of simulation per instance.

## 9    Experimental Evaluation

The primary goals of our experiments are to understand the additional value that our approach can bring to existing search methods, and how its performance is influenced by the characteristics of the problem.

**Experimental Setup.** As baseline search method, we implemented a state-of-the-art variant of conflict-based search, called DG [13], however our implementation does not include the run time reduction techniques and MDD merging technique proposed in [13]. When determining whether a pair of robots should share an edge in the conflict graph constructed in DG, we applied a two-agent MAPF solver to check this condition. As branching rule, we implement the rule proposed for ICBS [6], i.e., prioritizing cardinal over semi-cardinal over non-cardinal conflicts. We will refer to our implementation of DG as DG*. Our LRC approach implements DG with $\hat{f}_2(\cdot)$ as node evaluation function. We selected $\hat{f}_2(\cdot)$ over $\hat{f}_1(\cdot)$, because Eq. (21) implies $\hat{f}_2(sn) \geq \hat{f}_1(sn)$. We denote the combination of DG* with $\hat{f}_2(\cdot)$ by LR-DG*.

All experiments in this paper were carried out on an Intel 4 core i7-4790 processor running at 3.6 GHz with 16 GB RAM, and the program was written in C++. All our experiments were conducted on 30 × 30 4-connectivity grids, where some % of the locations on the grid are randomly chosen and blocked. The start and end locations for the robots on the grid are also randomly assigned. The makespan constraint **T** was set to 3 more than the shortest time it took for all robots to reach their goal from start when CCs are omitted.

In all our experiments for comparing DG* with LR-DG*, we allocated a time limit of 30 min for both algorithms. The 30 min allocated to LR-DG* is further split as follows. A time limit of 10 min was allocated beyond which cuts are not added in Algorithm 2 and MAX_CUT_ITER was set to 1000. Optimizing the Lagrangian multipliers using the accelerated step size update rule, in practice takes ≈2–4 min for the 100 robot instances. The remaining time was spent in performing conflict based search with $\hat{f}_2$ as evaluation function.

As all experiments were conducted on a 4-connectivity grid, we could pre-compute a template library. Our template library consisted of 64 templates with each template parameterized by at most 3 robots, time horizon (cf. $T(S)$) of at most 5 time units, and $|L_i^t(S)|$ varied between 6–9 nodes. This translates roughly to 0.2 million arcs in the size of the DD per template.

**Experimental Results.** To study the impact of problem characteristics on our algorithm performance, we considered problem scenarios with different blocked locations (%) and number of robots. For each such scenario, we record the progress of our solution method's lower bound with time. In Fig. 3, we graphically show the difference between LR-DG* and DG* lower bounds with time for different scenarios, and the results after 30 min are summarized in the table.

We first explain why the difference in bounds between the 2 methods looks like a step size function. Initially, the LR-DG* bound lags behind the DG* bound as it performs the cut addition phase of Algorithm 2. In the cut addition phase, the LR-DG* lower bound is not improving much. On the other hand the MVC heuristic in DG* is able to quickly identify pairs of robots that are in some sense constraining one another, and by branching on their conflicts it is able to make rapid progress initially. Once LR-DG* enters into the Lagrangian multipliers optimization phase with accelerated step size update rule, we see a marked improvement in the LR-DG* bound.

From the table shown in Fig. 3, one can observe the following trends in the gap between the lower bounds of the two methods. For a fixed number of robots, we see that as the block % increases, LR-DG* dominates over the DG* bound. Also, for a fixed block %, we observe that the gap between the 2 methods increases as the number of robots increases. The lower bound to the MAPF problem computed just after Lagrangian Relax and Cut phase i.e after Algorithm 2 in many cases dominates the bound obtained from DG* after 30 min. For 100 robot problems and 20% blocked cells, on average (over 10 instances) we observed this gap to be 4.89, and for 30% blocked cells the gap was 11.7. These results clearly indicate that the cuts generated in the LRC phase are strong, and capable of generating strong lower bounds for the given objective.

We explain the results observed. It is clear that when not many collisions are expected between robots, then it is unlikely that the robots have to wait for another robot or take a longer route. Trying to raise the lower bound with cuts is unlikely to result in an improvement of the lower bound, which is why we see that when the number of robots is few and/or blocked cells are also few, LR-DG* is unable to do any better than DG*. However, when the expected number of collisions is large, we see that LR-DG* generates strong bounds which reflects the strength of the cuts generated during the LRC phase. We explain this performance using an analogy. In environments which contain a lot of spatio-temporal bottleneck regions, i.e., local regions in time where many robots need to pass through to reach their goal, then by simply analyzing paths of robots within the bottleneck region, we may be able to infer facts such as, at least one robot must wait or take a longer route in order to pass through the bottleneck region without colliding. The strength of the inference improves as more robots are included in the analysis. Through the use of templates, our approach essentially focuses on a localized spatio-temporal neighbourhood. Our cut generating templates are able analyze all feasible paths through the neighbourhood at once for the robots parameterizing the template, thereby able to output strong cuts.

Despite LR-DG* producing stronger lower bounds than DG*, in general we observed that LR-DG* was unable to prove optimality for any problem that DG* also could not. The results indicates a need for stronger lower bounds and a better primal heuristic than the one used in this work for proving optimality. For problems that were solved to optimality by both DG* and LR-DG*, we compared the number of search tree nodes expanded. For 30 robot problems, LR-DG* on average expanded 37% fewer nodes, however the % reduction in nodes across instances displayed high variance. On many problems, the fact that LR-DG* proved optimality during LRC phase itself has skewed the results. In general however, observe that since the cuts in the procedure have been generated at the root node, their utility diminishes as the depth of the search node increases.

## 10    Conclusions

We proposed a new polyhedral approach for MAPF based on lower-dimensional polytopes called 'templates', which allows us to simultaneously analyze the paths of a number of robots within a spatio-temporal neighbourhood. We used decision diagrams to represent these templates and developed a cut generation scheme. The templates are translated spatio-temporally over the input graph to generate cuts for paths with conflicts. To obtain a lower bound, we embedded the cut generation into a Lagrangian Relax-and-Cut procedure. We incorporated the lower bound as a node evaluation function in a conflict-based search procedure. Our experimental results demonstrated that our lower bounds are particularly effective when the MAPF problem is very constrained due to large number of agents and\ or fewer traversable paths on the input graph.

**Acknowledgements.** Willem-Jan van Hoeve was partially supported by Office of Naval Research Grant No. N00014-18-1-2129 and National Science Foundation Award #1918102. The authors thank Viraj Parimi for his help with generating the plots.

## References

1. Baker, B.M., Sheasby, J.: Accelerating the convergence of subgradient optimisation. Eur. J. Oper. Res. **117**(1), 136–144 (1999)
2. Becker, B., Behle, M., Eisenbrand, F., Wimmer, R.: BDDs in a branch and cut framework. In: Nikoletseas, S.E. (ed.) WEA 2005. LNCS, vol. 3503, pp. 452–463. Springer, Heidelberg (2005). https://doi.org/10.1007/11427186_39
3. Benoist, T., Laburthe, F., Rottembourg, B.: Lagrange relaxation and constraint programming collaborative schemes for travelling tournament problems. In: CPAIOR, vol. 1, pp. 15–26 (2001)
4. Bergman, D., Cire, A.A., van Hoeve, W.-J.: Improved constraint propagation via Lagrangian decomposition. In: Pesant, G. (ed.) CP 2015. LNCS, vol. 9255, pp. 30–38. Springer, Cham (2015). https://doi.org/10.1007/978-3-319-23219-5_3

5. Bergman, D., Cire, A.A., Van Hoeve, W.J., Hooker, J.: Decision Diagrams for Optimization, vol. 1. Springer, Heidelberg (2016). https://doi.org/10.1007/978-3-319-42849-9
6. Boyarski, E., et al.: ICBS: improved conflict-based search algorithm for multi-agent pathfinding. In: Twenty-Fourth International Joint Conference on Artificial Intelligence (2015)
7. Davarnia, D., van Hoeve, W.-J.: Outer approximation for integer nonlinear programs via decision diagrams. Math. Program., 1–40 (2020). https://doi.org/10.1007/s10107-020-01475-4
8. Escudero, L.F., Guignard, M., Malik, K.: A Lagrangian relax-and-cut approach for the sequential ordering problem with precedence relationships. Ann. Oper. Res. **50**(1), 219–237 (1994). https://doi.org/10.1007/BF02085641
9. Felner, A., et al.: Adding heuristics to conflict-based search for multi-agent path finding. In: Twenty-Eighth International Conference on Automated Planning and Scheduling (2018)
10. Geoffrion, A.M.: Lagrangean relaxation for integer programming. In: Balinski, M.L. (ed.) Approaches to integer programming. Mathematical Programming Studies, vol. 2, pp. 82–114. Springer, Heidelberg (1974). https://doi.org/10.1007/BFb0120690
11. Khemmoudj, M.O.I., Bennaceur, H., Nagih, A.: Combining Arc-consistency and dual Lagrangean relaxation for filtering CSPs. In: Barták, R., Milano, M. (eds.) CPAIOR 2005. LNCS, vol. 3524, pp. 258–272. Springer, Heidelberg (2005). https://doi.org/10.1007/11493853_20
12. Lam, E., Le Bodic, P., Harabor, D., Stuckey, P.J.: Branch-and-cut-and-price for multi-agent pathfinding. In: Proceedings of the Twenty-Eighth International Joint Conference on Artificial Intelligence (IJCAI-19), pp. 1289–1296. International Joint Conferences on Artificial Intelligence Organization (2019)
13. Li, J., Boyarski, E., Felner, A., Ma, H., Koenig, S.: Improved heuristics for multi-agent path finding with conflict-based search. In: International Joint Conference on Artificial Intelligence, pp. 442–449 (2019)
14. Lucena, A.: Non delayed relax-and-cut algorithms. Ann. Oper. Res. **140**(1), 375–410 (2005). https://doi.org/10.1007/s10479-005-3977-1
15. Sharon, G., Stern, R., Felner, A., Sturtevant, N.R.: Conflict-based search for optimal multi-agent pathfinding. Artif. Intell. **219**, 40–66 (2015)
16. Tjandraatmadja, C., van Hoeve, W.J.: Target cuts from relaxed decision diagrams. INFORMS J. Comput. **31**(2), 285–301 (2019)
17. Wagner, G., Choset, H.: M*: a complete multirobot path planning algorithm with performance bounds. In: 2011 IEEE/RSJ International Conference on Intelligent Robots and Systems, pp. 3260–3267. IEEE (2011)
18. Wang, J., Li, J., Ma, H., Koenig, S., Kumar, T.: A new constraint satisfaction perspective on multi-agent path finding: preliminary results. In: Proceedings of the 18th International Conference on Autonomous Agents and MultiAgent Systems, pp. 2253–2255. International Foundation for Autonomous Agents and Multiagent Systems (2019)
19. Wurman, P.R., D'Andrea, R., Mountz, M.: Coordinating hundreds of cooperative, autonomous vehicles in warehouses. AI Mag. **29**(1), 9–9 (2008)
20. Yu, J., LaValle, S.M.: Planning optimal paths for multiple robots on graphs. In: 2013 IEEE International Conference on Robotics and Automation, pp. 3612–3617. IEEE (2013)
21. Yu, J., LaValle, S.M.: Structure and intractability of optimal multi-robot path planning on graphs. In: Twenty-Seventh AAAI Conference on Artificial Intelligence (2013)

# Hybrid Classification and Reasoning for Image-Based Constraint Solving

Maxime Mulamba$^{(\boxtimes)}$ ⓘ, Jayanta Mandi$^{(\boxtimes)}$ ⓘ, Rocsildes Canoy$^{(\boxtimes)}$ ⓘ, and Tias Guns$^{(\boxtimes)}$ ⓘ

Data Analytics Laboratory, Vrije Universiteit Brussel, Brussels, Belgium
{maxime.mulamba,jayanta.mandi,rocsildes.canoy,tias.guns}@vub.be

**Abstract.** There is an increased interest in solving complex constrained problems where part of the input is not given as facts, but received as raw sensor data such as images or speech. We will use 'visual sudoku' as a prototype problem, where the given cell digits are handwritten and provided as an image thereof. In this case, one first has to train and use a classifier to label the images, so that the labels can be used for solving the problem. In this paper, we explore the hybridisation of classifying the images with the reasoning of a constraint solver. We show that pure constraint reasoning on predictions does not give satisfactory results. Instead, we explore the possibilities of a tighter integration, by exposing the probabilistic estimates of the classifier to the constraint solver. This allows joint inference on these probabilistic estimates, where we use the solver to find the maximum likelihood solution. We explore the trade-off between the power of the classifier and the power of the constraint reasoning, as well as further integration through the additional use of structural knowledge. Furthermore, we investigate the effect of calibration of the probabilistic estimates on the reasoning. Our results show that such hybrid approaches vastly outperform a separate approach, which encourages a further integration of prediction (probabilities) and constraint solving.

**Keywords:** Constraint reasoning · Visual sudoku · Joint inference · Prediction and optimisation

## 1 Introduction

Artificial intelligence (AI) is defined as "systems that display intelligent behaviour by analysing their environment and taking actions - with some degree of autonomy - to achieve specific goals." [28]. In that regard, recent advancements in deep neural network (DNN) architectures have achieved highly accurate performance in object and speech recognition and classification. However, many real life problems are relational, where inference on one instance is related to another through various constraints and logical reasoning. Attaining good performance in tasks which require reasoning over constraints and relations still remains elusive. The DNN architectures rely heavily on learning latent representation from

© Springer Nature Switzerland AG 2020
E. Hebrard and N. Musliu (Eds.): CPAIOR 2020, LNCS 12296, pp. 364–380, 2020.
https://doi.org/10.1007/978-3-030-58942-4_24

the training datasets [18]. The main reason deep architectures struggle in constraint reasoning is that the nuances of the relationship between entities are often lost in the latent representation. For instance, when solving a sudoku, a DNN model would take the partially filled sudoku as an input and would then be expected to produce the solved sudoku as output. In this process, the model fails to comprehend the interactions among different cells.

Moreover, the high quality performance of DNNs at complex tasks comes at a cost. As DNN models fail to comprehend the logical reasoning, they have to adjust to gradual feedback of the error signals. As a consequence, to be proficient in any simple task, a DNN needs an enormous amount of data. As an example, to be an efficient video-gamer, a DNN model has to play a game for more than 900 h [9]. Motivated by such deficiencies, integrating logical and relational reasoning into DNN architecture has increasingly gained more attention.

In trying to bridge deep learning and logical reasoning, Wang et al. [30] propose SATNet, a differentiable satisfiability solver that can be used to learn both constraints and image classification through backpropagation. Internally, it uses a quadratic SDP relaxation of a MaxSAT model, and hence learns a relaxed representation of the constraints. We argue that in many cases, there is no need to learn everything end-to-end. Indeed, in a visual sudoku setting, while the constraints are easy to specify in a formal language, the image classification task is difficult for a machine to capture. Hence, we seek to bridge deep learning and logical reasoning by directly plugging the (probabilistic) output of the deep learning into a constraint solver that reasons over the relevant hard constraints.

In this work, we present a framework where we perform *joint inference* [24–26] over the different predictions, by integrating machine learning inference with first and second order logic. Specifically, instead of solving a constraint programming (*CP*) problem over a set of independently predicted values, we use CP to do joint inference over a set of probability vectors. The training of the DNN happens on individual image instances, as is typically done. Effectively, our framework can be considered as a *forward-only layer* on top of the predictions of a pre-trained network.

Specifically, we consider the "visual sudoku" problem where images of digits of some cells in the sudoku grid are fed as input. We first predict the digits using a DNN model and then use a CP solver to solve the sudoku puzzle. A conventional approach would use the predictions of the DNN as inputs to the CP. As the DNN model is not aware of the constraints of the sudoku problem, it misses the opportunity to improve its prediction by taking the constraints into account. When the predictions of the DNN are directly fed into the CP solver, in case of any error, the CP model is bound to fail. Note that in this case, even one prediction error will result in the failure of the whole problem.

We improve the process by considering the predicted class probabilities instead of directly using the *arg max* prediction. The advantage of our approach is that by avoiding hard assignments prior to the CP solver, we enable the CP solver to *correct the errors* of the DNN model. In this way, we use CP to do joint inference, which ensures that the predictions will respect the constraints of the problem.

The contributions of the paper are as follows:

- We explore hybridisation of classification and constraint reasoning on the visual sudoku problem;
- We show that constraint reasoning over the probabilistic predictions outperforms a pure reasoning approach, and that we can further improve by taking higher-order relations into account;
- We investigate the increased computational cost of reasoning over the probabilities, and the trade-offs possible when limiting the reasoning to the top-k probabilities.
- We experimentally explore the interaction of predictive power with the power of discrete reasoning, showing correction factors of 10% and more, as well as the effect of using *calibrated* probabilistic classifiers.

## 2   Related Work

**Predict-and-Optimize.** Our work is closely related to the growing body of research at the intersection of machine learning (ML) and combinatorial optimization [7,8,17] where the predictions of an ML model is fed into a downstream optimization oracle. In most applications, feeding machine learning predictions directly into a combinatorial optimization problem may not be the most suitable approach. Bengio [2] compared two ML approaches for optimizing stock returns—one uses a neural network model for predicting financial prices, and the second model makes use of a task-based loss function. Experimental results show that the second model delivers better optimized return. The results also suggest a closer integration of ML and optimization.

In this regard, Wilder et al. [32] propose a framework which trains the weight of the ML model directly from the task-loss of the downstream combinatorial problem from its continuous relaxation. The end-to-end model of [30] learns the constraints of a satisfiability problem by considering a differentiable SDP relaxation of the problem. A similar work [14] trains an ML model by considering a convex surrogate of the task-loss.

Our work differs from these as we do not focus on end-to-end learning. Rather, we enhance the predictions of an ML model by using CP to do joint inference over the raw probability vectors. In this way, we are taking the constraint interaction of the combinatorial problem into account.

**Joint Inference.** Our work is also aligned with the research in joint inference. For example, Poon and Domingos [24] have shown its advantage for information extraction in the context of citation matching. Recent work in linguistic semantic analysis of Wang et al. [31] forms a factor graph from the DNN output by encoding it into logical predicates and performs a joint inference over the factor graph. Several other works [3,11,12] focus on leveraging joint inference in DNN architecture for relation extraction from natural language. Our work differs from these, as we perform probabilistic inference on combinatorial constraint solving problem where one inference is linked with another by hard constraints.

**Training with Constraints.** Various works have introduced methods to enforce constraints on the outputs of an NN. One of the earlier work [23] does this by optimizing the Lagrangian coefficients of the constraints at every parameter update of the network. But this would not be feasible in the context of deep neural network as very large dimension matrices must be numerically solved for each parameter update [16]. Pathak et al. [20] introduce CCNN for image segmentation with size constraints where they introduce latent probability distributions over the labels and impose constraints on the latent distribution enabling efficient Lagrangian dual optimization. However, one drawback is, this involves solving an optimization problem at each iteration. Márquez-Neila et al. [16] use a Lagrangian based Krylov subspace approach to enforce linear equality constraints on the output of an NN. But this approach is not found to be scalable to large problem instances. The proposed framework of [13] quantifies inconsistencies of the NN output with respect to the logic constraints and is able to significantly reduce inconsistent constraint violating outcomes by training the model to minimize inconsistency loss.

The closest work to ours is [25], where Punyakanok et al. train a multiclass classifier to identify the label of an argument in the context of semantic role labeling and then feed the prediction scores of each argument to an Integer Linear Programming solver so that the final inferences abide by some predefined linguistic constraints.

## 3   Preliminaries

**CSP and COP.** The concept of a constraint satisfaction problem (CSP) is fundamental in constraint programming [27]. A CSP is formulated as a triplet $(V, D, C)$, where $V$ is a set of decision variables, each of which has its possible values in a domain contained in the set $D$, and $C$ is a set of constraints that need to be satisfied over the variables in $V$. In most cases, we are not only interested in knowing whether a constrained problem is solvable, but we want the *best* possible solution according to an objective.

A Constraint Optimization Problem $COP(V, D, C, o)$ finds a feasible solution of optimum value with respect to an objective function $o$ over the variables. In case of a minimisation problem, we have: $S \in COP(V, D, C, o)$ iff $S \in CSP(V, D, C)$ and $\nexists T \in CSP(V, D, C)$ with $o(T) < o(S)$.

**Sudoku.** In our work we consider a prototype CSP, namely the sudoku. Sudoku is a number puzzle, played on a partially filled $9 \times 9$ grid. The goal is to find the unique solution by filling in the empty grid cells with numbers from 1 to 9 in such a way that each row, each column and each of the nine $3 \times 3$ subgrids contain all the numbers from 1 to 9 once and only once.

Formally, the sudoku is a $CSP(V, D, C)$ where $V$ is the set of variables $v_{ij}$ $(i, j \in \{1, ..., 9\})$ for every cell in the grid, and $D(v_{ij}) = \{1, ..., 9\}$ for each $v_{ij} \in V$. We separate the sudoku constraints into two parts: the set of constraints $C_{given}$ defining the assignment of numbers in the filled cells (hereinafter referred to as the *givens*) of the grid and the set of constraints $C_{rules}$ defined by the rules of sudoku.

Formally, $C_{rules}$ consists of the following constraints:

$$\forall i \in \{1, ..., 9\} \quad \texttt{alldifferent}\{v_{i1}, ..., v_{i9}\}$$
$$\forall j \in \{1, ..., 9\} \quad \texttt{alldifferent}\{v_{1j}, ..., v_{9j}\}$$
$$\forall i, j \in \{1, 4, 7\} \quad \texttt{alldifferent}\{v_{ij}, ..., v_{(i+2)j}, \; v_{i(j+1)}, ..., v_{(i+2)(j+1)}, \quad (1)$$
$$v_{i(j+2)}, ..., v_{(i+2)(j+2)}\}$$
$$\forall i, j \in \{1, ..., 9\} \quad v_{ij} \in \{1, ..., 9\}$$

For the given cells, $C_{given}$ is simply an assignment: $D(v_{ij}) = y_{ij}, \; \forall v_{ij} \in \{v_{ij}\}^{given} \subset V$ where the $\{y_{ij}\}^{given}$ are known. Because $V$ and $D$ are obvious from the constraints, we will write $CSP(C_{rules} \wedge C_{given})$ or alternatively $CSP(C_{rules}, \{y_{ij}\}^{given})$ to represent a solution of a sudoku specification.

Sudoku has one additional property, namely that for a set of givens, the solution is unique: $S \in CSP(C_{rules} \wedge C_{given})$, $\nexists T \in CSP(C_{rules} \wedge C_{given})$, with $T \neq S$.

**ML Classifier.** We will consider the visual sudoku problem, where the given cells are not provided as facts, but each given cell will be an image of a hand-written digit. We will hence first use Machine Learning (ML) to classify what digit each of the images represents.

Given a dataset of size $n$, $\{(X_i, y_i)\}_{i=1}^n$ with $X_i \in \mathbb{R}^d$ (denoting that each element is a feature vector of $d$ real numbers) and $y_i$ the corresponding class label, the goal of an ML classifier is to learn a function approximator $f_\theta(X_i)$ (with $\theta$ the trainable parameters of the learning function), such that $f_\theta(X_i) \approx y_i$ for all $(X_i, y_i)$ pairs. In case of a probabilistic classifier, the predicted class label is $\hat{y}_i = f_\theta(X_i) = \arg\max_k P_\theta(y_i = k|X_i)$ with $P_\theta(y_i = k|X_i)$ the predicted probability that $X_i$ belongs to class $k$ [4].

Formally, the goal of training is to compute $\arg\min_\theta \mathcal{L}(f_\theta(X_i), y_i)$, where $\mathcal{L}(., .)$ is a *loss function* measuring how well the function approximates the target. An example of a loss function for probabilistic classifiers with $C$ possible classes is the *cross-entropy loss*, defined as:

$$\mathcal{L} = -\frac{1}{n} \sum_{i=1}^n \sum_{k=1}^C \mathbb{1}[y_i = k] \log P_\theta(y_i = k|X_i), \quad (2)$$

where $\mathbb{1}[y_i = k]$ is the indicator function having the value 1 only when $y_i$ has value $k$, i.e., belongs to class $k$.

# 4 Visual Sudoku and Solution Methods

We first introduce the visual sudoku problem as an example of an image-based constraint solving problem, and then propose three different approaches to solving it by combining classification and reasoning.

**Visual Sudoku.** In visual sudoku, the given cells of the sudoku are provided as unlabeled images of handwritten digits. We are also given a large dataset of labeled handwritten digits (the MNIST dataset [10]). It is inspired by an experiment in [30], although we consider the case where the constraints are known and can be used for reasoning.

Formally, $\texttt{VizSudoku}(C_{rules}, \{X_{ij}\}^{given})$ consists of the rules of sudoku (Eq. 1), and a set of given images $\{X_{ij}\}^{given}$ each one consisting of a pixel representation of the handwritten digit. The goal is to use a classifier $f_\theta$ on $\{X_{ij}\}^{given}$ such that the predicted labels $\{\hat{y}_{ij}\}^{given} = \{f_\theta(X_{ij}) | X_{ij} \in \{X_{ij}\}^{given}\}$ lead to the solution of the sudoku, that is: $CSP(C_{rules}, \{\hat{y}_{ij}\}^{given}) = CSP(C_{rules}, \{y_{ij}\}^{given})$ with $y_{ij}$ the true labels of the given images if known.

## 4.1 Separate Classification and Reasoning

The most straightforward approach to solving the visual sudoku problem is to consider the classification and reasoning problems separately. In this approach, first, the most likely digit for each of the given cells are predicted, after which the puzzle is solved using the resulting grid. This will be our baseline approach.

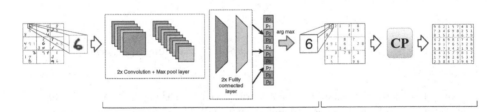

**Fig. 1.** Architecture of separate classification and reasoning approach

The baseline approach, explained on Fig. 1, is composed of a separate convolutional neural network and a CP solver. The process begins with the training of the DNN on the MNIST training set $\{(X, y)\}$ to obtain a handwritten digit classifier $f_\theta$. Then for each visual sudoku instance, we use the classifier to predict the value of each given cell's image. This takes us from a visual to a purely digital representation of the problem, which is then fed into the CP sudoku solver. Note, that training is separate from the concept of sudoku, and done on individual images as is standard in image recognition tasks.

Once the model is trained, we use it to solve $\texttt{VizSudoku}(C_{rules}, \{X_{ij}\}^{given})$. For that, we first predict the digit for each of the given images $\{X_{ij}\}^{given}$. For each $X_{ij}$ given, the trained DNN computes a class probability for each digit $k$ $P_\theta(y_{ij} = k|X_{ij})$ and predicts the value with the highest probability:

$$\hat{y}_{ij} = f_\theta(X_{ij}) = \underset{k \in \{0,..,9\}}{\arg\max} P_\theta(y_{ij} = k|X_{ij}), \qquad (3)$$

Once all the given images are predicted, the CP component finds a solution $S \in \mathrm{CSP}(C_{rules}, \{\hat{y}_{ij}\}^{given})$ as visualised in Fig. 1.

From an inference standpoint, the above approach commits to the independent predictions made by the classifier and tries to use them as best as possible.

## 4.2   Hybrid1: Reasoning over Class Probabilities

In this approach, we will use the same DNN architecture for digit classification as before. However, instead of using the hard labels from the DNN model, we will make use of the class probabilities of each of the given cells. Hence the outputs of the DNN, i.e., the inputs to the CP solver for each of the given cells, are 9 probabilities – one for each digit that can appear in a sudoku cell. The idea is to completely solve a sudoku grid by solving a COP. See Fig. 2 for a visual representation of the architecture.

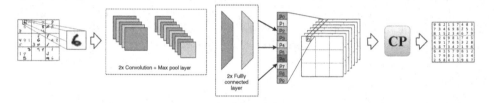

**Fig. 2.** Architecture of class-probability reasoning approach

Note that here, we make a joint inference over all the predictions, including their effect and relation to the a-priori empty cells. In the resulting solution, the digits of both given and non-given cells are obtained at once, while satisfying all the sudoku constraints.

First, the DNN is trained on images of single handwritten digits as before. After training, we store the DNN computed probabilities $P_\theta(y_{ij} = k|X_{ij})$ for each of the given $X_{ij}$. We wish to make the CP solver reason (do *inference*) over these probabilities directly, hence the sudoku problem formulation of Eq. 1 needs to be modified to accommodate the probabilities. Instead of only satisfying the regular sudoku constraints of Eq. 1, we seek to find a solution which *optimizes* the likelihood of the solution, given the probabilities obtained from the classifier.

More specifically, as each image is predicted on its own, we assume each to be an observation of an independent random variable, and hence the most likely solution is the one that maximizes the joint probability over the given images $\max \prod_{given(i,j)} \prod_{k \in \{1,..,9\}} (P_\theta(y_{ij} = k|X_{ij}))^{\mathbb{1}[s_{ij}=k]}$ for a solution $s$. We would like to find the most likely solution that also satisfies all constraints. After a log-transform, we can write the joint probability as a weighted sum objective function as follows:

$$min \sum_{\substack{(i,j)\in \\ given}} \sum_{\substack{k\in \\ \{1,..,9\}}} -log(P_\theta(y_{ij} = k|X_{ij})) * \mathbb{1}[s_{ij} = k] \tag{4}$$

Treating $-log(P_{ij})$ as a $k$-dimensional vector, one can see that the inner sum could be formulated with a traditional *element* constraint in a CP solver. We must emphasize that the log-likelihood is maximized only over the given cells and not for the whole grid due to the fact that we have the classifier provided probability vector only for these cells with given images.

Note that in this approach, the CP solver has to solve a more complex problem with larger domains for the given cells, and hence a larger search space. Contrary to the approach in Sect. 4.1 where the problem was a CSP, here the problem is a COP. The advantage of this approach is that it makes use of the constraint relationships of the sudoku problem. Moreover, it improves the prediction of the ML classifier by reasoning over these constraint relationships.

### 4.3  Hybrid2: Higher-Order Knowledge Exploitation

As mentioned before, a sudoku must have a unique solution for a set of givens. For traditional sudoku puzzles this is the case by construction, as otherwise, a human solver would be faced with having to choose among two or more options, rather than reasoning up to a full solution.

In the approach of Sect. 4.2, we simply find one solution and treat that as *the* solution, without verifying whether it is unique *with respect to the set of givens*. When projecting the solution of the entire sudoku back to only the assignment to the 'given' cells, e.g. those for which an image is given, then this assignment to the givens should have one and only one unique solution. If not, this assignment to the givens, and hence the entire sudoku solution, can not be the intended solution.

Therefore, we can use the (non) existence of a unique solution as an additional relational property that can steer the joint inference. The pseudo-code of this approach is shown in Algorithm 1. We start with finding the most likely solution *sol* as in the hybrid1 approach described in the previous section. We will write $\{sol_{ij}\}^{given}$ to represent the projected part of the solution, that is, only the part of the assignment of the cells with an image given.

Instead of *counting all* solutions given $\{sol_{ij}\}^{given}$, it is sufficient (and computationally cheaper) to only check whether *any* other solution exists. Hence, we will search for *any* sudoku solution (line 3) that is different from the *sol* solution that we already know exists (line 2).

If there does not exist such other solution, i.e. the assignment is an empty set (line 4), then the solution is unique and there is nothing more we can infer. If there is another solution, we reject $\{sol_{ij}\}^{given}$ for not being unique. That is, we

---

**Algorithm 1:** Higher-order COP of VizSudoku($C_{rules}, \{X_{ij}\}^{given}$) using a trained DNN $f_\theta(X)$

---

1   $sol \leftarrow$ VizSudoku($C_{rules}, \{X_{ij}\}^{given}$)    // as in hybrid1
2   $C'_{rules} \leftarrow C_{rules} \wedge \neg(V = sol))$        // temporarily forbid this solution
3   $sol' \leftarrow$ CSP($C'_{rules}, \{sol_{ij}\}^{given}$)    // check for other solutions having these
     givens
4   **while** $sol' \neq \emptyset$ **do**
5      $C_{rules} \leftarrow C_{rules} \wedge \neg(V^{given} = sol^{given})$    // add nogood on givens
6      $sol \leftarrow$ VizSudoku($C_{rules}, \{X_{ij}\}^{given}$)    // as in hybrid1
7      $C'_{rules} \leftarrow C_{rules} \wedge \neg(V = sol))$    // temporarily forbid this solution
8      $sol' \leftarrow$ CSP($C'_{rules}, \{sol_{ij}\}^{given}$)
9   **end**
10 **return** $sol$

---

add a *nogood* ensuring that no completion of $\{sol_{ij}\}^{given}$ will be found anymore (line 5), and repeat the procedure.

This use of a nogood, or a blocking clause, is common in solving such second-order logic problems. It can be seen as an instantiation of *solution dominance* [5].

## 5   Class Probability Calibration

In a machine learning context, *calibration* is the process of modifying the predicted probabilities so that they match the expected distribution of probabilities for each class [6]. We will investigate the effect of calibration on our joint inference approach. Our method reasons over all 9 probability estimates $\{(P_\theta(y = 1|X), \ldots, P_\theta(y = p|X)\}, pos\}$ and actively trades-off the probability of a prediction of one image to the prediction of another image in its objective function. Hence, it is not just a method of getting the top-predicted value right, but rather of getting all predicted probabilities correctly. Our reasoning approach hence assumes real (calibrated) probabilities and could be hampered by over- or under-confident class probability estimations.

In a multi-class setting, for a given handwritten digit a neural probabilistic classifier computes a vector $z$ containing raw scores for each class (i. e. a digit value), $z_k$ being the score assigned to class $k$. The SoftMax function is then applied to convert these raw scores into probabilities:

$$\sigma_{\text{SoftMax}}(z_k, z) = \frac{\exp(z_k)}{\sum_i \exp(z_i)}.$$

such that $P_\theta(y = k|X) = \sigma_{\text{SoftMax}}(z_k, z)$ is the output of the neural network.

While this output is normalized across classes to sum up to 1, the values are not real probabilities. More specifically, it has been shown that especially neural networks tend to overestimate the probability that an item belongs to its maximum likelihood class [6].

Post-processing methods such as Platt scaling [22] aim at calibrating the probabilistic output of a pre-trained classifier. Guo et al. [6] describe three variants of Platt scaling in the multi-class setting. In *matrix scaling*, a weight matrix $W$ and a bias vector $b$ apply a linear transform to the input vector of the softmax layer $z_i$ such that the calibrated probabilities become:

$$\widetilde{P}_\theta(y_i = k|X_i) = \sigma_{\text{SoftMax}}\left(W_k z_k + b_k, W z + b\right) \tag{5}$$

where $W$ and $b$ are parameters, learned by minimizing the Negative Log Likelihood loss on a validation set. *Vector scaling* applies the same linear transform, except that $W$ is a diagonal matrix, that is, only the diagonal is non-zero. Finally, *Temperature scaling* considers a single scalar value $T$ to calibrate the probability such that:

$$\widetilde{P}_\theta(y_i = k|X_i) = \sigma_{\text{SoftMax}}\left(\frac{z_k}{T}, \frac{z}{T}\right) \tag{6}$$

To *calibrate* the predictions, we train a model $f_{\theta,W,b}(X)$ where $\{(\widetilde{P}_\theta(y = 1|X), \ldots, \widetilde{P}_\theta(y = p|X))\}$ is calibrated on a validation set $\{X_i, y_i\}_{validation}$. More specifically, we will do calibration on top of a pre-trained neural network, so $\theta$ is pre-trained and the calibration learns the best $W, b$.

We will evaluate whether better calibrated probabilities lead to better joint inference reasoning in the experiments.

## 6   Experiments

Numerical experiments were done on a subset of the Visual Sudoku Dataset used in [30]. The subset contains 3000 sudoku boards whose givens are represented by MNIST digits. The average number of givens per sudoku grid is 36.2. Unless stated otherwise, the MNIST train data was split into 80%–20% train and validation set.

The DNN architecture for the digit classification task is the LeNet architecture [10] which uses two convolutional layers followed by two fully connected layers. The network is trained for 10 epochs to minimize cross-entropy loss, and is optimized via Adam with a learning rate of $10^{-5}$. Once trained on the MNIST train data, we use the same model for both separate and hybrid approaches. The neural network and CP model were implemented using PyTorch 1.3.0 [19] and OR-tools 7.4.7247 [21], respectively. All experiments were run on a laptop with $8 \times$ Intel® Core™ i7-8565U CPU @ 1.80GHz and 16 Gb of RAM.

To test the performance of our proposed frameworks, we define the following evaluation measures:

img accuracy = percentage of givens correctly labeled by the classifier
cell accuracy = percentage of cells matching the true solution
grid accuracy = percentage of correctly solved sudokus. A sudoku is correctly solved if its true solution was found. That is, if

$$s_1 \in \texttt{VizSudoku}(C_{rules}, \{X_{ij}\}^{given})$$

$$s_2 \in \text{CSP}(C_{rules}, \{y_{ij}\}^{given}) \implies s_1 \equiv s_2$$

`failure rate grid` = percentage of sudokus without a solution. A sudoku has no solution if $\text{VizSudoku}(C_{rules}, \{X_{ij}\}^{given}) = \emptyset$

In the subsequent experiments, we denote as `baseline` the separated classification and reasoning approach, whereas we refer to our proposed approaches as `hybrid1` and `hybrid2`.

**Table 1.** Comparison of hybrid solving approaches

|          | Accuracy |        |        | Failure rate | Time        |
|----------|----------|--------|--------|--------------|-------------|
|          | img      | cell   | grid   | grid         | average (s) |
| baseline | 94.75%   | 15.51% | 14.67% | 84.43%       | 0.01        |
| hybrid1  | 99.69%   | 99.38% | 92.33% | 0%           | 0.79        |
| hybrid2  | 99.72%   | 99.44% | 92.93% | 0%           | 0.83        |

## 6.1   Separate vs Hybrid Approaches

First we compare the result of the three approaches described in Sect. 4. As displayed on Table 1, the ability of the `baseline` approach to handle the image classification task with an accuracy of 94.75% translates to a meagre success rate of only 14.67% at the level of sudoku grids correctly solved. This is because the constraints relationships are not translated to the DNN model. As a consequence there is no way to ensure that the predictions would respect the constraints. Even a single mistake in predictions out of all the given images may result in an unsolvable puzzle. As an example, if one prediction error makes the same number appear twice in a row then the whole puzzle will be unsolvable even if the rest of the predictions are accurate.

On the other hand the hybrid approaches do not consider the model predictions as final and by using the constraints relationships, `hybrid2`, for instance, brings the classifier to correctly label 5361 additional images. As a result we observed an increase in overall accuracy of the predictions. The advantage of our frameworks is more prominent from the `grid` perspective, where we can see that more than 92% of the sudokus are now correctly solved. This is a huge improvement from the `baseline` approach which solves only 14.67% of the grids.

In terms of final performance `hybrid2` is more accurate as it exploits one more sudoku property; namely that sudoku must have a unique solution. By this mechanism we are able to further rectify more predictions and 18 additional puzzles are solved accurately.

However, from a computational standpoint, our hybrid approaches solve a COP instead of a CSP in the pure reasoning case. Hence they are almost a 100 times more time consuming (only the average per sudoku is shown). The average computation time is slightly higher for `hybrid2` as we need to prove that predicted givens only have a unique solution, or optimize again with a forbidden assignments if that is not the case; this situation happens 18 times in our experiments.

## 6.2   Reasoning over Top-$k$ Probable Digits

We are curious to know how the hybrid approaches outperform the separate approach. So we investigate when a digit is chosen by the hybrid approaches, how, on average, it is ranked by the ML classifier when ranking by probability.

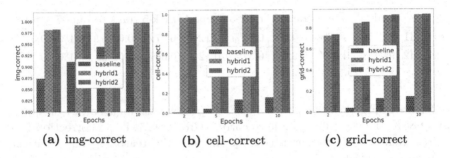

(a) img-correct          (b) cell-correct          (c) grid-correct

**Fig. 3.** Strength of hybrid with less accurate predictions

Table 2 reveals, among the instances where we find the correct solution, that the top-ranked value is chosen in most cases, with a quick decline in how often the other values are chosen. Remarkably, in 42 cases (i.e. 0.02% of predictions) `hybrid2` actually uses a digit which is ranked 8 or lower by the classifier.

From a combinatorial optimisation perspective, one can also consider that this allows to trade-off the size of the search space with the accuracy of the resulting solutions by only taking the $k$ highest probable digits into account and removing the others from the domains. In this regard the experiment in the previous section considered two extremes: the baseline uses only the maximum probable digit, and the hybrid approaches use all 9 digits.

Therefore, we investigate the effect of considering the top-k probability ranked digits on computational time and accuracy. Table 3 shows the effect of using only reasoning over the top-$k$ predicted values of the classifier:

When considering top-1 to top-4 values, we see that the image accuracy steadily goes up as does the grid correctness, and grid failure reaches 0 for top-4. As we consider 4 or more digits, both `grid` and `image` values slowly increase, with the best results obtained using all possible values; which makes the difference for 8 sudoku instances when using `hybrid2`.

This shows that there is indeed a trade-off between computational time of the joint inference and accuracy of the result, with runtime performance gains possible at low accuracy cost if needed.

## 6.3  Classifier Strength Versus Reasoning Strength

So far, we have used a fairly accurate model. We have also seen that joint inference by constraint solving could indeed correct many of the wrong predictions. In this experiment, we investigate the limits of this 'correcting' power of the reasoning. That is, for increasingly worse predictive models, we compare the accuracy of the baseline with our hybrid approaches.

**Table 2.** Rank distribution for cell values in correctly solved sudokus

|         | rank-0 | rank-1 | rank-2 | rank-3 | rank-4 | rank-5 | rank-6 | rank-7 | rank-8 |
|---------|--------|--------|--------|--------|--------|--------|--------|--------|--------|
| hybrid1 | 94.85% | 3.68%  | 0.93%  | 0.32%  | 0.12%  | 0.07%  | 0.02%  | 0.01%  | 0.01%  |
| hybrid2 | 94.84% | 3.68%  | 0.92%  | 0.33%  | 0.12%  | 0.06%  | 0.02%  | 0.01%  | 0.01%  |

Results in Fig. 3 show that even after 2 epochs, with an accuracy of approximately 88%, the reasoning is able to correct this to 98%, i.e., a correction factor of 10%. Hence, with weaker predictive models, the reasoning has even more potential for correcting.

Results on Table 4 show that this trend remains true even with a stronger classifier, obtained by considering a learning rate of $2 \times 10^{-3}$. In the stronger classifier case, `hybrid2` correctly classifies 654 more images than the `baseline`.

Also noteworthy is that the average runtime goes up by a significant factor, e.g., it is 10 times slower as the predictions become less accurate. Further investigation shows that the predicted values are less skewed at lower accuracy levels, e.g., the softmax probabilities are more similar and hence the branch-and-bound search takes more time in finding and proving optimality.

## 6.4  Effect of Calibration

As the joint inference reasons over the probabilities, we will investigate the effect of calibration on the reasoning. The first step towards that goal is to compare the different calibration methods we presented in Sect. 5, namely *Matrix scaling*, *Vector scaling*, and *Temperature scaling*. As described earlier, for each of these methods, calibration parameters are learned by minimizing the Negative Log Likelihood loss on the validation set (while remaining parameters of the network are fixed). Table 5 shows the validation NLL and the test accuracy before and after calibrating of the network. This table suggests that *Matrix scaling* produces the most calibrated classifier. Figure 4 shows how the classifier, although already quite well calibrated, is brought closer to a perfectly calibrated model.

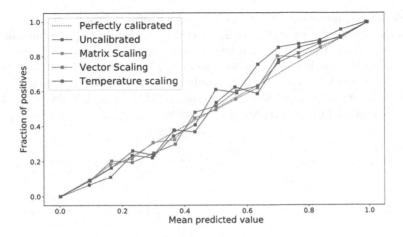

**Fig. 4.** Calibration curve, mean of probabilities over 15 equally-sized intervals

**Table 3.** Rank experiment using hybrid2 for joint inference

| top-$k$ | Accuracy | | | Failure rate | Time |
|---|---|---|---|---|---|
| | img | cell | grid | grid | average (s) |
| top-1 | 94.75% | 15.36% | 14.67% | 84.60% | 0.03 |
| top-2 | 96.15% | 63.63% | 55.43% | 34.20% | 0.03 |
| top-3 | 96.63% | 94.73% | 77.17% | 0.20% | 0.06 |
| top-4 | 98.78% | 98.04% | 86.33% | 0% | 0.12 |
| top-5 | 99.35% | 98.86% | 89.67% | 0% | 0.26 |
| top-6 | 99.57% | 99.21% | 91.60% | 0% | 0.38 |
| top-7 | 99.67% | 99.36% | 92.33% | 0% | 0.55 |
| top-8 | 99.69% | 99.40% | 92.63% | 0% | 0.66 |
| top-9 | 99.71% | 99.43% | 92.90% | 0% | 0.80 |

**Table 4.** Comparison of separate and hybrid approach with a stronger classifier

| | Accuracy | | | Failure rate |
|---|---|---|---|---|
| | img | cell | grid | grid |
| baseline | 99.384% | 80.380% | 80.100% | 19.6% |
| hybrid1 | 99.984% | 99.966% | 99.500% | 0% |
| hybrid2 | 99.986% | 99.972% | 99.600% | 0% |

**Table 5.** NLL loss on validation set and test accuracy for Platt scaling variants

| | Uncalibrated | Temp. scaling | Vector scaling | Matrix scaling |
|---|---|---|---|---|
| NLL | 12.07 | 11.61 | 11.38 | **10.12** |
| Test acc. | 96.75% | 96.75% | 96.70% | **96.93%** |

Figure 5 displays the effect of using a more calibrated model by running the top-k experiment with the `hybrid2` framework, with calibrated and uncalibrated classifiers. It shows that calibration *improves the accuracy* of our framework. This is true when considering not only less accurate, but also more accurate, neural networks, as reasoning over all 9 probabilities leads a calibrated classifier used within the `hybrid2` framework to an `img` rate of 99.80%, an `accuracy cell` rate of 99.62% and 94.30% of correctly solved grids.

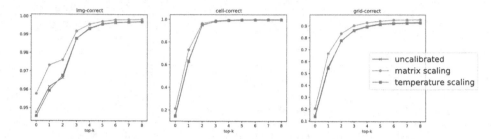

**Fig. 5.** Performance measures for joint inference from calibrated classifier and comparison with uncalibrated counterpart

## 7   Conclusions

In this paper we study a prototype application of hybrid prediction and constraint optimisation, namely the visual sudoku. Although deep neural networks have achieved unprecedented success in classification and reinforcement learning, they still fail at directly predicting the result of a combinatorial optimisation problem, due to the hard constraints and combinatorial optimisation aspect.

We propose a framework for solving challenging combinatorial problems like this, by adding a constraint programming layer on top of a neural network, which does joint inference over a set of predictions. We argue that reasoning over the actual predictions is limited as it ignores the probabilistic nature of the classification task, as confirmed by the experimental results. Instead, we can optimize the most likely joint solution over the classification probabilities which respects the hard constraints. Higher-order relations, such as that a solution must be unique, can also be taken into account to further improve the results.

Our proposed approach always finds a solution that satisfies the constraints, and *corrects* the underlying neural network output up to 10% in accuracy, for example transforming the output of a 94.8% accurate classifier into a 99.7% accurate joint inference classifier.

More broadly, we believe that this work is a notable path to incorporate domain-specific expertise in ML models. Practitioners often feel that they can help to make a ML model better by infusing their expertise into the model. However, incorporating such structured knowledge is often not feasible in a DNN setting. Our work proposes one way to impart human knowledge, namely on top of the neural network architecture and independent of the learning.

An interesting direction for future work is to look at differential classification+optimisation techniques, such as OptNet [1], and investigate whether it is possible to train better models end-to-end for this kind of hard constrained problems. In this respect, there is also a link with probabilistic programming techniques, which often use knowledge compilation to embed (typically simpler) constraints in a satisfaction setting [15]. Finally, we are keen to apply this technique on applications involving classification tasks, such as manhole maintenance [29] and more.

**Acknowledgements.** This research received funding from the Flemish Government under the "Onderzoeksprogramma Artificiële Intelligentie (AI) Vlaanderen" programme.

# References

1. Amos, B., Kolter, J.Z.: Optnet: differentiable optimization as a layer in neural networks. In: Proceedings of the 34th International Conference on Machine Learning, vol. 70, pp. 136–145. JMLR.org (2017)
2. Bengio, Y.: Using a financial training criterion rather than a prediction criterion. Int. J. Neural Syst. **8**(04), 433–443 (1997)
3. Chen, L., Feng, Y., Huang, S., Luo, B., Zhao, D.: Encoding implicit relation requirements for relation extraction: a joint inference approach. Artif. Intell. **265**, 45–66 (2018)
4. Goodfellow, I., Bengio, Y., Courville, A.: Deep Learning. MIT Press, Cambridge (2016)
5. Guns, T., Stuckey, P.J., Tack, G.: Solution dominance over constraint satisfaction problems. CoRR abs/1812.09207 (2018)
6. Guo, C., Pleiss, G., Sun, Y., Weinberger, K.Q.: On calibration of modern neural networks (2017)
7. Ifrim, G., O'Sullivan, B., Simonis, H.: Properties of energy-price forecasts for scheduling. In: Milano, M. (ed.) CP 2012. LNCS, pp. 957–972. Springer, Heidelberg (2012). https://doi.org/10.1007/978-3-642-33558-7_68
8. Kool, W., van Hoof, H., Welling, M.: Attention, learn to solve routing problems! In: ICLR 2019: 7th International Conference on Learning Representations (2019)
9. Lake, B.M., Ullman, T.D., Tenenbaum, J.B., Gershman, S.J.: Building machines that learn and think like people. Behav. Brain Sci. **40** (2017)
10. LeCun, Y., Bottou, L., Bengio, Y., Haffner, P., et al.: Gradient-based learning applied to document recognition. Proc. IEEE **86**(11), 2278–2324 (1998)
11. Li, Q., Anzaroot, S., Lin, W.P., Li, X., Ji, H.: Joint inference for cross-document information extraction. In: Proceedings of the 20th ACM International Conference on Information and Knowledge Management, pp. 2225–2228. ACM (2011)
12. Li, Q., Ji, H., Huang, L.: Joint event extraction via structured prediction with global features. In: Proceedings of the 51st Annual Meeting of the Association for Computational Linguistics (Volume 1: Long Papers), pp. 73–82 (2013)
13. Li, T., Gupta, V., Mehta, M., Srikumar, V.: A logic-driven framework for consistency of neural models. arXiv preprint arXiv:1909.00126 (2019)
14. Mandi, J., Demirović, E., Stuckey, P., Guns, T., et al.: Smart predict-and-optimize for hard combinatorial optimization problems. arXiv preprint arXiv:1911.10092 (2019)

15. Manhaeve, R., Dumancic, S., Kimmig, A., Demeester, T., De Raedt, L.: Deep-problog: neural probabilistic logic programming. In: Advances in Neural Information Processing Systems, pp. 3749–3759 (2018)
16. Márquez-Neila, P., Salzmann, M., Fua, P.: Imposing hard constraints on deep networks: promises and limitations. arXiv preprint arXiv:1706.02025 (2017)
17. Mukhopadhyay, A., Vorobeychik, Y., Dubey, A., Biswas, G.: Prioritized allocation of emergency responders based on a continuous-time incident prediction model. In: Adaptive Agents and Multi Agents Systems, pp. 168–177 (2017)
18. van den Oord, A., Vinyals, O., et al.: Neural discrete representation learning. In: Advances in Neural Information Processing Systems, pp. 6306–6315 (2017)
19. Paszke, A., et al.: Automatic differentiation in PyTorch. In: NeurIPS Autodiff Workshop (2017)
20. Pathak, D., Krahenbuhl, P., Darrell, T.: Constrained convolutional neural networks for weakly supervised segmentation. In: Proceedings of the IEEE International Conference on Computer Vision, pp. 1796–1804 (2015)
21. Perron, L., team: Google's or-tools
22. Platt, J.C.: Probabilistic outputs for support vector machines and comparisons to regularized likelihood methods. In: Advances in Large Margin Classifiers, pp. 61–74. MIT Press (1999)
23. Platt, J.C., Barr, A.H.: Constrained differential optimization. In: Neural Information Processing Systems, pp. 612–621 (1988)
24. Poon, H., Domingos, P.: Joint inference in information extraction. In: AAAI, vol. 7, pp. 913–918 (2007)
25. Punyakanok, V., Roth, D., Yih, W.T., Zimak, D.: Semantic role labeling via integer linear programming inference. In: Proceedings of the 20th International Conference on Computational Linguistics, COLING 2004, p. 1346. Association for Computational Linguistics, USA (2004)
26. Riedel, S.: Improving the accuracy and efficiency of map inference for Markov logic. arXiv preprint arXiv:1206.3282 (2012)
27. Rossi, F., Van Beek, P., Walsh, T.: Handbook of Constraint Programming. Elsevier, Amsterdam (2006)
28. The High-Level Expert Group on Artificial Intelligence (AI HLEG): A definition of AI (2017)
29. Tulabandhula, T., Rudin, C.: Machine learning with operational costs. J. Mach. Learn. Res. 14(1), 1989–2028 (2013)
30. Wang, P.W., Donti, P., Wilder, B., Kolter, Z.: Satnet: bridging deep learning and logical reasoning using a differentiable satisfiability solver. In: ICML 2019: Thirty-Sixth International Conference on Machine Learning, pp. 6545–6554 (2019)
31. Wang, Y., Chen, Q., Ahmed, M., Li, Z., Pan, W., Liu, H.: Joint inference for aspect-level sentiment analysis by deep neural networks and linguistic hints. IEEE Trans. Knowl. Data Eng. (2019)
32. Wilder, B.: Melding the data-decisions pipeline: decision-focused learning for combinatorial optimization. In: Proceedings of the 33rd AAAI Conference on Artificial Intelligence (2019)

# Multi-speed Gearbox Synthesis Using Global Search and Non-convex Optimization

Chiara Piacentini[2]([✉]) [iD], Hyunmin Cheong[1]([✉]), Mehran Ebrahimi[1],
and Adrian Butscher[1]

[1] Autodesk Research, 661 University Avenue, Toronto, ON M5G 1M1, Canada
{hyunmin.cheong,mehran.ebrahimi,adrian.butscher}@autodesk.com
[2] Augmenta Inc., 106 Front St E, Toronto, ON M5A 1E1, Canada
chiara@augmenta.ai

**Abstract.** We consider the synthesis problem of a multi-speed gearbox, a mechanical system that receives an input speed and transmits it to an outlet through a series of connected gears, decreasing or increasing the speed according to predetermined transmission ratios. Here we formulate this as a bi-level optimization problem, where the inner problem involves non-convex optimization over continuous parameters of the components, and the outer task explores different configurations of the system. The outer problem is decomposed into sub-tasks and optimized by a variety of global search methods, namely simulated annealing, best-first search and estimation of distribution algorithm. Our experiments show that a three-stage decomposition coupled with a best-first search performs well on small-size problems, and it outmatches other techniques on larger problems when coupled with an estimation of distribution algorithm.

**Keywords:** Global search · Best-first search · Stochastic search · Evolutionary algorithms · Non-convex optimization.

## 1 Introduction

As demonstrated by the rich literature available [7,13,23,25,33,40], the ability to generate optimal designs is of crucial importance in various engineering applications, since it can lead to significant cost reduction or increased quality of the designed product. In this paper, we focus on the configuration optimization of a specific type of multi-component mechanical system, namely a multi-speed gearbox, which is particularly challenging due to its discrete and bi-level nature.

A gearbox is a mechanical system that transmits and converts an input speed to one or multiple output speeds by means of a series of rotating elements such as shafts and gears. Depending on the arrangements of the gears, we have two different models: the planetary gearbox, used to achieve automatic transmission, and the gear-pairs model, used in manual transmission. Here we only consider the manual transmission, often used in heavy-duty systems due to its higher

© Springer Nature Switzerland AG 2020
E. Hebrard and N. Musliu (Eds.): CPAIOR 2020, LNCS 12296, pp. 381–398, 2020.
https://doi.org/10.1007/978-3-030-58942-4_25

durability. A power source rotates the input shaft with an input velocity, which in turn rotates the next connected shaft, propagating the speed throughout the gearbox and towards the output. Pairs of shafts are connected through gear-pairs with different radii. The size of the gears' radii determines the speed change between the connected shafts. Typically, a gearbox is designed to convert the input speed into multiple output speeds according to several transmission ratios. This is achieved by having multiple shafts and multiple gear-pairs connecting those shafts. When a desired transmission ratio is selected, the associated gear-pairs are activated via clutches. Objectives considered for gearbox design can be the deviation of the transmission ratios from the nominal values, the power capacity, and the volume or the mass of the gearbox. The gearbox synthesis involves choosing the optimal configuration of shafts and gear-pairs, as well as their parameters such as gears' radii, with respect to the objectives.

The current work formalizes the gearbox synthesis as a bi-level optimization problem. The inner problem (or parameter optimization) takes a gearbox configuration as input and finds the position and size of the gears and shafts within given boundaries. This problem can be cast onto a non-convex continuous optimization problem and solved using state-of-the-art solvers. The outer optimization generates different configurations and we can solve it as a single step or as a decomposed set of sub-tasks. For both these approaches, we implement a variety of algorithms: simulated annealing (SA) and best-first search (BFS), which exploit a formulation of the configuration problem as a state transition system, and an estimation of distribution algorithm (EDA), from the evolutionary algorithms family. While in literature the gearbox synthesis is often modelled as a state transition system, in this paper we explore novel models and adapt them to better suit the search algorithms used. In summary, the contributions of this paper are:

1. formalization of multi-speed gearbox synthesis as a bi-level optimization problem;
2. presentation of two different decomposition approaches of the problem;
3. modelling of different sub-problems as state transition models;
4. application of BFS for multi-speed gearbox synthesis considering both lower and upper bounds;
5. development of an EDA for multi-speed gearbox synthesis.

The paper is organized as follows. Section 2 provides a literature review on gearbox synthesis, while Sect. 3 presents the notation and the problem definition. In Sect. 4 we describe two ways in which we can decompose the problem and we summarize each sub-problems. In Sect. 5 we formulate the sub-problems as state transmission models. Global search algorithms used in this work are described in Sect. 6. The experimental evaluation of our approach is shown in Sect. 7. Section 8 concludes our paper.

## 2    Literature Review

Gearbox design has been extensively studied in the mechanical engineering literature and different variations of the problem have been considered. A large body of work deals with the parameter optimization problem, with the assumption that a fixed configuration is given as input. The problem can be formulated as a constraint satisfaction problem [26], and many works attempt to solve it using heuristic and meta-heuristic methods, such as local search [29,44], simulated annealing [17], genetic algorithms [4,11,30,43] and particle swarm optimization [31]. Multi-objective versions of the problem can also be found [10,15,27,41].

Gearbox configurations are often represented using graphs [42] and configuration synthesis is typically performed using *formal grammars*, i.e. a sets of rules that combine a finite set of elements to obtain a potentially infinite set of entities [9,24]. In the context of design synthesis, grammar rules offer a way to define how a design can be modified [6] and they have been widely used for gearbox generation [20,21,32,34,36]. Tsai et al. consider the configuration optimization of a planetary gearbox, where configurations are generated manually and duplicates are detected using graph isomorphism [36]. Schmidt & Chase developed a set of grammar rules that can be applied to generate several configurations of a planetary gearbox [32], while Li et al. propose a software that generates sketches of planetary gearboxes by modifying an initial design [20]. The set of rules developed by Lin et al. act on a graph representing the configuration of a manual transmission gearbox [21]. The graph is labelled to capture the relative position of gears and SA is used to generate candidate configurations and to fix some of the components' parameters. When a candidate is generated, collisions between elements are checked according to a set of constraints. The constraints identified by the authors, however, do not guarantee to avoid all possible collisions. Swantner & Campbell use an exhaustive tree search algorithm to generate a gearbox that can have a single transmission ratio, resulting in a fairly small system. The parameter optimization problem includes bending and stress constraints [34]. In the context of computational design synthesis, a complete search is used to study the quality of grammar rules [18]. Departing from this approach, Pomrehn & Papalambros consider the optimization of a gear train that outputs a single velocity as a mixed-integer non-linear programming model [28]. Berx et al. study a manual multi-speed gearbox and generate different configurations using constraint programming. A clustering procedure identifies promising candidate configurations, for which feasibility and objective value are calculated [2].

# 3    Problem Description

We consider the generation of manual multi-speed gearboxes and impose geometric constraints on gears and shafts, as proposed by Berx et al. [2].

## 3.1    Notation

Consider a gearbox consisting of a set of shafts $\mathbf{S}$ and a set of gear-pairs, i.e. gears connected with each other, $\mathbf{P}$. Each pair $p \in \mathbf{P}$ is comprised of an input gear $g_p^i$ and an output gear $g_p^o$. We call $\mathbf{G}$ the union of all the input and output gears and we denote with $s_g$ the shaft in which a gear $g$ is situated, and $s_p^i$

and $s_p^o$ the input and the output shafts of a gear-pair $p$, respectively. We call $\mathbf{C}$ the union of all the shafts and gears. We assume that all the shafts and gears are cylinders aligned along the $z$-axis. Each component $c \in \mathbf{C}$ is defined by the coordinates of the center of the bottom face of the cylinder $(x_c, y_c, z_c)$, a radius $\rho_c$ and a length (or thickness) $l_c$. We have set of transmission ratios $\Omega$, where a transmission ratio $\omega$ is a real number indicating the desired ratio of the input and output speed.

A gearbox layout can be represented as an $s$-$t$ multi-graph $\mathcal{G} = \langle \mathbf{S}, \mathbf{P} \rangle$ where the nodes $\mathbf{S}$ are the shafts and edges $\mathbf{P}$ are the gear-pairs. A source node $s$ is identified as the input shaft, while a sink $t$ is the output shaft of the system.

The graph is required to contain at least $k = |\Omega|$ simple paths from $s$ to $t$. Each simple path $\pi$ is a sequence of edges $\pi = (p_1, ..., p_n)$, with $p_i \in \mathbf{P}$ and all the vertices are distinct, producing a transmission ratio:

$$\omega_\pi = \prod_{j=1}^{n} \frac{\rho_{g_{p_j}^i}}{\rho_{g_{p_j}^o}} \tag{1}$$

where $\rho_{g_{p_j}^i}$ and $\rho_{g_{p_j}^o}$ are the radii of the input and output gears for gear-pair $p_j$.

Given the set of transmission ratios $\Omega$, we call assignment $\Pi^\Omega$ the set of 2-tuples, containing a transmission ratio and a path $\Pi^\Omega = \{(\omega, \pi), \forall \omega \in \Omega\}$. We call configuration $\mathcal{C} = (\mathcal{G}, \Pi^\Omega)$ of a gearbox the 2-tuple consisting of the gearbox layout $\mathcal{G}$ and an assignment $\Pi^\Omega$.

Figure 1 shows an example of a gearbox with an input shaft $s^i$, an output shaft $s^o$ and an intermediate shaft $s$. Two gear-pairs are connected between $s^i$ and $s$ ($p_1$ and $p_2$) and two between $s$ and $s^o$ ($p_3$ and $p_4$). There are four possible paths between $s^i$ and $s^o$: $\pi_1 = (p_1, p_3)$, $\pi_2 = (p_1, p_4)$, $\pi_3 = (p_2, p_3)$, and $\pi_4 = (p_2, p_4)$, corresponding to four different transmission ratios $\Omega = \{\omega_1, \omega_2, \omega_3, \omega_4\}$. An example of assignment can be $\Pi^\Omega = \{(\omega_i, \pi_i), \forall i = 1, ..., 4\}$.

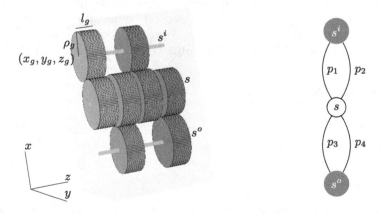

**Fig. 1.** Graph representation of the gearbox

## 3.2  Parameter Optimization

Let $\mathcal{C} = (\mathcal{G}, \Pi^\Omega)$ be a given configuration, where shafts, gear-pairs and their input and the output are known. We want to find the components' parameters $(x_c, y_c, z_c, \rho_c, l_c)$ that minimize the total mass of the gears and produce the desired transmission ratios. If we assume that all the gears have a constant thickness $(l_g)$ and are made of same material, the objective function is:

$$J_{\mathcal{C}}(\rho) = \sum_{g \in \mathbf{G}} \rho_g^2 \qquad (2)$$

The system must satisfy several constraints based on design requirements and physical feasibility. For each pair in the assignment $(\omega, \pi) \in \Pi^\Omega$ the transmission ratio produced by $\pi = (p_1, ..., p_n)$ is within a given $\epsilon$ to the desired value $\omega$:

$$\left| \prod_{j=1}^{n} \frac{\rho_{g_{p_j}^i}}{\rho_{g_{p_j}^o}} - \omega \right| \le \epsilon \qquad \forall (\omega, \pi) \in \Pi^\Omega, \qquad (3)$$

Every component $c \in \mathbf{C}$ must be contained inside a parallelepiped with maximum sizes $(x_{\max}, y_{\max}, z_{\max})$ and its radius has $\rho_{\min,c}, \rho_{\max,c}$ as limits:

$$\rho_c \le x_c \le x_{\max} - \rho_c \qquad \forall c \in \mathbf{C} \qquad (4)$$
$$\rho_c \le y_c \le y_{\max} - \rho_c \qquad \forall c \in \mathbf{C} \qquad (5)$$
$$0 \le z_c \le z_{\max} - l_c \qquad \forall c \in \mathbf{C} \qquad (6)$$
$$\rho_{\min,c} \le \rho_c \le \rho_{\max,c} \qquad \forall c \in \mathbf{C} \qquad (7)$$

A gear $g$ must be placed on its connected shaft $s_g$:

$$x_g = x_{s_g} \qquad \forall g \in \mathbf{G} \qquad (8)$$
$$y_g = y_{s_g} \qquad \forall g \in \mathbf{G} \qquad (9)$$
$$z_g \le z_{s_g} \qquad \forall g \in \mathbf{G} \qquad (10)$$
$$z_g + l_g \le z_{s_g} + l_{s_g} \qquad \forall g \in \mathbf{G} \qquad (11)$$

In addition, we impose that the gears of each gear-pair must touch:

$$z_{g_p^i} = z_{g_p^o} \qquad \forall p \in \mathbf{P} \qquad (12)$$
$$\rho_{g_p^i} + \rho_{g_p^o} = \sqrt{(x_{s_g^i} - x_{s_g^o})^2 + (y_{s_g^i} - y_{s_g^o})^2} \qquad \forall p \in \mathbf{P} \qquad (13)$$

Finally, we require that two elements cannot occupy the same position in space. For two components $c, d \in \mathbf{C}$ that are not a gear and the shaft in which the gear is placed on, there exist two parameters $\lambda_{c,d}$ and $\alpha_{c,d}$, such that:

$$\lambda_{c,d}(z_c - z_d - l_d) + (\alpha_{c,d} - \lambda_{c,d})(z_d - z_c - l_c) +$$
$$(1 - \alpha_{c,d})[(x_c - x_d)^2 + (y_c - y_d)^2 - (\rho_c + \rho_d)^2] > 0 \qquad (14)$$
$$0 \le \lambda_{c,d} \le \alpha_{c,d} \le 1 \qquad (15)$$

This is the continuous formulation of the non-overlapping constraint obtained from the Lagrange duality applied to the distance determination problem [2,3].

Given a configuration $\mathcal{C}$, we define the parametric optimization problem as:

$$\min_{\rho} J_{\mathcal{C}}(\rho) \qquad\qquad (\mathcal{I}(\mathcal{C}))$$

$$s.t.\ \text{Constraints (3)–(15)}$$

### 3.3   Configuration Optimization

When designing a gearbox configuration, we can impose a maximum limit on the number of shafts and the number of gears connecting any two shafts, $N_\mathbf{S}$ and $N_\mathbf{P}$, respectively. The configuration optimization problem is then:

$$\min_{\mathcal{C}} \mathcal{I}(\mathcal{C}) \qquad\qquad (\mathcal{O})$$

$$s.t.\ |\mathbf{S}| \leq N_\mathbf{S} \qquad\qquad\qquad\qquad\qquad (16)$$

$$|\mathbf{P}_{s_i,s_j}| \leq N_\mathbf{P} \qquad\qquad \forall s_i, s_j \in \mathbf{S} \qquad (17)$$

where $\mathbf{P}_{s_i,s_j}$ is the set of gear-pairs connecting shafts $s_i$ and $s_j$. Similarly, we define $\mathbf{P}_s$ as the set of gear-pairs connected to $s$.

## 4   Problem Decomposition

This section shows two decomposition approaches. In the **two-stage decomposition approach**, the problem is divided into two sub-tasks:

A1. **Transmission ratio path assignment:** given a complete multi-graph $\mathcal{G}^{N_s,N_p} = (\mathbf{S}, \mathbf{P})$ with $N_s$ shafts and $N_p$ gear-pairs between every two shafts, and a set of transmission ratios $\Omega$, generate an assignment $\Pi^\Omega$ using the $s$-$t$ simple paths in $\mathcal{G}^{N_s,N_p}$. The layout $\mathcal{G}$ is the union of paths in the $\Pi^\Omega$;

A2. **Parameter optimization:** given the layout $\mathcal{G}$ and the assignment $\Pi^\Omega$, solve $\mathcal{I}(\mathcal{G}, \Pi^\Omega)$.

Since working with the complete multi-graph $\mathcal{G}^{N_\mathbf{S},N_\mathbf{P}}$ may result in a prohibitive large number of possible assignments, we consider an alternative **three-stage decomposition approach**, where we first select a sub-graph of $\mathcal{G}^{N_s,N_p}$ with at least $k$ distinct $s$-$t$ simple paths:

B1. **Graph generation:** generate a $s$-$t$ multi-graph $\mathcal{G} = (\mathbf{S}, \mathbf{P})$ with at least $k$ distinct simple paths from the input and output shafts;

B2. **Transmission ratio path assignment:** given a graph $\mathcal{G}$, find an assignment $\Pi^\Omega$ using the $s$-$t$ simple paths in $\mathcal{G}$;

B3. **Parameter optimization:** given a graph $\mathcal{G}$ and an assignment $\Pi^\Omega$, solve problem $\mathcal{I}(\mathcal{G}, \Pi^\Omega)$.

## 4.1    Graph Generation

Graph generation entails finding a sub-graph $\mathcal{G}$ of $\tilde{\mathcal{G}}$ with at least $k$ $s$-$t$ simple paths. Given an edge-weighted graph $\tilde{\mathcal{G}} = (V, E)$, a source node $s \in V$, a sink node $t \in V$ and a positive number $k$, we call *min-k-simple paths* (mkSP) the problem of finding the minimum weighted edges $E' \subseteq E$, such that there exists $k$ distinct simple paths $\pi_i$ from $s$ to $t$, whose edges belong to $E'$: $\pi_i \subseteq E'$ for $i = 1, \ldots, k$. In the gearbox synthesis, the input is a complete multi-graph with $\mathcal{G}^{N_S, N_P}$ and the solution can provide a lower bound on the number of gears necessary to add to the system.

The problem can be seen as a particular type of coverage problem, namely *min-k-union* (mkU) [38]: given a set $\mathcal{U}$, a collection $\mathcal{S}$ of subsets of $\mathcal{U}$, and an integer number $k$, select $k$ subsets of $\mathcal{S}$ to minimize the number of covered elements in $\mathcal{U}$. Min-k-union is the minimization version of the classical maximum coverage problem. While for maximum coverage, a greedy solution has a $(1-1/e)$ approximation guarantee, mkU is harder to approximate [8].

The reduction of mkSP to mkU is trivial: the set of all the edges $E$ in mkSP is $\mathcal{U}$ in mkU and the set of all paths corresponds to $\mathcal{S}$. This, however, requires the identification of all the $s$-$t$ simple paths in $\mathcal{G}$, which is a #P-COMPLETE task in the general case [37].

When generating graphs, we can leverage *graph isomorphism* to avoid the repeated evaluation of equivalent graphs. While it is not known if the detection of isomorphic graphs can be solved in polynomial time or if it is a NP-COMPLETE problem, efficient solvers are available [22].

## 4.2    Transmission Ratio Path Assignment

Given a graph $\mathcal{G}$ with $n$ $s$-$t$ simple paths and a set of transmission ratios $\Omega$, the task requires to find an assignment $\Pi^\omega$. If we have $k$ transmission ratios, with $k \leq n$, we have $\frac{n!}{(n-k)!}$ assignments. To reduce the number of assignments that we need to check, we can identify those that lead to the same parameter optimization problem.

Consider the example in Fig. 2 with three transmission ratios $\omega_1, \omega_2, \omega_3$. The configuration contains three simple paths: $\pi_1 = (p_1, p_4), \pi_2 = (p_2, p_4), \pi_1 = (p_3)$ and six assignments can be made: $\{(\omega_1, \pi_1), (\omega_2, \pi_2), (\omega_3, \pi_3)\}, \{(\omega_1, \pi_1),$

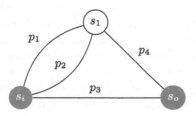

| | $\omega$ | $\pi$ | edge-seq | labeling |
|---|---|---|---|---|
| $\Pi^\Omega$ | $\omega_1$ | $\pi_1$ | $(p_1, p_4)$ | $(l_1, l_2)$ |
| | $\omega_2$ | $\pi_2$ | $(p_2, p_4)$ | $(l_3, l_2)$ |
| | $\omega_3$ | $\pi_3$ | $(p_3)$ | $(l_4)$ |
| $\Pi'^\Omega$ | $\omega_1$ | $\pi_2$ | $(p_2, p_4)$ | $(l_1, l_2)$ |
| | $\omega_2$ | $\pi_1$ | $(p_1, p_4)$ | $(l_3, l_2)$ |
| | $\omega_3$ | $\pi_3$ | $(p_3)$ | $(l_4)$ |

**Fig. 2.** Example of transmission ratio path assignments and their labeling

$(\omega_2, \pi_3)$, $(\omega_3, \pi_2)\}$, $\{(\omega_1, \pi_2)$, $(\omega_2, \pi_1)$, $(\omega_3, \pi_3)\}$, $\{(\omega_1, \pi_2)$, $(\omega_2, \pi_3)$, $(\omega_3, \pi_1)\}$, $\{(\omega_1, \pi_3)$, $(\omega_2, \pi_2)$, $(\omega_3, \pi_1)\}$, and $\{(\omega_1, \pi_3)$, $(\omega_2, \pi_1)$, $(\omega_3, \pi_2)\}$. If we have only a single transmission ratio, paths with the same lengths will lead to the same parameters. In our particular example, this allows us to check only three assignments: $\{(\omega_1, \pi_1)$, $(\omega_2, \pi_2)$, $(\omega_3, \pi_3)\}$, $\{(\omega_1, \pi_1)$, $(\omega_2, \pi_3)$, $(\omega_3, \pi_2)\}$, and $\{(\omega_1, \pi_3)$, $(\omega_2, \pi_2)$, $(\omega_3, \pi_1)\}$.

Formally, given a gearbox problem, a layout $\mathcal{G}$ and two assignments $\Pi^\Omega$ and $\Pi'^\Omega$ over a set of $s$-$t$ simple paths in $\mathcal{G}$, we say that the two assignments are equivalent $\Pi^\Omega \equiv \Pi'^\Omega$ if problems $\mathcal{I}((\mathcal{G}, \Pi^\Omega))$ and $\mathcal{I}((\mathcal{G}, \Pi'^\Omega))$ produce the same optimal solution. To detect equivalent assignments, we fix an arbitrary order of the transmission ratios and label the edges of the associated paths in their order of appearance. Thus, two assignments with the same labeling are equivalent.

### 4.3 Parameter Optimization

The parameter optimization $\mathcal{I}(\mathcal{C})$ is a non-convex optimization problem. We solve it using the solver IPOPT [39]. IPOPT is a software specialized in large scale continuous non-linear optimization problems, based on primal-dual interior point method. For non-convex problems, it does not guarantee that the solution found is globally optimal nor that the problem is globally infeasible. As shown in previous work [2], Constraints (14)–(15) are hard to enforce and can cause the solver to get stuck in a region of local infeasibility. To alleviate this problem, we first consider a relaxation of the problem containing only Constraint (3). If the relaxation is infeasible, we also consider the original problem to be infeasible. Otherwise, we use its solution to initialize the values of the radii of the gears. If the solver finds that $\mathcal{I}(\mathcal{C})$ is infeasible, we restart the solver with a different random initialization of the parameters, up to a fixed maximum number of times, or until a feasible solution is found.

## 5    State Transmission Models

In this section, we present different transition systems that can be used to solve the sub-tasks of our problem. Similar to grammar rules [18,21], a transition system can be used to describe how an algorithm moves from one partial solution (a state) to another, using applicable actions.

More formally, a state transition system is defined by a set of states $\mathcal{S}$, a set of possible actions $\mathcal{A}$ and a transition function $\mathcal{T} : \mathcal{S} \times \mathcal{A} \to \mathcal{S}$. An action can be applied to a state if the state satisfies some *preconditions*. Each transition is characterized by a cost function $\mathcal{Q}_\mathcal{T} : \mathcal{S} \times \mathcal{A} \to \mathbb{R}$. Given an initial state $\sigma_i \in \mathcal{S}$ and a set of goal states $\mathcal{S}_g \subseteq \mathcal{S}$, we want to find a sequence of applicable actions $\pi = (\alpha_1, ..., \alpha_n)$, such that after applying the actions from $\sigma_i$, we obtain a goal state $\sigma_g = \mathcal{T}(\mathcal{T}(\mathcal{T}(\sigma_i, \alpha_1), ...), \alpha_n) \in \mathcal{S}_g$, using a sequence of actions with minimal cost. In all the sub-tasks considered, the cost does not depend on the transition, but only on the state.

## 5.1    Graph Generation

A state is an $s$-$t$ multi-graph $\mathcal{G} = (\mathbf{S}, \mathbf{P})$ with nodes $\mathbf{S}$ and edges $\mathbf{P}$. The source of the graph is the input shaft $s^i \in \mathbf{S}$ while the sink is the output shaft $s^o \in \mathbf{S}$. A goal state is defined as a graph containing $k = |\Omega|$ simple paths from $s_i$ to $s_o$. We assign cost 0 to non-goal states, while the cost of a goal state is determined by the solution of the nested problem. We report here the primitive actions for this task. However, depending on the algorithm, we use a subset of such actions, or we define new actions as a sequence of these primitive actions. In the following, every action is identified by a name, where the subscript indicates the involved objects (gears, shafts, paths), its preconditions $\Phi$, and the transition function $\mathcal{T}$ from a state $\mathcal{G} = (\mathbf{S}, \mathbf{P})$:

- $a_{p, s_j, s_k}$: add a gear-pair $p$ between shafts $s_j, s_k \in \mathbf{S}$. $\Phi : |\mathbf{P}_{s_j, s_k}| < N_p$, $\mathcal{T} : (\mathbf{S}, \mathbf{P} \cup \{p\})$
- $a_s$: add a new shaft $s$. $\Phi : |\mathbf{S}| < N_s$. $\mathcal{T} : (\mathbf{S} \cup \{s\}, \mathbf{P})$
- $d_p$: delete edge $p \in \mathbf{P}$. $\Phi : p \in \mathbf{P}$. $\mathcal{T} : (\mathbf{S}, \mathbf{P}/\{p\})$
- $d_s$: delete shaft $s \in \mathbf{S}$. $\Phi : s \in \mathbf{S}$, $s \neq s^i$, $s \neq s^o$, $\mathcal{T} : (\mathbf{S}/\{s\}, \mathbf{P}/\mathbf{P}_s)$

## 5.2    Transmission Ratio Path Assignment

We consider two distinct models for the transmission ratio path assignment task. The first model, called **paths generation model**, incrementally builds paths adding edges, while the second, **assignment model**, finds all the $s$-$t$ simple paths in a graph and tries to assign each transmission ratio to one of them.

**Paths Generation Model.** In path generation, a state is defined as a tuple $(\pi_1, ..., \pi_k)$, where $\pi_i = (p_1, ..., p_{n_i})$ is the sequence of gear-pairs associated with the transmission ratio $\omega_i$, starting with shaft $s^i$. The sequence requires the gear-pairs to be connected: i.e. $s^i_{p_j} = s^o_{p_{j-1}}, \forall j = 2, ..., n_i$. A sequence may be empty. We say that a sequence is a *complete path* if it terminates with shaft $s^o$. A goal state is a state where all the sequences of gear-pairs are complete, distinct paths. A goal state uniquely identifies a transmission ratio path assignment. For this task, we only add or delete gear-pairs from a state $(\pi_1, ..., \pi_k)$ (shafts are automatically added if they are the input or the output of the gear-pair added):

- $a_{p, \pi_i}$: add gear-pair $p$ at the end of $\pi_i$. $\Phi : s^i_p = s^o_{p_{n_i}}$, $|\mathbf{P}_{s^i_p, s^o_p}| < N_p$, $|\mathbf{S}| < N_s$, $\pi_i \cup p$ is a simple path. $\mathcal{T} : (\pi_1, ..., \pi_i \cup p, ..., \pi_k)$,
- $d_{\pi_i}$: delete the last element of $\pi_i$. $\Phi : \pi_i$ is not empty. $\mathcal{T} : (\pi_1, ..., \pi_i/p_{n_i}, ..., \pi_k)$

The cost of a non-goal state is determined by the solution of the relaxed version of the parameter optimization problem, which considers only the transmission ratios associated with complete paths. The cost of goal states is the cost of the full parameter optimization problem.

**Assignment Model.** For this model, given a graph $\mathcal{G}$ containing $n$ simple paths from $s^i$ to $s^o$, a state is a partial assignment $\tilde{\Pi}^\Omega$, i.e. a subset of $\Pi^\Omega$. Actions assign a simple path to a transmission ratio, i.e. add an element $(\omega, \pi)$ to the partial assignment. A goal state is defined as a state where all the transmission ratios are assigned. The cost of a non-goal state is determined by the solution of the relaxed version of the parameter optimization problem over the partial assignment, while for goal states, the cost is the objective value of the full parameter optimization problem.

# 6    Global Search

This section describes three search algorithms used to solve the gearbox synthesis problem. For each search, both the two-stage and three-stage decomposition approaches are considered (see Fig. 3). Each sub-task is solved using transition systems described in the previous section and its solution is the input of the next sub-task. The first algorithm presented, our baseline, is SA [16]. Next, we consider a BFS [12] that keeps track of both the upper bound and lower bound of solutions, similarly to branch-and-bound [19]. Finally, an EDA, a population-based meta-heuristics approach is used.

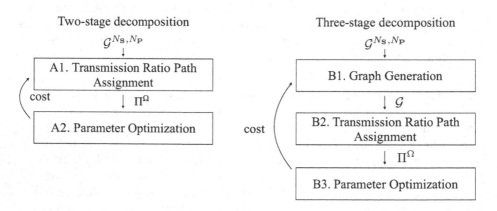

**Fig. 3.** Flowchart of the decomposition approaches

## 6.1    Simulated Annealing

In SA, states are feasible solutions and their cost corresponds to the cost of a solution. A new state is generated by randomly selecting one transition of the transition model and solving the subsequent sub-tasks. If this results in a non-feasible solution, the state is automatically rejected.

**Two-Stage Decomposition.** The search is initialized with randomly generated paths and the neighborhood is defined by the paths generation model in Sect. 5.2. We consider the following actions, obtained by concatenating the actions in Sect. 5.2:

- delete gear-pair $p$ at position $i$ in $\pi$: $d_{p,i,\pi} = d_{p_n,\pi}, d_{p_{n-1},\pi}, ..., d_{p_{i-1},\pi},$
$a_{(s^i_{p_{i-1}}, s^i_{p_{i+1}}),\pi}, a_{p_{i+2},\pi}, ..., a_{p_n,\pi}$
- add gear-pair $p$ at position $i$ in $\pi$: $a_{p,i,\pi} = d_{p_n,\pi}, d_{p_{n-1},\pi}, ..., d_{p_{i-1},\pi},$
$a_{(s^i_{p_{i-1}}, s^i_p),\pi}, a_{p,\pi}, a_{(s^o_p, s^o_{p_i}),\pi}, a_{p_i,\pi}, ..., a_{p_n,\pi}$
- replace gear-pair $p_j$ with $p_k$ in $\pi$: $r_{p_j,p_k,i,\pi} = d_{p_j,i,\pi}, a_{p_k,i,\pi}$

**Three-Stage Decomposition.** For this approach, we first generate a graph using the graph generation model in Sect. 5.1 with the actions devised by Konigseder et al. [18], which are a combination of actions in Sect. 5.1:

- create a new gear-pair between two existing shafts: $a_{p,s_j,s_k}$
- delete an existing gear-pair $p$: $d_p$
- create a new shaft and connect to two existing ones: $a_s, a_{p_i,s^i_{p_j},s}, a_{p_j,s,s^o_{p_j}}$
- delete a shaft: $d_s$
- replace a gear-pair: $d_p, a_s, a_{p_j,s^i_p,s}, a_{p_k,s,s^o_p}$

Starting from a randomly generated graph, the SA randomly selects one of the actions and finds a new graph. The transmission ratio-path assignment problem is solved heuristically by selecting a limited number of assignments, which are then used to solve the parameter optimization problem. The cost of the graph is the best objective value found among the sub-problems considered.

## 6.2    Best-First Search

In BFS, starting from the initial state, the algorithm selects a node for expansion based on an evaluation function $f$, which represents an estimation of the cost of the best solution. During *expansion* all the possible successor states are generated, *evaluated* and inserted into a priority queue. The process is repeated, selecting each time the state in the queue with the lowest $f$, until a goal state is retrieved from the queue. We consider here a tweaked version of BFS that also keeps track of an upper bound [5], which is simply the cost of the best incumbent solution found during the evaluation stage of the algorithm. This allows us to return the such feasible solution when we limit the running time of the algorithm. Our evaluation function $f$ is defined as a lower bound of the cheapest configuration that can be achieved from a state $\sigma$ and it is set to 0 at initialization. The optimality of the algorithm cannot be guaranteed because our parameter optimization problem is not solved to (global) optimality.

**Two-Stage Decomposition.** Similarly to SA, we solve the transmission ratio path assignment problem using the path generation model in Sect. 5.2. We start a search with a state with empty paths and incrementally add all possible edges to create the first path. When the path is complete, we solve the inner optimization problem to determine the cost of the state and we can start adding edges to create the next path. Notice that the only action that we need is the addition of a gear-pair. The evaluation function is calculated by running the relaxed solver on paths that are completed. In addition, we add a lower bound related to the number of gear-pairs that we need to add to the graph to have $k$ paths. This is calculated by building the graph representing the gearbox layout as the union of the paths and estimating the number of edges necessary to have $k$ $s$-$t$ simple paths: if the graph has $k$ $s$-$t$ simple paths, this is the number of edges in the graph, otherwise, we take the number of edges plus one. We calculate the lower bound on the cost by multiplying such number by two (every edge is a pair of gears) and the square of the minimum radius of the gears.

**Three-Stage Decomposition.** We set the initial state to be a graph containing only the input $s^i$ and the output $s^o$ shafts and consider the actions in Sect. 5.1:

- create a new gear-pair between two existing shafts: $a_{p,s_j,s_k}$
- create a new gear-pair $p$ between an existing and a new shafts $s$: $a_s, a_{p,s_j,s}$

Since we insert all the states explored in the queue and we start with an empty graph, we do not need actions that delete gear-pairs or shafts. We use graph isomorphism to detect duplicated states. When a goal state (a graph with $k$ simple paths from $s^i$ to $s^o$) is generated, we calculate the minimal cost of the graph. This cost is calculated by solving the transmission ratio path assignment problem. The evaluation function is calculated by estimating the minimum number of edges that we need to have a graph with at least $k$ simple paths, similarly to the 2-stage decomposition approach. Since the cost of a graph is defined by $\mathcal{I}$, which is a non-convex problem, adding a gear-pair does not necessarily increase the cost of a configuration. For this reason, the evaluation function of a goal state is not the cost of the configuration but is the lower bound defined above. Our algorithm does not terminate when a goal state is found, but when the gap between lower and upper bound is 0 or all the states have been explored.

To solve the transmission ratio path assignment problem, we run another BFS using the assignment model in Sect. 5.2, where we start from a state where none of the transmission ratios is assigned and actions correspond to the assignment of a simple path to a transmission ratio.

### 6.3   Estimation of Distribution Algorithm

The last type of global search method is EDA, a class of meta-heuristic approaches based on the evolution of populations [14]. While typical evolutionary algorithms, such as genetic algorithms, use variation operators such as cross-over and mutation, EDAs use an explicit probability model to generate new solutions, showing advantages in terms of performance, theoretical convergence, and capturing the structure of the problem space [14]. The probabilities are computed by directly using the frequency statistics of the selected top individuals from a population and indicate the likelihood of a particular solution being included in a set of top quality solutions based on prior observations. EDA has also been successfully applied to the configuration design of vehicle suspension systems [7]. As with the other algorithms, both two and three-stage decomposition approaches are implemented and EDA is used to solve the top-level task. The overall framework for the both approaches is the same except for the details in each of the steps, as detailed below.

**i) Generation of Initial Population:** As a first step, an initial population of individuals $\mathcal{P}$ is randomly generated.

In the two-stage decomposition approach, an individual is a tuple of $k$ simple paths, each corresponding to a transmission ratio path assignment. Each path is constructed by randomly choosing an edge at a given node from a uniform distribution defined over all possible outgoing edges of the node, starting from the input shaft node $s^i$ and ending at the output shaft node $s^o$.

In the three-stage decomposition approach, an individual is a graph $\mathcal{G}$. An individual is generated by randomly selecting an edge from a uniform distribution defined over all possible edges, and incrementally adding the selected edge to a null graph until the number of simple paths $n$ is such that $k \leq n \leq n_{max}$. We impose an upper bound on the number of paths to limit the number of transmission ratio path assignments. If the last edge added results in $n > n_{max}$, we backtrack one step and add another edge until $k \leq n \leq n_{max}$.

**ii) Evaluation and Selection:** Each individual in the population is evaluated by solving the parameter optimization problem. For the three-stage decomposition approach, several transmission ratio path assignments are created exhaustively before parameter optimization. A subset of the population, $\mathcal{P}' \subseteq \mathcal{P}$, representing top $t$ individuals, is selected. $|\mathcal{P}'|/|\mathcal{P}|$ is called truncation rate.

**iii) Estimation of Probability Distribution:** From $\mathcal{P}'$, the probability distributions of the edges in the individuals are estimated.

The probability model used for the two-stage decomposition approach is:

$$\mathbb{P}_{path,\pi_j}(\mathbf{P}) = \prod_{p \in \mathbf{P}} \mathbb{P}_{path,\pi_j}(p|s_p^i) \qquad \forall j = 1, ..., k \qquad (18)$$

Here, a conditional probability distribution is assumed and the probability of a gear-pair $p$ is dependent on its input shaft, $s_p^i$.

The probability model used for the three-stage decomposition approach is:

$$\mathbb{P}_{graph}(\mathbf{P}) = \prod_{p \in \mathbf{P}} \mathbb{P}_{graph}(p) \qquad (19)$$

In other words, a univariate probability distribution is assumed and the probability of an edge $p$ is determined independently from other edges.

**iv) New Population Generation:** A new population of individuals is generated using the same techniques as in Step i) with probability distributions estimated in Step iii). That is, instead of randomly selecting edges from uniform distributions to construct a graph or a path, the edges are sampled from the probability distributions in Eq. (19) or Eq. (18), depending on the approach.

**v) Iterate Steps ii)–iv):** Steps ii) to iv) are repeated with the newly generated population until an allocated number of iterations is reached.

## 7    Experimental Evaluation

Experiments are run on a single desktop computer with two Intel Xeon CPUs E5-2650 v2 2.60 GHz and 32G of RAM.

### 7.1    Experimental Setup

**Datasets.** We generate a first dataset with 45 synthetic problem instances (*synthetic dataset*). Instances have 4, 7 and 10 transmission ratios. We fix the minimum transmission ratio to 1, and the maximum to 2, 6 and 10, respectively. The values of the transmission ratios are generated randomly between the minimum and the maximum. The second dataset (*tremec dataset*) contains nine realistic problems, five with five transmission ratios and four with ten transmission ratios. The transmission ratios and the size of the gearbox are taken from the specifications found on the TREMEC website [35].

**Implementation Details.** The inner problem $\mathcal{I}(\mathcal{C})$ is solved with IPOPT [39], using the modeling language CASADI [1]. Isomorphic graphs are detected using the NAUTY library [22]. For EDAs, evaluations are run in parallel across 32 CPU threads available in the computer used. For the two-stage decomposition approach, we use population sizes of 192 for the five-ratio problems and 384 for the ten-ratio problems in the *tremec* dataset. For the three-stage decomposition approach, we use population size of 64, while using different time limits for evaluating each individual: three minutes for the five-ratio problems and six minutes for the ten-ratio problems. For all EDAs, the truncation rate and the number of iterations are set to 0.2 and 10, respectively. Both SA and BFS are run on a single thread. The temperature parameter in SA is set to 2000 and decremented every 1000 evaluations with a step size of 5. Time limits vary depending on the size of the problem and are reported in the next section. They are assumed to be tolerable waiting times from a design process perspective.

**Table 1.** Results on the *synthetic dataset*. We report the average and the standard deviation of objective value for problems that are solved by all algorithms (excluding those that cannot solve any problem). We report also the number of problems solved by each algorithm.

| k |  | 4 |  |  |  |  |  | 7 |  |  |  |  |  | 10 |  |  |  |  |  |
|---|---|---|---|---|---|---|---|---|---|---|---|---|---|---|---|---|---|---|---|
| max ω |  | 2 |  | 6 |  | 10 |  | 2 |  | 6 |  | 10 |  | 2 |  | 6 |  | 10 |  |
|  |  | μ | σ | μ | σ | μ | σ | μ | σ | μ | σ | μ | σ | μ | σ | μ | σ | μ | σ |
| SA2 | obj | 1349 | 92 | 2868 | 336 | 3377 | 105 | 6821 | 581 | 13389 | 2083 | 22805 | 1449 | - | - | - | - | - | - |
|  | # | 5 | - | 5 | - | 5 | - | 5 | - | 5 | - | 5 | - | 0 | - | 0 | - | 0 | - |
| BFS2 | obj | **1168** | 99 | **2015** | 97 | **2596** | 121 | 1715 | 45 | - | - | - | - | - | - | - | - | - | - |
|  | # | 5 | - | 5 | - | 5 | - | 0 | - | 0 | - | 0 | - |  |  |  |  |  |  |
| SA3 | obj | 1168 | 100 | 2033 | 120 | 2644 | 145 | 1764 | 116 | **3456** | 268 | **4427** | 285 | 3161 | 608 | **5337** | - | **6787** | - |
|  | # | 5 | - | 5 | - | 5 | - | 5 | - | 5 | - | 5 | - | 5 | - | 4 | - | 5 | - |
| BFS3 | obj | 1168 | 99 | **2015** | 97 | 2597 | 121 | **1640** | 22 | 3858 | 307 | 6018 | 499 | **2258** | 364 | 7565 | - | 24206 | - |
|  | # | 5 | - | 5 | - | 5 | - | 5 | - | 5 | - | 5 | - | 5 | - | 1 | - | 1 | - |

**Table 2.** Results on the *tremec dataset*. We report the average and the standard deviation of the objective value of each problem. The value marked with * is the objective value of the only solution found in all runs of the algorithm.

| problem id | 5-transmission ratios |  |  |  |  |  |  |  |  |  | 10-transmission ratios |  |  |  |  |  |  |  |
|---|---|---|---|---|---|---|---|---|---|---|---|---|---|---|---|---|---|---|
|  | es42-5a |  | es52-5a |  | es60-5a |  | es60-5c |  | tr-3550 |  | tr-t-10d |  | tr-t-10v |  | tr-to-10s |  | tr-to-10v |  |
|  | μ | σ | μ | σ | μ | σ | μ | σ | μ | σ | μ | σ | μ | σ | μ | σ | μ | σ |
| EDA2 | 23281 | 1307 | 23733 | 957 | 24294 | 1522 | 24468 | 1538 | 12670 | 761 | - | - | - | - | - | - | - | - |
| SA2 | 28523 | 2102 | 28114 | 2363 | 27468 | 4187 | 28863 | 3966 | 14501 | 1797 | - | - | - | - | - | - | 466813* | 0 |
| BFS2 | - | - | - | - | 23797 | 265 | 24651 | 696 | **10952** | 200 | - | - | - | - | - | - | - | - |
| EDA3 | **21644** | 473 | **22014** | 414 | **21199** | 0 | **22674** | 155 | 11067 | 69 | **22891** | 1247 | **25936** | 1343 | **21844** | 711 | **22698** | 631 |
| SA3 | 25866 | 1957 | 26401 | 1773 | 25401 | 1993 | 24232 | 2618 | 11976 | 800 | 44105 | 10130 | 46084 | 20198 | 33093 | 9432 | 40640 | 10993 |
| BFS3 | 31022 | 573 | 31370 | 552 | 30799 | 438 | 30792 | 538 | 10995 | 162 | 55458 | 1049 | 69725 | 1571 | 47989 | 1291 | 51338 | 1127 |

## 7.2 Results

We use the *synthetic dataset* to test the behaviour of the algorithms running on a single thread: BFS and SA. In Table 1 we report the average solution qualities and the number of problems solved for different groups of problems, using a time-limit of 1 h. The name of the algorithm is followed by 2 or 3, indicating the two-stage or three-stage decomposition approach, respectively. Every algorithm is run 3 times to account for the randomization in the algorithms. Both SA and BFS perform better when using the three-stage decomposition approach. The two-stage approach fails to find feasible solutions to medium-size problems. In terms of solution quality, BFS outperforms SA for small size problems, while SA generally performs better on problems with more transmission ratios.

All algorithms are tested on the *tremec dataset* and results are in Table 2. The time limit is 1 and 2 h for problems with 5 and 10 transmission ratios, respectively. Each algorithm is run 10 times for each problem. Results on this dataset confirms that the three-stage decomposition generally outperforms the two-stage. Among the algorithms, EDA is the best performing, mainly attributed to the larger number of solutions that can be evaluated in parallel.

## 8    Conclusion

In this paper, we consider the multi-speed gearbox synthesis problem formulated as a bi-level optimization problem. The inner task, or parameter optimization, is a non-convex continuous optimization problem, solved with a state-of-the-art solver. For the outer problem, we used two approaches to search over gearbox configurations: the two-stage decomposition approach searches over transmission ratio path assignments, while the three-stage decomposition approach first selects a sub-graph, and then performs the transmission ratio path assignment. We found that the latter consistently outperforms the first in all test problems and search algorithms. We investigated a variety of global search algorithms for solving the sub-tasks of the outer problem. While best-first-search usually performs well on small-size problems, the estimation of distribution algorithm produces better quality solutions for realistic instances. This work demonstrates the value of integrating methods from both the artificial intelligence and optimization fields applied to configuration design problems in mechanical engineering.

## References

1. Andersson, J.A.E., Gillis, J., Horn, G., Rawlings, J.B., Diehl, M.: CasADi: a software framework for nonlinear optimization and optimal control. Math. Program. Comput. **11**(1), 1–36 (2019). https://doi.org/10.1007/s12532-018-0139-4
2. Berx, K., Gadeyne, K., Dhadamus, M., Pipeleers, G., Pinte, G.: Model-based gearbox synthesis. In: Mechatronics Forum International Conference, pp. 599–605 (2014)
3. Boyd, S., Vandenberghe, L.: Convex Optimization. Cambridge University Press, Cambridge (2004)
4. Buiga, O., Tudose, L.: Optimal mass minimization design of a two-stage coaxial helical speed reducer with genetic algorithms. Adv. Eng. Softw. **68**, 25–32 (2014)
5. Castro, M.P., Piacentini, C., Cire, A.A., Beck, J.C.: Relaxed decision diagrams for cost-optimal classical planning. In: Workshop HSDIP, ICAPS, pp. 50–58 (2018)
6. Chakrabarti, A., et al.: Computer-based design synthesis research: an overview. J. Comput. Inf. Sci. Eng. **11**(2), 021003 (2011)
7. Cheong, H., Ebrahimi, M., Butscher, A., Iorio, F.: Configuration design of mechanical assemblies using an estimation of distribution algorithm and constraint programming. In: 2019 IEEE Congress on Evolutionary Computation (CEC), pp. 2339–2346. IEEE (2019)
8. Chlamtáč, E., Dinitz, M., Makarychev, Y.: Minimizing the union: tight approximations for small set bipartite vertex expansion. In: Proceedings of the Twenty-Eighth Annual ACM-SIAM Symposium on Discrete Algorithms, pp. 881–899. SIAM (2017)
9. Chomsky, N., Lightfoot, D.W.: Syntactic structures. Walter de Gruyter (1957)
10. Chong, T.H., Bae, I., Kubo, A.: Multiobjective optimal design of cylindrical gear pairs for the reduction of gear size and meshing vibration. JSME Int J., Ser. C **44**(1), 291–298 (2001)
11. Deb, K., Jain, S.: Multi-speed gearbox design using multi-objective evolutionary algorithms. J. Mech. Des. **125**(3), 609–619 (2003)

12. Dechter, R., Pearl, J.: Generalized best-first search strategies and the optimality of A. J. ACM (JACM) **32**(3), 505–536 (1985)
13. Eschenauer, H., Koski, J., Osyczka, A.: Multicriteria Design Optimization: Procedures and Applications. Springer, Heidelberg (2012)
14. Hauschild, M., Pelikan, M.: An introduction and survey of estimation of distribution algorithms. Swarm Evol. Comput. **1**(3), 111–128 (2011)
15. Huang, H.Z., Tian, Z.G., Zuo, M.J.: Multiobjective optimization of three-stage spur gear reduction units using interactive physical programming. J. Mech. Sci. Technol. **19**(5), 1080–1086 (2005)
16. Hwang, C.R.: Simulated annealing: theory and applications. Acta Applicandae Mathematicae **12**(1), 108–111 (1988)
17. Jain, P., Agogino, A.M.: Theory of design: an optimization perspective. Mech. Mach. Theory **25**(3), 287–303 (1990)
18. Königseder, C., Shea, K.: Comparing strategies for topologic and parametric rule application in automated computational design synthesis. J. Mech. Des. **138**(1), 011102 (2016)
19. Land, A., Doig, A.: An automatic method of solving discrete programming problems. Econometrica **28**(3), 497–520 (1960)
20. Li, X., Schmidt, L.: Grammar-based designer assistance tool for epicyclic gear trains. J. Mech. Des. **126**(5), 895–902 (2004)
21. Lin, Y.S., Shea, K., Johnson, A., Coultate, J., Pears, J.: A method and software tool for automated gearbox synthesis. In: ASME 2009 International Design Engineering Technical Conferences and Computers and Information in Engineering Conference, pp. 111–121. American Society of Mechanical Engineers Digital Collection (2009)
22. McKay, B.D., Piperno, A.: Practical graph isomorphism, II. J. Symb. Comput. **60**, 94–112 (2014)
23. Mittal, S., Frayman, F.: Towards a generic model of configuraton tasks. In: IJCAI, vol. 89, pp. 1395–1401. Citeseer (1989)
24. Mullins, S., Rinderle, J.R.: Grammatical approaches to engineering design, part i: an introduction and commentary. Res. Eng. Design **2**(3), 121–135 (1991)
25. Murthy, S.S., Addanki, S.: PROMPT: an innovative design tool. IBM Thomas J, Watson Research Division (1987)
26. Nadel, B.A., Lin, J.: Automobile transmission design as a constraint satisfaction problem: modelling the kinematic level. AI EDAM **5**(3), 137–171 (1991)
27. Osyczka, A.: An approach to multicriterion optimization problems for engineering design. Comput. Methods Appl. Mech. Eng. **15**(3), 309–333 (1978)
28. Pomrehn, L., Papalambros, P.: Discrete optimal design formulations with application to gear train design. J. Mech. Des. **117**(3), 419–424 (1995)
29. Prayoonrat, S., Walton, D.: Practical approach to optimum gear train design. Comput. Aided Des. **20**(2), 83–92 (1988)
30. Rai, P., Agrawal, A., Saini, M.L., Jodder, C., Barman, A.G.: Volume optimization of helical gear with profile shift using real coded genetic algorithm. Procedia Comput. Sci. **133**, 718–724 (2018)
31. Savsani, V., Rao, R., Vakharia, D.: Optimal weight design of a gear train using particle swarm optimization and simulated annealing algorithms. Mech. Mach. Theory **45**(3), 531–541 (2010)
32. Schmidt, L.C., Shetty, H., Chase, S.C.: A graph grammar approach for structure synthesis of mechanisms. J. Mech. Des. **122**(4), 371–376 (1999)
33. Sobieszczanski-Sobieski, J., Haftka, R.T.: Multidisciplinary aerospace design optimization: survey of recent developments. Struct. Optim. **14**(1), 1–23 (1997)

34. Swantner, A., Campbell, M.I.: Topological and parametric optimization of gear trains. Eng. Optim. **44**(11), 1351–1368 (2012)
35. TREMEC: Tremec. http://www.tremec.com. Accessed 28 Nov 2019
36. Tsai, L.W., Maki, E., Liu, T., Kapil, N.: The categorization of planetary gear trains for automatic transmissions according to kinematic topology. Technical report, SAE Technical Paper (1988)
37. Valiant, L.G.: The complexity of enumeration and reliability problems. SIAM J. Comput. **8**(3), 410–421 (1979)
38. Vinterbo, S.A.: A note on the hardness of the k-ambiguity problem. Technical Report DSG-T R-2002-006 (2002)
39. Wächter, A., Biegler, L.T.: On the implementation of an interior-point filter line-search algorithm for large-scale nonlinear programming. Math. Program. **106**(1), 25–57 (2006)
40. Wang, G.G., Shan, S.: Review of metamodeling techniques in support of engineering design optimization. J. Mech. Des. **129**(4), 370–380 (2006)
41. Wang, H., Wang, H.P.: Optimal engineering design of spur gear sets. Mech. Mach. Theory **29**(7), 1071–1080 (1994)
42. Wojnarowski, J., Kopeć, J., Zawiślak, S.: Gears and graphs. J. Theor. Appl. Mech. **44**(1), 139–162 (2006)
43. Yokota, T., Taguchi, T., Gen, M.: A solution method for optimal weight design problem of the gear using genetic algorithms. Comput. Ind. Eng. **35**(3–4), 523–526 (1998)
44. Zarefar, H., Muthukrishnan, S.: Computer-aided optimal design via modified adaptive random-search algorithm. Comput. Aided Des. **25**(4), 240–248 (1993)

# Enumerative Branching with Less Repetition

Thiago Serra[(⊠)]

Bucknell University, Lewisburg, USA
thiago.serra@bucknell.edu

**Abstract.** We can compactly represent large sets of solutions for problems with discrete decision variables by using decision diagrams. With them, we can efficiently identify optimal solutions for different objective functions. In fact, a decision diagram naturally arises from the branch-and-bound tree that we could use to enumerate these solutions if we merge nodes from which the same solutions are obtained on the remaining variables. However, we would like to avoid the repetitive work of finding the same solutions from branching on different nodes at the same level of that tree. Instead, we would like to explore just one of these equivalent nodes and then infer that the same solutions would have been found if we explored other nodes. In this work, we show how to identify such equivalences—and thus directly construct a reduced decision diagram—in integer programs where the left-hand sides of all constraints consist of additively separable functions. First, we extend an existing result regarding problems with a single linear constraint and integer coefficients. Second, we show necessary conditions with which we can isolate a single explored node as the only candidate to be equivalent to each unexplored node in problems with multiple constraints. Third, we present a sufficient condition that confirms if such a pair of nodes is indeed equivalent, and we demonstrate how to induce that condition through preprocessing. Finally, we report computational results on integer linear programming problems from the MIPLIB benchmark. Our approach often constructs smaller decision diagrams faster and with less branching.

**Keywords:** Branch-and-Bound · Decision Diagrams · Depth-First Search · Integer Programming · Solution Enumeration

## 1 Introduction

The enumeration of near-optimal solutions is a feature that is present in commercial MILP solvers such as CPLEX [33] and Gurobi [26] as well as algebraic modelling environments such as GAMS [25]. This feature is important because some users need to qualitatively compare the solutions of a mathematical model. However, those solutions are often a small set collected along the way towards solving for the optimal solution. While the option for complete enumeration exists in some solvers, it comes with observations like the following in CPLEX [33]:

© Springer Nature Switzerland AG 2020
E. Hebrard and N. Musliu (Eds.): CPAIOR 2020, LNCS 12296, pp. 399–416, 2020.
https://doi.org/10.1007/978-3-030-58942-4_26

Beware, however, that, even for small models, the number of possible solutions is likely to be huge. Consequently, enumerating all of them will take time and consume a large quantity of memory.

In fact, the problem of enumerating integer solutions is #P-complete [49,50].

In practice, the enumeration of solutions has been an extension of the same methods used for optimization. When searching for an optimal solution of a problem with linear constraints and integer variables, the first step is often to solve a relaxation of this problem: a linear program in which we ignore that the variables should have integer values [18]. If the resulting solution has fractional values for some of the variables, then we may resort to *branching*: exploring two or more subproblems in which we fix or restrict the domain of these variables to exclude their corresponding fractional values. We can ignore some of the resulting subproblems if they are provably suboptimal, for which reason this process is known as *branch-and-bound* [36]. However, if we are interested in enumerating some or all the solutions, then we may need to keep branching even if no value is fractional. In such a case, we continue while the domains of the variables have multiple values, the relaxation remains feasible, and the objective function value is within a desired optimality gap. That is the case of the one-tree approach [19], which has been used to populate the solution pool of the CPLEX solver [32].

The branch-and-bound process is often represented by a directed tree, the *branch-and-bound tree*, in which a root node $r$ corresponds to the problem of interest and the children of each node are the subproblems defined by branching. If we branch by assigning the value of one decision variable at a time to each possible value, then a path from the root to a leaf defining a feasible subproblem corresponds to a set of assignments leading to a unique solution. Such explicit representation of the solution set emerging from branch-and-bound, which may reach an exponential number of nodes with respect to the number of variables, could be naturally transformed into a *decision diagram* by merging the leaves of the tree as a single terminal node $t$ [44]. Likewise, we can merge any other nodes from which the same solutions are obtained on the remaining variables while preserving a correspondence between solutions of the problem and $r$–$t$ paths in the diagram. That results in a representation that is substantially more compact and, in some cases, can have less nodes than the number of solutions represented. If every path assigns a value to every variable, then we can find optimal solutions for varied linear objective functions by assigning the corresponding weights to the arcs and computing a minimum weight $r$–$t$ path, which is usually faster than resolving the problem with different objective functions [44].

Figure 1 illustrates these different graphic representations for the solutions of inequality $2x_1 - 2x_2 - 3x_3 \leq -1$ on a vector of binary variables $x \in \{0,1\}^3$: (a) is a tree in which every path from the root to a different leaf assigns a distinct set of values to the variables; (b) is the decision diagram produced by merging the leaves of the tree; and (c) is a decision diagram in which we merge the three nodes in (b) from which the only arc towards $t$ corresponds to $x_3 = 1$. We use thin arcs for assignments of value 0 and bold arcs for assignments of value 1. In the example, we always assign a value to variable $x_1$ first, then to variable $x_2$,

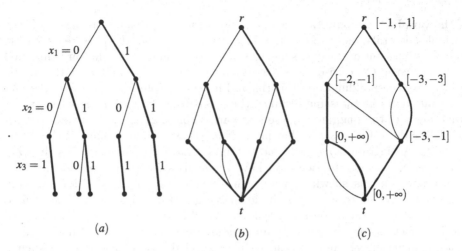

**Fig. 1.** (a) Tree representing the solutions of $\{x \in \{0,1\}^3 : 2x_1 - 2x_2 - 3x_3 \leq -1\}$. (b) Decision diagram from merging the leaf nodes. (c) Reduced decision diagram with interval of equivalent right-hand side values for the inequality on each node.

and finally to variable $x_3$. When the same order of variables is used in every $r$–$t$ path, then nodes at the same arc-distance from $r$ define a *layer* with the same *selector variable* for the next assignment, and we say that we have an *ordered* decision diagram. In fact, Fig. 1(c) has the smallest ordered decision diagram for that sequence of variable selectors, and we call it a *reduced* decision diagram.

In this paper, we discuss how to compare the formulation of subproblems involving one or more inequalities to determine the equivalence of branch-and-bound nodes, and hence directly construct reduced decision diagrams.

## 1.1  Contribution

We generalize prior work on identifying equivalent subproblems with a single inequality [1,2,6] and introduce a variant for the case of multiple inequalities. First, we show how to compute the interval of equivalent Right-Hand Side (RHS) values for any inequality on finite domains with additively separable Left-Hand Side (LHS) and fractional RHS. Second, we discuss why the same idea cannot be directly applied to problems with multiple inequalities. In that case, we show how to eliminate candidates for equivalence among the explored nodes in such a way that we are left with at most one potentially equivalent node. Finally, we present a sufficient condition achievable by bottom-up preprocessing to determine if such a remaining node is indeed equivalent.

## 2  Decision Diagrams

A *decision diagram* is a directed acyclic graph with root node $r$. Arcs leaving each node denote assignments to a variable, which we denote its *selector variable*.

If the variables can only take two possible values, the diagram is *binary*. If the variables can take more values, the diagram is *multi-valued*. When representing only feasible solutions, there is one terminal node $t$ and each solution is mapped to an $r$–$t$ path. Otherwise, there are two terminal nodes $\mathcal{T}$ and $\mathcal{F}$, with feasible solutions corresponding to $r$–$\mathcal{T}$ paths and infeasible solutions to $r$–$\mathcal{F}$ paths. We say that nodes at the same distance (in arcs) from the root node $r$ are in the same *layer* of the diagram. The sets of paths from a given node $u$ toward $t$ (or $\mathcal{T}$) define the *solution set* of that node. Nodes have the same *state* if those sets coincide. A diagram is *reduced* if no two nodes have the same state. A diagram is *ordered* if all nodes in each layer have the same selector variable.

Bryant [14] has shown that we can efficiently reduce a decision diagram through a single bottom-up pass by identifying and merging nodes with equivalent sets of assignments on the remaining variables.

Two factors may help us obtain smaller decision diagrams. First, equivalences are intuitively more frequent in ordered decision diagrams. In such a case, all nodes in each layer have solution sets defined on the same set of variables, which is a necessary condition to merge such nodes. Second, it is easier to identify equivalences if the problem is defined by inequalities in which the *Left-Hand Side* (LHS) is an *additively separable function*. A function $f(x)$ on an $n$-dimensional vector of variables $x$ is additively separable if we can decompose it as $f(x) = \sum_{i=1}^{n} f_i(x_i)$, i.e., a sum of univariate functions on each of those variables. In such a case, all the nodes in a given layer define subproblems with inequalities having the same LHS. For example, if we have a problem $\{x : \sum_{i=1}^{n} f_i(x_i) \leq \rho\}$, then each node obtained by assigning a distinct value $\bar{x}_1$ to variable $x_1$ defines a subproblem $\{x : \sum_{i=2}^{n} f_i(x_i) \leq \rho - f_1(\bar{x}_1)\}$ on the same $n - 1$ variables and with the same LHS for the inequality.

Following those conditions, we can anticipate that two nodes have the same solutions if their subproblems have inequalities with the same *Right-Hand Side* (RHS) values. In such a case, we can save time if we just branch on one of those nodes to enumerate its solutions and then merge it with the second node. For example, in the problem of Fig. 1 we obtain the same subproblem $\{x_3 \in \{0,1\} : -3x_3 \leq -1\}$ with assignments $x_1 = 0$ and $x_2 = 0$ or $x_1 = 1$ and $x_2 = 1$.

More generally, we say that two subproblems have *equivalent formulations* if they have the same solutions, even if the formulations are different. For example, in Fig. 1 we obtain subproblem $\{x_3 \in \{0,1\} : -3x_3 \leq -3\}$ by assigning $x_1 = 1$ and $x_2 = 0$, which has the same solution set as the previously mentioned assignments leading to $\{x_3 \in \{0,1\} : -3x_3 \leq -1\}$. A method to compute such equivalent RHS values for nodes in the same layer of ordered decision diagrams representing a linear inequality with integer coefficients has been independently proposed twice on binary domains [1,6] and later extended to multi-valued domains [2]. This method is based on search for all the solutions for some nodes and then inferring if a new node would have the same solutions as one of those nodes. We will use the following definition for those previous nodes.

**Definition 1 (Explored node).** *A node is said to be explored if all the solutions for the subproblem rooted at that node are known.*

Once a node has been explored, we can compute the minimum RHS value that would produce the same solutions and the minimum RHS value, if any, that would produce more solutions. In Fig. 1(c), the interval of equivalent integer values for the RHS of the inequalities on the remaining variables is shown next to each node. For example, in the penultimate layer we have the intervals of RHS values for $-3x_3$ as LHS, which includes $[-3, -1]$ for the node that is reached by the three assignments to $x_1$ and $x_2$ that we discussed previously.

However, we cannot directly apply the same method to problems with multiple inequalities and expect to find all the nodes that are equivalent. For example, in Fig. 2 we have three equivalent formulations for the solution set $\{(0,0), (0,1), (1,0)\}$. The inequality with LHS $5x_1 + 4x_2$ has equivalent RHS values in $[5, 8]$. The inequality with LHS $6x_1 + 10x_2$ has equivalent RHS values in $[10, 15]$. But there are equivalent RHS values beyond $[5, 8] \times [10, 15]$: if the RHS of only one inequality is made larger, such as in the first two examples of Fig. 2, the other inequality prevents the inclusion of solution $(1, 1)$. In other words, not every inequality needs to separate every infeasible solution.

$$5x_1 + 4\,x_2 \leq 6 \qquad \leq 10 \qquad \leq 5$$
$$6x_1 + 10\,x_2 \leq 17 \qquad \leq 11 \qquad \leq 10$$
$$x_1, x_2 \in \{0,1\}$$

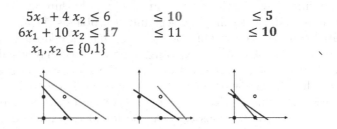

**Fig. 2.** Equivalent formulations in which the inequalities have the same functions in the Left-Hand Side (LHS) but different values in the Right-Hand Side (RHS).

## 3   Related Work

Put in context, our work may also improve exhaustive search. Even in cases where the goal is not to construct a decision diagram for the set of solutions, the characterization of a *state* for the nodes of the corresponding decision diagram based on their solution sets can be exploited to avoid redundant work during the branch-and-bound process.

Beyond the case in which all RHS values are the same, prior work on identifying equivalent branch-and-bound nodes has focused on detecting inequalities that are always satisfied. One example is the detection of *unrestricted subtrees*, in which any assignment to the remaining variables is feasible [3]. The same principle was later used to ignore all unrestricted inequalities and only compare the RHS values of inequalities that separate at least one solution [44]. That can be particularly helpful with set covering constraints, in which it suffices to have one variable assigned to 1. This line of work relies on computing the minimum RHS value after which each inequality does not separate any solution, and therefore any RHS value larger than that is deemed equivalent. For example, in Fig. 1(c) the first node in the penultimate layer allows both possible values to $x_3$, which for the LHS $-3x_3$ implies a RHS of 0 or more. In contrast to [44], the present paper aims to identify all nodes that can be merged during the top-down construction of a decision diagram corresponding to an ILP formulation. Hence, we also compare different RHS values that exclude at least one solution.

While the search effort may decrease, it can still be substantial and depend on other factors. First, the size of a reduced decision diagram for a single constraint can be exponential on the number of variables for any order of variable selectors [30]. Second, finding a better order of variable selectors for binary decision diagrams is NP-complete [13], hence implying that it is NP-hard to find an order of variable selections that minimizes the size of the decision diagram [6,22]. Nevertheless, some ordering heuristics have been found to work well in practice [6,7], while strong bounds for the size of the decision diagrams according to the ordering of variable selectors have been shown for certain classes of problems [28]. In other cases, a convenient order may be deduced from the context of the problem. For example, by identifying a temporal relationship among the decision variables, such as in sequencing and scheduling problems [17,42].

A related approach consists of analyzing *dominance relations* among branch-and-bound nodes [31,35]. When exploring a node $u$, we may wonder if there is another node $v$ that can be reached by fixing the same variables with different values such that (i) the assignments leading to $v$ are preferable according to the objective function; and (ii) $v$ defines a relaxation of the subproblem defined by $u$, in which case all solutions that can be found by exploring $u$ can also be found by exploring $v$ [16,23,24]. If such a node $v$ exists and we only want to find an optimal solution, then we can safely prune the subtree rooted at $u$ and only explore the subtree rooted at $v$. However, such an approach is not fully compatible with enumerating solutions because the subproblems on $u$ and $v$ may not have the same set of solutions. Furthermore, a change in the objective function could make the discarded node $u$ preferable to node $v$ in applications of reoptimization. Finally, finding such a node $v$ may entail solving an optimization problem on the variables that are fixed to reach node $u$, and thus it only pays off closer to the root node because not many variables have been assigned yet.

In contrast, our approach pays off closer to the bottom of the decision diagram representation. For a node at distance $k$ from $r$, there are at most $2^k$ states if all variables have binary domains, which are the distinct assignments to the first $k$ variables. For a node at distance $k$ from $t$, there are at most $2^{2^k} - 1$ states, which are the non-empty sets of solutions for the last $k$ variables. Hence, we should generally expect that equivalences occur more often when the number of top-down states exceeds the number of bottom-up states.

In the context of integer linear programming, a small set of solutions is often obtained through the application of inequalities separating previous solutions [5,15]. Recent work has also focused on the generation of a diverse set of solutions [20,41,48], and there are also methods to obtain exact upper bounds [34] and probabilistic lower bounds [46] on the size of the solution set.

Decision diagrams and some extensions have also been widely used to solve combinatorial optimization problems [4,8,9,29,38,40,43,47,51–53], and more recently stochastic [27,37,45] and nonlinear optimization problems [10,21].

## 4    Prior Result

We begin our analysis with a prior result about the direct construction of reduced decision diagrams, which concerns the decision diagram rooted at a node $v$ and defined by an inequality of the form $a_1 x_1 + \ldots + a_n x_n \leq a_0$ with integer coefficients and each decision variable $x_i$ having a discrete domain of the form $\{0, 1, \ldots, d_i\}$. For such a node $v$ and each of the other nodes in the decision diagram, we want to compute a corresponding interval $[\beta, \gamma]$ or $[\beta, \infty)$, where $\beta$ is the smallest integer RHS value that could replace $a_0$ and still yield the same solutions. Similarly, $\gamma$ is the largest integer RHS value that would still yield the same solutions, and $\gamma$ only exists if at least one solution is infeasible. We recursively compute such intervals after exploring a node and all of its descendants.

**Theorem 1 (Abío & Stuckey[1] [2]).** *Let $\mathcal{M}$ be the multi-valued decision diagram of a linear integer constraint $a_1 x_1 + \ldots + a_n x_n \leq a_0$. Then, the following holds:*

- *The interval of the true node $\mathcal{T}$ is $[0, \infty)$.*
- *The interval of the false node $\mathcal{F}$ is $(-\infty, -1]$.*
- *Let $v$ be a node with selector variable $x_i$ and children $\{v_0, v_1, \ldots, v_{d_i}\}$. Let $[\beta_j, \gamma_j]$ be the interval of $v_j$. Then, the interval of $v$ is $[\beta, \gamma]$, with $\beta = \max\{\beta_s + sa_i \mid s \in \{0, \ldots, d_i\}\}$, $\gamma = \min\{\gamma_s + sa_i \mid s \in \{0, \ldots, d_i\}\}$.*

The interval of the explored nodes can then be compared with the RHS of the inequalities defining each of the unexplored nodes in the same layer to identify equivalences. In order to fully avoid redundant work, the branch-and-bound algorithm should perform a *depth-first search* (DFS), where the unexplored node with fewer unfixed variables is explored next. In such a case, we do not risk branching on two nodes that would later be found to have the same state.

In this paper, we consider decision diagrams with a single terminal node $t$, hence disregarding infeasible solutions and nodes in which roughly $\beta = -\infty$.

## 5    The Case of One Inequality

In this section, we discuss how Theorem 1 can be further generalized when applied to a single inequality. Our contribution is evidencing that the sequence of work culminating in the result by Abío and Stuckey [2] can be extended to

---

[1] Following the recommendation of one of the anonymous reviewers, there is small correction in comparison to [2]: we use $s \in \{0, \ldots, d_i\}$ instead of $0 \leq s \leq d_i$.

account for the case of linear inequalities with fractional coefficients, and to the slightly more general case of inequalities involving additively separable functions.

We begin with an intuitive argument for generalizing that result for the case of fractional RHS values. In the interval $[\beta, \gamma]$ used in Theorem 1, both $\beta$ and $\gamma$ are integer values corresponding to the smallest and largest integer RHS values defining the same solutions for the inequality. If the LHS coefficients and the decision variables are integer, then it follows that any right-hand side value larger than $\gamma$ but smaller than $\gamma + 1$ would also define the same solutions. Hence, $[\beta, \gamma+1)$ is the maximal interval of equivalent right-hand side values if we allow a fractional right-hand side. Now the upper end value $\gamma + 1$ is the smallest RHS value yielding a proper superset of solutions. Furthermore, with such half-closed intervals, both ends may also become fractional if the LHS coefficients are fractional and assigning a variable changes the RHS by a fractional amount.

In addition, the only reason to represent infeasible solutions is to compute those upper ends of the RHS intervals. But for each node that has missing arcs for some values of its selector variable, which if represented would only reach the terminal $\mathcal{F}$, the corresponding upper end is the sum of the impact of that assignment with the smallest RHS associated with a solution on the remaining variables. But as we will see next, that value can be calculated by inspection.

In summary, we can ignore infeasible solutions by using a single terminal node $t$, lift the integrality of all coefficients, and consequently allow the LHS to be additively separable. That leads us to the following result:

**Theorem 2.** *Let $\mathcal{D}$ be a decision diagram with variable ordering $x_1, x_2, \ldots, x_n$ of $f(x) = f_1(x_1) + \ldots + f_n(x_n) \leq \rho$ on finite domains $D_1$ to $D_n$. Then, the following holds:*

- *The interval of the terminal node $t$ is $[0, \infty)$.*
- *Let $v$ be a node with selector variable $x_i$ and non-empty multi-set[2] of descendants $\{v_j \mid j \in D^v\}$, $D^v \subseteq D_i$, where $v_j$ is reached by an arc denoting $x_i = j$. Let $[\beta_j, \gamma_j)$ be the interval of $v_j$. The interval of $v$ is $[\beta, \gamma)$, with*

$$\beta = \max\{\beta_j + f_i(j) \mid j \in D^v\}$$

*and*

$$\gamma = \min\left\{ \min\{\gamma_j + f_i(j) \mid j \in D^v\},\ \min\{\xi_i + f_i(j) \mid j \in D_i \setminus D^v\}\right\},$$

*where*

$$\xi_i = \sum_{l=i+1}^{n} \min\{f_l(d) \mid d \in D_l\}.$$

Since we can prove a stronger result regarding the lower end $\beta$, which also applies to the case of multiple inequalities that will be discussed in Sect. 6, we split the proof of Theorem 2 into two lemmas.

---

[2] We use a multi-set because two nodes might be connected through multiple arcs for different variable assignments.

**Lemma 1.** *Let $\mathcal{D}$ be a decision diagram with variable ordering $x_1, x_2, \ldots, x_n$ of $\{(x_1, x_2, \ldots, x_n) \in D_1 \times D_2 \times \ldots D_n | f^k(x) = f_1^k(x_1) + \ldots + f_n^k(x_n) \leq \rho^k \; \forall k = 1, \ldots, m\}$. Then, computing $\beta^k$ for each inequality $f^k(x) \leq \rho^k$ as in Theorem 2 yields the smallest RHS not affecting the set of solutions satisfying all inequalities on each node.*

*Proof.* We proceed by induction on the layers, starting from the bottom. The base case holds since $0 \leq \rho^k$ is only valid if $\rho^k \geq \beta^k = 0$ for each inequality $k$ at the terminal node $t$. Now suppose, for contradiction, that Lemma 1 holds for the $n - i$ bottom layers of $\mathcal{D}$ and not for the one above. Hence, there would be a node $v$ with selector variable $x_i$ such that $v$ has the same solutions if we replace the RHS of the $k$-th inequality with some $\delta < \beta^k \leq \rho^k$, thereby obtaining $\sum_{l=i}^{n} f_l^k(x_l) \leq \delta$. Let $j \in \arg\max\{\beta_j^k + f_i^k(j) \mid j \in D^v\}$, i.e., $v_j$ would be one of the children maximizing the expression with which we calculate $\beta^k$ in the statement of Lemma 1 and $\beta_j^k$ is the smallest RHS value not affecting the set of solutions of $v_j$ with respect to the $k$-th inequality. Consequently, node $v_j$ would have the same solutions if the $k$-th inequality for $v_j$ becomes $\sum_{l=i+1}^{n} f_l^k(x_l) \leq \delta - f_i(j)$. However, $\beta_j^k = \beta^k - f_i^k(j) > \delta - f_i^k(j)$, and we have a contradiction since by induction hypothesis any value smaller than $\beta_j^k$ would yield proper subset of the solution set of $v_j$. $\square$

**Lemma 2.** *Computing $\gamma$ for one inequality as in Theorem 2 yields the smallest RHS value that would yield a proper superset of the solutions of node $v$.*

*Proof.* We proceed by induction on the layers, starting from the bottom. The base case holds since the terminal node $t$ has a set with one empty solution and we denote $\gamma = +\infty$.

By induction hypothesis, we assume that Lemma 2 holds for the $n - i$ lower layers of $\mathcal{D}$, and we show next that it consequently holds for the $i$-th layer of $\mathcal{D}$.

For any node $v$ in the $i$-th layer with interval $[\beta, \gamma)$ for a finite $\gamma$ such that $\sum_{l=i}^{n} f_l(x_l) \leq \gamma$, let $j \in \arg\min\left\{\min\{\gamma_j + f_i(j) \mid j \in D^v\}, \min\{\xi_i + f_i(j) \mid j \in D_i \setminus D^v\}\right\}$. If (I) $j \in \arg\min\{\gamma_j + f_i(j) \mid j \in D^v\}$, then there is a child node $v_j$ that minimizes the expression with which we calculate $\gamma$ in the statement of Lemma 2. By induction hypothesis, that implies that there is a solution that is not feasible for node $v_j$ and such that $\sum_{l=i+1}^{n} f_l(\bar{x}_j) = \gamma_j$. Consequently, solution $(\bar{x}_i = j, \bar{x}_{i+1}, \ldots, \bar{x}_n)$ is not feasible for node $v$ and $\sum_{l=i}^{n} f_l(\bar{x}_j) = \gamma$. If (II) $j \in \arg\min\{\xi_i + f_i(j) \mid j \in D_i \setminus D^v\}$, then $\gamma = f_i(j) + \xi_i$ and node $v$ does not have any solution in which $x_i = j$. Consequently, there is an infeasible solution $(\bar{x}_i = j, \bar{x}_{i+1}, \ldots, \bar{x}_n)$ such that $\sum_{l=i+1}^{n} f_l(\bar{x}_j) = \xi_i$ and $\sum_{l=i+1}^{n} f_l(\bar{x}_j) = \gamma$. In either case, a RHS of $\gamma$ yields a proper superset of the solutions of $v$. Furthermore, a smaller RHS value yielding a proper superset of the solutions of $v$ would either contradict the choice of $j$ in (I) if there is at least one solution of $v$ such that $x_i = j$ or in (II) if there is no solution of $v$ such that $x_i = j$. $\square$

We are now able to prove the main result of this section.

*Proof (Theorem 2).* Lemma 1 implies that $\beta$ is the smallest RHS value yielding the same solution set as $v$ and Lemma 2 implies that $\gamma$ is the smallest RHS value yielding at least one more solution than node $v$. Since there is a single inequality, then a RHS of $\gamma$ yields a proper superset of the solutions of $v$.                                   □

## 6  The Case of Multiple Inequalities

For variables $x_1$ to $x_n$ with finite domains $D_1$ to $D_n$, we now consider constructing a decision diagram for a problem defined by $m$ inequalities in the following form:

$$f^i(x) = f_1^i(x_1) + \ldots + f_n^i(x_n) \le \rho^i \qquad \forall i = 1, \ldots, m$$

We will exploit the fact that we still can apply Lemma 1 to a problem with multiple inequalities. Theorem 2 is not as helpful because the equivalent upper ends for one inequality may depend on the RHS values of other inequalities. That prevents us from immediately identifying if two nodes are equivalent by comparing the intervals of the explored node with the RHS values of the unexplored node. Nevertheless, we can characterize and distinguish nodes having different solution sets by their lower ends as in Lemma 1. We show that such comparison is enough to exclude all but one of the explored nodes as potentially equivalent, to which we describe a sufficient condition to guarantee equivalence.

### 6.1  Necessary Conditions

With multiple inequalities, we cannot simply use the intervals of RHS values associated with each of the inequalities independently. We have previously observed that with the example in Fig. 2, which we now revisit with half-closed intervals. For the inequality with LHS $5x_1 + 4x_2$, the solution set $\{(0,0), (0,1), (1,0)\}$ corresponds to RHS values in $[5,9)$. For the inequality with LHS $6x_1 + 10x_2$, that same solution set corresponds to RHS values in $[10,16)$. Hence, $[5,9) \times [10,16)$ defines a valid combination of RHS values yielding the same solutions. However, we can relax one inequality if the other still separates the remaining infeasible solution $(1,1)$. Consequently, the solution set is actually characterized by following combination of RHS values: $[5,9) \times [10,+\infty) \cup [5,+\infty) \times [10,16)$.

Note that the *lower ends* in $\beta$ are nevertheless the same. In fact, Lemma 1 implies that we can characterize node states by the smallest RHS value of each inequality that would yield the same solutions. The key difference is that pushing any RHS lower than $\beta$ restricts the solution set, whereas increasing some RHS to $\gamma$ or above may not enlarge the solution set if another inequality separates the solutions that would otherwise be included. Hence, we ignore the upper ends in what follows and focus on the consequences of Lemma 1.

In what follows, let $v$ be an unexplored node with RHS values $\rho^1$ to $\rho^m$ and let $\bar{v}$ and $\bar{\bar{v}}$ be explored nodes with lower ends $\bar{\beta}^1$ to $\bar{\beta}^m$ and $\bar{\bar{\beta}}^1$ to $\bar{\bar{\beta}}^m$, respectively.

**Corollary 1 (Main necessary condition).** *Node $v$ is equivalent to node $\bar{v}$ only if $\rho^k \geq \bar{\beta}^k$ for $k = 1, \ldots, m$.*

*Proof.* Lemma 1 implies that $\rho^k < \bar{\beta}^k$ for any inequality $k$ would make a solution of $\bar{v}$ infeasible for $v$. Conversely, if $\rho^k \geq \bar{\beta}^k$ for $k = 1, \ldots, m$, then $v$ has at least the same solutions as $\bar{v}$, a necessary condition for equivalence.     □

**Corollary 2 (Dominance elimination).** *Node $v$ is equivalent to an explored node $\bar{v}$ only if no other explored node $\bar{\bar{v}}$ for which $v$ satisfies the necessary condition has a strictly larger lower end for any of the inequalities, i.e., there is no such $\bar{\bar{v}}$ for which $\bar{\bar{\beta}}^k > \bar{\beta}^k$ for any $k = 1, \ldots, m$.*

*Proof.* If two nodes $\bar{v}$ and $\bar{\bar{v}}$ have different lower ends for inequality $k$ and they are such that $\bar{\bar{\beta}}^k > \bar{\beta}^k$, then $\bar{\bar{v}}$ has a solution requiring a larger RHS value on inequality $k$ to be feasible. Thus, $\bar{v}$ does not have such a solution. However, if both $\bar{v}$ and $\bar{\bar{v}}$ have a subset of the solutions of node $v$, then $v$ also has that solution and thus cannot be equivalent to $\bar{v}$.     □

**Corollary 3 (One or none).** *Node $v$ has at most one explored node satisfying both the necessary condition and the dominance elimination in a reduced decision diagram.*

*Proof.* Since the lower ends characterize the state of a node and no two nodes have the same state in a reduced decision diagram, then any pair of explored nodes will differ in at least one lower end value. Consequently, no more than one explored node can satisfy both conditions for node $v$.     □

When multiple nodes meet the necessary condition of having at least as many solutions as node $v$, we can eliminate some by dominance. If they differ in the lower end of some inequality, that implies that node $v$ and one of them have a solution that the other does not have, hence allowing us to discard the latter. In fact, explored nodes can eliminate one another in different inequalities. Since no two nodes have all lower ends matching in a reduced decision diagram, no more than one node is left as candidate, but possibly none is.

## 6.2   A Sufficient Condition

From Corollary 3, we are left with at most one explored node $\bar{v}$ that could be equivalent to a given unexplored node $v$. If there is no such node, then the solution set of $v$ is distinct from all the solution sets of explored nodes in the layer. Otherwise, the solution set of such node $\bar{v}$ is different from that of $v$ only if the solutions of node $v$ are a proper superset of the solutions of $\bar{v}$. If the layer has explored nodes corresponding to every possible solution set, then nodes $v$ and $\bar{v}$ would be equivalent. However, having all such nodes would be prohibitive.

Alternatively, we can individually consider each of the solutions that could be missing from $\bar{v}$. If a given layer has explored nodes containing each one of them, then node $\bar{v}$ is always equivalent. We use the following definition for sufficiency:

**Definition 2 (Populated layer).** *A layer is said to be populated if it has explored nodes corresponding to the minimal solution set containing each of the solutions on the remaining variables.*

For a solution $x$ in a populated layer, there is a node $v_x$ with lower ends $\beta(x)$ corresponding to the tightest RHS values for which $x$ is feasible. In other words, all the inequalities are active for $x$ when the RHS is replaced with $\beta(x)$. Node $v_x$ may also have any other solution $y$ such that $\beta(y) \leq \beta(x)$, in which case solution $x$ is only feasible when $y$ is. Note that we only need $O(2^k)$ nodes with distinct states to populate a layer instead of $O(2^{2^k})$ to cover all possible states. Figure 3 shows the tightest RHS values associated with each solution for the inequalities of the problem illustrated in Fig. 2. The next result formalizes the condition.

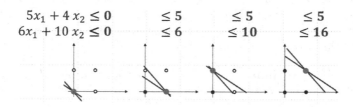

$$5x_1 + 4\,x_2 \leq 0 \qquad \leq 5 \qquad \leq 5 \qquad \leq 5$$
$$6x_1 + 10\,x_2 \leq 0 \qquad \leq 6 \qquad \leq 10 \qquad \leq 16$$

**Fig. 3.** Smallest (right-hand side) RHS values for which each solution in $\{0,1\}^2$ is feasible for the inequalities with left-hand side (LHS) $5x_1 + 4x_2$ and $6x_1 + 10x_2$.

**Theorem 3.** *Let $v$ belong to a populated layer. If there is a node $\bar{v}$ satisfying the necessary condition and the dominance elimination, then node $\bar{v}$ is equivalent to $v$. If no node satisfies the necessary condition, then $v$ has no solution.*

*Proof.* Let us suppose, for contradiction, that there is an explored node $\bar{v}$ that satisfies both conditions but is not equivalent to $v$. In such a case, node $v$ has a solution $x$ that $\bar{v}$ does not have. However, a populated layer containing node $v$ would also contain a node $v_x$ corresponding to the lower ends in $\beta(x)$. Node $v_x$ satisfies the necessary condition since $v$ contains $x$ and hence any solution that is always feasible when $x$ is feasible. Therefore, either $v_x$ is the node left after both conditions or else $\bar{v}$ contains all solutions of $v_x$, thereby contradicting that $v$ has a solution that $\bar{v}$ does not have.

Now let us suppose, for contradiction, that node $v$ has a solution $x$ but no explored node satisfies the necessary condition. That would imply that the layer does not contain a node $v_x$ corresponding to $\beta(x)$, a contradiction.     $\square$

One way to guarantee that a layer is populated is through bottom-up generation of nodes corresponding to the smallest RHS values yielding each solution.

# 7    Computational Experiments

We evaluated the impact of identifying equivalent search nodes when constructing reduced decision diagrams for integer linear programming problems. The construction of these diagrams mimics the branch-and-bound tree that emerges from a mathematical optimization solver as it enumerates the solutions of a problem, which we captured through callback functions when a solution is found or a branching decision is about to be made. We use pure 0–1 problems that are small enough to have a reasonable runtime, since enumerating all solutions takes much longer than solving the problem to optimality [19].

For each problem, we defined a gap with respect to the optimal value to limit the enumeration. The gap was chosen as large as possible to either enumerate all solutions or to avoid a considerable solver slowdown, for example from storing search nodes in disk. Nevertheless, we tried to push the gap to the largest possible value since more equivalences can be identified as the solution set gets denser. All problems are either directly obtained or adapted from MIPLIB [11,12,44]. When enumerating near-optimal solutions, we add an inequality to limit the objective function value. The order of the selector variables follows the indexing of the decision variables for the corresponding problem. We did not consider the possibility of changing the order of the selector variables, since any such change could have an unrelated effect on the number of branches and runtime.

For each problem, we constructed decision diagrams with preprocessing in the bottom $k$ layers above the terminal node, where $k \subset \{3, 6, 9\}$. We use $k = 0$ as the baseline, which is the case in which we cannot always determine if two nodes are equivalent by inspecting the RHS. For $k > 0$, we generated the smallest RHS values for each feasible solution on the remaining $k$ variables. We created marked nodes corresponding to each of such RHS vectors, which are explored when first matched with an unexplored node. Finally, we kept track of the solutions found before fixing all variables to avoid recounting them at the bottom of the diagram.

However, the number of equivalences that can be identified decays significantly as we move further away from the bottom of the decision diagram. Since the number of possible states for the last $k$ levels is $O(2^{2^k})$, it is rather expected that these equivalences will only be identified closer to the bottom—even in the cases that we can significantly reduce the size of the decision diagram. In practice, we did not observe significant gains with a value of $k$ larger than 10.

In our experiments, the code is written in C++ (gcc 4.8.4) using CPLEX Studio 12.8 as a solver and ran in Ubuntu 14.04.4 on a machine with 40 Intel(R) Xeon(R) CPU E5-2640 v4 @ 2.40GHz processors and 132 GB of RAM.

Table 1 reports the number of branches, runtime, and solutions found for each problem. We do not report results for stein9 with $k = 9$ because that problem has exactly 9 variables. Hence, we exclude the results for stein9 from the geometric mean in all cases, although we note that the number of branches and the runtime for stein9 reduced as $k$ increased up to 6. For the rest of the problems, we found a reduction of over 27.7% in number of branches and 7.7% in runtimes when comparing the geometric mean of the baseline with the geometric mean for $k = 9$. We observed a consistent reduction in the number of branches

**Table 1.** Number of branches and runtime to construct a decision diagram that enumerates near-optimal solutions using the baseline approach and bottom-up preprocessing in the bottom $k$ layers for $k \in \{3, 6, 9\}$ to identify equivalent nodes.

| Problem | Variables | Gap | Solutions | Branches | | | | Runtime (s) | | | |
|---|---|---|---|---|---|---|---|---|---|---|---|
| | | | | Baseline | $k = 3$ | $k = 6$ | $k = 9$ | Baseline | $k = 3$ | $k = 6$ | $k = 9$ |
| air01 | 771 | 5,400 | 117,997 | 9,402,967 | 9,403,025 | 9,398,697 | 9,393,475 | 3,142.1 | 3,178.8 | 3,188 | 3,184.4 |
| bm23 | 27 | 60 | 2,168 | 17,931 | 17,873 | 17,573 | 16,054 | 4.2 | 4 | 4.1 | 4.1 |
| enigma | 100 | 1 | 4 | 3,772 | 3,772 | 3,772 | 3,772 | 0.4 | 0.5 | 0.5 | 0.5 |
| l152lav | 1,989 | 30 | 37,741 | 7,576,394 | 7,576,394 | 7,576,394 | 7,576,394 | 7,976.5 | 9,274.3 | 9,360.4 | 8,036.5 |
| lp4l | 1,086 | 100 | 91,672 | 6,641,504 | 6,641,504 | 6,641,504 | 6,641,504 | 3,631.6 | 3,534.3 | 3,518.3 | 3,555 |
| lseu | 89 | 340 | 659,207 | 12,934,695 | 12,935,416 | 12,887,431 | 12,662,594 | 14,223 | 14,455.3 | 14,367.8 | 14,133.8 |
| misc01b | 82 | 250 | 600 | 4,302 | 4,264 | 3,963 | 3,760 | 0.9 | 0.9 | 0.7 | 0.8 |
| misc02b | 58 | 1,500 | 1,168 | 3,998 | 3,936 | 3,564 | 3,185 | 0.8 | 0.7 | 0.7 | 0.6 |
| misc03b | 159 | 1,600 | 4,320 | 75,730 | 75,651 | 74,309 | 72,019 | 14.8 | 14.6 | 14.1 | 13.9 |
| misc07b | 259 | 400 | 16,272 | 805,012 | 804,545 | 801,219 | 798,062 | 248.6 | 238.6 | 234.6 | 228.5 |
| p0033 | 33 | 2,200 | 10,746 | 17,174 | 16,368 | 11,783 | 7,498 | 2.6 | 2.5 | 2 | 1.5 |
| p0040 | 40 | 7,200 | 519,216 | 851,356 | 645,703 | 328,415 | 144,066 | 138 | 84.5 | 42.6 | 19.1 |
| p0201 | 201 | 600 | 864,128 | 27,675,260 | 27,661,111 | 27,501,029 | 27,140,363 | 9,017.5 | 8,564.5 | 8,865 | 8,418.1 |
| p0291 | 291 | 2 | 214 | 685,472 | 685,472 | 685,472 | 685,472 | 1,174.8 | 2,124.4 | 4,926 | 5,469 |
| pipex | 48 | 200 | 12,266 | 47,420 | 47,420 | 47,420 | 47,420 | 9 | 9.1 | 9 | 9.1 |
| sentoy | 60 | 270 | 67,820 | 6,587,574 | 6,504,322 | 5,919,396 | 5,219,636 | 8,052.8 | 7,743.3 | 7,054.3 | 6,732.1 |
| stein9 | 9 | 4 | 172 | 101 | 73 | 73 | — | 0.05 | 0.04 | 0.04 | — |
| stein15 | 15 | 6 | 2,809 | 1,739 | 1,233 | 842 | 841 | 0.5 | 0.3 | 0.3 | 0.4 |
| stein27 | 27 | 9 | 367,525 | 301,488 | 247,169 | 114,290 | 44,497 | 79.2 | 56.6 | 52.9 | 61.9 |
| stein45 | 45 | 2 | 795,064 | 3,970,424 | 3,955,623 | 3,833,862 | 3,575,399 | 896.9 | 882.4 | 1,014 | 1,973.9 |
| Geometric mean[a] | | | | $2.9 \times 10^5$ | $2.8 \times 10^5$ | $2.4 \times 10^5$ | $2.1 \times 10^5$ | 97.0 | 92.8 | 91.2 | 89.5 |

[a]The geometric mean excludes results for stein9 in all columns.

for most cases, which is often not offset by the number of corresponding bottom-up states generated: 15 for $k = 3$, 127 for $k = 6$, and 1,023 for $k = 9$. While generating these additional nodes in advance is cheaper than branching, the extra time to check equivalence might explain the lesser impact on runtime.

# 8    Conclusion

This paper discussed the connection between redundant work in branch-and-bound and the direct construction of reduced decision diagrams. In both cases, such redundancy may manifest as nodes defining equivalent subproblems that are repetitively explored. That connection is particularly stronger if we want to generate a pool of feasible or near-optimal solutions of a discrete optimization problem, which requires substantially more branching than finding an optimal solution. The enumeration of solutions is a relatively unexplored topic, especially in integer linear programming. Nevertheless, alternate solutions are important in practice and generating them in smaller problems is now technically feasible due to the continuous advances in hardware and algorithms. Furthermore, decision diagrams provide a compact representation of solution sets, with which we can more efficiently manipulate to solve the same problem with different objectives.

Our first contribution is a simple but useful extension of prior work on identifying equivalent problems with a single inequality [1,2,6], which can be mainly useful for integer linear programs with fractional coefficients.

Our second contribution, which we believe is the most important, is the theoretical distinction between the case of equivalence involving one inequality and multiple inequalities. For problems defined by inequalities with additively separable LHS, we have seen that explored nodes can be uniquely identified by the smallest RHS values yielding the same solutions. When the nodes are explored in depth-first search, we showed that it is possible to isolate a single explored node as potentially equivalent to each unexplored node. This fact alone simplifies considerably the identification of potentially equivalent search nodes.

Our third contribution is a first sufficient condition to confirm the equivalence among such pairs of nodes, which consists of a bottom-up preprocessing technique based on fixing $k$ among the $n$ variables to branch last. Note that we can reasonably expect that equivalences will be more frequent if there are fewer variables left. For $k \ll n$, this would only marginally affect the search behavior and, in fact, our experiments showed a positive impact with small values of $k$.

For large solution sets, we found some instances in which our approach reduced branching and runtime. If the solution set of the decision diagram is not sufficiently dense, the effectiveness of our method would depend on identifying hidden structure in the problems. One potential example are problems with variables that produce the same effect when assigned, which has been previously exploited with orbital branching [39]: if we leave such variables for last in the decision diagram, we can anticipate that many nodes will have equivalent states.

We believe that there is further room to improve on runtime based on the reduction on the number of branches. Another topic for future work would be identifying simpler sufficient conditions for equivalence, which would allow us to directly construct decision diagrams for much larger problems.

**Acknowledgements.** I would like to thank John Hooker for bringing this topic to my attention and the anonymous reviewers for their detailed feedback.

# References

1. Abío, I., Nieuwenhuis, R., Oliveras, A., Rodríguez-Carbonell, E., Mayer-Eichberger, V.: A new look at BDDs for pseudo-Boolean constraints. J. Artif. Intell. Res. **45**, 443–480 (2012)
2. Abío, I., Stuckey, P.J.: Encoding linear constraints into SAT. In: O'Sullivan, B. (ed.) CP 2014. LNCS, vol. 8656, pp. 75–91. Springer, Cham (2014). https://doi.org/10.1007/978-3-319-10428-7_9
3. Achterberg, T., Heinz, S., Koch, T.: Counting solutions of integer programs using unrestricted subtree detection. In: Perron, L., Trick, M.A. (eds.) CPAIOR 2008. LNCS, vol. 5015, pp. 278–282. Springer, Heidelberg (2008). https://doi.org/10.1007/978-3-540-68155-7_22
4. Andersen, H.R., Hadzic, T., Hooker, J.N., Tiedemann, P.: A constraint store based on multivalued decision diagrams. In: Bessière, C. (ed.) CP 2007. LNCS, vol. 4741, pp. 118–132. Springer, Heidelberg (2007). https://doi.org/10.1007/978-3-540-74970-7_11
5. Balas, E., Jeroslow, R.G.: Canonical cuts on the unit hypercube. SIAM J. Appl. Math. **23**, 61–69 (1972)
6. Behle, M.: Binary decision diagrams and integer programming. Ph.D. thesis, Universität des Saarlandes (2007)
7. Bergman, D., Cire, A.A., van Hoeve, W.-J., Hooker, J.N.: Variable ordering for the application of BDDs to the maximum independent set problem. In: Beldiceanu, N., Jussien, N., Pinson, É. (eds.) CPAIOR 2012. LNCS, vol. 7298, pp. 34–49. Springer, Heidelberg (2012). https://doi.org/10.1007/978-3-642-29828-8_3
8. Bergman, D., Cire, A.A., van Hoeve, W.J., Hooker, J.N.: Discrete optimization with decision diagrams. INFORMS J. Comput. **28**(1), 47–66 (2016)
9. Bergman, D., Cire, A., van Hoeve, W.J., Hooker, J.: Decision Diagrams for Optimization. Springer, Heidelberg (2016). https://doi.org/10.1007/978-3-319-42849-9
10. Bergman, D., Cire, A.A.: Discrete nonlinear optimization by state-space decompositions. Manag. Sci. **64**(10), 4700–4720 (2018)
11. Bixby, R.E., Boyd, E.A., Indovina, R.R.: MIPLIB: a test set of mixed integer programming problems. SIAM News **25**, 16 (1992)
12. Bixby, R.E., Ceria, S., McZeal, C.M., Savelsbergh, M.W.P.: An updated mixed integer programming library: MIPLIB 3.0. Optima **58**, 12–15 (1998)
13. Bollig, B., Wegener, I.: Improving the variable ordering of OBDDs is NP-complete. IEEE Trans. Comput. **45**(9), 993–1002 (1996)
14. Bryant, R.: Graph-based algorithms for Boolean function manipulation. IEEE Trans. Comput. **C–35**(8), 677–691 (1986)
15. Camm, J.D.: ASP, the art and science of practice: a (very) short course in suboptimization. INFORMS J. Appl. Anal. **44**(4), 428–431 (2014)
16. Chu, G., de la Banda, M.G., Stuckey, P.J.: Exploiting subproblem dominance in constraint programming. Constraints **17**(1), 1–38 (2012). https://doi.org/10.1007/s10601-011-9112-9

17. Cire, A.A., van Hoeve, W.J.: Multivalued decision diagrams for sequencing problems. Oper. Res. **61**(6), 1411–1428 (2013)
18. Conforti, M., Cornuéjols, G., Zambelli, G.: Integer Programming. Springer, Heidelberg (2014). https://doi.org/10.1007/978-3-319-11008-0
19. Danna, E., Fenelon, M., Gu, Z., Wunderling, R.: Generating multiple solutions for mixed integer programming problems. In: Fischetti, M., Williamson, D.P. (eds.) IPCO 2007. LNCS, vol. 4513, pp. 280–294. Springer, Heidelberg (2007). https://doi.org/10.1007/978-3-540-72792-7_22
20. Danna, E., Woodruff, D.L.: How to select a small set of diverse solutions to mixed integer programming problems. Oper. Res. Lett. **37**, 255–260 (2009)
21. Davarnia, D., van Hoeve, W.J.: Outer approximation for integer nonlinear programs via decision diagrams (2018). https://doi.org/10.1007/s10107-020-01475-4
22. Ebendt, R., Gunther, W., Drechsler, R.: An improved branch and bound algorithm for exact bdd minimization. IEEE Trans. Comput. Aided Des. Integr. Circuits Syst. **22**(12), 1657–1663 (2003)
23. Fischetti, M., Salvagnin, D.: Pruning moves. INFORMS J. Comput. **22**(1), 108–119 (2010)
24. Fischetti, M., Toth, P.: A new dominance procedure for combinatorial optimization problems. Oper. Res. Lett. **7**(4), 181–187 (1988)
25. GAMS Software GmbH: Getting a list of best integer solutions of my MIP (2017). https://support.gams.com/solver:getting_a_list_of_best_integer_solutions_of_my_mip_model. Accessed 29 Nov 2019
26. Gurobi Optimization, LLC: Finding multiple solutions (2019). https://www.gurobi.com/documentation/8.1/refman/finding_multiple_solutions.html. Accessed 29 Nov 2019
27. Haus, U.U., Michini, C., Laumanns, M.: Scenario aggregation using binary decision diagrams for stochastic programs with endogenous uncertainty. CoRR abs/1701.04055 (2017)
28. Haus, U.U., Michini, C.: Representations of all solutions of Boolean programming problems. In: International Symposium on Artificial Intelligence and Mathematics (ISAIM) (2014)
29. Hooker, J.N.: Decision diagrams and dynamic programming. In: Gomes, C., Sellmann, M. (eds.) CPAIOR 2013. LNCS, vol. 7874, pp. 94–110. Springer, Heidelberg (2013). https://doi.org/10.1007/978-3-642-38171-3_7
30. Hosaka, K., Takenaga, Y., Kaneda, T., Yajima, S.: Size of ordered binary decision diagrams representing threshold functions. Theor. Comput. Sci. **180**, 47–60 (1997)
31. Ibaraki, T.: The power of dominance relations in branch-and-bound algorithms. J. Assoc. Comput. Mach. **24**(2), 264–279 (1977)
32. IBM Corp.: IBM ILOG CPLEX Optimization Studio Getting Started with CPLEX Version 12 Release 8 (2017)
33. IBM Corp.: How to enumerate all solutions (2019). https://www.ibm.com/support/knowledgecenter/SSSA5P_12.9.0/ilog.odms.cplex.help/CPLEX/UsrMan/topics/discr_optim/soln_pool/18_howTo.html. Accessed 29 Nov 2019
34. Jain, S., Kadioglu, S., Sellmann, M.: upper bounds on the number of solutions of binary integer programs. In: Lodi, A., Milano, M., Toth, P. (eds.) CPAIOR 2010. LNCS, vol. 6140, pp. 203–218. Springer, Heidelberg (2010). https://doi.org/10.1007/978-3-642-13520-0_24
35. Kohler, W., Steiglitz, K.: Characterization and theoretical comparison of branch-and-bound algorithms for permutation problems. J. Assoc. Comput. Mach. **21**(1), 140–156 (1974)

36. Land, A.H., Doig, A.G.: An automatic method of solving discrete programming problems. Econometrica **28**(3), 497–520 (1960)
37. Lozano, L., Smith, J.C.: A binary decision diagram based algorithm for solving a class of binary two-stage stochastic programs. Math. Program. (2018). https://doi.org/10.1007/s10107-018-1315-z
38. Morrison, D., Sewell, E., Jacobson, S.: Solving the pricing problem in a branch-and-price algorithm for graph coloring using zero-suppressed binary decision diagrams. INFORMS J. Comput. **28**(1), 67–82 (2016)
39. Ostrowski, J., Linderoth, J., Rossi, F., Smriglio, S.: Orbital branching. Math. Program. **126**, 147–178 (2011). https://doi.org/10.1007/s10107-009-0273-x
40. Perez, G., Régin, J.C.: Efficient operations on MDDs for building constraint programming models. In: International Joint Conference on Artificial Intelligence (IJCAI) (2015)
41. Petit, T., Trapp, A.C.: Enriching solutions to combinatorial problems via solution engineering. INFORMS J. Comput. **31**(3), 429–444 (2019)
42. Raghunathan, A., Bergman, D., Hooker, J., Serra, T., Kobori, S.: Seamless multimodal transportation scheduling. CoRR abs/1807.09676 (2018)
43. Sanner, S., Uther, W., Delgado, K.V.: Approximate dynamic programming with affine ADDs. In: AAMAS (2010)
44. Serra, T., Hooker, J.: Compact representation of near-optimal integer programming solutions. Math. Program. **182**, 199–232 (2019). https://doi.org/10.1007/s10107-019-01390-3
45. Serra, T., Raghunathan, A.U., Bergman, D., Hooker, J., Kobori, S.: Last-mile scheduling under uncertainty. In: Rousseau, L.-M., Stergiou, K. (eds.) CPAIOR 2019. LNCS, vol. 11494, pp. 519–528. Springer, Cham (2019). https://doi.org/10.1007/978-3-030-19212-9_34
46. Serra, T., Ramalingam, S.: Empirical bounds on linear regions of deep rectifier networks. CoRR abs/1810.03370 (2018)
47. Tjandraatmadja, C., van Hoeve, W.J.: Target cuts from relaxed decision diagrams. INFORMS J. Comput. **31**(2), 285–301 (2019)
48. Trapp, A.C., Konrad, R.A.: Finding diverse optima and near-optima to binary integer programs. IIE Trans. **47**, 1300–1312 (2015)
49. Valiant, L.G.: The complexity of computing the permanent. Theoret. Comput. Sci. **8**, 189–201 (1979)
50. Valiant, L.G.: The complexity of enumeration and reliability problems. SIAM J. Comput. 8(3), 410–421 (1979)
51. Verhaeghe, H., Lecoutre, C., Schaus, P.: Compact-MDD: efficiently filtering (s) MDD constraints with reversible sparse bit-sets. In: IJCAI (2018)
52. Verhaeghe, H., Lecoutre, C., Schaus, P.: Extending compact-diagram to basic smart multi-valued variable diagrams. In: Rousseau, L.-M., Stergiou, K. (eds.) CPAIOR 2019. LNCS, vol. 11494, pp. 581–598. Springer, Cham (2019). https://doi.org/10.1007/978-3-030-19212-9_39
53. Ye, Z., Say, B., Sanner, S.: Symbolic bucket elimination for piecewise continuous constrained optimization. In: van Hoeve, W.-J. (ed.) CPAIOR 2018. LNCS, vol. 10848, pp. 585–594. Springer, Cham (2018). https://doi.org/10.1007/978-3-319-93031-2_42

# Lossless Compression of Deep Neural Networks

Thiago Serra[1]($\boxtimes$), Abhinav Kumar[2], and Srikumar Ramalingam[2]

[1] Bucknell University, Lewisburg, USA
thiago.serra@bucknell.edu
[2] The University of Utah, Salt Lake City, USA
abhinav.kumar@utah.edu, srikumar@cs.utah.edu

**Abstract.** Deep neural networks have been successful in many predictive modeling tasks, such as image and language recognition, where large neural networks are often used to obtain good accuracy. Consequently, it is challenging to deploy these networks under limited computational resources, such as in mobile devices. In this work, we introduce an algorithm that removes units and layers of a neural network while not changing the output that is produced, which thus implies a lossless compression. This algorithm, which we denote as LEO (Lossless Expressiveness Optimization), relies on Mixed-Integer Linear Programming (MILP) to identify Rectified Linear Units (ReLUs) with linear behavior over the input domain. By using $\ell_1$ regularization to induce such behavior, we can benefit from training over a larger architecture than we would later use in the environment where the trained neural network is deployed.

**Keywords:** Deep learning · Mixed-Integer Linear Programming · Neural network pruning · Neuron stability · Rectified Linear Unit

## 1 Introduction

Deep Neural Networks (DNNs) have achieved unprecedented success in many domains of predictive modeling, such as computer vision [17,36,43,60,91], natural language processing [90], and speech [45]. While complex architectures are usually behind such feats, it is not fully known if these results depend on such DNNs being as wide or as deep as they currently are for some applications.

In this paper, we are interested in the compression of DNNs, especially to reduce their size and depth. More generally, that relates to the following question of wide interest about neural networks: given a neural network $\mathcal{N}_1$, can we find an *equivalent* neural network $\mathcal{N}_2$ with a different architecture? Since a trained DNN corresponds to a function mapping its inputs to outputs, we can formalize the equivalence among neural networks as follows [78]:

**Definition 1 (Equivalence).** *Two deep neural networks $\mathcal{N}_1$ and $\mathcal{N}_2$ with associated functions $f_1 : \mathbb{R}^{n_0} \to \mathbb{R}^m$ and $f_2 : \mathbb{R}^{n_0} \to \mathbb{R}^m$, respectively, are equivalent if $f_1(x) = f_2(x) \ \forall \ x \in \mathbb{R}^{n_0}$.*

© Springer Nature Switzerland AG 2020
E. Hebrard and N. Musliu (Eds.): CPAIOR 2020, LNCS 12296, pp. 417–430, 2020.
https://doi.org/10.1007/978-3-030-58942-4_27

In other words, our goal is to start from a trained neural network and identify neural networks with fewer layers or smaller layer widths that would produce the exact same outputs. Since the typical input for certain applications is bounded along each dimension, such as $x \in [0,1]^{n_0}$ for the MNIST dataset [63], we can consider a broader family of neural networks that would be regarded as equivalent in practice. We formalize that idea with the concept of local equivalence:

**Definition 2 (Local Equivalence).** *Two deep neural networks $\mathcal{N}_1$ and $\mathcal{N}_2$ with associated functions $f_1 : \mathbb{R}^{n_0} \to \mathbb{R}^m$ and $f_2 : \mathbb{R}^{n_0} \to \mathbb{R}^m$, respectively, are local equivalent with respect to a domain $\mathbb{D} \subseteq \mathbb{R}^{n_0}$ if $f_1(x) = f_2(x) \ \forall \ x \in \mathbb{D}$.*

For a given application, local equivalence with respect to the domain of possible inputs suffices to guarantee that two networks have the same accuracy in any test. Hence, finding a smaller network that is local equivalent to the original network implies a compression in which there is no loss. In this paper, we show that simple operations such as removing or merging units and folding layers of a DNN can yield such lossless compression under certain conditions. We denote as *folding* the removal of a layer by directly connecting the adjacent layers, which is accompanied by adjusting the weights and biases of those layers accordingly.

## 2 Background

We study feedforward DNNs with Rectified Linear Unit (ReLU) activations [38], which are comparatively simpler than other types of activations. Nevertheless, ReLUs are currently the type of unit that is most commonly used [64], which is in part due to landmark results showing their competitive performance [35,77].

Every network has input $x = [x_1 \ x_2 \ \dots \ x_{n_0}]^T$ from a bounded domain $\mathbb{X}$ and corresponding output $y = [y_1 \ y_2 \ \dots \ y_m]^T$, and each hidden layer $l \in \mathbb{L} = \{1, 2, \dots, L\}$ has output $h^l = [h_1^l \ h_2^l \dots h_{n_l}^l]^T$ from ReLUs indexed by $i \in \mathbb{N}_l = \{1, 2, \dots, n_l\}$. Let $W^l$ be the $n_l \times n_{l-1}$ matrix where each row corresponds to the weights of a neuron of layer $l$, and let $b^l$ be vector of biases associated with the units in layer $l$. With $h^0$ for $x$ and $h^{L+1}$ for $y$, the output of each unit $i$ in layer $l$ consists of an affine function $g_i^l = W_i^l h^{l-1} + b_i^l$ followed by the ReLU activation $h_i^l = \max\{0, g_i^l\}$. The unit $i$ in layer $l$ is denoted *active* when $h_i^l > 0$ and *inactive* otherwise. DNNs consisting solely of ReLUs are denoted *rectifier* networks, and their associated functions are always piecewise linear [7].

### 2.1 Mixed-Integer Linear Programming

Our work is primarily based on the fast growing literature on applications of Mixed-Integer Linear Programming (MILP) to deep learning. MILP can be used to map inputs to outputs of each ReLU and consequently of rectifier networks. Such formulations have been used to produce the image [27,71] and estimate the number of pieces [86,87] of the piecewise linear function associated with the network, generate adversarial perturbations to test the network robustness [6, 16,31,89,96,106], and implement controllers based on DNN models [85,109].

For each unit $i$ in layer $l$, we can map $h^{l-1}$ to $g_i^l$ and $h_i^l$ with a formulation that also includes a binary variable $z_i$ denoting if the unit is active or not, a variable $\bar{h}_i^l$ denoting the output of a complementary fictitious unit $\bar{h}_i^l = max\{0, -g_i^l\}$, and constants $H_i^l$ and $\bar{H}_i^l$ that are positive and as large as $h_i^l$ and $\bar{h}_i^l$ can be:

$$W_i^l h^{l-1} + b_i^l = g_i^l, \qquad g_i^l = h_i^l - \bar{h}_i^l \tag{1}$$

$$h_i^l \leq H_i^l z_i^l, \qquad \bar{h}_i^l \leq \bar{H}_i^l(1 - z_i^l) \tag{2}$$

$$h_i^l \geq 0, \qquad \bar{h}_i^l \geq 0, \qquad z_i^l \in \{0,1\} \tag{3}$$

This formulation can be strengthened by using the smallest possible values for $H_i^l$ and $\bar{H}_i^l$ [31,96] and valid inequalities to avoid fractional values of $z_i^l$ [6,86].

The largest possible value of $g_i^l$, which we denote $\mathcal{G}_i^l$, can be obtained as

$$\mathcal{G}_i^{l'} = \max \qquad W_i^{l'} h^{l'-1} + b_i^{l'} \tag{4}$$

$$\text{s.t.} \qquad (1)\text{-}(3) \qquad \forall l \in \{1,\ldots,l'-1\},\ i \in \mathbb{N}_l \tag{5}$$

$$x \in \mathbb{X} \tag{6}$$

If $\mathcal{G}_i^l > 0$, then $\mathcal{G}_i^l$ is also the largest value of $h_i^l$ and it can be used for constant $H_i^l$. Otherwise, $h_i^l = 0$ for any input $x \in \mathbb{X}$. We can also minimize $W_i^l h^{l-1} + b_i^l$ to obtain $\overline{\mathcal{G}}_i^l$, the smallest possible value of $g_i^l$, and use $-\overline{\mathcal{G}}_i^l$ for constant $\bar{H}_i^l$ if $\overline{\mathcal{G}}_i^l < 0$; whereas $\overline{\mathcal{G}}_i^l > 0$ implies that $h_i^l > 0$ for any input $x \in \mathbb{X}$. By solving these formulations from the first to the last layer, we have the tightest values for $H_i^l$ and $\bar{H}_i^l$ for $l \in \{1,\ldots l'-1\}$ when we reach layer $l'$. Units with only zero or positive outputs were first identified using MILP in [96], where they are denoted as *stably inactive* and *stably active*, and their incidence was induced with $\ell_1$ regularization. That was later exploited to accelerate the verification of robustness by making the corresponding MILP formulation easier to solve [106].

In this paper, we use the stability of units to either remove or merge them while preserving the outputs produced by the DNN. The same idea could be extended to other architectures with MILP mappings, such as Binarized Neural Networks (BNNs) [18,78]. MILP has been used in BNNs for adversarial testing [56] and along with Constraint Programming (CP) for training [52]. BNNs have characteristics that also make them suitable under limited resources [78].

## 2.2   Related Work

Our work relates to the literature on neural network compression, and more specifically to methods that simplify a trained DNN. Such literature includes low-rank decomposition [22,26,53,81,101,112], quantization [18,83,97,105], architecture design [48,49,51,91,94], non-structured pruning [66], structured pruning [1,39,40,65,72,74,110], sparse learning [4,68,102,114], automatic discarding of layers in ResNets [44,98,111], variational methods [113], and the recent Lottery Ticket Hypothesis [32] by which training only certain subnetworks in the

DNN—the *lottery tickets*—might be good enough. However, network compression is often achieved with side effects to the function associated with the DNN.

In contrast to many lossy pruning methods that typically focus on removing unimportant neurons and connections, our approach focuses on developing lossless transformations that exactly preserve the expressiveness during the compression. A necessary criterion for equivalent transformation is that the resulting network is as expressive as the original one. Methods to study neural network expressiveness include universal approximation theory [19], VC dimension [9], trajectory length [82], and linear regions [7, 41, 42, 75, 76, 79, 82, 86, 87].

We can also consider our approach as a form of reparameterization, or equivalent transformation, in graphical models [59, 100, 103]. If two parameter vectors $\theta$ and $\theta'$ define the same energy function (i.e., $E(\boldsymbol{x}|\theta)) = E(\boldsymbol{x}|\theta'), \forall \boldsymbol{x}$), then $\theta'$ is called a reparameterization of $\theta$. Reparameterization has played a key role in several inference problems such as belief propagation [100], tree-weighted message passing [99], and graph cuts [57]. The idea is also associated with characterizing the functions that can be represented by DNNs [7, 19, 46, 62, 67, 73, 95].

Finally, our work complements the vast literature at the intersection of mathematical optimization and Machine Learning (ML). General-purpose methods have been applied to train ML models to optimality [12, 13, 52, 84]. Conversely, ML models have been extensively applied in optimization [11, 34]. To mention a few lines of work, ML has been used to find feasible solutions [24, 33] and predict good solutions [21, 25]; determine how to branch on [3, 8, 47, 55, 69] or add cuts [93], when to linearize [14], or when to decompose [61] an optimization problem; how to adapt algorithms for each problem [10, 20, 23, 50, 54, 58, 88]; obtain better optimization bounds [15]; embed the structure of the problem as a layer of a neural network [2, 5, 29, 108]; and predict the resolution by a time-limit [30], the feasibility of the problem [107], and the problem itself [28, 70, 92].

## 3  LEO: Lossless Expressiveness Optimization

Algorithm 1, which we denote LEO (Lossless Expressiveness Optimization), loops over the layers to remove units with constant outputs regardless of the input, some of the stable units, and any layers with constant output due to those two types of units. These modifications of the network architecture are followed by changes to the weights and biases of the remaining units in the network to preserve the outputs produced. The term *expressiveness* is commonly used to refer to the ability of a network architecture to represent complex functions [86].

First, LEO checks the weights and stability of each unit and decides whether to immediately remove them. A unit with constant output, which is either stably inactive or has zero input weights, is removed as long as there are other units left in the layer. A stably active unit with varying output is removed if the column of weights of that unit is linearly dependent on the column of weights of stably active units with varying outputs that have been previously inspected in that layer. We consider the removal of such stably active units as a *merging* operation since the output weights of other stable units need to be adjusted as well.

Second, LEO checks if layers can be removed in case the units left in the layer are all stable or have constant output. If they are all stably active with varying output, then the layer is removed by directly joining the layers before and after it, which we denote as a *folding* operation. In the particular case that only one unit is left with constant output, be it stably inactive or not, then all hidden layers are removed because the network has an associated function that is constant in $\mathbb{D}$. We denote the latter operation as *collapsing* the neural network.

Figure 1 shows examples of units being removed and merged on the left as well as of a layer being folded on the right. Although possible, folding or collapsing a trained neural network is not something that we would expect to achieve in practice unless we are compressing with respect to a very small domain $\mathbb{D} \subset \mathbb{X}$.

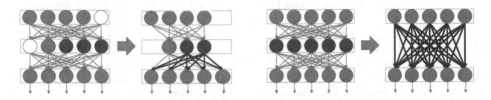

**Fig. 1.** Examples of output-preserving neural network compression obtained with LEO. On the left, two units in white are stably inactive and three units indexed by set $S$ in darker blue are stably active, where rank($\boldsymbol{W}_S^2$)=2. In such a case, we can remove the stably inactive units and merge the stably active units to produce the same input to the next layer using only two units. On the right, an entire layer is stably active. In such a case, we can fold the layer by directly connecting the layers before and after it. In both cases, the red arcs correspond to coefficients that need to be adjusted accordingly.

**Theorem 1.** *For a neural network $\mathcal{N}_1$, Algorithm 1 produces a neural network $\mathcal{N}_2$ such that $\mathcal{N}_1$ and $\mathcal{N}_2$ are local equivalent with respect to an input domain $\mathbb{D}$.*

*Proof.* If $\overline{\mathcal{G}}_i^l < 0$, then $h_i^l = 0$ for any input in $\mathbb{D}$ and unit $i$ in layer $l$ can be regarded as stably inactive. Otherwise, if $\boldsymbol{W}_i^l = \boldsymbol{0}$, then the output of the unit is positive but constant. Those two types of units are analyzed by the block starting at line 5. If there are other units left in the layer, which are either not stable or stably active but not removed, then removing unit a stably inactive unit $i$ does not affect the output of subsequent units since the output of the unit is always 0 in $\mathbb{D}$. Likewise, in the case that $\boldsymbol{W}_i^l = \boldsymbol{0}$ and $b_i^l > 0$, then the output of the network remains the same after removing that unit if such removal of $h_i^l$ from each unit $j$ in layer $l + 1$ is followed by adding $w_{ji}^{l+1}b_i^l$ to $b_j^{l+1}$.

If $\overline{\mathcal{G}}_i^l > 0$, then $h_i^l = \boldsymbol{W}_i^l h^{l-1} + b_i^l$ for any input in $\mathbb{D}$ and unit $i$ in layer $l$ can be regarded as stably active. Those units are analyzed by the block starting at line 14. If the rank of the submatrix $\boldsymbol{W}_S^l$ consisting of the weights of stably active units in set $S$ is the same as that of $\boldsymbol{W}_{S \cup \{i\}}^l$ and given that $\boldsymbol{W}_i^l \neq \boldsymbol{0}$ for every $i \in S$, then $S \neq \emptyset$ and $h_i^l = \sum_{k \in S} \alpha_k w_k^l h^{l-1} + b_i^l = \sum_{k \in S} \alpha_k (h_k^l - b_k^l) + b_i^l$. Since

**Algorithm 1.** LEO produces a smaller equivalent neural network with respect to a domain $\mathbb{D}$ by removing units and layers while adjusting weights and biases

```
1:  for l ← 1,...,L do
2:      S ← {}                                          ▷ Set of stable units left in layer l
3:      Unstable ← False                                ▷ If there are unstable units in layer l
4:      for i ← 1,...,n_l do
5:          if 𝒢_i^l < 0 for x ∈ 𝔻 or W_i^l =0 then     ▷ Unit i has constant output
6:              if i < n_l or |S| > 0 or Unstable then   ▷ Layer l still has other units
7:                  if W_i^l = 0 and b_i^l > 0 then
8:                      for j ← 1,...,n_{l+1} do          ▷ Adjust activations in layer l + 1
9:                          b_j^{l+1} ← b_j^{l+1} + w_{ji}^{l+1}b_i^l
10:                     end for
11:                 end if
12:                 Remove unit i from layer l            ▷ Unit i is not necessary
13:             end if
14:         else if 𝒢̄_i^l > 0 for x ∈ 𝔻 then            ▷ Unit i is stably active
15:             if rank(W_{S∪{i}}^l) > |S| then           ▷ w_i^l is linearly independent
16:                 S ← S ∪ {i}                           ▷ Keep unit in the network
17:             else                                      ▷ Output of unit i is linearly dependent
18:                 Find {α_k}_{k∈S} such that w_i^l = ∑_{k∈S} α_k w_k^l
19:                 for j ← 1,...,n_{l+1} do              ▷ Adjust activations in layer l + 1
20:                     w_{jk}^{l+1} ← w_{jk}^{l+1} + ∑_{k∈S} α_k w_{ji}^{l+1}
21:                     b_j^{l+1} ← b_j^{l+1} + w_{ji}^{l+1}(b_i^l + ∑_{k∈S} α_k b_k^l)
22:                 end for
23:                 Remove unit i from layer l            ▷ Unit i is no longer necessary
24:             end if
25:         else
26:             Unstable ← True                           ▷ At least one unit is not stable
27:         end if
28:     end for
29:     if not Unstable then                             ▷ All units left in layer l are stable
30:         if |S| > 0 then                              ▷ The units left have varying outputs
31:             Create matrix W̄ ∈ ℝ^{n_{l-1}×n_{l+1}} and vector b̄ ∈ ℝ^{n_{l+1}}
32:             for i ← 1,...,n_{l+1} do                  ▷ Directly connect layers l − 1 and l + 1
33:                 b̄_i ← b_i^{l+1} + ∑_{k∈S} w_{ik}^{l+1}b_k^l
34:                 for j ← 1,...,n_{l-1} do
35:                     w̄_{ij} ← ∑_{k∈S} w_{kj}^l w_{ik}^{l+1}
36:                 end for
37:             end for
38:             Remove layer l; replace parameters in next layer with W̄ and b̄
39:         else                                         ▷ Only unit left in layer l has constant output
40:             Compute output Υ for any input χ ∈ 𝔻     ▷ Function is constant
41:             (W^{L+1}, b^{L+1}) ← (0, Υ)               ▷ Set constant values in output layer
42:             Remove layers 1 to L and break            ▷ Remove all hidden layers and leave
43:         end if
44:     end if
45: end for
```

there would be other units left in the layer, the output of the network remains the same after removing the unit if such removal of $h_i^l$ from each unit $j$ in layer $l + 1$ is followed by adding $\alpha_k w_{ji}^{l+1}$ to $w_{jk}^{l+1}$ and $w_{ji}^{l+1} \left( b_i^l - \sum_{k \in A} \alpha_k b_k^l \right)$ to $b_j^{l+1}$.

If all units left in layer $l$ are stably active and $|S| > 0$, then layer $l$ is equivalent to an affine transformation and it is possible to directly connect layers $l - 1$ and $l + 1$, as in the block starting at line 30. Since $h_k^l = W_j^l h^{l-1} + b_k^l$ for each stably active unit $k$ in layer $l$, then $h_i^{l+1} = W_i^{l+1} h^l + b_i^{l+1} = W_i^{l+1} \left( \sum_{k=1}^{n_l} W_k^l h^{l-1} + b_k^l \right) +$

$b_i^{l+1} = \sum_{j \in n_{l-1}} \left( \sum_{k \in S} w_{kj}^l w_{ik}^{l+1} \right) h_j^{l-1} + b_i^{l+1} + \left( \sum_{k \in S} w_{ik}^{l+1} b_k^l \right).$

If the only unit $i$ left in layer $l$ is stably inactive or stably active but has zero weights, then any input in $\mathbb{D}$ results in $h_i^l = \max\{0, b_i^l\}$, and consequently the neural network is associated with a constant function $f : x \to \Upsilon$ in $\mathbb{D}$. Therefore, it is possible to remove all hidden layers and replace the output layer with a constant function mapping to $\Upsilon$ as in the block starting at line 39. $\square$

*Implementation.* We do not need to solve (4)–(6) to optimality to determine if $\underline{\mathcal{G}}_i^l < 0$: it suffices to find a negative upper bound to guarantee that, or a solution with positive value to refute that. A similar reasoning applies to $\bar{\mathcal{G}}_i^l > 0$.

## 4  Experiments

We conducted experiments to evaluate the potential for network compression using LEO. In these experiments, we trained rectifier networks on the MNIST dataset [63] with input size 784, two hidden layers of same width, and 10 outputs. The widths of the hidden layers are 25, 50, and 100. For each width, we identified in preliminary experiments a weight for $\ell_1$ regularization on layer weights that improved the network accuracy in comparison with no regularization: 0.0002, 0.0002, and 0.0001, respectively. We trained 31 networks with that regularization weight, with 5 times the same weight to induce more stability, and with zero weight as a benchmark. We use the negative log-likelihood as the loss function after taking a softmax operation on the output layer, a batch size of 64 and SGD with a learning rate of 0.01, and momentum of 0.9 for training the model to 120 epochs. The learning rate is decayed by a factor of 0.1 after every 50 epochs. The weights of the network were initialized with the Kaiming initialization [43] and the biases were initialized to zero. The models were trained using Pytorch [80] on a machine with 40 Intel(R) Xeon(R) CPU E5-2640 v4 @ 2.40 GHz processors and 132 GB of RAM. The MILPs were solved using Gurobi 8.1.1 [37] on a machine with Intel(R) Core(TM) i5-6200U CPU @ 2.30 GHz and 16 GB of RAM. We used callbacks to check bounds and solutions and then interrupt the solver after determining unit stability and bounds for each MILP.

Tables 1 and 2 summarize our experiments with mean and standard error. We note that the compression grows with the size of the network and the weight of $\ell_1$ regularization, which induces the weights of each unit to be orders of magnitude

**Table 1.** Compression of 2-hidden-layer rectifier networks trained on MNIST. Each line summarizes tests on 31 networks. Depending on how the network is trained, the higher incidence of stable units allows for more compression while preserving the trained network accuracy. For example, training with $\ell_1$ regularization induces such stability and then inactive units can be removed. Interestingly, the small amount of regularization that improves accuracy during training also helps compressing the network later.

| Layer width | $\ell_1$ weight | Accuracy (%) | Units removed | | Network compression (%) |
|---|---|---|---|---|---|
| | | | 1st layer | 2nd layer | |
| 25 | 0.001 | 95.76 ± 0.05 | 5.7 ± 0.3 | 5.1 ± 0.3 | 22 ± 1 |
| 25 | 0.0002 | 97.24 ± 0.02 | 1.2 ± 0.1 | 3.0 ± 0.4 | 8.3 ± 0.7 |
| 25 | 0 | 96.68 ± 0.03 | 0 ± 0 | 0 ± 0 | 0 ± 0 |
| 50 | 0.001 | 96.05 ± 0.04 | 16.9 ± 0.6 | 12.5 ± 0.6 | 29.4 ± 0.7 |
| 50 | 0.0002 | 97.81 ± 0.02 | 7.6 ± 0.4 | 7.5 ± 0.5 | 15.1 ± 0.6 |
| 50 | 0 | 97.62 ± 0.02 | 0 ± 0 | 0 ± 0 | 0 ± 0 |
| 100 | 0.0005 | 97.14 ± 0.02 | 36.7 ± 0.7 | 24.9 ± 0.6 | 30.8 ± 0.5 |
| 100 | 0.0001 | 98.14 ± 0.01 | 18.6 ± 0.5 | 11.1 ± 0.7 | 14.9 ± 0.4 |
| 100 | 0 | 98.00 ± 0.01 | 0 ± 0 | 0 ± 0 | 0 ± 0 |

**Table 2.** Additional details about the experiments for each type of network, including runtime per test, incidence of stably active units, and overall network stability.

| Layer width | $\ell_1$ weight | Runtime (s) | Stably active units | | Network Stability (%) |
|---|---|---|---|---|---|
| | | | 1st layer | 2nd layer | |
| 25 | 0.001 | 27.9 ± 0.3 | 2.5 ± 0.3 | 7.4 ± 0.4 | 41.3 ± 0.6 |
| 25 | 0.0002 | 29 ± 1 | 0 ± 0 | 1.0 ± 0.2 | 10.4 ± 0.7 |
| 25 | 0 | 28.4 ± 0.3 | 0 ± 0 | 0 ± 0 | 0 ± 0 |
| 50 | 0.001 | 103 ± 2 | 15 ± 0.5 | 24.9 ± 0.6 | 69.3 ± 0.4 |
| 50 | 0.0002 | 106 ± 3 | 2.7 ± 0.3 | 8.8 ± 0.5 | 26.6 ± 0.6 |
| 50 | 0 | 112 ± 3 | 0 ± 0 | 0 ± 0 | 0 ± 0 |
| 100 | 0.0005 | 421 ± 4 | 35.7 ± 0.6 | 57.7 ± 0.7 | 77.5 ± 0.2 |
| 100 | 0.0001 | 456 ± 8 | 11.1 ± 0.5 | 18 ± 0.7 | 29.4 ± 0.5 |
| 100 | 0 | 385 ± 2 | 0 ± 0 | 0 ± 0 | 0 ± 0 |

smaller than its bias. The compression identified is all due to removing stably inactive units. Most of the runtime is due to solving MILPs for the second hidden layer. Given the incidence of stably active units, we conjecture that inducing rank deficiency in the weights or negative values in the biases could also be beneficial.

# 5  Conclusion

We introduced a lossless neural network compression algorithm, LEO, which relies on MILP to identify parts of the neural network that can be safely removed after reparameterization. We found that networks trained with $\ell_1$ regularization are particularly amenable to such compression. In a sense, we could interpret $\ell_1$ regularization as inducing a subnetwork to represent the function associated with the DNN. Future work may explore the connection between these subnetworks identified by LEO and lottery tickets, bounding techniques such as in [104] to help efficiently identifying stable units, and other forms of inducing posterior compressibility while training. Concomitantly, we have shown another form in which discrete optimization can play a key role in deep learning applications.

# References

1. Aghasi, A., Abdi, A., Nguyen, N., Romberg, J.: Net-trim: convex pruning of deep neural networks with performance guarantee. In: NeurIPS (2017)
2. Agrawal, A., Amos, B., Barratt, S., Boyd, S., Diamond, S., Kolter, Z.: Differentiable convex optimization layers. In: NeurIPS (2019)
3. Alvarez, A., Louveaux, Q., Wehenkel, L.: A machine learning-based approximation of strong branching. INFORMS J. Comput. (2017)
4. Alvarez, J., Salzmann, M.: Learning the number of neurons in deep networks. In: NeurIPS (2016)
5. Amos, B., Kolter, Z.: OptNet: differentiable optimization as a layer in neural networks. In: ICML (2017)
6. Anderson, R., Huchette, J., Tjandraatmadja, C., Vielma, J.: Strong mixed-integer programming formulations for trained neural networks. In: IPCO (2019)
7. Arora, R., Basu, A., Mianjy, P., Mukherjee, A.: Understanding deep neural networks with rectified linear units. In: ICLR (2018)
8. Balcan, M.F., Dick, T., Sandholm, T., Vitercik, E.: Learning to branch. In: ICML (2018)
9. Bartlett, P., Maiorov, V., Meir, R.: Almost linear VC-dimension bounds for piecewise polynomial networks. Neural Comput. **10**, 2159–2173 (1998)
10. Bello, I., Pham, H., Le, Q.V., Norouzi, M., Bengio, S.: Neural combinatorial optimization with reinforcement learning. In: ICLR (2017)
11. Bengio, Y., Lodi, A., Prouvost, A.: Machine learning for combinatorial optimization: a methodological tour d'horizon. CoRR abs/1811.06128 (2018)
12. Bertsimas, D., Dunn, J.: Optimal classification trees. Mach. Learn. **106**(7), 1039–1082 (2017)
13. Bienstock, D., Muñoz, G., Pokutta, S.: Principled deep neural network training through linear programming. CoRR abs/1810.03218 (2018)
14. Bonami, P., Lodi, A., Zarpellon, G.: Learning a classification of mixed-integer quadratic programming problems. In: CPAIOR (2018)
15. Cappart, Q., Goutierre, E., Bergman, D., Rousseau, L.M.: Improving optimization bounds using machine learning: decision diagrams meet deep reinforcement learning. In: AAAI (2019)
16. Cheng, C., Nührenberg, G., Ruess, H.: Maximum resilience of artificial neural networks. In: ATVA (2017)

17. Ciresan, D., Meier, U., Masci, J., Schmidhuber, J.: Multi column deep neural network for traffic sign classification. Neural Netw. **32**, 333–338 (2012)
18. Courbariaux, M., Hubara, I., Soudry, D., El-Yaniv, R., Bengio, Y.: Binarized neural networks: training deep neural networks with weights and activations constrained to +1 or -1. In: NeurIPS (2016)
19. Cybenko, G.: Approximation by superpositions of a sigmoidal function. Mathematics of Control, Signals and Systems (1989)
20. Dai, H., Khalil, E.B., Zhang, Y., Dilkina, B., Song, L.: Learning combinatorial optimization algorithms over graphs. In: NeurIPS (2017)
21. Demirović, E., et al.: An investigation into prediction + optimisation for the knapsack problem. In: CPAIOR (2019)
22. Denton, E., Zaremba, W., Bruna, J., LeCun, Y., Fergus, R.: Exploiting linear structure within convolutional networks for efficient evaluation. In: NeurIPS (2014)
23. Deudon, M., Cournut, P., Lacoste, A., Adulyasak, Y., Rousseau, L.M.: Learning heuristics for the TSP by policy gradient. In: CPAIOR (2018)
24. Ding, J.Y., et al.: Accelerating primal solution findings for mixed integer programs based on solution prediction. CoRR abs/1906.09575 (2019)
25. Donti, P., Amos, B., Kolter, Z.: Task-based end-to-end model learning in stochastic optimization. In: NeurIPS (2017)
26. Dubey, A., Chatterjee, M., Ahuja, N.: Coreset-based neural network compression. In: Ferrari, V., Hebert, M., Sminchisescu, C., Weiss, Y. (eds.) ECCV 2018. LNCS, vol. 11211, pp. 469–486. Springer, Cham (2018). https://doi.org/10.1007/978-3-030-01234-2_28
27. Dutta, S., Jha, S., Sankaranarayanan, S., Tiwari, A.: Output range analysis for deep feedforward networks. In: NFM (2018)
28. Elmachtoub, A., Grigas, P.: Smart predict, then optimize. CoRR abs/1710.08005 (2017)
29. Ferber, A., Wilder, B., Dilkina, B., Tambe, M.: MIPaaL: mixed integer program as a layer. In: AAAI (2020)
30. Fischetti, M., Lodi, A., Zarpellon, G.: Learning MILP resolution outcomes before reaching time-limit. In: CPAIOR (2019)
31. Fischetti, M., Jo, J.: Deep neural networks and mixed integer linear optimization. Constraints (2018)
32. Frankle, J., Carbin, M.: The lottery ticket hypothesis: Finding sparse, trainable neural networks. In: ICLR (2019)
33. Galassi, A., Lombardi, M., Mello, P., Milano, M.: Model agnostic solution of CSPs via deep learning: a preliminary study. In: CPAIOR (2018)
34. Gambella, C., Ghaddar, B., Naoum-Sawaya, J.: Optimization models for machine learning: a survey. CoRR abs/1901.05331 (2019)
35. Glorot, X., Bordes, A., Bengio, Y.: Deep sparse rectifier neural networks. In: AISTATS (2011)
36. Goodfellow, I., Warde-Farley, D., Mirza, M., Courville, A., Bengio, Y.: Maxout networks. In: ICML (2013)
37. Gurobi Optimization, L.: Gurobi optimizer reference manual (2018). http://www.gurobi.com
38. Hahnloser, R., Sarpeshkar, R., Mahowald, M., Douglas, R., Seung, S.: Digital selection and analogue amplification coexist in a cortex-inspired silicon circuit. Nature **405**, 947–951 (2000)
39. Han, S., et al.: DSD: regularizing deep neural networks with dense-sparse-dense training flow. arXiv preprint arXiv:1607.04381 (2016)

40. Han, S., Pool, J., Tran, J., Dally, W.: Learning both weights and connections for efficient neural network. In: NeurIPS (2015)
41. Hanin, B., Rolnick, D.: Complexity of linear regions in deep networks. In: ICML (2019)
42. Hanin, B., Rolnick, D.: Deep relu networks have surprisingly few activation patterns. In: NeurIPS (2019)
43. He, K., Zhang, X., Ren, S., Sun, J.: Deep residual learning for image recognition. In: CVPR (2016)
44. Herrmann, C., Bowen, R., Zabih, R.: Deep networks with probabilistic gates. CoRR abs/1812.04180 (2018)
45. Hinton, G., et al.: Deep neural networks for acoustic modeling in speech recognition. IEEE Sig. Process. Mag. **29**, 82–97 (2012)
46. Hornik, K., Stinchcombe, M., White, H.: Multilayer feed-forward networks are universal approximators. Neural Net. **2**(5), 359–366 (1989)
47. Hottung, A., Tanaka, S., Tierney, K.: Deep learning assisted heuristic tree search for the container pre-marshalling problem. Comput. Oper. Res. (2020)
48. Howard, A., et al.: Mobilenets: efficient convolutional neural networks for mobile vision applications. arXiv preprint arXiv:1704.04861 (2017)
49. Huang, G., Liu, Z., Maaten, L.V.D., Weinberger, K.: Densely connected convolutional networks. In: CVPR (2017)
50. Hutter, F., Hoos, H.H., Leyton-Brown, K.: Sequential model-based optimization for general algorithm configuration. In: LIOn (2011)
51. Iandola, F., Han, S., Moskewicz, M., Ashraf, K., Dally, W., Keutzer, K.: Squeezenet: alexnet-level accuracy with 50x fewer parameters and $< 0.5$ MB model size. arXiv preprint arXiv:1602.07360 (2016)
52. Icarte, R., Illanes, L., Castro, M., Cire, A., McIlraith, S., Beck, C.: Training binarized neural networks using MIP and CP. In: International Conference on Principles and Practice of Constraint Programming (CP) (2019)
53. Jaderberg, M., Vedaldi, A., Zisserman, A.: Speeding up convolutional neural networks with low rank expansions. In: BMVC (2014)
54. Kadioglu, S., Malitsky, Y., Sellmann, M., Tierney, K.: ISAC – Instance-Specific Algorithm Configuration. In: ECAI (2010)
55. Khalil, E., Bodic, P., Song, L., Nemhauser, G., Dilkina, B.: Learning to branch in mixed integer programming. In: AAAI (2016)
56. Khalil, E., Gupta, A., Dilkina, B.: Combinatorial attacks on binarized neural networks. In: ICLR (2019)
57. Kolmogorov, V., Rother, C.: Minimizing nonsubmodular functions with graph cuts-a review. In: TPAMI (2007)
58. Kotthoff, L.: Algorithm selection for combinatorial search problems: a survey. AI Mag. **35**(3) (2014)
59. Koval, V., Schlesinger, M.: Two-dimensional programming in image analysis problems. USSR Academy of Science, Automatics and Telemechanics (1976)
60. Krizhevsky, A., Sutskever, I., Hinton, G.: Imagenet classification with deep convolutional neural networks. In: NeurIPS (2012)
61. Kruber, M., Lübbecke, M., Parmentier, A.: Learning when to use a decomposition. In: CPAIOR (2017)
62. Kumar, A., Serra, T., Ramalingam, S.: Equivalent and approximate transformations of deep neural networks. arXiv preprint arXiv:1905.11428 (2019)
63. LeCun, Y., Bottou, L., Bengio, Y., Haffner, P.: Gradient-based learning applied to document recognition. Proc. IEEE **86**(11), 2278–2324 (1998)

64. LeCun, Y., Bengio, Y., Hinton, G.: Deep learning. Nature **521**, 436–444 (2015)
65. Li, H., Kadav, A., Durdanovic, I., Samet, H., Graf, H.: Pruning filters for efficient convnets. arXiv preprint arXiv:1608.08710 (2016)
66. Lin, C., Zhong, Z., Wei, W., Yan, J.: Synaptic strength for convolutional neural network. In: NeurIPS (2018)
67. Lin, H., Jegelka, S.: Resnet with one-neuron hidden layers is a universal approximator. In: NeurIPS (2018)
68. Liu, B., Wang, M., Foroosh, H., Tappen, M., Pensky, M.: Sparse convolutional neural networks. In: CVPR (2015)
69. Lodi, A., Zarpellon, G.: On learning and branching: a survey. Top **25**(2), 207–236 (2017)
70. Lombardi, M., Milano, M.: Boosting combinatorial problem modeling with machine learning. In: IJCAI (2018)
71. Lomuscio, A., Maganti, L.: An approach to reachability analysis for feed-forward ReLU neural networks. CoRR abs/1706.07351 (2017)
72. Luo, J.H., Wu, J., Lin, W.: Thinet: A filter level pruning method for deep neural network compression. In: ICCV (2017)
73. Mhaskar, H., Poggio, T.: Function approximation by deep networks. CoRR abs/1905.12882 (2019)
74. Molchanov, P., Tyree, S., Karras, T., Aila, T., Kautz, J.: Pruning convolutional neural networks for resource efficient transfer learning. arXiv preprint arXiv:1611.06440 (2016)
75. Montúfar, G.: Notes on the number of linear regions of deep neural networks. In: SampTA (2017)
76. Montúfar, G., Pascanu, R., Cho, K., Bengio, Y.: On the number of linear regions of deep neural networks. In: NeurIPS (2014)
77. Nair, V., Hinton, G.: Rectified linear units improve restricted boltzmann machines. In: ICML (2010)
78. Narodytska, N., Kasiviswanathan, S., Ryzhyk, L., Sagiv, M., Walsh, T.: Verifying properties of binarized deep neural networks. In: AAAI (2018)
79. Pascanu, R., Montúfar, G., Bengio, Y.: On the number of response regions of deep feedforward networks with piecewise linear activations. In: ICLR (2014)
80. Paszke, A., et al.: Automatic differentiation in pytorch. In: NeurIPS Workshops (2017)
81. Peng, B., Tan, W., Li, Z., Zhang, S., Xie, D., Pu, S.: Extreme network compression via filter group approximation. In: Ferrari, V., Hebert, M., Sminchisescu, C., Weiss, Y. (eds.) ECCV 2018. LNCS, vol. 11212, pp. 307–323. Springer, Cham (2018). https://doi.org/10.1007/978-3-030-01237-3_19
82. Raghu, M., Poole, B., Kleinberg, J., Ganguli, S., Dickstein, J.: On the expressive power of deep neural networks. In: ICML (2017)
83. Rastegari, M., Ordonez, V., Redmon, J., Farhadi, A.: XNOR-Net: Imagenet classification using binary convolutional neural networks. In: Leibe, B., Matas, J., Sebe, N., Welling, M. (eds.) ECCV 2016. LNCS, vol. 9908, pp. 525–542. Springer, Cham (2016). https://doi.org/10.1007/978-3-319-46493-0_32
84. Ryu, M., Chow, Y., Anderson, R., Tjandraatmadja, C., Boutilier, C.: CAQL: Continuous action Q-learning. CoRR abs/1909.12397 (2019)
85. Say, B., Wu, G., Zhou, Y.Q., Sanner, S.: Nonlinear hybrid planning with deep net learned transition models and mixed-integer linear programming. In: IJCAI (2017)
86. Serra, T., Ramalingam, S.: Empirical bounds on linear regions of deep rectifier networks. In: AAAI (2020)

87. Serra, T., Tjandraatmadja, C., Ramalingam, S.: Bounding and counting linear regions of deep neural networks. In: ICML (2018)
88. Serra, T.: On defining design patterns to generalize and leverage automated constraint solving (2012)
89. Singh, G., Gehr, T., Püschel, M., Vechev, M.: Robustness certification with refinement. In: ICLR (2019)
90. Sutskever, I., Vinyals, O., Le, Q.: Sequence to sequence learning with neural networks. In: NeurIPS (2014)
91. Szegedy, C., Liu, W., Jia, Y., Sermanet, P., Reed, S., Anguelov, D., Erhan, D., Vanhoucke, V., Rabinovich, A.: Going deeper with convolutions. In: CVPR (2015)
92. Tan, Y., Delong, A., Terekhov, D.: Deep inverse optimization. In: CPAIOR (2019)
93. Tang, Y., Agrawal, S., Faenza, Y.: Reinforcement learning for integer programming: learning to cut. CoRR abs/1906.04859 (2019)
94. Tang, Z., Peng, X., Li, K., Metaxas, D.: Towards efficient u-nets: a coupled and quantized approach. In: TPAMI (2019)
95. Telgarsky, M.: Benefits of depth in neural networks. In: COLT (2016)
96. Tjeng, V., Xiao, K., Tedrake, R.: Evaluating robustness of neural networks with mixed integer programming. In: ICLR (2019)
97. Tung, F., Mori, G.: Clip-q: Deep network compression learning by in-parallel pruning-quantization. In: CVPR (2018)
98. Veit, A., Belongie, S.: Convolutional networks with adaptive computation graphs. CoRR abs/1711.11503 (2017)
99. Wainwright, M., Jaakkola, T., Willsky, A.: Map estimation via agreement on (hyper)trees: Message-passing and linear-programming approaches. IEEE Trans. Inf. Theory 51(11), 3697–3717 (2005)
100. Wainwright, M., Jaakkola, T., Willsky, A.: Tree consistency and bounds on the performance of the max-product algorithm and its generalizations. Stat. Comput. 14, 143–166 (2004). https://doi.org/10.1023/B:STCO.0000021412.33763.d5
101. Wang, W., Sun, Y., Eriksson, B., Wang, W., Aggarwal, V.: Wide compression: tensor ring nets. In: CVPR (2018)
102. Wen, W., Wu, C., Wang, Y., Chen, Y., Li, H.: Learning structured sparsity in deep neural networks. In: NeurIPS (2016)
103. Werner, T.: A linear programming approach to max-sum problem: a review. Technical Report CTU-CMP-2005-25, Center for Machine Perception (2005)
104. Wong, E., Kolter, J.Z.: Provable defenses against adversarial examples via the convex outer adversarial polytope. In: ICML (2018)
105. Wu, J., Leng, C., Wang, Y., Hu, Q., Cheng, J.: Quantized convolutional neural networks for mobile devices. In: CVPR (2016)
106. Xiao, K., Tjeng, V., Shafiullah, N., Madry, A.: Training for faster adversarial robustness verification via inducing ReLU stability. ICLR (2019)
107. Xu, H., Koenig, S., Kumar, T.S.: Towards effective deep learning for constraint satisfaction problems. In: CP (2018)
108. Xue, Y., van Hoeve, W.J.: Embedding decision diagrams into generative adversarial networks. In: CPAIOR (2019)
109. Ye, Z., Say, B., Sanner, S.: Symbolic bucket elimination for piecewise continuous constrained optimization. In: CPAIOR (2018)
110. Yu, R., et al.: NISP: pruning networks using neuron importance score propagation. In: CVPR (2018)
111. Yu, X., Yu, Z., Ramalingam, S.: Learning strict identity mappings in deep residual networks. In: CVPR (2018)

112. Zhang, X., Zou, J., Ming, X., He, K., Sun, J.: Efficient and accurate approximations of nonlinear convolutional networks. In: CVPR (2015)
113. Zhao, C., Ni, B., Zhang, J., Zhao, Q., Zhang, W., Tian, Q.: Variational convolutional neural network pruning. In: CVPR (2019)
114. Zhou, H., Alvarez, J.M., Porikli, F.: Less is more: towards compact CNNs. In: Leibe, B., Matas, J., Sebe, N., Welling, M. (eds.) ECCV 2016. LNCS, vol. 9908, pp. 662–677. Springer, Cham (2016). https://doi.org/10.1007/978-3-319-46493-0_40

# CP and Hybrid Models for Two-Stage Batching and Scheduling

Tanya Y. Tang$^{(\boxtimes)}$ and J. Christopher Beck

Department of Mechanical and Industrial Engineering,
University of Toronto, Toronto, ON M5S 3G8, Canada
{tytang,jcb}@mie.utoronto.ca

**Abstract.** Batch scheduling is a common problem faced in industrial scheduling when groups of related jobs must be processed consecutively or simultaneously on the same resource. Motivated by the composites manufacturing industry, we present a complex batch scheduling problem combining two-stage bin packing with hybrid flowshop scheduling. We propose five solution approaches: a constraint programming model, a three-phase logic-based Benders decomposition model, an earliest due date heuristic, and two hybrid heuristic/constraint programming approaches. We then computationally test these approaches on generated problem instances modelled on real-world instances. Numeric results show that the heuristic approaches perform as well as or better than the exact models, especially on large instances. The relative success of a simple heuristic suggests that such problems pose an interesting challenge for further research in mathematical and constraint programming.

## 1 Introduction

Batch scheduling arises when it is desirable that a set of jobs that share common characteristics are processed together either consecutively or simultaneously on the same resource. Many motivating examples for batch scheduling come from semiconductor manufacturing, where batching can be modelled as a one-stage bin packing problem [12]. In contrast, we study a two-stage batching and scheduling problem motivated by composites manufacturing.

In the first stage, called layup, multiple parts are created by layering alternating sheets of raw materials and epoxy resin in a mould tool. The mould tool with its sheets of materials is then cured in the second stage by a high-temperature and pressure autoclave oven. The extreme conditions of the autoclave compress and heat the material sheets to create parts made of a new composite material. The parts are then delivered to downstream machines for further processing. Because of the high expense incurred by the autoclave stage, it is desirable to group multiple tools and cure them together as an autoclave batch. This hierarchical two-stage process requires parts, hereby referred to as jobs, to be batched onto mould tools and then the tools themselves to be batched in the autoclave.

The one-stage bin packing and scheduling problem, as appears in semiconductor production, is well-studied for both exact [3,10] and heuristic [11,14]

© Springer Nature Switzerland AG 2020
E. Hebrard and N. Musliu (Eds.): CPAIOR 2020, LNCS 12296, pp. 431–446, 2020.
https://doi.org/10.1007/978-3-030-58942-4_28

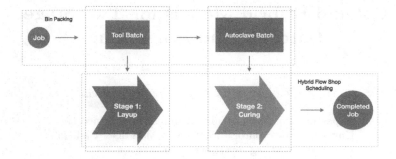

**Fig. 1.** Overview of the flow through the 2BPHFSP.

methods. Existing literature on scheduling for composites manufacturing [1,7] does not consider batching jobs onto tools, and hence is also one-stage. The presence of tool batches changes the batching aspect into a *Two-Stage Bin Packing Problem* [5]. In addition, the two-stage scheduling aspect of this problem is a *Two-Stage Hybrid Flow Shop Scheduling Problem* [6]. The actual composites manufacturing process is complex and multi-layered; in this paper, we study an abstraction that captures its core complexities. We denote this abstracted problem as the *Two-Stage Bin Packing and Hybrid Flow Shop Scheduling Problem* (2BPHFSP), and formally define it in the next section.

While our motivation comes from an industrial problem, and we will use the terms of composite manufacturing in our problem description, the problem we study abstracts away the application-specific details. Our goal is to investigate the core complexities of the hybridization of hierarchical bin packing with scheduling. Figure 1 provides an overview of the 2BPHFSP.

## 2    Problem Definition

In the layup stage, jobs are grouped onto mould tools. Each tool has a one-dimensional capacity that cannot be exceeded. We will refer to these groups of jobs on tools as tool batches. Each job has a layup processing time; as the jobs are laid up sequentially, the processing time to assemble a tool batch is the sum of its job processing times. The layup stage is thus an example of the *Family Batch Scheduling* model [13]. There are multiple unary capacity machines at stage 1; the process to form a tool batch requires one such machine.

After a tool batch is created in stage 1, it is cured in an autoclave in stage 2. Autoclaves are capable of curing more than one tool at a time, so multiple tool batches can be aggregated into an autoclave batch. Autoclaves are constant-capacity and identical, therefore, all autoclave batches have the same capacity and the sum of tool volumes within each autoclave batch cannot exceed that capacity. Similar to stage 1, there are multiple parallel identical autoclave machines; each machine is capable of processing one autoclave batch at a time. However, each autoclave batch is processed for the same length of time regardless of the contained tool batches. Therefore, autoclave batches fall under the

*Batching Machines* model [13]. Lastly, there is an upper limit on the time a laid-up tool batch can wait before entering an autoclave due to the epoxy that is layered between the material sheets. We will refer to this time limit as the restricted waiting time.

Now, let us formally present the parameters that define the 2BPHFSP. We are given a set of jobs $\mathcal{J}$, a set of empty tool batches $\mathcal{B}^1$, and a set of empty autoclave batches $\mathcal{B}^2$. There exists three sets of resources: an unlimited set of tools $\mathcal{T}$, a set of unary-capacity stage 1 machines $\mathcal{M}^1$, and a set of single-batch-capacity stage 2 machines $\mathcal{M}^2$. Each job $j \in \mathcal{J}$ is associated with a one-dimensional volume parameter $s_j$, a due date $d_j$, and a layup processing time $p_j$, and each tool $t \in \mathcal{T}$ is associated with a one-dimensional volume parameter $v_t$. The volume of a tool batch $k \in \mathcal{B}^1$ is therefore the volume of its assigned tool. Each job $j \in \mathcal{J}$ needs to be assigned to a tool batch $k \in \mathcal{B}^1$, and each tool batch $k \in \mathcal{B}^1$ needs to be assigned to an autoclave batch $i \in \mathcal{B}^2$. Each tool batch $k \in \mathcal{B}^1$ is scheduled on one machine from $\mathcal{M}^1$ for a length of time equal to $\sum_{j \in k} p_j$, and each autoclave batch is scheduled on one machine from $\mathcal{M}^2$ for a length of time equal to $P^2$. There are precedence constraints and a restricted waiting time between the end of tool batches and the start of their assigned autoclave batches.

The upper bound on the number of tool batches and autoclave batches is the number of jobs (i.e. each job assigned to its own tool batch and each tool batch assigned to its own autoclave batch). Thus, we create $|\mathcal{J}|$ empty batches in $\mathcal{B}^1$ and $\mathcal{B}^2$. However, we almost always find solutions that use fewer batches, so we denote any non-empty batch as *open*.

The objective of the 2BPHFSP is to minimize a weighted sum of the number of open autoclave batches and job tardiness. We minimize the number of open autoclave batches to decrease autoclave operational costs and we minimize job tardiness to ensure parts are delivered to downstream machines on time.

## 3    Mathematical Programming Approaches

We developed two mathematical programming approaches to solve the 2BPHFSP, a monolithic constraint programming (CP) and a hybrid CP/mixed integer programming (MIP) logic-based Benders decomposition (LBBD).[1]

### 3.1    Constraint Programming

This formulation uses two sets of interval decision variables for bin packing and two sets of interval decision variables for scheduling. The variables and parameters are shown in Table 1.

The first set of bin packing variables are optional two-indexed interval variables, $x_{j,k}$. We assign job $j$ to a tool batch $k$ by enforcing one variable from

---

[1] We also developed a monolithic MIP model, which we do not report on here as it was unable to solve even the smallest instances in our experiments. The main bottleneck was the model size as time-indexed variables were used to handle the scheduling decisions.

**Table 1.** CP model variables and parameters.

| Variable | Description |
|---|---|
| $x_{j,k}$ | Interval variable for job $j$ in tool batch $k$ |
| $y_{k,i}$ | Interval variable for tool batch $k$ in autoclave batch $i$ |
| $\gamma_k$ | Interval variable for tool batch $k$ |
| $\sigma_i$ | Interval variable for autoclave batch $i$ |
| $t_k$ | Tool assigned to tool batch $k$ |
| $\tau_k$ | Volume of tool assigned to tool batch $k$ |
| $l_j$ | Tardiness of job $j$ |

| Parameter | Description | Parameter | Description |
|---|---|---|---|
| $j \in \mathcal{J}$ | Set of jobs | $k \in \mathcal{B}^1$ | Set of tool batches |
| $i \in \mathcal{B}^2$ | Set of autoclave batches | $t \in \mathcal{T}$ | Set of tools |
| $s_j$ | Volume of job $j$ | $p_j$ | Processing time of job $j$ |
| $d_j$ | Due date of job $j$ | $v_t$ | Volume of tool $t$ |
| $V^1$ | Max tool batch capacity | $V^2$ | Autoclave batch capacity |
| $P^2$ | Autoclave processing time | $C^1$ | Number of stage 1 machines |
| $C^2$ | Number of stage 2 machines | $W$ | Restricted waiting time |
| $\alpha_j$ | Weighting for tardiness of job $j$ in the objective function | $\beta$ | Weighting for an open autoclave batch in the objective function |

$\{x_{j,k}; k \in \mathcal{B}^1\}$ to be present for each job $j$. The same concept is applied to the second set of bin packing variables, $y_{k,i}$, where each present variable indicates tool batch $k$ is assigned to autoclave batch $i$. We use two sets of single-index interval variables for scheduling batches on the horizon $\mathcal{H}$, $\gamma_k$ for tool batches and $\sigma_i$ for autoclave batches. We also defined a set of variables $t_k$ that assigns tool batch $k$ to a specific tool so we can calculate the volume of $k$, $\tau_k$.

We present this model in two sections, focusing on bin packing and scheduling, respectively.

$$\min \quad \beta \sum_{i \in \mathcal{B}^2} \mathtt{presenceOf}(\sigma_i) + \sum_{j \in \mathcal{J}} \alpha_j l_j \tag{1}$$

$$\text{s.t.} \quad \sum_{k \in \mathcal{B}^1} \mathtt{presenceOf}(x_{j,k}) = 1 \qquad \forall j \in \mathcal{J} \tag{2}$$

$$\sum_{i \in \mathcal{B}^2} \mathtt{presenceOf}(y_{k,i}) = \mathtt{presenceOf}(\gamma_k) \qquad \forall k \in \mathcal{B}^1 \tag{3}$$

$$\mathtt{presenceOf}(\gamma_k) \geq \mathtt{presenceOf}(\gamma_{k+1}) \qquad \forall k \in \{0, ..., |\mathcal{B}^1| - 1\} \tag{4}$$

$$\mathtt{presenceOf}(\sigma_i) \geq \mathtt{presenceOf}(\sigma_{i+1}) \qquad \forall i \in \{0, ..., |\mathcal{B}^2| - 1\} \tag{5}$$

$$\tau_k = \mathtt{element}(\{v_t; t \in \mathcal{T}\}, t_k) \qquad \forall k \in \mathcal{B}^1 \tag{6}$$

$$\sum_{j \in \mathcal{J}} s_j \times \mathtt{presenceOf}(x_{j,k}) \leq \tau_k \qquad \forall k \in \mathcal{B}^1 \tag{7}$$

$$\sum_{k \in \mathcal{B}^1} \tau_k \times \mathtt{presenceOf}(y_{k,i}) \leq V^2 \qquad \forall i \in \mathcal{B}^2 \tag{8}$$

$$t_k \in \{0, ...\mathcal{T}\}; \tau_k \in \{0, ...V^1\}$$

Constraint (2) assigns each job to one tool batch. Constraint (3) assigns each open tool batch to one autoclave batch. Constraints (4) and (5) enforce that batches are opened in order of index, to reduce symmetry. Constraint (6) is an element constraint that assigns the tool capacity associated with tool $t_k \in \mathcal{T}$ to $\tau_k$.[2] Constraint (7) makes sure the sum of job volumes in each tool batch is less than its tool volume and constraint (8) makes sure the sum of tool volumes in each autoclave batch is less than the autoclave capacity.

The standard global *pack* constraint does not allow the size of an item to be variable, and therefore cannot be used for two-stage bin packing. Thus, we choose to use interval variables for the packing stage to easily connect them with the corresponding scheduling variables via existing global constraints.

Next, we present the scheduling constraints.

$$\mathtt{sizeOf}(x_{j,k}) = p_j \qquad\qquad \forall j \in \mathcal{J} \quad \forall k \in \mathcal{B}^1 \quad (9)$$

$$\mathtt{span}\left(\gamma_k, \{x_{j,k}; j \in \mathcal{J}\}\right) \qquad\qquad \forall k \in \mathcal{B}^1 \quad (10)$$

$$\mathtt{sizeOf}(\gamma_k) = \sum_{j \in \mathcal{J}} p_j \times \mathtt{presenceOf}(x_{j,k}) \qquad\qquad \forall k \in \mathcal{B}^1 \quad (11)$$

$$\mathtt{sizeOf}(y_{k,i}) = P^2 \qquad\qquad \forall k \in \mathcal{B}^1 \quad \forall i \in \mathcal{B}^2 \quad (12)$$

$$\mathtt{synchronize}(\sigma_i, \{y_{k,i}; i \in \mathcal{B}^2\}) \qquad\qquad \forall i \in \mathcal{B}^2 \quad (13)$$

$$\mathtt{alwaysIn}\left(\sum_{k \in \mathcal{B}^1} \mathtt{pulse}(\gamma_k, 1), 0, |\mathcal{H}|, 0, C^1\right) \qquad\qquad (14)$$

$$\mathtt{alwaysIn}\left(\sum_{i \in \mathcal{B}^2} \mathtt{pulse}(\sigma_i, 1), 0, |\mathcal{H}|, 0, C^2\right) \qquad\qquad (15)$$

$$\mathtt{endBeforeStart}(\gamma_k, y_{k,i}) \qquad\qquad \forall k \in \mathcal{B}^1 \quad \forall i \in \mathcal{B}^2 \quad (16)$$

$$\mathtt{startBeforeEnd}(y_{k,i}, \gamma_k, -W) \qquad\qquad \forall k \in \mathcal{B}^1 \quad \forall i \in \mathcal{B}^2 \quad (17)$$

$$l_j \geq \mathtt{endOf}(\sigma_i) - d_j \qquad\qquad \forall j \in \mathcal{J} | \ j \text{ assigned to } i \quad (18)$$

$$l_j \in \{0, ... |\mathcal{H}|\}$$

Constraint (9) sets the processing time of each job. Constraints (10) and (11) make sure jobs in a tool batch are processed sequentially without overlapping. The actual sequence of jobs in a tool batch does not matter because their processing times are not sequence-dependent and jobs cannot leave the first stage until the entire tool batch is processed. Thus, a span constraint is sufficient. Constraint (12) sets the processing time of each tool batch. Constraint (13) makes sure all tool batches in the same autoclave batch are processed simultaneously. Constraints (14) and (15) enforce that the total number of tool batches or autoclave batches processed at one time is at most the respective stage's capacity. Constraint (16) says that each tool batch must finish processing before its associated autoclave batch can begin processing. Constraint (17) restricts the waiting

---

[2] An alternative approach is to create tool batches with predefined tools. However, this approach expands the number of possible tool batches from $|\mathcal{J}|$ to $|\mathcal{J}| \times |\mathcal{T}|$.

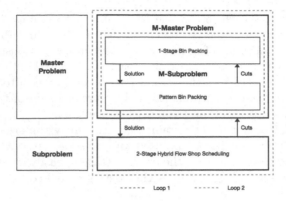

**Fig. 2.** Decomposition flow between problems, each iteration of loop 2 finds a feasible solution to the entire problem.

time between tool batches and their associated autoclave batches. Constraint (18) defines job tardiness.

## 3.2   Logic-Based Benders Decomposition

As the 2BPHFSP is clearly composed of packing and scheduling problems, we are motivated to investigate a decomposition approach. Thus, in this section, we present a three-stage decomposition, shown in Fig. 2, that separates the 2BPHFSP into a one-stage bin packing problem, a pattern bin packing problem, and a two-stage hybrid flow shop scheduling problem, denoted as the *m-master problem*, the *m-subproblem*, and the *subproblem*, respectively. Each of these problems is much smaller and simpler to solve than the 2BPHFSP in its entirety, so a decomposition approach may scale better than a monolithic approach.

The two loops shown in Fig. 2 show the order in which the problems are solved. The m-master problem is a one-stage bin packing problem which packs jobs into capacity-constrained autoclave batches, while minimizing the number of open autoclave batches. Autoclave batches have a constant capacity, so the sum of job volumes in each batch must be less than the capacity. We chose to assign jobs to autoclave batches directly because part of the objective is to minimize the number of open autoclave batches and the job-to-autoclave batching is a relaxation of the two-stage batching requirement.

A feasible packing of jobs in an autoclave batch may not be feasibly partitionable into tool batches as jobs may not fit perfectly on tools. The m-subproblem attempts to find such a feasible partitioning for each autoclave batch. We denote this constrained partitioning problem as the *Pattern Bin Packing Problem*. If we find a feasible packing for all autoclave batches, we schedule the tool and autoclave batches in the subproblem. If no feasible packing exists for an autoclave batch, a cut is added to the m-master problem.

**Table 2.** M-Master problem variables.

| Variable | Description |
|---|---|
| $x_j$ | Autoclave batch containing job $j$ |
| $\theta_i$ | Volume of jobs in autoclave batch $i$ |
| $q$ | Number of open autoclave batches |

**Table 3.** M-subproblem variables.

| Variable | Description |
|---|---|
| $y_{j,j'}$ | Binary variable, 1 if job $j$ is in tool batch defined by job $j'$, 0 otherwise |
| $z_{j',t}$ | Binary variable, 1 if tool batch defined by job $j'$ is on tool $t$, 0 otherwise |
| $\tau_{j'}$ | Volume of tool for the tool batch defined by job $j'$ |
| $\rho_{j'}$ | Processing time of tool batch define by job $j'$ |

*M-Master Problem.* The m-master problem was modelled using CP. The objective is to minimize the number of open autoclave batches. The variables in the m-master problem are shown in Table 2; see Table 1 for parameter descriptions.

$$\min \quad q \tag{19}$$
$$\text{s.t.} \quad \texttt{pack}\left(\{\theta_i; i \in \mathcal{B}^2\}, \{x_j; j \in \mathcal{J}\}, \{s_j; j \in \mathcal{J}\}\right) \tag{20}$$
$$q = \texttt{max}\left(\{x_j; j \in \mathcal{J}\}\right) \tag{21}$$
$$x_j \in \{0, ... |\mathcal{B}^2|\}; \ \theta_i \in \{0, ... V^2\}; \ q \in \{0, ... |\mathcal{B}^2|\}$$

Constraint (20) is a global constraint which packs jobs in set $\mathcal{J}$ into autoclave batches in set $\mathcal{B}^2$ while keeping the sum of job volumes in each autoclave batch below the autoclave capacity. Constraint (21) defines the number of open bins.

*M-Subproblem.* Inspired by approaches to related problems [4,9], we modelled the m-subproblem using MIP with two sets of binary decision variables. The first set creates tool batches from jobs and the second set assigns each tool batch to a tool. The objective function minimizes the total sum of tool volumes.

Consider a tool batch $k$ that contains the set of jobs $\mathcal{J}^*$, let $j'$ be the job with the lowest index in $\mathcal{J}^*$. We denote job $j'$ as the representative job of tool batch $k$ and create decision variables $y_{j,j'}$ that assign jobs to representative jobs instead of directly to tool batches. If job $j$ is assigned to the tool batch with representative job $j'$, then $y_{j,j'} = 1$, otherwise, $y_{j,j'} = 0$. Therefore, each $y_{j',j'} = 1$ indicates an open tool batch. Only jobs in the same autoclave batch can be assigned to the same tool batch, so we define a $|\mathcal{J}| \times |\mathcal{J}|$ matrix, where entry $A_{j,j'}$ is equal to 1 if job $j$ and job $j'$ are in the same autoclave batch in the m-master problem solution, and equal to 0 otherwise. The variables in the m-master problem are presented in Table 3; see Table 1 for parameter descriptions.

$$\min \quad \sum_{j' \in \mathcal{J}} \tau_{j'} \tag{22}$$

$$\text{s.t.} \quad \sum_{j' \in \mathcal{J}} y_{j,j'} = 1 \qquad \forall j \in \mathcal{J} \tag{23}$$

$$y_{j,j'} \leq A_{j,j'} \qquad \forall j, j' \in \mathcal{J} \tag{24}$$

$$y_{j,j'} = 0 \qquad \forall j, j' \in \mathcal{J} : j < j' \tag{25}$$

$$y_{j,j'} \leq y_{j',j'} \qquad \forall j, j' \in \mathcal{J} \tag{26}$$

$$\sum_{t \in \mathcal{T}} z_{j',t} = y_{j',j'} \qquad \forall j' \in \mathcal{J} \tag{27}$$

$$\tau_{j'} = \sum_{t \in \mathcal{T}} v_t z_{j',t} \qquad \forall j' \in \mathcal{J} \tag{28}$$

$$\tau_{j'} \geq \sum_{j \in \mathcal{J}} s_j y_{j,j'} - \sum_{j \in \mathcal{J}} s_j (1 - y_{j',j'}) \qquad \forall j' \in \mathcal{J} \tag{29}$$

$$\rho_{j'} = \sum_{j \in \mathcal{J}} p_j y_{j,j'} \qquad \forall j' \in \mathcal{J} \tag{30}$$

$$y_{j,j'} \in \{0,1\}; \ z_{j',t} \in \{0,1\}; \ \tau_{j'} \in \{0,...V^1\}; \ \rho_{j'} \in \{0,...,\textstyle\sum_{j \in \mathcal{J}} p_j\}$$

Constraint (23) makes sure each job is assigned to one tool batch. Constraints (24) and (25) enforce assignment restrictions. If job $j'$ defines a tool batch, it is forced by constraint (25) to be the lowest indexed job in that tool batch, so constraint (26) tightens the linear relaxation. Each open tool batch is associated with a job $j'$, and only jobs with a higher index than $j'$ can be assigned to the tool batch associated with $j'$. Constraint (27) makes sure each open tool batch is assigned to a specific tool. Constraints (28) and (29) define and enforce the tool volume for each tool batch to be larger than the sum of job volumes. Constraint (30) defines the processing time for each tool batch.

*Feasibility Cuts.* Once the m-subproblem is solved to optimality, if the sum of tool volumes for autoclave batch $i^*$ is larger than the autoclave capacity then $i^*$ is an infeasible batch. Cuts are added to prevent the subset of jobs in each infeasible autoclave batch from being packed together again. Let $\mathcal{B}^2$ be the set of infeasible autoclave batches from the incumbent m-master problem solution, and let $\mathcal{J}^{i^*}$ be the set of jobs assigned to batch $i^* \in \hat{\mathcal{B}}^2$.

These cuts are written as a global cardinality constraint (GCC) for each infeasible autoclave batch. The standard GCC format is $\text{gcc}(\{cards\}, \{vals\}, \{vars\})$, where sets $\{cards\}$ and $\{vals\}$ are the same size. Over the set of variables in $\{vars\}$, each value $val[i]$ should only be taken $cards[i]$ times; $cards[i]$ can be a single value or a range of values. Using GCC cuts also removes symmetrical solutions arising from batch indexing. Constraint (31) forms the feasibility cut.

$$\text{gcc}\left(\{[0,...,|\mathcal{J}^{i^*}| - 1], ...\}, \{1, ..., |\mathcal{B}^2|\}, \{x_j; j \in \mathcal{J}^{i^*}\}\right) \qquad \forall i^* \in \hat{\mathcal{B}}^2 \tag{31}$$

**Table 4.** Subproblem variables and parameters.

| Variable | Description | Parameter | Description |
|---|---|---|---|
| $\gamma_k$ | Interval variable for tool batch $k$ | $k \in \mathcal{B}^{1*}$ | Set of open tool batches |
| $\sigma_i$ | Interval variable for autoclave batch $i$ | $i \in \mathcal{B}^{2*}$ | Set of open autoclave batches |
| $l_j$ | Tardiness of job $j$ | $(k,i) \in \mathcal{E}^*$ | Associated tool and autoclave batches |
| | | $(j,i) \in \mathcal{F}^*$ | Associated jobs and autoclave batches |
| | | $\rho_k^*$ | Processing time of tool batch $k$ |

*Subproblem.* The subproblem is modelled using CP and has two sets of interval decision variables, one to schedule tool batches and one to schedule autoclave batches. The objective is to minimize the sum of job tardiness. The subproblem variables and model-specific parameters are presented in Table 4; see Table 1 for other parameter descriptions.

$$\min \quad \sum_{j \in \mathcal{J}} \alpha_j l_j \tag{32}$$

$$\text{s.t.} \quad \texttt{sizeOf}(\gamma_k) = \rho_k^* \qquad \forall k \in \mathcal{B}^{1*} \tag{33}$$

$$\texttt{sizeOf}(\sigma_i) = P^2 \qquad \forall i \in \mathcal{B}^{2*} \tag{34}$$

$$\texttt{endBeforeStart}\,(\gamma_k, \sigma_i) \qquad \forall (k,i) \in \mathcal{E}^* \tag{35}$$

$$\texttt{startBeforeEnd}\,(\sigma_i, \gamma_k, -W) \qquad \forall (k,i) \in \mathcal{E}^* \tag{36}$$

$$\texttt{alwaysIn}\left( \sum_{k \in \mathcal{B}^1} \texttt{pulse}(\gamma_k, 1), 0, |\mathcal{H}|, 0, C^1 \right) \tag{37}$$

$$\texttt{alwaysIn}\left( \sum_{i \in \mathcal{B}^2} \texttt{pulse}(\sigma_i, 1), 0, |\mathcal{H}|, 0, C^2 \right) \tag{38}$$

$$l_j = \texttt{endOf}(\sigma_i) - d_j \qquad \forall (j,i) \in \mathcal{F}^* \tag{39}$$

$$l_j \in \{0, ... |\mathcal{H}|\}$$

Constraints (33) and (34) define the interval variable lengths for tool and autoclave batches, respectively. Constraint (35) makes sure any autoclave batch starts after all its associated tool batches have finished processing. Constraint (36) restricts the waiting time between stages. Constraints (37) and (38) enforce resource capacities for each stage. Constraint (39) defines job tardiness.

*Optimality Cuts.* Once the subproblem is solved after a loop 2 cycle, a feasible solution to the entire problem has been found. Thus, to start the cycle again, the incumbent solution must be removed from the search space. Due to the fact that only one autoclave batch needs to be changed to cut off the incumbent solution, we need to add a disjunction to require at least one autoclave batch to be different. GCC constraints cannot be disjoined in the solver we are using [8]. Thus, we introduce a new set of binary variables, $\omega_{i^*} \in \{0,1\}$, where $i^* \in \mathcal{B}^{2*}$ is the set of open autoclave batches.

The binary variable $\omega_{i^*}$ is set to be equal to 1 if autoclave batch $i^*$ is required to be different in subsequent solutions, and equal to 0 otherwise. If autoclave batch $i^*$ is forced to be different, then constraints must be added such that only a strict subset of jobs in that batch can be assigned to the same autoclave batch. We add a set of counting constraints that limit how many times each autoclave batch index can be assigned in the set of decision variables belonging to jobs in that batch. Then, a final constraint needs to be added to enforce that at least one binary variable out of the set $\{\omega_{i^*}; i^* \in \mathcal{B}^{2^*}\}$ is equal to 1, forcing the incumbent solution to change. Constraints (40) and (41) form the optimality cuts.

$$(\omega_{i^*} == 1) \rightarrow \left(\text{count}(\{x_j; j \in \mathcal{J}^{i^*}\}, i) \leq |\mathcal{J}^{i^*}| - 1\right) \quad i \in \mathcal{B}^2 \quad i^* \in \mathcal{B}^{2^*} \quad (40)$$

$$\sum_{i^* \in \mathcal{B}^{2^*}} \omega_{i^*} \geq 1 \tag{41}$$

*Theoretical Results.* The proof of the validity of the m-subproblem cut is straightforward and, following Chu and Xia [2], can be sketched as follows. Given a solution with an infeasible autoclave batch $i^*$:

1. Cut (31) removes $i^*$ from the search space. Suppose we have an infeasible autoclave batch $i^*$ containing the set of jobs $\mathcal{J}^{i^*}$. The decision variables associated with jobs in $i^*$, $\{x_j; j \in \mathcal{J}^{i^*}\}$ take on values ranging from $\{1, ..., |\mathcal{B}^2|\}$, where $|\mathcal{B}^2|$ is the total number of autoclave batches available. If $\{x_j; j \in \mathcal{J}^{i^*}\}$ are assigned to the same autoclave batch, then in any solution where these jobs are grouped into an autoclave batch one of the values from $\{1, ..., |\mathcal{B}^2|\}$ appears $|\mathcal{J}^{i^*}|$ times. Cut (31) explicitly removes all such solutions.
2. Cut (31) does not remove any globally optimally solutions from the search space. Suppose we have a globally optimal solution $\mathcal{S}$ to the 2BPHFSP containing autoclave batch $i'$, which contains the same jobs as infeasible autoclave batch $i^*$. As $i^*$ is infeasible, the minimum sum of tool volumes that can contain all the jobs in $i^*$ is larger than the autoclave machine capacity. The m-subproblem finds the set of tools with minimal volume that can hold all jobs within each autoclave batch. Since $i'$ contains the same jobs as $i^*$, it holds that $i'$ is also infeasible. Therefore, $\mathcal{S}$ is infeasible and cannot be a globally optimal solutions.

Similarly, the proof of the validity of the subproblem cut is sketched as follows. Given a feasible solution, cuts (40) to (41) are sufficient to remove the incumbent solution and all symmetrical solutions from the search space. Suppose the incumbent solution has $|\mathcal{B}^{2^*}|$ open autoclave batches, where each autoclave batch $i \in \{1, ..., |\mathcal{B}^{2^*}|\}$ is associated with binary variable $\omega_i$. If $\omega_{i^*} = 1$, autoclave batch $i^*$ cannot appear in any subsequent solutions. Let the sum of $\omega_i$ over all $i \in \mathcal{B}^{2^*}$ be equal to $\pi$. If $\pi = 0$, then all $\omega_i = 0$ and no autoclave batch is enforced to be different, which would lead to the algorithm finding the same solution as the incumbent. Note that it is possible that exactly one autoclave batch from the incumbent solution can be changed to form a new distinct solution. Therefore, we forbid at least one autoclave batch from appearing again to remove the incumbent solution from the search space.

# 4  Heuristic Approaches

In addition to the CP and MIP models, we developed a greedy heuristic based on earliest due date (EDD) and hybridizations of the heuristic with CP.

## 4.1  Earliest Due Date Heuristic

The pseudocode for the heuristic is shown in Algorithm 1. See Table 1 for parameter descriptions.

This heuristic algorithm flows through three distinct sections. The first section orders jobs by due date then greedily assigns them one by one to the first available tool batch. If no tool batch is available, then a new tool batch is opened and randomly assigned to a tool. Next, all open tool batches are sorted by due date and the algorithm greedily assigns them one by one to the first available autoclave batch. If no autoclave batch is available, a new batch is opened. Lastly, the autoclave batches are sorted by due date. For each autoclave batch, the algorithm assigns each of its associated tool batches to start as soon as possible on the first available stage 1 machine. Then, the autoclave batch itself is assigned to start, after all its associated tool batches have ended, on the first available stage 2 machine.[3]

We also incorporated a simple randomization technique where each time an item is picked from a sorted list, i.e. picking the next job or tool batch to be batched or picking the next autoclave batch to schedule, a random item from the first $L$ items in the list is selected. This method allows us to apply the algorithm on an instance multiple times and use the best solution found.

## 4.2  Hybrid Heuristic Approaches

Once EDD finds a feasible solution to the 2BPHFSP, we can try to improve the solution using CP. Thus, we formulated two hybrid heuristic-CP approaches. First, the heuristic solution can be used to warm-start the monolithic CP model, and second, packing decisions from the heuristic solution can be fixed and scheduling decisions can be improved using the subproblem from LBBD. These two approaches will be referred to as WCP and EDD-CP, respectively.

# 5  Numerical Results

Our five approaches were tested on 11 sets of 30 randomly generated problem instances, ranging from 5 jobs to 100 jobs per instance. All CP and MIP models were implemented in Java using CPLEX Optimization Studio 12.9. Each approach was given a sixty minute total time limit to solve each instance. For the LBBD model, each component was given a ten minute time limit to prevent the model from timing out globally within a single component.

---

[3] Note that EDD is not guaranteed to find a feasible solution even if one exists. For example, if the restricted waiting time constraints are too tight some search may be required to find a solution.

---

**Algorithm 1:** EDD heuristic

---

**for** *each iteration* **do**
    sort $\mathcal{J}$ by increasing due date; // `pack in tool batches`
    **while** $\mathcal{J}$ *not empty* **do**
        $j \leftarrow$ randomly selected job from first $L$ items of $\mathcal{J}$;
        **if** $\mathcal{B}^1$ *empty or j does not fit in any open batch* **then**
            create new tool batch $k$ and add to $\mathcal{B}^1$, assign $j$ to $k$;
            select a $t$ from $\mathcal{T}$ with larger volume than $j$ and assign to $k$;
        **else**
            assign $j$ to first $k \in \mathcal{B}^1$ with enough space;
        **end**
    **end**
    sort $\mathcal{B}^1$ by increasing due date; // `pack in autoclave batches`
    **while** $\mathcal{B}^1$ *not empty* **do**
        $k \leftarrow$ randomly selected tool batch from first $L$ items of $\mathcal{B}^1$;
        **if** $\mathcal{B}^2$ *empty or k does not fit in any open batch* **then**
            create new autoclave batch $i$ and add to $\mathcal{B}^2$, assign $k$ to $i$;
        **else**
             assign $k$ to first $i \in \mathcal{B}^2$ with enough space;
        **end**
    **end**
    sort $\mathcal{B}^2$ by increasing due date;
    **while** $\mathcal{B}^2$ *not empty* **do**
        $i \leftarrow$ randomly selected autoclave batch from first $L$ items of $\mathcal{B}^2$;
        sort tool batches in $i$ by decreasing processing time; // `schedule batches`
        **for** $k \in$ *sorted tool batches of i* **do**
            $m_1 \leftarrow$ `GetNextFreeMachine`($\mathcal{M}^1$);
            schedule $k$ next on $m_1$;
        **end**
        $m_2 \leftarrow$ `GetNextFreeMachine`($\mathcal{M}^2$);
        schedule $i$ next on $m_2$ after end of all $k$ in $i$;
    **end**
    add solution to solution list;
**end**
pick best solution from solution list;

---

*Overall Performance.* Figure 3 shows the number of instances for which each approach was able to find feasible solution(s) and the average time to find the best solution within one hour. The monolithic CP model is clearly the worst performing model: it was not able to solve any instances with more than 25 jobs. The LBBD model starts to show signs of degraded performance at 50 jobs per instance, and can only solve around 50% of instances with 100 jobs. These two results imply that we will not be able to use these non-heuristic methods to solve real-world instances, which contain at least 1000 jobs. EDD, EDD-CP, and WCP were able to find feasible solutions for all instances. Both CP and WCP are very unlikely to scale, but LBBD's mediocre performance still warrants some exploration with the full problem. EDD and EDD-CP show strong potential, but Fig. 3b indicates that EDD-CP does not use much more time than EDD, meaning we can solely pursue EDD-CP.

*Solution Quality.* We compared solution qualities of each approach by calculating optimality gaps using equation (42), where $z(n)$ is the objective value found for instance $n$ and $z(\text{RSP}(n))$ is a lower bound for instance $n$. The CP and LBBD

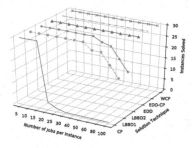

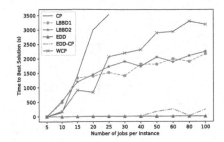

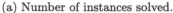

(a) Number of instances solved.

(b) Average time to best solution within one hour.

**Fig. 3.** Performance of all solution techniques.

models did not find feasible solutions for some instances; the optimality gap for any unsolved instance $n$ is not included in the associated averages in Fig. 4.

$$\text{Optimality Gap of Approach } n = \frac{z(n) - z(\text{RSP}(n))}{z(\text{RSP}(n))} \tag{42}$$

The lower bound is obtained by solving the *Relaxed Scheduling Problem* (RSP). The RSP is a version of the LBBD subproblem with all batching removed: each job is processed once in the first stage and once in the second stage. We model the first stage as a multi-capacity machine and each job occupies one unit of capacity during processing. The second stage is also a multi-capacity machine, but each job occupies $s_j$ units of space during processing and the second stage machine has a constant capacity of $C^2 \times V^2$. There are precedence constraints between the stages of each job. The RSP is a relaxation of the 2BPHFSP, so the optimal objective of the RSP is a lower bound on the 2BPHFSP. There is a small subset of instances where the RSP is unable to prove optimality within one hour. These instances are not included in the averages shown in Fig. 4. This lower bound is very weak and does not serve any purpose beside providing a comparison basis for solution qualities across the different approaches.

Figure 4 shows that, other than CP, the techniques do not show a large range of solution qualities when they are all able to find feasible solutions. From 15 to 60 jobs per instance, EDD and EDD-CP find consistently worse solutions than LBBD and WCP, both of which can change packing decisions to find better schedules. However, the four techniques start to converge at 80 jobs per instance. Given the censoring of the data for instances which LBBD could not find a feasible solution, the true performance of the heuristic techniques is likely better than LBBD for the 100-job instances. We can conclude that heuristic techniques have comparable or better performance than the sophisticated mathematical models for large problem instances.

*Hybrid Heuristic Approaches.* We can calculate how much improvement CP was able to give to a heuristic solution in the two hybrid heuristic approaches by

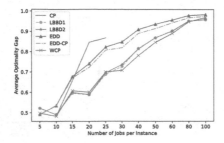

**Fig. 4.** Comparing the optimality gaps of all solution techniques tested.

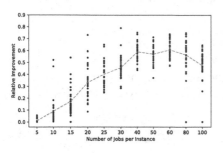

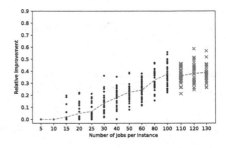

(a) Warm-started CP model.

(b) EDD packing plus CP scheduling.

**Fig. 5.** Relative decrease in objective value after using CP to improve the heuristic solution; average decrease is shown by the dotted line. The blue data points show results from larger instances that WCP was not able to solve. (Color figure online)

using equation (43), where $z(\text{EDD})$ is the heuristically found solution objective value and $z(\text{CP})$ is the improved objective value.

$$\text{Relative Decrease} = \frac{z(\text{EDD}) - z(\text{CP})}{z(\text{EDD})} \tag{43}$$

Figure 5 allows us to see the convergence trend shown by Fig. 4 more clearly. WCP can make large improvements to the EDD solution up to 60 jobs per instance, where its performance starts to degrade, whereas EDD-CP shows gradually improving performance from 5 to 100 jobs per instance. We generated larger problem instances of size 110, 120, and 130 to further test WCP and EDD-CP. Figure 5b shows that EDD-CP's performance plateaus after 100 jobs per instance. WCP was not able to solve any problems larger than 100 jobs per instance due to high memory usage. We can see that EDD-CP is clearly the most robust and promising approach to solving the full problem.

## 6    Conclusions and Next Steps

We introduced the 2BPHFSP and developed three primary solution approaches, constraint programming (CP), logic-based Benders decomposition (LBBD), and an earliest due date heuristic (EDD). Each approach was tested on 11 sets of 30 instances, with each set containing 5 to 100 jobs per instance. CP was able to solve small instances, but scaled poorly. LBBD scaled better than CP but did not find feasible solutions for some instances with 50 or more jobs. EDD was able to find a feasible solution within seconds for every instance and had comparable quality to LBBD solutions for larger instances.

Two hybrid heuristic-CP approaches were tested on the same sets of problem instances. First, the EDD solution was used to warm-start the CP model (WCP), and second, the packing from the EDD solution was fixed and the scheduling subproblem from LBBD was used to improve the EDD solution schedule (EDD-CP). Both methods were able to find feasible solutions for every instance. WCP performed better within the range of problem instances tested, but showed signs of decreasing performance at around 50 to 60 jobs per instance. EDD-CP has consistent performance up to 100 jobs per instance, implying scalability.

Overall, the most interesting conclusion is the strong performance of heuristic techniques compared to the mathematical formulations. The complexity of the problem lends itself to CP, but scaling is an issue for complete approaches. Future work includes considering some complexities of the original problem that were abstracted away, and increasing instance sizes to match the real world. The positive heuristic results point us towards focusing future work on developing meta-heuristic, constraint-based local search, and hybrid heuristic-CP approaches. Ultimately, the success shown by EDD and the hybrid heuristic-CP approaches suggests that complex problems, such as two-stage hierarchical batching and scheduling, pose a challenge for the development of robust and efficient exact methods.

**Acknowledgements.** This research was supported by the Natural Sciences and Engineering Research Council of Canada and Visual Thinking International Ltd (Visual8).

# References

1. Azami, A., Demirli, K., Bhuiyan, N.: Scheduling in aerospace composite manufacturing systems: a two-stage hybrid flow shop problem. Int. J. Adv. Manuf. Technol. **95**, 3259–3274 (2018)
2. Chu, Y., Xia, Q.: A hybrid algorithm for a class of resource constrained scheduling problems. In: Barták, R., Milano, M. (eds.) CPAIOR 2005. LNCS, vol. 3524, pp. 110–124. Springer, Heidelberg (2005). https://doi.org/10.1007/11493853_10
3. Dupont, L., Dhaenens-Flipo, C.: Minimizing the makespan on a batch machine with non-dentical job sizes: an exact procedure. Comput. Oper. Res. **29**, 807–819 (2002)
4. Emde, S., Polten, L., Gendreau, M.: Logic-based benders decomposition for scheduling a batching machine. Discussion paper, CIRRELT, Montreal, Canada (2018)
5. Gilmore, P.C., Gomory, R.E.: Multistage cutting stock problems of two and more dimensions. Oper. Res. **13**, 94–120 (1965)
6. Gupta, J.N.D.: Two-stage, hybrid flowshop scheduling problem. J. Oper. Res. Soc. **39**, 359–364 (1988)

7. Hueber, C., Fischer, G., Schwingshandl, N., Schledjewski, R.: Production planning optimization for composite aerospace manufacturing. Int. J. Prod. Res. **57**, 5857–5873 (2018)
8. IBM: Class IloDistribute, see note. https://www.ibm.com/support/knowledgecenter/SSSA5P_12.9.0/ilog.odms.cpo.help/refcppcpoptimizer/html/classes/IloDistribute.html. Accessed 11 Sept 2019
9. Kosch, S., Beck, J.C.: A new MIP model for parallel-batch scheduling with non-identical job sizes. In: Simonis, H. (ed.) CPAIOR 2014. LNCS, vol. 8451, pp. 55–70. Springer, Cham (2014). https://doi.org/10.1007/978-3-319-07046-9_5
10. Malapert, A., Gueret, C., Rousseau, L.M.: A constraint programming approach for a batch processing problem with non-identical job sizes. Eur. J. Oper. Res. **221**, 533–545 (2012)
11. Melouk, S., Damodaran, P., Chang, P.Y.: Minimizing makespan for single machine batch processing with non-identical job sizes using simulated annealing. Int. J. Prod. Econ. **87**, 141–147 (2004)
12. Monch, L., Fowler, J.W., Dauzere-Peres, S., Mason, S.J., Rose, O.: Scheduling semiconductor manufacturing operations: problems, solution techniques, and future challenges. J. Sched. **14**, 583–599 (2011)
13. Potts, C.N., Kovalyov, M.Y.: Scheduling with batching: a review. Eur. J. Oper. Res. **120**, 228–249 (2000)
14. Uzsoy, R.: Scheduling a single batch processing machine with non-identical job sizes. Int. J. Prod. Res. **32**, 1615–1635 (1994)

# Improving a Branch-and-Bound Approach for the Degree-Constrained Minimum Spanning Tree Problem with LKH

Maximilian Thiessen[1,2(✉)], Luis Quesada[3], and Kenneth N. Brown[3]

[1] Department of Computer Science, University of Bonn, Bonn, Germany
thiessen@cs.uni-bonn.de
[2] Fraunhofer IAIS, Sankt Augustin, Germany
[3] Insight Centre for Data Analytics, School of Computer Science and IT,
University College Cork, Cork, Ireland
{luis.quesada,ken.brown}@insight-centre.org

**Abstract.** The degree-constrained minimum spanning tree problem, which involves finding a minimum spanning tree of a given graph with upper bounds on the vertex degrees, has found multiple applications in several domains. In this paper, we propose a novel CP approach to tackle this problem where we extend a recent branch-and-bound approach with an adaptation of the LKH local search heuristic to deal with trees instead of tours. Every time a solution is found, it is locally optimised by our new heuristic, thus yielding a tightened cut. Our experimental evaluation shows that this significantly speeds up the branch-and-bound search and hence closes the performance gap to the state-of-the-art bottom-up CP approach.

**Keywords:** Degree-constrained minimum spanning tree · Branch-and-bound · Local search · LKH

## 1 Introduction

The degree constrained minimum spanning tree problem (DCMSTP) involves finding a minimum spanning tree (MST) of a given graph where the degree of every vertex is bounded. Minimum spanning trees are commonly used in the design of protocols for wireless sensor networks [30]. Having upper bounds on the degree of the vertices of a tree is a very common constraint in the design of such protocols due to many factors (e.g., bounded number of radios per vertex, limited capacity to store routing tables, etc.) [14].

The data collection process (convergecasting) in wireless sensor networks is commonly accomplished by using a routing tree between the sensors [1,5,20],

This work is supported by the Chist-ERA project Dyposit with funding from IRC, by Science Foundation Ireland under Grant No. 12/RC/2289 P2 and 16/SP/3804, by the EU and Enterprise Ireland under grant IR 2017 0041/ EU 737422-2 SCOTT, and by the German Academic Scholarship Foundation (Studienstiftung).

© Springer Nature Switzerland AG 2020
E. Hebrard and N. Musliu (Eds.): CPAIOR 2020, LNCS 12296, pp. 447–456, 2020.
https://doi.org/10.1007/978-3-030-58942-4_29

mostly due to time and energy efficiency reasons. The authors of [5] showed that the network topology is one of the main efficiency bottlenecks and significantly improved the practical performance by enforcing degree constraints on the vertices of the routing tree. Additionally, degree constraints are crucial in situations with battery-driven sensors in inaccessible terrains [20]. Similarly, these constraints can play an important role in the diversion of the flow to avoid interdicted links that result from cyber attacks in cyber-physical networks [25].

Another area of application is Software Defined Networks (SDNs). An SDN attempts to centralise network intelligence in one network component by disassociating the forwarding process of network packets (data plane) from the routing process (control plane). The control plane consists of one or more controllers which are considered as the brain of the SDN where the whole intelligence is incorporated [8]. The OpenFlow protocol is an open standard and is the main and most widespread enabling technology of the SDN architecture. An Open-Flow switch is equipped with a Forwarding Information Base (FIB) table, storing matching rules for the incoming packets, one or more actions (e.g., forward to a port, drop the packet, or modify its header) and counters. If an incoming packet matches a rule in the FIB, the corresponding action is taken and the counters are updated [26]. Reducing the energy impact of SDNs is an important challenge nowadays. Researchers have proposed protocols based on MST to address this challenge [23,26]. The limited space for storing FIB tables seems to motivate the degree constraint on the vertices naturally.

DCMSTP subsumes the path version of the Traveling Salesman Problem (TSP), where one is interested in finding a Hamiltonian path of minimum cost. If we set the degree bound to 2 for each vertex, the Hamiltonian path problem reduces to the DCMSTP, and thus it is NP-hard [13]. While several CP approaches have managed to push the state-of-the-art of the TSP by primarily taking advantage of relaxations of the problem [2,10–12,17], we are only aware of one CP approach that has managed to do the same for DCMSTP [6].

A common feature in the applications we have mentioned is that the user is not necessarily interested in finding an optimal solution. In reality, the user is much more concerned about the time spent in the computation of the solution and is usually satisfied with a solution that is close to the optimal one. For instance, consider the case where a wireless network has to be restored after a link failure to a, possibly non-optimal, acceptable working state as fast as possible and only afterwards make further adjustments to save costs. This is certainly an issue with the recent bottom-up approach proposed in [6] since it only produces a satisfiable solution (the optimal one) at the end of the search process, besides the typically bad initial one. Branch-and-bound approaches do not have this drawback but suffer from poor performance due to the lack of good upper bounds.

In this paper, we propose a novel CP approach to tackle the DCMSTP where we extend a branch-and-bound approach with a local search heuristic inspired by LKH [16], which is the most widely used heuristic method for TSP, to combine the benefits of a branch-and-bound approach with an acceptable runtime. We apply the heuristic to every found solution to improve its cut value, which not only skips a lot of intermediate solutions, but also strengthens the filtering of all used propagators. Our experiments show that our approach is competitive with respect to the state-of-the-art when it comes to CP [6] and is preferable when it comes to finding close-to-optimal solutions.

## 2   Background

Let $G = (V, E)$ be an undirected graph with integer edge costs $c : E \rightarrow \mathbb{Z}$ and a degree upper bound for each vertex, given by the mapping $d : V \rightarrow \mathbb{N}$. For all vertices $v \in V$ let $\delta(v)$ denote the set of incident edges of $v$. We want to find a spanning tree, satisfying the degree constraints for all vertices $v \in V$: $|\delta(v)| \leq d(v)$, of minimum total cost, which is the sum of all edge costs in the tree.

A typical CP formulation is:

$$
\begin{align}
\text{Minimize:} & \quad Z & (1) \\
\text{s.t.} & \quad \text{WST}(G, Z, c) & (2) \\
& \quad |\delta(v)| = D_v \leq d(v) \qquad \forall v \in V & (3)
\end{align}
$$

It consists of a graph variable $G$, an integer variable $Z$ representing an upper bound on the total cost, and integer degree variables $D$. A graph variable is an abstraction over the edge set of the graph, where the lower bound forms the set of already fixed (mandatory) edges and the upper bound is a superset of the lower bound. The difference between the upper bound and the lower bound is the set of optional edges [4]. The weighted spanning tree constraint $\text{WST}(G, Z, w)$ (2) forces $G$ to contain a spanning tree of cost at most $Z$ [28]. Additionally, we have the degree upper bound constraints (3). Due to the minimization $Z$ will become the total cost of a spanning tree eventually.

The current state-of-the-art CP approach [6] combines (2) and (3) into the powerful $\text{DCWST}(G, Z, w, D)$ constraint. It is a direct generalization of the WST constraint and uses an adapted sub-gradient method of [15] to provide a lower bound on the total cost, typically much tighter than a normal minimum spanning tree would yield. On top of this model, the authors of [6] additionally adapted the Last-Conflict search strategy [18] to graph variables, which significantly improves the performance.

As a search procedure, they suggest to first greedily find any feasible solution, which is needed for the DCWST constraint, and continue with a simple bottom-up approach. They fix the total cost variable $Z$ to a lower bound obtained via the pruning performed by the DCWST constraint and then branch on the graph variable. If a feasible solution is found it is guaranteed to be optimal. Otherwise, the lower bound is increased by one and the process is repeated.

Due to the lack of a good upper bounding heuristic on the cost, a basic branch-and-bound search is typically much slower than the bottom-up approach, as can be seen in our experiments.

Therefore, we suggest an improved branch-and-bound approach, which tackles this problem. Our ideas are inspired by the local search technique $k$-Opt [21] for the TSP and especially its popular generalization the Lin-Kernighan algorithm [22]. $k$-Opt has recently been applied to the DCMSTP [19], but using an incomplete local search approach, and without using LKH. At each step, they test all possible k-Opt swaps, including those which would change the vertex degree. In contrast, our approach is complete, and motivated by LKH, we check only a subset of possible moves.

The idea of $k$-Opt is to iteratively perform improving $k$-Opt swaps, which consist of swapping $k$ edges of the current spanning tree (or tour in the TSP case) with $k$ new ones to reduce the total cost of the graph until no more swaps can be found. As the original $k$-Opt swaps for tours, we consider only moves that do not change the vertex degrees. The case $k = 2$ is particularly simple because for any spanning tree there is only one correct way to reconnect two removed edges without violating the connectivity of the tree and changing the vertex degrees (see Fig. 1):

**Proposition 1.** *Two non-incident edges $\{v, w\}$ and $\{x, y\}$ of a tree can be exchanged with the edges $\{v, y\}$ and $\{w, x\}$ without violating the tree property of the graph and changing the vertex degrees if and only if the (unique) connecting path between $v$ and $x$ is not using $w$ or $y$.*

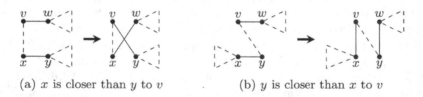

(a) $x$ is closer than $y$ to $v$          (b) $y$ is closer than $x$ to $v$

**Fig. 1.** If we remove two edges $\{v, w\}$ and $\{x, y\}$ from a tree, either (a) $\{v, y\}$, $\{w, x\}$ or (b) $\{v, x\}$, $\{w, y\}$ can be used to reconnect the tree without changing the vertex degrees. Depending on whether $x$ or $y$ is closer to $v$ in the tree without using $w$ determines the valid swap. The dashed parts represent the rest of the tree.

Indeed the tree property is maintained as no cycle is introduced with the addition of edges $\{v, y\}$ and $\{w, x\}$ since these edges are connecting the disconnected components that were created by the removal of the edges $\{v, w\}$ and $\{x, y\}$. Similarly, the degrees are unmodified as the only vertices affected by the change are $v$, $y$, $w$ and $x$, which trivially maintain their original degrees.

The quality of this local optimisation is strongly influenced by $k$. Typically, $k = 2$ is not enough and therefore $k \geq 3$ is used to produce satisfactory results. However, this is also the main drawback of the $k$-Opt approach, since to perform one single edge swap the algorithm has to check all possible edge sets of size $k$, resulting in a runtime of $\mathcal{O}(|V|^k)$, which is too slow for typical real-world graphs and an appropriate $k$.

This problem motivates the Lin-Kernighan algorithm [22] for the TSP unifying $k$-Opt swaps by dynamically choosing $k$ during runtime. The main idea is to start with 2-Opt swaps and only consider larger ones if needed. The currently available implementation of the Lin-Kernighan algorithm [16], called LKH, is one of the best-performing methods to heuristically solve the TSP in practice (see e.g. [24, 29] for recent developments).

One of the main difference to $k$-Opt is that LKH focuses on a certain subset of possible moves called *sequential* $k$-Opt swaps, being a $k$-Opt swap, which can be decomposed into $k - 1$ many 2-Opt swaps, applied one after another each sharing an edge with the previous one. Additionally, the set of potential edge candidates is typically restricted to heuristically promising ones, for example only edges connecting nearest neighbours (see [16] for more details).

Motivated by the impressive results for the TSP, we propose to adapt LKH to spanning trees and use it on all found solutions of a branch-and-bound CP approach, in the sense of [9], to tighten the cuts, while still keeping the search complete.

## 3   Improving Solutions with LKH

As it is not possible to apply the state-of-the-art LKH implementation [16] to our problem due to its inherent design for tours, we decided to re-implement a barebones version of it using the mentioned sequential $k$-Opt swaps and the nearest-neighbour heuristic to use it directly in our CP setting. To further simplify the implementation we set an upper bound on the largest allowed $k$.

Since sequential $k$-Opt swaps can be decomposed into 2-Opt swaps (see [16]) we have to only find an efficient way to perform 2-Opt swaps. The main challenge here is to decide which one of the two possible swapping moves (see Fig. 1) has to be performed to keep connectivity and the vertex degrees. Using Proposition 1 the problem simplifies to checking the vertex distances in the tree. Given a first fixed edge $\{v, w\}$ of an edge pair candidate $\{v, w\}$ and $\{x, y\}$, the question is whether $x$ or $y$ is closer to $v$ (see Fig. 2).

**Fig. 2.** Fixing the edge $\{v, w\}$ as the first part of an edge pair determines the correct way of swapping it together with any non-incident second edge. For all the (dashed) edges, which are reachable from $v$ without using $w$, the correct way of swapping is to connect $w$ with the vertex $x$ reached first and $v$ to the vertex $y$ after it. For all the (dotted) edges reachable from $w$ without using $v$, the opposite swap has to be performed using $\{w, y'\}$ and $\{v, x'\}$, connecting $v$ to the vertex $x'$ reached first.

To efficiently perform this check, we can use any graph traversal algorithm (like BFS or DFS) starting from $v$ without visiting $w$. By Proposition 1 every traversed edge $\{x, y\}$, where $x$ is reached before $y$, can be swapped using $\{v, y\}$ and $\{w, x\}$. The same procedure for $w$ is reversed, resulting in the opposite swap $\{v, x\}$ and $\{w, y\}$. In total, this yields a linear runtime to identify all valid swaps for a fixed edge.

As mentioned above, instead of trying all these edge pairs we only perform promising ones using the following nearest-neighbour heuristic. We first sort the neighbours of each vertex by increasing distance, which can be performed once at the beginning of the branch-and-bound approach. Then we only apply an edge swap, if one of the new edges, say $\{v, x\}$, connects two *close* vertices $v$ and $x$, where close means being one of the nearest neighbours, typically restricted to 3–5 neighbours. To prevent checking an edge pair twice, we can enforce $v < x$ with some fixed order on the vertices as a condition for any edge pair candidate. We want to emphasize that this local optimisation does not change the vertex degrees.

To sum up, our proposed approach is the following. Exactly as [6] we start with an initial greedy solution, but then apply a branch-and-bound search. Every time a solution is found, including the greedy one, it is locally optimised with LKH to reduce its total cost, hence yielding a better cut.

This significantly speeds up the branch-and-bound search, as can be seen in the experiments, since on the one hand, we can skip a lot of intermediate solutions and on the other hand, the improved cuts strengthen the filtering of the used propagators.

## 4    Experiments

In the following section, we compare our new approach with the state-of-the-art bottom-up CP approach from [6] and to a branch-and-bound search without our additional local optimisation on two benchmark datasets DE and ANDINST [3]. We implemented our approach in Java 8 on top of the CP library Choco 4.0.6 [27] with Choco-graph 4.2.4 [7] and made it publicly available[1]. All experiments are run on a Debian Linux 9 workstation with an Intel® Xeon® X5675 CPU and 128 GB of total RAM. A time limit of 3 h is used.

In our approach, we set the highest $k$ for any $k$-Opt swap, as well as the number of nearest neighbours to 3. These parameters are a tradeoff between runtime and efficiency of the local optimisation. We determined these empirically, as they seemed to be the most appropriate for our test datasets. Due to space limitations, we omit a discussion about the parameter selection.

Motivated by our applications we also record the time to reach a solution of cost 1% close to the optimum, i.e. smaller than or equal to 1.01 times the optimal value. For the bottom-up approach, this almost always coincides with the overall runtime, because the bottom-up approach does not generate any intermediate solutions, besides the typically bad initial one.

---

[1] https://github.com/mthiessen/CP-LKH-DCMST.

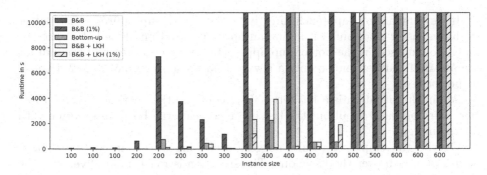

**Fig. 3.** Runtimes on DE instances

In Fig. 3 we depict the runtimes on the DE instances. The striped parts of the branch-and-bound approaches indicate the time to reach a solution value 1% close to optimal. Our adapted approach is much faster than the basic branch-and-bound. We also observe that the time to reach 1% optimality almost coincides with the overall runtime for the basic branch-and-bound, while drops significantly for our adapted approach.

On most of the instances, the runtime of the bottom-up approach is comparable to our approach. In three instances the bottom-up approach is significantly better, but on the other hand on three other instances, our approach performs better. If we look at the runtime to reach 1% the situation improves. On all but two instances our approach finds such solutions faster than the bottom-up approach. This indicates that our approach is preferable over the bottom-up approach in a dynamic environment, such as in our mentioned applications.

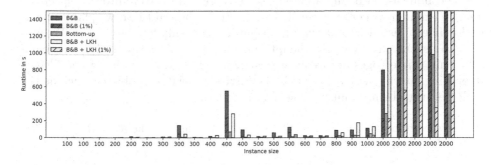

**Fig. 4.** Runtimes on ANDINST instances

On the ANDINST dataset (see Fig. 4), our approach is better than the basic branch-and-bound on most instances. Unfortunately, our approach is slower than the basic branch-and-bound on some larger instances. We are convinced that a more sophisticated implementation of LKH on trees would resolve this issue.

If we compare our approach directly to the bottom-up approach, the latter is always faster than ours in finding an optimum solution. This is because the lower bound used in the bottom-up approach is very close to the optimum for the ANDINST dataset and so only a few bottom-up steps, typically 1–2, have to be performed.

Again the situation is a lot better if we compare the runtimes to reach 1%. Here our approach on all but the last instance beats the bottom-up, often with a large margin.

Overall the empirical results suggest preferring our adapted branch-and-bound approach over the basic one. The situation is much more diverse in the comparison of the bottom-up approach to ours, where most of the time the results are either similar or in favour of the bottom-up approach. Nevertheless, taking the time to reach 1% into account shows the benefits of using our adapted approach.

## 5    Conclusion

We have extended a CP based branch-and-bound approach to the degree constrained minimum spanning tree problem with an adaptation of the LKH local search heuristic. In this adaptation, the heuristic deals with trees instead of tours. The branch-and-bound search is significantly enhanced by this adaptation as shown in the experiments, which makes our approach competitive with the state-of-the-art bottom-up CP approach on some well-known benchmarks. We have discussed and shown empirically that our branch-and-bound approach is preferable in a dynamic environment, where close to optimal solutions are of interest.

This study is a proof-of-concept and indicates the high potential of an adapted branch-and-bound approach since there is a lot to improve. Especially, apart from improving the implementation of the adapted LKH, we plan to focus on Euclidean instances and build propagators especially for these cases. We also plan to generate nogoods and adapt our approach to deal with dynamic environments where edges are frequently changing. Lastly, generalising the idea of improving intermediate solutions by heuristics to related problems, such as the TSP, seems to be a promising research direction.

**Acknowledgement.** The authors thank Keld Helsgaun from the Roskilde University, Denmark for helpful discussions.

## References

1. Annamalai, V., Gupta, S.K.S., Schwiebert, L.: On tree-based convergecasting in wireless sensor networks. In: 2003 IEEE Wireless Communications and Networking, WCNC 2003, vol. 3, pp. 1942–1947, March 2003
2. Benchimol, P., van Hoeve, W.J., Régin, J.-C., Rousseau, L.-M., Rueher, M.: Improved filtering for weighted circuit constraints. Constraints **17**(3), 205–233 (2012)

3. da Cunha, A.S., Lucena, A.: Lower and upper bounds for the degree-constrained minimum spanning tree problem. Netw.: Int. J. **50**(1), 55–66 (2017)
4. Dooms, G., Deville, Y., Dupont, P.: CP(graph): introducing a graph computation domain in constraint programming. In: van Beek, P. (ed.) CP 2005. LNCS, vol. 3709, pp. 211–225. Springer, Heidelberg (2005). https://doi.org/10.1007/11564751_18
5. Durmaz Incel, O., Ghosh, A., Krishnamachari, B., Chintalapudi, K.: Fast data collection in tree-based wireless sensor networks. IEEE Trans. Mobile Comput. **11**(1), 86–99 (2012)
6. Fages, J.-G., Lorca, X., Rousseau, L.-M.: The salesman and the tree: the importance of search in CP. Constraints **21**(2), 145–162 (2014). https://doi.org/10.1007/s10601-014-9178-2
7. Fages, J.-G., Prud'homme, C., Lorca, X.: Choco graph documentation. COSLING S.A.S, IMT Atlantique (2018)
8. Farhadi, H., Lee, H.Y., Nakao, A.: Software-defined networking: a survey. Comput. Netw. **81**, 79–95 (2015)
9. Pesant, G., Gendreau, M.: A view of local search in constraint programming. In: Freuder, E.C. (ed.) CP 1996. LNCS, vol. 1118, pp. 353–366. Springer, Heidelberg (1996). https://doi.org/10.1007/3-540-61551-2_86
10. Focacci, F., Lodi, A., Milano, M.: Cost-based domain filtering. In: Jaffar, J. (ed.) CP 1999. LNCS, vol. 1713, pp. 189–203. Springer, Heidelberg (1999). https://doi.org/10.1007/978-3-540-48085-3_14
11. Focacci, F., Lodi, A., Milano, M.: Embedding relaxations in global constraints for solving TSP and TSPTW. Ann. Math. Artif. Intell. **34**(4), 291–311 (2002)
12. Focacci, F., Lodi, A., Milano, M.: Optimization-oriented global constraints. Constraints **7**(3–4), 351–365 (2002)
13. Garey, M.R., Johnson, D.S.: Computers and Intractability: A Guide to the Theory of NP-Completeness. W. H. Freeman & Co., New York (1979)
14. Gouveia, L., Moura, P.: Spanning trees with node degree dependent costs and knapsack reformulations. Electron. Notes Discrete Math. **36**, 985–992 (2010)
15. Held, M., Karp, R.M.: The traveling-salesman problem and minimum spanning trees: part ii. Math. Program. **1**(1), 6–25 (1971)
16. Helsgaun, K.: An effective implementation of the lin-kernighan traveling salesman heuristic. Eur. J. Oper. Res. **126**(1), 106–130 (2000)
17. Isoart, N., Régin, J.-C.: Integration of structural constraints into TSP models. In: Schiex, T., de Givry, S. (eds.) CP 2019. LNCS, vol. 11802, pp. 284–299. Springer, Cham (2019). https://doi.org/10.1007/978-3-030-30048-7_17
18. Lecoutre, C., Saïs, L., Tabary, S., Vidal, V.: Reasoning from last conflict(s) in constraint programming. Artif. Intell. **173**(18), 1592–1614 (2009)
19. Lee, S.-U.: A degree-constrained minimum spanning tree algorithm using k-opt. J. Korea Soc. Comput. Inf. **20**(5), 31–39 (2015)
20. Liang, C., Huang, Y., Lin, J.: An energy efficient routing scheme in wireless sensor networks. In: 22nd International Conference on Advanced Information Networking and Applications - Workshops (AINA Workshops 2008), pp. 916–921, March 2008
21. Lin, S.: Computer solutions of the traveling salesman problem. Bell Syst. Tech. J. **44**(10), 2245–2269 (1965)
22. Lin, S., Kernighan, B.W.: An effective heuristic algorithm for the traveling-salesman problem. Oper. Res. **21**(2), 498–516 (1973)
23. Liu, H., Li, J., Yang, K.: A dynamic method of constructing MST based on openflow protocol. Int. J. Comput. Commun. Eng. **5**(6), 398–408 (2016)

24. McMenemy, P., Veerapen, N., Adair, J., Ochoa, G.: Rigorous performance analysis of state-of-the-art TSP heuristic solvers. In: Liefooghe, A., Paquete, L. (eds.) EvoCOP 2019. LNCS, vol. 11452, pp. 99–114. Springer, Cham (2019). https://doi.org/10.1007/978-3-030-16711-0_7
25. Perelman, L., Amin, S.: A network interdiction model for analyzing the vulnerability of water distribution systems. In: Proceedings of the 3rd International Conference on High Confidence Networked Systems, HiCoNS 2014, New York, NY, USA, pp. 135–144. ACM (2014)
26. Prete, L., Farina, F., Campanella, M., Biancini, A.: Energy efficient minimum spanning tree in openflow networks. In: 2012 European Workshop on Software Defined Networking, pp. 36–41, October 2012
27. Prud'homme, C., Fages, J.-G., Lorca, X.: Choco Documentation. TASC - LS2N CNRS UMR 6241, COSLING S.A.S. (2017)
28. Régin, J.-C., Rousseau, L.-M., Rueher, M., van Hoeve, W.-J.: The weighted spanning tree constraint revisited. In: Lodi, A., Milano, M., Toth, P. (eds.) CPAIOR 2010. LNCS, vol. 6140, pp. 287–291. Springer, Heidelberg (2010). https://doi.org/10.1007/978-3-642-13520-0_31
29. Tinós, R., Helsgaun, K., Whitley, D.: Efficient recombination in the Lin-Kernighan-Helsgaun traveling salesman heuristic. In: Auger, A., Fonseca, C.M., Lourenço, N., Machado, P., Paquete, L., Whitley, D. (eds.) PPSN 2018. LNCS, vol. 11101, pp. 95–107. Springer, Cham (2018). https://doi.org/10.1007/978-3-319-99253-2_8
30. Vashist, R., Dutt, S.: Minimum spanning tree based improved routing protocol for heterogeneous wireless sensor network. Int. J. Comput. Appl. **103**, 29–33 (2014)

# Insertion Sequence Variables for Hybrid Routing and Scheduling Problems

Charles Thomas[1]($\boxtimes$) (ID), Roger Kameugne[2] (ID), and Pierre Schaus[1] (ID)

[1] UCLouvain, Ottignies-Louvain-la-Neuve, Belgium
{charles.thomas,pierre.schaus}@uclouvain.be
[2] Faculty of Sciences, University of Maroua, Maroua, Cameroon
rkameugne@gmail.com

**Abstract.** The Dial a Ride family of Problems (DARP) consists in routing a fleet of vehicles to satisfy transportation requests with time-windows. This problem is at the frontier between routing and scheduling. The most successful approaches in dealing with DARP are often tailored to specific variants. A generic state-of-the-art constraint programming model consists in using a sequence variable to represent the ordering of visits in a route. We introduce a possible representation for the domain called *Insertion Sequence Variable* that naturally extends the standard subset bound for set variables with an additional insertion operator after any element already sequenced. We describe the important constraints on the sequence variable and their filtering algorithms required to model the classical DARP and one of its variants called the Patient Transportation Problem (PTP). Our experimental results on a large variety of instances show that the proposed approach is competitive with existing sequence based approaches.

## 1 Introduction

Door-to-door transportation services and on demand public transport are increasingly important due to the flexibility it offers to the customers. Two such problems are the Dial a Ride Problem (DARP) [7] and the Patient Transportation Problem (PTP) [3] which consist in transporting a maximum of patients to and from medical appointments. These problems often involve large number of requests and thus require efficient algorithmic solutions. Many approaches have been proposed and applied successfully to different variants of the DARP [11]. However, these solutions are often tailored towards specific use-cases and are difficult to adapt to other variants of the problem. It is thus crucial to develop approaches to model and solve efficiently such problems while remaining generic and easily adaptable.

We describe an *Insertion Sequence Variable* (ISV) for modeling and solving DARPs. Our domain representation includes the subset bound domain [9] for set variables. This allows to represent optional elements in the domain and prevents a repetition of the same element at different positions in the sequence. The set domain is extended with an internal sequence that can be grown with arbitrary

© Springer Nature Switzerland AG 2020
E. Hebrard and N. Musliu (Eds.): CPAIOR 2020, LNCS 12296, pp. 457–474, 2020.
https://doi.org/10.1007/978-3-030-58942-4_30

insertions available from a set of possible insertions. By letting the constraints remove impossible insertions, the search space is pruned by restricting the set of possible sequences. We describe two important global constraints on the Insertion Sequence Variable for modeling the DARP and PTP: 1) The `TransitionTimes` constraint links a sequence variable with time interval variables to take into account a transition time matrix between consecutive elements in the sequence. 2) The `Cumulative` constraint ensures that the load profile does not exceed a fixed capacity when pairs of elements in the sequence represent the load and discharge on a vehicle. We experimentally test the performances of the Insertion Sequence Variable on the two problems and show that it is competitive with the state-of-the-art CP approaches.

## 2    Related Work

In [14], the authors propose a constraint-based approach called *LNS-FFPA* to solve DARPs with a cost objective and show that it outperforms other state-of-the-art approaches. While highly efficient, the LNS-FFPA algorithm is difficult to adapt to other variants of the DARP such as the PTP. Indeed, the approach is not declarative since some constraints are enforced with the search. Furthermore, the model is not able to deal with optional visits that occur in the PTP and similar problems.

Two recent approaches for solving the PTP are [3] and [20]. The approach proposed in [3] consists in representing the problem with a scheduling model where trips are represented by activities. The approach of [20] is based on IBM ILOG CP Optimizer solver [19]. It makes use of the sequence variables from CP Optimizer to decide the order of visits in each vehicle.

The high level functionalities and constraints related to sequence variables of CP Optimizer have been briefly described in [17,18]. Unfortunately, no details are given on the implementation of such variables and the filtering algorithms of the constraints in the literature. According to the API and documentation available at [12,13], the sequence variable of CP Optimizer is based on a Head-Tail Sequence Graph structure. It consists of maintaining separate growing head and tail sub-sequences. Interval variables not yet sequenced can be added either at the end of the head or at the beginning of the tail. When no more interval variable can be added and all members of the head and tail are decided, both sub-sequences are joined to form the final sequence. Google OR-Tools [22] also propose sequence variables [23] with the same approach as CP Optimizer. The approach proposed in this paper differs from the one of CP Optimizer in the following ways: 1) the insertion sequence variable is generic and usable in a large variety of problems. In particular, the variable is independent of the notion of time intervals; 2) insertions are allowed at any point in the sequence which allows flexible modeling and search; 3) the variable proposed keeps track of the possible insertions for each element inside its domain which allows advanced propagation techniques.

In [10], the authors discuss the usage of a path variable in the context of Segment Routing Problems. Their implementation is based on a growing prefix to which candidates elements can be appended.

## 3   Preliminary

Let $X = \{0, ..., n\}$ be a finite set and $\mathcal{P}(X)$ the set of subsets (power set) of $X$. The inclusion $\subseteq$ relation defines a partial order over $\mathcal{P}(X)$ and the structure $(\mathcal{P}(X), \subseteq)$ is a lattice generally used to represent the domain of a finite set variable. To avoid explicit exhaustive enumeration of set domain, three disjoint subsets of $X$ are used to represent the current state of the set domain (see [9]). The domain is defined as $\langle P, R, E \rangle \equiv \{S' \mid S' \subseteq X \wedge R \subseteq S' \subseteq R \cup P\}$ where $P$, $R$, and $E$ denote respectively the set of Possible, Required and Excluded elements of $X$. At any time we have that $P, R$ and $E$ form a partition of $X$. The variable $S$ with domain $\langle P, R, E \rangle$ is bound if $P$ is empty. Table 1 contains the supported operations on a set variable $S$ of domain $\langle P, R, E \rangle$ with their complexity.

**Table 1.** Operations supported by set variables

| Operation | Description | Complexity |
|---|---|---|
| requires$(S, e)$ | move $e$ to $R$, fails if $e \in E$ | $\Theta(1)$ |
| excludes$(S, e)$ | move $e$ to $E$, fails if $e \in R$ | $\Theta(1)$ |
| isBound$(S)$ | return true iff $S$ is bound | $\Theta(1)$ |
| is{Possible/Required/Excluded}$(S, e)$ | return true iff $e \in \{P/R/E\}$ | $\Theta(1)$ |
| all{Possible/Required/Excluded}$(S)$ | enumerate $\{P/R/E\}$ | $\Theta(|\{P/R/E\}|)$ |

We denote by $\overrightarrow{S}$ a *sequence* without duplicates over $X$ ($S \subseteq X$). The sequence $\overrightarrow{S}$ defines an order over the elements of $S$. Each element of $X$ is unique and can appear only once in $S$. The set of all sequences of $X$ is denoted by $\overrightarrow{\mathcal{P}}(X)$. Let $a$ and $b$ be two elements of $S$. The relation $a$ *precedes* $b$ *in* $\overrightarrow{S}$ is noted $a \overset{\overrightarrow{S}}{\prec} b$ or $a \prec b$ when it is clear from the context that the relation applies in regards to $S$. The relation $a$ *directly precedes* $b$ *in* $\overrightarrow{S}$ is noted $a \overset{\overrightarrow{S}}{\longrightarrow} b$ or $a \rightarrow b$ when clear from the context. In this case, $b$ is called the *successor* of $a$ and $a$ is called the *predecessor* of $b$ in $\overrightarrow{S}$. A sequence $\overrightarrow{S}'$ is a *super-sequence* of $\overrightarrow{S}$ if $S \subseteq S'$ and $\forall a, b \in S, a \overset{\overrightarrow{S}}{\prec} b \implies a \overset{\overrightarrow{S}'}{\prec} b$. This relationship is noted $\overrightarrow{S} \subseteq \overrightarrow{S}'$. Conversely, $\overrightarrow{S}$ is a *sub-sequence* of $\overrightarrow{S}'$.

The *insertion* operation $insert(\overrightarrow{S}, e, p)$ consists in inserting the element $e$ in the sequence $\overrightarrow{S}$ after the element $p$ where $e \in X \setminus S$ and $p \in S$. Performing this operation results in a super-sequence $\overrightarrow{S}'$ of $\overrightarrow{S}$ such that $S' = S \cup \{e\}$ and $p \overset{\overrightarrow{S}'}{\longrightarrow} e$. The operation is also noted $\overrightarrow{S} \underset{(e,p)}{\Longrightarrow} \overrightarrow{S}'$.

The insertion of an element $e$ at the beginning of a sequence or in an empty sequence is defined as $insert(\vec{S}, e, \bot)$. An insertion in a sequence $\vec{S}$ is thus characterized by a tuple $(e, p)$ where $e \notin S$ and $p \in S \lor p = \bot$.

Given $I$, a set of tuples, each corresponding to a potential insertion in $\vec{S}$, the one-step derivation $\vec{S} \underset{I}{\Longrightarrow} \vec{S}'$ between a sequence $\vec{S}$ and its super-sequence $S'$ is defined as $\vec{S} \underset{I}{\Longrightarrow} \vec{S}' \iff \exists i = (e, p) \in I \mid \vec{S} \underset{i=(e,p)}{\Longrightarrow} \vec{S}'$. In other words, the sequence $S$ is transformed into $S'$ by applying one possible insertion from $I$. More generally *zero or more steps derivation* is defined as $\vec{S} \underset{I}{\overset{*}{\Longrightarrow}} \vec{S}' \equiv \vec{S} = \vec{S}' \lor \left( \exists i \in I \mid \vec{S} \underset{i}{\Longrightarrow} \vec{S}'' \land \vec{S}'' \underset{I \setminus \{i\}}{\overset{*}{\Longrightarrow}} \vec{S}' \right)$. Note that $I$ may contain tuples that do not correspond to a possible insertion in $\vec{S}$ but instead to a possible insertion in a super-sequence of $\vec{S}$. Also note that several sequences of insertions in $I$ may lead to a same super-sequence.

*Example 1.* Let us consider the sequence $\vec{S} = (1, 2, 3)$ and the set of insertions $I = \{(4, 2), (5, 2), (5, 4)\}$. We have that $\vec{S} \underset{I}{\overset{*}{\Longrightarrow}} \vec{S}' = (1, 2, 4, 5, 3)$ since it can be obtained with consecutive derivations over $I$: $(1, 2, 3) \underset{(4,2)}{\Longrightarrow} (1, 2, 4, 3) \underset{(5,4)}{\Longrightarrow} (1, 2, 4, 5, 3)$.

## 4    Insertion Sequence Variable

**Definition 1.** *An insertion sequence variable $Sq$ on a set $X$ is a variable whose domain is represented by a tuple $\langle \vec{S}, I, P, R, E \rangle$ where $\langle P, R, E \rangle$ is the domain of a set variable on $X$, $\vec{S}$ is a sequence $\in \vec{\mathcal{P}}(R)$ and $I$ is a set of tuples $(e, p)$, each corresponding to a possible insertion. The domain of $Sq$, also noted $D(Sq)$, is defined as*

$$\langle \vec{S}, I, P, R, E \rangle \equiv \left\{ \vec{S}' \in \vec{\mathcal{P}}(P \cup R) \mid R \subseteq S' \land \vec{S} \underset{I}{\overset{*}{\Longrightarrow}} \vec{S}' \right\} \tag{1}$$

$Sq$ is bound if $P$ is empty and $|S| = |R|$. Initially, all elements of the domain are optional ($\in P$). During the search, elements can be set as mandatory or excluded (moved to $R$ or $E$) and possible insertions can be removed from $I$.

**Lemma 1.** *Checking the consistency of the domain $\langle \vec{S}, I, P, R, E \rangle$ is NP-complete.*

*Proof.* It requires verifying the following properties: $\exists \vec{S}' \mid \vec{S} \underset{I}{\overset{*}{\Longrightarrow}} \vec{S}' \land R \subseteq S'$ and $\forall e \in P, \exists S' \mid \vec{S} \underset{I}{\overset{*}{\Longrightarrow}} \vec{S}' \land R \cup \{e\} \subseteq S'$. The Hamiltonian path problem for a directed graph $G = (\mathcal{V}, \mathcal{E})$ can be reduced to checking the consistency of the domain $D(Sq) = \langle \vec{S} = (), I = \mathcal{E}_{reverse} \cup \{(v, \bot) \mid \forall v \in \mathcal{V}\}, P = \emptyset, R = \mathcal{V}, E = \emptyset \rangle$ where $\mathcal{E}_{reverse}$ is the result of applying the *reverse* operation on each edge $(i, j) \in \mathcal{E}$ defined as $(i, j)_{reverse} = (j, i)$. □

Consequently, instead of checking the full domain consistency at each change in the domain, the following invariant is maintained internally by the sequence variable:

$$P \cup R \cup E = X \wedge P \cap R = R \cap E = P \cap E = \phi \qquad (2)$$

$$S \subseteq R \qquad (3)$$

$$\forall (e, p) \in I, e \notin S \wedge e \notin E \wedge p \notin E \qquad (4)$$

$$\forall p \in S, \nexists(e, p) \in I \implies e \in E \qquad (5)$$

At any moment: $P \cup R \cup E$ form a partition of $X$ (2); any member of $\overrightarrow{S}$ is required (3); any member of $\overrightarrow{S}$ cannot be inserted in $\overrightarrow{S}$ ; any excluded element cannot be inserted in $\overrightarrow{S}$ and is not a valid predecessor (4); any element that cannot be inserted at any position in $\overrightarrow{S}$ is excluded (5).

*Example 2.* Let us consider $X = \{a, b, c, d, e, f\}$, the variable $Sq$ of domain $\langle \overrightarrow{S} = (f, b), I = \{(c, \bot), (c, e), (c, f), (e, c), (e, f)\}, P = \{c\}, R = \{b, e, f\}, E = \{a, d\}\rangle$ corresponds to the sequences $\{(f, e, b), (c, f, e, b), (f, c, e, b), (f, e, c, b)\}$. The sequences $\{(f, b), (c, f, b), (f, c, b)\}$ are not valid as they do not contain $e$ which is required.

The insertion sequence variable inherits all the operations defined on the set variable (see Table 1) and supports the additional operations summarized in Table 2.

**Table 2.** Operations supported by insertion sequence variables

| Operation | Description | Complexity |
|---|---|---|
| isBound$(Sq)$ | return true iff $Sq$ is bound | $\Theta(1)$ |
| isMember$(Sq, e)$ | return true iff $e$ is present in $\overrightarrow{S}$ | $\Theta(1)$ |
| allMembers$(Sq)$ | enumerate $\overrightarrow{S}$ | $\Theta(|S|)$ |
| allCurrentInserts$(Sq)$ | enumerate $\{(e, p) \in I \mid p \in S\}$ | $\mathcal{O}(\min(|I|, |S|))$ |
| nextMember$(Sq, e)$ | return the successor of $e$ in $\overrightarrow{S}$ | $\Theta(1)$ |
| insert$(Sq, e, p)$ | insert $e$ in $\overrightarrow{S}$ after $p$, update $P$, $R$ and $I$, fail if $e \in E \vee p \notin S$ | $\Theta(1)$ |
| canInsert$(Sq, e, p)$ | return true iff $(e, p) \in I$ | $\Theta(1)$ |
| allInserts$(Sq)$ | enumerate $I$ | $\Theta(|I|)$ |
| remInsert$(Sq, e, p)$ | remove $(e, p)$ from $I$ | $\Theta(1)$ |

## 4.1   Implementation

The implementation of the internal set variable $\langle P, R, E \rangle$ uses array-based sparse sets as in [24] to ensure efficient update and reversibility during a backtracking depth-first-search. It consists of an array of length $|X|$ called elems and two reversible integers: r and p. The position of the elements of $X$ in elems indicates in which subset the element is. Elements before the position r are part of $R$ while elements starting from position p are part of $E$. Elements in between are part of $P$. An array called elemPos maps each element of $X$ with its position in elems, allowing access in $\Theta(1)$.

The internal partial sequence $\vec{s}$ is implemented using a reversible chained structure. An array of reversible integers called succ indicates for each element its successor in the partial sequence. An element which is not part of the partial sequence points towards itself. An additional dummy element $\perp$ marks the start and end of the partial sequence. It can be specified as predecessor in the insertion operation to insert an element at the beginning of the sequence or in an empty sequence. Inserting an element $e$ in the partial sequence after $p$ consists in modifying the successor of $e$ to point to the previous successor of $p$ and modifying the successor of $p$ to point to $e$.

The set of possible insertions $I$ is implemented using an array of sparse sets called posPreds. For each element, the corresponding sparse set contains all the possible predecessors after which the element can be inserted. If the element is a member of the sequence $\vec{s}$ or excluded, its set is empty. The sparse sets are initialized with the following domain: $R \cup P \cup \{\perp\}$. Constraints may remove possible insertions during their propagation. If doing so results in an empty set, the corresponding element is excluded according to the invariant (5).

An illustration of the domain representation for the variable $Sq$ with a domain of $\langle \vec{s} = (f,b), I = \{(c,\perp), (c,e), (c,f), (e,c), (e,f)\}, P = \{c\}, R = \{b,e,f\}, E = \{a,d\}\rangle$ is given in Fig. 1.

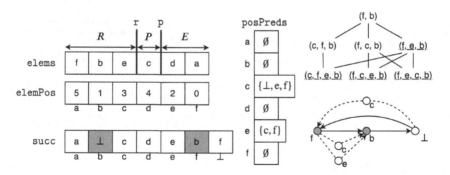

**Fig. 1.** The insertion sequence variable domain $\langle\langle \vec{s} = (f,b), I = \{(c,\perp), (c,f), (c,e), (e,c), (e,f)\}, P = \{c\}, R = \{b,e,f\}, E = \{a,d\}\rangle$ (left and middle) and the corresponding lattice (with valid sequences underlined) and graphical representation (right)

# 5    Global Constraints on Insertion Sequence Variables

## 5.1    Transition Times Constraint

In a scheduling context, the elements to sequence correspond to activities performed over time, each associated with a time window and requiring a minimum transition time to move to the next that depends on the pair of consecutive activities. The `Transition Times` constraint links the sequenced elements with their time window to make sure that transition time constraints are satisfied between any two consecutive elements of the sequence. More formally, each element $i \in X$ is associated with an activity defined by a start $start_i$ and a duration variable $dur_i$. A matrix $trans_{i,j}$, satisfying the triangle inequality, specifies transition times associated to each couple of activities $(i, j)$. The `TransitionTimes` constraint is then defined as

$$\texttt{TransitionTimes}(Sq, [start], [dur], [[trans]]) \equiv$$
$$\left\{ \overrightarrow{s}' \in D(Sq) \mid \forall a, b \in S', a \overset{\overrightarrow{s}'}{\prec} b \implies start_b \geq start_a + dur_a + trans_{a,b} \right\} \quad (6)$$

**Filtering.** The filtering algorithm is triggered whenever an element is either inserted in $\overrightarrow{s}'$ or required or if one of the bounds of a time window changes. The algorithm is split into three parts: *time windows update, insertion update* and *feasible path checking and filtering.*

*Time Window Update.* This filtering algorithm is used to adjust the start and duration of the activities already present in $\overrightarrow{s}$. This update is done in linear time by iterating over the elements of the sequence and updating their time windows depending on the time needed to transition from the previous element and to the next element. If the time window of an element is shrunk outside its domain, this leads to a failure.

*Insertion Update.* This filtering algorithm is used to filter out the invalid insertions in $I$ based on the current state of $\overrightarrow{s}$ and the transition times of the activities. The algorithm is linear and consists in iterating over $I$. For each possible insertion, if the transition times between the inserted activity and its predecessor and successor lead to a violation of a time window, the insertion is removed.

*Feasible Path Checking and Filtering.* The problem of verifying that there exists at least one *transition time feasible* extension of the current sequence composed of the required activities not yet inserted is NP-Complete [8] by a reduction from the TSP. Algorithm 1 is a recursive depth first search used to check that there exists at least one feasible extension of the current sequence composed of the required activities not yet inserted (i.e. in the set $R \setminus S$). Given the current sequence $\overrightarrow{s}$, the recursive call $\texttt{feasiblePath}(\ell, p, \Omega, t, d)$ checks that it is possible to build a sequence starting from $\ell$ at time $t$ that contains at least $d$ elements of $\Omega$ and is a super-sequence of the sub-sequence of $\overrightarrow{s}$ starting in $p$.

The parameter $\ell$ indicates the last element visited at time $t$ whereas the parameter $p$ indicates the last element of $S$ that has been visited (possibly several steps before $\ell$). The initial call $\texttt{feasiblePath}(\ell = \bot, p = \bot, \Omega = R \setminus S, t = 0, d)$ thus checks that there exists a super-sequence of $\overrightarrow{S}$ containing $d$ elements of $R \setminus S$.

---

**Algorithm 1:** $\texttt{feasiblePath}(\ell, p, \Omega, t, d)$

---

**Input:** $\ell$: last element reached, $p$: previous element reached in $\overrightarrow{S}$, $\Omega$: set of elements to reach, $t$: departure time from $\ell$, $d$: depth, $Sq = \langle \overrightarrow{S}, I, P, R, E \rangle$: Sequence variable, cache: memoization map

1  $n \leftarrow \texttt{nextMember}(Sq, p)$ ;
2  **if** $n \neq \bot$ *and* $t + trans_{\ell,n} > \max(start_n)$ **then**
3  $\quad$ **return** false;
4  **if** $\Omega = \emptyset$ **then**
5  $\quad$ **return** true;
6  $(t_f, t_i) \leftarrow \texttt{cache.getOrElse}((\ell, p, \Omega), (-\infty, +\infty))$ ;
7  **if** $t \leq t_f$ **then return** true;
8  **if** $t \geq t_i$ **then return** false;
9  **for** $a \in \Omega$ **do**
10 $\quad$ **if** $t + trans_{\ell,a} > \max(start_a)$ **then**
11 $\quad\quad$ $\texttt{cache.update}((\ell, p, \Omega), (t_f, \min(t_i, t)))$ ;
12 $\quad\quad$ **return** false ; // pruning (infeasible sequence)
13 **if** $d \leq 0$ **then**
14 $\quad$ **return** true ; // pruning (maximum depth reached)
15 **else**
16 $\quad$ **for** $a \in \Omega \mid (a, \ell) \in I$, *sorted in increasing* $(\min(start_a) + \min(dur_a))$ **do**
17 $\quad\quad$ **if** $feasiblePath(a, p, \Omega \setminus \{a\}, \max(t + trans_{\ell,a}, \min(start_a) + \min(dur_a)), d - 1)$ **then**
18 $\quad\quad\quad$ $\texttt{cache.update}((\ell, p, \Omega), (\max(t_f, t), t_i))$ ;
19 $\quad\quad\quad$ **return** true;
20 $\quad$ **if** $n \neq \bot$ *and* $feasiblePath(n, n, \Omega, \max(t + trans_{\ell,n}, \min(start_n) + \min(dur_a)), d)$ **then**
21 $\quad\quad$ $\texttt{cache.update}((\ell, p, \Omega), (\max(t_f, t), t_i))$ ;
22 $\quad\quad$ **return** true;
23 $\quad$ **return** false;

---

At each node, the algorithm either explores the insertion of a new element after $\ell$ which corresponds to branching over an element of $\Omega$ (line 16) or follows the current sequence $\overrightarrow{S}$ which consists in branching over the successor of $p$ in $\overrightarrow{S}$ (line 20). A pruning is done at lines 2 and 9 if one realizes that at least one activity cannot be reached. By the triangle inequality assumption of the transition times, if either the successor of $p$ or at least one activity of $\Omega$ cannot be reached directly after $\ell$, then it can surely not be reached later in time if some activities were visited in between. Therefore $\texttt{false}$ is returned in such case which corresponds to the infeasibility pruning. The possible extensions considered recursively at line 16 are based on the current state of $I$ and the value of $p$. The maximum depth is controlled by the parameter $d$ to avoid prohibitive computation. The algorithm can thus return a false positive result by returning $\texttt{true}$ at line 14 if this limit is reached.

The time complexity of Algorithm 1 is $\mathcal{O}(|S| \cdot |\Omega|^d)$ in worse case as it corresponds to an iteration over $\overrightarrow{s}$ with a depth-first search of depth $d$ and branching factor $|\Omega|$ at each step. In practice, as the branching is based on $I$, the search tree will often be smaller. In order to reduce the time complexity of the successive calls to `feasiblePath`, a cache is used to avoid exploring several times a partial extension that can be proven infeasible or feasible based on previous executions. A global map called `cache` is assumed to contain keys composed of the arguments of the function, that is a tuple with $(\Omega, \ell, p)$. At each key, the map associates a couple of integer values $(t_f, t_i)$ where $t_f$ is the latest known time at which it is possible to depart from $\ell$ and find a feasible path among the sub-sequence starting after $p$ and the activities of $\Omega$ and $t_i$ is the earliest known time at which the departure from $\ell$ is too late and there exists no feasible path. Line 6 is called to find if a corresponding entry exists in the map. If it is the case, the departure time $t$ is compared to the couple $(t_f, t_i)$ of the map. If $t \leq t_f$, the value *true* is immediately returned. If $t \geq t_i$, *false* is returned. If $t_f < t < t_i$, the algorithm continues its exploration. The cache is updated at lines 11, 18 and 21 depending on the result found. Usage of the cache is highlighted in gray in Algorithm 1.

This checking algorithm can be used in a shaving-like fashion into Algorithm 2. A value is filtered out from the possible set if its requirement made the sequencing infeasible according to the transition times. This `TransitionTimesFiltering` algorithm executes in $\mathcal{O}(|P| \cdot (|S| \cdot |R \setminus S|)^d)$. Notice that the cache is shared and reused along the calls in order to avoid many subtree explorations. Due to the extensive nature of the algorithm, a parameter $\rho$ defines a threshold for the size of $P$ above which the `feasiblePath` algorithm is not executed for each element of $P$ (line 4).

---

**Algorithm 2:** `TransitionTimesFiltering`$(Sq, d, \rho)$

---

**Input:** $d$: maximum depth, $\rho$: filtering threshold, $Sq = \langle \overrightarrow{s}, I, P, R, E \rangle$: seq. variable

1  $cache \leftarrow map()$ ; // initializing memoization map
2  **if** $!feasiblePath(\bot, \bot, R \setminus S, 0, d)$ **then**
3  $\quad$ **return** failure ;
4  **if** $|P| \leq \rho$ **then**
5  $\quad$ **forall** $a \in P$ **do**
6  $\quad\quad$ **if** $!feasiblePath(\bot, \bot, (R \setminus S) \cup \{a\}, 0, d)$ **then**
7  $\quad\quad\quad$ `excludes`$(Sq, a)$;

---

*Example 3.* Let us consider the following example where $X = \{a, b, c, d\}$ is the set of activities. The transition times between activities are given in Table (a) of Fig. 2 and the initial time windows (column start) in Table (b) of Fig. 2. We consider the sequence variable $Sq$ of domain $\langle \overrightarrow{s} = (a, d), I = \{(b, a), (b, d), (c, \bot), (c, d)\}, P = \{c\}, R = \{a, b, d\}, E = \emptyset \rangle$. The duration of each activity is fixed at 2. Let us apply the propagation of `TransitionTimes` on this example:

1. *Time window update* is applied. The time windows of $a$ and $d$ are reduced. The updated time windows are displayed in Table (b) (column start').

2. *Insertion update* is applied. The insertion $(b, a)$ is removed from $I$ as $b$ cannot be inserted after $a$ without violation ($b$ would end at the earliest at 9 which implies that $d$ would start at the earliest at 16, outside its time window).
3. *Transition Time Filtering* (Algorithm 2) is applied. The search trees for the checker (c) and the filter (d) are displayed in Fig. 2. Failures are denoted with × and successes with √. The initial value of the parameter $d$ is 3. The domain of $Sq$ after propagation is $\langle (a, d), \{(b, d)\}, \emptyset, \{a, b, d\}, \{c\} \rangle$ as the filter excludes $c$.

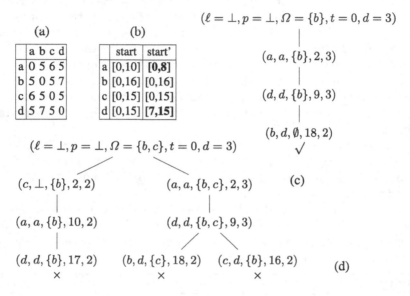

**Fig. 2.** Propagation on $Sq = \langle (a, d), \{(b, a), (b, d), (c, \bot), (c, d)\}, \{c\}, \{a, b, d\}, \emptyset \rangle$

## 5.2 Cumulative Constraint

In both the DARP and PTP, one has to satisfy requests that correspond to embarking and disembarking a person in a vehicle. The activities of transport are modeled as pairs of elements in an insertion sequence variable that must occur in this specific order: embarking before disembarking. Also this pair of elements must both be present or absent from the sequence. During the trip, the person occupies some load in the vehicle. By analogy to scheduling problems, a request is called an activity $A_i$ and is composed of the two elements $(start_i, end_i)$ corresponding to the embarking and disembarking. This activity will consume a load $load_i$ while it is on the board of the vehicle. The set of activities is denoted $A$. Also by analogy to scheduling [1], we call Cumulative the constraint that ensures that the capacity $C$ of the resource is respected at any point in the ordering defined by the sequence $Sq$ over $X$ where $\forall i \in A, start_i, end_i \in X$.

More formally

$$\texttt{Cumulative}(Sq, [start], [end], [load], C) \equiv$$

$$\left\{ \vec{s}' \in D(Sq) \mid \forall e \in S', \sum_{i \in A \mid start_i \preceq e \preceq end_i} load_i \leq C \right\}. \quad (7)$$

**Filtering.** The propagation is triggered when new elements are inserted in $\vec{s}$. It consists in filtering insertions in the current sequence $\vec{s}$ by checking if they are supported. An insertion for the element corresponding to one extremity of an activity is supported if there exists at least one possible insertion for the other extremity of the activity such that the activity load does not overloads the capacity between both insertion positions.

The first step of the propagation algorithm is to build a minimum load profile that maps each element $e$ of the sequence to the minimal load at this point in the sequence based on the activities that are part of $\vec{s}$. These can be either fully inserted (both the start and end of the activity $\in \vec{s}$) or partially inserted (only the start or end $\in \vec{s}$). For partially inserted activities, the position for the element not yet inserted is chosen among the possible insertions in $I$ as the closest one to the inserted element. Note that a violation of the capacity at this point would trigger a failure.

Once the cumulative profile is built, possible insertions for activities that are partially inserted are filtered. The algorithm used consists in iterating over $\vec{s}$ starting from the inserted element. Possible insertions for the missing element are considered and allowed as long as the load of the activity can be added to the minimal load profile without overloading the capacity. If the capacity is overloaded at some point, the current insertion as well as the insertions not yet reached are removed.

Finally, Algorithm 3 is used to check activities for which neither element is inserted. The loop at line 5 iterates over $\vec{s}$ starting from the dummy element $\perp$. When a potential start predecessor is encountered, it is added to the `activeStarts` set which maintains potential valid predecessors for the start element that have been encountered so far (line 7). The boolean `canClose` indicates if there exists at least one possible insertion position for the start of the activity that would be valid if the end is inserted at this point. It is set to true whenever a start predecessor is added to `activeStarts`. If adding the activity to the load profile for the current element violates the capacity, `canClose` is set to false and `activeStarts` is emptied as the potential start predecessors will not be matched to a valid insertion for the end element. When a valid predecessor for the end element is encountered, the end predecessor and all the start predecessors in `activeStarts` are validated (lines 13 and 14). The possible predecessors that have not been validated at the end of the loop are removed at line 18.

---

**Algorithm 3:** CumulFiltering($Sq$, $start$, $end$, $load$, $C$, $profile$)

**Input:**  $start$, $end$, $load$: starts, ends and loads of activities, $C$: capacity,
         $Sq = \langle \overrightarrow{s}, I, P, R, E \rangle$: Sequence variable, $profile$: minimum load profile

1   **forall** $i \mid start_i \notin S \wedge end_i \notin S$ **do**
2      $activeStarts \leftarrow \emptyset$ ;
3      $current \leftarrow \perp$ ;
4      $canClose \leftarrow false$ ;
5      **do**
6          **if** canInsert($Sq$, $start_i$, $current$) **then**
7              $activeStarts \leftarrow activeStarts \cup \{current\}$ ;
8              $canClose \leftarrow true$ ;
9          **if** $profile(current) + load_i > C$ **then**
10             $activeStarts \leftarrow \emptyset$ ;
11             $canClose \leftarrow false$ ;
12          **if** canInsert($Sq$, $end_i$, $current$) $\wedge$ $canClose$ **then**
13             $current$ is valid predecessor for $end_i$ ;
14             $\forall p \in activeStarts$, $p$ is valid predecessor for $start_i$ ;
15             $activeStarts \leftarrow \emptyset$ ;
16          $current \leftarrow$ nextMember($Sq$, $current$)
17      **while** $current \neq \perp$ ;
18      remove predecessors for $start_i$ and $end_i$ that have not been validated ;

---

The complexity to build the minimum load profile is linear. The complexity to check all the activities $\in A$ is $\mathcal{O}(|A| \cdot |S|)$.

*Example 4.* Let us consider four activities: $A_0 = [a, e]$, $A_1 = [b, f]$, $A_2 = [c, g]$ and $A_3 = [d, h]$. Each activity $A_i$ has a load of 1. The capacity is $C = 3$. The current partial sequence is $\overrightarrow{s} = (a, b, c, e, f)$. Before propagation, the current possible insertions in $I$ are: $\{(d, \perp), (d, a), (d, b), (d, g), (g, d), (g, e), (g, f),$ $(g, h), (h, a), (h, c), (h, e), (h, d), (h, g)\}$. Note that the possible insertions that are not in the current sequence $((d, g), (g, d), (g, h), (h, d), (h, g))$ will be ignored by the filtering algorithm. Let us propagate the Cumulative constraint:

1. The minimal load profile is built based on $A_0 = [a, e]$ and $A_1 = [b, f]$ which are both fully inserted and $A_2 = [c, g]$ which is partially inserted (only $c$ is member in $\overrightarrow{s}$). The possible insertion for the end of $A_2$ ($g$) that is the closest to its start ($c$) is $(g, e)$. Thus, $A_2$ is considered ending after $e$ to compute the minimum load profile which is $\{\perp : 0, a : 1, b : 2, c : 3, e : 2, f : 0\}$.
2. The possible insertions for the partially inserted activity $A_2$ are filtered. The sequence is iterated over starting from $c$. As $(g, e)$ is part of the minimal load profile, it is validated. The remaining possible insertion $(g, f)$ is reached without overloading the capacity and thus validated.
3. The possible insertions for non-inserted activity $A_3 = [d, h]$ are filtered. To do so, Algorithm 3 iterates over the elements in $\overrightarrow{s}$, starting from $\perp$. Both $\perp$ and $a$ are added to activeStart and canClose is set to *true*. When considering $a$ as possible predecessor for $h$, as canClose is true, the insertions $(h, a), (d, \perp)$ and $(d, a)$ are validated. Afterwards $b$ is added to activeStart. When considering $c$, adding the activity $A_3$ at this point would overload the

capacity $C$. Thus, canClose is set to *false* and activeStart is emptied. $c$ and $d$ are not validated as possible predecessors, as canClose is *false* when they are considered.

At the end of the propagation, the validated insertions are $(g, e), (g, f), (d, \perp)$, $(d, a)$ and $(h, a)$. The possible insertions $(d, b), (h, c)$ and $(h, e)$ are removed from $I$.

# 6   Applications of the Insertion Sequence Variable

This section presents the application of the insertion sequence variable on two variants of the Dial a Ride Problem.

## 6.1   Dial a Ride Problem

The Dial a Ride Problem (DARP) consists at routing a fleet of vehicles in order to transport clients from one place to another. The variant experimented in this paper was proposed by Cordeau and Laporte [5]. The objective is to minimize the total routing cost of the vehicles (defined as the total distance traveled by the vehicles) under various constraints such a maximum trip duration and time-windows. This problem is modeled with one insertion sequence variable for each route. Each request is modeled by two stops (its pickup and drop) that must be part of the same sequence. A Cumulative constraint ensures the capacity of the vehicle is satisfied. The time-window and time constraints are enforced with the help of the TransitionTimes constraints.

**Search.** A Large Neighborhood Search (LNS) [25] is used. The relaxation procedure randomly selects a subset of requests that must be reinserted into the sequences. If the search tree is completely explored during a given number of consecutive iterations given by a stagnation threshold $s$, the relaxation size is increased. Two different search heuristics are considered: 1) **A generic First Fail search**. Similarly as in [14], at each step of the search, it selects the element (the stop) not yet decided with the minimal number of possible insertions in all compatible sequences. Then, it branches in a random order over the possible insertions for the element. 2) **A problem specific heuristic** called Cost Driven search. It uses a similar approach to the first fail heuristic to select a stop with a minimal number of possible insertions. The cost metric used in [14] for their LNS-FFPA algorithm is used to improve the heuristic. The minimum cost between all possible insertions for a stop is used as a tie breaker for the selection of the next stop to insert. Additionally, the branching decisions, each corresponding to a possible insertion for the stop selected, are explored by increasing order of cost.

## 6.2  Patient Transportation Problem (PTP)

The Patient Transportation Problem (PTP) [3] is a variation of the classical DARP where clients are patients that must be transported to medical appointments and possibly brought back to a specified location after their care. This implies that some pairs of requests are dependent from each other. The objective consists in maximizing the number of requests served instead of minimizing the total distance. Additionally, the problem introduces additional constraints such as categories of patients that can only be taken in charge by specific vehicles. The fleet of vehicles is heterogeneous, each has its own capacity, can only serve some types of patients and departure from different points. Also each vehicle is available in a given time window only.

**Search.** Such as for the DARP, LNS is used and Two search heuristics are considered: 1) The same **generic First Fail heuristic** as the one described in Sect. 6.1; 2) **A problem specific heuristic** called `Slack Driven` search. It is similar to the `Cost Driven` heuristic described in Sect. 6.1. The cost metric is replaced by a *slack difference* metric which is defined as the total size difference of the time windows of the predecessor and successor of the stop to insert before and after insertion. The intuition is to minimize this difference in order to keep the sequences as flexible as possible and maximizing potential future insertions.

# 7  Experimental Results

This section reports the comparison of the models presented in Sect. 6 with state-of-the-art CP approaches for the DARP and PTP. The models based on insertion sequences variables are referred as the *Insertion Sequence* (ISEQ) approaches. The generic *First Fail* heuristic is referred as FF. The *Cost Driven* and *Slack Driven* heuristics are referred as CDS and SDS respectively.

For the DARP, the insertion sequence approach was compared with 1) the *LNS with First Feasible Probabilistic Acceptance*(LNS-FFPA) model and heuristic proposed in [14]; 2) an implementation of our model with the sequence variables and interval variables of *CP Optimizer* is referred as DARP_CPO. The approaches were run on 68 DARP instances from [4,6] and are available at [2].

For the PTP, The insertion sequence model was compared with 1) the model proposed in [3], referred as *Scheduling with Maximum Selection Search* (SCHED+MSS); 2) the model proposed in [20], referred as *Liu CP Optimizer model* (LIU_CPO). A greedy approach referred as (GREEDY) was used to compute the initial PTP solutions given to the compared models in the LNS setting. Tests were performed on the benchmark of instances used in [3]. It contains both real exploitation instances and randomly generated instances which are available at [26].

For the `TransitionTimes` constraint, the maximum depth was fixed to 3 and the filtering threshold to 10. The LNS used an initial relaxation of 20% of the requests, a failure limit of 500, a stagnation threshold of 50 and an increase factor of 20%.

Each approach was run 10 times on each instance, with a time limit of 600 s. The system used for the experiments is a PowerEdge R630 server (128GB, 2 proc. Intel E5264 6c/12t) running on Linux. The approaches using CP Optimizer were implemented using the Java API of CPLEX Optimization Studio V12.8 [19]. The other models were implemented on OscaR [21] running on Scala 2.12.4.

In order to compare the anytime behavior of the approaches, we define the *relative distance* of an approach at a time $t$ as the current distance from the best known objective (BKO) divided by the distance to the worse initial objective (WSO): $(objective(t) - BKO)/(WSO - BKO)$. If an approach has not found an initial solution, the worse initial objective (WSO) is used as objective value. A relative distance of 1 thus indicates that the approach has not found an initial solution or is stuck at the initial solution while a relative distance of 0 indicates that the best known solution has been reached.

**Results.** Figure 3 shows the evolution of the average relative distance during the search. The DARP results are shown on the left. For the PTP, the approaches are compared in two different settings: 1) in the same experimental setting as in [3] (with a LNS search starting from an initial solution given by a greedy approach) (middle); 2) in a DFS starting without an initial solution (right).

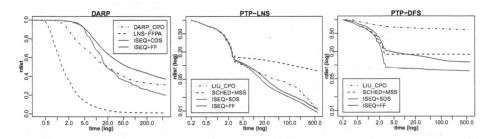

**Fig. 3.** Average relative distance in function of time

These results suggest that, on the DARP, the sequence based approaches are not able to compete with the dedicated LNS-FFPA algorithm. However, they are able to successfully outperform the dedicated SCHED+MSS approach on the PTP. As can be observed, the approaches using the insertion sequence variable obtain slightly better result than the approach using the state-of-the-art CP Optimizer. Note that the comparison with CP Optimizer is not straightforward as it is mostly black box and its interface does not offer much control over its behavior. However, despite the adaptive LNS search [16] and the advanced techniques (failure directed search [28], objective landscapes [15]) used by CP Optimizer, our approach is competitive in a LNS setting. The experiment in a DFS setting where the advanced search of CP Optimizer is not used suggests that the difference is mainly due to the modeling and propagation as even our generic search outperforms CP Optimizer in this setting.

**Constraint Parameters.** Several values were tested for the parameters of the `TransitionTimes` constraint by using the methodology proposed in [27]. It consists in storing the search tree obtained with the weakest filtering and replaying it with the constraints and parameters to test. The impact of the `Cumulative` constraint was also tested by comparing it to a simple checker.

Table 3 presents the results of this experiment on 3 medium sized PTP instances in a DFS setting. The instances are expressed in terms of the number of hospitals (h), number of available vehicles (v) and number of patients (p). The values are displayed in terms of percentage compared to the base case (the parameter value for the first column). The first row corresponds to the percentage of size (in terms of the number of nodes) of the new search tree compared to the base case. The second row consists in the percentage of time taken to explore the new search tree. For example, on the Hard instance, for a depth d of 2, the search tree is 76.26% smaller which results in an exploration 63.67% faster. Each parameter was tested independently with the others set to their default values.

**Table 3.** Number of nodes explored (top) and time taken (bottom) with various parameter values

| Instance | | | | $\rho$ | | | | $d$ | | | | | cache | | Cumul. | |
|---|---|---|---|---|---|---|---|---|---|---|---|---|---|---|---|---|
| Set | h | v | p | 0 | 10 | 20 | ∞ | 1 | 2 | 3 | 6 | ∞ | × | √ | × | √ |
| Easy | 24 | 9 | 96 | 100 | 100 | 100 | 100 | 100 | 100 | 100 | 100 | 100 | 100 | 100 | 100 | 100 |
| | | | | 100 | 96.27 | 92.27 | 101 | 100 | 102.53 | 96.68 | 93.87 | 91.73 | 100 | 95.39 | 100 | 126.61 |
| Medium | 48 | 5 | 96 | 100 | 100 | 100 | 100 | 100 | 0.01 | 0.01 | 0.01 | 0.01 | 100 | 100 | 100 | 0.01 |
| | | | | 100 | 80.79 | 55.67 | 53.07 | 100 | 0.05 | 0.06 | 0.05 | 0.06 | 100 | 77.24 | 100 | 0.01 |
| Hard | 96 | 5 | 96 | 100 | 100 | 100 | 55.35 | 100 | 76.26 | 70.31 | 64.48 | 59.86 | 100 | 100 | 100 | 0.01 |
| | | | | 100 | 85.06 | 56.47 | 22.74 | 100 | 63.67 | 53.8 | 38.8 | 51.93 | 100 | 91.44 | 100 | 0.03 |

As can be observed, the constraints have an important impact on both the size of the search tree and the search time for the medium and difficult instances. The easy instance search tree was not affected by the constraints. Note that an increase in depth may result in a faster search despite having the same search tree size (such as for the Easy instance). This is most likely due to the cache that is filled faster in the first calls to Algorithm 1 and thus allows smaller searches in the subsequent calls which results in a gain of time over the whole propagation.

## 8   Conclusion

In this paper, we propose a new variable called *Insertion Sequence Variable* (ISV) to provide a flexible and efficient model for the DARP and its variant the PTP. The ISV domain extends the set domain variable with the possibility to insert an element after any sequenced element. Experimental results show that the proposed approach is competitive with existing sequence based approaches, outperforms dedicated approaches for the PTP and confirm the effectiveness of the new filtering algorithms proposed.

While used only in the context of the Dial-a-Ride problem in this paper, sequence variables could be used to model a large variety of Routing and Scheduling problems. As future work, it would be interesting to study the usage of the ISV on other problems as well as developing new global constraints and filtering algorithms.

**Acknowledgments.** This research is financed by the Walloon Region (Belgium) as part of PRESupply Project. We thank Siddhartha Jain and Pascal Van Hentenryck for sharing with us their implementation of the LNS-FFPA algorithm.

# References

1. Aggoun, A., Beldiceanu, N.: Extending chip in order to solve complex scheduling and placement problems. Math. Comput. Modell. **17**(7), 57–73 (1993)
2. Braekers, K.: Dial-a-Ride Problems Instances. http://alpha.uhasselt.be/kris. braekers/ (2019). Accessed 2 Dec 2019
3. Cappart, Q., Thomas, C., Schaus, P., Rousseau, L.-M.: A constraint programming approach for solving patient transportation problems. In: Hooker, J. (ed.) CP 2018. LNCS, vol. 11008, pp. 490–506. Springer, Cham (2018). https://doi.org/10.1007/978-3-319-98334-9_32
4. Cordeau, J.F.: A branch-and-cut algorithm for the dial-a-ride problem. Oper. Res. **54**(3), 573–586 (2006)
5. Cordeau, J., Laporte, G.: The dial-a-ride problem (DARP): variants, modeling issues and algorithms. 4OR **1**(2), 89–101 (2003). https://doi.org/10.1007/s10288-002-0009-8
6. Cordeau, J.F., Laporte, G.: A Tabu search heuristic for the static multi-vehicle dial-a-ride problem. Transp. Res. Part B: Methodol. **37**(6), 579–594 (2003)
7. Cordeau, J.F., Laporte, G.: The dial-a-ride problem: models and algorithms. Ann. Oper. Res. **153**(1), 29–46 (2007)
8. Garey, M.R., Johnson, D.S.: Computers and Intractability: A Guide to the Theory of NP-Completeness. W. H. Freeman (1979)
9. Gervet, C.: Interval propagation to reason about sets: definition and implementation of a practical language. Constraints **1**(3), 191–244 (1997)
10. Hartert, R., Schaus, P., Vissicchio, S., Bonaventure, O.: Solving segment routing problems with hybrid constraint programming techniques. In: Pesant, G. (ed.) CP 2015. LNCS, vol. 9255, pp. 592–608. Springer, Cham (2015). https://doi.org/10.1007/978-3-319-23219-5_41
11. Ho, S.C., Szeto, W., Kuo, Y.H., Leung, J.M., Petering, M., Tou, T.W.: A survey of dial-a-ride problems: Literature review and recent developments. Transp. Res. Part B: Methodol. **111**, 395–421 (2018)
12. IBM Knowledge Center: Interval variable sequencing in CP Optimizer. https://www.ibm.com/support/knowledgecenter/SSSA5P_12.9.0/ilog.odms.ide.help/refcppopl/html/interval_sequence.html (2019). Accessed 22 Nov 2019
13. IBM Knowledge Center: Search API for scheduling in CP Optimizer. https://www.ibm.com/support/knowledgecenter/SSSA5P_12.9.0/ilog.odms.cpo.help/refcppcpoptimizer/html/sched_search_api.html?view=kc#85 (2019). Accessed 22 Nov 2019
14. Jain, S., Van Hentenryck, P.: Large neighborhood search for dial-a-ride problems. In: Lee, J. (ed.) CP 2011. LNCS, vol. 6876, pp. 400–413. Springer, Heidelberg (2011). https://doi.org/10.1007/978-3-642-23786-7_31

15. Laborie, P.: Objective landscapes for constraint programming. In: van Hoeve, W.-J. (ed.) CPAIOR 2018. LNCS, vol. 10848, pp. 387–402. Springer, Cham (2018). https://doi.org/10.1007/978-3-319-93031-2_28

16. Laborie, P., Godard, D.: Self-adapting large neighborhood search: application to single-mode scheduling problems. In: Proceedings MISTA-07, Paris, vol. 8 (2007)

17. Laborie, P., Rogerie, J.: Reasoning with conditional time-intervals. In: FLAIRS conference. pp. 555–560 (2008)

18. Laborie, P., Rogerie, J., Shaw, P., Vilím, P.: Reasoning with conditional time-intervals. part ii: an algebraical model for resources. In: Twenty-Second International FLAIRS Conference (2009)

19. Laborie, P., Rogerie, J., Shaw, P., Vilím, P.: IBM ILOG CP optimizer for scheduling. Constraints 23(2), 210–250 (2018)

20. Liu, C., Aleman, D.M., Beck, J.C.: Modelling and solving the senior transportation problem. In: van Hoeve, W.-J. (ed.) CPAIOR 2018. LNCS, vol. 10848, pp. 412–428. Springer, Cham (2018). https://doi.org/10.1007/978-3-319-93031-2_30

21. OscaR Team: OscaR: Scala in OR (2012). https://bitbucket.org/oscarlib/oscar

22. Perron, L., Furnon, V.: Or-tools (2019). https://developers.google.com/optimization/

23. Perron, L., Furnon, V.: OR-Tools Sequence Var. https://developers.google.com/optimization/reference/constraint_solver/constraint_solver/SequenceVar (2019). Accessed 22 Nov 2019

24. de Saint-Marcq, V.l.C., Schaus, P., Solnon, C., Lecoutre, C.: Sparse-sets for domain implementation. In: CP Workshop on Techniques for Implementing Constraint Programming Systems (TRICS), pp. 1–10 (2013)

25. Shaw, P.: Using constraint programming and local search methods to solve vehicle routing problems. In: Maher, M., Puget, J.-F. (eds.) CP 1998. LNCS, vol. 1520, pp. 417–431. Springer, Heidelberg (1998). https://doi.org/10.1007/3-540-49481-2_30

26. Thomas, C., Cappart, Q., Schaus, P., Rousseau, L.M.: CSPLib problem 082: patient transportation problem (2018). http://www.csplib.org/Problems/prob082

27. Cauwelaert, S.V., Lombardi, M., Schaus, P.: How efficient is a global constraint in practice? Constraints 23(1), 87–122 (2017). https://doi.org/10.1007/s10601-017-9277-y

28. Vilím, P., Laborie, P., Shaw, P.: Failure-directed search for constraint-based scheduling. In: Michel, L. (ed.) CPAIOR 2015. LNCS, vol. 9075, pp. 437–453. Springer, Cham (2015). https://doi.org/10.1007/978-3-319-18008-3_30

# Relaxation-Aware Heuristics for Exact Optimization in Graphical Models

Fulya Trösser[1], Simon de Givry[1]([✉])(iD), and George Katsirelos[2]([✉])(iD)

[1] MIAT, UR-875, INRAE, 31320 Castanet Tolosan, France
{fulya.ural,simon.de-givry}@inrae.fr
[2] UMR MIA-Paris, INRAE, AgroParisTech, Univ. Paris-Saclay, 75005 Paris, France
gkatsi@gmail.com

**Abstract.** Exact solvers for optimization problems on graphical models, such as Cost Function Networks and Markov Random Fields, typically use branch-and-bound. The efficiency of the search relies mainly on two factors: the quality of the bound computed at each node of the branch-and-bound tree and the branching heuristics. In this respect, there is a trade-off between quality of the bound and computational cost. In particular, the Virtual Arc Consistency (VAC) algorithm computes high quality bounds but at a significant cost, so it is mostly used in preprocessing, rather than in every node of the search tree.

In this work, we identify a weakness in the use of VAC in branch-and-bound solvers, namely that they ignore the information that VAC produces on the linear relaxation of the problem, except for the dual bound. In particular, the branching heuristic may make decisions that are clearly ineffective in light of this information. By eliminating these ineffective decisions, we significantly reduce the size of the branch-and-bound tree. Moreover, we can optimistically assume that the relaxation is mostly correct in the assignments it makes, which helps find high quality solutions quickly. The combination of these methods shows great performance in some families of instances, outperforming the previous state of the art.

**Keywords:** Graphical model · Cost Function Network · Weighted Constraint Satisfaction Problem · Virtual Arc Consistency · Branch-and-bound · Linear relaxation · Local polytope · Variable ordering heuristic

## 1 Introduction

Undirected graphical models like Cost Function Networks, aka Weighted Constraint Satisfaction Problems (WCSP), and Markov Random Fields (MRF) can be used to give a factorized representation of a function, in which vertices of a

---

Some supplementary figures available at http://genoweb.toulouse.inra.fr/~degivry/evalgm/TrosserCPAIOR20supp.pdf.

© Springer Nature Switzerland AG 2020
E. Hebrard and N. Musliu (Eds.): CPAIOR 2020, LNCS 12296, pp. 475–491, 2020.
https://doi.org/10.1007/978-3-030-58942-4_31

graph represent variables of the function and (hyper)edges represent factors. The factors can be, for example, cost functions, in which case the graphical model represents a factorization of a cost function, or local probability tables, in which case the model represents a non-normalized joint probability distribution [17].

The two models, WCSP and MRF, are equivalent under a − log transformation, hence the NP-complete cost minimization query in WCSP is equivalent to the maximum a posteriori (MAP) assignment query in MRF. This optimization problem has applications in many areas, such as image analysis, speech recognition, bioinformatics, and ecology.

Exact solution methods for this problem are mostly based on branch-and-bound. For example, one can express WCSP optimization as an integer linear program (ILP) and use a solver for that problem. However, ILP solvers need to solve the linear relaxation of the instance exactly to obtain a bound at each node of the branch-and-bound tree, an operation that is too expensive for the scale of problems encountered in many applications. Instead, the most successful dedicated solvers use algorithms that in effect solve the linear relaxation approximately and therefore potentially suboptimally. Specifically, algorithms like EDAC [9], VAC [6], TRWS [18] and others, produce feasible solutions to the dual of the linear relaxation of the WCSP, which can be used as lower bounds. For the loss of precision that they give up, these algorithms gain significantly in computational efficiency. In stark contrast to integer programming, not only is exact LP solving not used, but the preferred method for branch-and-bound, EDAC, is by far the weakest, while VAC or TRWS are most often used only in preprocessing.

There are a few exceptions to the norm of using branch-and-bound for this problem: core-guided MaxSAT solvers [21], logic-based Benders decomposition [8], cut generation [24], to name a few. Here, we are interested in the COMBILP method [13], which solves the linear relaxation and decomposes the problem into two parts: the "easy" part which corresponds to the set of integral variables in the linear relaxation and a combinatorial part which contains the variables assigned fractional values. They then proceed to solve the combinatorial subset exactly and if that solution can be combined with the easy part without incurring extra cost, it reports optimality. Otherwise, it moves some variables from the easy part to the combinatorial part and iterates. Crucially, they identify integral variables by identifying a condition called Strict Arc Consistency (Strict AC) on the dual solution produced, and can therefore be used with approximate dual LP solvers, like VAC and TRWS, which may produce a suboptimal dual LP solution for which no corresponding primal solution exists.

We make several contributions here. First, we relax Strict AC, the condition that COMBILP uses to detect integrality. We show in Sect. 3 that the relaxed condition admits larger sets of integral variables. Second, we show that a class of fixpoints of an LP solver like VAC implies a specific set of integral variables regardless of the dual solution it finds, even when those variables do not satisfy the Strict AC condition in this solution. This avoids the need to bias the LP solver towards solutions that contain Strict AC variables. On the practical side, we introduce two simple techniques that exploit this property within a branch-

and-bound solver. The first, given in Sect. 5 modifies the branching heuristic to avoid branching on Strict AC variables, as that is unlikely to be informative. The second, given in Sect. 6, is a variant of the well-known RINS heuristic in integer programming [7], which optimistically assumes that the set of Strict AC variables assigned their integral values actually appear in the optimal solution and solves a restricted sub-problem to help quickly identify high quality solutions. In Sect. 7, we show that integrating these techniques in the TOULBAR2 solver [6] improves performance significantly over the state of the art in some families of instances.

## 2    Preliminaries

**Definition 1.** A *Constraint Satisfaction Problem (CSP)* [6] is a triple $\langle X, D, C \rangle$. $X$ is a set of $n$ variables $X = \{1, \ldots, n\}$. Each *variable* $i \in X$ has a domain of values $D_i \in D$ and can be assigned any value $a \in D_i$, also noted $(i, a)$. $C$ is a set of constraints. Each *constraint* $c_S \in C$ is defined over a set of variables $S \subseteq X$ (called the *scope* of the constraint) by a subset of the Cartesian product $\prod_{i \in S} D_i$ which defines all consistent tuples of values.

We assume, without loss of generality, that at most one constraint is defined over a given set of variables. The unary constraint on variable $i$ will be denoted $c_i$, and binary constraints $c_{ij}$. The cardinality $|S|$ is the *arity* of $c_S$. For $J \subseteq X$, $\ell(J)$ denotes the set of all possible tuples for $J$, i.e., $\ell(J) = \prod_{i \in J} D_i$. Let $S \subseteq X$, and $t \in \ell(S)$, the projection of $t$ onto $V \subseteq S$ is denoted by $t[V]$. A tuple $t$ satisfies a constraint $c_S$ if $t[S] \in c_S$. A tuple $t \in \ell(X)$ is a *solution* iff it satisfies all the constraints in $C$. Finding a solution is NP-complete.

**Definition 2.** A *Weighted Constraint Satisfaction Problem (WCSP)* [6] is a quadruple $\langle X, D, C, k \rangle$ where $X$ is a set of $n$ variables $X = \{1, \ldots, n\}$, each variable $i \in X$ has a domain of possible values $D_i \in D$, as in CSP. $C$ is a set of cost functions, and $k$ is a positive integer or infinity serving as the upper bound. Each cost function $\langle S, c_S \rangle \in C$ is defined over a set of variables $S \subseteq X$ (its *scope*) and $c_S$ maps each assignment to the variables in $S$ to non-negative integer costs.

WCSPs generalize CSPs as they can represent the same set of feasible solutions with infinite cost $c_S(t) = k$ for forbidden tuples $t$, but additionally define a cost for feasible assignments. We assume all WCSPs contain a unary cost function for each variable and a cost function $c_\emptyset$, which represents a constant in the objective function. Since all costs are non-negative, $c_\emptyset$ is a lower bound on the cost of feasible solutions of the WCSP.

If the largest arity of any cost function in a WCSP is 2, then we say this is a *binary WCSP*. We focus on binary WCSPs here, both for simplicity and because of technical limitations of the implementation. However, all definitions and properties we present can easily be generalized to higher arities.

A binary WCSP can be graphically represented as shown in Fig. 1a. Each variable $i \in X$ corresponds to a cell. Each value $a \in D_i$ corresponds to a dot in the cell. A unary cost $c_i(a)$ is written next to the dot only if it is non-zero. If there is a non-zero binary cost between $(i, a)$ and $(j, b)$, then an edge is drawn between the dots corresponding to these assignments.

The problem is to find a solution $t \in \ell(X)$ which minimizes the sum of all cost functions, denoted as $c_P(t) = c_\emptyset + \sum_{i \in X} c_i(t[i]) + \sum_{c_{ij} \in C} c_{ij}(t[i], t[j])$, and such that $c_P(t) < k$. This is denoted $opt(P)$. This problem is NP-hard.

Given two WCSPs $P$, $P'$ with the same set of variables and scopes, we say they have the same structure. If $c_P(t) = c_{P'}(t)$ for all $t \in l(X)$ then $P$ and $P'$ are equivalent and they are reparameterizations of each other. It has been shown [6] that the optimal reparameterization, which maximizes the constant factor in the objective, is given by the dual of the following linear program (LP), called the local polytope of the WCSP:

$$\min c_\emptyset + \sum_{i \in X, a \in D_i} c_i(a) x_{ia} + \sum_{c_{ij} \in C, a \in D_i, b \in D_j} c_{ij}(a, b) y_{iajb}$$

s.t.

$$\sum_{a \in D_i} x_{ia} = 1 \qquad\qquad \forall i \in X$$

$$x_{ia} = \sum_{b \in D_j} y_{iajb} \qquad\qquad \forall i \in X, a \in D_i, c_{ij} \in C$$

$$0 \leq x_{ia} \leq 1 \qquad\qquad \forall i \in X, a \in D_i$$

$$0 \leq y_{iajb} \leq 1 \qquad\qquad \forall c_{ij} \in C, a \in D_i, b \in D_j$$

$$(1)$$

From the optimal solution of the above LP, the reparameterization is extracted from the *reduced costs* $r(x_{ia})$ and $r(y_{iajb})$ of each variable and binary cost function, respectively, by setting $c_i(a)$ to $r(x_{ia})$ and $c_{ij}(a, b)$ to $r(y_{iajb})$ and setting $c_\emptyset$ to the optimum of the LP (1). Because of the correspondence between reparameterizations and solutions of this LP, we use the two interchangeably.

In this paper, we work with algorithms that do not solve this LP exactly but compute a feasible solution of its dual. In particular, the VAC algorithm computes a reformulation which is virtual arc consistent (VAC), as defined below.

**Definition 3.** A CSP $P$ is *arc consistent* (AC) if for all $c_{ij} \in C$, $\forall a \in D_i$, $\exists b \in D_j$ with $\{(i, a), (j, b)\} \in c_{ij}$, and $\forall b \in D_j$, $\exists a \in D_i$ with $\{(i, a), (j, b)\} \in c_{ij}$. The arc consistent closure $AC(P)$ is the unique CSP which results from removing values from domains that violate the arc consistency property.

The CSP $AC(P)$ is equivalent to $P$, i.e., it has exactly the same set of solutions. In particular, if $AC(P)$ is empty (has empty domains), $P$ is unsatisfiable.

**Definition 4** ([6]). Let $P = \langle X, D, C, k \rangle$ be a WCSP. Then $Bool(P) = \langle X, \overline{D}, \overline{C} \rangle$ ($Bool_\theta(P) = \langle X, \overline{D_\theta}, \overline{C_\theta} \rangle$) is the CSP where, for all $i \in X$, $a \in \overline{D_i}$ (resp. $\overline{D_{\theta i}}$) if and only if $c_i(a) = 0$ (resp. $c_i(a) < \theta$) and for all $i, j \in X^2$,

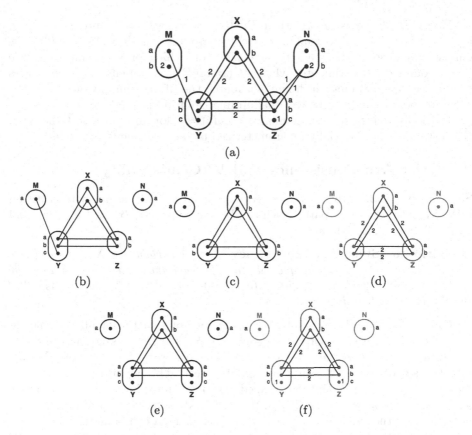

**Fig. 1.** Stratified VAC and RASPS examples for two different thresholds. Blue variables are VAC-integral. (a) A WCSP instance $P$ with 5 variables $\{M, X, Y, Z, N\}$. (b) $Bool_1(P) = Bool(P)$ (c) $AC(Bool_1(P))$. (d) WCSP instance $P_1$ constructed by RASPS from $AC(Bool_1(P))$. Optimal solution is $\{a, a, b, a, a\}$ with cost 2. (e) $Bool_2(P) = AC(Bool_2(P))$. (f) WCSP instance $P_2$ constructed by RASPS from $AC(Bool_2(P))$. Optimal solution is $\{a, a, b, c, a\}$ with cost 1, optimum of $P$. (Color figure online)

$\langle \{i, j\}, R_{ij} \rangle \in \overline{C}$ (resp. $\overline{C_\theta}$) iff $\exists c_{ij} \in C$, where $R_{ij}$ is the relation $\forall a \in \overline{D_i}$ (resp. $\overline{D_{\theta i}}$), $\forall b \in \overline{D_j}$ (resp. $\overline{D_{\theta j}}$), $\{(i, a), (j, b)\} \in R_{ij} \Leftrightarrow c_{ij}(a, b) = 0$ (resp. $c_{ij}(a, b) < \theta$).

By construction, $Bool(P)$ admits exactly the solutions of $P$ with cost $c_\emptyset$, since all assignments that have non-zero cost in any cost function of $P$ are mapped to forbidden tuples in $Bool(P)$. Thus, if $Bool(P)$ is unsatisfiable, $c_\emptyset < opt(P)$. There is no such clear result for $Bool_\theta(P)$, but it is useful in practice. Examples of $Bool_\theta(P)$ are shown in Fig. 1b, 1e where edges correspond to forbidden tuples.

**Definition 5.** A WCSP $P$ is *virtual arc consistent* (VAC) if the arc consistency closure of the CSP $Bool(P)$ is non-empty [6].

If $AC(Bool(P))$ is empty, then $Bool(P)$ is unsatisfiable and hence $c_\emptyset <$
$Opt(P)$. The VAC algorithm iteratively computes whether $AC(Bool(P))$ is
empty and if so extracts a reparameterization which provably improves the
lower bound $c_\emptyset$. It terminates when $AC(Bool(P))$ is non-empty. It converges
to a non-unique fixpoint which may not match the LP optimum. Conversely, the
reparameterization given by the dual optimal solution is VAC.

In the following, we assume $\min_{t\in\ell(S)} c_S(t) = 0$ for all scopes $S$. Otherwise,
the instance can be trivially reparameterized to increase the lower bound.

# 3   Strict Arc Consistency and VAC-integrality

Savchynskyy et al. introduced Strict Arc Consistency ([23]) as a way to partition
a WCSP into an "easy" part, which can be solved exactly by an LP solver and
a "hard" combinatorial part.

**Definition 6 (Strict Arc Consistency [23]).** *A variable $i \in X$ is Strictly Arc
Consistent if there exists a unique value $a \in D_i$ such that $c_i(a) = 0$ and a unique
tuple $\{(i, a), (j, b)\}$ which satisfies $c_{ij}(a, b) = 0 \quad \forall c_{ij} \in C$. The value $a$ is called
the Strict AC value of $i$.*

Given a WCSP $P$ and a subset $S$ of its variables such that all variables in
$S$ are Strict AC, we can solve $P$ restricted to $S$ exactly by assigning the Strict
AC value to each variable. This property gives a natural partition of a WCSP
into the set of Strict AC variables and the rest. This partition was used by
Savchynskyy et al. [23] and in a refined algorithm introduced later [13]. These
algorithms exploit the solvability of the Strict AC subset of variables and only
need to solve the smaller non-Strict-AC subset using a combinatorial solver.

Our first contribution here is to note that the Strict AC property is stronger
than necessary[1]. In particular, we can weaken the second condition as follows:

**Definition 7 (VAC-integrality).** *A variable $i \in X$ is VAC-integral if there
exists a unique value $a \in D_i$ such that $c_i(a) = 0$ and at least one tuple
$\{(i, a), (j, b)\}$ which satisfies $c_{ij}(a, b) + c_j(b) = 0 \quad \forall c_{ij} \in C$. The value $a$ is
the VAC-integral value of $x$.*

The difference between VAC-integrality and Strict AC is that in
VAC-integrality, the second condition requires that the witness value appears
in at least one rather than exactly one 0-cost tuple in each incident constraint.
The VAC-integral subset of a WCSP maintains the main property of Strict AC,
namely that it is exactly solvable by inspection and its optimal solution has cost
0. The optimal solution, as in Strict AC, simply assigns to each VAC-integral
variable its VAC-integral value. By definition, this has cost 0.

Since VAC-integrality is a relaxation of Strict AC, every Strict AC set of
variables is also VAC-integral. The inverse does not hold (see Supplementary
Fig. 1). However, this only holds for instances that are at a VAC fixpoint.

---

[1] We also change the name of the property from "consistency", which implies an
algorithm that achieves said consistency, to "integrality". Adding the VAC term will
become clear after Proposition 3.

**Proposition 1.** *If a WCSP instance P is VAC and a variable i is strict AC then it is also VAC-integral.*

As with Strict AC, VAC-integrality implies integrality of the corresponding primal solution by complementary slackness.

**Proposition 2.** *The VAC-integral variables in an optimal dual solution of* (1) *correspond to the variables i for which there exists a unique a with $x_{ia} = 1$ and $x_{ib} = 0$ for $b \neq a$ in the corresponding optimal primal solution.*

*Proof (Sketch).* Given an optimal dual solution of the local polytope LP, for each VAC-integral variable $i$ with VAC-integral value $a$, the primal solution must have $x_{ib} = 0$ for all $b \neq a$ by complementary slackness and hence $x_{ia} = 1$.   □

Note that in the case of approximate dual LP solvers like VAC and TRWS, this observation does not hold: if the dual solution is not optimal, there is no primal solution with the same cost. Rather, we use Strict AC and VAC-integrality as proxies for conditions which would lead to integrality in optimal solutions, while maintaining the property that they admit zero cost solutions.

One complication with both Strict AC and VAC-integrality is that any lower bound given by a dual solution can in fact be produced by several dual solutions, but they do not all give the same VAC-integrality subset. One way to deal with this is to bias the LP solver towards solutions that maximize the VAC-integral subset [13]. Here we propose another method, given by the following observation.

**Proposition 3.** *Given a WCSP instance P which is VAC and a variable i, if in $AC(Bool(P))$ it holds that $\overline{D}_i = \{a\}$ then i is VAC-integral with value a.*

*Proof (Sketch).* Since $AC(Bool(P))$ is arc consistent, if a value remains in the domain of $i$ in $Bool(P)$, it has unary cost 0 in $P$ and is supported by tuples and values of cost 0 in all incident constraints. Conversely, if a value is removed in $Bool(P)$, either it has non-zero unary cost in $P$ or some non-zero amount of cost can be moved onto it [6].   □

The effect of Proposition 3 is that the class of dual feasible solutions which have the same $AC(Bool(P))$ produce the same set of VAC-integral variables, even though most of these solutions do not satisfy Definition 7. This is shown in in the WCSP of Fig. 1a which is VAC but only variable $N$ is VAC-integral. However, by applying Proposition 3, we get from $AC(Bool(P))$ that $M$ is also VAC-integral. Therefore, this observation allows us to construct a larger VAC-integral subset than that given by collecting the variables that satisfy the VAC-integrality property given a dual solution. This has the advantage that we do not need to modify the LP solver to be biased towards specific dual solutions and is easy to use with VAC, which explicitly maintains $Bool(P)$. We can apply the same reasoning to find Strict AC sets of variables, using the following condition.

**Proposition 4.** *If a WCSP instance P is VAC, then a VAC-integral variable i is strict AC if and only if all its neighbors are VAC-integral.*

*Complexity.* It is natural to ask whether the presence of a large VAC-integral subset makes the problem easier to solve, in the sense of fixed parameter tractability [11]. Unfortunately, this turns out not to be the case. Let ALMOST-INTEGRAL-WCSP be the class of WCSPs which are VAC with $n - 1$ VAC-integral variables. We show that it is NP-complete, which implies that WCSP is para-NP-complete for the parameter of number of non-VAC-integral variables.

**Theorem 1.** ALMOST-INTEGRAL-WCSP *is NP-Complete.*

*Proof.* Membership in NP is obvious since this is a subclass of WCSP. For hardness, we reduce from binary WCSP. Let $P = \langle X, D, C, k \rangle$ be an arbitrary WCSP instance and assume that it is VAC. We construct $P' = \langle X \cup X' \cup \{q\}, D, C', k + |C| \rangle$, where $X'$ is a copy of the variables in $X$, including the unary cost functions and $q$ has domain $\{a, b\}$ with $c_q(a) = c_q(b) = 0$. For each cost function with scope $\{i, j\}$ in $C$, $P'$ has two cost functions with scopes $\{i, j, q\}$ and $\{i', j', q\}$.

Each variable in $P$ has at least one value with unary cost 0, since it is VAC. Let this value be $a$ for all variables. We define the ternary cost functions to be $c_{ijq}(a, a, a) = 0$, $c_{ijq}(u, v, a) = k$ when $u \neq a$ or $v \neq a$, $c_{ijq}(u, v, b) = c_{ij}(u, v) + 1$ for all $u, v$. Similarly, $c_{i'j'q}(a, a, b) = 0$, $c_{i'j'q}(u, v, b) = k$ when $u \neq a$ or $v \neq a$, $c_{i'j'q}(u, v, a) = c_{ij}(u, v) + 1$ for all $u, v$.

$P'$ is an instance of ALMOST-INTEGRAL-WCSP with $q$ the non-VAC-integral variable. Indeed, $c_i(a) = 0$ for all variables $i \in X \cup X'$ and it is supported by the zero cost tuple $(a, a, a)$ in each ternary constraint. All other values appear only in ternary tuples with non-zero cost hence will be pruned in $AC(Bool(P'))$.

$P$ has optimum solution of cost $c$ if and only if $P'$ has optimum of cost $c + |C|$. Indeed, when we assign $q$ to $a$ or $b$, the problem is decomposed into independent binary WCSPs on $X$ and $X'$. One of these admits the all-$a$, 0-cost assignment and the other is identical to $P$ with an extra cost of 1 per cost function.    □

Although this construction uses ternary cost functions, we can convert them to binary using the hidden encoding [4]. This preserves arc consistency, hence it also preserves VAC, so the result holds also for binary WCSPs.

## 4    Stratified VAC

The foundation of all the heuristics we present in this paper is the implementation of VAC in TOULBAR2 [6], which is restricted to binary WCSPs. In this implementation, the non-zero binary costs $c_{ij}(a, b)$ are stratified. Specifically, they are sorted in decreasing order and placed in a fixed number $l$ of buckets. The minimum cost $\theta_i$ of each bucket $i \in \{1, \ldots, l\}$ defines a sequence of thresholds $(\theta_1, \ldots, \theta_l)$. At each $\theta_i$ for $i$ from 1 to $l$, it constructs the $Bool_{\theta_i}(P)$ and iterates on it until no domain wipe-out occurs. After $\theta_l$, it follows a geometric schedule $\theta_{i+1} = \frac{\theta_i}{2}$ until $\theta_i = 1$. The reader is referred to [6] for more details.

For a smaller $\theta_i$, $Bool_{\theta_i}(P)$ is more restricted, i.e. the domain sizes are reduced. The overall, informal observation is that the set of VAC-integral variables expands as $\theta_i$ gets smaller, and saturates at some point, which is usually before $\theta_i = 1$. However, even after this saturation, the domain sizes of non-VAC-integral variables do not necessarily cease shrinking. For the heuristic we present in Sect. 6, we aim to choose a threshold $\theta$ where we have a good compromise between the number of VAC-integral variables and the domain sizes of non-VAC-integral variables. This way, we increase the size of the easy (VAC-integral) part and decrease the complexity of the difficult part, while hopefully keeping most (may-be all) values belonging to the optimal solution (see Fig. 1d and 1f). In an informal sense, we consider those VAC-integral variables that were present with a higher $\theta$ to be more informative, and more likely to appear in an optimal solution. For example, if we assign against the VAC-integral value for a high value $\theta_i$, the cost of the best possible solution is at least $c_\emptyset + \theta_i$, whereas for $\theta_{i'} = 1$, the cost of the best possible solution that disagrees with the VAC-integral value can only be shown to be $c_\emptyset + 1$. Thus, the higher the $\theta$ for which a variable is VAC-integral, the less tight the relaxation needs to be for the corresponding VAC-integral value to appear in an optimal solution.

## 5    Branching Heuristics Based on VAC-integrality

For a branch-and-bound algorithm, the order in which variables are assigned has a crucial impact on the performance. In general, a branching decision should help the solver quickly prune sub-trees which contain no improving solutions, by creating sub-problems with increased dual bound in all branches [1].

Based on this observation and the connection of VAC-integrality to integrality explained in Sect. 3, we observe that branching on a VAC-integral variable $x$ will create a branch which must contain the VAC-integral value $a$ of $x$. Since $a$ is the only value in the domain of $x$ in $Bool(P)$ and its unary cost does not change by branching, $Bool(P \mid_{x=a})$ is identical to $Bool(P)$, so the dual bound is not improved in this branch. Therefore, it makes sense to avoid branching on VAC-integral variables[2].

To implement this, we find the set of VAC-integral variables implied by Proposition 3, i.e., those that have singleton domain in $Bool(P)$ and only allow branching on the rest. The choice among the rest of the variables is made using whatever branching heuristic the solver uses normally. In the case of TOULBAR2, which we use in our implementation, that is DOM+WDEG [5] together with the last conflict heuristic [19].

When only VAC-integral variables remain, we assign them all at the same time and check that the lower bound did not increase (see premature termination of VAC in [6]). If so, we update the upper bound if a better solution was found, unassign VAC-integral variables, and keep branching with the default heuristic.

For efficiency reasons, during search, EDAC [9] is established before enforcing VAC in TOULBAR2. If the VAC property (Definition 5) cannot be enforced at a

---

[2] Although, as a heuristic we cannot expect this to always be the best choice.

given search node due to premature termination of VAC, then VAC-integrality is unavailable for that node and again we rely on the default branching heuristic[3].

## 5.1  Exploiting Larger Zero-Cost Partial Assignments

Definition 7 requires there is a unique VAC-integral value $a \in D_i$ for each VAC-integral variable. The partial assignment of unique values to their corresponding variables implies a zero-cost lower-bound increase as said before. Thus, our branching heuristic will avoid branching on these variables. We could search for other potentially-larger assignments with the same zero-cost property. A simple way to do that is to test a particular value assignment and keep the variables not in *conflict* with it, *i.e.*, with no cost violations related to them or with their neighbors. We choose first to test the assignment based on *EAC values*, which are maintained by EDAC [14]. An EAC value is defined like a VAC-integral value but it is not required to be unique in the domain. If a variable is kept,

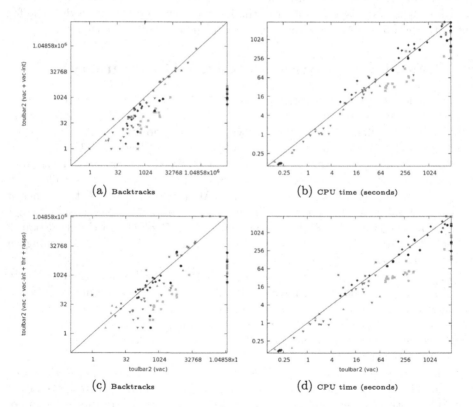

**Fig. 2.** Comparison with TOULBAR2 using VAC during search. CELAR: ∗, CPD: ●, ProteinDesign: ▲, ProteinFolding: ▼, Warehouse: ◆, Worms: ■. (Color figure online)

---

[3] We also tried to exploit the last valid VAC-integrality information collected along the current search branch, but it did not improve the results.

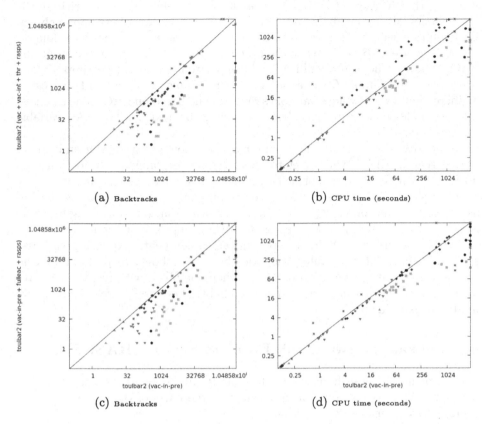

**Fig. 3.** Comparison with default TOULBAR2. CELAR: ✳, CPD: ●, ProteinDesign: ▲, ProteinFolding: ▼, Warehouse: ◆, Worms: ▥. (Color figure online)

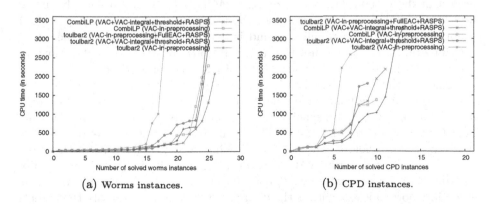

**Fig. 4.** Comparison with COMBILP on Worms and CPD instances.

it means that its EAC value is fully compatible, *i.e.*, has zero-cost, with all the EAC values of its neighbors. We call it a *Full EAC value*. This approach can be combined with VAC-integrality. By restricting the EAC values to belong to the AC closure of $Bool(P)$ when the problem is made VAC, we ensure that any VAC-integral value is also Full EAC. The opposite is not true (see Supp. Fig. 1). Thus, the set of Full EAC values can be larger. We perform the Full EAC test in an incremental way (using a variable's revise queue based on EAC value changes) at every node of the search tree, before choosing the next non Full EAC variable to branch on.

In order to improve this approach further, as soon as a new solution is found during search, EDAC will prefer to select the corresponding solution value as its EAC value for each unassigned variable if this value belongs to the current set of feasible EAC values. By doing so, we may exploit larger zero-cost partial assignments found previously during search. Notice that our branching heuristic is related to the min-conflicts repair algorithm [20] as it will only branch on variables in conflict with respect to a given assignment. Exploiting the best solution found so far for value heuristics has been shown to perform well on several constraint optimization problems when combined with restarts [10]. We use such a value ordering heuristic inside an hybrid best-first search algorithm [3] in all our experiments.

# 6   Relaxation-Aware Sub-Problem Search (RASPS)

One problem that branch-and-bound faces, especially in depth-first order, is that without a good upper bound it may explore large parts of the search tree that contain only poor quality solutions.

Here, we propose to use integrality information to try to quickly generate solutions that are close to the optimum. We describe a primal heuristic that we call Relaxation-Aware Sub-Problem Search (RASPS), which runs in preprocessing. We simply fix all VAC-integral variables to their values, prune values from the rest of the variables that are pruned in $AC(Bool(P))$, and then solve using the EDAC lower bound the resulting subproblem (see examples in Fig. 1d, 1f) to optimality or until a resource bound is met[4]. In order to choose the set of VAC-integral variables, we use the dual solutions constructed in iterations of VAC before the last, hence examine $Bool_\theta(P)$ for an appropriate $\theta$.

Although the idea of the heuristic is pretty straightforward, the key issue is to choose the threshold value (the $\theta$) (recall Sect. 4) to construct $Bool_\theta(P)$, as it has an impact on the quality of the upper bound produced and the time spent for this. To determine the threshold value for the RASPS, we observe the curves of the threshold $\theta_i$, the ratio of VAC-integral variables $r_i$, and the value $\alpha_i = r_i/\theta_i$, collected during VAC iterations. The idea is that, once the ratio of VAC-integral variables saturates, $\theta_i$ continues to decrease. As a result, $\alpha_i$ starts increasing more quickly which is the desired cutoff point. To identify that point,

---

[4] In our implementation, we set an upper bound of 1000 backtracks for solving the subproblem.

we track of the curve of $\alpha_i$ over the VAC iterations and choose the threshold value when the angle of the curve reaches 10° (see Supplementary Figure 2).

This idea is related to the COMBILP method of Haller et al. [13], described earlier. Compared to COMBILP, RASPS solves a simpler combinatorial subproblem because of the larger VAC-integral set and the remaining pruned domains. Then, it only aims to produce a good initial upper bound and leaves proving optimality to the branch-and-bound solver.

Even more closely related is the RINS heuristic of Danna et al. [7]. It also searches for primal bounds by extending the integral part of the relaxation. In contrast to RASPS, it permits values of the incumbent solution and may be invoked in nodes other than the root. However, it has no way of distinguishing among integral variables as RASPS does with its choice of $\theta > 1$. We have experimented with RASPS during search but have so far not found it worthwhile.

## 7   Experimental Results

We have implemented VAC-integrality and RASPS inside TOULBAR2, an open-source exact branch-and-bound WCSP solver in C++[5]. All computations were performed on a single core of Intel Xeon E5-2680 v3 2.50 GHz and 256 GB of RAM with a 1-hour CPU time limit. No initial upper bounds were used, as is the default of the solver.

### 7.1   Benchmark Description

We performed experiments on probabilistic and deterministic graphical models coming from different communities [15]. We considered a large set of 431 instances[6] which are all binary. It includes 251 instances (170 Auction, 16 CELAR, 10 ProteinDesign, 55 Warehouse) from the Cost Function Library[7], 129 instances (108 DBN, 21 ProteinFolding) from the *Probabilistic Inference Challenge* (PIC 2011)[8], 30 "Worms" instances [16] where COMBILP is state-of-the-art [13], and 21 Computational Protein Design (CPD) large instances for which TOULBAR2 is state-of-the-art [2,22]. We discarded Max-CSP, Max-SAT, Constraint Programming (CP), and Computer Vision (CVPR) instances which are either unweighted (all costs equal to 1), or non-binary, or being too easy (small search tree) or unsolved by all the tested approaches including MRF and ILP solvers [15].

### 7.2   Comparison with VAC

First, we compared our new heuristics with default VAC maintained during search (option $-A = 999999$ for all tested methods). We skipped Auction and DBN as they do not have VAC-integral variables.

[5] https://github.com/toulbar2/toulbar2, version 1.1.0.
[6] http://genoweb.toulouse.inra.fr/~degivry/evalgm/.
[7] forgemia.inra.fr/thomas.schiex/cost-function-library.
[8] www.cs.huji.ac.il/project/PASCAL.

In Fig. 2a, we show a scatter plot comparing the number of backtracks between VAC and VAC exploiting VAC-integral variable heuristic. The size of the search tree is significantly reduced thanks to VAC-integrality for most instance families. Notice the logarithmic axes. The improvement in terms of CPU time (Fig. 2b) is less important but still significant for CPD, ProteinFolding, Worms, and some CELAR instances. However, we found several Warehouse instances where it was significantly slower using VAC-integrality. In this case, we found the explanation was a larger number of VAC iterations per search node (8 times more in average) corresponding to small lower bound improvements at small threshold values ($\theta$ near 1) that did not reduce the search tree sufficiently (only by a mean factor 2.2 on difficult Warehouse instances).

In order to avoid such pathological cases, we placed a bound on the minimum threshold value $\theta$ for VAC iterations during search. We selected the same limit as for RASPS (e.g., $\theta_{30}$ for Worms/cnd1threeL1_1228061). We found that using this threshold mechanism alone speeds up Warehouse resolution and does not significantly deteriorate the results in the other families (see Supp. Fig. 4). Furthermore we obtained consistent results when combining with VAC-integrality, reducing the number of backtracks and CPU time for several families while being equivalent for the others (see Supp. Fig. 5).

Next, we analyzed the impact of applying the RASPS upper-bounding procedure in preprocessing. We limit RASPS to 1000 backtracks. Again, our new heuristic RASPS significantly reduces the search effort in terms of backtracks and time, except for Warehouse and some CELAR. For Warehouse, the upper bounds found did not reduce the total number of backtracks. For CELAR scen06_r, it reduces backtracks by 3.4 and solving time by 4.2. For Worms, it was more than 10 times faster for some instances (see Supp. Fig. 6).

Finally, we combine the two heuristics, VAC-integrality and RASPS, with VAC threshold limit and show the results compared to VAC alone in Fig. 2c and 2d. We keep this best configuration in the rest of the paper.

## 7.3  Comparison with VAC-in-preprocessing and COMBILP

One might expect using VAC only in preprocessing to be the fastest option, as it is the default for TOULBAR2 and significantly outperforms VAC during search in most cases [15]. For certain instance families, we manage to outperform it.

When VAC is used only in the preprocessing, using RASPS in addition considerably improves runtimes except for Warehouse and some CELAR (see Supp. Fig. 8). If we add VAC-integrality and RASPS when using VAC during search, we manage to outperform VAC in preprocessing for all families except CELAR and Warehouse, where the overhead of VAC is too high (see Fig. 3a and 3b).

Moreover, if we compare methods using VAC in preprocessing only, then exploiting our simpler Full EAC branching heuristic and RASPS performs even better in most cases, being as good as default TOULBAR2 on Warehouse instances (55 instances solved in average in 128 s) and comparable on CELAR (our approach solved graph13 and scen06 one-order-of-magnitude faster, but could not solve graph11 compared to default VAC in preprocessing, see Fig. 3c, 3d).

Next, we compare TOULBAR2 and COMBILP (which uses the same TOULBAR2 as its internal ILP solver) with different lower bound techniques, showing the advantages of exploiting VAC-integrality or Full EAC and RASPS extensions.

In Fig. 4a, we see a cactus plot[9] for the Worms benchmark where there are 30 instances. We solved these instances with different combinations of solvers and heuristics with a CPU time limit of 1 h. COMBILP was reported to solve 25 of these instances in [13] within 1 h CPU time. Here, we compare COMBILP with parameters used in [13] (VAC in preprocessing and EDAC during search), as well as our version of TOULBAR2 plugged in it. In addition to those, we have standalone TOULBAR2 either with VAC in preprocessing and EDAC during search, with or without Full EAC, or VAC-integrality-aware branching, and RASPS options. TOULBAR2 alone can go up to 25 instances. However, by plugging our version of TOULBAR2 in COMBILP, we manage to solve 26 of these instances, which makes 1 more than [13]. Another important detail is that, although it is costly to use VAC throughout the search tree, it becomes better with VAC-integrality and RASPS. Still, it was slightly dominated by Full EAC.

This simpler heuristic performed even better on the CPD benchmark (Fig. 4b). Our Full EAC heuristic with RASPS got the best results, solving 13 instances, compared to VAC-integrality and RASPS which solves 9, and only 8 by default TOULBAR2. COMBILP using VAC during search with VAC-integrality and RASPS solved 11 instances, instead of 10 without these options and VAC in preprocessing.

# 8    Conclusions

We revisited the Strict Arc Consistency property which was recently used in an iterative relaxation solver. We identified properties that make it easier to use within a branch-and-bound solver and in particular in conjunction with the VAC algorithm. This property allows us to integrate information about the relaxation that VAC computes to be used in heuristics. We presented three new heuristics that exploit this information, two for branching and the other for finding good quality upper bounds. In an experimental evaluation, these heuristics showed great performance in some families of instances, improving on the previous state of the art. VAC-integrality identifies a single zero-cost satisfiable partial assignment in a particular CSP $Bool(P)$ of the original problem $P$. Other CSP techniques such as neighborhood substitutability [12] could be used to detect larger tractable sub-problems. The integral subproblem can also be viewed as a particularly easy tractable class, where each variable has a single value. Therefore, another possible direction is to detect subproblems that are tractable for more sophisticated reasons.

**Acknowledgements.** This work has been partly funded by the "Agence nationale de la Recherche" (ANR-16-CE40-0028).

---

[9] It shows on the x-axis the number of instances solved for a time limit given in y-axis.

# References

1. Achterberg, T.: SCIP: solving constraint integer programs. Math. Program. Comput. **1**(1), 1–41 (2009)
2. Allouche, D., et al.: Computational protein design as an optimization problem. Artif. Intell. **212**, 59–79 (2014)
3. Allouche, D., de Givry, S., Katsirelos, G., Schiex, T., Zytnicki, M.: Anytime hybrid best-first search with tree decomposition for weighted CSP. In: Pesant, G. (ed.) CP 2015. LNCS, vol. 9255, pp. 12–29. Springer, Cham (2015). https://doi.org/10.1007/978-3-319-23219-5_2
4. Bacchus, F., Chen, X., van Beek, P., Walsh, T.: Binary vs. non-binary constraints. Artif. Intell. **140**(1/2), 1–37 (2002)
5. Boussemart, F., Hemery, F., Lecoutre, C., Sais, L.: Boosting systematic search by weighting constraints. In: Proceedings of ECAI 2004, pp. 146–150, Valencia (2004)
6. Cooper, M., De Givry, S., Sánchez, M., Schiex, T., Zytnicki, M., Werner, T.: Soft arc consistency revisited. Artif. Intell. **174**(7–8), 449–478 (2010)
7. Danna, E., Rothberg, E., Le Pape, C.: Exploring relaxation induced neighborhoods to improve MIP solutions. Math. Program. **102**(1), 71–90 (2005)
8. Davies, J., Bacchus, F.: Solving MAXSAT by solving a sequence of simpler SAT instances. In: Lee, J. (ed.) CP 2011. LNCS, vol. 6876, pp. 225–239. Springer, Heidelberg (2011). https://doi.org/10.1007/978-3-642-23786-7_19
9. de Givry, S., Zytnicki, M., Heras, F., Larrosa, J.: Existential arc consistency: getting closer to full arc consistency in weighted CSPs. In: Proceedings of IJCAI 2005, pp. 84–89, Edinburgh (2005)
10. Demirović, E., Chu, G., Stuckey, P.J.: Solution-based phase saving for CP: a value-selection heuristic to simulate local search behavior in complete solvers. In: Hooker, J. (ed.) CP 2018. LNCS, vol. 11008, pp. 99–108. Springer, Cham (2018). https://doi.org/10.1007/978-3-319-98334-9_7
11. Downey, R.G., Fellows, M.R.: Fundamentals of Parameterized Complexity. Texts in Computer Science. Springer, London (2013). https://doi.org/10.1007/978-1-4471-5559-1
12. Freuder, E.C.: Eliminating interchangeable values in constraint satisfaction problems. In: Proceedings of AAAI 1991, pp. 227–233, Anaheim (1991)
13. Haller, S., Swoboda, P., Savchynskyy, B.: Exact map-inference by confining combinatorial search with LP relaxation. In: Proceedings of AAAI 2018, pp. 6581–6588, New Orleans (2018)
14. Heras, F., Larrosa, J.: New inference rules for efficient max-SAT solving. In: Proceedings of the National Conference on Artificial Intelligence, AAAI-2006 (2006)
15. Hurley, B., O'Sullivan, B., Allouche, D., Katsirelos, G., Schiex, T., Zytnicki, M., de Givry, S.: Multi-language evaluation of exact solvers in graphical model discrete optimization. Constraints **21**(3), 413–434 (2016)
16. Kainmueller, D., Jug, F., Rother, C., Myers, G.: Active graph matching for automatic joint segmentation and annotation of *C. elegans*. In: Golland, P., Hata, N., Barillot, C., Hornegger, J., Howe, R. (eds.) MICCAI 2014. LNCS, vol. 8673, pp. 81–88. Springer, Cham (2014). https://doi.org/10.1007/978-3-319-10404-1_11
17. Koller, D., Friedman, N.: Probabilistic Graphical Models: Principles and Techniques. The MIT Press, Cambridge (2009)
18. Kolmogorov, V.: Convergent tree-reweighted message passing for energy minimization. IEEE Trans. Pattern Anal. Mach. Intell. **28**(10), 1568–1583 (2006)

19. Lecoutre, C., Sais, L., Tabary, S., Vidal, V.: Reasoning from last conflict(s) in constraint programming. Artif. Intell. **173**(18), 1592–1614 (2009)
20. Minton, S., Johnston, M., Philips, A., Laird, P.: Minimizing conflicts: a heuristic repair method for constraint satisfaction and scheduling problems. Artif. Intell. **58**, 160–205 (1992)
21. Morgado, A., Ignatiev, A., Marques-Silva, J.: MSCG: robust core-guided MaxSAT solving. JSAT **9**, 129–134 (2014)
22. Ouali, A., et al.: Variable neighborhood search for graphical model energy minimization. Artif. Intell. **278**(103194), 22p. (2020)
23. Savchynskyy, B., Kappes, J.H., Swoboda, P., Schnörr, C.: Global map-optimality by shrinking the combinatorial search area with convex relaxation. In: Proceedings of NIPS 2013, pp. 1950–1958, Lake Tahoe (2013)
24. Sontag, D., Meltzer, T., Globerson, A., Weiss, Y., Jaakkola, T.: Tightening LP relaxations for MAP using message-passing. In: Proceedings of UAI, pp. 503–510, Helsinki (2008)

# Exact Method Approaches for the Differential Harvest Problem

Gabriel Volte[1]([✉]), Eric Bourreau[1]([✉]), Rodolphe Giroudeau[1]([✉]),
and Olivier Naud[2]([✉])

[1] LIRMM, University of Montpellier, CNRS, Montpellier, France
{volte,bourreau,giroudeau}@lirmm.fr
[2] ITAP, Univ Montpellier, INRAE, Institut Agro, Montpellier, France
olivier.naud@inrae.fr

**Abstract.** The trend towards a precise, numerical, and data-intensive agriculture brings forward the need to design and combine optimization techniques to obtain decision support methodologies that are efficient, interactive, robust and adaptable. In this paper, we consider the Differential Harvest Problem (DHP) in precision viticulture. To tackle this problem, we dedicated a specific column generation approach with enumeration techniques and a constraint programming model. Therefore, a set of simulated instances (which differ in field shape, zone shape, and size) was created to perform a parametric study on our different approaches. The specific column generation approach presented in this paper is preliminary work in the development path of more sophisticated resolution methods such as robust optimization and column generation/constraint programming hybridization.

**Keywords:** Column generation · Enumeration technique · Constraints programming · Exact method · Precision agriculture

## 1 Problem Description

The Differential Harvest Problem, introduced by [6], consists of optimizing harvests of different grape qualities in vineyards so that we obtain a certain quantity, denoted by $Rmin$, of good quality grapes. In the problem, there are only two types of grape quality: A-grapes and B-grapes (let us assume that A-grapes are of better quality than B-grapes).

Thanks to agronomic information obtained a priori, it is possible to map (see Fig. 1) a vineyard by distinguishing areas according to the quality of the grapes. Meanwhile, geolocated harvesting machines equipped with two hoppers (harvest tanks with a maximum load of $CapaMax$) can use such a map. When one of the two hoppers is full, both must be emptied into a bin located at the edge of the plot. These machines have two harvesting modes, that can be changed only

Supported by organization French National Research Agency under the Investments for the Future Program, referred as ANR-16-CONV-0004.

**Fig. 1.** Illustration of the agronomic map with different zones of grapes quality (real data from the Gruissan vineyard).

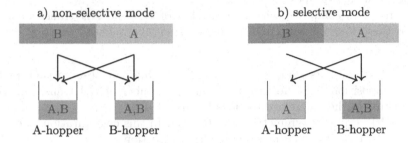

**Fig. 2.** Illustration of the different harvesting modes.

when they are emptying the hoppers at the bin, the selective mode, and the non-selective mode. The selective mode corresponds to sorting good grapes quality in one hopper and other grapes in the second hopper. When the $Rmin$ quantity of A-grapes is harvested in the selective mode the machine can change to the non-selective mode where the loading capacity is perfectly balanced between both hoppers (cf. Fig. 2).

There is a technical issue with the harvesting machine which is related to the longitudinal size of its picking head of the harvester. This size is almost the length of the wheelbase, approximately 5 m. The problem occurs during zone changes and more particularly during the change from a zone with $B$-grapes to a zone with $A$-grapes. Let us note this change $BA$ ($AB$ for the change from zone $A$ to $B$).

When a $BA$ transition occurs, three cases can be isolated (see Fig. 3):

- case a), the machine head has just entered zone A. The harvesting hopper (which is hopper B) cannot be changed to hopper A because it is still harvesting B-grapes and therefore hopper A would be corrupted with B-grapes.
- In case b), the grape picker is overlapping both zones. The reasoning is the same as in case a) because it is still harvesting $B$-grapes.
- In case c), the harvester is completely in zone $A$, we no longer harvest $B$-grapes. The harvesting hopper can, therefore, be changed from hopper $B$ to hopper $A$ to harvest the $A$-grapes in hopper $A$.

a)                    b)                    c)

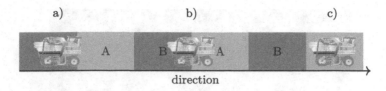

direction

**Fig. 3.** Latency illustration.

When changing from $AB$, the harvesting hopper is changed from hopper $A$ to hopper $B$ as soon as the area transition has occurred, avoiding the corruption of $A$-grapes with $B$-grapes.

Any row $r$ is composed with a succession of $BA$ and $AB$ transitions, this leads to these four following row representations (consecutive $A$ or $B$ can be merged):

- $r = (AB)^*$ and its symmetric $r = (BA)^*$. One can observe that in this row composition the difference between the number of $BA$ transition when harvesting in the different directions is exactly one.
- $r = (AB)^*A$ and its symmetric $r = (BA)^*B$. In this configuration, the number of $BA$ transitions is the same in either direction.

Depending on the row configuration and the row direction the latency causes an asymmetry in the harvested quantities, therefore in the optimization part, we will need to determine the direction of each row to obtain the $Rmin$ quantity. For the sake of simplicity, the vineyard rows extremities are oriented upper to bottom, the upper extremity can be chosen arbitrarily.

The article is organized as follows: the Sect. 1 gives the context of this work and provides the problem definition. The Sect. 2 refers to the previous and related works on the problem. In Sect. 3 we introduce a new graph model and provide models for two exact methods; first a column generation approach (see Subsect. 3.1) then a constraint programming model (see Subsect. 3.2). In the Sect. 4 we outline the results obtained with our different approaches.

## 2   Related Works

Precision agriculture is a principle of agricultural parcel management that aims to optimize yields and investments, by seeking to take better account of the variability of environments. In [19], the authors asserted that precision farming was the future of crop nutrition offering benefits in crop quality, sustainability, food safety, etc. A few years later, [13] reviewed some precision agriculture advances and proposed new directions of research. [17] analyzed the adoption of precision agriculture technologies for several actors in the agricultural industry.

In the last decades, a lot of works has been done in digital and precision agriculture using operation research techniques involved in the resolution of vehicle routing problems, scheduling problems or stochastic problems.

The surveys [2] and [3] assert that most of the agricultural applications involve the motion of machines hence they first classify the agricultural field operations which can be modeled as a vehicle routing problem. They present an approach to represent the planning and scheduling of moving machines as a vehicle routing problem with time windows where the machine has to process deterministic, stochastic or dynamic requests.

VRP-specific optimization methods are addressed in [20] to solve their fleet routing problem, they have reduced the operating time by at most 17.3% using their approach rather than a human-made solution.

One can find some works on-field coverage problem [7,14], where the goal is to cover a field under technical constraints. The purpose of the work presented in [14] is to reduce the soil compaction. Smaller vehicles are used to cover the fields but those vehicles have smaller storage capacity. Thus the full field coverage in a single run is no longer possible and a path planning strategy is used to partially cover the field under *compacted area minimization constraints*. In comparison, the article [7] focuses both on finding minimal operating costs and balancing the vehicle's workload.

The paper [18] explores the consequence of groundwater resource use under climate change scenarios. The problem is modeled as a dynamic stochastic problem with several temporal decision stages with multiple sources of risk that should impact farmer decisions.

Investment behavior under different policy and price scenarios was studied in [22]. They employ a dynamic multi-objective farm-household integer programming model on a northern Italy case study application highlighting the potentialities and the limits of the methodology applied.

A column generation approach is used to solve a crop rotation scheduling problem to produce a pre-determined demand for crops while respecting some ecological production constraints, this problem is called the sustainable vegetable crop demand-supply problem [10].

In few available literatures, the DHP has been solved with methods based on Constraint Programming (CP), first, a step model was proposed in [5] where they were able to optimally solve a 16 rows instance in 6 days, then a precedence model was introduced [6] which gives better results yet instances with 14 rows reach the 2 h time limit, and Cost-Optimal Reachability Analysis [24] (CORA, which is a model-checking techniques branch) finding optimal solutions in vineyards with up to 12 rows in a few minutes.

## 3  Models

We propose a new graph model of the problem that we call the "flatten" representation of the vineyard (extremities are merged to consider only the rows) in opposition with the "physical" representation of the vineyard (rows are split into two extremities) used by [6] and [24]. Let $n$ be the number of rows, $CapaMax$ be the hopper's capacity, $Rmin$ the minimum A-grapes quality needed to be harvested, $b$ be the bin/depot, an undirected weighted complete

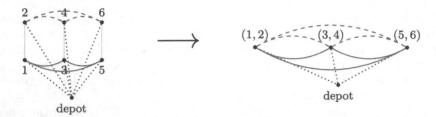

**Fig. 4.** The flattened graph (right) and the "physical" graph (left) representation of the vineyard. Solid edges correspond to direction 0 and dashed edges refer to direction 1. Dotted edges bear no direction.

graph $G = (V, E)$ with labeled edges, $V = \{0, \ldots, n\} \cup \{b\}$ and $E = \{(i, j, k) | i, j \in V^2, \ k \in \{0, 1\}\}$ ($k$ refers to the row harvest direction see Fig. 4). Taking the edge $(i, j, k)$ means that the vehicle moves from row $i$ to row $j$ harvesting row $i$ with direction $k$, if $k = 0$ the row is harvested from top to bottom, otherwise the row is harvested from bottom to top. Let say $D$ is the weight of $G$ computed from a distance matrix. It must be noted that because the vehicle has an important turning radius, the distance between rows is not exactly euclidean. $qA/qB$ are the two harvesting functions: $(i, k) \mapsto \mathbb{N}, i \in V, k \in \{0, 1\}$.

The goal is to find a set of disjoint routes (except for the bin $b$), respecting the $CapaMax$ capacity, covering $V$ and minimizing the harvesting time (here the distance) while harvesting at least $Rmin$ quantity of A-grapes.

This problem can be seen as a Heterogeneous Vehicle Routing Problem [21] with two resource constraints (the grapes quality, A and B) and two vehicle types (one selective and the other non-selective) with different capacities. If the vehicle is selective the harvested grapes are sorted according to their quality, thus the vehicle has two hoppers each with a capacity equal to $CapaMax$, otherwise, both grapes quality are mixed in both hoppers, therefore, the vehicle hoppers can be merged into only one hopper with a capacity $2 * CapaMax$. Let us consider $\gamma$ the number of selective vehicles and $\lambda$ the total number of vehicles, $Qtot$ refers to the total amount of grapes present in the vineyard. The following bounds are easy to verify:

$$\left\lceil \frac{Rmin}{CapaMax} \right\rceil \leq \gamma \leq n \text{ and } \left\lceil \frac{Qtot}{2 * CapaMax} \right\rceil \leq \lambda \leq n$$

Previous works on the problem tackled it with constraint programming and model checking. We would like to investigate integer linear programming and more precisely column generation to observe the effectiveness of these techniques on the DHP. As early preliminary work, we tried to solve this problem with integer linear programming models and out-of-the-box solving methods. Because this was unsuccessful, and as our problem is a VRP variant, we concluded that column generation techniques may be appropriate here. The following Subsect. 3.1 explains the use of these techniques on the DHP. We identified that the principal weakness of the constraint programming approach used previously was to introduce a new global constraint that specifically fits the DHP. To avoid this, we decided to focus on designing constraint programming models and decision strategies that suit the most to our problem while using state-of-the-art generic constraints. We dedicate the Subsect. 3.2 to highlight our constraint programming approaches.

## 3.1   The Column Generation Approach

We use Dantzig-Wolfe decomposition [9] on the DHP to obtain a set partitioning master problem with a side constraint which is the $Rmin$ constraint and an elementary shortest path with resource constraints pricing problem.

Let us denote $\Omega$ the set of all feasible routes. The master problem selects routes in $\Omega$ to obtain at least the $Rmin$ quantity. For the sake of memory usage, only solving the master problem, restricted to a subset $R$ of $\Omega$, is worth considering because the number of routes in $\Omega$ grows exponentially with the number of rows. Let's consider $K = \{s, \bar{s}\}$ as the set of vehicle types. Assume $R \subset \Omega$ is the set of feasible routes $R = (R^{\bar{s}} \cup R^{s})$, with $R^{\bar{s}}$ be the set of non-selective routes and $R^{s}$ be the set of selective routes. Selective routes must be generated independently of the non-selective routes, they use vehicle types with different capacities and resources consumption, therefore two pricing problems will be needed, one to generate selective routes and the other to generate non-selective routes.

Despite the classical branch-and-price scheme to VRP, we decided to opt for an enumeration technique [1]. There are several works on enumeration (see [1,8,15,23] for more details). The enumeration technique operates as follows. First, the relaxation of the restricted master problem is solved, like if we were solving the root node with branch-and-price, to obtain the linear relaxation optimal value $z_{MP}^{*}$. With the last solved master problem special dual costs are obtained, denoted by $CR^{*}$. Then any integer solution gives an upper bound $z_{IP}$. These bounds and the dual costs are provided to a specific pricing problem which will generate all columns with a reduced cost of $0 < z_{IP} - z_{MP}^{*} = \epsilon$. If we solve the new restricted master problem, this time without the linear relaxation, the solution obtained is optimal for the restricted master problem and thus for the master problem. The principal drawback of this technique is that too many columns may be generated if the linear relaxation of the restricted master problem is of poor quality, the gap with any integer solution would be too large.

Let us define the restricted master problem decision variables:

$$
y_r^k = \begin{cases} 1 \text{ if route } r \text{ is used by vehicle of type k} \\ 0 \text{ otherwise} \end{cases} , \; \forall k \in K, \; \forall r \in R^k
$$

Note $a_{ir} = 1$ if row $i$ is harvested in the route $r$, 0 otherwise. Let $q_r^t$ be the quantity of $t$-grapes collected in the route $r$ and $c_r$ the cost of the route $r$ (total traveled distance computed from the weight matrix $D$). The route cost is the total traveled distance on the route. A non-selective route does not harvest A-grapes, all grapes are mixed in B-grapes.

---

Model M1: Restricted Master Problem

---

$$\text{Min} \sum_{k \in K} \sum_{r \in R^k} y_r^k c_r \tag{1}$$

$$\sum_{k \in K} \sum_{r \in R^k} y_r^k a_{ir} = 1 \qquad \forall i \in V \qquad\qquad [\pi_i] \tag{2}$$

$$\sum_{r \in R^s} q_r^A y_r^s \geq Rmin \qquad\qquad\qquad\quad [\theta] \tag{3}$$

$$y_r^k \in \{0, 1\} \qquad\qquad \forall k \in K, \ \forall r \in R^k \tag{4}$$

---

The aim is to minimize the cost of each route (1) while the entire field is harvested (2). The constraint (3) verifies that at least the $Rmin$ quantity of good quality grapes is harvested. And finally, the constraints (4) ensures the integrality of the $y$ variables.

**Pricing Sub-problem.** The pricing problem generates improving routes for the master problem based on the value of the dual variables associated with the constraints of the restricted master problem.

We solve the pricing problem, the shortest path problems with resource constraints, with dynamic programming using a labeling algorithm, introduced first by [12], enhanced by [11]. We adjusted this algorithm to our two pricing problems. Because extension and dominance rules are straightforward from the original, we decided not to reintroduce them in this paper. The only originality in our label algorithm is that two labels lists are used. One list, $L^0$, for labels with direction 0 and the other, $L^1$, for labels with direction 1. Therefore as the direction changes for every row in a route, the label obtained from the extension of the label $L^0$ in the labels list with direction 0 is added to the labels list with direction 1.

For each constraint (2), we obtain a dual cost $\pi_i$ and each constraint (3) gives the $\theta$ dual cost. Using these dual costs we compute a reduced cost $\hat{c}_r^{\bar{s}}$ for any route $r$ in $R^{\bar{s}}$ and $\hat{c}_r^s$ for any route $r$ in $R^s$ in the master problem. It also important when generating new selective routes to take the $\theta$ dual cost into account at each label extension to avoid interesting routes from being dominated:

$$\hat{c}_r^s = c_r - \sum_{i \in r} a_{ir} \pi_i - q A_r \theta = c_r - \sum_{i \in r} a_{ir} (\pi_i + q A_i \theta)$$

$$\hat{c}_r^{\bar{s}} = c_r - \sum_{i \in r} a_{ir} \pi_i$$

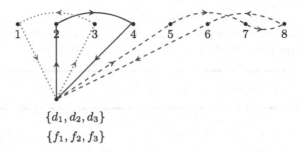

$$\{d_1, d_2, d_3\}$$
$$\{f_1, f_2, f_3\}$$

**Fig. 5.** Illustration of the optimal solution obtained with constraint programming for an instance with 8 rows. Plain arcs (resp. dotted and dashed arcs) representing vehicle one (resp. two and three), for the sake of clarity only the used vehicles are represented.

In the column generation part, routes with positive reduced cost can be discarded because they cannot belong to any optimal solution. Nevertheless, in the enumeration procedure, a new pricing problem is solved which brings out the routes with a reduced cost strictly less than $\epsilon$.

### 3.2 Constraint Programming

We introduce in this section our constraint programming approach (CP). First, we describe a classical precedence model mainly used in VRP (partially used by [6], fully explained in [4]) enhanced by specific bin packing capacity constraints. Finally, a constructive search strategy is detailed.

Let $V$ be the set of at most $n$ vehicles($|V| = n$). For each vehicle, two dummy depots are created: an initial depot and a final depot. Note $V^d$ (resp. $V^f$) the set of initial (resp. final) depots. The set of nodes $N$ in the graph, such that $|N| = n + 2 * |V|$ (one node per row and two nodes per vehicle), is ordered as follows:

$$N = \{\overbrace{1, \ldots, n}^{rows}, \overbrace{n+1, \ldots, n+|V|}^{V^d}, \overbrace{n+|V|+1, \ldots, n+2*|V|}^{V^f}\}$$

Let us define the decision variables (see Fig. 5 and Table 1, the distance matrix are displayed in the Fig. 11 and Fig. 12):

- $\forall i \in N, successor_i \in N$ (resp. $predecessor_i \in N$) variables giving the successor (resp. predecessor) of the row $i$ in the route.
- $\forall i \in N, position_i \in N$, variables providing the position of the row $i$ in the route, initial depots are in position 0.
- $\forall i \in N, assignment_i \in V$ variables indicating the vehicle assignment of the row $i$.
- $\forall i \in N, direction_i \in \{0, 1\}$, variables showing the direction of row $i$.
- $\forall v \in V, selective_v \in \{0, 1\}$, variables indicating whether a vehicle is selective.

- $\forall i \in N$, $\forall u \in \{A, B\}$, $CapaSumq_i^u \in [0, CapaMax]$, variables measuring the cumulative sum of the harvested $u$-grapes quantity before harvesting row $i$.
- $Obj$ is the objective variable which computes the total traveled distance.

---

Model M2: Constraint Programming

---

$$\forall f_i \in V^f \qquad successor_{f_i} = n + i \tag{5}$$

$$\forall d_i \in V^d \qquad assignment_{d_i} = i \tag{6}$$

$$\forall f_i \in V^f \qquad assignment_{f_i} = i \tag{7}$$

$$\forall i \in V^d \qquad position_i = 0 \tag{8}$$

$$AllDifferent(successor_1, \ldots, successor_{|N|}) \tag{9}$$

$$\forall i \in N \qquad direction_{successor_i} = 1 - direction_i \tag{10}$$

$$\forall i \in N \qquad position_{successor_i} = 1 + position_i \tag{11}$$

$$\forall i \in N \qquad assignment_{successor_i} = assignment_i \tag{12}$$

$$\sum_{i \in [1,n]} (qA_i * selective_{assignment_i}) \geq Rmin \tag{13}$$

$$\forall i \in N \qquad CapaSumqA_{successor_i} = CapaSumqA_i + Q_i^A \tag{14}$$

$$\forall i \in N \qquad CapaSumqB_{successor_i} = CapaSumqB_i + Q_i^B \tag{15}$$

$$\forall d_i \in V^d \qquad s_{d_i} < s_{d_{i+1}} \tag{16}$$

$$\forall i \in N \qquad predecessor_k = i \Leftrightarrow successor_i = k \tag{17}$$

$$\forall u \in \{A, B\} \quad diffN(assignment, CapaSumq^u, (1, \ldots, 1)^T, Q^u) \tag{18}$$

$$Obj = \sum_{i \in N} D_{i, successor_i, direction_i} \tag{19}$$

$$\min Obj \tag{20}$$

---

Constraints (5), (6), (7) and (8) assign the successor of a final depot to the corresponding initial depot, the same vehicle is affected to initial and final depot and initial depots are in position 0. The $AllDifferent$ constraint (9) ensures that routes are disjointed and that they cover all the rows, every successor is distinct implies that a row is taken exactly once (there are $n$ values for $n$ variables). The direction, position and assignment variables are updated depending on their successors (10), (11) and (12). Constraint (13) ensures that the $Rmin$ A-grapes quantity is collected. The cumulative amount of already harvested A-grapes (resp. B-grapes) is computed with (14) (resp. (15)), these constraints ensure that no vehicle exceeds its hoppers loading due to the maximum value of the

**Table 1.** Variables affectation for an eight rows instance, $CapaMax$ is set to 200 and $Rmin$ to 174, illustrated in Fig. 5. The harvesting quantity qA/qB is displayed for the direction 1, but for the opposite direction, the harvesting quantity is computed by adding 5 (the latency) to qB and to subtract 5 to qA.

| Node | 1 | 2 | 3 | 4 | 5 | 6 | 7 | 8 | $d_1$ | $d_2$ | $d_3$ | $f_1$ | $f_2$ | $f_3$ |
|------|---|---|---|---|---|---|---|---|-------|-------|-------|-------|-------|-------|
| qA | 87 | 75 | 62 | 50 | 37 | 25 | 12 | 0 | 0 | 0 | 0 | 0 | 0 | 0 |
| qB | 13 | 25 | 38 | 50 | 63 | 75 | 88 | 100 | 0 | 0 | 0 | 0 | 0 | 0 |
| successor | $f_2$ | 4 | 1 | $f_1$ | 7 | $f_3$ | 8 | 6 | 2 | 3 | 5 | $d_1$ | $d_2$ | $d_3$ |
| position | 2 | 1 | 1 | 2 | 1 | 4 | 2 | 3 | 0 | 0 | 0 | 3 | 3 | 5 |
| assignment | 2 | 1 | 2 | 1 | 3 | 3 | 3 | 3 | 1 | 2 | 3 | 1 | 2 | 3 |
| direction | 0 | 1 | 1 | 0 | 1 | 0 | 0 | 1 | 0 | 0 | 0 | 1 | 1 | 1 |
| selective | 1 | 1 | 1 | 1 | 0 | 0 | 0 | 0 | 1 | 1 | 0 | 1 | 1 | 0 |
| CapaSumqA | 57 | 0 | 0 | 70 | 0 | 150 | 50 | 100 | 0 | 0 | 0 | 120 | 144 | 200 |
| CapaSumqB | 43 | 0 | 0 | 30 | 0 | 150 | 50 | 100 | 0 | 0 | 0 | 80 | 56 | 200 |

$CapaSum$ domain. The constraints (19) and (20) minimize the total traveled distance.

We added redundant constraints to improve the performance of this model: first, symmetry breaking constraints (16) between vehicles are used then a channeling constraint (17) between *successor* and *predecessor* variables. The *diffN* constraints (18) are global constraints which are used to ensure that each task $T_i$ (the harvesting of row $i$), represented as a rectangle with coordinates $(assignement_i, CapaSumq_i^u)$ and with width of 1 and height of $Q_i^A$, do not overlap each other [16]. These constraints are also redundant and handle the vehicle capacity taking account of the vehicle filling and the object positioning in the vehicle, this positioning depends on both the $X$ and $Y$ axis (see Fig. 6). We create two tasks/rectangles per row: one indicating the A-grape quantity harvested in the rows and the other indicating the B-grapes quantity harvested. Tasks are affected by vehicle assignment ($X$ axis) and ordered depending on their position in the route with the $CapaSumq^u$ variables $\forall u \in \{A, B\}$ ($Y$ axis) which link row with its direct successor. The height of a task, $Q_i^u \forall u \in \{A, B\}$ , depends on the type of the node $i \in N$: $q_{i,selective_i,direction_i}^u$ for rows, 0 for initial depots and $CapaMax - CapaSum_{v_f}^u$ for final depots. The height of the final depots thereby defined is used to make the *diffN* constraint even more compact, all vehicles will have their maximum capacity reached due to the height of the final depots which fill the remaining space (see Fig. 6).

**The Snake Constructive Search Strategy.** One of the advantages of constraint programming is the possibility to choose the search strategy in the decision tree. The strategy is based on two axes: variable choice and value choice. The variable choice is a determining factor in a good strategy. Indeed, if the first

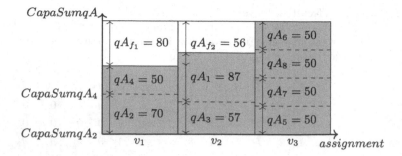

**Fig. 6.** Representation of the *diffN* constraint for the A-grapes for the instance shown in the Fig. 5.

variables used have a large domain or are weakly constrained then the decision tree is larger.

With this in mind, we wanted to create a strategy that mimics a greedy method, which we call: the Snake. The Snake strategy builds a route and then constructs entirely the next one and so on. The purpose is to find a good solution in a few decision nodes. Other strategies (as DomOverWDeg) generally build route fragments, and, at the bottom of the decision tree, try to assemble these fragments to obtain a feasible route. This assembly is very combinatorial due to the many possible arrangements of the fragments, see Fig. 7.

The Snake strategy behaves as the nearest neighbor heuristic: the first variable is the first available initial depot. The choice of the new variable is determined by the choice of the value of the previously instantiated variable. The first available value for the variable is selected, the values are increasingly sorted (we want to minimize the total distance). Once a final depot is instantiated, the next variable selected is the next initial depot. The variables that calculate the distance to the successor are used. Besides, these variables are sorted so that optimization begins with selective routes.

## 4    Experimental Results

In this section, we present the experimental results, first, we introduce the data sets used for the experimental study, then we outline the results obtained with our models. For the parametric study, we were focusing on exact methods and thus on finding optimal solutions.

### 4.1    Description of Data Sets

We have tested our models on different instances, some were artificially generated based on real data observed in a vineyard, at INRA Pech-Rouge (Gruissan), located in the southern France (cf. Fig. 1).

As [6,24] we decompose this instance to create a set of smaller instances, denoted by *G*-instances. We create instances with 10 consecutive rows and up to

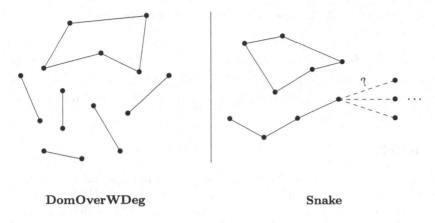

**DomOverWDeg**                              **Snake**

**Fig. 7.** The Snake strategy vs the DomOverWDeg strategy illustration.

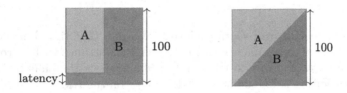

**Fig. 8.** Vineyard shape representation of the square instances (left side) and the triangle instances (right side).

24 rows (the original instances), with a hopper capacity of 900 and 1800 and the $Rmin$ value takes value in $\{0, 0.25, 0.5, 0.7, 0.9, 1\}$ of the total available A-grapes that can be effectively be harvested in the vineyard taking into account minimum loses due to the latency effect.

Artificial instances were generated based on two vineyard shapes, the square instances ($S$-instances) and the triangle instances ($T$-instances) see Fig. 8, that seems to be theoretically interesting. The study of these shapes is sufficient because the vineyard shape is hidden in the harvested A-grapes and B-grapes in each row and the difference between the number of latency in each row depending on the direction is at most one. We assume that the triangle instances must be the worst-case instances because everything is symmetric. The square instances were determined to see the influence of the increase of the $Rmin$ value on the objective function.

The same construction as the original instance is used to construct artificial instances. The Table 2 summarizes the parameters variation of each instance type.

**Table 2.** Parameters variation for each data set.

|              | #rows                  | $CapaMax$             | $Rmin$                        | Total |
|--------------|------------------------|-----------------------|-------------------------------|-------|
| $G$-instances | $\{10, 11, \ldots, 24\}$ | $\{900, 1800\}$        | $\{0, 0.25, 0.5, 0.7, 0.9, 1\}$ | 1140  |
| $T$-instances | $\{10, 11, \ldots, 24\}$ | $\{100, 200, \ldots, 500\}$ | $\{0, 0.25, 0.5, 0.7, 0.9, 1\}$ | 540   |
| $S$-instances | $\{10, 11, \ldots, 24\}$ | $\{100, 200, \ldots, 500\}$ | $\{0, 0.25, 0.5, 0.7, 0.9, 1\}$ | 540   |

## 4.2   Results

In this section, we will detail the results obtained for our approaches. The results were performed on an Optiplex 5055 computer with an AMD RYZEN 7 Pro 1700 (8 cores/4 Mo/16 Threads/3 GHz) processor. For the column generation approach, we use CPLEX 12.8.0.0 and for the constraint programming, we use Choco 4.10.1.

**Results for the Column Generation Methods.** The time limit for all the instances was fixed to 10 min, except for the original instance with 24 rows where the time limit was 1 h, due to the number of instances that has to be solved. The integer solution is found using a MIP solver on the RMP.

To the best of our knowledge, we are the first who were able to solve the real instance with 24 rows, see Fig. 1, in 2210 s and almost all the computation time is used to generated routes and to find an integer solution, the enumeration technique took less than a few minutes and generates only 18 columns.

In the Table 5, a comparison is made between the results obtained with the variation of the $CapaMax$ parameter for the S and T instances and the same results for the G instances are shown in the Table 3. The results displayed are the average computation time and the average integrality gap of the $Rmin$ values for every row's number and $CapaMax$ value. When the $CapaMax$ parameter increases there is a huge increase in the computation time since the pricing problem is more difficult to solve (the number of non-dominated routes generated is larger). For the G instances, the gap is really small (less than 0.002 %) however almost all the large instances reach the time limit.

The Fig. 9 and Fig. 10 indicate the results obtained for the G instances according to the $Rmin$ parameter variation and each point represents the average value of the $CapaMax$ values. One can see that the objective value and the computation time appear not to increase from 0% to 90% $Rmin$ but when demanding to harvest all the A-grapes the objective value grows due to the latency, all the rows have to be harvested in the direction that maximizes the A-grapes harvested, yet the computation time is low even for the biggest instances.

**Table 3.** Results for the G instances with the variation of the *CapaMax* parameter.

| #rows | CapaMax | | | |
|---|---|---|---|---|
| | 900 | | 1800 | |
| | CPU(s) | gap(%) | CPU(s) | gap(%) |
| 10 | 1.87 | $\sim 0$ | 10.71 | $\sim 0$ |
| 11 | 5.98 | $\sim 0$ | 38.43 | 0.016 |
| 12 | **6.35** | $\sim 0$ | **81.78** | $\sim 0$ |
| 13 | **118.84** | $\sim 0$ | **344.90** | 0.01 |
| 14 | **6.46** | $\sim 0$ | **217.08** | 0 |
| 15 | 85.04 | $\sim 0$ | 462.85 | 0.02 |
| 16 | 11.15 | $\sim 0$ | 429.40 | $\sim 0$ |
| 17 | 227.66 | $\sim 0$ | 538.73 | 0.02 |
| 18 | 19.27 | $\sim 0$ | 536.99 | 0.0003 |
| 19 | 313.19 | $\sim 0$ | 535.27 | $\sim 0$ |
| 20 | **97.11** | $\sim 0$ | **534.45** | $\sim 0$ |
| 21 | 331.28 | $\sim 0$ | 528.05 | 0.03 |
| 22 | **165.95** | $\sim 0$ | **539.18** | 0.04 |
| 23 | 545.24 | $\sim 0$ | 530.19 | 0.08 |

**Results for the Constraint Programming Approaches.** Precedent works using constraint programming to solve the DHP were able to solve optimally a 16 rows instances in 6 days, then optimally solve 10 and 12 rows instances with a 2 h time limit. We propose to compare our approach to its ability to obtain a good feasible solution in a short time.

With this in mind, we only show the results for a few instances (one with 10 rows, 13 rows, and 16 rows), with a 10 min time limit. The results for the constraint programming approach are summarized in the Table 4, we are looking for statistics for the first, the best solution (whether the optimal solution is found according to the column generation solution) and for optimality proof. For the 10 rows instance, with the *DomOverWDeg* strategy a poor quality first solution has been found. Moreover, the last solution is not improving so much the first solution (4379 to 4358). Meanwhile, the Snake strategy gives a good first solution in a few decision nodes (27 nodes). This solution is improved to optimality in under 2 min. The *diffN* constraint adds some reasoning to the model because the nodes number decreases but the total time to compute an optimal solution is higher. For the instance with 13 rows, the analysis is even worse for the *DomOverWDeg* strategy cannot find any solution within the 10 min time limit even with the *diffN* constraint. Nevertheless, with the *Snake* strategy a solution is found in less than 1 s but at the end of the time limit, no optimal solution has been proved. For the 16 rows instance, the Snake strategy also finds a solution in less than 1 s but is not able to improve it to optimality in less than 10 min.

These results are interesting for a hybridization approach because the feasible solution found by the constraint programming in less than a few minutes can be used in the enumeration part for the column generation approach.

**Table 4.** Constraint programming results for the instance with 10, 13 and 16 rows with a 600 s time limit.

| | | 10 rows | | | | 13 rows | | | | 16 rows | | | |
|---|---|---|---|---|---|---|---|---|---|---|---|---|---|
| | | V1 | V2 | V3 | V4 | V1 | V2 | V3 | V4 | V1 | V2 | V3 | V4 |
| DOWDeg | | ✓ | ✓ | | | ✓ | ✓ | | ✓ | ✓ | | | |
| diffN | | | ✓ | | ✓ | | ✓ | | ✓ | | ✓ | | ✓ |
| Snake | | | | ✓ | ✓ | | | ✓ | ✓ | | ✓ | ✓ | |
| First Sol | obj | 1379 | 4370 | 4222 | 4222 | ∅ | ∅ | 5241 | 5241 | ∅ | ∅ | 6603 | 6603 |
| | Nodes | 1669 | 1306 | 27 | 27 | \ | \ | 31 | 31 | \ | \ | 39 | 39 |
| | CPU(s) | 3 | 2 | 0 | 0 | \ | \ | 0 | 0 | \ | \ | 0 | 0 |
| Best Sol | obj | ∅ | ∅ | **4066** | **4066** | ∅ | ∅ | **5234** | **5234** | ∅ | ∅ | ∅ | ∅ |
| | Nodes | \ | \ | 91k | 81k | \ | \ | 466k | 452k | \ | \ | \ | \ |
| | CPU(s) | \ | \ | 26 | 36 | \ | \ | 159 | 231 | \ | \ | \ | \ |
| | #sol | \ | \ | 10 | 10 | \ | \ | 6 | 6 | \ | \ | \ | \ |
| Proof | obj | 4358 | 4 206 | **4066** | **4066** | ∅ | ∅ | **5234** | **5234** | ∅ | ∅ | 6447 | 6447 |
| | Nodes | 547k | 525k | 398k | 339k | 492k | 433k | 1767k | 1171k | 399k | 308k | 2033k | 1281k |
| | CPU(s) | 600 | 600 | 115 | 148 | 600 | 600 | 600 | 600 | 600 | 600 | 600 | 600 |

## 5 Conclusion

In this paper, we have proposed both a column generation, with enumeration technique, approach based on a new graph representation of the DHP and a constraint programming model using global constraints. We have performed a parametric study on various instance sets, some based on real data and other purposely generated. The main difficulty is that computing the improving routes in the pricing problem may be awful with the growth of the $CapaMax$ parameter. The small integrality gap obtained gives us good hope to solve larger instances with a bigger time limit, and validate the efficiency of the column generation approach to tackle the DHP, the real instance with 24 rows was optimally solved. The constraint programming results are promising for a hybridization approach, computing all the $\epsilon$-improving routes with the constraint programming integer solution (found in a few minutes) for the enumeration part.

**Acknowledgements.** This work was supported by the French National Research Agency under the Investments for the Future Program, referred to as ANR-16-CONV-0004.

# Appendix

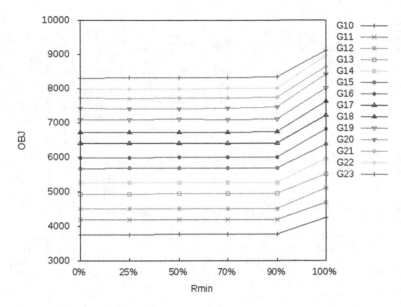

**Fig. 9.** The evolution of the objective function value when the *Rmin* parameters increase for the G instances.

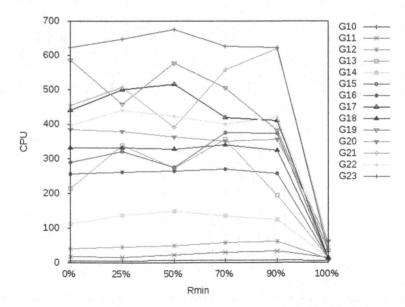

**Fig. 10.** The evolution of the computation time when the *Rmin* parameters increase for the G instances.

**Table 5.** Results for T and S instance type with the variation of the $CapaMax$ parameter.

| #rows | $CapaMax$ | T CPU(s) | T gap(%) | S CPU(s) | S gap(%) | #rows | $CapaMax$ | T CPU(s) | T gap(%) | S CPU(s) | S gap(%) |
|---|---|---|---|---|---|---|---|---|---|---|---|
| | 100 | 0.29 | 4.25 | 0.2 | 3.37 | | 100 | 0.82 | 0.13 | 47.66 | 3.70 |
| | 200 | 0.49 | 0.11 | 0.94 | 0.41 | | 200 | 16.52 | 4.55 | 216.75 | 5.25 |
| 10 | 300 | 2.36 | 0.04 | 2.64 | 0.64 | 17 | 300 | 529.01 | 5.12 | 603.39 | 5.72 |
| | 400 | 6.52 | 0.41 | 9.05 | 0.87 | | 400 | 521.21 | 1.77 | 525.95 | 2.48 |
| | 500 | 9.40 | 1.10 | 16.40 | 1.67 | | 500 | 542.55 | 1.80 | 624.74 | 4.15 |
| | 100 | 1.32 | 4.95 | 0.15 | 2.26 | | 100 | 0.71 | 0.99 | 2.78 | 3.94 |
| | 200 | 14.96 | 7.03 | 0.91 | 8.16 | | 200 | 6.55 | 0.22 | 7.11 | 0.54 |
| 11 | 300 | 48.99 | 7.28 | 79.15 | 8.45 | 18 | 300 | 441.58 | 0.52 | 512.01 | 1.04 |
| | 400 | 87.25 | 7.30 | 52.64 | 8.24 | | 400 | 518.07 | 1.59 | 549.86 | 0.27 |
| | 500 | 59.52 | 7.53 | 98.07 | 8.35 | | 500 | 550.98 | 2.15 | 620.04 | 3.87 |
| | 100 | 0.23 | 1.22 | 0.24 | **2.56** | | 100 | 0.55 | 1.67 | 0.08 | 0.64 |
| | 200 | 0.90 | 0.58 | 0.93 | **0.50** | | 200 | 310.40 | 3.79 | 311.51 | 4.63 |
| 12 | 300 | 8.96 | 0.38 | 10.50 | 0.27 | 19 | 300 | 521.54 | 4.20 | 522.08 | 4.93 |
| | 400 | 15.42 | 0.68 | 20.34 | 0.58 | | 400 | 517.21 | 4.35 | 626.03 | 7.15 |
| | 500 | 42.56 | 0.50 | 26.88 | 1.12 | | 500 | 512.78 | 2.23 | 613.55 | 3.85 |
| | 100 | 1.35 | 3.46 | 2.34 | **3.32** | | 100 | 0.047 | 0 | 106.80 | 1.91 |
| | 200 | **20.37** | 6.07 | 159.77 | **7.017** | | 200 | 10.36 | 0.26 | 16.08 | 0.31 |
| 13 | 300 | **318.57** | 5.95 | 419.87 | 7.16 | 20 | 300 | 500.71 | 0.53 | 513.86 | 0.73 |
| | 400 | 523.88 | 6.56 | 518.98 | 7.51 | | 400 | 504.41 | 0.76 | 620.26 | 2.10 |
| | 500 | 455.87 | 6.83 | 267.43 | 7.39 | | 500 | 512.33 | 2.53 | 607.29 | 3.03 |
| | 100 | 0.52 | 1.93 | 0.33 | 2.73 | | 100 | 0.14 | 1.67 | 0.08 | 0.53 |
| | 200 | **2.87** | 0.82 | 3.09 | 0.48 | | 200 | 415.4 | 3.71 | 515.31 | 4.24 |
| 14 | 300 | **21.14** | 1.07 | 25.82 | 0.75 | 21 | 300 | 520.53 | 3.83 | 623.61 | 4.34 |
| | 400 | 125.38 | 0.44 | 92.45 | 0.60 | | 400 | 506.63 | 3.95 | 608.38 | 4.78 |
| | 500 | 249.35 | 0.65 | 245.66 | 0.39 | | 500 | 519.92 | 2.30 | 607.89 | 3.07 |
| | 100 | **1.97** | 2.19 | **0.591** | 2.04 | | 100 | 1.08 | 2.04 | 0.18 | 1.09 |
| | 200 | **307.76** | 4.83 | **454.34** | 5.88 | | 200 | 48.26 | 0.87 | 38.33 | 0.58 |
| 15 | 300 | 428.17 | 5.32 | 524.62 | 6.29 | 22 | 300 | 515.78 | 2.96 | 522.21 | 1.34 |
| | 400 | 526.90 | 5.54 | 545.35 | 6.36 | | 400 | 519.62 | 0.70 | 626.18 | 2.26 |
| | 500 | 515.87 | 3.29 | 530.25 | 6.18 | | 500 | 549.81 | 2.48 | 640.07 | 2.57 |
| | 100 | 0.36 | 1.10 | 0.21 | 0.36 | | 100 | 0.22 | 0.98 | 0.10 | 1.34 |
| | 200 | 3.65 | 1.34 | 7.00 | 0.77 | | 200 | 418.62 | 3.47 | 418.97 | 3.78 |
| 16 | 300 | 156.90 | 0.24 | 136.39 | 1.07 | 23 | 300 | 515.7 | 2.51 | 542.45 | 4.02 |
| | 400 | 526.57 | 0.44 | 513.76 | 0.26 | | 400 | 524.80 | 2.57 | 621.25 | 4.65 |
| | 500 | 517.73 | 0.99 | 550.97 | 2.74 | | 500 | 521.29 | 2.32 | 621.82 | 2.63 |

$$
\begin{pmatrix}
0 & 5 & 4 & 6 & 8 & 10 & 12 & 14 & 16 \\
5 & 0 & 5 & 4 & 6 & 8 & 10 & 12 & 14 \\
4 & 5 & 0 & 5 & 4 & 6 & 8 & 10 & 12 \\
6 & 4 & 5 & 0 & 5 & 4 & 6 & 8 & 10 \\
8 & 6 & 4 & 5 & 0 & 5 & 4 & 6 & 8 \\
10 & 8 & 6 & 4 & 5 & 0 & 5 & 4 & 6 \\
12 & 10 & 8 & 6 & 4 & 5 & 0 & 5 & 4 \\
14 & 12 & 10 & 8 & 6 & 4 & 5 & 0 & 5 \\
16 & 14 & 12 & 10 & 8 & 6 & 4 & 5 & 0
\end{pmatrix}
$$

**Fig. 11.** Distance matrix between rows and the depot with direction 0, for the example instance Fig. 5.

$$\begin{pmatrix} 0 & 105 & 104 & 106 & 108 & 110 & 112 & 114 & 116 \\ 105 & 0 & 5 & 4 & 6 & 8 & 10 & 12 & 14 \\ 104 & 5 & 0 & 5 & 4 & 6 & 8 & 10 & 12 \\ 106 & 4 & 5 & 0 & 5 & 4 & 6 & 8 & 10 \\ 108 & 6 & 4 & 5 & 0 & 5 & 4 & 6 & 8 \\ 110 & 8 & 6 & 4 & 5 & 0 & 5 & 4 & 6 \\ 112 & 10 & 8 & 6 & 4 & 5 & 0 & 5 & 4 \\ 114 & 12 & 10 & 8 & 6 & 4 & 5 & 0 & 5 \\ 116 & 14 & 12 & 10 & 8 & 6 & 4 & 5 & 0 \end{pmatrix}$$

**Fig. 12.** Distance matrix between rows and the depot with direction 1, for the example instance Fig. 5.

# References

1. Baldacci, R., Christofides, N., Mingozzi, A.: An exact algorithm for the vehicle routing problem based on the set partitioning formulation with additional cuts. Math. Program. **115**(2), 351–385 (2008). https://doi.org/10.1007/s10107-007-0178-5
2. Bochtis, D.D., Sørensen, C.G.: The vehicle routing problem in field logistics part I. Biosyst. Eng. **104**(4), 447–457 (2009)
3. Bochtis, D.D., Sørensen, C.G.: The vehicle routing problem in field logistics: part II. Biosyst. Eng. **105**(2), 180–188 (2010)
4. Bourreau, E., Gondran, M., Lacomme, P., Vinot, M.: De la Programmation Linéaire à la Programmation Par Contraintes (2019)
5. Briot, N., Bessiere, C., Tisseyre, B., Vismara, P.: Integration of operational constraints to optimize differential harvest in viticulture. In: Precision Agriculture 2015, pp. 111–129. Wageningen Academic Publishers (2015)
6. Briot, N., Bessiere, C., Vismara, P.: A constraint-based approach to the differential harvest problem. In: Pesant, G. (ed.) CP 2015. LNCS, vol. 9255, pp. 541–556. Springer, Cham (2015). https://doi.org/10.1007/978-3-319-23219-5_38
7. Burger, M., Huiskamp, M., Keviczky, T.: Complete field coverage as a multi-vehicle routing problem. IFAC Proc. Vol. **46**(18), 97–102 (2013)
8. Contardo, C., Martinelli, R.: A new exact algorithm for the multi-depot vehicle routing problem under capacity and route length constraints. Discrete Optim. **12**, 129–146 (2014)
9. Dantzig, G., Wolfe, P.: Decomposition principle for linear programs. Oper. Res. **8**(1), 101–111 (1960)
10. dos Santos, L.M.R., Costa, A.M., Arenales, M.N., Santos, R.H.S.: Sustainable vegetable crop supply problem. Eur. J. Oper. Res. **204**(3), 639–647 (2010)
11. Feillet, D., Dejax, P., Gendreau, M., Gueguen, C.: An exact algorithm for the elementary shortest path problem with resource constraints: application to some vehicle routing problems. Networks **44**(3), 216–229 (2004)
12. Irnich, S., Desaulniers, G.: Shortest path problems with resource constraints. In: Desaulniers, G., Desrosiers, J., Solomon, M.M. (eds.) Column Generation, pp. 33–65. Springer, Boston (2005). https://doi.org/10.1007/0-387-25486-2_2
13. McBratney, A., Whelan, B., Ancev, T., Bouma, J.: Future directions of precision agriculture. Precision Agric. **6**(1), 7–23 (2005). https://doi.org/10.1007/s11119-005-0681-8

14. Plessen, M.: Partial field coverage based on two path planning patterns. Biosyst. Eng. **171**, 16–29 (2018)
15. Poullet, J., Parmentier, A.: Ground staff shift planning under delay uncertainty at air france. arXiv preprint arXiv:1811.00171 (2018)
16. Prud'homme, C., Fages, J.-G., Lorca, X.: Choco documentation. TASC, INRIA Rennes, LINA CNRS UMR **6241**, 64–70 (2014)
17. Reichardt, M., Jürgens, C.: Adoption and future perspective of precision farming in Germany: results of several surveys among different agricultural target groups. Precision Agric. **10**(1), 73–94 (2009)
18. Robert, M., Bergez, J.-E., Thomas, A.: A stochastic dynamic programming approach to analyze adaptation to climate change-application to groundwater irrigation in India. Eur. J. Oper. Res. **265**(3), 1033–1045 (2018)
19. Robert, P.C.: Precision agriculture: a challenge for crop nutrition management. In: Horst, W.J., et al. (eds.) Progress in Plant Nutrition: Plenary Lectures of the XIV International Plant Nutrition Colloquium. Developments in Plant and Soil Sciences, vol. 98, pp. 143–149. Springer, Dordrecht (2002). https://doi.org/10.1007/978-94-017-2789-1_11
20. Seyyedhasani, H., Dvorak, J.: Reducing field work time using fleet routing optimization. Biosyst. Eng. **169**, 1–10 (2018)
21. Toth, P., Vigo, D.: The Vehicle Routing Problem. SIAM, Philadelphia (2002)
22. Viaggi, D., Raggi, M., y Paloma, S.: An integer programming dynamic farm-household model to evaluate the impact of agricultural policy reforms on farm investment behaviour. Eur. J. Oper. Res. **207**(2), 1130–1139 (2010)
23. Wolsey, L.A., Nemhauser, G.L.: Integer and Combinatorial Optimization. Wiley, Hoboken (2014)
24. Yagoubi, R.S., Naud, O., Dejean, K.G., Crestani, D.: New approach for differential harvest problem: the model checking way. IFAC-PapersOnLine **51**(7), 57–63 (2018)

# Scheduling of Dual-Arm Multi-tool Assembly Robots and Workspace Layout Optimization

Johan Wessén[1,2]([⊠]) [iD], Mats Carlsson[3] [iD], and Christian Schulte[2] [iD]

[1] ABB Corporate Research, Västerås, Sweden
johan.wessen@se.abb.com
[2] KTH Royal Institute of Technology, Stockholm, Sweden
cschulte@kth.se
[3] RISE Research Institutes of Sweden, Uppsala, Sweden
mats.carlsson@ri.se

**Abstract.** The profitability of any assembly robot installation depends on the production throughput, and to an even greater extent on incurred costs. Most of the cost comes from manually designing the layout and programming the robot as well as production downtime. With ever smaller production series, fewer products share this cost. In this work, we present the dual arm assembly program as an integrated routing and scheduling problem with complex arm-to-arm collision avoidance. We also present a set of high-level layout decisions, and we propose a unified CP model to solve the joint problem. The model is evaluated on realistic instances and real data. The model finds high-quality solutions in short time, and proves optimality for all evaluated problem instances, which demonstrates the potential of the approach.

**Keywords:** Assembly manufacturing · Constraint programming · Robot planning and scheduling

## 1 Introduction

Consumer products are being manufactured in ever smaller series, often using assembly manufacturing. This means that complex sub-parts, such as cameras and circuit boards, are put together, akin to a 3D puzzle. These products are mainly manufactured by hand, in countries with competitive labor cost.

At the same time, lightweight industrial robots have been introduced, such as YuMi by ABB [1] and Sawyer by Rethink Robotics [17]. Many of these robots are designed to take over repetitive and tedious tasks from humans, occupying similar floor area and have dual arms with similar arm reach. Some even target small parts assembly manufacturing [1].

However, robot based assembly manufacturing is at a price point out of reach for many products. It *is* important to obtain good enough throughput of the robots, as it determines how much value the robots create. However, cost

© Springer Nature Switzerland AG 2020
E. Hebrard and N. Musliu (Eds.): CPAIOR 2020, LNCS 12296, pp. 511–520, 2020.
https://doi.org/10.1007/978-3-030-58942-4_33

has the greatest impact on net value of the robot installation, and the cost is largely determined by the time it takes to deploying the robot, as it often costs significantly more than the robot itself, plus the cost of production downtime. The inspiration for this work comes from a real world installation of a YuMi, where a highly skilled programmer and researcher from ABB spent six weeks to design the layout and program the robot. At deployment, unforeseen obstacles were present in the workspace, and the layout design and program had to be changed, which took another week. This work builds on [4] and [10].

Deploying industrial robots is a time-consuming, iterative process, and requires trained developers. The process entails **A)** layout design of the robot workspace, and robot programming. Layout design means deciding where to put locations of interest, such as pick-locations, as well as deciding what apparatus to use when multiple are available, based on their locations. These decisions greatly affect the throughput of the robots. The robot programming can be subdivided into **B)** task allocation and **C)** task sequencing, **D)** (inter-arm) scheduling, all while efficiently utilizing the multiple tools of each robot hand. It is slow and error-prone, this is especially true for multiple arms, where the programmer must coordinate the arms such that they do not collide during **E)** execution. This coordination greatly adds to the complexity of the problem. The two-dimensional case is NP-hard [19], and is an active field of research, with CP solutions at the forefront [11]. All sub-problems **A)**–**E)** are tightly interconnected.

Each task, such as pick or press, is in itself a small robot program. However, for the level of abstraction of this and related work, each task can be abstracted as a constraint on the duration of said task. Similarly, travel between task locations is a small program, which, in this and most of the related work, is abstracted as a constraint on the travel time between task pairs.

The AI community has a long history of addressing model-based robot programming, both Automated Ground Vehicles (AGV:s) using planning [6] and first-order logic [3] to achieve **C)**. AI for robotics is a long-standing topic [16], and robotics is an active part of the AI community [15]. Task planning, including temporal planning, is a field of active research [8], and there is a sub-field called integrated task and motion planning [5,7,9,12–14,18] addressing **C)** while also generating the travel motion between tasks. These works address single arms, while one recent CP approach includes a second arm [2], thus addressing **B)**, **D)**, **E)**. For that work, however, **C)** was given.

To our knowledge, no published work dealing with **A)** in this context exists, and no work approaching **B)**, **C)**, **D)** **E)** in unison. Some prior works include motion generation in their problem formulation, however due to our approach to arm coordination, these motions can be generated separately, under weak assumptions, and still result in collision-free robot programs. Thus, this work takes a larger scope on the robot deployment process and addresses all aspects **A)**–**E)**. We present the following contributions:

1. The first optimization-based model integrating **A)** layout optimization, **B)** task allocation, **C)** task sequencing, **D)** scheduling and **E)** collision avoidance for a dual-arm robot, to the best of our knowledge. The model has features

from vehicle routing and flexible job-shop, with variable locations and travel time that depends on location and performing agent.

2. An evaluation showing that the model is efficient, quickly delivering good solutions as well as reaching optimality for all problem instances.

3. An efficient way of handling arm-to-arm collisions while generating robot programs, based on dividing the workspace into two sides, and the preliminary observation that this does not affect the makespan.

## 2  The Assembly - Application, Cell and Robot Program

Assembly manufacturing is a type of production, where several *components* are combined into an *assembly*. Components could be of any kind such as electronic boards, motors, and sensors. Finished assemblies are forwarded to further processing steps. The assembly process is then repeated. In industrial robotics, a *robot workspace* is the space used by one or more robots and the equipment relating to the application, if the robot is included it is called a *robot cell*, and if it is used for assembly manufacturing it is called a *assembly cell*. Figure 1 shows the assembly cell used in this paper.

**Cell and Application:** In our example, there are five *tray locations* for holding *component trays*. Each component tray holds multiple components of one particular component type. At most one component tray can be put at any of the tray locations. Each component tray is paired with a camera facing the tray, to keep track of available components. However, the focal point of the cameras is located between the camera and the tray. The reason for this is explained later.

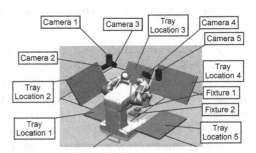

**Fig. 1.** An assembly cell with an ABB YuMi, two fixtures, five tray locations and a camera facing each tray location

In the assembly cell there are two *fixtures*. These are constructions in front of the robot, and are designed to hold one specific component each. When that component is placed, the next component is mounted on top of the first one, and so on, until we finish that fixture's *sub-assembly*. Subsequently, one sub-assembly is picked up from its fixture, and merged with the other sub-assembly, to form the final assembly, using each component type once. The final assembly is then picked up from its fixture and put on a output tray for further processing, and the assembly process starts over. The output tray is placed at the right side of the robot, not shown in Fig. 1.

One should note that it is possible to use one fixture to finish the entire assembly, and this is probably how humans would do it. However, since the robot arms work in parallel, and only one robot hand can work at each fixture at a time, a second fixture is often used to shorten the makespan. So, in this study we use two fixtures. It is also worth noting that there could be scenarios where the component mounting orders are not fixed, however this study is restricted to cases where the ordering *is* fixed.

**The YuMi Robot:** Our particular assembly cell contains a YuMi [1], a robot from ABB with two *arms* having one *hand* each. Each hand, shown in Fig. 2, features one *gripper* tool and two *suction cup* tools. These tools are designed to *pick*, move and then *place* objects. An object being a component, or a (sub-)assembly. The (empty) gripper tool is also used to *press* components after placing, to ensure good fit. A gripper tool *picks* objects with its two parallel "fingers". A suction cup tool picks objects by adhering to a flat surface. The object, the application and the robot programmer determine what tool to use for each object. When holding an object with the gripper, the position and orientation, or *pose*, are sufficiently well known to place it. This is not true for the suction cup tool, however the object pose can be determined by processing an image of the object in the tool. Thus, when using suction cup tool, a *camera* task must be performed at any of the five camera focal point, introduced previously.

Note that one tool can hold one object, independently of the other tools. This means that one robot hand can hold up to three objects simultaneously. This fact is often used by experienced programmers to improve the makespan.

**Robot Programming.** The software controlling the robot and cameras is called the *robot program*. It consists of two sequences of tasks, one per arm. Each arm program alternates between performing a task and moving to the next task. The tasks mentioned thus far are *pick, camera, place, press*. Each task is itself a program, provided by the robot programmer. However, for the abstraction level of this work, a task is a constraint on the duration of said task. Similarly, travel between task locations is abstracted as a constraint on the travel time between said locations.

Fig. 2. A robot hand with one gripper tool and two suction cup tools

In summary, the robot program must obey task orders at the fixtures. The program must be realizable. That is, no more than one component per tool, and perform pick-(camera)-place with the same arm, in that order. Finally, the program must avoid arm-to-arm collisions. A robot programmer often spends considerable effort to keep arms apart.

To allow the arms to work independently of each other, the optimization model of this work relies on the following assumptions on tasks and travel. (1) Parallel work: the model assumes that any task at one fixture, performed by one arm, does not interfere with any task at the other fixture, performed by the other arm. (2) Elbows out: if the left robot hand is located to the left of the right robot hand, then no part of the arms are colliding. (3) No zig-zag: travel between fixtures and other locations does not pass through volumes to the left (right) of that location when using the right (left) arm.

There is a small (implicit) volume around each fixture, dedicated to the arm working at the fixture. When a task is done, the robot arm leaves this volume, *as a part of the task implementation* (and task duration). This makes sure no collisions occur when tasks are executed by different arms at the same fixture.

## 3   Modeling

Our CP model captures the dual robot arm program for manufacturing one assembly. It captures variable task locations, thus optimizing tray layout and which camera to use while avoiding arm collisions. It is a scheduling model that borrows from vehicle routing models, augmented with variable locations.

The model features a set $\mathbb{T} = \{0, .., T - 1\}$ of tasks, a set $\mathbb{L} = \{0, .., L - 1\}$ of locations, a set $\mathbb{W} = \{0, .., W - 1\}$ of time slots, and a set $\mathbb{H} = \{Gripper, SuctionCup\}$ of robot hand tools.

For task $t$, we have decision variables $arm_t \in \{0, 1\}$, the arm performing $t$; $succ_t \in \mathbb{T}$, the successor task of $t$; $location_t \in \mathbb{L}$, the location of $t$; $load_t^\tau \in \mathbb{N}$, the load of tool $\tau$ when arriving at $t$; $travel\_cost_t \in \mathbb{W}$, the travel time from $t$ to its successor; $duration_t \in \mathbb{W}$, the duration of processing $t$; $arrival_t \in \mathbb{W}$, the time at which the arm is ready to perform $t$; $start_t \in \mathbb{W}$, the time at which $t$ starts processing; and $end_t \in \mathbb{W}$, the time at which $t$ finishes processing. For many tasks, the location is fixed up front. There can be a waiting period between the arrival time and the start time.

We use successor variables to capture task order, i.e., $succ_i = j$ means that task $j$ is the successor of task $i$. The model includes $T - 4$ actual tasks and four dummy tasks, with the convention that $T-4$ and $T-2$ are the dummy start and end tasks of arm 0, $T-3$ and $T-1$ are the dummy start and end tasks of arm 1, the start task of arm 1 is the successor of the task of arm 0, and vice versa. The total set of tasks thus form a Hamiltonian circuit. Let $\mathbb{T}_{start} = \{T - 4, T - 3\}$ and $\mathbb{T}_{end} = \{T - 2, T - 1\}$ denote the start and end tasks, respectively. Let $\mathbb{T}_{TrayPick} \subset \mathbb{T}$ denote the set of tasks that occur at a tray location.

The application requires both robot arms to return to the original position to be ready for the next assembly, so we must account for the travel time from the last actual task back to the original position. To achieve this, we constrain the locations of the end task, the start task, and the first actual task to coincide.

The objective is to minimize the makespan, i.e., the maximal completion time of the two arms. Instance data are encoded in part by a set of functions:

$W_L(a, s, g) \in \mathbb{Z} =$ travel time from source location $s$ to target location $g$
using arm $a$, or -1 if infeasible

$W_D(a, t) \in \mathbb{N} =$ duration of task $t$ for arm $a$

$P_L(l) \in \mathbb{Z} =$ relative position of location $l$ in robot workspace, left to right,
$\geq 0$ indicating relative positions, -1 indicating shared space

$$O(t, t') = \begin{cases} 2, & \text{if task } t \text{ must precede task } t' \text{ using the same arm,} \\ 1, & \text{if task } t \text{ must precede task } t', \text{ but } t, t' \text{ may use different arms,} \\ 0, & \text{otherwise} \end{cases}$$

$$\Delta(t, \tau) = \begin{cases} 1, & \text{if task } t \text{ uses tool } \tau \text{ to pick a component or (sub)-assembly,} \\ -1, & \text{if task } t \text{ uses tool } \tau \text{ to place it,} \\ 0, & \text{otherwise} \end{cases}$$

$Cap(t, \tau) \in \mathbb{N} =$ the max possible load for tool $\tau$ when arriving to task $t$

The model's constraints are grouped as follows:

**Successors.** Recall that the tasks should form a Hamiltonian circuit, which can be seen as arm 0's task sequence followed by arm 1's task sequence, linked via the dummy tasks. Of course, the same arm must be used within the given task sequence. Finally, the robot arms must return to their original position. This all is encoded by the following constraints:

$$\begin{aligned} \textbf{Circuit}([succ_t \mid t \in \mathbb{T}]), & \\ arm_t = arm_{succ_t} \quad \forall t \in \mathbb{T} \setminus \mathbb{T}_{\text{end}}, & \\ succ_{T-2} = T - 3, succ_{T-1} = T - 4, & \\ arm_{T-4} = arm_{T-2} = 0, arm_{T-3} = arm_{T-1} = 1, & \\ location_{succ_{T-4}} = location_{T-4} = location_{T-2}, & \\ location_{succ_{T-3}} = location_{T-3} = location_{T-1} & \end{aligned} \quad (1)$$

**Time Constraints.** Travel costs depend on locations and arm, and durations depend on task and arm:

$$\begin{aligned} travel\_cost_t = W_L(arm_t, location_t, location_{succ_t}) \quad \forall t \in \mathbb{T} \setminus \mathbb{T}_{\text{start}} \setminus \mathbb{T}_{\text{end}}, & \\ duration_t = W_D(arm_t, t) \quad \forall t \in \mathbb{T} \setminus \mathbb{T}_{\text{start}} \setminus \mathbb{T}_{\text{end}} & \end{aligned} \quad (2)$$

The following relates each actual and dummy task's time variables. Note that this forms the link between the routing and scheduling parts of the model. Together with the routing constraints, this prevents tasks, including the arms' travel time, from overlapping:

$$arrival_t \leq start_t \quad \forall t \in \mathbb{T},$$
$$start_t + duration_t = end_t \quad \forall t \in \mathbb{T},$$
$$end_t + travel\_cost_t = arrival_{succ_t} \quad \forall t \in \mathbb{T},$$
$$duration_t = travel\_cost_t = 0 \quad \forall t \in \mathbb{T}_{start} \cup \mathbb{T}_{end},$$
$$arrival_t = start_t = end_t = arrival_{succ_t} \quad \forall t \in \mathbb{T}_{start} \cup \mathbb{T}_{end},$$
$$arrival_t = 0 \quad \forall t \in \mathbb{T}_{start} \tag{3}$$

**Precedence Constraints.** Precedences at fixtures are imposed by constraints on the time intervals of tasks. Precedences among pick-(camera)-(other)-place tasks are handled similarly. Note that the same arm is required for each such task, and so the inequation can be stricter and include all time spent on a task:

$$end_t \leq start_{t'} \quad \forall t, t' \in \mathbb{T} \text{ where } O(t, t') = 1,$$
$$arrival_{succ_t} \leq arrival_{t'} \wedge arm_t = arm_{t'} \quad \forall t, t' \in \mathbb{T} \text{ where } O(t, t') = 2 \tag{4}$$

**Capacity Constraints.** The $load_t^\tau$ variables are related via the $\Delta(t, \tau)$ values:

$$load_t^\tau = 0 \quad \forall \tau \in \mathbb{H} \quad \forall t \in \mathbb{T}_{start} \cup \mathbb{T}_{end},$$
$$load_{succ_t}^\tau = load_t^\tau + \Delta(t, \tau) \quad \forall t \in \mathbb{T} \quad \forall \tau \in \mathbb{H} \tag{5}$$

Tools cannot exceed their physical capacity, and for some tasks, they must not be occupied:

$$load_t^\tau \leq Cap(t, \tau) \quad \forall t \in \mathbb{T} \quad \forall \tau \in \mathbb{H} \tag{6}$$

**Location Constraints.** Each $location_t$ is initialized with a given subset of $\mathbb{L}$. For this model, all task locations are fixed, except tray locations and camera locations. There is only room for one tray at each tray location:

$$\textbf{AllDifferent}([location_t \mid t \in \mathbb{T}_{TrayPick}]) \tag{7}$$

Finally, to prevent the arms from colliding, we split the robot workspace into non-shareable zones and shareable zones. For every pair of tasks that use different arms and placed in non-shareable zones, we require the left arm (arm 0) to be left of the right arm (arm 1):

$$arm_p = 0 \wedge arm_q = 1 \implies$$
$$P_L(location_p) = -1 \vee P_L(location_q) = -1 \vee$$
$$P_L(location_p) < P_L(location_q) \quad \forall p, q \in \mathbb{T} \tag{8}$$

# 4    Evaluation

We evaluate dual arm, multi tool, assembly programs using real world data. The real world data consists of the possible locations of all task types, the travel times of manually generated paths between all locations, for a workspace of a real case, as well as task durations for each arm. The problem instances are solved by the model presented in Sect. 3. All instances are solved using single thread Gecode 6.2.0 on Intel Core i7-6850 3.6 GHz CPU. The primary decisions of our search are load balanced task assignment to arms, using a mix of static and dynamic variable order, and a mix of static and random value selection.

The number of tasks of a problem instance is mainly dependent on a) number of components, and b) whether a component is picked using suction cup or gripper. We evaluated instances of 3, 4 and 5 components, using suction cup vs. gripper for different components, and equipping the robot with 1 or 2 suction cup tools per arm. This results in 11 problem instances of 17 to 26 tasks.

For each problem instance we generated 8 variants with nominal travel times randomly perturbed by ±25%, each variant solved 10 times by using different randomization seeds for value selection in the search. The problem instances are named e.g. 5C4S-2 meaning 5 components, out of which 4 are handled by suction cup with 2 suction cups per arm.

We present the runtime distribution solving each instance to optimality as a box plot extending from the first and third quartile ($Q1,Q3$), with a line indicating the median. The whiskers extends to at most the range $[Q3 + 1.5 \cdot (Q3 - Q1), Q1 - 1.5 \cdot (Q3 - Q1)]$. Data outside this range are shown as outliers.

As a complementary perspective on the optimization progress, we report runtime distributions to reach optimality gaps 0.05 and 0.10 as well, i.e., the time at which the makespan is 5% and 10% away from the optimal value. This information can be used for selecting timeouts in a real-world scenario.

Most runs finish within seconds to a few minutes, with the notable exception of the case this work was inspired by, and also the largest case: 5C4S-2. Having two suction cup tools per arm gives more flexibility and opens up for better throughput, however, inspecting instances *−1 vs *−2 in Fig. 3 it is clear that this incurs an increased optimization time.

We also repeated all the experiments relaxing constraint (8), thus ignoring arm-to-arm collisions. Very much against the working hypothesis, for every problem instance the optimal makespan was unaffected by whether or not constraint (8) was active. Note that, if some or all locations become fixed, this is definitely not the case.

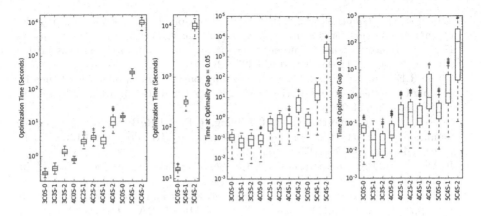

**Fig. 3.** Optimization time and time to reach optimality gap 0.05 and 0.10.

# 5  Conclusions and Future Work

The contributions of this work include: a first optimization-based model that integrates all aspects of the problem, an evaluation that shows that the model delivers good solutions quickly and is able to solve all evaluated instances to optimality, and an approach, based on spatial separation, to handling arm-to-arm collisions without any observed degradation of solution quality.

Current results are promising. Future work includes more than two robot arms, and explore how this affects arm-to-arm collision avoidance. It would be interesting to involve other characteristics, scale up to larger problem instances and optimize over the number of fixtures used.

**Acknowledgements.** This work was partially supported by the Wallenberg AI, Autonomous Systems and Software Program (WASP).

# References

1. ABB AB: The ABB YuMi industrial robot. http://new.abb.com/products/robotics/industrial-robots/yumi
2. Behrens, J.K., Lange, R., Mansouri, M.: A constraint programming approach to simultaneous task allocation and motion scheduling for industrial dual-arm manipulation tasks. In: International Conference on Robotics and Automation (ICRA) (2019). https://doi.org/10.1109/ICRA.2019.8794022
3. Burgard, W., et al.: The interactive museum tour-guide robot, pp. 11–18 (1998). https://doi.org/10.5555/295240.295249
4. Ejenstam, J.: Implementing a time optimal task sequence for robot assembly using constraint programming. Master's thesis, Department of Information Technology, Uppsala University, Sweden (2014). http://urn.kb.se/resolve?urn=urn:nbn:se:uu:diva-229581

5. Lagriffoul, F., Dimitrov, D., Saffiotti, A., Karlsson, L.: Constraint propagation on interval bounds for dealing with geometric backtracking. In: Proceedings of IEEE/RSJ International Conference on Intelligent Robots and Systems (2012). https://doi.org/10.1109/IROS.2012.6385972
6. Fikes, R.E., Nilsson, N.J.: STRIPS: a new approach to the application of theorem proving to problem solving. Artif. Intell. **2**(3), 189–208 (1971). https://doi.org/10.1016/0004-3702(71)90010-5
7. Garrett, C.R., Lozano-Pérez, T., Kaelbling, L.P.: FFRoB: leveraging symbolic planning for efficient task and motion planning. CoRR abs/1608.01335 (2016). https://doi.org/10.1177/0278364917739114
8. ICAPS: ICAPS competitions. http://icaps-conference.org/index.php/Main/Competitions
9. Kaelbling, L.P., Lozano-Perez, T.: Hierarchical task and motion planning in the now, pp. 1470–1477 (2011). https://doi.org/10.1109/ICRA.2011.5980391
10. Knutar, F.: Automatic generation of assembly schedules for a dual-arm robot using constraint programming. Master's thesis, Department of Information Technology, Uppsala University, Sweden (2015). http://urn.kb.se/resolve?urn=urn:nbn:se:uu:diva-276340
11. Li, J., Harabor, D., Stuckey, P., Ma, H., Koenig, S.: Symmetry-breaking constraints for grid-based multi-agent path finding. In: Proceedings of the AAAI Conference on Artificial Intelligence, vol. 33, pp. 6087–6095, July 2019. https://doi.org/10.1609/aaai.v33i01.33016087
12. Lozano-Pérez, T., Kaelbling, L.P.: A constraint-based method for solving sequential manipulation planning problems. In: IEEE/RSJ International Conference on Intelligent Robots and Systems (2014). https://doi.org/10.1109/IROS.2014.6943079
13. Mösenlechner, L., Beetz, M.: Parameterizing actions to have the appropriate effects. In: Proceedings of IEEE/RSJ International Conference on Intelligent Robots and Systems (2011). https://doi.org/10.1109/IROS.2011.6048523
14. Mösenlechner, L., Beetz, M.: Fast temporal projection using accurate physics-based geometric reasoning. In: Proceedings of IEEE International Conference on Robotics and Automation (2013). https://doi.org/10.1109/ICRA.2013.6630817
15. Pecora, F., Mansouri, M., Hawes, N., Kunze, L.: Special issue on reintegrating artificial intelligence and robotics. Künstliche Intelligenz **33**(4), 315–317 (2019). https://doi.org/10.1007/s13218-019-00625-x
16. Pecora, F.: Is model-based robot programming a mirage? A brief survey of AI reasoning in robotics. KI - Künstliche Intelligenz **28**(4), 255–261 (2014). https://doi.org/10.1007/s13218-014-0325-0
17. Rethink Robotics: The Rethink Robotics Sawyer robot. https://www.rethinkrobotics.com/sawyer/
18. Toussaint, M.: Logic-geometric programming: an optimization-based approach to combined task and motion planning. In: In Proceedings of the International Joint Conference on Artificial Intelligence (IJCAI) (2015). https://dl.acm.org/doi/10.5555/2832415.2832517
19. Yu, J., LaValle, S.M.: Structure and intractability of optimal multi-robot path planning on graphs. In: Proceedings of the AAAI Conference on Artificial Intelligence (2013). https://doi.org/10.1609/aaai.v33i01.33016087

# Conflict-Free Learning for Mixed Integer Programming

Jakob Witzig[1](✉) and Timo Berthold[2]

[1] Zuse Institute Berlin, Takustr. 7, 14195 Berlin, Germany
witzig@zib.de
[2] Fair Isaac Germany GmbH, Stubenwald-Allee 19, 64625 Bensheim, Germany
timoberthold@fico.com

**Abstract.** Conflict learning plays an important role in solving mixed integer programs (MIPs) and is implemented in most major MIP solvers. A major step for MIP conflict learning is to aggregate the LP relaxation of an infeasible subproblem to a single globally valid constraint, the dual proof, that proves infeasibility within the local bounds. Among others, one way of learning is to add these constraints to the problem formulation for the remainder of the search.

We suggest to not restrict this procedure to infeasible subproblems, but to also use global proof constraints from subproblems that are not (yet) infeasible, but can be expected to be pruned soon. As a special case, we also consider learning from integer feasible LP solutions. First experiments of this *conflict-free* learning strategy show promising results on the MIPLIB2017 benchmark set.

## 1 Introduction

In this paper, we consider *mixed integer programs (MIPs)* of the form

$$\min\{c^\mathsf{T} x \,|\, Ax \geq b, \ \ell \leq x \leq u, \ x_j \in \mathbb{Z} \ \forall j \in \mathcal{I}\}, \tag{1}$$

with objective coefficient vector $c \in \mathbb{R}^n$, constraint coefficient matrix $A \in \mathbb{R}^{m \times n}$, constraint right-hand side $b \in \mathbb{R}^m$, and variable bounds $\ell, u \in \overline{\mathbb{R}}^n$, where $\overline{\mathbb{R}} := \mathbb{R} \cup \{\pm\infty\}$. Furthermore, let $\mathcal{N} = \{1, \dots, n\}$ be the index set of all variables and let $\mathcal{I} \subseteq \mathcal{N}$ be the set of variables that need to be integral in every feasible solution. Moreover, we allow that the constraint set might change during the course of the search. More specifically, we allow that the right-hand side $b$ can be tightened at any point in time to $\tilde{b} \in \mathbb{R}^m$ with $\overline{b} \geq \tilde{b} \geq b$ and $\overline{b} \in \overline{\mathbb{R}}^m$. As a consequence, the set of feasible solutions to both, the MIP and its LP relaxation might only be further restricted and parts of the search space that have been discarded due to infeasibility will still be infeasible. Note, we say that $\tilde{b}$ is greater or equal to $b$, if $\tilde{b}_i \geq b_i$ for all $i = 1, \dots, m$.

An important special case of this general setting is the tightening of the so-called *cutoff bound* $\overline{c}$ during the MIP search. The cutoff bound is either defined by the objective value of current incumbent, i.e., the currently best known, solution

© Springer Nature Switzerland AG 2020
E. Hebrard and N. Musliu (Eds.): CPAIOR 2020, LNCS 12296, pp. 521–530, 2020.
https://doi.org/10.1007/978-3-030-58942-4_34

$\bar{x}$ or $+\infty$ if no solution has been found yet. It gives rise to the *objective cutoff constraint*

$$-c^\mathsf{T} x \geq -\bar{c}. \tag{2}$$

The objective cutoff constraint (2) models an upper bound on all MIP solutions found in the remainder of the search. In the following, we assume that the objective cutoff constraint (2) is explicitly contained in $Ax \geq b$. The computational experiments of this paper will focus on the case that the objective cutoff constraint is the only constraint being tightened during the search, the theoretic background, however, will be given for the general case $\bar{b} \geq \tilde{b} \geq b$. A lower bound on the MIP solution is given by the *linear programming (LP) relaxation* which omits the integrality conditions of (1). The optimal objective value of the LP relaxation provides a lower bound on the optimal solution value of the MIP (1).

In LP-based branch-and-bound [7,16], the most commonly used method to solve MIPs, the LP relaxation is used for bounding. Branch-and-bound is a divide-and-conquer method which splits the search space sequentially into smaller subproblems that are expected to be easier to solve. During this procedure, we may encounter infeasible subproblems. Infeasibility can be detected by contradicting implications, e.g., derived by domain propagation, by an infeasible LP relaxation, or an LP relaxation that exceeds the objective value of the current incumbent solution. Following our assumption that the objective cutoff constraint is part of the constraint matrix, the latter is just a special case of an infeasible LP relaxation.

## 1.1   Conflict Analysis in MIP

Modern MIP solvers try to 'learn' from infeasible subproblems, e.g., by applying *conflict graph analysis* or *dual proof analysis*. Conflict graph analysis for MIP has its origin in solving satisfiability problems (SAT) and goes back to [20]. Similar ideas are used in constraint programming, e.g., see [10,14,24]. First integration of these techniques into MIP were independently suggested by [2,8,23]. Dual proof analysis and its combination with conflict graph analysis has been recently studied for both MIPs [21,25] and mixed integer nonlinear programs (MINLPs) [18,27]. While conflict graph analysis is based on combinatorial arguments, dual proof analysis is a purely LP-based approach. We will briefly describe both concepts in the remainder of this section.

Assume we are given an infeasible node of the branch-and-bound tree defined by the subproblem

$$\min\{c^\mathsf{T} x \mid Ax \geq b, \ \ell' \leq x \leq u', \ x_j \in \mathbb{Z} \ \forall j \in \mathcal{I}\} \tag{3}$$

with local bounds $\ell \leq \ell' \leq u' \leq u$. In LP-based branch-and-bound, the infeasibility of a node/subproblem is either detected by an infeasible LP relaxation or by contradicting implications in domain propagation.

In the latter case, a *conflict graph* gets constructed which represents the logic of how the set of branching decisions led to the detection of infeasibility. More precisely, the conflict graph is a directed acyclic graph in which the vertices represent bound changes of variables and the arcs $(v, w)$ correspond to bound changes implied by propagation, i.e., the bound change corresponding to $w$ is based (besides others) on the bound change represented by $v$. In addition to these inner vertices which represent the bound changes from domain propagation, the graph features source vertices for the bound changes that correspond to branching decisions and an artificial sink vertex representing the infeasibility. Then, each cut that separates the branching decisions from the artificial infeasibility vertex gives rise to a valid *conflict constraint*. A conflict constraint consists of a set of variables with associated bounds, requiring that in each feasible solution at least one of the variables has to take a value outside these bounds. This corresponds to no-good learning in CP. A variant of this procedure is implemented in SCIP, the solver in which we will conduct our computational experiments.

## 1.2   Deriving Dual Proofs for Infeasible LP Relaxations

If infeasibility is proven by the LP relaxation, however, the proof of infeasibility is given by a ray in the dual space. Consider a node of the branch-and-bound tree and the corresponding subproblem of type (3) with local bounds $\ell \leq \ell' \leq u' \leq u$. The *dual LP* of the corresponding LP relaxation of (3) is given by

$$\max\{y^\mathsf{T} b + r^\mathsf{T}\{\ell', u'\} \mid y^\mathsf{T} A + r^\mathsf{T} = c^\mathsf{T},\ y \in \mathbb{R}_+^m,\ r \in \mathbb{R}^n\}, \qquad (4)$$

$$r^\mathsf{T}\{\ell', u'\} := \sum_{j \in \mathcal{N}:\, r_j^\ell > 0} r_j^\ell \ell'_j - \sum_{j \in \mathcal{N}:\, -r_j^u < 0} r_j^u u'_j \qquad (5)$$

with $r^\ell, r^u \in \mathbb{R}_+^n$ representing the dual variables on the finite bound constraints, see, e.g. [6]. Note, variable $x_j$ can only be tight in at most one bound constraint, thus, $r_j^\ell$ and $r_j^u$ cannot be non-zero at the same time and it holds that $r = r^\ell - r^u$. For every variable $x_j$ it holds that $r_j = c_j - y^\mathsf{T} A_{.j}$, where $A_{.j}$ denotes the $j$-th column of $A$. By LP theory, each ray $(y, r) \in \mathbb{R}^{m+n}$ in the dual space that satisfies

$$\begin{aligned} y^\mathsf{T} A + r^\mathsf{T} &= 0 \\ y^\mathsf{T} b + r^\mathsf{T}\{\ell', u'\} &> 0 \end{aligned} \qquad (6)$$

proves infeasibility of (the LP relaxation of) (3), which is a direct consequence of the Farkas Lemma [9]. Hence, there exists a solution $(y, r)$ of (6) with

$$\Delta_{\max}(y^\mathsf{T} A, \ell', u') < y^\mathsf{T} b,$$

where $\Delta_{\max}(y^\mathsf{T} A, \ell', u') := \sum_{y^\mathsf{T} A < 0}(y^\mathsf{T} A)\ell' + \sum_{y^\mathsf{T} A > 0}(y^\mathsf{T} A)u'$ is called the *maximal activity* of $y^\mathsf{T} A$ w.r.t. the local bounds $\ell'$ and $u'$. Consequently, the inequality

$$y^\mathsf{T} A x \geq y^\mathsf{T} b \qquad (7)$$

has to be fulfilled by every feasible solution of the MIP. In the following, this type of constraint will be called *dual proof constraint*. If locally valid constraints are present in subproblem (3), e.g., due to the separation of local cutting planes, the corresponding dual multipliers are assumed to be zero, thereby leaving those constraints out of aggregation (7). Otherwise, the resulting dual proof constraint might not be globally valid anymore. For an approach how to deal with local cutting planes we refer to [27].

**Observation 1.** *Let $b \in \mathbb{R}^m$ be the right-hand side vector and $y^\mathsf{T} Ax \geq y^\mathsf{T} b$ be a dual proof constraint that was derived from an infeasible subproblem. After tightening the (global) right-hand side to $\tilde{b} \in \mathbb{R}^m$, with $\bar{b} \geq \tilde{b} \geq b$, the following holds.*

*(i) $y^\mathsf{T} Ax \geq y^\mathsf{T} b$ is still globally valid.*

*(ii) The dual proof can be strengthened to $y^\mathsf{T} Ax \geq y^\mathsf{T} \tilde{b}$, while preserving global validity.*

The contribution of the paper is twofold. Firstly, it describes LP-based solution learning, and approach to generate feasible, conflict-like constraints from feasible primal MIP solutions. Secondly, it describes conflict-free learning, a technique to generate conflict-like constraints from feasible dual LP solutions. Both approaches are evaluated in a computational study.

## 2   LP-Based Solution Learning

Conflict-driven learning or no-good learning [22,28], is a fundamental concept in SAT and CP. Besides learning from infeasibility, the methodology of solution-driven learning or good-learning [5,11], i.e., learning from feasibility, has been applied in SAT and CP. Recently, good learning has been successfully applied to nested constraint programming [4]. Generally, algorithms for infeasibility learning can be extended to solution learning by pretending that the corresponding cutoff constraint with the updated incumbent was already present for the current subproblem and would prove it to be infeasible (after the incumbent update).

To the best of our knowledge, solution learning has not yet been studied for MIP. Every LP that yields an optimal solution that is MIP-feasible, i.e., feasible for (1), can be used to apply *LP-based solution learning*.

Consider a subproblem (3) with local bounds $\ell'$ and $u'$. Moreover, let $x^\star_{\mathrm{LP}}$ be an optimal solution of its LP relaxation that is feasible for MIP (1). If $x^\star_{\mathrm{LP}}$ is an improving solution, i.e., $c^\mathsf{T} x^\star_{\mathrm{LP}} < \bar{c}$, $x^\star_{\mathrm{LP}}$ defines the new incumbent solution. Consequently, the cutoff bound can be updated to

$$\bar{c} = c^\mathsf{T} x^\star_{\mathrm{LP}} - \epsilon$$

with $\epsilon > 0$. Note that MIP solvers using floating point arithmetic typically subtract a small epsilon in the order or magnitude of the used tolerances, e.g., SCIP uses $\epsilon = 10^{-6}$, to enforce strict improvement during the search. If all variables with a non-zero coefficient in the objective function are integral, the minimal improvement in the objective value can be computed by a GCD-like algorithm and used as an epsilon. For example, the objective value always improves by a multiple of 1 if $c_j \in \{-1, 0, 1\}$ with $j \in \mathcal{I}$ and $c_j = 0$ with $j \in \mathcal{N} \setminus \mathcal{I}$. Thus, $\epsilon = 1$ could be used in this case.

Since the objective cutoff constraint (2) is part of the constraint matrix $A$, the right-hand side vector $b$ changes when updating the cutoff bound. Assume that row index $m$ represents the matrix row associated to the objective cutoff constraint, i.e., $A_{m.}x \geq b$ with $-c^\mathsf{T} = A_{m.}$ and $b_m = -\bar{c}$. After an incumbent update, the right-hand side vector changes to $\tilde{b}$ with $\tilde{b}_m = b_m + \bar{c} - c^\mathsf{T}x^\star_{\mathrm{LP}}$ and $\tilde{b}_i = b_i$ for all $i = 1, \ldots, m-1$. Thus, the feasible LP relaxation defined by the local bounds $\ell'$ and $u'$ turns infeasible after the update. Henceforth, we can apply both conflict graph analysis and dual proof analysis to learn from LP relaxations that yield integer feasible solutions.

## 2.1 Implementation

In our implementation, LP-based solution learning is applied whenever the LP relaxation yields a feasible solution, i.e., all integrality conditions are satisfied, that improved the incumbent solution. Note, in principle LP-based solution learning could also be applied for all improving solutions, e.g., found within a heuristic, with an objective value equal to the objective value of the LP relaxation. However, in this publication we only apply solution whenever LP information are immediately available, e.g., during diving heuristics.

Solution learning can immediately apply both *conflict graph analysis* and *dual proof analysis* [26] when the feasible LP relaxation turns into a bound exceeding LP without introducing much computational overhead.

## 3   Conflict-Free Dual Proofs

State-of-the-art MIP solvers like SCIP and FICO Xpress do not actively steer the tree search towards the exploration of infeasible subproblems. Thus, learning from infeasibility information can be considered to be a "byproduct".

Here, we will discuss how the concept of conflict analysis can be extended to learn from subproblems that are not (yet) infeasible. Therefore, we consider dual proofs of form (7) that are *conflict-free*.

**Definition 2 (Conflict-Free Dual Proof).** *Let $\ell \leq \ell' \leq u' \leq u$ be a set of local bounds and $y^\mathsf{T}Ax \geq y^\mathsf{T}b$ be an aggregation of globally valid constraint weighted by $y \in \mathbb{R}^m_+$. The inequality $y^\mathsf{T}Ax \geq y^\mathsf{T}b$ is called conflict-free dual proof with respect to $\ell'$ and $u'$ if*

*i)* $\Delta_{\max}(y^\mathsf{T}A, \ell', u') \geq y^\mathsf{T}b$ *and*
*ii)* $\exists\, \tilde{b} \in \mathbb{R}$ *with* $\bar{b} \geq \tilde{b} \geq b$ *such that* $\Delta_{\max}(y^\mathsf{T}A, \ell', u') < y^\mathsf{T}\tilde{b}$.

Within a black-box MIP solver (e.g., SCIP, FICO Xpress, Gurobi, and CPLEX) that considers the cutoff bound for pruning subproblems, the concept of conflict-free dual proofs simplifies as follows. W.l.o.g. let $A_{m.}x \geq b_m$ be the row assigned to the cutoff bound. Moreover, let $\hat{A} \in \mathbb{R}^{n \times (m-1)}$ be the coefficient matrix without the row assigned to the objective cutoff constraint (2) and $\hat{b} \in \mathbb{R}^{m-1}$ the corresponding constraint right-hand side, i.e.,

$$A := \begin{bmatrix} \hat{A} \\ -c \end{bmatrix} \quad \text{and} \quad b := \begin{bmatrix} \hat{b} \\ -\bar{c} \end{bmatrix}.$$

Let (3) be feasible with respect to the local bounds $\ell'$, $u'$, and the current solution $\bar{x}$. Moreover, let $(y, r)$ be a dual feasible solution for (4). By complementary slackness, it follows that $y_m = 0$. Thus, it holds that

$$c^\mathsf{T} \bar{x} \geq y^\mathsf{T} b + r^\mathsf{T} \{\ell', u'\}$$
$$\Leftrightarrow \qquad \bar{c} \geq y^\mathsf{T} b + (c - y^\mathsf{T} A)\{\ell', u'\}$$
$$\Leftrightarrow \qquad \bar{c} \geq \hat{y}^\mathsf{T} \hat{b} + y_m \bar{c} + (c - (\hat{y}^\mathsf{T} \hat{A} + y_m c))\{\ell', u'\}$$
$$\Leftrightarrow \qquad \bar{c} \geq \hat{y}^\mathsf{T} \hat{b} + (c - \hat{y}^\mathsf{T} \hat{A})\{\ell', u'\}$$
$$\Leftrightarrow \qquad (\hat{y}^\mathsf{T} \hat{A} - c)\{\ell', u'\} \geq \hat{y}^\mathsf{T} \hat{b} - \bar{c}$$
$$\Leftrightarrow \qquad y^\mathsf{T} A\{\ell', u'\} \geq y^\mathsf{T} b \quad \text{with } y_m = 1.$$

Consequently, from every dual feasible solution $(y, r) \in \mathbb{R}^{m+n}$ a globally valid constraint

$$y^\mathsf{T} A x \geq y^\mathsf{T} b \tag{8}$$

can be derived. This constraint is generally not violated with respect to $\ell, u$ and will not be violated with respect to $\ell'$ and $u'$ either, when the local LP relaxation is feasible. Moreover, let $\underline{c} \in \bar{\mathbb{R}}$ be the current dual bound, i.e., the global lower bound on the MIP solution value. If there exists a $\tilde{c} \in \mathbb{R}$ with $\underline{c} < \tilde{c} < \bar{c}$ such that

$$(\hat{y}^\mathsf{T} \hat{A} - c)\{\ell', u'\} < \hat{y}^\mathsf{T} \hat{b} - \tilde{c},$$

then (8) is a *conflict-free dual proof*. The new global right-hand side is defined by $\tilde{b} := \begin{bmatrix} b & -\tilde{c} \end{bmatrix}^\mathsf{T}$.

## 3.1   Implementation

In our implementation, we maintain a storage of conflict-free dual proofs which is restricted to at most 200 entries. For every conflict-free dual proof we calculate the *primal target bound* $\tilde{c} := \bar{c} + (\Delta_{\max}(y^\mathsf{T} A, \ell', u') - \hat{y}^\mathsf{T} b)$. The decision whether a conflict-free dual proof is added to the storage for later considerations only depends on the primal target bound. If the storage maintains less than 200 entries, a conflict-free dual proof is accepted if its primal target bound is at least the current global dual bound. In case of a completely filled storage, the newly derived conflict-free dual proof is immediately rejected if its primal target bound is smaller (i.e., worse) than the smallest target bound among all maintained conflict-free dual proofs. Otherwise, the conflict-free dual proof is accepted if it has a larger (i.e., better) target bound or less nonzero coefficients. With this strategy we aim to prefer short conflict-free dual proofs that tend to propagate earlier with respect to the cutoff bound, i.e., the improvements on the primal side. Whenever a new conflict-free dual proof is derived, all maintained conflict-free dual proofs whose primal target bound become worse than the global dual bound are immediately removed from the storage.

If a new incumbent solution $\bar{x}$ is found, we add at most 10 conflict-free dual proofs for which $\tilde{c} \geq c^\mathsf{T}\bar{x}$ holds to the actual solving process, i.e., these (conflict-free) dual proofs become "active" and are considered during the remainder of the search for, e.g., variable bound propagation. Moreover, we allow for slight relaxed primal target bounds. Thus, every conflict-free dual proof for which $\tilde{c} \geq (1 + \alpha)c^\mathsf{T}\bar{x}$, with $\alpha \geq 0$, holds is considered to become active. In our computational experiments we used $\alpha = 0.1$.

## 4   Computational Experiments

This section presents a first computational study of solution learning and conflict-free learning for MIP. Our preliminary implementation covers the main features, but is still missing some fine-tuning, as we will see in the following.

We implemented the techniques presented in this paper within the academic MIP solver SCIP 6.0.2, using SoPlex 4.0.2 as LP solver [12]. In the following, we will refer to SCIP with default settings as default and to SCIP with enabled LP-based solution learning and enabled conflict-free learning as sollearning and conffree, respectively. To SCIP using both techniques simultaneously, we will refer to as combined. Our experiments were run on a cluster of identical machines equipped with Intel Xeon E5-2690 CPUs with 2.6 GHz and 128 GB of RAM. A time limit of 7200 s was set.

As test set we used the benchmark set of MIPLIB 2017 [13] which consists of 240 MIP problems. To account for the effect of performance variability [15,17] all experiments were performed with three different global random seeds [19]. Every pair of MIP problem and seed is treated as an individual observation, effectively resulting in a test set of 720 *instances*. Instances where at least one setting finished with numerical violations are not considered in the following.

Aggregated results on MIPLIB 2017 comparing all three configurations to SCIP with default settings as baseline are shown in Table 1. For every set of instances the group of affected and hard instances is shown. We denote an instance to be hard, when at least one setting takes more than 100 s and as affected, if it could be solved by at least one setting and the number of nodes differs among settings. The columns of Table 1 show the number of instances in every groups (#) and the number of solved instances (S). For the baseline (default) the shifted geometric mean [2] of solving times in seconds (T, shift = 1) and explored search tree nodes (N, shift = 100) is shown. For conffree, sollearning, and combined relative solving times (T_Q) and nodes (N_Q) compared to default are shown. Relative numbers less than 1 indicate improvements.

**Table 1.** Aggregated computational results on MIPLIB 2017 benchmark over three random seeds. Improvements by at least 5% are highlighted in bold and blue.

| | # | default S | T | N | conffree S | $T_Q$ | $N_Q$ | sollearning S | $T_Q$ | $N_Q$ | combined S | $T_Q$ | $N_Q$ |
|---|---|---|---|---|---|---|---|---|---|---|---|---|---|
| MIPLIB 2017 | | | | | | | | | | | | | |
| All | 716 | 369 | 1124 | 6069 | 370 | 1.001 | 0.976 | 368 | 0.995 | 0.991 | 368 | 1.000 | 0.973 |
| Affected | 127 | 122 | 567 | 30265 | 123 | 0.993 | **0.927** | 121 | 0.975 | 0.952 | 121 | 0.987 | **0.902** |
| ≥100 s | 107 | 102 | 969 | 45972 | 103 | 0.987 | **0.919** | 101 | 0.970 | **0.944** | 101 | 0.979 | **0.887** |
| MIXED BINARY | | | | | | | | | | | | | |
| All | 523 | 272 | 1153 | 5639 | 273 | 1.000 | 0.976 | 272 | 0.991 | 0.983 | 273 | 0.996 | 0.967 |
| Affected | 83 | 81 | 517 | 26406 | 82 | 0.983 | **0.913** | 81 | **0.943** | **0.901** | 82 | 0.958 | **0.857** |
| ≥100 s | 72 | 70 | 817 | 42289 | 71 | 0.973 | **0.899** | 70 | **0.934** | **0.886** | 71 | **0.942** | **0.834** |

Our computational experiments indicate that both individual techniques and the combination of them are superior compared to `default` on affected instances. There, we observe an overall speed-up of up to 2.5% (`sollearning`). At the same time, the tree size reduced by up to 10% (`combined`). Regarding solving time and tree size, `sollearning` alone is superior to `conffree`. `combined` is superior to both individual settings regarding nodes and almost identical to `sollearning` regarding solving time. For the set of all instances, the number of nodes reduces for all settings, while the impact on running time is almost neutral.

A reason why `conffree` seems to be less powerful than `sollearning` might be the fact that dual proof constraints are known to work better in the neighborhood of the subproblem where they were derived from, which is usually controlled by maintaining a small pool of around 100 dual proof constraints [3, 25, 26]. In our implementation, the origin of conflict-free dual proofs is not yet considered; this is a direction of future research. Also, conflict-free learning is applied much more frequently than solution learning (every feasible LP relaxation versus every integral LP relaxation), leading to a larger overhead. While `sollearning` only increases the time SCIP spends during conflict analysis by marginal 2.4%, `conffree` learning increases it by a *factor* of 3.4. This shows the need to better choose at which nodes to run conflict-free learning in future implementations.

In our computational study we observed that both techniques perform poorly on instances with general integer variables. One reason for the deterioration might be that for such instances, conflict graph analysis will generate bound disjunction constraints [1] which are generally weaker than conflict constraints on binary variables. Table 1 also presents results when applying the techniques only to (mixed) binary problems. In this case, improvements of over 5% (`sollearning`) with respect to running time and 14% (`combined`) with respect to the number of nodes can be observed on affected individual.

## 5    Conclusion and Outlook

In this paper, we discussed how conflict analysis techniques can be applied to learn from subproblems that are not (yet) proven to be infeasible. For our computational study, we implemented two conflict-free learning techniques, namely conflict-free dual proofs and LP-based solution learning, within the academic MIP solver SCIP. The results of our study indicate promising results on the benchmark set of MIPLIB 2017 when applying conflict-free learning techniques within SCIP. In particular, our experiments indicate that solution learning seems to work best on mixed binary instances.

For future research, we plan to consider the locality of derived proofs to increase the efficiency and we plan to predict, e.g., by ML techniques, from which subproblems conflict-free dual proofs should be derived to reduce the overhead.

**Acknowledgments.** The work for this article has been conducted within the Research Campus Modal funded by the German Federal Ministry of Education and Research (fund number 05M14ZAM).

# References

1. Achterberg, T.: Conflict analysis in mixed integer programming. Discrete Optim. **4**(1), 4–20 (2007)
2. Achterberg, T.: Constraint integer programming (2007)
3. Achterberg, T., Bixby, R.E., Gu, Z., Rothberg, E., Weninger, D.: Presolve reductions in mixed integer programming. INFORMS J. Comput. **32**, 473–506 (2019)
4. Chu, G., Stuckey, P.J.: Nested constraint programs. In: O'Sullivan, B. (ed.) CP 2014. LNCS, vol. 8656, pp. 240–255. Springer, Cham (2014). https://doi.org/10.1007/978-3-319-10428-7_19
5. Chu, G., Stuckey, P.J.: Goods and nogoods for nested constraint programming (2019)
6. Conforti, M., Cornuéjols, G., Zambelli, G.: Integer Programming. GTM, vol. 271. Springer, Cham (2014). https://doi.org/10.1007/978-3-319-11008-0
7. Dakin, R.J.: A tree-search algorithm for mixed integer programming problems. Comput. J. **8**(3), 250–255 (1965)
8. Davey, B., Boland, N., Stuckey, P.J.: Efficient intelligent backtracking using linear programming. INFORMS J. Comput. **14**(4), 373–386 (2002)
9. Farkas, J.: Theorie der einfachen Ungleichungen. J. für die reine und angewandte Mathematik **124**, 1–27 (1902)
10. Ginsberg, M.L.: Dynamic backtracking. J. Artif. Intell. Res. **1**, 25–46 (1993)
11. Giunchiglia, E., Narizzano, M., Tacchella, A.: Backjumping for quantified boolean logic satisfiability. Artif. Intell. **145**(1), 99–120 (2003)
12. Gleixner, A., et al.: The SCIP optimization suite 5.0. Technical report, 17–61, ZIB, Takustr. 7, 14195 Berlin (2017)
13. Gleixner, A., et al.: MIPLIB 2017: data-driven compilation of the 6th mixed-integer programming library. Technical report, Technical report, Optimization Online (2019)
14. Jiang, Y., Richards, T., Richards, B.: Nogood backmarking with min-conflict repair in constraint satisfaction and optimization. In: Borning, A. (ed.) PPCP 1994. LNCS, vol. 874, pp. 21–39. Springer, Heidelberg (1994). https://doi.org/10.1007/3-540-58601-6_87
15. Koch, T., et al.: MIPLIB 2010. Math. Program. Comput. **3**(2), 103–163 (2011)
16. Land, A.H., Doig, A.G.: An automatic method of solving discrete programming problems. Econometrica **28**(3), 497–520 (1960)

17. Lodi, A., Tramontani, A.: Performance variability in mixed-integer programming. In: Theory Driven by Influential Applications, pp. 1–12. INFORMS (2013)
18. Lubin, M., Yamangil, E., Bent, R., Vielma, J.P.: Extended formulations in mixed-integer convex programming. In: Louveaux, Q., Skutella, M. (eds.) IPCO 2016. LNCS, vol. 9682, pp. 102–113. Springer, Cham (2016). https://doi.org/10.1007/978-3-319-33461-5_9
19. Maher, S.J., et al.: The SCIP optimization suite 4.0. Technical report 17–12, ZIB, Takustr. 7, 14195 Berlin (2017)
20. Marques-Silva, J.P., Sakallah, K.: GRASP: a search algorithm for propositional satisfiability. IEEE Trans. Comput. **48**(5), 506–521 (1999)
21. Pólik, I.: Some more ways to use dual information in MILP. In: International Symposium on Mathematical Programming, Pittsburgh, PA (2015)
22. Prosser, P.: Hybrid algorithms for the constraint satisfaction problem. Comput. Intell. **9**(3), 268–299 (1993)
23. Sandholm, T., Shields, R.: Nogood learning for mixed integer programming. In: Workshop on Hybrid Methods and Branching Rules in Combinatorial Optimization, Montréal (2006)
24. Stallman, R.M., Sussman, G.J.: Forward reasoning and dependency-directed backtracking in a system for computer-aided circuit analysis. Artif. Intell. **9**(2), 135–196 (1977)
25. Witzig, J., Berthold, T., Heinz, S.: Experiments with conflict analysis in mixed integer programming. In: Salvagnin, D., Lombardi, M. (eds.) CPAIOR 2017. LNCS, vol. 10335, pp. 211–220. Springer, Cham (2017). https://doi.org/10.1007/978-3-319-59776-8_17
26. Witzig, J., Berthold, T., Heinz, S.: Computational aspects of infeasibility analysis in mixed integer programming. Technical report 19–54, ZIB, Takustr. 7, 14195 Berlin (2019)
27. Witzig, J., Berthold, T., Heinz, S.: A status report on conflict analysis in mixed integer nonlinear programming. In: Rousseau, L.-M., Stergiou, K. (eds.) CPAIOR 2019. LNCS, vol. 11494, pp. 84–94. Springer, Cham (2019). https://doi.org/10.1007/978-3-030-19212-9_6
28. Zhang, L., Madigan, C.F., Moskewicz, M.H., Malik, S.: Efficient conflict driven learning in a Boolean satisfiability solver. In: Proceedings of the 2001 IEEE/ACM International Conference on Computer-aided Design, pp. 279–285. IEEE Press (2001)

# Author Index